11th International Conference on Mesh Methods for Boundary-Value Problems and Applications 2016

IOP Conference Series: Materials Science and Engineering
Volume 158

Kazan, Russia
20 - 25 October 2016

ISBN: 978-1-5108-3524-5
ISSN: 1757-8981

Printed from e-media with permission by:

Curran Associates, Inc.
57 Morehouse Lane
Red Hook, NY 12571

Some format issues inherent in the e-media version may also appear in this print version.

Printed by Curran Associates, Inc. (2017)

For permission requests, please contact the Institute of Physics at the address below.

Institute of Physics
Dirac House, Temple Back
Bristol BS1 6BE UK

Phone: 44 1 17 929 7481
Fax: 44 1 17 920 0979

techtracking@iop.org

Additional copies of this publication are available from:

Curran Associates, Inc.
57 Morehouse Lane
Red Hook, NY 12571 USA
Phone: 845-758-0400
Fax: 845-758-2633
Email: curran@proceedings.com
Web: www.proceedings.com

Table of contents

Volume 158

11th International Conference on "Mesh methods for boundary-value problems and applications"

20–25 October 2016, Kazan, Russia

Accepted papers received: 14 November 2016
Published online: 19 December 2016

Preface

Papers

This volume contains papers from the 11th International Conference "Mesh methods for boundary-value problems and applications" which took place in Kazan, October 20 - 25, 2016. The organizers of the workshop were Kazan Federal University, Keldysh Institute of Applied Mathematics (Russian Academy of Sciences), Tatarstan Academy of Sciences and Lomonosov Moscow State University.

The scientific program of the conference was focused primarily on the theory mesh methods for boundary value problems of mathematical physics. Considerable attention is also paid to the mathematical models of complex nonlinear processes of mechanics and physics, and on the construction of efficient algorithms for solving finite-dimensional approximations of nonlinear equations and variational inequalities.

The following topics were presented:

1. Theory of mesh methods for boundary-value problems of mathematical physics: Methods for constructing and studying mesh schemes for problems with non-smooth solutions; mesh methods of domain decomposition; mixed finite element methods, mesh methods for solving nonlinear degenerate elliptic and parabolic equations; mesh methods for solving nonlinear spectral problems; multigrid methods.
2. Mathematical models of mechanics and physics: Methods for studying equations and inequalities arising in the nonlinear theory of liquid and gas filtration, the nonlinear theory of thin elastic shells, the theory of biological membranes, plasma physics, and hydrodynamic lubrication theory; methods for constructing self-similar solutions; numerical methods; methods for solving inverse problems.
3. Methods for solving variational inequalities: Construction of mesh approximations of stationary and evolutionary variational inequalities; accuracy estimation; construction and study of efficient iterative methods; techniques for improved approximation of a free boundary; domain decomposition methods for variational inequalities.
4. Numerical methods in plasma physics Numerical models and methods in low-temperature plasma physics: kinetic, hydrodynamic, and hybrid models. Models of interaction between plasma and surfaces: plasma spraying, activation, plasma etching.
5. Numerical methods in filtration theory: Numerical methods for solving seepage problems in porous media. Numerical solution of filtration consolidation problems. Numerical methods of multi-phase filtration. Numerical simulation of enhanced oil recovery processes. Numerical methods for solving problems of rational and environmentally friendly mining. Numerical simulation of filtration processes of heavy oils.
6. Computer simulation and training systems: Information and communication technologies; computational experiment; numerical modeling of technological processes using modern supercomputers; principles of training systems; simulation of random processes.
7. Numerical simulation of dynamic processes in multiphase media: Development of models and numerical techniques for computing dynamic processes in multiphase media and their applications. Development of models and numerical techniques for computing dynamics of single gas and vapor bubbles, dynamics of bubbles in clusters and their applications. Development of models and numerical techniques for computing cavitation impact on the surfaces of solid bodies and their applications.

Peer review statement

All papers published in this volume of *IOP Conference Series: Materials Science and Engineering* have been peer reviewed through processes administered by the proceedings Editors. Reviews were conducted by expert referees to the professional and scientific standards expected of a proceedings journal published by IOP Publishing.

Method of investigation of deformations of solids of incompressible materials

A I Abdrakhmanova, I R Garifullin and L U Sultanov

Kazan Federal University, 18 Kremlyovskaya Street, Kazan, 420008, Russian Federation

E-mail: A061093@mail.ru

Abstract. The aim of this work is development mathematical models, algorithm for the investigation stress-strain state of elastic solids, taking into account the incompressibility materials. The constitutive equations are received using a potential energy of deformations. The system of the linear algebraic equations is received by linearization of a resolving equation. The penalty method is applied for a modelling of the incompressibility of the material. The finite element method is used for numerical solution of the problems.

1. Introduction

The problem of mathematical modelling of elastic incompressible materials, which are referred to rubber and the other elastomers, is of great significance in the modern solid mechanics. From the point of view of mechanics, such problems are the objects of researches of non-linear theory of elasticity. This is due to their property that they allow large deformations, while maintaining the elastic properties. The authors often use numerical methods to solve non-linear problems [1–18]. The objective of our work is development of mathematical model and numerical realization of algorithm of research of stress-strained state of elastic bodies in terms of incompressibility of material and large deformations.

2. Kinematics. Constitutive equations

Kinematics of finite deformation is described using [11–14, 19]: deformation gradient tensor $(\boldsymbol{F})$ left Cauchy–Green tensor (Finger tensor) $(\boldsymbol{B})$ spatial velocity gradient tensor $(\boldsymbol{h})$; the deformation rate tensor $(\boldsymbol{d})$.

A stored strain energy function or elastic potential W, which depends on the Finger tensor can be determined as

$$W = W\left(I_{1B}, I_{2B}, I_{3B}\right),$$

where I_{1B}, I_{2B}, I_{3B} – invariants of the tensor $(\boldsymbol{B})$.

The stress state is characterized by a Cauchy–Euler stress tensor, which is defined in the current state [20–27]:

$$(\sigma) = \sigma_{ij}\left(\boldsymbol{e}_i \boldsymbol{e}_j\right).$$

Constitutive relationships are written in the following form:

$$\left(\sigma^{Tr}\right)=(c)\cdot\cdot(d),$$

where $\left(\sigma^{Tr}\right)=(\dot\sigma)+(h)\cdot(\sigma)+(\sigma)\cdot(h)^T-I_{1d}(\sigma),(\dot\sigma)=(c)\cdot\cdot(d)+(h)\cdot(\sigma)+(\sigma)\cdot(h)^T-(\sigma)I_{1d}$, $\mathbf{c}=\frac{4}{J}(B)\cdot\frac{\partial^2 W}{\partial B\partial B}\cdot(B)$ – elasticity tensor, $J=\frac{d\Omega}{d\Omega_0}$ – a changing of volume, Ω – current volume, Ω_0 – initial volume.

We applied the penalty method for solving of problems with incompressibility. The material is considered as nearly incompressible, and the pressure introduced into the equation as an independent variable:

$$P=k(J-1).$$

When the penalty number (bulk modulus) $k\to\infty$, $J\to1$, which means that the condition of incompressible material $J=1$.

For an isotropic nearly compressible material, the modified deformation gradient and left Cauchy-Green tensor are considered:

$$\hat{\mathbf{F}}=J^{-1/3}\mathbf{F},\quad \hat{\mathbf{B}}=J^{-2/3}\mathbf{B}.$$

For these measures $J=I_{3\hat{\mathbf{B}}}=1$.

Thus, the strain energy density is represented in the form of a sum of two terms; the first depends on the changes in volume, and the second – on the invariants of the measures of deformations:

$$W=W\left(I_{1\hat B},I_{2\hat B}\right)+W(J)=W\left(I_{1\hat B},I_{2\hat B}\right)+\frac{k}{2}(J-1)^2.$$

3. Variational formulation

The algorithm of investigation is based on an Update Lagrange formulation. The equation of principle of virtual work in terms of the virtual velocity in the current configuration is used [27–29]:

$$\int_{\Omega}(\sigma)\cdot\cdot(\delta d)d\Omega=\int_{S^\sigma}t_n\cdot\delta v ds+\int_{\Omega}f\cdot\delta v d\Omega,$$

where $\vec v$ – velocity vector of a material point; S^σ – part of the surface on which are defined forces; $\vec t_n$, $\vec f$ – vectors of surface and body forces respectively.

After linearization, we received the resolving system of the linear algebraic equations:

$$\int_{\Omega_k}\{\left({}^k d\right)\cdot\cdot\left({}^k c\right)\cdot\cdot(\delta d)+\frac{1}{2}\left({}^k\sigma\right)\cdot\cdot\left[(\delta h)^T\cdot\left({}^k h\right)+\left({}^k h\right)^T\cdot(\delta h)\right]-$$

$$-\left[{}^k\Delta_y\cdot{}^k v\right]{}^k f\cdot\delta v\}d\Omega_k+\int_{S_k^\sigma}\left\{{}^k t_n\cdot\left({}^k h\right)^T-\left[{}^k\Delta_y\cdot{}^k v\right]{}^k t_n\right\}\delta v dS_k=$$

$$=\int_{S_k^\sigma}{}^k\vec t_n\cdot\delta\vec v dS_k+\int_{\Omega_k}{}^k\vec f\cdot\delta\vec v d\Omega_k-\frac{1}{\Delta t}\left\{\int_{\Omega_k}\left({}^k\sigma\right)\cdot\cdot\left(\delta^k d\right)d\Omega_k-\int_{S_k^\sigma}{}^k\vec t_n\cdot\delta\vec v dS_k-\int_{\Omega_k}{}^k\vec f\cdot\delta\vec v d\Omega_k\right\}.$$

For static problems take ${}^k v=\frac{\Delta^k u}{\Delta t}$. Since the argument t is arbitrary, allowable to take $\Delta t=1$. As a result, we get the resolving equation for the displacement increments $\Delta^k\vec u$.

4. Numerical example

The numerical implementation is based on the finite element method [14]. An 8-node brick element is used. As an example the strain energy density of the neo-Hookean material is considered:

$$W=\frac{\mu}{2}\left(tr(B)-3\right),$$

where μ – shear modulus.

Elasticity tensor $\mathbf{c}$ is written as a sum of three terms:

$$\mathbf{c} = \hat{\mathbf{c}} + \mathbf{c}_{\mathrm{p}} + \mathbf{c}_{\mathrm{k}},$$

where

$$\hat{\mathbf{c}} = 2\mu J^{-\frac{5}{3}}\left(-\frac{1}{3}(\boldsymbol{B})\cdot(\boldsymbol{I}) + \frac{1}{9}I_{1B}\cdot(\boldsymbol{I})\cdot(\boldsymbol{I}) + \frac{1}{3}(\boldsymbol{C}_{II})\cdot I_{1B} - \frac{1}{3}(\boldsymbol{I})\cdot(\boldsymbol{B})\right),$$

$$\mathbf{c}_{\mathrm{p}} = pJ\left((\boldsymbol{I})\cdot(\boldsymbol{I}) - 2(\boldsymbol{C}_{II})\right).$$

Here we have introduced the tensor of the fourth rank $(\boldsymbol{C}_{II}) = (\vec{e}_i\vec{e}_j\vec{e}_i\vec{e}_j)$.

The Cauchy-Euler stress tensor then takes the following form:

$$(\sigma) = \frac{2}{J}(\boldsymbol{B})\cdot\left(\frac{\partial W}{\partial \boldsymbol{B}}\right) = \mu J^{-\frac{5}{3}}\left((\boldsymbol{B}) - \frac{1}{3}(\boldsymbol{I})\cdot I_{1B}\right) + p(\boldsymbol{I}).$$

The plane stress of the strip 6.5×6.5×0.079 mm^3 with a hole 0.5 mm in diameter is considered [5]. We use neo-Hookean material with μ=0.4225 N/mm^2.The strip was stretched in the horizontal direction in six time (Figure 1). In the Figure 2 the finite element model is shown and in the Figure 3 – the final mesh.

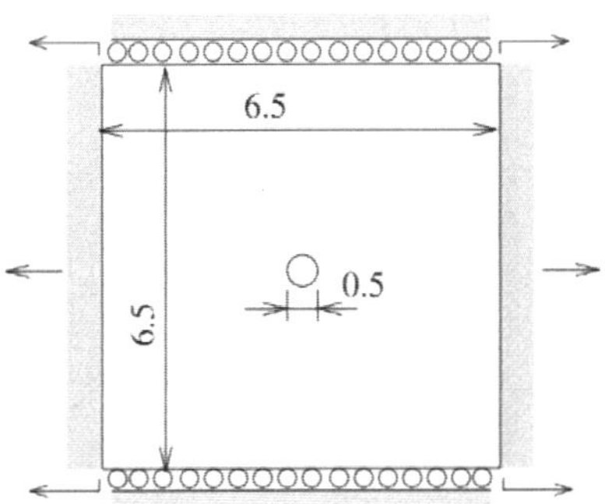

Figure 1. Strip with hole.

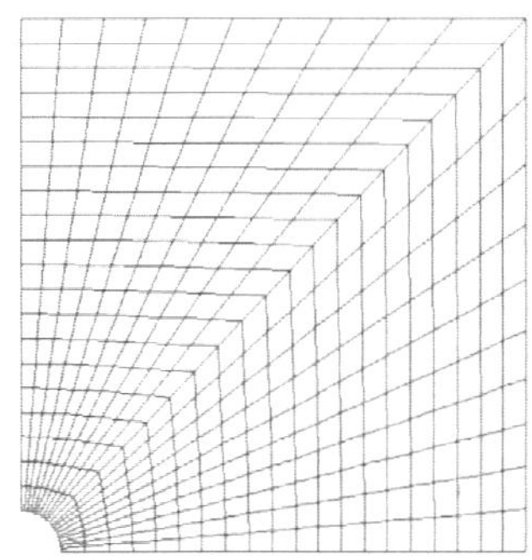

Figure 2. Finite element model.

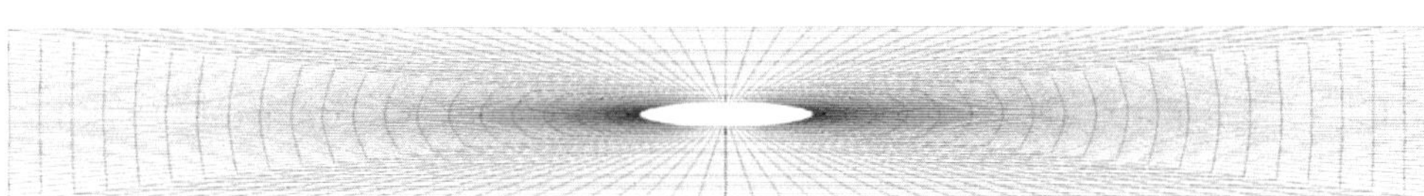

Figure 3. Final mesh.

5. Conclusion

Thus, in the paper we construct a method of numerical investigation of solids that physical relationships are specified using the strain energy density. The constitutive equations and resolving equation are obtained. The problem of plane deformation of the strip is solved. The calculation results show efficiency of this method of investigation of nonlinear problems.

Acknowledgments

This work was supported by the Russian Science Foundation (project 16-11-10299).

References

[1] Paimushin V N 2008 *Journal of Applied Mathematics and Mechanics* **72 (5)** 597–610
[2] Simo J S 1988 *Computer Methods in Applied Mechanics and Engineering.* **66 (2)** 199–219
[3] Basar Y and Itskov M 1999 Computational mechanics **23 (5)** 466–481.
[4] Meyers A, Schievbe P and Bruhns O T 2000 Acta Mech. **139 (1)** 91–103.
[5] Sagdatullin M and Berezhnoi D 2014 *Applied Mathematical Sciences* **8 (35)** 1731–1738
[6] Sagdatullin M and Berezhnoi D 2014 *Applied Mathematical Sciences* **8 (60)** 2965–2972
[7] Badriev I B, Banderov V V, Garipova G Z, Makarov M V and Shagidullin R R 2015 *Applied Mathematical Sciences* **9 (82)** 4095–4102
[8] Badriev I B and Shagidullin R R 1995 *Journal of Mathematical Sciences* **73 (5)** 519–525
[9] Sachenkov O, Kharislamova L, Shamsutdinova N, Kirillova E and Konoplev Yu 2015 *IOP Conference Series: Materials Science and Engineering* **98**
[10] Shigapova F A, Mustakimova R F, Saleeva G T and Sachenkov O A *International Journal of Applied Engineering Research* **10(24)** 44711–44714
[11] Golovanov A I, Konoplev Yu G and Sultanov L U 2008 *Uchenye Zapiski Kazanskogo Universiteta. Seriya Fiziko-Matematicheskie Nauki* **150 (1)** 29–37
[12] Golovanov A I, Konoplev Yu G and Sultanov L U 2008 *Uchenye Zapiski Kazanskogo Universiteta. Seriya Fiziko-Matematicheskie Nauki* **150 (3)** 122–132
[13] Golovanov A I, Konoplev Yu G and Sultanov L U 2010 *Uchenye Zapiski Kazanskogo Universiteta. Seriya Fiziko-Matematicheskie Nauki* **151 (3)** 108–120
[14] Golovanov A I, Konoplev Yu G and Sultanov L U 2010 *Uchenye Zapiski Kazanskogo Universiteta. Seriya Fiziko-Matematicheskie Nauki* **152 (4)** 115–126
[15] Sachenkov O A, Mitryaikin V I, Zaitseva T A and Konoplev Yu G 2014 *Applied Mathematical Sciences* **8 (159)** 7889–7897
[16] Berezhnoi D V, Sachenkov A A and Sagdatullin M K 2014 *Applied Mathematical Sciences* **8 (127)** 6341–6348
[17] Berezhnoi D V, Sachenkov A A and Sagdatullin M K 2014 *Applied Mathematical Sciences* **8 (143)** 7107–7115
[18] Sultanov L U 2015 *Procedia Earth and Planetary Science* **15** 119–124
[19] Bonet J and Wood R D 1997 *Nonlinear continuum mechanics for finite element analysis* (Cambridge University Press) 283
[20] Davydov R L and Sultanov L U 2013 *PNRPU Mechanics Bulletin* **1** 81–93
[21] Golovanov A I and Sultanov L U 2005 *International Applied Mechanics* **41 (6)** 614–620
[22] Sultanov L U 2014 *Applied Mathematical Sciences* **8 (143)** 7117–7124
[23] Sultanov L U and Davydov R L and 2014 *Applied Mathematical Sciences* **8 (60)** 2991–2996
[24] Abdrakhmanova A I and Sultanov L U 2016 *Materials Physics and Mechanics* **26 (1)** 30–32
[25] Davydov R and Sultanov L 2013 *Sixth International Conference on Nonlinear Mechanics (ICNM-VI)* 64–67
[26] Davydov R L and Sultanov L U 2015 *Journal of Engineering Physics and Thermophysics* **88 (5)** 1280–1288
[27] Sultanov L U and Fakhrutdinov L R 2013 *Magazine of Civil Engineering* **44(9)** 69–74
[28] Sultanov L U and Davydov R L 2013 *Magazine of Civil Engineering* **44(9)** 64–68
[29] Sultanov L U 2016 *Lobachevskii Journal of Mathematics* **37 (6)** 787–793

A variant of the method of quadratures for solving integral equations with fractional integral of Weyl in the main part

J R Agachev and A F Galimyanov

N.I. Lobachevsky Institute of Mathematics and Mechanics, Kazan (Volga region) Federal University, Kremlevskaya 35, Kazan, 420008, Russian Federation

E-mail: jagachev@gmail.com

Abstract. In this paper the method of mechanical quadrature solutions fractional integral equation. Computational scheme quadrature method is based on the quadrature formula of rectangles with equidistant nodes, which is the formula of the highest trigonometric degree of accuracy, using a regularizing parameter. This decision is taken for the approximate trigonometric interpolation polynomial constructed from the values that make up the solution of the quadrature method. The substantiation of the method in Holder spaces.

1. Introduction

The work is a natural extension of [1]. Here we remove conditions of zero mean value for the interval of integration coefficients of the equation

$$K\varphi \equiv \frac{1}{2\pi}\int_0^{2\pi}\Psi_{\pm}^{\alpha}(x-t)\varphi(t)dt + \int_0^{2\pi}v(x,t)\varphi(t)dt = f(x), \quad -\infty < x < \infty, \tag{1}$$

which was imposed in [2], where $\alpha\ (0<\alpha<1)$ – predetermined number, $v(x,t), f(t), \varphi(t)$ – given and the sought-for function respectively; function

$$\Psi_{\pm}^{\alpha}(t) = 2\sum_{k=1}^{\infty}\frac{\cos(kt \mp \alpha\pi/2)}{k^{\alpha}} \tag{2}$$

sets known fractional integral operator G, defined by fractional Weyl integral (see., eg., in [2, p.~264]) of order α:

$$G\varphi \equiv \frac{1}{2\pi}\int_0^{2\pi}\Psi_{\pm}^{\alpha}(x-t)\varphi(t)dt.$$

Equation (1) is seen in a pair of holder spaces $(H_{\beta}, H_{\alpha+\beta})$, where the parameter β is subject to the condition $0<\beta<1-\alpha$.

Note that for this equation in [1] given justification general polynomial projection method, in particular, of the collocation. Unfortunately, the practical realization of the circuit of the projection method requires the calculation of integrals.

This shortage is deprived of the method of mechanical quadrature. However, the justification of the quadrature method requires much more effort. Computational scheme quadrature method is based on the quadrature formula of rectangles with equidistant nodes, which is the formula of the highest trigonometric degree of accuracy, using a regularizing parameter. This decision is taken for the approximate trigonometric interpolation polynomial constructed from the values that make up the solution of the quadrature method.

2. Definitions and auxiliary results

Let $C \equiv C_{2\pi}$ be the space of continuous 2π-periodic functions with the usual $\max$-norm $\|\cdot\|_C$. We denote by H_δ $(0 < \delta \le 1)$ space of $2\pi-$ periodic functions satisfying a Holder condition with exponent $\delta > 0$, with the norm $\|\varphi\|_\delta = \|\varphi\|_C + H(\varphi;\delta)$, where $H(\varphi;\delta)$ is the smallest Holder constant. Through $H_{0,\delta}$ we will denote the subspace of space H_δ of functions $f(x)$, for which $\int_0^{2\pi} f(x)dx = 0$. We have the following (see, eg., in [1], p. 275, the corollary Theorem 19.8)

Lemma 1. Let $0 < \beta < 1-\alpha$. Then the operator G isomorphically reflects $H_{0,\beta}$ to $H_{0,\alpha+\beta}$.

Lemma 1 means that the operator G, the operator of the $H_{0,\beta}$ in $H_{0,\alpha+\beta}$, is continuously invertible if $\beta \in (0,1-\alpha)$.

For arbitrarily fixed number $\beta, 0 < \beta < 1-\alpha$, let's fix two spaces: $\Phi = H_\beta \times \mathrm{R}^1, F = H_{\alpha+\beta}$. Norm in Φ is given by the formula $\|\overline{\varphi}\|_\Phi = |\gamma| + \|\varphi\|_\beta$, where $\overline{\varphi} = (\varphi;\gamma) \in \Phi$.

Consider the operator $\overline{G}$, defined by following formula

$$\overline{G}\overline{\varphi} \equiv \gamma + G\varphi, \quad \overline{\varphi} = (\varphi;\gamma) \in \Phi. \tag{3}$$

It is clear that $\overline{G} : \Phi \to F$. Let $\tilde{\Phi} \subset \Phi$ be composed of elements of the form $(\tilde{\varphi};\gamma)$, where $\tilde{\varphi} \in H_{0,\beta}$.

Lemma 2. For any $\beta, 0 < \beta < 1-\alpha$, $\overline{G} : \tilde{\Phi} \to F$ is continuously invertible.

In the future we will need the following (see in [3], as well as in [4, p.~102, Lemma~3])

Lemma 3. Let $0 < \beta < \mu < \infty$, the function $f \in C$ and a polynomial $f_n \in H_n^T$ are such that

$$\| f - f_n \|_C \le c_1 n^{-\mu},$$

where c_1 is an absolute positive constant.

Then we have the inequality

$$H(f - f_n;\beta) \le c_2 n^{\beta-\mu},$$

where $c_2 = c_2(\beta,\mu)$ is positive constant independent from n.

From Lemma 3 immediately ensues the following result.

Corollary. The conditions of Lemma 3 we have the estimate

$$\| f - f_n \|_\beta \le c_3 n^{\beta-\mu}, c_3 = c_3(\beta,\mu), 0 < \beta \le 1, \beta < \mu.$$

Consider the situation where a polynomial f_n in Lemma 3 is constructed using the operator P_n of projection method. In that case takes the place

Lemma 4. Let $0 < \beta < \mu < \infty$, the function $f \in C$ and the operator $P_n : C \to H_n^T$ are such that

$$\| f - P_n f \|_C = O\left(\frac{\ln n}{n^\mu}\right).$$

Then for $\beta \leq 1$

$$\| f - P_n f \|_\beta = O\left(\frac{\ln n}{n^{\mu-\beta}}\right).$$

Corollary. If $f \in H_\mu, 0 < \beta < \mu \leq 1$ and operator $P_n : C \to H_n^T$ have the properties

$$P_n^2 = P_n, \| P_n \|_{C\to C} = O(\ln n), n \to \infty,$$

then

$$\| f - P_n f \|_\beta = O(\ln n \cdot n^{\beta-\mu}), n \to \infty.$$

3. Alternative method of mechanical quadratures

Let's include the equation (1) to a wider class of equations of the form

$$\overline{K\varphi} \equiv \gamma + \frac{1}{2\pi}\int_0^{2\pi} \Psi_\pm^\alpha(xt)\varphi(t)dt + (V\varphi)(x) = f(x), \quad -\infty < x < \infty,$$

where $\gamma \in \mathrm{R}^1, \overline{\varphi} = (\varphi;\gamma)$. and the integral operator V is defined by the formula

$$(V\varphi)(x) \equiv (Vv\varphi)(x) = \int_0^{2\pi} v(x,t)\varphi(t)dt$$

Considering this equation in two spaces (Φ, F), introduced above, we write it in operator form

$$\overline{K\varphi} \equiv \overline{G\varphi} + V\varphi = f \quad (\overline{\varphi} = (\varphi;\gamma) \in \Phi, f \in F) \tag{1'}$$

Let P_n further is the operator of trigonometric Lagrange interpolation with equally spaced points $t_j = 2j\pi/(2n+1), j = \overline{0,2n}$, projecting F on $F_n = H_n^T$, where H_n^T - set of trigonometric polynomials of order at most n, and $\Phi_n = \{\overline{\varphi}_n = (\varphi_n;\gamma_n) \in F_n \times R^1\}$. Then the approximate solution $\overline{\varphi}_n$ of (1') going to find the exact solution of the equation

$$\overline{K}_n\overline{\varphi}_n \equiv \overline{G\varphi_n} + P_n V(P_n^t(v\varphi_n)) = P_n f \quad (\overline{\varphi}_n = (\varphi_n;\gamma_n) \in \Phi_n,) \tag{4}$$

Here P_n^t denotes the operator P_n, applied for the integration variable t.

Note that equation (4) with respect to γ_n and the coefficients of the polynomial φ_n is a system of linear algebraic equations.

Theorem 1. Suppose that the assumptions are true:

1) function $f \in H_\lambda, \alpha + \beta < \lambda \leq 1$;
2) $v \in H_\lambda$ in the variable x and t uniformly with respect to another variable;
3) equation (1') is uniquely solvable for any right part of the F

Then the equation (4) for all sufficiently large n is uniquely solvable. The approximate solutions found by the method (4), agree to the the exact solution of (1') in the space of Φ with speed

$$\| \overline{\varphi} - \overline{\varphi}_n \|_\Phi = O(n^{\alpha+\beta-\lambda} \ln n). \tag{5}$$

Proof. First, we note that under the conditions of theorem for an arbitrary function $\varphi \in H_\beta$, we have equality: $G\varphi = G\tilde{\varphi}$, where $\tilde{\varphi} = \varphi - \frac{1}{2\pi}\int_0^{2\pi}\varphi dt$. Therefore, the equation $(1')$ and quadrature method equation $(4')$ are of the form:

$$\overline{K}\tilde{\overline{\varphi}} \equiv \overline{G}\tilde{\overline{\varphi}} + V\varphi = f \quad (\tilde{\overline{\varphi}} \in \tilde{\Phi}) \tag{1''}$$

$$\overline{K}_n\tilde{\overline{\varphi}}_n \equiv \overline{G}\tilde{\overline{\varphi}}_n + P_n V(P_n^t(v\varphi))_n = P_n f \quad (\tilde{\overline{\varphi}}_n \in \tilde{\Phi}_n). \tag{4'}$$

Since, by Lemma 2, the operator $\overline{G}:\tilde{\Phi} \to F$ is continuously invertible, and $V: H_\beta \to F$ is compact, the existence of unique solution of the equation (1′) for any right of F is equivalent to a invertibility of continuous operator $\overline{K}:\tilde{\Phi} \to F$.

Next, the operator P_n trigonometric Lagrange interpolation at equidistant points of the estimate (see., eg., in [4])

$$\| f - P_n f \|_C \le 2\| P_n \|_{C\to C} \cdot E_n^T(f) = O\{n^{-\lambda} \ln n\},$$

where $E_n^T(f)$ is the best uniform approximation of a function $f(x)$ by trigonometric polynomials of order at most n. Then, by Corollary to Lemma 4 we can easily deduce proximity of right-hand sides of equations (1'') and (4')

$$\delta_n \equiv \| f - P_n f \|_F = O\{n^{\alpha+\beta-\lambda} \ln n\} \to 0, n \to \infty.$$

Let's now determine the proximity of operators of (1'') and (4'). For this we will consider for an arbitrary $\tilde{\overline{\varphi}}_n \in \tilde{\Phi}_n$ difference $\overline{K\tilde{\varphi}}_n - \overline{K}_n\tilde{\overline{\varphi}}_n$. We have

$$\| \overline{K\tilde{\varphi}}_n - \overline{K}_n\tilde{\overline{\varphi}}_n \|_F \le \| V\varphi_n - P_n V\varphi_n \| + \| = P_n V\varphi_n - P_n V P_n^t(v\varphi_n) \| \equiv I_1 + I_2 \tag{6}$$

For the first term on the right side in the inequality (6), we have:

$$I_1 = O\{\ln n \cdot E_n^T(V\varphi_n)\} = O\{\ln n \cdot E_n^{Tx}(v)_{C\times L_1} \| \varphi_n \|_C\} =$$

$$= O\{n^{\alpha+\beta-\lambda} \ln n\} \| \varphi_n \|_\beta = O\{n^{\alpha+\beta-\lambda} \ln n\} \| \tilde{\varphi}_n \|_\beta, \tag{7}$$

where $E_n^{Tx}(v)_{C\times L_1}$ is the quotient of the best approximation of kernel $v(x,t)$ on the variable x by trigonometric polynomials of order at most n in a mixed metric.

We now estimate the second term in (6). Given trigonometric the degree of precision interpolation quadrature formula for equidistant nodes $\{t_j\}_0^{2n}$ it is easy to see that

$$I_2 = \| P_n V[(v - P_n^t v)\varphi_n] \|_F = O\{\ln n \cdot E_n^{Tt}(v)_{F\times C_{2\pi}} \| \varphi_n \|_{C_{2\pi}}\} =$$
$$O\{\ln n \cdot n^{\alpha+\beta-\lambda}\} \| \varphi_n \|_\beta = O\{\ln n \cdot n^{\alpha+\beta-\lambda}\} \| \tilde{\varphi}_n \|_\beta .$$

Thus, considering the ratio of the latter (7) and (6) we derive

$$\varepsilon_n \equiv \| \overline{K} - \overline{K}_n \|_{\tilde{\Phi}\to F} = O\{n^{\alpha+\beta-\lambda} \ln n\}.$$

The rest is obvious, considering the Theorem 7 of Chapter I of monograph [4].

Let us consider the case where the operator $V: H_\beta \to H_{\alpha+\beta}$ is continuous and a small norm.

Theorem 2. Suppose that the assumptions are true:

1) function $f \in H_\lambda, \alpha+\beta < \lambda \le 1$;

2) operator $V:\Phi \to H_\lambda$ is continuous, and

$$\| V \|_{H_\beta\to F} < \| G^{-1} \|^{-1}_{H_{0,\alpha+\beta}\to H_{0,\beta}};$$

Then equation (4) has a unique solution, at least for all sufficiently large n. Approximate solutions have been found by (4) converge to the exact solution of the equation $(1')$ in the space Φ with speed

$$\| \overline{\varphi} - \overline{\varphi}_n \|_{\Phi} = O(n^{\alpha+\beta-\lambda} \ln n).$$

Proof. Given that for any $\varphi \in H_\beta$ $G\varphi = G\tilde{\varphi}$, where $\tilde{\varphi} = \varphi - \frac{1}{2\pi}\int_0^{2\pi} \varphi dt$, equation $(1')$ can be written as

$$\overline{K}\overline{\varphi} \equiv \overline{G}\overline{\tilde{\varphi}} + V\varphi = f \quad (\overline{\tilde{\varphi}} \in \tilde{\Phi}).$$

By Lemma 2, the operator $\overline{G} : \tilde{\Phi} \to F$ has a continuous inverse $\overline{G}^{-1} : F \to \tilde{\Phi}$. Applying it to the last equation, we obtain the following equation of the second kind:

$$\overline{K}_1\overline{\varphi} \equiv \overline{\tilde{\varphi}} + \overline{G}^{-1} V\varphi = \overline{G}^{-1} f \quad (\overline{\tilde{\varphi}} \in \tilde{\Phi}). \tag{8}$$

Since

$$\| \overline{G}^{-1} V \|_{H_\beta \to \Phi} \le \| \overline{G}^{-1} \|_{F \to \tilde{\Phi}} \cdot \| V \|_{H_\beta \to F} < 1,$$

Banach's theorem on the perturbation of the identity operator with a small norm (see, eg., in [5, p.~140]), it follows that the equation (8) has a unique solution $\overline{\varphi}^* = (\varphi^*; \gamma^*) \in \Phi$. Last means that equivalent to (8), equation $(1')$ also has a unique solution $\overline{\varphi}^*$. Moreover, the operator $\overline{K} = \overline{G}\overline{K}_1 : \Phi \to F$ is continuously invertible:

$$\| \overline{K}^{-1} \|_{F \to \Phi} \le \| \overline{G}^{-1} \|_{F \to \tilde{\Phi}} \cdot \| \overline{K}_1^{-1} \|_{\tilde{\Phi} \to \Phi} < \infty.$$

We now show that the equation of the projection method (4) will also have a unique solution, at least for sufficiently large n. With For this purpose, we will verify that the conditions of Theorem 7 Chapter I monograph [4], which gives sufficient conditions for convergence of direct method of solution of the equation $(1')$.

We have (see proof of Theorem 1)

$$\delta_n \equiv \| f - P_n f \|_F = O\{n^{\alpha+\beta-\lambda} \ln n\} \to 0, n \to \infty. \tag{9}$$

Furthermore, for any element $\overline{\varphi}_n \in \Phi_n, \overline{\varphi}_n \ne 0$

$$\| \overline{K}\overline{\varphi}_n - \overline{K}_n \overline{\varphi}_n \|_F = \| V\varphi_n - P_n V\varphi_n \| + \| P_n V\varphi_n - P_n V(P_n^t(v\varphi_n)) \| =$$

$$= \| \varphi_n \|_\Phi \cdot \{ \| V\psi_n - P_n V\psi_n \| + \| P_n V\psi_n - P_n V(P_n^t(v\psi_n)) \| \}, \psi_n = \varphi_n / \| \varphi_n \|_\Phi .$$

It implies that

$$\| \overline{K}\varphi_n - \overline{K}_n \varphi_n \|_F \le \| \varphi_n \|_\Phi \cdot \{ \sup_{g \in VS(0,1)} \| g - P_n g \|_F + \| P_n V(v - P_n^t v)\psi_n \|_F \} \equiv \| \varphi_n \|_\Phi \{ I_3 + I_4 \},$$

(10)

where $S(0,1)$ – the unit ball centered at the origin of space Φ. Thus, taking into account the limit relation (9) we obtain (see. also our work [2])

$$I_3 \le \sup_{g \in VS(0,1)} \| g - P_n g \|_F \to 0, n \to \infty \tag{11}$$

For the value of I_4 implies the estimate

$$I_4 = O(\ln n \cdot \| v - P_n^t v \|_{F \times L_2} \| \psi_n \|_{L_2}) = O(\ln n \cdot E_n^{Tt}(v)_{F \times C_{2\pi}} \| \psi_n \|_{C_{2\pi}}) = O(\ln n \cdot E_n^{Tt}(v)_{F \times C_{2\pi}})$$

The rest follows from this theorem 7 and theorem 6 about representation of error of the approximate solutions from monograph [4, p.~17].

References

[1] Agachev J R and Galimjanov A F 2015 On Justification of General Polynomial Projection Method for Solving Periodic Fractional Integral Equations *Lobachevskii Journal of Mathematic* **36 (2)** pp 97–102

[2] Samko S G, Kilbas A A and Marichev O I 1993 *Fractional integrals and derivatives: theory and applications* (Gordon and Breach Science Publishers)

[3] Gabdulkhaev B G 1970 *Proceedings of the Intern. conf. By constructive theory of functions* (Varna) pp 35–49

[4] Gabdulkhaev B G 1980 *Optimal approximations of linear problems solutions* (Kazan: in Kazan. University Press)

[5] Trenogin V A 1980 *Functional analysis* (Moscow: Nauka)

The calculation of weakly non-spherical cavitation bubble impact on a solid

A A Aganin, T S Guseva, L A Kosolapova and N A Khismatullina

Institute of Mechanics and Engineering, Kazan Science Center, Russian Academy of Sciences, 2/31 Lobachevsky str., Kazan 420111, Russia

E-mail: nailya_hism@mail.ru

Abstract. The effect of small spheroidal non-sphericity of a cavitation bubble touching a solid at the beginning of its collapse on its impact on the solid of a copper-nickel alloy is investigated. The impact on the solid is realized by means of a high-speed liquid jet arising at collapse on the bubble surface. The shape of the jet, its velocity and pressure are calculated by the boundary element method. The spatial and temporal characteristics of the pressure pulses on the solid surface are determined by the CIP-CUP method on dynamically adaptive grids without explicitly separating the gas-liquid interface. The solid surface layer dynamics is evaluated by the Godunov method. The results are analyzed in dimensionless variables obtained with using the water hammer pressure, the time moment and the jet-solid contact area radius at which the jet begins to spread on the solid surface. It is shown that in those dimensionless variables, the dependence of the spatial and temporal characteristics of the solid surface pressure pulses on the initial bubble shape non-sphericity is relatively small. The non-sphericity also slightly influences the main qualitative features of the dynamic processes inside the solid, whereas its effect on their quantitative characteristics can be significant.

1. Introduction

It is widely known that the solid surfaces contacting with a liquid with the variable pressure are often subject to cavitation erosion resulting from the action of cavitation bubbles [1]. One of the main mechanisms of cavitation erosion is considered the impact of the high-speed liquid jet arising on the surface of the bubble at its asymmetrical collapse near a solid [2]. It is shown in [3] that the parameters of such a jet (the velocity and the pressure, the geometry of its end) are strongly dependent on the shape of the bubble at the beginning of its collapse.

The present paper is devoted to investigating the effect of a small non-sphericity of the bubble at the beginning of its collapse on the surface layer of a solid in the case of a cavitation bubble touching the solid surface. The solid material is a copper-nickel alloy (Monel 400), which has high strength, ductility and corrosion resistance in various media, including sea water, hydrofluoric acid, sulfuric acid and alkalis. Such materials are used in ship manufacturing, chemical industry, etc.

2. Problem statement and the main points of the numerical technique

The impact of a cavitation bubble at its collapse in liquid is considered. The liquid is water (the density ρ_L = 1000 kg / m^3), the solid material is Monel 400 (a copper-nickel alloy with the Young modulus E = 173 GPa, the Poisson ratio υ = 0.3, the density ρ = 7850 kg / m^3, the yield strength

Y_0 = 250 MPa), which is used for corrosion protection. The bubble collapses due to a large difference between the initial pressure in its cavity (0.023 bar, which corresponds to the saturated water vapor at room temperature) and the pressure in the surrounding liquid (4 bar). The bubble touches the solid whose dimensions are much larger than those of the bubble so that the solid surface is considered plane. At the beginning of collapse the bubble and the liquid are at rest, the bubble is spheroidal. The contour of the bubble in its axial cross-section is an ellipse with the semi-axes a and b, the former being orthogonal to the solid surface. The initial non-sphericity of the bubble is characterized by a parameter $\varepsilon = 1 - e$ where $e = a / b$. The influence of the small initial bubble non-sphericity in the range $-0.02 \le \varepsilon \le 0.02$ is studied. At that, $\varepsilon < 0$, $\varepsilon > 0$, and $\varepsilon = 0$ respectively correspond to the slightly elongated, slightly flattened, and purely spherical bubbles.

The study is conducted with the help of a numerical technique, in which the process of computing the impact of a bubble on a solid is divided into three stages, according to the features of the impact process (figure 1).

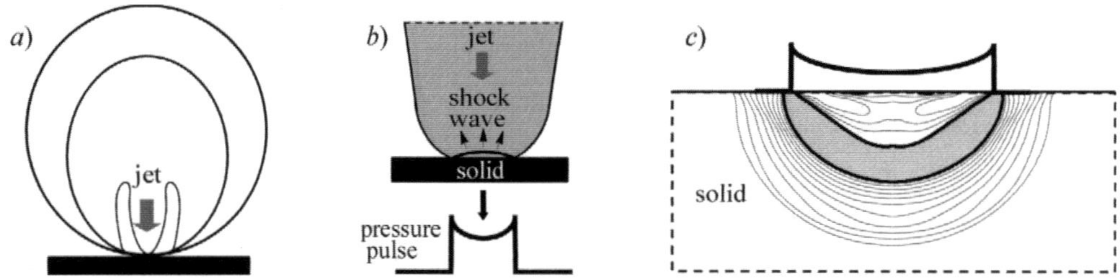

Figure1. Three stages of the impact of a bubble on a solid according to which the numerical technique of the present work is divided into three stages: I – formation of a jet hitting the solid surface (a), II – the impact of the jet on the solid surface resulting in a surface pressure pulse (b), III – the solid surface layer response to the pressure pulse arising on the solid surface (c)

The first stage is devoted to computing the relatively long-lasting bubble collapse until the moment of contact of the end of the jet arising on the bubble surface with the solid surface, i.e. until the beginning of impact on the solid surface. The main results of the first stage are the liquid jet characteristics at the beginning of its impact on the solid, namely, the jet velocity, the jet pressure, the shape of the jet end, which are required for the next stage. At the first stage the effects of the liquid viscosity and compressibility, the inhomogeneity of the vapor in the bubble, the solid surface layer dynamics are not taken into account due to their smallness. The computation is carried out by the numerical technique [4] based on the boundary element method.

The second stage is devoted to computing the spatial and temporal characteristics of the pressure pulse occurring on the solid surface. At this stage, shock waves appear in the jet, the jet surface is strongly deformed. At the same time, in the considered range of the small initial bubble non-sphericity most intensive jet impact on the solid surface is realized in the region of their contact the radius of which is significantly less than the transverse dimension of the bubble. Because of this, the influence of the liquid outside the jet is not taken into account, so that an impact of a cylindrical liquid jet, surrounded by a vapor, with the hemispherical end on a plane rigid solid surface is actually considered at the second stage. The radius of the hemispherical end is taken equal to the radius of the end of the bubble jet. The backward effect of the solid dynamics on the liquid and vapor dynamics is here, as in the previous stage, not taken into account. The computation is carried out by the numerical technique [5] based on the CIP-CUP method on dynamically adaptive grids without explicit separation of the gas-liquid interface [6].

At the third stage, the dynamics of the solid surface layer is determined. To this end, the solid surface load is calculated by approximating the spatial and temporal characteristics of the pressure computed earlier at the second stage. The solid is modeled by an elastic-plastic isotropic semi-space,

IOP Conf. Series: Materials Science and Engineering **158** (2016) 012003 doi:10.1088/1757-899X/158/1/012003

in which the deformations and the displacements are assumed small. The Mises yield condition is used in the plastic zones. The solution is found numerically by the numerical technique [7] based on the classic Godunov method [8]. In that technique the non-reflective conditions are posed on the artificial boundaries of the computational domain.

3. Characteristics of the jet at the beginning of its impact

Due to the presence of a solid, the bubble initially close to the spherical collapses strongly non-spherically (figure 1a). In particular, a high-speed liquid jet directed to the solid arises on the bubble surface.

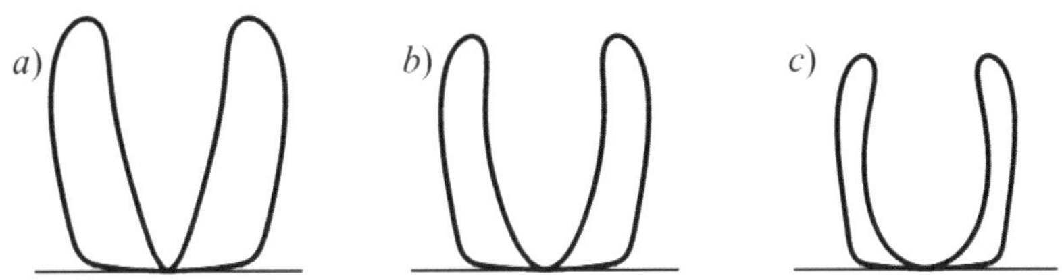

Figure 2. Contours of the bubbles in their axial cross-sections at the beginning of impact on the solid surface, the initial bubble non-sphericity $\varepsilon = 0.02$ (*a*), $\varepsilon = 0$ (*b*), and $\varepsilon = -0.02$ (*c*)

Figure 2 illustrates the shape of a bubble at the beginning of its impact on the solid surface in the cases of the bubble initially slightly flattened ($\varepsilon = 0.02$), purely spherical ($\varepsilon = 0$) and slightly elongated ($\varepsilon = -0.02$). One can see that the jet markedly thickens as the non-sphericity of the bubble at the beginning of its collapse varies from 0.02 to -0.02. In particular, the radius of its end curvature R increases from 15 μm to 116 μm, the jet end velocity V, with which it hits the solid, decreases from 270 m / s to 220 m / s, and the jet end pressure increases from 4.4 bar to 17 bar. The jet end pressure is determined by the vapor pressure in the bubble cavity so that its increase with decreasing ε is due to the decrease in the volume of the bubble. It should be noted that because of the relatively rapid collapse of the bubble the vapor in its cavity is modeled as a perfect gas.

4. Characteristics of the pressure pulses on the solid surface

As mentioned in the previous section, for all values of the initial bubble non-sphericity ε the action of the bubble on the solid is simulated in the present technique as impact of a jet with the hemispherical end on a rigid surface. Upon the impact, a shock wave forms in the jet, propagating away from the solid. At the beginning of the impact the pressure behind the front of this wave is approximately equal to the water hammer pressure $p_{wh} = \rho_L V D$ where D is the shock wave velocity (for water $D \approx C_L + 2V$). In the initial phase of the impact the shock wave remains attached to the solid surface by its edge on the periphery of the circular rapidly-expanding jet-solid contact area. The jet-solid contact area expansion velocity rapidly decreases so that at $t \approx \tau_j = RV / (2D^2)$ (hereinafter, the time t is measured from the beginning of the jet impact) the shock wave edge leaves the solid surface and goes away from it along the jet boundary. At this moment the jet begins to spread radially on the body surface. The radius of the jet-solid contact area at $t = \tau_j$ is $R_j = RV / D$. The mentioned features of the jet impact in the numerical simulations are consistent with those known in the literature [9].

Figure 3 illustrates the spatial and temporal characteristics of the pressure pulses that occur on the solid surface at impact by the jets arising in the bubbles with an initial bubble non-sphericity in the considered range $-0.02 \leq \varepsilon \leq 0.02$. In this figure, the pressure is referred to the water hammer pressure p_{wh}, the time t and the radial coordinate r to the time moment τ_j and the radius R_j, respectively, at which the jet begins to spread on the solid surface. One can see that all three solid lines 1 and all three solid lines 2 are in fact graphically coincident. Two solid lines 3 and two solid lines 4 corresponding to the initially non-spherical bubbles are different from those corresponding to the spherical ones by no more than 5% and 10%, respectively. The difference between three solid curves 5

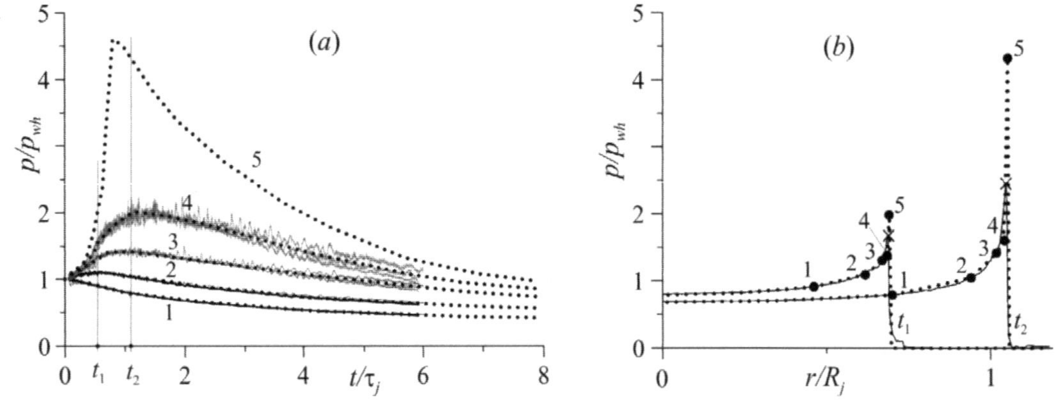

Figure 3.(*a*) – the time-dependences of the pressure (lines 1-5) at 5 points (shown in (*b*)) of its radial profiles on the solid surface, corresponding to the 5 values of r / R_{max} (t) where R_{max} (t) is the radius at which the solid surface pressure at a moment t is maximum; (*b*) - the radial solid surface pressure profiles at two moments $t_{1,2}$ shown in (*a*) by the vertical lines. In (*a*): solid lines 1-4 in the interval $0 \leq t / \tau_j \leq 6$ are the numerical simulation results with $\varepsilon = -0.02$, 0 and 0.02, dashed lines 1-5 are the approximations of the numerical simulation results. In (*b*): solid lines (their maxima are marked by crosses) are the numerical simulation results with $\varepsilon = 0.02$, dotted lines are the approximations of the numerical simulation results.

is much larger (to avoid encumbering they are not given in figure 3). But the large difference takes place in only a very small spatial vicinity of the maximum pressure. The smallness of this vicinity allows one to assume that the influence of the large difference on the solid dynamics is small. Taking this into account, we conclude that in terms of dimensionless variables of figure 3 the difference of the solid surface pressure pulse characteristics in the spherical and non-spherical cases is not significant. This allows us to approximate all the numerical simulation results in the range $-0.02 \leq \varepsilon \leq 0.02$ by the dotted curves of figure 3. Such approximations are utilized in the next stage.

5. Dynamics of the solid surface layer

As mentioned in the previous section, the solid surface layer dynamics is studied with using dimensionless variables obtained with the pressure p_{wh}, the time moment τ_j, and the radius R_j. In such variables, the spatial and temporal characteristics of the solid surface pressure pulses used in the numerical simulations are the same (in figure 3 they are presented by the dotted lines) for any initial bubble non-sphericity ε in the range under consideration. However, the dimensionless plasticity value defined by the ratio Y_0 / p_{wh}, becomes variable due to the dependence of p_{wh} on the velocity V, which, in its turn, depends on ε. For example, $p_{wh} \approx 5.6$, 4.7, and 4.2 kbar for $\varepsilon = 0.02$, 0, and -0.02, respectively. Note that at one-dimensional impact with these pressures the yield strength of the considered alloy ($Y_0 = 250$ MPa) gets exceeded only when $p_{wh} = 5.6$ and 4.7 kbar.

Numerical experiments show that the dynamic processes realizing in the copper-nickel alloy solid surface layer at impact on the surface by the jets arising in a bubble with an initial non-sphericity in the range $-0.02 \leq \varepsilon \leq 0.02$ have the same qualitative features. Figure 4 illustrates the change of the stress intensity in the solid surface layer in a small vicinity of the point of impact in the case of the initial bubble non-sphericity $\varepsilon = 0.02$. With decreasing ε, the region in which the stress intensity is equal to the yield strength, becomes, in the dimensionless coordinates, smaller, and disappears, in the dimensionless time, faster.

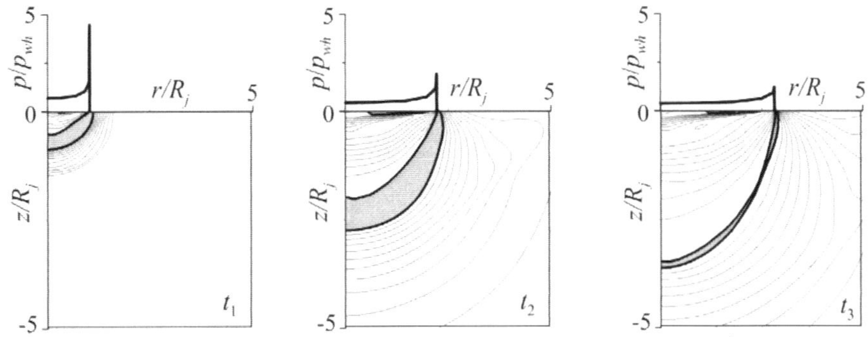

Figure 4. Contours of the stress intensity σ_i in the solid in a small vicinity of the jet impact point at the three consecutive time moments $t_{1\text{-}3}$ and the corresponding radial pressure profiles in the case of the initial bubble non-sphericity $\varepsilon = 0.02$. The bold contours and the shaded regions corresponds to $\sigma_i = Y_0$. At the moment $t_1 = \tau_j$ the pressure on the periphery of the loaded area is maximum (figure 3).

Quantitative changes in the solid surface layer dynamics with varying the initial bubble non-sphericity in the range $-0.02 \le \varepsilon \le 0.02$ is illustrated in figure 5 where the increase in the depth of the micropit on the solid surface during the jet impact is shown for $\varepsilon = -0.02$, 0, and 0.02. One can see that for $\varepsilon = 0.02$ ($\varepsilon = -0.02$) the micropit is deeper (shallower) than for $\varepsilon = 0$ by 37.5% (17.5%), though in the former case the solid surface pressure pulse amplitude is larger (smaller) than in the latter case by only 17% (9%). The difference in the pressure pulse amplitude is determined by the difference in the corresponding values of the water hammer pressure p_{wh}.

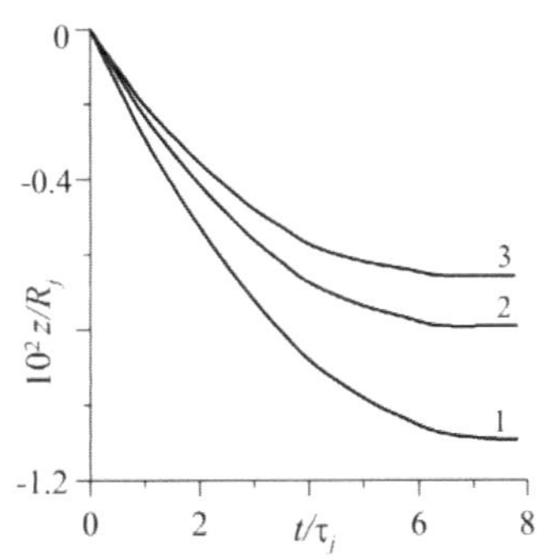

Figure 5. Time-dependences of the axial coordinates of the center of the circular micropit arising on the solid surface at the jet impact, the initial bubble non-sphericity $\varepsilon = 0.02$ (curve 1). 0 (curve 2), and -0.02 (curve 3)

6. Conclusion

A numerical study of the influence of small spheroidal non-sphericity of a cavitation bubble at the beginning of its collapse on impact of this bubble on the body of copper-nickel alloy has been performed. The velocity, the pressure and the shape of the jet hitting the solid are determined by the boundary element method. The spatial and temporal characteristics of the solid surface pressure pulses are evaluated by the CIP-CUP method on dynamically adaptive grids without explicitly separating the gas-liquid interface. The dynamics of the solid surface layer in the region of impact is calculated by the Godunov method. The study has been performed in dimensionless variables obtained with using the water hammer pressure, the time moment and the jet-solid contact area radius at which the jet begins to spread on the solid surface. It has been shown that in those dimensionless variables the spatial and temporal characteristics of the solid surface pressure pulses are slightly dependent on the initial bubble non-sphericity. The initial bubble non-sphericity also slightly influence the main features of the dynamic processes inside the solid (for example, the shape of the plastic zone). However, its effect on the quantitative characteristics of those processes can be significant. In particular, the difference in the depth of the micropits arising on the solid surface due to the jet impact can differ by more than 35%.

References

[1] Bowden F P and Brunton J H 1961 *Proc. R. Soc. Lond.* A. **263** 433
[2] Kornfeld M and Suvorov L 1944 *J. Appl. Phys.* **15** 495
[3] Voinov O V and Voinov V V 1976 *Soviet Physics. Doclady* **21** 133
[4] Aganin A A, Ilgamov M A, Kosolapova L A and Malakhov V G 2013 *Vestnik Bashkirckogo universiteta* **18** Issue 1 15
[5] Aganin A A and Guseva T S 2014 *Uchenye zapiski Kazanskogo universiteta Seriya fiziko-matematicheskiye nauki* **156** Book 2 55
[6] Takizawa K, Yabe T, Tsugawa Y, Tezduyar T E and Mizoe H 2007 *Comput. Mech.* **40** 167
[7] Aganin A A, Guseva T S and Khismatullina N A 2015 *Uchenye zapiski Kazanskogo universiteta Seriya fiziko-matematicheskiye nauki* **157** Book 1 75
[8] Godunov S K, Zabrodin A V, Ivanov M Y, Krayko A N and Prokopov G P 1976 *Numerical solving of gas dynamics multidimensional problem* (Moscow: Nauka) 400
[9] Field J E 1999 *Wear* **233-235** 1

Numerical simulation of liquid jet impact on a rigid wall

A A Aganin and T S Guseva

Institute of Mechanics and Engineering, Kazan Science Center, Russian Academy of Sciences, 420111 Kazan, Russian Federation

E-mail: ts.guseva@mail.ru

Abstract. Basic points of a numerical technique for computing high-speed liquid jet impact on a rigid wall are presented. In the technique the flows of the liquid and the surrounding gas are governed by the equations of gas dynamics in the density, velocity, and pressure, which are integrated by the CIP-CUP method on dynamically adaptive grids without explicitly tracking the gas-liquid interface. The efficiency of the technique is demonstrated by the results of computing the problems of impact of the liquid cone and the liquid wedge on a wall in the mode with the shockwave touching the wall by its edge. Numerical solutions of these problems are compared with the analytical solution of the problem of impact of the plane liquid flow on a wall. Applicability of the technique to the problems of the high-speed liquid jet impact on a wall is illustrated by the results of computing a problem of impact of a cylindrical liquid jet with the hemispherical end on a wall covered by a layer of the same liquid.

1. Introduction

The impact of high-speed liquid jets (or drops) on a wall is of considerable interest to applications as it can lead to the wall damages. Such problems can occur, for example, with aircrafts flying in a rain, with operating steam turbines [1, 2]. High-speed microjets directed to a wall can arise on the surface of cavitation bubbles at their non-spherical collapse near the wall. The impact of such jets is considered to be one possible cause of destruction of the surfaces of bodies operating in the conditions of cavitation [3, 4].

Important features of the high-speed jet impact on a wall are shockwaves which can arise in the liquid and the surrounding gas, the strong deformations of the liquid-gas interface. These features need to be allowed for while choosing a mathematical model as well as a numerical method, and they were taken into account in realizing the numerical technique of the present work.

2. Basic points of the numerical technique

The numerical technique is based on the approach without explicit separation of the interface. Similar to [5], an identifier-function φ is used to identify a fluid: $\varphi = 1$ in the liquid and $\varphi = 0$ in the gas. In a small vicinity of the gas-liquid interface the identifier φ is continuous and monotonically changing from 0 to 1. The identifier φ is governed by the advection equation

$$\varphi_t + \mathbf{u} \cdot \nabla\varphi = 0, \tag{1}$$

where $\mathbf{u}$ is the advection velocity.

Dynamics of the liquid and the gas is computed by the CIP-CUP method [6]. It is based on equations of motion of a compressible medium in primitive variables (density, velocity, pressure), which allows one to perform calculation without explicit separation of the interface [7]. Without allowing for the effects of viscosity and heat conductivity those equations can be written in the form

$$\mathbf{f}_t + \mathbf{u} \cdot \nabla \mathbf{f} = \mathbf{G} \tag{2}$$

where $\mathbf{f} = (\rho, \mathbf{u}, p)^T$, $\mathbf{G} = (-\rho\nabla\cdot\mathbf{u}, -\rho^{-1}\nabla p, -\rho C_S^2\nabla\cdot\mathbf{u})^T$, ρ is the density, p is the pressure, $C_S = \varphi C_{S1}+(1-\varphi) C_{S2}$ is the sound speed, $C_{Si} = [\Gamma_i(p+B_i)/\rho]^{1/2}$, $i = 1$ corresponds to the liquid, $i = 2$ to the gas, Γ_1, B_1 are the constants of the Tait equation of the liquid state, $\Gamma_2 = \gamma$, $B_2 = \mathbf{0}$, γ is the isentropic exponent of gas.

System (2) is numerically integrated by splitting into the advection and non-advection parts.

$$\frac{\mathbf{f}^* - \mathbf{f}^n}{\Delta t^n} + \mathbf{u} \cdot \nabla \mathbf{f} = 0, \tag{3}$$

$$\frac{\mathbf{f}^{n+1} - \mathbf{f}^*}{\Delta t^n} = \mathbf{G}. \tag{4}$$

Semi-Lagrangian method CIP (Constrained Interpolation Profile) [5] is applied to compute the group of advection equations (3). The solution of the advection equation for a function f at a grid point $\mathbf{x}$ at a time moment $t^{n+1} = t^n + \Delta t^n$ can be approximately written as $f(\mathbf{x}-\mathbf{u}\Delta t^n)$ where the argument is a departure point meaning the location of the "Lagrangian particle" which in time interval Δt^n will arrive at the grid point $\mathbf{x}$. The value of the function f at the departure point is determined by CIP-interpolation using known values of f and its spatial derivatives at the grid points at the moment t^n.

Non-advection part (4) is reduced to the following pressure equation

$$\frac{p^{n+1} - p^*}{\rho^* C_S^{*2} \Delta t^{n2}} = \nabla \cdot \left(\frac{\nabla p^{n+1}}{\rho^*} \right) - \frac{\nabla \cdot \mathbf{u}^*}{\Delta t^n},$$

which is integrated by the well-known SOR method. Then $\mathbf{u}^{n+1}$ and ρ^{n+1} are calculated. In solving problems with strong shockwaves the CIP-CUP method is used in combination with the artificial viscosity.

Calculations are carried out with the use of dynamically adaptive grids [8]. In the two-dimensional case, the grid is a set of points located on parallel straight lines. At each time step a new grid adapted to the solution and independent of the old grid is constructed. In doing so, the number and the relative position of the grid lines and the grid points on those lines can change.

In creating a grid, a monitoring function $M(x, y, t)$ defined as

$$M(x,y,t) = \left(1 + \alpha\left(f_x^2 + f_y^2\right)\right)^{1/2} + \beta\left(|f_{xx}| + |f_{yy}|\right)$$

is first calculated. Here f is some parameter of the solution, α, β are some positive coefficients. The lines and the points of the grid are condensed in the zones with large values of the monitoring function and are rarefied where the values of the function are small. In the finite-difference approximations to the derivatives, the necessary values of the parameters at the points of the stencil are determined by CIP interpolation [8].

3. Impact of a liquid cone and a liquid wedge on a wall

To estimate the efficiency of a numerical technique of computing the jet impact on a solid surface it is possible to use a model problem of a liquid cone impact on a wall. For this purpose the case of impact of a liquid blunt cone on a plane wall is considered in the present paper (Fig. 1*a*). The slope of the cone surface to the wall is sufficiently small so that a shockwave with the edge attached to the wall is formed in the liquid. The circular area $r \le R_L$ ($R_L = Vt \operatorname{ctg} \theta$, V is the impact velocity, θ is the angle of the slope of the cone surface, r is the distance to the axis of symmetry) of contact of the liquid and the wall quickly expands. The sideway spreading of the liquid on the wall does not arise.

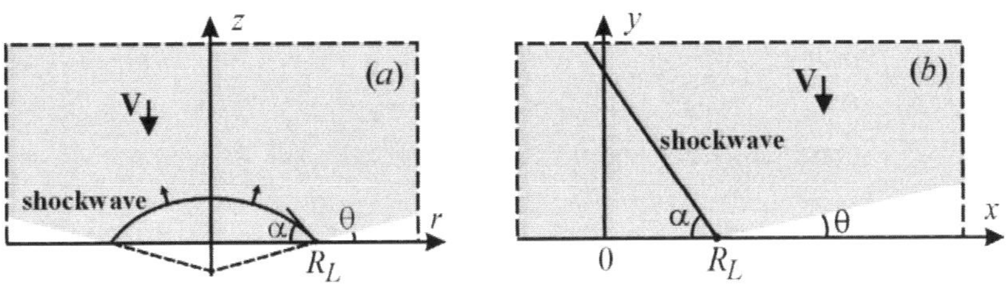

Figure 1. The schematic of the impact of a liquid blunt cone (*a*) and the plane semi-infinite liquid flux with the same slope of the boundary (*b*) on the horizontal fixed wall.

Following [9], we estimate the value of the pressure of the compressed liquid in the vicinity of the shockwave edge ($r \approx R_L$) in the problem of the liquid cone impact (Fig. 1*a*) by the value of the pressure of the everywhere-equally compressed liquid behind the shockwave in the problem of impact of a plane semi-infinite liquid flux with the same liquid velocity and the same slope of the flux boundary (Fig. 1*b*). In the latter problem, the pressure of the compressed liquid behind the shockwave is determined by

$$p = p_0 - \rho_{L0}\frac{V^2}{\sin^2\theta}\sin(\alpha+\theta)\left[\sin(\alpha+\theta) - \cos(\alpha+\theta)\,\mathrm{tg}\,\alpha\right], \qquad (5)$$

where the shockwave slope angle α is found from the equation

$$\mathrm{tg}^3(\alpha+\theta)\left(\frac{\Gamma-1}{2}M_1^2+1\right) - \theta\left(M_1^2-1\right)\mathrm{tg}^2(\alpha+\theta) + \left(\frac{\Gamma+1}{2}M_1^2+1\right)\mathrm{tg}(\alpha+\theta) + \mathrm{ctg}\,\theta = 0\,.$$

Here Γ is the constant of the Tait equation of the liquid state, $M_1 = V/(C\sin\theta)$, p_0, ρ_{L0}, C are the pressure, the density, and the sound speed in the undisturbed liquid.

Some differences between the values of the pressure of the compressed liquid in the vicinity of the shockwave edge in the former problem and that behind the shockwave in the latter problem can be caused by the influence of the axial symmetry. To estimate this influence the numerical solution of the similar plane problem of the liquid wedge impact is utilized.

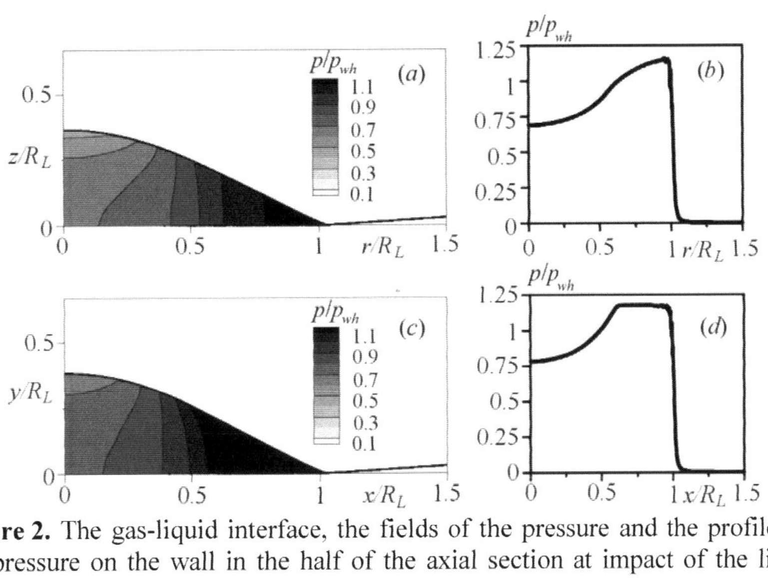

Figure 2. The gas-liquid interface, the fields of the pressure and the profiles of the pressure on the wall in the half of the axial section at impact of the liquid cone (*a*, *b*) and the liquid wedge (*c*, *d*), $\theta = 0.06$, $V = 250$ m/s ($p_{wh} = 5$ kbar).

Typical numerical results on impact of the liquid cone and the liquid wedge on the wall in case of small values of θ are illustrated in Fig. 2. The liquid ($\Gamma_l = 7.15$, $B_l = 3072$ bar) is assumed to be surrounded by a gas (γ=1.4). At the initial moment $\varphi = 1$, $\rho = 10^3$ kg/m^3 in the liquid, $\varphi = 0$, $\rho = 1$ kg/m^3 in the gas, and $p = 1$ bar everywhere. In Fig. 2 the pressure p is referred to the waterhammer pressure p_{wh}, which can be estimated approximately as $p_{wh} = \rho_{L0}DV$ where D is the shockwave velocity, $D \approx C + 2V$ for water [9]. One can see in Fig. 2 that the main features of the impact (the geometry of the shockwave front, the distribution and the maximum and minimum levels of the compressed liquid pressure) are almost infependent of the axial symmetry. At the same time, in the presence of the axial symmetry the distribution of the liquid pressure on the wall is everywhere nonuniform with the clear maximum at the imact area edge $r = R_L$, while in the absence of the axial symmetry this distibution has, near the imact area edge $x = R_L$, quite a large plateau at the level of the maximum pressure.

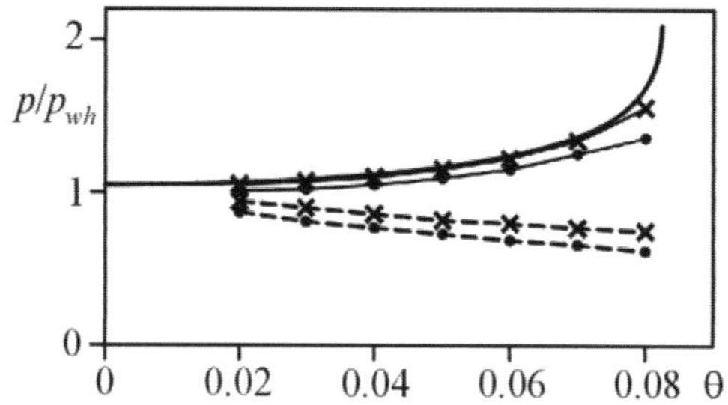

Figure 3. Dependences of the liquid pressure on the wall on the slope angle θ in the cases of impact by (i) the plane semi-infinite liquid flux (curve without symbols, calculated by (5)), (ii) the liquid cone (curves with •), and (iii) the liquid wedge (curves with ×). The dashed (solid) curves with symbols correspond to the centers $r = 0$, $x = 0$ (edges $r \approx R_L$, $x \approx R_L$) of the loaded area. In the case (i) the liquid pressure behind the shockwave is everywhere the same.

According to Fig. 3, for the small slope angles $\theta \leq 0.07$ the pressure values, computed by the technique of the present work in the case of impact by the liquid wedge, in the vicinity of the shockwave edge $x = R_L$ are in very good agreement with those, calculated by (5), behind the shockwave in the case of impact by the plane semi-infinite liquid flux. The influence of the axial symmetry manifests itself in only slightly reducing the wall pressure both in the center and at the edge of the loaded area.

4. Impact of a jet with hemispherical end on a liquid layer on a wall

Impact of an axisymmetric liquid ($\Gamma_l = 7.15$, $B_l = 3072$ bar) jet (surrounded by the air, $\gamma = 1.4$) with the hemispherical end on a rigid wall covered by a thin uniform layer of the same liquid is considered (Fig. 4*a*). Initially $\varphi = 1$, $\rho = 10^3$ kg/m^3 in the liquid, $\varphi = 0$, $\rho = 1$ kg/m^3, in the gas, and $p = 1$ bar everywhere. The jet velocity and radius are $V = 350$ m/s and $R = 10$ μm, the thickness of the liquid layer is equal to 0.04 R.

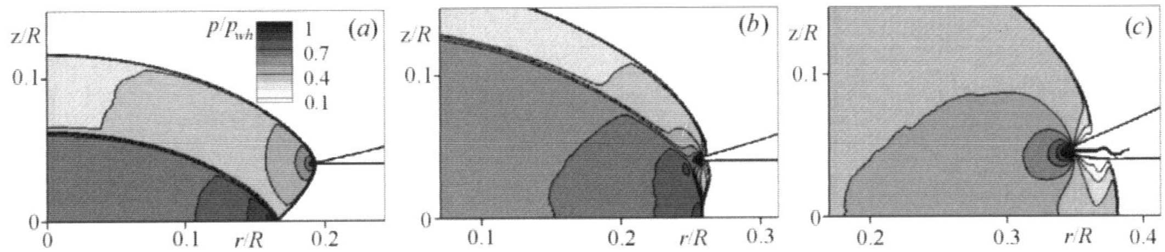

Figure 4. Impact of a liquid jet on a thin liquid layer on a wall: the gas-liquid interface, the fields of pressure in the half of the axial section in the vicinity of the jet end, p_{wh} = 7.7 kbar.

Liquid jet impact on a liquid layer on a wall results in appearance of two shockwaves, the pressure at the place of impact being approximately equal to half of the waterhammer pressure. One of the shockwaves propagates up the jet, while the other first moves in the layer to the wall and then reflects from it (Fig. 4*a*). Interaction of the second shockwave with the wall is divided into the stages of regular (Fig. 4*a*) and irregular (Fig. 4*b*) reflection. At the stage of regular reflection and in the beginning of the stage of irregular reflection the maximum pressure on the wall is attained at the edge of the reflection area. The levels of that maximum pressure are comparable with the waterhammer pressure. At the stage of irregular reflection the fronts of the oncoming and reflected shockwaves merge to produce the resultant shockwave in the vicinity of the wall. The resultant shockwave begins to propagate in the liquid layer and interact with its surface. The interaction leads to creation of a rarefaction wave which soon acts on the wall (Fig. 4*c*). A thin ring-like liquid splash arises in the space between the jet and the layer (Fig. 4*c*).

5. Conclusion

Main points of a numerical technique for computing problems of the high-speed liquid jet impact on a rigid wall are presented. In the technique, the liquid and gas flows are governed by the gas dynamics equations in the density, velocity and pressure. Their numerical integration is carried out by the CIP-CUP method with application of dynamically adaptive grids, the gas-liquid interface is calculated without explicit tracking.

To estimate the efficiency of the technique problems of impact on a wall of a liquid cone and a liquid wedge in the mode with the shockwave attached to the wall by their edges are considered. The pressure values computed by the presented technique are found to be in the vicinity of the edge of the shockwave attached to the wall in good agreement with the analytically determined pressure of the compressed liquid in a similar problem of impact on the pressure wall of a flat semi-infinite flux. The influence of the axial symmetry of the liquid cone impact on the liquid pressure in the vicinity of the shockwave edge is shown to be small.

Applicability of the presented technique to numerically simulating the high-speed liquid jet impact on a wall is illustrated by computing impact of a cylindrical liquid jet with the hemispherical end on a wall covered by a layer of the same liquid.

References

[1] Bowden F P and Brunton J H 1961 *Proc. R. Soc. Lond. A* **263** 433–50
[2] Bourne N K 2005 *Proc. R. Soc. Lond. A* **461** 1129–45
[3] Kornfeld M and Suvorov L 1944 *J. Appl. Phys.* **15** 495–506
[4] Hawker N A and Ventikos Y 2012 *J. Fluid Mech.* **701** 59–97
[5] Yabe T, Xiao F and Utsumi T 2001 *J. Comput. Phys.* **169** 556–93

[6] Yabe T and Wang PY 1991 *J. Phys. Soc. Japan* **60** 2105–08
[7] Abgrall R and Karni S 2001 *J. Comp. Phys.* 1**69** 594–623
[8] Takizawa K, Yabe T, Tsugawa Y, Tezduyar T E and Mizoe H 2007 *Comput. Mech.* **40** 167–83
[9] Heymann F J 1969 *J. Appl. Phys.* **40** 5113–22

One algorithm for branch and bound method for solving concave optimization problem

A A Andrianova[1], A A Korepanova[2] and I F Halilova[2]

[1]Department of System Analysis and Information Technologies, Kazan (Volga Region) Federal University, Kazan, Kremlevskaya st. 18, Russia
[2]Master program Fundamental Informatics and Information Technologies, Kazan (Volga Region) Federal University, Kazan, Kremlevskaya st. 18, Russia

E-mail: Anastasiya.Andrianova@kpfu.ru

Abstract. The article describes the algorithm for branch and bound method for solving the concave programming problem, which is based on the idea of similarity the necessary and sufficient conditions of optimum for the original problem and for a convex programming problem with another feasible set and reverse the sign of the objective function. To find the feasible set of the equivalent convex programming problem we construct an algorithm using the idea of the branch and bound method. We formulate various branching techniques and discusses the construction of the lower objective function evaluations for the node of the decision tree. The article discusses the results of experiments of this algorithm for some famous test problems of a particular form.

1. Introduction

The concave programming problem has many applications in various fields of human activity (medicine, biology, technology, economics). Very often a model of concave programming problem encountered in the decision theory, concave programming problems applied for solving equilibrium problems, complementarity problems and others. Thus, it can be stated a serious interest in the construction an effective methods [1-4].

The main methods for the concave programming problem are based on the ideas of the cutting-plane methods, element methods and techniques of branches and borders based on the decomposition of the admissible set. However, most of these methods do not provide a solution in a reasonable time for practical problems [1,4].

In this paper we propose a different approach of branch and border method. The idea of this approach is based on the theoretical result - the similarity of the necessary and sufficient conditions for optimum some convex programming problem and original concave programming problem. Thus, the branch and bound method is only necessary in order to build this equivalent auxiliary convex programming problem. This theoretical result for the concave programming problem with linear constraints has been proposed and substantiated in [5,6].

So, we suggest a generalization of this theoretical result to the case of the feasible set defined by a system of inequalities with concave functions, formulate a general scheme of branch and bound method for solving the concave programming problem, an one algorithm for the special form of a concave programming problem and shows the results of experiments on a series of well-known test problems [7].

2. Theoretical aspects

Suppose that in Euclidean space R_n we consider the problem

$$\min_{\boldsymbol{x}\in D} \to f(\boldsymbol{x}), \tag{1}$$

where $D=\{\boldsymbol{x}\in R_n \mid f_i(\boldsymbol{x})\le 0, i\in I\}$, $I=\{1,2...m\}$, $f(\cdot), f_i(\cdot)$ $i\in I$ are smooth concave functions. As known, this problem can be several local minima. The global minimum is among them.

We formulate the necessary conditions for a local minimum the problem (1) known as Lagrange principle (for example, [8]). Each point which satisfies the necessary conditions for a local minimum is called a stationary point.

Theorem 1. *If the point* $\boldsymbol{x}^*\in D$ *is a local minimum for problem (1), then there are exists numbers* $y_i^*\ge 0$, $i=0,1...m$, *such that:*

$$y_0^* f(\boldsymbol{x})+\sum_{i=1}^{m} y_i^* f_i(\boldsymbol{x}^*)=\boldsymbol{0}, \tag{2}$$

$$y_i^* f_i(\boldsymbol{x}^*)=0, \quad i=1...m. \tag{3}$$

It is obvious that the conditions (2) - (3) are equivalent to the following condition

$$y_0^* f(\boldsymbol{x})+\sum_{i\in I(x^*)} y_i^* f_i(\boldsymbol{x}^*)=\boldsymbol{0}, \tag{4}$$

where $I(\boldsymbol{x}^*)=\{\mathrm{i}\in \mathrm{I} \mid f_i(\boldsymbol{x}^*)=0\}$ is the active indices set in point $\boldsymbol{x}^*$. Note that finding the number y_0^* can be neglected, if the problem (1) satisfies the regularity conditions: gradients of functions in stationary point $\left\{f_i'(\boldsymbol{x}^*)\right\}_{i=1}^{m}$ is a linearly independent system..

Choose a subset of indices $I_0\subset I$. Define an auxiliary convex programming problem

$$\max_{\boldsymbol{x}\in D_0} \to f(\boldsymbol{x}), \tag{5}$$

where $D_0=\{\boldsymbol{x}\in R_n \mid f_i(\boldsymbol{x})\ge 0, i\in I_0\}$.

Theorem 2. *The solution of problem (1) is the solution of problem (5) with* $I_0=I(\boldsymbol{x}^*)$.

Prof. The problem (5) is a convex programming problem, therefore, any of its local minimum points is a global minimum point. For convex programming problem stationary point conditions are also sufficient conditions for the global optimum ([8]):

$$y_0^* f(\boldsymbol{x}^*)+\sum_{i\in I_0} y_i^* f_i(\boldsymbol{x}^*)=\boldsymbol{0}, \tag{6}$$

$$y_i^* f_i(\boldsymbol{x}^*)=0, \quad i\in I_0. \tag{7}$$

So, for $I_0=I(\boldsymbol{x}^*)$ conditions (2)-(3) and (6)-(7) are the same. It is obvious that any local minimum point of problem (1) is a solution some problem of form (5). Conversely, if the point $\boldsymbol{x}^*$ solves the problem (5) and $\boldsymbol{x}^*\in D$, then $\boldsymbol{x}^*$ is the stationary point of the problem (1).

Consequently, the solution of the problem (1) can be found as a point of local minimum obtained as the solution of the problem (5) for some index set, with a minimum value of the objective function. Therefore, it can be found by trying all possible subsets of indices and solution for their auxiliary problem (5).

It should be noted that the problem (5) for some subset $I_0 \subset I$ may have empty feasible set. Consequently, the method of solving the auxiliary problem should have a way to determine this fact.

An effective way to solve the auxiliary problem (5) can be a solution of its dual problem:

$$\min_{y \in Y} \to \varphi(\boldsymbol{y}), \tag{8}$$

where objective function defines as $\varphi(\boldsymbol{y}) = \sup_{x \in R_n} L(\boldsymbol{x}, \boldsymbol{y})$ with $L(\boldsymbol{x}, \boldsymbol{y}) = y_0 f(\boldsymbol{x}) + \sum_{i \in I_0} y_i f_i(\boldsymbol{x})$ and feasible set has the form $Y = \{\boldsymbol{y} \in R_{m+1} \mid y_i \geq 0, i \in I_0, y_i = 0, i \notin I_0\}$. A simple form of this set may allow the use of less sophisticated computational methods for solving the problem (8). In addition, by the solving of the dual problem (8) we can conclude about the incompatibility of the restrictions of problem (5): if objective function of problem (8) is unbounded from bottom in its feasible set, then the problem (5) has no solutions.

3. The general scheme of branch and bound algorithm

So, as noted in the previous section, the solution of problem (1) can be found by searching all possible subsets $I_0 \subseteq I$. Thus, it is possible to organize this search with using the branch and bound method is constructed as follows.

General scheme. We denote $I_0 = \varnothing$, $f_{rec} = +\infty$. In each node of the decision tree following actions are performed:

1. The convex programming problem (5) is solved (for example, using the dual problem (8)). Note this solution as $\boldsymbol{x}'$. If found that the feasible set of the problem (5) is empty, the branching is not performed and returns to the parent node of the decision tree. Note that in the root node of the decition tree point $\boldsymbol{x}'$ will be the point of the absolute maximum of the function $f(\cdot)$ if such maximum exists.
2. Calculate the lower bound evaluation f'_{low} of objective function for node and all its child subtree.
3. If $f'_{low} > f_{rec}$, the branching is not performed and returns to the parent node of the decision tree.
4. If $\boldsymbol{x}' \in D$, then $\boldsymbol{x}'$ is the stationary point pf the problem (1). If $f_{rec} > f(\boldsymbol{x}')$, then $f_{rec} = f(\boldsymbol{x}')$ and we memorize a point $\boldsymbol{x}'$ as a possible solution of problem (1).
5. We perform the branching. Choosing indexes are not yet included in the feasible set $J = \{i \mid i \in I, i \notin I_0\}$. For each index $j \in J$ we construct child node with $I_0 = I_0 \cup \{j\}$. So, a count of child nodes for current neode is equal to the cardinality of the set of indices J. The order of enumeration indices form set J is not significant.

Note that the general scheme is principled. To implement it, we need a way for solving the problem (5), the ability to determine that the feasible set is inconsistent, the method of calculating the lower bound evaluation of the objective function for the node and its child subtree. The process of branching as described in step 5 is very general and does not preclude re-examination of the same set under a different order enabling the restrictions in it. However, this method guarantees branching viewing subsets of this.

We should also discuss the construction of the lower bound evaluation of the objective function for the node. Traditionally, for construction of the branch and bound methods is expected that the optimum value of the auxiliary problem solved in a node can be used as an evaluation for the node. But in this case it is not.

Indeed, let the auxiliary convex programming problem is solved with its dual problem. Let the fixed node of the tree is constructed for a certain set of indices $I_1 \neq \varnothing$ and to branching some index $t \notin I_1$ is added. So, $I_2 = I_1 \cup \{t\}$ and we have two duals problems $\min\limits_{y \in Y_1} \varphi_1(\boldsymbol{y})$ and $\min\limits_{y \in Y_2} \varphi_2(\boldsymbol{y})$ where

$$\varphi_1(\boldsymbol{y}) = \sup_{x \in R_n} L_1(\boldsymbol{x}, \boldsymbol{y}),\ \varphi_2(\boldsymbol{y}) = \sup_{x \in R_n} L_2(\boldsymbol{x}, \boldsymbol{y}),$$

$$L_1(\boldsymbol{x}, \boldsymbol{y}) = y_0 f(\boldsymbol{x}) + \sum_{i \in I_1} y_i f_i(\boldsymbol{x}),$$

$$L_2(\boldsymbol{x}, \boldsymbol{y}) = y_0 f(\boldsymbol{x}) + \sum_{i \in I_2} y_i f_i(\boldsymbol{x}) = y_0 f(\boldsymbol{x}) + \sum_{i \in I_1} y_i f_i(\boldsymbol{x}) + y_t f_t(\boldsymbol{x}),$$

$$Y_1 = \{\boldsymbol{y} \in R_{m+1} \mid y_i \geq 0, i \in I_1, y_i = 0, i \notin I_1\},\ Y_2 = \{\boldsymbol{y} \in R_{m+1} \mid y_i \geq 0, i \in I_2, y_i = 0, i \notin I_2\}.$$

Obviously, $Y_1 \subseteq Y_2$. So, for $y \in Y_1$ we have $\varphi_1(\boldsymbol{y}) = \sup\limits_{x \in R_n} L(\boldsymbol{x}, \boldsymbol{y})$ and for $y \in Y_2$ - $\varphi_2(\boldsymbol{y}) = \sup\limits_{x \in R_n} L(\boldsymbol{x}, \boldsymbol{y})$, where the Lagrangian $L(\boldsymbol{x}, \boldsymbol{y}) = y_o f(\boldsymbol{x}) + \sum\limits_{i \in I} y_1 f_1(\boldsymbol{x})$ is constructed for full set of indeces of the problem (1). Consequently, $\min\limits_{y \in Y_2} \varphi_2(\boldsymbol{y}) \leq \min\limits_{y \in Y_1} \varphi_1(\boldsymbol{y})$. The same happens with the values of the objective function of the primal problem (5). Thus, the addition of new constraint can reduce the objective function value at the optimum auxiliary problem in the node.

4. An algorithm of branch and bound method

Consider the simplest form of the problem (1):

$$\min_{x \in D} \rightarrow f(\boldsymbol{x}) \tag{9}$$

where $f(\boldsymbol{x}) = 0.5\langle Q\boldsymbol{x}, \boldsymbol{x}\rangle + \langle d, \boldsymbol{x}\rangle$ - quadratic function with symmetric negative definite $n \times n$ matrix Q, $d \in R_n$ and feasible set has a form $D = \{\boldsymbol{x} \in R_n \mid \alpha_i \leq x_i \leq \beta_i, i = 1..n\}$. It should be noted that despite its simplicity, this type of problem is computationally difficult to solve, because it has 2^n local minima. Nevertheless, this type of problem allows us to formulate specific rules for the branch and bound method.

1. An easy way to solve the auxiliary problem (5) in the node tree. The dual problem to (5) also has the form of quadratic programming problem and there is an explicit formula for obtaining the solution of the direct problem (5) $\boldsymbol{x}'$ with solution of dual problem $\boldsymbol{y}'$: $\boldsymbol{x}' = -Q^{-1}[d + A^{\mathrm{T}}\boldsymbol{y}']$ ([5]).
2. Another rule for branching. The decision tree will be in binary tree form. In each node of this tree we choose the next variable for branching, for example x_t. In one of child nodes we solve the problem with additional constraint $x_t \leq \alpha_t$ and in another child node - $x_t \geq \beta_t$. So, in the step 5 of general scheme we choose a variable, but not constraint. Moreover, because the variable is not found in other limitations, there is no possibility of constructing a problem (5) incompatible. So, step 1 of general scheme will be easier.
3. The rule for calculate of lower bound evaluation. This rule is based on using known bound values of each variable. Let we have the solution of problem (5) $\boldsymbol{x}'$ in some node Write the objective function:

$$f(\boldsymbol{x}) = 0.5\sum_{i=1}^{n}\sum_{j=1}^{n} q_{ij} x_i x_j + \sum_{i=1}^{n} b_i x_i$$

It can be seen that the lower value of the function in stationary points obtained in the child subtree, we can estimate the sum of lower bound evaluation of each term. Thus, the lower bound evaluation is the same sum, in which coordinates uncommitted variables are replaced with the bound values that give a minimum term. For example, if the only uncommitted variable x_t evaluation will be as follows:

$$f_{low} = 0.5\sum_{i=1,i\neq t}^{n}\sum_{j=1,j\neq t}^{n} q_{ij}x_i'x_j' + \sum_{i=1,i\neq t}^{n} b_i x_i' +$$

$$+\sum_{j=1,j\neq t}^{n} \min\{q_{tj}x_j'\alpha_t, q_{tj}x_j'\beta_t\} + 0.5\min\{q_{tt}\alpha_t^2, q_{tt}\beta_t^2\} + \min\{b_t\alpha_t, b_t\beta_t\}.$$

5. Results of experiments

We provide the experiment on some known test problems in form (9) which often used in research of concave optimization methods [7]:

Type 1. Q is diagonal matrix, $d = \mathbf{0}$, $q_{jj} = -(n-1-0.1j)$, $\alpha_j = -1-j$, $\beta_j = 1+5j$, $j = 1..n$;

Type 2. Q is diagonal matrix, $d_j = 1$, $q_{jj} = -1$, $\alpha_j = -j$, $\beta_j = [n/j]$, $j = 1..n$;

Type 3. Q is diagonal matrix, $d = \mathbf{0}$, $q_{jj} = -1$, $\alpha_j = -(n-j+1)$, $\beta_j = n + j/2$, $j = 1..n$;

Тип 4. Q is diagonal matrix, $d_j = -j$, $q_{jj} = -1$, $\alpha_j = 1-j$, $\beta_j = 2j$, $j = 1..n$.

Each problem is solved with $n = 3...12$. In solving was measured the time of operation process, the percentage of nodes in the tree, in which auxiliary convex programming problem (5) did not solved ($P_{nosolve}$), the percentage of nodes in obtaining the global optimum (P_{rec}).The Table 1 will be given the maximum, minimum and average values of parameters $P_{nosolve}$ and P_{rec} for each type of problem.

We considered several rules for selecting the order of the variables:

Rule 1. Direct order variables, i.e. first fixed variable number is $i = 1$, then – with number $i = 2$. The last variable will be with number $i = n$. So, the left child node is corresponded the constraint $x_i \le \alpha_i$ and the right child node – $x_i \ge \beta_i$.

Rule 2. Reverse order variables, i.e. first fixed variable number is $i = n$, then – with number $i = n-1$. The lasr variable will be with number $i = 1$. So, the left child node is corresponded the constraint $x_i \ge \beta_i$ and the right child node – $x_i \le \alpha_i$.

Rule 3. More intellectual case based on the analysis of the deviation of the boundary values of each variable from its value at the absolute maximum point of function $f(\cdot)$. Note x_{abs} is the absolute maximum point of function $f(\cdot)$. The order for variables is defined by sorted order of values $\rho_i = \max\{|x^i_{abs} - \alpha_i|, |x^i_{abs} - \beta_i|\}$. In this case left child node will be corresponded to those constraint $x_i \le \alpha_i$ or $x_i \ge \beta_i$, which bound has maximum deviation of value of i variable form absolute maximum point x^i_{abs}.

A detailed analysis of the overall process time showed the presence of proportion to the percentage of nodes in the tree, in which auxiliary convex programming problem (5) are solved $(1 - P_{nosolve})$. In general, the calculation time ranged from a few seconds for $n = 3..6$ up to 8 hours at $n = 12$. The sharp increase in overall process time observed since $n = 9$. Obviously, this was due to an increase in the computational complexity of solving the auxiliary convex programming problem in the tree nodes.

The results of experiments were carried out in the Table 1. So, we make the following conclusions on the basis of the experiment:

1. The solution of Type 1 and Type 4 problems is primarily for upper bounds of variables. So, value of P_{rec} for Rule 1 is large, while for Rule 2 it is sufficiently small. Thus, in this case, Rule 2 of the order of enumeration variables is preferable.
2. The solution of Type 2 problems is at the lower boundaries of variables. Therefore, Rule 1 of the order of enumeration variables is preferable.
3. The Rule 1 and Rule 2 of the order of enumeration variables do not account for problems specific. Therefore, the values P_{rec} are very different depending on the type of problem. When using Rule 3 is visible that indicator P_{rec} is stabilized regardless of the type of problem. On average for the optimal solution of problem (1) is sufficient to look a little less than 10% of the decision tree.
4. Dynamics of the indicator $P_{nosolve}$ does not allow to make qualitative conclusions about the effectiveness of Rule 3. For example, in the problems of Type 2 and Type 4, percentage of nodes in the tree, in which auxiliary convex programming problem (5) is not solved, is large enough, and for problems of Type 3 - is very small. This is due to low lower bound evaluation.

Table 1. A results of experiments.

	$P_{nosolve}$			P_{rec}		
	min	avg	max	min	avg	max
Type 1:						
Rule 1	0	0,04393	0,0967	0,8895	0,9636	1
Rule 2	0,258	0,3851	0,4211	0,0053	0,0803	0,0952
Rule 3	0,258	0,3851	0,4211	0,0053	0,0803	0,0952
Type 2:						
Rule 1	0,2666	0,798	0,988	0,0053	0,108	0,266
Rule 2	0	0	0	0,933	0,9838	1
Rule 3	0,5333	0,7251	0,8725	0,0053	0,0847	0,266
Type 3:						
Rule 1	0	0,0119	0,0293	0,4904	0,5333	0,5056
Rule 2	0,1333	0,165	0,1935	0,0097	0,091	0,266
Rule 3	0,1333	0,1502	0,18	0,0097	0,091	0,266
Type 4:						
Rule 1	0	0	0	1	1	1
Rule 2	0,4	0,729	0,6132	0,0097	0,091	0,266
Rule 3	0,4	0,729	0,6132	0,0097	0,091	0,266

6. Conclusions

In general, the experiments leads to the conclusion that the approach used in the article in the construction algorithm of branch and bound method, can be quite competitive in solving practical problems of concave programming. Prospects for the development of this method are related to the construction of good lower bounds in the tree nodes and the possible use of parallel computing.

References

[1] Strekalovsky A S 2003 *Principles of Nonconvex Optimization* (Novosibirsk, Nauka)

[2] Strekalovsky A S 2014 Modern Methods for Solving Nonconvex Optimal Control Problems *IIGU Ser. Matematika* **8** 141–163

[3] Gruzdeva T V, Strekalovsky A S, Orlov A V, Druzinina O V 2011 Nonsmooth minimization problems for the difference of two convex functions *Vychisl.Metody Programm* **12(4)** 384-396

[4] Audet C, Hansen P, Savard G 2005 *Essays and Surveys in Global Optimization* (New York, Springer) 312

[5] Konnov I V 2010 Sign reversion approach to concave minimization problems *Optim.Lett.* **4** 491-500

[6] Andrianova A A, Konnov I V 2014 The branch and bound method for concave optimization problem in *Problem of Theoretical Cybernetics* (Kazan, Otechestvo) 23-26

[7] Chinchuluun A, Pardalos P M, Enkhbat R 2005 *Global minimization algorithms for concave quadratic programming problems* Optimization **54** 627-639

[8] Sukharev A G, Timokhov A V, Fedorov V V 2005 *A course in optimization methods* (Moscow, Nauka) chapter 4 147-160

About decomposition approach for solving the classification problem

A A Andrianova

Department of System Analysis and Information Technologies, Kazan (Volga Region) Federal University, Kazan, Kremlevskaya st. 18, Russia

E-mail: Anastasiya.Andrianova@kpfu.ru

Abstract. This article describes the features of the application of an algorithm with using of decomposition methods for solving the binary classification problem of constructing a linear classifier based on Support Vector Machine method. Application of decomposition reduces the volume of calculations, in particular, due to the emerging possibilities to build parallel versions of the algorithm, which is a very important advantage for the solution of problems with big data. The analysis of the results of computational experiments conducted using the decomposition approach. The experiment use known data set for binary classification problem.

1. Introduction

The classification problem is one of the most popular big data problems. Often their solution reduces to optimization problems. However, the traditional computational approaches for solving optimization problems is not effective because of the large number of variables and constraints. For such problems, the calculation itself is gradient of function is a time-consuming task. Even worse is placed when the methods of the second order are used. Therefore, decomposition approach for solving such problems is promising, as it allows to reduce the solution to a simple computational procedures, as well as makes it possible to build distributed and parallel algorithms for solving big data problems.

In [1] was considered decomposition approach to the construction of coordinate descent algorithms for problem $\min_{x \in X} \mu(\boldsymbol{x})$ where objective function define as sum $\mu(\boldsymbol{x}) = f(\boldsymbol{x}) + h(\boldsymbol{x})$, $f(\cdot)$ is a smooth, but not necessary convex function in Euclidean space R_n, $h(\cdot)$ is convex, but not necessary smooth function what is more $h(\boldsymbol{x}) = \sum_{i=1}^{m} h_i(\boldsymbol{x}^i)$, $X = X_1 \times X_2 \times \ldots \times X_m$, $\boldsymbol{x}^i \in X_i$ $i = 1..m$. The partition of vector $\boldsymbol{x} \in R_n$ into components $\{\boldsymbol{x}^i\}_{i=1}^m$ determines the possibility of using the decomposition approach.

In this paper, the algorithm will be reformulated for the case of classification problem solving method of Support Vector Machine and held on the known problems of the analysis of computational experiment ([2]).

2. Algorithm for Support Vector Machine method with decomposition

Let us formulate the optimization problem of Support Vector Machines for construction of linear binary classifier ([3]).

Suppose we have a training set of $K>0$ examples $(\boldsymbol{x}_i, y_i)$ $i=1..K$ of two known classes. There $\boldsymbol{x}_i \in R_L$ - L-dimensional vector characteristics, $y_i \in \{-1;1\}$ is a class label. The binary classification problem is a problem of constructing a linear classifier as a hyperplane $\langle \boldsymbol{w}, \boldsymbol{x} \rangle = 1$ where $\boldsymbol{w} \in R_L$.

Traditionally ([3]) Support Vector Machine method determines the separating hyperplane as a solution to the following optimization problem:

$$\min_{w} \to 0.5\|\boldsymbol{w}\|^2 + C\sum_{i=1}^{K}\xi_i \tag{1}$$

with constraints

$$y_i\langle \boldsymbol{w}, \boldsymbol{x}_i \rangle \geq 1-\xi_i, \quad i=1..K \tag{2}$$

$$\xi_i \geq 0, \quad i=1..K. \tag{3}$$

The variables ξ_i $i=1..\mathrm{K}$ allows to determine the penalty for the error in determining the well-known i-th example to another class, $C>0$ - fixed constant.

Instead, the problem (1) - (3) usually its dual problem solves:

$$\max_{0\leq\lambda_i\leq C,\, i=1..K} \to \sum_{i=1}^{K}\lambda_i - 0.5\sum_{s=1}^{K}\sum_{t=1}^{K}\lambda_s\lambda_t y_s y_t \langle \boldsymbol{x}_s, \boldsymbol{x}_t \rangle, \tag{4}$$

where λ_i, $i=1..\mathrm{K}$ are dual variables, whereby their optimum values the vector coefficients hyperplane $\boldsymbol{w}$ can be obtained:

$$\boldsymbol{w} = \sum_{i=1}^{K}\lambda_i y_i \boldsymbol{x}_i. \tag{5}$$

Thus, we can solve instead of the problem (1) - (3) with a large number of constraints its dual problem (4) a large number of variables, but with simple constraints, and then to obtain the solution of primary problem (1)-(3) using formula (5).

The problem (4) has a form suitable for decomposition. The set $X=\{\Lambda=(\lambda_1,\lambda_2...\lambda_K)\in R_K \mid 0\leq\lambda_i\leq C, i=1..K\}$ allows decomposed of the vector Λ on $m>0$ disjoint groups of variables Λ^i $i=1..m$ and their corresponding sets of X_i. Thus, we have

$$h(\Lambda) = -\sum_{i=1}^{K}\lambda_i, \qquad f(\Lambda) = -0.5\sum_{i=1}^{K}\sum_{j=1}^{K}\lambda_i\lambda_j y_i y_j < \boldsymbol{x}_i, \boldsymbol{x}_j >.$$

In paper [1] it was proved that the solution of problem решение задачи (4) Λ^* is a solution a mixed variational inequality $\sum_{i=1}^{m}\langle \boldsymbol{g}^i(\Lambda^{i*}), \boldsymbol{Y}^i - \Lambda^{i*}\rangle + \sum_{i=1}^{m}\left[h_i(\boldsymbol{Y}^i) - h_i(\Lambda^{i*})\right] \geq 0$ for all vector $\boldsymbol{Y}^i \in X_i$, $i=1..m$. Here $\boldsymbol{g}^i(\Lambda^i)$ is a vector with components of the gradient of the function $f(\cdot)$ corresponding group of variables Λ^i.

After apply the methods for mixed variational inequalities problem we can using as a the direction of decrease of the function $\mu(\Lambda) = f(\Lambda) + h(\Lambda)$ in point Λ the vector $\boldsymbol{d} \in R_K$ which components are follows:

$$d_i = \begin{cases} Y_i(\Lambda) - \Lambda_i, & \text{при } i = i_0 \\ 0, & \text{при } i \neq i_0 \end{cases}, \tag{6}$$

Here $\boldsymbol{Y}(\Lambda)$ is a solution with a predetermined Λ for a problem:

$$\min_{\Lambda^i\in X_i, i=1..m} \to \sum_{i=1}^{m}\left(\langle \boldsymbol{g}^i(\Lambda), \boldsymbol{Y}^i \rangle + 0.5\alpha^{-1}\left\|\Lambda^i - \boldsymbol{Y}^i\right\|^2 + h_i(\boldsymbol{Y}^i)\right). \tag{7}$$

Obviously, the problem (7) can be decomposed into the m problems of follow form:

$$\min_{\Lambda^i \in X_i} \rightarrow \left\langle \boldsymbol{g}^i(\Lambda), \boldsymbol{Y}^i \right\rangle + 0.5\alpha^{-1} \left\| \Lambda^i - \boldsymbol{Y}^i \right\|^2 + h_i(\boldsymbol{Y}^i) \quad i = 1..m. \tag{8}$$

These problems may be solved independently including their parallel solution.

For determining the direction of decrease of function $\mu(\Lambda)$ according formula (6) the index i_0 should also be specified. For this index the following condition must be performed:

$$\left| Y_{i_0}(\Lambda) - \Lambda_{i_0} \right| > \delta . \tag{9}$$

Here $\delta > 0$ is the parameter that provides a choice of reasonably good direction of decrease. If such an index does not exist, the value of parameter decreases (for example, values of parameter δ we can choose from sequence $\{\delta_i\}$: $\lim_{i\to\infty} \delta_i = 0$).

The search of index i_0 can be carried out from sequentially view the decomposed set of variable (we call it decomposed block). The origin of this search can be random. Then we must to solve for decomposed block the problem (8) and to check the condition (9). After obtaining the index i_0 all remaining unsolved problems (8) for other decomposed blocks will not be further considered.

Then we must calculate the step $\tau > O$ on the found direction $\boldsymbol{d} \in R_K$. In paper [1] it was proved that is sufficient to provide a reduce of the objective function is not less than an amount proportional to the $\left| Y_{i_0}(\Lambda) - \Lambda_{i_0} \right|^2$. Such step can be easily found using finite procedures. Note that the increase of value $\mu(\Lambda)$ is possible only at i_0 variable component from vector Λ. Therefore, the step on descent direction can be found if minimized only those terms of function $\mu(\Lambda)$ which depends on the variable Λ_{i_0}. It also reduces the computational complexity. So, we move on to the next iteration point:

$$\Lambda^{new} = \Lambda + \tau \boldsymbol{d} .$$

In paper [1] it was proved that this process converges to the solution of the problem (4). Calculations can be stopped according to heuristic conditions about values of the objective function in neighboring iterative points.

3. Experimental study of the effectiveness of decomposition

The experiment contains several typical problems generated by well-known learning sets from samples (a1a, a2a, a3a, a4a, a5a, a6a, a7a, a8a, a9a, [2]). These samples containing an examples with the vectors of characteristics with size $L = 14$.

The solution was carried out with the help of three variants of the algorithm - sequential version without decomposition, sequential version with decomposition and parallel version using OpenMP.

The experiment was provided by a series of random samples from the data set with sizes $K = 100, 250, 500, 700, 1000, 1500$. We solved more than 100 different problems. The auxiliary problems (8) was solved using a gradient projection method.

The tests looked at various options of decomposed blocks of contiguous variables. Size of decomposed blocks took up the same (except for the last block, which contains all the remaining variables).

The following table shows the average time (in seconds) for solving the problem (4) by sequential version without decomposition (T_1) and with decomposition (T_2). Also in Table 1 the count of example in sample (K) and the size of decomposed block (S) are listed.

As seen from Table 1 for small amounts examples in the sample the decomposition approach has a little effect. When the sample size $K = 700$ difference becomes noticeable and in the future it begins to grow very quickly. Thus, the decomposition approach can give a great effect for serious practical problems.

Table 1. The time without decomposition and with it.

K	S	T_1	T_2
100	2,10,50	<<1 sec	<<1 sec
250	100	12 sec	4 sec
500	250	75 sec	15 sec
700	400	298 sec	48 sec
1000	500	30 min	140 sec
1500	500	1 hour 45 min.	346 sec

Then give the solution time performance when using the same decomposition in sequential (T_3) and parallel (T_4) versions of the algorithm. We note immediately that the size $K \leq 500$ of the training set for the parallel version provides the benefits only for a small amount of decomposed block. In other cases, there was even the same time sequential and parallel versions of the algorithm. This is probably due to overhead costs of the time for parallelization problems.

Table 2. The time with different sizes of decomposed blocks.

K	S	T_3	T_4
1000	10	60 sec	18 sec
1000	50	21 sec	5 sec
1000	100	13 sec	8 sec
1000	200	21 sec	10 sec
1000	300	32 sec	28 sec
1500	100	39 sec	18 sec
1500	200	37 sec	14 sec
1500	300	48 sec	38 c sec
1500	500	254 sec	204 sec

As can be seen from Table 2, the use of parallel version of the algorithm have advantages in almost all values of the decomposed block sizes, but it becomes sensible when the size of at least 5 times smaller than the training set size. Another interesting conclusion is that with decrease of the decomposed block size the time is decreased for both versions of the algorithm only to a certain moment. Further reduction of the size decomposed block increases the computational time. So, it is clear for $K = 1000$ in the sequential version of the algorithm between 50 and 100 variables in the block. The computational time is worsened in 1.6 times. Parallel version worsened the time later – between 10 and 50 variables in a block. For $K = 1500$ we observe the same behavior between 100 and 200 variables in a decomposed block.

4. Conclusions

In general, the experiment allows us to conclude that the decomposition approach for solving the Support Vector Machine optimization problem (1)-(3) can provide significant effect. Note that the implementation of the algorithm is independent of the way of partition variables into decomposed blocks. Using of decomposition, compute the direction of descent the objective function (6) saves computing resources. However, an excessive reduction in the size of the decomposed block can lead to

IOP Conf. Series: Materials Science and Engineering **158** (2016) 012006 doi:10.1088/1757-899X/158/1/012006

additional computational time. Therefore, the size of the decomposed block is recommended in 5-10% of the training sets size.

Acknowledgments
This work is performed by financial support of RFBR (project 16-01-00109).

References

[1] Konnov I V 2015 Sequantial threshold control in descent splitting methods for decomposable optimization problems *Optimization Methods and Software* **30(6)** 1238-1254
[2] Chang C C, Lin C J *LIBSVMdata: Classification, regression and multi-label* (http://www.csie.ntu.edu.tw/cjlin/libsvmtools/datasets/)
[3] Burges C J 1998 A Tutorial on Support Vector Machines for Pattern Recognition *Data Mining and Knowledge Discovery* **2(2)** 121–167

Dynamic generation of supercritical water fluid in a strong electrical discharge in a liquid

V Antonov[1], N Kalinin[2] and A Kovalenko[2]

[1] Department of Mathematics, Peter the Great Saint-Petersburg Polytechnic University, Saint-Petersburg, Polytechnicheskaja Str. 29, Russia
[2]Joffe Physical-Technical Institution Russian Academy of Science, Saint-Petersburg, Polytechnicheskaja Str. 26, Russia

E-mail: antonovvi@mail.ru

Abstract. A new impetus for the development of electro physics is associated with using different types of electrical discharges in biology and medicine. These applications are based on their energetic and non-toxic factors affecting the medium on a cellular level. For the study of such processes, a mathematical model of a high-current low-temperature Z-discharge in a liquid, forming by the electrical explosion of a thin-walled metal shell, connected to a pulsed high-voltage generator, has been developed. High efficiency energy conversion, introduced into the plasma discharge to the energy of fluid motion, provides various bio chemical applications of such physical processes. The investigation is conducted through numerical solution of one-dimensional single-temperature non-stationary equations of radiation magneto hydrodynamics, one way describing the evolution of hydrodynamic, thermal and electrical characteristics of the medium throughout the area under consideration. The electrical approximation based on the assumption that the electric field in the discharge has a uniform distribution. The results are presented as a function of the electric current and the plasma channel length of time, as well as the temperature and pressure distributions at different time points along the radius of the cylindrical region in which the explosion occurs.

1. Introduction

Powerful electrical discharges in the plasma channels produced in liquid media (eg water), are widely used in scientific research, technics and development of modern advanced technologies. A large part of applied research, which examines the powerful electric discharges in the liquid, are directed to the use of hydrodynamic energy pulses for the treatment of materials. High efficiency energy conversion, introduced into the plasma discharge to the energy of fluid motion and implementation of programming pulses of hydrodynamic characteristics, provides a variety of technological applications of this type of electrical discharges. In this regard, a series of simplified discharge models useful for solving a given applied problem have been developed. But the most correct description of the discharge may frame MRGD models to consistently analyze the change in the physical condition of plasma and liquid based on the current pulses generated by the discharge circuit.

A new impetus for the development of this area of electro physics is associated with recent study of using different types of electrical discharges in biology and medicine. The most successful to date is the theoretical rationale for the use of impact of electric explosion on living organisms. These applications are based on their unique energetic and highly non-toxic factors affecting the medium on a cellular level. In particular, the theoretical studies substantiate the possibility of the application of these factors to cell destruction.

The high-power nanosecond discharges in the electrode gap work a large number of free electrons that have a relatively high energy. These electrons can initiate and be the main actors of plasma chemical reactions. Their interaction with water molecules and biological substrates results in the formation of reactive species such as O, O_3, OH, O_2 and H_2, which are active oxygen species. They are the first to appear in the chain reactions of cell metabolism, and are involved in the biological processes of the organism and the cells.

As a first step in the direction indicated by the above, we carried out numerical simulations of the nanosecond high-current electrical discharge in the water, which is one of the important components of biological media. Emphasis is placed on MRGD simulation of changes in the physical state of water associated with the development of a powerful electric discharge. This special interest to a powerful electrical discharge in the water is due to the idea of getting supercritical water, in which it acquires the properties of an effective solvent in a sufficiently wide area, adjacent to the discharge. It is also necessary to study the dynamics of this area in the process of discharge development, and to determine the optimal operating conditions which will permit to obtain the physical and chemical state of the supercritical fluid of water required for the implementation of technological processes.

2. Mathematical model of discharge

In this paper we study the dynamics of discharge into the water conducted through numerical solution of one-dimensional single-temperature non-stationary equations of radiation magneto hydrodynamics, one way describing the evolution of hydrodynamic, thermal and electrical characteristics of the medium (plasma and liquid) throughout the area under consideration from the axis to the cylindrical discharge chamber inner wall. The electrical approximation based on the assumption that the electric field in the discharge is uniform distribution, are taken into account transient processes in the electric circuit of the discharge, the work of the magnetic field generated by the discharge current, the Joule heating of the plasma, the formation and propagation of the shock wave, thermal conductivity of water and plasma and radiation transfer in the continuous spectrum.

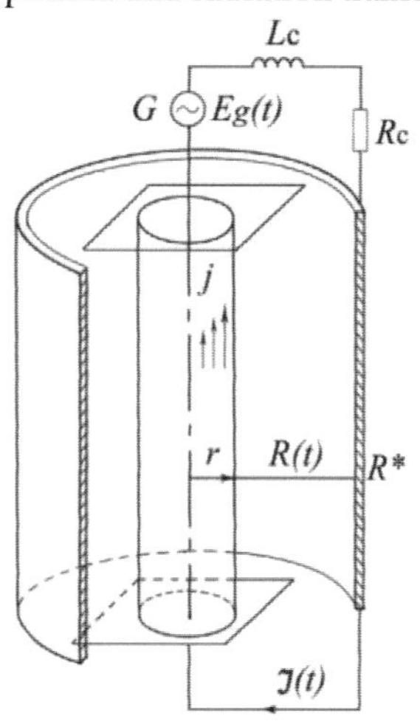

Figure 1. Discharge circuit.

Schematic diagram of the discharge is shown in Figure 1. Discharge power supply system consists of a high-voltage pulse generator *Eg* (*t*), exacerbating fuses, inductor *Lc* and an electric circuit of the transmission line, which is represented in Scheme by a wave resistance *Rc*. The current supplied to the discharge by a solid copper electrode. Assuming that the characteristic dimensions of the system are small (in this case, not more than 1 cm), compared with the diffusion length of the magnetic field for the discharge time, it may be considered the electrodes as equipotential surfaces, and the electric field between them uniform. In this case the plasma channel can be regarded as a conductor, having a variable radial conductivity and inductance, and its capacity can be neglected. Pulse load - high-current discharge - represented in the diagram variable as an impedance plasma channel *R*(*t*) and the inductance of the area enclosed between the plasma channel and the reverse current lead *Ld* (t).

The discharge is initiated by electrical breakdown of the discharge gap, thereby forming a plasma channel in water. The phenomenon of electrical breakdown of water in this paper is not considered. The radius of the channel and the resulting plasma state properties are considered as parameters of the problem and must be specified.

The plasma temperature in the discharge channel can reach tens of thousands of degrees, and heat radiation in the energy balance is beginning to play a significant role [1-5]. To calculate the characteristics of the radiation in the continuous spectrum used a simplified model based on the approach of "forward - backward" (Schuster-Schwarzschild approximation) [6, 7] with only two spectral bands.

The one-dimensional unsteady magnetic hydrodynamics equations, uniform way to describe the entire area from the axis of the level before the walls of the bit cell, have the form:

$$\frac{\partial}{\partial t}\rho+\frac{1}{r}\frac{\partial}{\partial r}(r\rho u)=0$$

$$\frac{\partial}{\partial t}(\rho u)+\frac{1}{r}\frac{\partial}{\partial r}[r(\rho u^2+\tau)]+\frac{\partial}{\partial r}(P+P_\omega)+jB=0$$

$$\frac{\partial}{\partial t}W+\frac{1}{r}\frac{\partial}{\partial r}\left[ru(\varepsilon+P+P_\omega+\tau)+r\chi\frac{\partial T}{\partial r}\right]=\sigma E^2-Q_R \tag{1}$$

$$\frac{1}{r}\frac{\partial}{\partial r}(rB)=\mu_0 j;\quad j=\sigma E$$

$$W=\rho\left(\varepsilon+\frac{1}{2}u^2\right);\quad \tau=\eta\left(\frac{3}{4}\frac{\partial u}{\partial r}-\frac{2}{3}\frac{u}{r}\right)$$

$$P=P(T,\rho),\quad \varepsilon=\varepsilon(T,\rho)$$

Here r, t-spatial and temporal coordinates; ρ- density; u-speed; P - pressure; T - temperature; ε - specific internal energy; χ, σ, η-thermal conductivity, electrical conductivity and viscosity respectively; Q_R-volumetric energy loss by radiation; E-electric field strength; B-induction of own magnetic field; j-current density; P_ω -pressure caused by artificial viscosity entered for "end-to-end" (i.e. without an explicit allocation of fronts) calculation of shock waves.

Time variation of the discharge current describes the electrical circuit equation

$$U_G(t)=L_0\frac{\mathrm{d}I}{\mathrm{d}t}+R_0 I+\frac{\mathrm{d}}{\mathrm{d}t}(L_D I)+R_\rho I+E_S \tag{2}$$

where $U_G(t)$-strength of the electric field generated by a high voltage pulse generator range bit (considered to be a function of time); L_0, R_0-inductance and resistance in an electrical circuit; R_ρ-impedance transporting line; L_D-inductance of discharge plasma- reverse conductor of electric current; E_S-electrical field strength on the border out of the discharge. All the characteristics of the electrical circuit in the formula attributed to the length of the bit period.

The resistance level is determined by the formula

$$R_D(t)=I(t)^{-2}\int_0^{r_{pl}(t)}\sigma E^2 r\mathrm{d}r \tag{3}$$

The equations of state of the water used in the work presented in [5]. There are restrictions on the temperature $T\leq 10^5$ K and pressure $P\leq 10^4$ MPa. However, these limitations make it possible to consider supercritical water conditions. Method for determining the plasma water composition, knowledge of which is necessary to calculate its thermodynamic, transport and optical properties described in [8]. Wide-range semi-empirical model is used to calculate the conductivity and thermal conductivity [9], generalized to the case of non-homogeneous elemental composition of plasma. Collisions of electrons with atoms and ions were recorded in accordance with [9].The boundary conditions are set on the axis and on the inner wall surface of the discharge chamber, taking into account the symmetry of the problem [10-12]. Due to the electrical breakdown of water, it forms a cylindrical plasma channel. Outside of this channel at the initial time the water is in the normal state. The radial pressure distribution at the initial time was set spatially uniform. The initial discharge channel radius on the basis of recommendations known [1-4] was assumed to be r_0~0.1 mm, the initial temperature of the plasma in the discharge channel after electrical breakdown, as usual ~ 10000 - 20000 K. The problem was solved in Lagrangian variables using a conservative difference scheme through computation with the introduction of artificial viscosity [11].

3. Calculation results

In numerical experiments Ramp Up front was set at 5 ns, the maximum electric field strength in the discharge gap - up to 10 MV/cm. Figure 2, 3 show the time dependence of the discharge current *I* (*t*) and the dimensionless radius of plasma channel *R*(*t*).

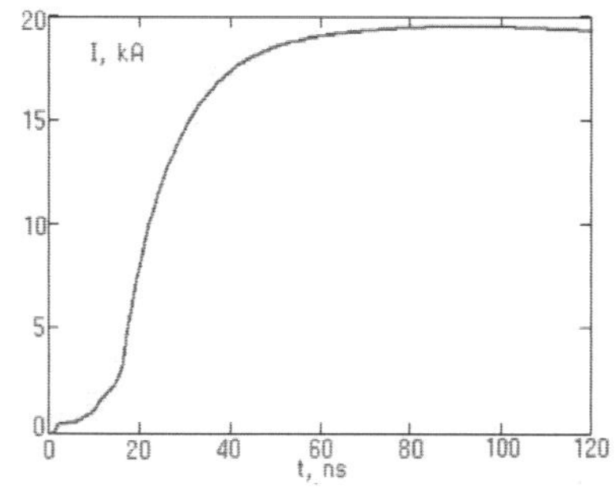

Figure 2. Discharge current

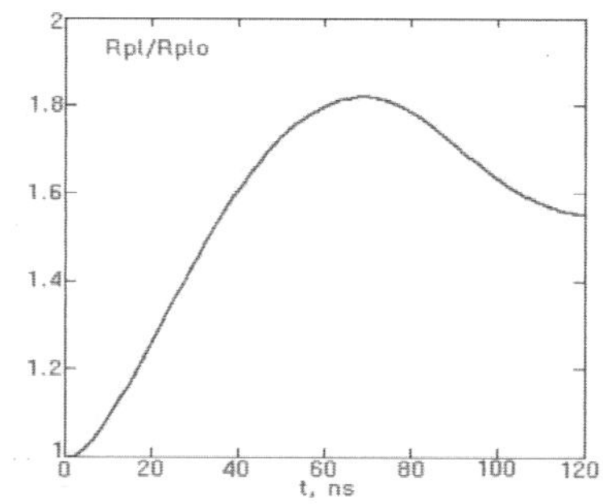

Figure 3. Plasma channel

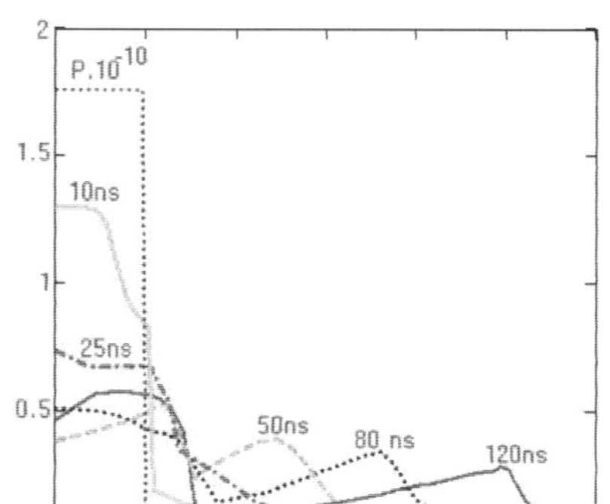

Figure 4. Pressure distribution

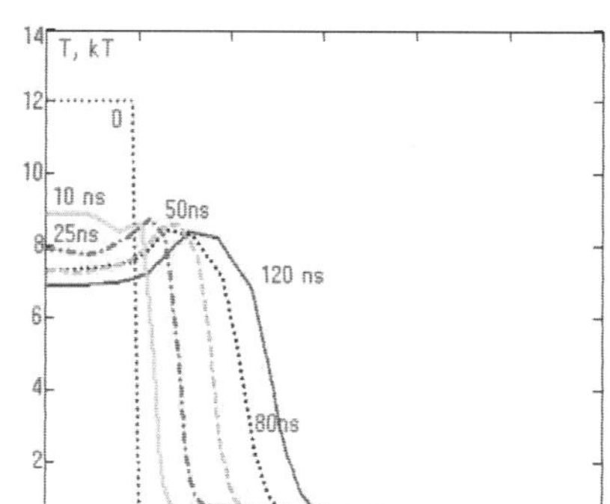

Figure 5. Temperature distribution

Under the influence of a pressure gradient formed in the initial moment of time due to an electrical breakdown of water begins expansion of the plasma in the discharge channel, which leads to the formation of a cavity of reduced density in the region adjacent to the axis, and the formation of a shock wave in the fluid moving at a rate of ~ $2{\cdot}10^3$ m/c in the direction of the discharge chamber wall. Joule heating of the plasma in the discharge leads to the formation of the second shock wave traveling in the discharge channel with a reduced density. This shock wave is reflected almost completely by the channel on which the density increases by almost an order of magnitude. The reflected shock wave causes a sharp increase in pressure in the discharge channel and compressing the conductive region, the formation of an intense flow of the medium and the cooling agent in the inner region of the discharge channel. In the inner regions of the discharge channel realized high temperatures. Transfer of radiation can be described in the approximation of radiant heat conduction, but temperature gradients are small, and therefore the radiation from this region hardly leaves. In the rest of the discharge can be applied the approach of the volume emission.

Carried MRGD - Calculations have shown in particular that the cavity of reduced density and boundaries of the discharge boundaries do not coincide. Inside the cavity there is a secondary shock wave associated with the energy input into the plasma. Interacting with the boundary of the cavity (reflected from it), this shock wave causes pressure pulsations noted earlier in [8], forms the radial no uniform distributions of hydrodynamic, thermal and electrical properties of the medium in the discharge, and has an impact on the integral characteristics of the discharge.

4. Conclusions

The designed computer program makes possible to study in detail the dynamics of the high-current Z-discharges in liquid media, receiving the radial distribution characteristics of the medium at different times, to analyze the interaction of shock waves, the impact of the magnitude and speed of input in energy environment in the discharge structure and to obtain the necessary high physical-chemical conditions of the medium. The reliability of the calculation results is confirmed by comparison with the data of other researchers, as well as a comparison with experimental data. An important step in the further development of this work should be to obtain more accurate data on the process of plasma generation by explosion of metallic conductors. Equally important to get a more detailed description of processes in the metal, water and plasma, in particular equation of state.

The results can be an important step in the study of the influence of high energy processes on the biological systems. However, any interference in the process of homeostasis must be carefully verified experimentally, and also obtained a statistical analysis of the results. We must remember that all living organisms depend on maintaining a complex set of interacting metabolic chemical reactions.

References

[1] Pozdeev V A 1980 *Applied fluid dynamics of the electric discharge in liquid* (Kiev: Naukova Dumka)
[2] Arsent'ev V V 1965 *J. Appl. Mech. Tech. Physics* **B 5** 51
[3] Naugolnykh K A and Roy N A 1971 *Electrical discharges in water* (Moscow: Nauka)
[4] Sinkevich O A and Shevchenko A L 1983 *Fluid Dynamics* **B 3** 104
[5] Kolgatin S N 1995 *J. Technical Physics* **B 65** (7) 10
[6] Zel'dovich I B and Raiser Y P 1968 *Physics of Shock Waves and High-Temperature Hydrodynamic Phenomena* (Moscow: Nauka)
[7] Wrekin A D. 1978 *Thermal physics high temperatures* **B16** (4) 811
[8] Romanov G S, Stanchits L K and Stepanov K L 1979 *Journal Applied Spectroscopy* B30 (1) 35
[9] Polishchuk A J 1990 *Thermal physics high temperatures* **B 28** (5) 877
[10] Sinkevich O A and Shevchenko A L 1983 *Fluid Dynamics* **B 3** 104
[11] Samarsky A A and Popov Y P 1980 *Difference methods for solving problems of gas dynamics* (Moscow: Nauka)
[12] Chetverushkin B N 1985 *Mathematical modeling of dynamic problems of radiating gas* (Moscow: Nauka).

Application of a second order accurate finite-difference method to problems of diffraction of elastic waves by gradient layers

A V Anufrieva, E V Rung, D N Tumakov

Department of Applied Mathematics, Institute of Computer Mathematics and Information Technologies, Kazan Federal University, 18 Kremlyovskaya Street, Kazan 420008, Russia

E-mail: dtumakov@kpfu.ru

Abstract. A generalized statement is formulated for the boundary-value problem describing diffraction of elastic waves by gradient isotropic and transversely isotropic layers. Numerical experiments are conducted for various types of materials filling the layer. A conclusion is drawn that the obtained finite-difference scheme is second-order accurate, when distributions of elastic parameters in the layer are described by smooth curves.

1. Introduction

Boundary value problems for the Lame equations are encountered in consideration of processes of propagation and diffraction of elastic waves. Passage of waves through media with continuously changing parameters represent a special interest here. One of the simplest approaches to analyzing diffraction of an elastic wave by layers of a varying structure is applying an isotropic model of the layer. A more general approach to studying layers of a varying structure is through applying a transversely isotropic model of the layer. In our previous studies, we investigated a problem of passage of a planar elastic wave through gradient isotropic [1], transversely isotropic [2] and anisotropic layers [4]. A disadvantage of those studies was that the used numerical methods were first-order accurate.

As an alternative method of solution of problems of this type, one can consider the finite-element method being widely used in recent years [5], which is particularly efficient for solving multidimensional problems in domains with complex boundaries [6]. However, in the one-dimensional case finite-difference methods can be competitive to the finite-element methods. One of such approaches, called the method of approximation of integral identities [7], is used in the present study.

The method of approximation of integral identities allows achieving the approximation error $O(h^2)$. This algorithm of constructing finite-difference schemes was applied by us earlier to solving the Helmholtz equation, to which a one-dimensional problem of diffraction of an elastic wave by a gradient layer is reduced [8], as well as to solving the saturated-unsaturated filtration consolidation problem [9].

In the present study, a concept of a generalized solution to the Cauchy problem describing the process of diffraction of elastic waves by gradient isotropic and transversely isotropic layers is introduced and its uniqueness conditions are formulated. A second-order accurate finite-difference scheme for solving the problem is presented. A series of numerical experiments demonstrating applicability of the method in terms of smoothness of curves of distribution of the layer's elastic parameters is provided.

2. Problem statement

Let an elastic harmonic wave u_0 fall on a uniform in the transverse direction and transversely isotropic layer of thickness L (medium 2 $\{0 < x < L\}$ having continuous density $\rho_2(x)$ and tensor (3×3) of elasticity modulus $K(x)$ from medium 1 $\{x > L\}$ at angle θ with respect to axis x (see Fig. 1). As a result of diffraction, there appear a wave reflected to medium 1, a wave passing to medium 3 $\{x < 0\}$ and a field in the layer. We will assume that the media 1 and 3 are uniform and isotropic.

The process of diffraction of a planar elastic wave by a gradient isotropic and transversely isotropic layer (at $x \in (0, L)$) is governed by the system of ordinary differential equations of the second order with respect to displacement components in the Cartesian coordinate system $u = (u_1(x), u_2(x))$, which has the form:

$$\frac{d}{dx}\left(c_{s1}\frac{du_s}{dx}\right) + c_{s2}u_s + c_{s3}\frac{du_k}{dx} + c_{s4}u_k = q_s, \qquad s,k = 1,2, \quad s \neq k, \tag{1}$$

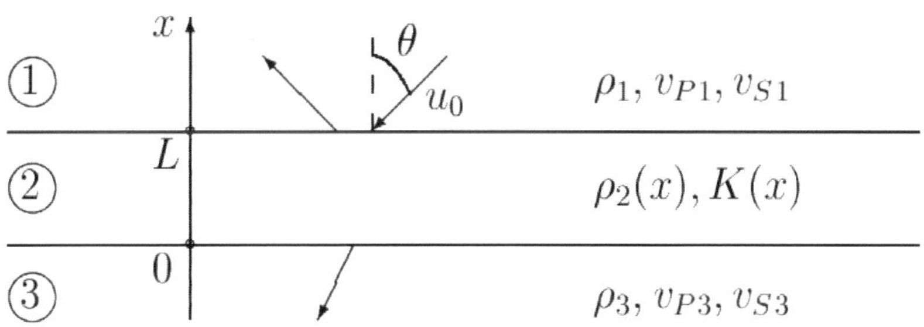

Figure 1. Geometry of the problem.

and boundary conditions have the form ($s, k = 1,2, \ s \neq k$):

$$a_{s1}\frac{du_s(0)}{dx} + a_{s2}u_s(0) + a_{s3}\frac{du_k(0)}{dx} + a_{s4}u_k(0) = f_s, \tag{2}$$

$$b_{s1}\frac{du_s(L)}{dx} + b_{s2}u_s(L) + b_{s3}\frac{du_k(L)}{dx} + b_{s4}u_k(L) = g_s. \tag{3}$$

Expressions (2), (3) correspond to conditions of continuity on boundaries $x = 0$ and $x = L$ of normal and tangential stresses as well as of both displacement components.

We will assume that functions $c_{si}(x)$, $q_s(x)$, $i = 1..4$ and $s = 1,2$ are continuous complex-valued functions on interval $[0, L]$, and the following relations are fulfilled:

$$c_{s1}(0) \neq 0, \quad c_{s1}(L) \neq 0, \quad s = 1,2, \tag{4}$$

and the numbers a_{si}, b_{si}, $i = 1,2,3$, $s = 1,2$ satisfy the conditions:

$$a_{11}\cdot a_{21} - a_{23}\cdot a_{13} \neq 0, \qquad b_{11}\cdot b_{21} - b_{23}\cdot b_{13} \neq 0. \tag{5}$$

Equation (4) has the following physical meaning. If the coefficients $c_{s1}(x)$ are not equal to zero, then it means that longitudinal and transverse velocities at points $x=0$ and $x=L$ are also not equal to zero for the case of equation (1) describing propagation of an elastic wave in the gradient isotropic layer. In addition, the latter means that, for the case of transversely isotropic media, the longitudinal and transverse velocities in the isotropy plane are not equal to zero either [3].

With regard to conditions (5), it is worth noting that if the boundary-value problem (1)–(3) describes diffraction by gradient layers, then conditions (5) imply lack of excitation of surface modes (Rayleigh waves) in the adjacent half planes.

Let us construct a second-order accurate finite-difference scheme for solving the problem (1)–(3) and estimate the error of the obtained solution at various distributions of the layer's elastic parameters. At first, we will proceed from the classical solution to the generalized solution.

3. Generalized solution to the problem

If conditions (4), (5) are fulfilled, then system (1)–(3) can be rewritten in the following equivalent operator form [10]:

$$-\frac{d}{dx}\left(A\frac{du}{dx}\right)+B\frac{du}{dx}+Du=q,\quad x\in(0,L), \tag{6}$$

$$\left.\left(A\frac{du}{dx}\right)\right|_{x=0}=G_0u(0)+\varphi^0, \tag{7}$$

$$-\left.\left(A\frac{du}{dx}\right)\right|_{x=L}=G_Lu(L)+\varphi^L, \tag{8}$$

where $u=(u_1(x),u_2(x))$ is a vector function; $\frac{du}{dx}=\left(\frac{du_1}{dx},\frac{du_2}{dx}\right)$, A, B, D are linear operators acting from a two-dimensional complex plane C^2 into space C^2; $q=(q_1,q_2)$; $\varphi^0=(\varphi_1^0,\varphi_2^0)$; $\varphi^L=(\varphi_1^L,\varphi_2^L)$; matrices G_0 and G_L are given in the form:

$$G_0=\begin{pmatrix}\mu_{11}^0 & \mu_{12}^0\\ \mu_{21}^0 & \mu_{22}^0\end{pmatrix},\quad G_L=\begin{pmatrix}\mu_{11}^L & \mu_{12}^L\\ \mu_{21}^L & \mu_{22}^L\end{pmatrix},$$

here complex numbers μ_{ij}^0, μ_{ij}^L, φ_i^0, φ_i^L are determined via coefficients of boundary conditions a_{ik}, b_{ik}, f_i, g_i, where $i,j=1,2$, $k=1..4$.

We define the space $V=W_2^{(1)}(0,L)\times W_2^{(1)}(0,L)$ having the norm

$$\|u\|=\int_0^L\left(|u|^2+\left|\frac{du}{dx}\right|^2\right)dx$$

and introduce the concept of a generalized solution to problem (6)–(8) in the form of a function $u=(u_1(x),u_2(x))\in V$. For function u, the following equality holds:

$$\int_0^L\left(A\frac{du}{dx}\cdot\frac{dv}{dx}+B\frac{du}{dx}\cdot v+Du\cdot v\right)dx+G_0u(0)\cdot v(0)+G_Lu(L)\cdot v(L)=$$

$$=\int_0^L q\cdot v\,dx-\varphi^0\cdot v(0)-\varphi^L\cdot v(L)\qquad\forall v=(v_1,v_2)\in V, \tag{9}$$

where $u\cdot v=u_1\bar{v}_1+u_2\bar{v}_2$ is a standard scalar product in C^2.

Next, we define sesquilinear form $a: V \times V \to C$ and linear functional $f: V \to C$ using the following relations:

$$a(u,v) \equiv \int_0^L \left(A\frac{du}{dx}\cdot\frac{dv}{dx} + B\frac{du}{dx}\cdot v + Du\cdot v \right)dx + G_0 u(0)\cdot v(0) + G_L u(L)\cdot v(L), \quad (10)$$

$$f(v) \equiv \int_0^L q\cdot v dx - \varphi^0 \cdot v(0) - \varphi^L \cdot v(L). \quad (11)$$

Then, seeking a generalized solution to problem (6)–(8) is equivalent to seeking a function $u \in V$, meeting the condition:

$$a(u,v) = f(v) \qquad \forall v \in V. \quad (12)$$

We set the following conditions for functions c_{ij}, q_i and matrices G_0, G_L:

A_1) Functions $c_{ij} \in L_\infty(0,L), q_i \in L_2(0,L), i = 1,2,\ j = 1..4$.

A_2) $\mathrm{Re}(c_{i1}(x)) \geq \alpha_{i1},\quad x \in [0,L],\quad i = 1,2$.

A_3) $\mathrm{Re}(G_0\xi\cdot\xi) \geq \alpha_0|\xi|^2$, $\mathrm{Re}(G_L\xi\cdot\xi) \geq \alpha_L|\xi|^2 \quad \forall \xi \in C^2$, where α_{11}, α_{21}, α_0, α_L are some positive constants.

We will assume that the conditions $A_1) - A_3)$ are met along with a condition providing V-ellipticity of form a, i.e. there exists a positive real number δ_3 such that for any $u \in V$, the following inequality is satisfied:

$$\mathrm{Re}(a(u,v)) \geq \delta_3\|u\|. \quad (13)$$

Then, problem (6)–(8) has a unique generalized solution $u \in V$ and the following inequality is true:

$$\|u\| \leq c\cdot\left(\sum_{i=0}^{2}\left(\|q_i\|_{L_2} + |\varphi_i^0| + |\varphi_i^L|\right)\right), \quad (14)$$

where c is a positive constant.

4. Approximation of the problem

For approximating the problem (6)–(8), we will use the method of approximation of integral identities. For that, on interval $[0,L]$ we will construct a uniform grid $\overline{\omega}$ with step size $h = L/N$:

$$\overline{\omega} = \{x_i = ih,\quad i = 0..N\}.$$

Using a classical approach [7], we will define a space of grid functions V_h on $\overline{\omega}$.

We call grid functions $(y_1, y_2) \in V_h$ as a solution to the finite-difference scheme of problem (6)–(8), if, for any grid functions $\eta_1, \eta_2 \in V_h$ and $s,k = 1,2,\ s \neq k$, the following inequalities are satisfied [10]:

$$-\frac{1}{2}\left[c_{s1}(y_s)_x, (\eta_s)_x\right) - \frac{1}{2}\left(c_{s1}(y_s)_{\bar{x}}, (\eta_s)_{\bar{x}}\right]$$

$$+\frac{1}{2}\left[c_{s3}(y_k)_x, \eta_s\right) + \frac{1}{2}\left(c_{s3}(y_k)_{\bar{x}}, \eta_s\right] + \left[c_{s2}y_s + c_{s4}y_k, \eta_s\right]$$

$$+\left(\mu_{s1}^0 y_{1,0} + \mu_{s2}^0 y_{2,0} + \varphi_s^0\right)\overline{\eta}_{s,0} + \left(\mu_{s1}^L y_{1,N} + \mu_{s2}^L y_{2,N} + \varphi_s^L\right)\overline{\eta}_{s,N} = [q_s, \eta_s]. \quad (15)$$

Let (u_1, u_2) be a solution to problem (6)–(8), and functions c_{s1} and u_s have three and four constrained derivatives in the neighborhood of nodes of grid $\overline{\omega}$, respectively. Then, the finite-difference scheme (15) has approximation error $\Psi_i^s = O(h^2),\ s = 1,2,\ i = 0..N$.

It is worth noting that the finite-difference scheme (15) represents a linear system of equations, for which one can apply the Thomas algorithm [7].

5. Numerical results

Let us estimate an error of solution of problem (1)–(3) by using the finite-difference scheme (15). For that purpose, we will consider some model problems having known solutions (u_1, u_2). We calculate functions $c_{si}(x)$ and coefficients a_{si}, b_{si} for certain materials filling the layer. By substitution of known functions (u_1, u_2) into (1)–(3), we find right-hand sides $q_s(x)$, f_s and g_s. Next, we solve numerically the Cauchy problem with coefficients and right-hand sides obtained. As a results, will obtain an exact (a-priori known) solution $u=(u_1,u_2)$ and approximated solution $y=(y_1,y_2)$. For determining the error of the obtained solution $(y_{1,j}, y_{2,j})$ in node x_j $(j=0..N)$, we calculate the norm:

$$\left\|y_j-u(x_j)\right\|=\left(\left|y_{1,j}-u_1(x_j)\right|^2+\left|y_{2,j}-u_2(x_j)\right|^2\right)^{1/2}. \tag{16}$$

An error of the approximated solution in the entire layer will be determined through the maximum norm:

$$\varepsilon=\max_{j=0..N}\left\|y_j-u(x_j)\right\|. \tag{17}$$

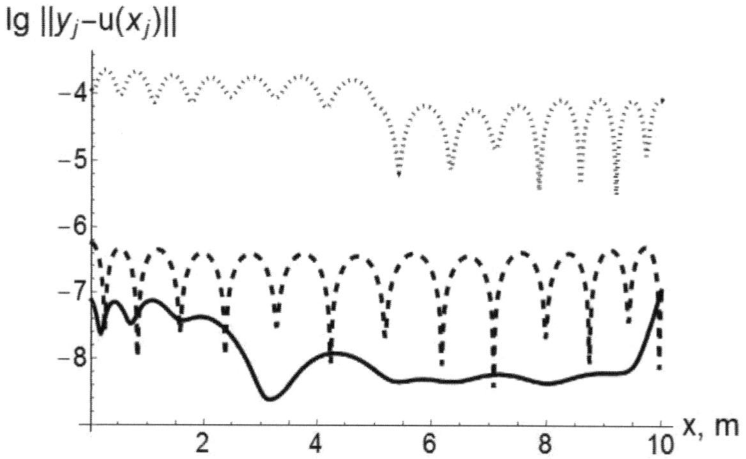

Figure 2. Values of logarithm of the error of the approximated solution in the layer's points. Solid line corresponds to a linear change of elastic parameters; dashed line corresponds to a parabolic profile; dotted line corresponds to a broken profile.

Let us consider a layer with transversely isotropic material located inside a sand-rock and having parameters: $\rho_1=\rho_3=2400$ kg/m^3; $v_{P1}=v_{P3}=3300\,m/\mathrm{s}$ and $v_{S1}=v_{S3}=2200\,m/\mathrm{s}$. We confine ourselves only to three numerical experiments. The first numerical experiment corresponds to the case of the layer's elastic parameters changing continuously from sand-rock (isotropic material) to silt-rock (transversely isotropic material). At $x=0$, we have a continuous transition from the external medium toward the layer. After that, the layer parameters as well as coefficients of the system of equations (1) also change continuously, obeying the linear law, toward parameters corresponding to a transversely isotropic material, i.e. silt-rock [11]: $E=56.8\,\mathrm{GPa}$, $E'=62.1\,\mathrm{GPa}$, $\nu=0.29$, $\nu'=0.26$, $G'=22.9\,\mathrm{GPa}$ (more details on connections between elastic parameters for a transversely isotropic material and elasticity tensor K can be found in [2]). At another point of contact of the layer with sand-rock, $x=10\,m$, parameters change discontinuously, by a leap.

The second numerical experiment is carried out for the case of elastic parameters changing continuously by the parabolic law from sand-rock at the beginning of the layer toward silt-rock at $x=5\,m$ followed by a symmetric change of parameters back toward sand-rock at $x=10\,m$. The third numerical experiment is carried out for the case of elastic parameters changing continuously and linearly from sand-rock toward silt-rock in the middle of the layer followed by a continuous and linear change toward sand-rock by the end of the layer. In the latter case, for elastic parameters we have a broken profile, i.e. a profile having discontinuity of the first derivative in the middle of the layer. We assume constant density throughout the layer for all considered cases: $\rho_2 = 2400\,kg/\mathrm{m}^3$.

System (1)–(3) is solved for the problem of a planar longitudinal wave of a single amplitude falling on the layer at angle $\theta = 20^0$.

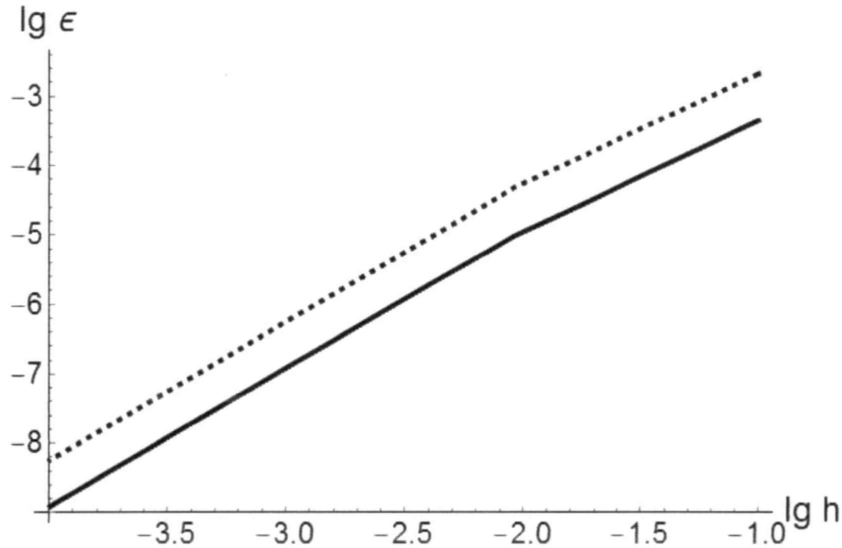

Figure 3. Dependence of logarithm of the approximated solution on logarithm of the grid step size. Solid line corresponds to a linear profile; dotted line corresponds to a parabolic profile.

Let us estimate an error of the approximated solution for each of the cases in the layer's points via formula (16) choosing the grid step size $h=10^{-3}$. An error for the linear profile, when compared with the other experiments, is smallest (solid line in Fig. 2). At the end point, there is slight worsening of the error, which is explained with discontinuity of elastic parameters at that point. The error for the parabolic profile (dashed line) is slightly worse, but it is still of order $O(h^2)$. In case of the broken profile (dotted line), the error becomes of the first order due to breakdown of smoothness of coefficients of the original equation (1) in the central point $x=5$ m.

Now we will consider dependence of the solution error in the entire layer ε (17) on the grid step size. One can see from Fig. 3 that, for experiments with smooth coefficients of the original equation, logarithm of the error is proportional to logarithm of the error. The solution error for the linear profile is smaller than that for the parabolic profile.

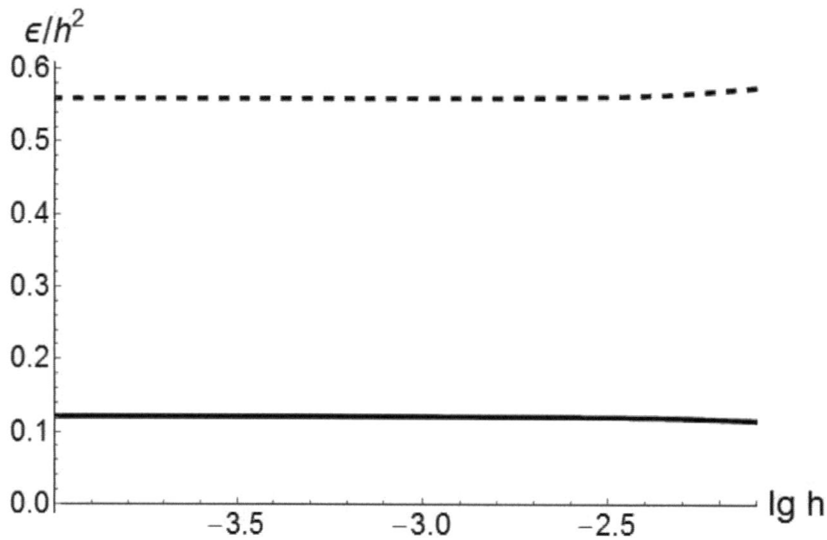

Figure 4. Graphs depicting values of the constant C in estimating $\varepsilon \le Ch^2$. Solid line corresponds to the linear profile; dashed line corresponds to the parabolic profile.

Thus, one can draw a conclusion that the numerical scheme has second-order accuracy for problems with smooth coefficients. For clarifying the coefficient in estimating $\varepsilon \le Ch^2$, let us present graphs for dependence of ε / h^2 on $\lg h$. We will obtain that $C \approx 0.12$ for the linear profile case and $C \approx 0.56$ for parabolic profile case. This also confirms our assumption that the problem with linear profile is solved more accurately. Note that the obtained estimations for C are valid at $h < 0.01$, and at $h > 0.01$, the values of C slightly increase.

6. Conclusions

A conclusion is drawn that for the problem of diffraction of an elastic wave by a gradient layer with elastic parameters having smooth distributions, the finite-difference scheme obtained by the method of approximation of integral identities is second-order accurate. It is shown that in case of breakdown of smoothness of curves of distribution of elastic parameters in the layer, the error of the scheme reduces to the first-order.

Acknowledgments
The work is performed according to the Russian Government Program of Competitive Growth of Kazan Federal University.

References

[1] Anufrieva A V, Tumakov D N and Kipot V L 2013 *Proc. Days on Diffraction (St. Petersburg, Russia)* 11–16
[2] Anufrieva A V and Tumakov D N 2013 Diffraction of a plane elastic wave by a gradient transversely isotropic layer *Advances in Acoustics and Vibration* 262067
[3] Anufrieva A V and Tumakov D N 2014 *Proc. Days on Diffraction (St. Petersburg, Russia)* 23–28
[4] Anufrieva A, Chikrin D and Tumakov D 2015 On Peculiarities of Propagation of a Plane Elastic Wave through a Gradient Anisotropic Layer *Advances in Acoustics and Vibration* 515263
[5] Khoei A R 2014 *Extended Finite Element Method : Theory and Applications* (New York: Wiley)

[6] Liu G R and Quek S S 2003 *The Finite Element Method: A Practical Course* (Butterworth-Heinemann)

[7] Samarskii A A 2001 *The Theory of Difference Schemes* (N.Y.: Marcel Dekker)

[8] Anufrieva A V, Tumakov D N and Kipot V L 2012 *Proc. Days on Diffraction (St. Petersburg, Russia)* 21-26

[9] Pavlova M F and Rung E V 2013 *Lobachevskii Journal of Mathematics* **34** **(4)** 392-405

[10] Anufrieva A V, Rung E V and Tumakov D N 2016 *Uchenye zapiski Kazanskogo universiteta. Seriya: Fiz.-mat. nauki* **158** **(1)** 26-39

[11] Batugin S A and Nirenburg R K 1972 *Fiziko-tekhnicheskie problemy razrabotki poleznykh iskopaemykh* **7** **(1)** 7–12

Capacitive coupled RF discharge: modelling at the local and not local statement of the problem

R M Askhatov[1], V Yu Chebakova[1] and V S Zheltukhin[2]

[1]Kazan Federal University, 18 Kremlevskaya Street, Kazan, 420008, Russia
[2]Kazan State Technological University, 68 Karl Marx Street, Kazan, 420015, Russia

E-mail: vchebakova@mail.ru

Abstract. The models provided in the research paper describe a capacitive coupled radio-frequency discharge in argon between two parallel plate electrodes, one of which is grounded, and the other is connected to the high-frequency capacitive generator. Herein we review various approaches to simulate a high-frequency capacitive discharge depending on modelled pressure rates. The model of a high-frequency capacitive discharge under low pressure is simulated in non-local approximation, and under high pressure is simulated in local approximation and is sensitive to dimers and molecular ions. We provide calculation data with respect to different pressures and make comparative analysis of data provided by other authors in particular, analysis of data obtained with real experiment.

1. Introduction

The Among the discharges applied to treat materials, significant part is assigned to capacitive coupled radio-frequency (CCRF) discharges [1-6]. CCRF-discharges differ under different pressures as by various values of discharge characteristics and by the mechanisms of internal processes. Nowadays the models of HFC-discharges of middle and low pressures are studied in details [5]. Recently the interest to CCRF-discharges of low pressure has increased due to emerged possibility to treat natural materials because of low atom temperature in such discharges, as well as interest to CCRF-discharges atmosphere pressure. Experimental calculation methods, which complement each other, are used to tie internal and external parameters of the discharge [7-11].

The models provided in this research paper describe a capacitive coupled radio-frequency discharge in argon between two parallel plate electrodes, one of which is grounded, and the other is connected to the high-frequency capacitive generator providing that distance between the electrodes is less then dimensions of the electrodes themselves. Under such conditions the electrical field is close to the potential one and the discharge is uniform along the electrodes, what allows using one-dimensional model [5]. Due to use of one dimensional model, we can review kinetics of the discharge with respect to diffusive drift approximation [12-13]. Comparison of time and distance needed for electron to lose the energy obtained from the field with the field period and dimensions of computational space revealed that the local and non-local approximations shall be used for modelling of CCRF -discharges under high and low pressures [3, 14].

IOP Conf. Series: Materials Science and Engineering **158** (2016) 012009 doi:10.1088/1757-899X/158/1/012009

2. Setting a problem in non-local approximation

The this section devoted to the pressure range from 13.3 Pa to 133 Pa describes the self-consistent mathematical model of CCRF-discharge in non-local approximation under low pressure in argon. This model includes process factors occurred under electron impacts, which are taken with account of electron-electron collision depending on electron temperature and ionization rate [15].

Provided herein model of CCRF-discharge under low pressure includes time-dependent equation of balances for electron gas, metastable atoms, atomic ions, Poisson's equation for potential of electrical field, time-dependent equation of electronic energy balance, as well steady state equation of thermal conductivity of atomic-ion gas at average parameters. At that here is used the simplified diagram of argon atom, where 4 lowest approximate electronically excited states are replaced with uniform rate resulted with efficient mixing due to their electronic impact [16]. The model includes plasma chemical reactions $Ar+e \rightarrow Ar^{+}+2e$; $Ar^{*}+Ar^{*} \rightarrow Ar+Ar^{+}+e$; $Ar^{*}+e \rightarrow Ar^{+}+2e$; $Ar^{+}+e \rightarrow Ar+h_{v}$; $Ar^{+}+2e \rightarrow Ar+e$; $Ar+e \rightarrow Ar^{*}+e$; $Ar^{*} \rightarrow Ar+h_{v}$; $Ar^{*}+Ar \rightarrow 2Ar$; $Ar^{*}+e \rightarrow Ar+e$. Velocity factors of the corresponding reactions are described in the work papers [17-23].

In order to solve the nonlinear system of boundary value problems and initial boundary value problems described herein, we used the approximation method, which is based on preliminary finite dimensional approximation of the problem using difference scheme including further use of iteration process applied for its implementation [24]. Moreover, the research papers [25-29] are devoted to the methods to solve nonlinear problems of continuum mechanics as well describe-layer iteration methods including methods lowering nonlinearity to a lower layer [25-29].

3. Calculation data

The calculation data of CCRF-discharge in plasmatron, which has electron spacing 22 mm, pressure P=13.3 Pa, applied voltage rate Va=65 V qualitatively coincide with the data of the research paper [30], however, ion concentration has specific bell-bottom shape with maximum located in the center of the discharge $\approx 10^{15}$ m^{-3}, while the experiments show value [30] $\approx 7 \cdot 10^{14}$ m^{-3}

Electron temperature reaches its maximum level, per the period 3 eV, at near-electrode layers, and maximum value at the center of the discharge was 2.5 eV. However, gas temperature stayed almost permanent: 305 K.

According to the research paper [30] the following temperature range was fixed under voltage Va=65 V and pressure P=13.3 Pa: maximum - 3.4 eV, minimum 1.7 eV providing that inaccuracy is 20%. As well the researchers made calculations under the pressure 133 Pa. Comparison of calculation data revealed that the distance, when an electron lose its energy under pressure 133 Pa decreased significantly (see Fig. 1 and Fig. 2).

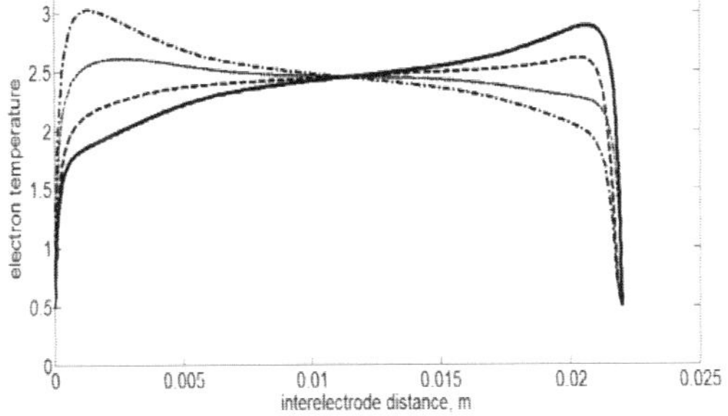

Figure 1. Change of electron temperature in time under pressure 13.3 Pa

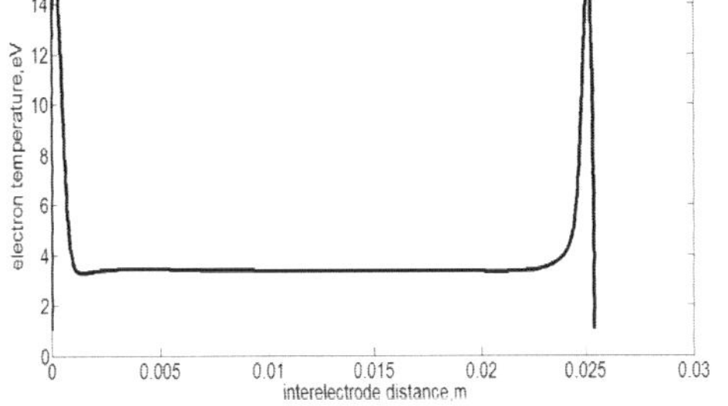

Figure 2. Change of electron temperature under pressure 13.3 Pa (Maximum value per the period is given at each point of interelectrode distance)

4. Setting a problem in local approximation

The research paper [31] shows dependency of concentration ratio of atomic and molecular ions depending on gas temperature in CCRF-discharges. Owing to nonuniform distribution of gas temperature in interelectrode spacing, the provided self-consistent model of CCRF-discharge under high pressure includes time-dependent equation of balance for electron gas, metastable atoms, molecular and atomic ions, Poisson's equation for potential of electrical field, steady state equation of thermal conductivity of atomic-ion gas under boundary conditions of heat exchange, which is calculated according to average parameters, and kinetic reactions include added: $Ar^*+Ar^*\to Ar_2^++e$; $2Ar+Ar^+\to Ar_2^++Ar$; $Ar_2^++e\to Ar+Ar^++e$; $Ar_2^++e\to Ar+Ar$; $Ar_2^*\to 2Ar+h_\nu$; $Ar^*+2Ar\to Ar_2^*+Ar$; $Ar_2^*+Ar_2^*\to e+2Ar+Ar_2^+$; $Ar_2^*+Ar^*\to e+Ar+Ar_2^+$; $e+Ar_2^*\to 2Ar^*+e$; $Ar^*+2Ar \to 3Ar+h_\nu$; $Ar_2^++Ar\to 2Ar+Ar^+$. Velocity factors of the processes being the subject for electron impact depend on local value of provided electrical field and ionization rate with account of electron-electron collision [15].

5. Calculation data

The work paper [20] states that electron density available from experiments is $5\cdot 10^{11}$ sm^{-3} under atmosphere pressure and interelectrode distance – 2mm, according to the calculation data provided at the same work paper, density of electrons and ions approximately is $\approx 7\cdot 10^{11} sm^{-3}$. Calculations made for our model under the similar conditions revealed that concentration maximum is $\approx 3.9\cdot 10^{11}$ sm^{-3} (Fig. 3).

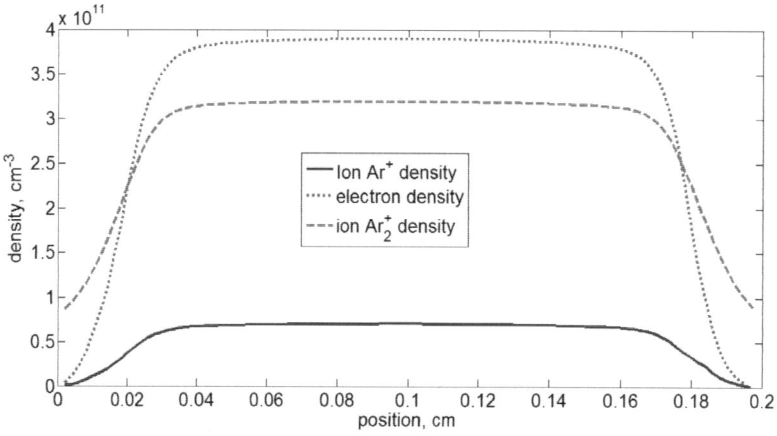

Figure 3. Distribution of average value per the period of charged-particle concentration

6. Conclusion

This research paper describes the self-consistent mathematical model of CCRF-discharge in local approximation under low pressure and non-local approximation under high pressure. Velocity factors of the processes being the subject for electron impact are taken with account of electron-electron collision as well as with account of space variations of gas temperature. The researches have developed the software package, which allows counting parameters of a CCRF-discharge under low pressure in non-local approximation. They performed qualitative and quantitative comparison of numerical calculation data with the data provided by other authors, particularly, with the results of real experiments.

Acknowledgments

The work is performed according to the Russian Government Program of Competitive Growth of Kazan Federal University and this work was supported by Russian Science Foundation, project no. 16-11-10299.

References

[1] Fridman A 2008 *Plasma chemistry* (Cambridge: Cambridge Univ. Press)
[2] Raiser Yu P 2009 *Physics of gas discharge* (Dolgoprudny: Publishing house "Intellect")
[3] Abdullin I Sh, Zheltukhin V S and Kashapov N Ph 2000 *Radio-frequency plasma jet treatment under low pressure. Theory and practice in the application* (Kazan: Publishing house affiliated with Kazan University)
[4] Savinov V P 2013 *Physics of a high-frequency capacitive discharge* (Moscow: PHYSMATLIT)
[5] Riser Yu P, Shneider M N and Yatsenko N A 1995 *Righ-frequency capacitive discharge. Physics. Experimental procedures. Attachments*(Moscow: Publishing house affiliated with MPhTI)
[6] Chebert P and Braithwaite N 2011 *Physics of radio-frequency plasmas* (Cambridge: Cambridge Univ. Press)
[7] Balcon N, Hagelaar G J M and Boeuf J P 2008 Numerical Model of an Argon Atmospheric Pressure RF Discharge *IEEE transactions on plasma science* **36** (5) 2782-7
[8] Karoulina E and Lebedev Yu 1992 Computer simulation of microwave and dc plasmas: comparative characterization of plasmas *J.Phys.D: Appl.Phys.* **25** 401–12
[9] Boeuf J P and Pitchford L C 1995 Two-dimensional model of a capacitively coupled rf discharge and comparisons with experiments in the Gaseous Electronics Conference reference reactor, *Physical Review E.* **51** (2) 1376–90
[10] Lebedev Y A, Tatarinov A V, Epstein I L and Averin K A 2015 Mathematical modeling of the gas bubbles in the microwave discharge of boiling n-heptane *Contemporary Engineering Sciences* **8** (21) 1057-65
[11] Zheltuhin V S, Solovyev P S and Chebakova V Y 2015 Boundary conditions for electron balance equation in the stationary High-Frequency Induction discharges *Research Journal of Applied Sciences* **10** (10) 658-62
[12] Chebakova V Ju, Zheltukhin V S and Dubrovin V T 2016 Problem of radio-frequency discharge at atmospheric pressure in local statement *Applied Mathematical Sciences* **10** (21) 1013-22
[13] Zheltukhin V S, Chebakova V J, Abaidullin R N and Kadyrov R F 2015 On solving of some nonlinear problems of plazma physics *Applied Mathematical Sciences* **9** (88) 4351-9.
[14] Kudryavtsev A A, Smirnov A S and Tsendin L D 2010 *Physics of glow-discharge* (Saint Petersburg: Publishing house "Lan'")
[15] Hagelaar G J M and Pitchford L C 2005 Solving the Boltzmann equation to obtain electron transport coefficients and rate coefficients for fluid models *Plasma Sources Sci. Techn.* **14** 722–33
[16] Ferreira C M, Loureiro J and Ricard A 1985 Populations in the metastable and the resonance levels of argon and stepwise ionization effects in a low-pressure argon positive column, *Journal of Applied Physics* **57** (82) 82–90
[17] Lymberopoulos D P and Economou D J 1993 Fluid simulations of glow discharge & Effect of metastable atoms in argon, *J. Appl. Phys.* **73** (8) 3668–79
[18] Moravej M, Yang X, Hicks R F, Penelon J and Babayan S E 2006 A radio-frequency nonequilibrium atmospheric pressure plasma operating with argon and oxygen *J. Appl. Phys.* **99** Art. 093305
[19] Bora B, Bhuyan H, Favre M, Wyndham E and Chuaqui H 2011 Diagnostic of Capacitively Coupled Low Pressure Radio Frequency Plasma: An Approach through Electrical Discharge Characteristic *International Journal of Applied Physics and Mathematics* **1** (2) 124–8
[20] Balcon N P, Aanesland A, Hagelaar G J M., Boswell R and Boeuf J P 2007 Atmospheric pressure RF discharge in argon: optical diagnostic, fluid model and applications, *28th ICPIG, July 15-20, 2007, Prague, Czech Republic* 957–60
[21] Diatko N A, Ionokh Yu Z, Meshanov A B and Napartovich A P 2005 Study of dark study of development of positive column of glow discharge in argon *Plasma Physics Reports* **31** (10) 939-53

[22] Epstein I L, Gavrilović M, Jovircević S, Konjević N, Lebedev Yu A and Tatarinov A V 2014 The study of a homogeneous column of argon plasma at a pressure of 0.5 torr, generated by means of the Beenakker's cavity *Eur. Phys. J. D* **68** (334)

[23] Dubrovin V T, Gabbasov F G and Chebakova V J 2016 Multidimensional central limit theorem for sums of functions of the trajectories of endomorphisms *Lobachevskii Journal of Mathematics* **37** (4) 409-17

[24] Dubrovin V T, Chebakova V Ju and Zheltukhin V S 2016 Radio-Frequency Discharge at Low Pressure: a Non-Local Problem Statement Approach *Procedia Engineering* **150** 1041–5

[25] Badriev I B, Zadvornov O A 2003 A decomposition method for variational inequalities of the second kind with strongly inverse-monotone operators *Differential equations.* **39** (7) 936-44

[26] Badriev I B; Zadvornov O A and Lyashko A D 2004 A study of variable step iterative methods for variational inequalities of the second kind *Differential equations.* **40** (7) 971-83

[27] Badriev I B, Banderov V V 2014 Numerical method for solving variation problems in mathematical physics *Mechanical components and control engineering III.* **668-669** 1094-7

[28] Badriev I B, Zadvornov O A 2002 Construction and convergence analysis of iterative methods for variational problems with a nondifferentiable functional *Differential equations* **38** (7) 985-91

[29] Badriev I B 1983 Difference-schemes for linear-problems of the filtration theory with discontinuous law *Izvestiya Vysshikh Uchebnykh Zavedenii Matematika.* **5** 3-12

[30] Lisovsky V A 1998 Features of α-γ – transition in RF discharge in argon at low pressure *Technical Physics Magazine* **68** (5) 52-60 (in Russian)

[31] Martinez E Castonos, Kabouzi Y, Makasheva K and Moisan M 2004 Modeling of microwave-sustained plasmas at atmospheric pressure with application to discharge contraction *Physical review E70* 066405

Numerical modeling of the initial fluctuation condensation stage with charge drops

T A Averina[1,2] and G I Zmievskaya[3,4]

[1] Institute of Computational Mathematics and Mathematical Geophysics SB RAS, prospect Akademika Lavrentjeva, 6, Novosibirsk, 630090, Russia
[2] Novosibirsk State University, Pirogova street, 2, Novosibirsk, 630090, Russia
[3] Keldysh Institute of Applied Mathematics, Moscow, Miusskaya Square, 4, 125047, Russia
[4] Moscow Institute of Physics and Technology, Moscow, Kerchenskaya street 1A, 117303, Russia

E-mail: ata@osmf.sscc.ru

Abstract. This paper deals with a mathematical model of the phase transition of the first kind at the initial stage of forming drops in a liquid or in melted state in a volume of steam with a fixed charge on drops. The model of the process is represented by superposition of random diffusion and jump stochastic processes. The algorithms for solving stochastic differential equations (SDEs) of the model of processes, which form the cluster size, allow one to calculate a distribution function of drops according to their size. The kinetic approach makes possible evaluate the role of the Rayleigh capillary instability at the initial condensation stage and to employ the analysis of electrodispersion mechanisms in the production of metal and semiconductor powders.

1. Introduction

Numerical investigation of phase transitions as processes of particle association in clusters is connected with many applications, such as the formation of aerosols in the atmosphere (gas and dust interplanetary clouds, comet tails, etc.), condensation in high-velocity gas streams issuing from a nozzle, polymerization and crystallization, metals vapor deposition and some others.

Vapor condensation phenomena at the nucleation step in terms of non-equilibrium kinetics of physical and chemical processes can be viewed as a continuous process of forming the germs clusters [1]. The numerical model of the initial fluctuation stage of condensation has been developed on the assumption that clustering germs of liquid drops in the vapor are represented by the diffusion in the phase space of the clusters size, the density of the transition probability of the process being described by the Fokker-Planck-Kolmogorov equation (FPK) [2, 3]. Statistical algorithms for solving the SDE, which are equivalent to the FPK were built based on of numerical methods for solving SDE [4, 5]. Also, efficient algorithms for modeling an inhomogeneous Poisson flow have been developed to simulate random jump processes [6].

Earlier the stochastic models and the kinetic approach to the simulation of non-equilibrium collisional processes in gases and plasmas [7] and, also, the formation of new-phase islands [8] and *3C–SiC* powders [9] at the initial stage of vapor condensation were generalized to the formation of porosity in solids. This problem was considered as specific features of the development of pores (as new-phase germs) in metal [10] and 3C–SiC silicon carbide [11]. In this case, at small times instants ranging from 10^{-10} to 10^{-4} s, the phase transition of the first kind occurs, which accompanies Xe^{++} implantation of ions. The above phenomena are caused by fluctuations of the thermodynamic parameters of the "gases or vacancies in a crystal lattice" [2] or "vapor and liquid (melt)" [9].

How does the fluctuation of the phase transition stage of the vapor condensation occur if the drops acquire a charge in the discharge plasma, or as a result of thermionic emission, or otherwise? According to the criterion of the Rayleigh capillary instability, beginning with a certain cluster size the drops start crushing. In the "open" physical system, the non-equilibrium distribution of the germs size becomes bimodal. We use the generalized asymptotically unbiased method of the Rosenbrock type for solving the SDE (for the evolution of the cluster size) at irregular time discretization comprising the moments of drops division. Numerical simulation can be claimed in the problems of dispersing a charge during the preparation of powders. For the composite of particles, it is necessary to include particles of increased strength and resistance to various kinds of damage such as crystal defects, so it is necessary to obtain amorphous powders. Models of the non-equilibrium phase transition are based on the methods developed in the plasma physics [12] as well as computational experiments, object-oriented plasma kinetics codes [13] and calculations of dusty plasma charging [11].

2. A clustering model and approximation used

The simulation of the fluctuation stage of phase transition (PT) of the first kind for the cluster formation process is carried out under conditions of an "open" physical system. Non-equilibrium processes, forming clusters, cause the PT fluctuation instability.

The process of forming clusters (aggregates of molecules formed in the course of the volume condensation or on the surface of a solid) is considered as consecutive reactions:

$$a_1 + a_1 \leftrightarrow a_2,$$
$$\cdots$$
$$a_{n-1} + a_1 \leftrightarrow a_n,$$

leading to diffusion in the clusters sizes space.

According to the ideas discussed in [1], the clusters are formed as a result of the fact that to the new phase germs a greater number of particles is sticked than vaporized per time unit. Fluctuations of thermodynamic parameters (such as temperature, supersaturation and vapor pressure) cause clustering the germs at a new thermodynamic phase. The growth of the cluster size g is described by a random diffusion process in the phase space of the cluster sizes G. The kinetic Kolmogorov-Feller equation for the distribution density of germs according to their size has the following form

$$\frac{\partial f(g,t)}{\partial t} = \frac{1}{kT}\frac{\partial}{\partial g}\left[D(g,t)f(g,t)\frac{\partial \Delta\Phi(g,t)}{\partial g}\right] + \frac{\partial}{\partial g}\left[D(g,t)\frac{\partial f(g,t)}{\partial g}\right] + S - Q, \quad (1)$$

$$f(g,0) = f_0, \quad \frac{df(g,t)}{dg}\Big|_{g=2} = 0, \quad f(g,t)\big|_{g<2} = 0,$$

where g is the number of atoms in the cluster, D is the diffusion coefficient; $\Delta\Phi(g,t)$ is free energy of nucleus formation (or Gibbs energy); S is the source of particles forming a nucleus (the source of "monomers"); Q is the "monomers" particles run-off. The diffusion coefficient is defined as follows:

$$D(g) = g^{2/3} \cdot n_1^2 \beta_k (4\pi)^{1/3} (3V_l)^{2/3} (kT)^{1/2} (2\pi m)^{-1/2},$$

where V_l is the volume per one particle of the liquid phase, $\beta_k = \beta_k(T)$ is the accommodation coefficient of particles on the cluster surface, n_1 is the density of the gaseous phase, m is the mass of vapor molecules. The value $f(g,t)dg$ determines the number of clusters in the size range (per volume unit of the medium). The range of variables: $g \in [2, g_{\max}]$, $t \in [0, \infty]$.

The function $f(g,t)$ is normalized so that the volume unit is, at least, one cluster comprising no less than two particles. Macroscopic characteristics of the gas-vapor mixture (such as the number of clusters per volume unit, etc.) can be calculated with allowance for the function $f(g,t)$. The computer implementation of mathematical models of clustering (1) requires the solution of quasilinear partial integro-differential equations of the second order. In this paper, this equation is solved with the help of the stochastic analog method [2], which is based on the theorems according to which the kinetic equations are uniquely associated with stochastic differential equations [3].

Let us write down the SDE for Markov random process (MP) g (t), whose density function is a solution of the corresponding kinetic equation (1).

Let us consider the case $S = Q = 0$. If from physical considerations the following expression holds:

$$\frac{\partial}{\partial g}\left[\frac{\partial D(g,t)}{\partial g} f(g,t)\right] = 0, \qquad (2)$$

then equation (1) corresponds to the SDE in the Stratonovich sense:

$$dg(t) = -\frac{D(g,t)}{kT}\frac{\partial \Delta\Phi(g,t)}{\partial g}dt - \frac{1}{2}\frac{\partial D(g,t)}{\partial g} + \sqrt{2D(g,t)} \circ dw(t), \quad g(t_0) = g_0. \qquad (3)$$

If (*) is not satisfied, then equation (1) will correspond to the SDE in the Stratonovich sense

$$dg(t) = -\frac{D(g,t)}{kT}\frac{\partial \Delta\Phi(g,t)}{\partial g}dt + \frac{1}{2}\frac{\partial D(g,t)}{\partial g} + \sqrt{2D(g,t)} \circ dw(t), \quad g(t_0) = g_0. \qquad (4)$$

An additional factor of non-disequilibrium of the condensation process is the instability of large drops, which is associated either with the fact that the vapor moves with a certain velocity (the Weber criterion), or condensation drops are in conditions where there is charge on the drop surface [14]. There is implemented a reaction $a_i \to a_k + a_l$, where indices mark the drops, whose specific choice occurs when the law of conservation of mass of clusters holds. In the Kolmogorov-Feller equation (1) one should take into account the type of a collision model (drops decay)

$$\int_G \int_\Gamma f(\xi,t)[\delta(g - \xi - c(\xi,t,\theta)) - \delta(g - \xi)]\pi(\theta,\xi,t)d\theta d\xi, \qquad (5)$$

where δ is the Dirac delta-function, and the functions c, π and a set $\Gamma \in R$ are determined by properties of the medium. The function π is a scalar non-negative function, which specifies probabilities of crushing the drops and also determines the decay intensity. Then equations (1), (5) under condition (2) correspond to the SDE, combining diffusive and the Poisson components:

$$dg(t) = -\frac{D(g,t)}{kT}\frac{\partial \Delta\Phi(g,t)}{\partial g}dt - \frac{1}{2}\frac{\partial Dg,t)}{\partial g} + \sqrt{2D(g,t)} \circ dw(t) + \int_\Gamma z(g(t^-),t^-,\theta)\nu(d\theta \times dt), \qquad (6)$$

where the function z is determined by the type of drop decay, ν is the Poisson random measure on the set $\Gamma \times [0, T_{finish}]$ with the characteristic measure

$$\Pi(A) = \int_A \pi(\theta)d\theta, \quad A \in \Gamma, \quad \Pi(\Gamma) < \infty, \qquad (7)$$

$z(t^-)$ is the value of the function at the point $(t-0)$. As is noted in [15], the introduction of the dependence of the function $\pi(\theta, g, t)$ on g and t makes the Poisson measure inhomogeneous, but it allows one to form a broader class of Markov processes than in the case when $\pi(\theta)$ is independent of g and t.

The next section describes the numerical method for solving stochastic differential equations used. Note that the numerical simulation of the SDE (6) requires not only the modeling of the Wiener component, but also an inhomogeneous Poisson measure, which is a special case of Poisson ensemble [6].

3. Numerical Methods for solving Stochastic Differential Equations

In general, the Cauchy problem for stochastic differential equations in the sense of Ito is written as

$$dY(t) = H_1(Y,t)dt + \tilde{\sigma}(Y,t)dW(t), \quad Y(t_0) = Y_0, t \in [0, T_{fihish}], \tag{8}$$

where $Y(t)$ is a continuous random process of dimension n; $W(t)$ is m-dimensional standard Wiener process; $H_1(Y,t)$ is n-dimensional vector function; $\tilde{\sigma}(Y,t)$ is the matrix function of the size $n \times m$. The initial state of the system is set by a random vector Y_0. If the matrix $\tilde{\sigma}(Y,t)$ is differentiable, then the Cauchy problem for the SDE (8) in the Ito sense corresponds to the Cauchy problem for the SDE in the Stratonovich sense

$$dY(t) = H(Y,t)dt + \tilde{\sigma}(Y,t) \circ dW(t), \quad Y(t_0) = Y_0, \tag{9}$$

where

$$H(Y,t) = H_1(Y,t) - \frac{1}{2}\frac{\partial \tilde{\sigma}(Y,t)}{\partial Y}\tilde{\sigma}(Y,t), \quad \left(\frac{\partial \tilde{\sigma}(Y,t)}{\partial Y}\tilde{\sigma}(Y,t)\right)_i = \sum_{k=1}^{n}\sum_{j=1}^{n}\frac{\partial \tilde{\sigma}_{ij}(Y,t)}{\partial y_k}\tilde{\sigma}_{kj}(Y,t).$$

Equations (8) and (9) are equivalent. The one-dimensional probability density function of the SDE solutions satisfies the direct Kolmogorov equation

$$\frac{\partial p(Y,t)}{\partial t} = -\sum_{i=1}^{n}\frac{\partial}{\partial y_i}\left[(H_1)_i(Y,t)p(Y,t)\right] + \frac{1}{2}\sum_{i=1}^{n}\sum_{j=1}^{n}\frac{\partial^2}{\partial y_i \partial y_j}\left[b_{ij}p(Y,t)\right] \quad p(Y,t_0) = \psi_0(Y_0), \tag{10}$$

where ψ_0 is the probability density function of the distribution of the phase coordinates at the initial time t_0, $B = \tilde{\sigma}\tilde{\sigma}^T$, $b_{ij} = \sum_{k=1}^{m}\tilde{\sigma}_{ik}\tilde{\sigma}_{jk}$ is the diffusion matrix.

For the numerical solution of the SDE in the Stratonovich sense (9) we use the generalized method of the Rosenbrock type

$$Y_{k+1} = Y_k + [I - \frac{h_k}{2}\frac{\partial H}{\partial Y}(Y_k)]^{-1}[h_k H(Y_k) + h_k H(Y_{k+1}^p)] + \sqrt{h_k}\sum_{j=1}^{n}(\tilde{\sigma}_{\bullet j_2}(Y_k) + \tilde{\sigma}_{\bullet j_2}(Y_{k+1}^p))\xi_{jk}], \tag{11}$$

$$Y_{k+1}^p = Y_k + \tilde{\sigma}(Y_k)\sqrt{h_k}\xi_k.$$

This method converges with second order for the SDE systems with single noise and is asymptotically unbiased. These properties of the method allow one to use it for solving unstable SDE systems with high accuracy.

Here Y_k is the numerical solution of (9) at a grid node t_k; $h_k = t_{k+1} - t_k$ is the integration step; $\xi_k = (\xi_{1k}, \ldots, \xi_{nk})^T$ is the vector of mutually independent normal random variables with zero mean and unit variance; $\tilde{\sigma}_{\bullet j}^{(l)}$ denotes the j-th column of the matrix $\tilde{\sigma}^{(l)}$; I is the identity matrix; $\partial a^{(l)} / \partial Y$ is the

Jacobi matrix. We supplement the stochastic differential equation (9) with a stochastic integral with respect to the Poisson measure

$$dY(t) = H(Y,t)dt + \tilde{\sigma}(Y,t) \circ dW(t) + \int_{\Gamma} z(Y(t^-),t^-,\theta)\nu(d\theta \times dt),\ Y(t_0) = Y_0, \tag{12}$$

$z(Y,t,\theta)$ is n-dimensional vector-function; π is the non-negative scalar function defining characteristic measure (7); the set $\Gamma \in R^n$.

Let us write down the *algorithm of the simulation of the SDE solution with a Poisson component (12) when the Poisson measure is homogeneous*:
previously, for each trajectory we simulate the time instants of the trajectory discontinuity points (for model (1) these are the time instants of decay) $\tau_1,\ldots,\tau_l < T_{fihish}$ by the formula $\tau_s = \tau_{s-1} + \zeta_s$, $\tau_0 = 0$, where ζ_s are independent random variables with the probability density $p(x) = \Pi(\Gamma)\exp(-\Pi(\Gamma)x)$ (according to the formula $\zeta_s = -\ln\alpha_s / \Pi(\Gamma)$, α_s are independent uniformly distributed random variables on the interval $(0,1)$);

0) k:=0; simulate according to the specified initial values;
1) solve equation (9) on the interval $[\tau_k, \tau_{k+1}]$ by numerical method (11) with the step h and find Y_{k+1} at the time instant τ_{k+1}; the step h should be coordinated with the transition intensity, for example, $h \le 0.1/\Pi(\Gamma)$;
2) $k := k + 1$;
3) simulate a random variable θ according to the density $p(x) = \pi(x)/\Pi(\Gamma)$, then $Y_k := Y_k + z(Y_k, \tau_k, \theta)$;
4) if $\tau_k < T$, then go to step 1, otherwise the trajectory simulation is completed.

Using the method of maximum cross-section, the algorithm can be generalized to the case of inhomogeneous Poisson measure [6].

4. Numerical Experiments

With the algorithm proposed let us carry out the numerical simulation, illustrating the formation of the condensation nuclei with a fixed charge on the drops with the diffusion coefficient and the Gibbs energy of the form

$$D = D_0 g^{2/3}, \tag{13}$$

$$\Delta\Phi = \Delta\Phi_{chem} + \Delta\Phi_{st} + \Delta\Phi_{ch} = -ag + bg^{2/3} + cg^{-1/3}, \tag{14}$$

where $\Delta\Phi_{chem} = -ag$, $a = \eta(\chi_\beta - \chi_\alpha)$, and $(\chi_\beta - \chi_\alpha)$ is the difference between the chemical potentials of phases, and $\eta = 2\pi/3V$ and V is the volume of a water molecule. The second term in (14) is $\Delta\Phi_{st} = bg^{2/3}$, where $b = (36\pi)^{1/3}V^{2/3}\sigma$ and σ being the surface tension at the boundary of a drop. The third term in (14), related to the drop charge, is $\Delta\Phi_{ch} = cg^{-1/3}$, where $c \sim eZ$ with Z being the drop charge measured in thousands of elementary charges. We assume that the drop charge is matched in accordance with [9, 11, 12]. For coefficients (13) - (14), the equation model (6) takes the form

$$dg(t) = H(g)dt + \tilde{\sigma}(g) \circ dw(t) + \int_{\Gamma} z(g(t^-),t^-,\theta)\nu(d\theta \times dt), \quad g(t_0) = g_0, \tag{15}$$

where

$$H=-\frac{D_0}{kT}\left(-ag^{2/3}+\frac{2}{3}bg^{1/3}-\frac{c}{3}cg^{-2/3}\right)-\frac{D_0}{3}g^{-1/3},\quad \tilde{\sigma}=\sqrt{2D(g,t)},\quad z=z_0 g. \tag{16}$$

According to the Rayleigh criterion for the charged drops of radius r_d with a constant charge Z the instability condition in dimensionless form looks like the following [14]:

$$W=\frac{Z^2}{16\pi\sigma r_d^3}\geq 1.$$

We have studied the fluctuation stage of the phase transition of the water vapor condensation with allowance for the Rayleigh instability criterion of the charged drop [14]. In this case the temperature $T=350K$, the pressure vapor $p=0.5M\Pi a$, the saturation $s=2$, the value of a charge is measured in $1.6\cdot 10^{-16}$ coulomb, the clusters concentration of the condensation vapor is $3\cdot 10^5\,cm^{-3}$. The parameters were taken from [16]. The parameters (figures 1-5) provide the value of the dimensionless time step $h=5.43\cdot 10^{-11}$ s. We have considered the following dimensionless parameters of the model:

$$a=0.24318,\ b=2.366613,\ c=25,\ D_0=0.25,\ kT=1,\ \Gamma=(0,1),\ \pi(\theta)\equiv 1,\ \Pi(\Gamma)=\int_\Gamma \pi(\theta)d\theta=1. \tag{17}$$

The Gibbs energy (14) for the model parameters (17) is shown in figure 1.

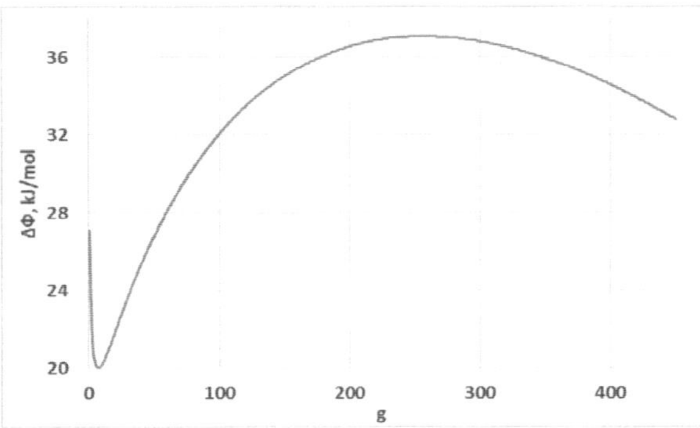

Figure 1. The Gibbs energy (14), (17) of forming drops (with the drop charge) as a function of the cluster size g.

For the numerical solution of SDE (15) with parameters (16)-(17) we used the algorithm developed. The grid on the interval [0, 30] is constructed for each trajectory of the numerical solution and includes a uniform grid with the step $h=0.5$ and all the time instants of the drop decay. The simulation of the drop decay time instants was carried out by the formula $s_0=0,\ \ s_k=s_{k-1}-\ln\alpha/\Pi(\Gamma),\ \ k=1,\ldots,k_T,\ \ s_{k_T}<T_{finish}=30$. The RAND pseudorandom generator was used for the simulation of uniform random variables α on the interval [0, 1]. In the numerical experiment, 10^6 trajectories of the solution were used.

The initial conditions were chosen from the "critical" area of energy values $\Delta\Phi-kT\sim 3kJ/mol$. The critical value g_{crit} is found from the condition $\partial\Delta\Phi/\partial g=0$. The initial cluster size was set larger than the critical size ($g_0=400,\ \ g_{\max}=420,\ \ g_{crit}=274$). Since the initial value was not random but the same for all simulated trajectories, then the distribution density of nuclei clusters at the initial time is the delta-function. The figures show the evolution of the distribution of clusters size of the water (figures 2, 4 without taking into account the drops charge; figures 3, 5 with the drops charge).

The drops decay as a result of capillary instability leads to a non-equilibrium distribution of the clusters size (figures 3, 5).

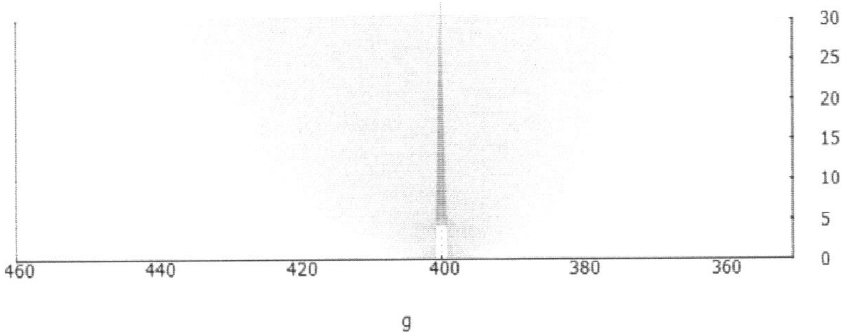

Figure 2. The evolution of the distribution of cluster size. (without drop charge).

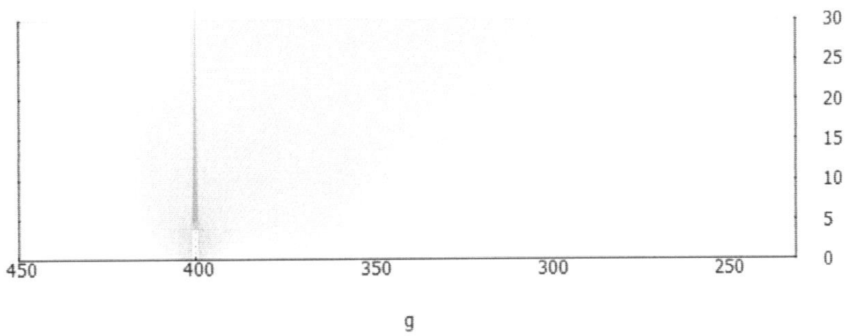

Figure 3. The evolution of the distribution of cluster size. (with the drop charge)

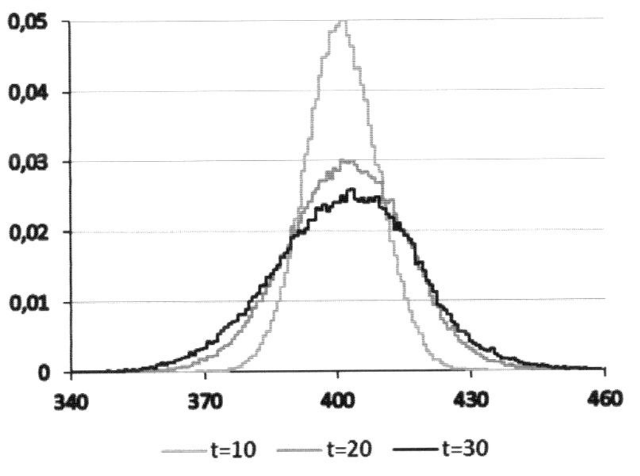

Figure 4. Histograms of the cluster size distribution (without drop charge) at three time instants.

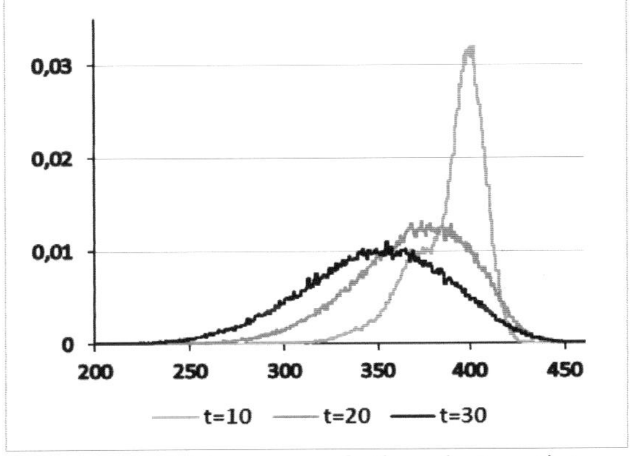

Figure 5. Histograms of the cluster size distribution (with the drop charge) at three time instants.

The algorithm developed allows one to simulate the phase transition of the first kind at the initial stage of the drops formation with a fixed charge on the drops. The numerical experiments show the fluctuation instability resulting in the bimodal distribution of the condensate drop size. The kinetic approach makes possible evaluate the role of the Rayleigh capillary instability at the initial condensation stage and to employ the analysis of electrodispersion mechanisms in the production of metal and semiconductor powders.

Acknowledgments
This work was partially supported by the Russian Foundation for Basic Research (Projects no. 14-01-00787, no. 15-01-05052) and the program "Leading Scientific Schools" (grant 5111.2014.1).

References

[1] Zeldovich Ya B 1942 *Eksp. Teor. Fiz.* **12** 525
[2] Zmievskaya G I, Averina T A and Bondareva A L 2015 *Appl. Numer. Math.* **93** 15–29
[3] Korolyuk V S, Portenko N I, Skorokhod A V and Turbin A F 1985 *A Handbook on Probability Theory and Mathematical Statistics* (Moscow: Nauka)
[4] Averina T A and Artemiev S S 1986 *Soviet Math. Dokl.* **33** 736-738
[5] Averina T A *Bul Buryat State Univer.* **2012** **9** 91-94
[6] Averina T A 2009 *Comp. Math. and Math. Phys.* **50** pp 12-18
[7] Zmievskaya G I 1997 *Plasma Phys. Rep.* **23** **(4)** 340
[8] Zmievskaya G I and Bondareva A. L. 2005 Dokl. Phys. **50** **(4)** 169
[9] Zmievskaya G I and Bondareva A L 2011 Plasma Phys. Reports **37** 87-95
[10] Zmievskaya G I, Bondareva A L, Levchenko V D and Levchenko T V 2007 J. Phys. D 40 4842
[11] Zmievskaya G I, In'kov L V, Levchenko V D and Levchenko T V 2004 Poverkhn. Rentgen. Sinkhrotr. Neitron. Issled. No. 5 106-112
[12] Sigov Yu S 2001 *Computer Simulation: The Link between the Past and Future of Plasma Physics (Selected Articles)* ed. G I Zmievskaya and V D Levchenko (Moscow: Fizmatlit)
[13] Levchenko V D 2002 *New in Synergetics: a Look into the Third Millenium* ed. G G Malinetskii and S P Kurdyumov (Moscow: Nauka) 159
[14] Taylor G.I. 1964 Disintegration of water drops in an electric field *Proc. R. Soc.* London **A280** 383–397
[15] Paraev Yu I 1976 *Introduction to Statistical dynamics of Processes of Control and filtering.* (Moscow: Sovradio)
[16] Zmievskaya G I 1996 *Dynamics of Transport in Plasmas and Charged Beams* ed G Maino and M Ottaviani (Singapore: World Scientific) 84-98

IOP Conf. Series: Materials Science and Engineering **158** (2016) 012011 doi:10.1088/1757-899X/158/1/012011

Longitudinal and transverse bending by a cylindrical shape of the sandwich plate stiffened in the end sections by rigid bodies

I B Badriev[1], **M V Makarov**[1, 2] **and V N Paimuhin**[1, 2]

[1]Kazan Federal University, 18 Kremlyovskaya Street, 420008, Kazan, Russia
[2]Kazan National Research Technical University, 10 K.Marks Street, 420111, Kazan, Russia

E-mail: ildar.badriev1@mail.ru

Abstract. We study the problems of deformation mechanics of sandwich constructions with taking into account the interaction with the contour reinforcing rods. To derive the basic equations of equilibrium, static boundary conditions for the shell and reinforcing rods, as well as conditions of the kinematic conjugation the carrier layers with a core, the carrier layers and a core with reinforcing rods we use a generalized variational Lagrange principle. We reduce the boundary value problem on the to the integral-algebraic system of Volterra equations of the second kind. To approximate the obtained integral equations of Volterra type a collocation method with Gaussian nodes and a method for constructing the integrating matrices are proposed. For the numerical realization of the proposed methods we have developed a software package. Numerical calculations were performed. Analyze the results of numerical experiments is carried out.

1. Introduction. Statement of the problem

Sandwich plates and shells with external (carrier) layers of rigid materials and cores relating to a transversely soft class (cores of cellular or folded structure, reinforced with polystyrene etc.) are one of the most popular structural elements for various purposes. The external layers of these elements in the contour are always reinforcing by the rods to ensure the transfer of the load on the carrier layers of interaction with other construction elements (Figure 1).

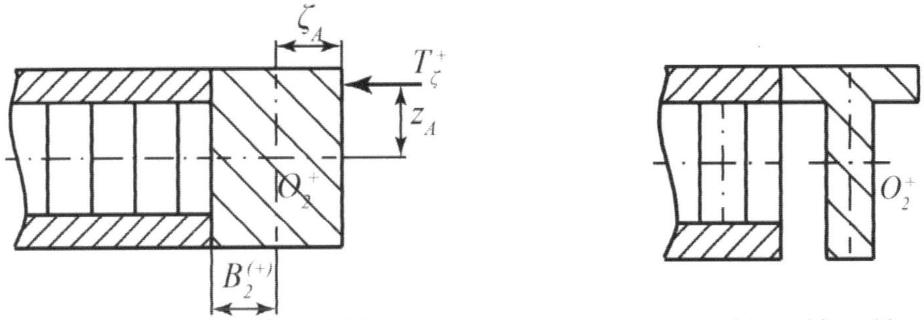

$a)\ \tau_1 \neq 0, Q_{11}^{(k)} \neq 0, Q_{13}^{(k)} \neq 0, L_{11}^{(k)} \neq 0$ $\qquad b)\ \tau_1 = Q_{11}^{(k)} = Q_{13}^{(k)} = L_{11}^{(k)} = 0$

Figure 1. Schemes of fixing the plate and reinforcing rods and loads.

Problems of deformation mechanics of sandwich constructions without taking into account the interaction with the contour reinforcing rods are studied in many of papers [1-13], from which we select articles [8-13]. In these works for sandwich plates and shells with transversal-soft core at small strains and middle displacements was built the refined geometrically nonlinear theory, which allows to describe the process of subcritical deformation and to identify all the possible buckling of carrier layers (in-phase, antiphase mixed bending and bending-shear mixed, as well as arbitrary, which include all of the above). Summarizing the results of these studies to the sandwich plates and shells with the contour reinforcing rods (Figure 2), we introduce as unknowns, as in [5-7], the contact forces of interaction the carrier layers with the a core, and the carrier layers, and a core with reinforcing rods at all points in their conjugation surfaces [14, 15].

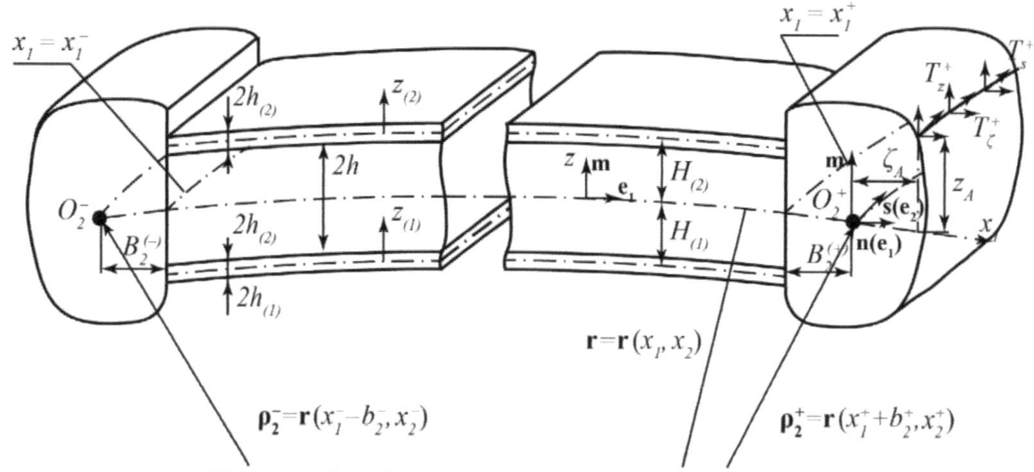

Figure 2. The scheme of the plate and the reinforcing rods.

To derive the basic equations of equilibrium, static boundary conditions for the shell and reinforcing rods, as well as conditions of the kinematic conjugation the carrier layers with a core, the carrier layers and a core with reinforcing rods, as in [14], we use the previously proposed a generalized variational Lagrange principle [16]. With regard to the sandwich plate experiencing a cylindrical bending, derived in this way the equilibrium equations will have the form (hereinafter $k=1,2$, is the layer number, $\delta_{(1)}=1,\ \delta_{(2)}=-1$):

$$\frac{dN_1^{(k)}}{dx}+\frac{\delta_{(k)}E_3}{2h}\left(w^{(2)}-w^{(1)}\right)+X_3^{(k)}=0 \tag{1}$$

$$\frac{dT_{11}^{(k)}}{dx}+\delta_{(k)}q_1+X_1^{(k)}=0, \tag{2}$$

$$\mu_1=u_1^{(1)}-u_1^{(2)}-H_{(1)}\omega^{(1)}-H_{(2)}\omega^{(2)}+\frac{2hq_1}{G_{13}}-\frac{2h^3}{3E_3}\frac{d^2q_1}{dx^2}=0\,. \tag{3}$$

In these equations, introduced in consideration of the tangential forces $T_{11}^{(k)}$ and generalized shear forces $N_1^{(k)}$ through the unknown axial displacements $u_1^{(k)}$, deflections $w^{(k)}$ of medial surface of the carrier layers and the tangential stresses q_1 in the core, constant over its thickness, are expressed by the formulas

$$N_1^{(k)} = Q_1^{(k)} + H_{(k)}q_1,\ Q_1^{(k)} = \frac{dM_{11}^{(k)}}{dx} + \left(T_{11}^{(k)}\omega^{(k)}\right),$$
$$M_{11}^{(k)} = -D_{11}^{(k)}\frac{d^2w^{(k)}}{dx^2},\quad T_{11}^{(k)} = B_{11}^{(k)}\left[\frac{du_1^{(k)}}{dx} + \frac{1}{2}\left(\omega^{(k)}\right)^2\right], \omega^{(k)} = \frac{dw^{(k)}}{dx}, \tag{4}$$

where $B_{(k)} = 2h_{(k)}E^{(k)}/(1-\nu_{12}^{(k)}\nu_{21}^{(k)})$, $D_{(k)} = B_{(k)}h_{(k)}^2/3$ are the tension and compression stiffness and bending stiffness of k-th layer having a thickness $2h_{(k)}$ and made of a material with a Young's modulus and Poisson's ratios $\nu_{12}^{(k)}$, $\nu_{21}^{(k)}$; $H_{(k)} = h_{(k)} + h$,, $2h$ is the thickness of a core having the elasticity modulus E_3 in the transverse direction and transverse shear modulus G_{13}; $M_{11}^{(k)}$ is the internal bending moment at k-th layer; $X_{(k)}^1$, $X_{(k)}^3$ are the surface load components actuated to the middle surface in k-th layer, $Q_1^{(k)}$ are the shear forces in k-th carrier layer excluding the tangential stresses in the core. If we introduce the unknown contact forces of the interaction the carrier layers and a core with reinforcing rods, the solutions of equations (1) - (4) must satisfy the boundary conditions

$$M_{11}^{(k)} = L_{11}^{(k)},\ \frac{dM_{11}^{(k)}}{dx} + T_{11}^{(k)}\omega^{(k)} + H_{(k)}q_1 = Q_{13}^{(k)} + h\tau_1,\ T_{11}^{(k)} = Q_{11}^{(k)},\ q_1 = \tau_1,\ \text{at } x_1 = x_1^{\pm} \tag{5}$$

where $Q_{11}^{(k)}, Q_{13}^{(k)}$, $L_{11}^{(k)}$ are the unknown Loop contact forces and the bending moments, τ_1 are the surface contact stresses that are formed at the points of coupling the reinforcing rod with a core.

We assume that the loading of plate forms in it the stress-strain state, which is symmetrical with respect to the cross section $x_1 = 0$. Then this cross section is necessary to satisfy the symmetry conditions of solutions of equations (1) - (4) of the following types

$$\omega^{(k)} = 0,\ N_1^{(k)} = 0,\ u_1^{(k)} = 0,\ q_1 = 0 \tag{6}$$

If there is a cylindrical bending sandwich plate, the rods, reinforcing it in the face sections $x_1 = x_1^-, x_1 = x_1^+$ are transformed into an absolutely solid body. Therefore, the equilibrium equations of the reinforcing rod in cross-section $x_1 = x_1^+$ will have the following form

$$f_u^+ = -\left(Q_{11}^{(1)} + Q_{11}^{(2)}\right) + T_\zeta^+ = 0, \tag{7}$$

$$f_w^+ = -\left(Q_{13}^{(1)} + Q_{13}^{(2)} + 2h\tau_1\right) + T_z^+ = 0, \tag{8}$$

$$f_\varphi^+ = -\left[L_{11}^{(1)} + L_{11}^{(2)} - Q_{11}^{(1)}H_{(1)} + Q_{11}^{(2)}H_{(2)} + B_2^{(+)}\left(Q_{13}^{(1)} + Q_{13}^{(2)} + 2h\tau_1\right)\right] + m_s^+ = 0, \tag{9}$$

where T_ζ^+, T_z^+ are the components of the external loop load нагрузки $\mathbf{T}^+ = T_\zeta^+\mathbf{n} + T_z^+\mathbf{m}$, components applied to the point of line $\mathbf{r}_A = \zeta_A\mathbf{n} + z_A\mathbf{m}$ (Fig. 2), $m_s^+ = T_\zeta^+ z_A - T_z^+\zeta_A$ is the loop bending moment. Emerging in the reinforcing rods the displacements U, W and rotation φ must satisfy the kinematic conditions of the coupling the rods with carrier layers

$$u_1^{(k)} - \left(U - \delta_{(k)}H_{(k)}\varphi\right) = 0,\ \text{at } \delta Q_{11}^{(k)} \neq 0 \tag{10}$$

$$w^{(k)} - \left(W + B_2^{(+)}\varphi\right) = 0,\ \text{at } \delta Q_{13}^{(k)} \neq 0 \tag{11}$$

$$w_{,1}^{(k)} + \varphi = 0,\ \text{at } \delta L_{11}^{(k)} \neq 0 \tag{12}$$

to which it is necessary to add the kinematic conditions

$$w^{(1)} + w^{(2)} + \frac{2h^2}{3E_3}q_{1,1} - 2\left(W + B_2^{(+)}\varphi\right) = 0, \tag{13}$$

taking place at $\delta\tau_1 \neq 0$. Note that in (10) - (13) the unknowns U, W and φ are the displacements of the point O_2^+ (Figure 2) and the rotation angle of the cross section of the reinforcing body.

Under the conditions of $Q_{11}^{(k)} = Q_{13}^{(k)} = L_{11}^{(k)} = \tau_1 = 0$ (at absence in the sections $x_1 = x_1^\pm$ the reinforcing rods, see, eg, Figure 1) to solve the problem (1) - (5) with respect to the vector functions of unknowns $X_1 = (w^{(1)}, w^{(2)}, u_1^{(1)}, u_1^{(2)}, q_1)$ in [8] a numerical method based on the method of summation identities is proposed. In this case, however, for the boundary value problem in the form of five differential equations (1) - (3) with the boundary conditions (5), (6) and ten algebraic equations (7) - (13) it is required to determine the vector function of unknowns $X_2 = (w^{(1)}, w^{(2)}, u_1^{(1)}, u_1^{(2)}, q_1, Q_{11}^{(1)}, Q_{11}^{(2)}, Q_{13}^{(1)}, Q_{13}^{(2)}, L_{11}^{(1)}, L_{11}^{(2)}, \tau_1, U, W, \varphi)$.

2. Reduction of the boundary value problem to the integro-algebraic form

We reduce the formulated boundary value problem (1) - (13) on the definition of the function $X_2(x)$ to the integral-algebraic system of Volterra equations of the second kind by using additional relationships to determine the unknown integration constants. Note that, in the original differential problem includes derivatives of order $2n$ of the unknown function, while the integral equation will contain only n-th derivatives. This reduction is carried out by integrating the equations (1)–(3), satisfying the conditions (5), (6) and using the following equations

$$u_1^{(k)}(x) = \int_0^x \frac{du_1^{(k)}}{ds} ds, \quad q_1(x) = \int_0^x \frac{dq_1}{ds} ds,$$

$$\frac{dw^{(k)}}{dx} = \int_0^x \frac{d^2 w^{(k)}}{ds^2} ds, \quad w^{(k)}(x) = w_a^{(k)} - \int_x^a \int_0^\xi \frac{d^2 w^{(k)}}{ds^2} ds d\xi,$$

where $w_a^{(k)} = w|_{x=x_1^+}$. As a result, relatively the vector function

$$X(x) = (w_{,11}^{(1)}, w_{,11}^{(2)}, u_{,1}^{(1)}, u_{,1}^{(2)}, q_{1,1}, w_a^{(1)}, w_a^{(2)}, Q_{11}^{(1)}, Q_{11}^{(2)}, Q_{13}^{(1)}, Q_{13}^{(2)}, L_{11}^{(1)}, L_{11}^{(2)}, \tau_1, U_A, \varphi), \quad x \in [0, x_1^+] = [0, a]$$

we obtain the following system of integro-algebraic equations:

$$\begin{aligned} &L_{11}^{(k)} - D_{11}^{(k)} \frac{d^2 w^{(k)}}{dx^2} + \int_x^a T_{11}^{(k)} \int_0^\xi \frac{d^2 w^{(k)}}{ds^2} ds d\xi + H_{(k)} \int_x^a \int_0^\xi \frac{dq_1}{ds^2} ds d\xi + \\ &+\delta_{(k)} \frac{E_3}{2h} \int_x^a \int_0^\xi \left[w_a^{(2)} - w_a^{(1)} - \int_s^a \int_0^\eta \left(\frac{d^2 w^{(2)}}{ds^2} - \frac{d^2 w^{(1)}}{ds^2} \right) dy d\eta \right] ds d\xi = -\int_x^a \int_0^\xi X_3^{(k)} ds d\xi, \end{aligned} \tag{14}$$

$$Q_{11}^{(k)} - B_{11}^{(k)} \left(\frac{du_1^{(k)}}{dx} + \frac{1}{2} \left(\int_0^x \frac{d^2 w^{(k)}}{ds^2} ds \right)^2 \right) + \delta_{(k)} \int_x^a \int_0^\xi \frac{dq_1}{ds} ds d\xi = 0 \tag{15}$$

$$\frac{2h^3}{3E_3} \frac{dq_1}{dx} + \frac{2h}{G_{13}} \int_x^a \int_0^\xi \frac{dq_1}{ds} ds d\xi + \int_x^a \int_0^\xi \left(\frac{du_1^{(1)}}{ds} - \frac{du_1^{(2)}}{ds} - H_{(1)} \frac{d^2 w^{(1)}}{ds^2} - H_{(2)} \frac{d^2 w^{(2)}}{ds^2} \right) ds d\xi = 0 \tag{16}$$

$$\int_0^a \frac{dq_1}{dx} dx - \tau_1 = 0, \tag{17}$$

$$\frac{\delta_{(k)}}{2h} \int_0^a \left\{ E_3 \left[w_a^{(2)} - w_a^{(1)} - \int_x^a \int_0^\xi \frac{d^2 w^{(2)}}{ds^2} - \frac{d^2 w^{(1)}}{ds^2} ds d\xi \right] \right\} dx + Q_{13}^{(k)} + h\tau_1 = -J_3 X_3^{(k)} \tag{18}$$

to which must be added the algebraic equilibrium equations (7)-(9) for the rod.

If the point of force application T_ζ^+ has a displacement W_A in the direction of the axis z equal to zero, and the displacement in the direction of the axis ζ equal is U_A, then by virtue of (1) $u_1^{(k)}\Big|_{x=a}=\int\limits_0^a \frac{du_1^{(k)}}{dx}dx,\ W=\zeta_A\varphi,\ U=U_A-z_A\varphi$ the coupling conditions (10) - (13) take the form

$$\int\limits_0^a \frac{du_1^{(k)}}{dx}dx-U_A+\left(z_A+\delta_{(k)}H_{(k)}\right)\varphi=0, \tag{19}$$

$$w_a^{(k)}-\left(\zeta_A+B_2^{(+)}\right)\varphi=0, \tag{20}$$

$$\int\limits_0^a \frac{d^2w^{(k)}}{dx}dx+\varphi=0, \tag{21}$$

Thus, to determine the introduced into consideration sixteen unknowns $X\in H^{\times 5}\times R^{\times 11}$ the system of sixteen resolution integro-algebraic equations (14) - (21) and the rod of equilibrium equations (7), (9) are composed, where $H=L_2(0,a)$, and a is the half-length of the plate.

3. Approximation of integral equations by the collocation method with Gaussian nodes

To approximate the obtained integral equations of Volterra type in [17] a collocation method with Gaussian nodes and a method for constructing the integrating matrices are proposed. We introduce the integral operators $\mathcal{J}(f)=\int\limits_0^x f(\xi)d\xi,\ \mathcal{J}^*(f)=\int\limits_x^a f(\xi)d\xi,\ \mathcal{J}_3(f)=\int\limits_0^a f(\xi)d\xi$ that are replaced by their finite-dimensional analogues in the form of integrating matrices J_1,J_2,J_3, respectively. соответственно. For this purpose, in the segment $[0,a]$ we introduce a grid $\omega=\{x_i,i=1,2,\dots,N\}$ according to the Gauss quadrature formula:

$$J_3(f)=\sum_{i=1}^N d_i f(x_i),$$

where $\{d_i\}$, $\{x_i\}$ are the weights and collocation nodes connected with the roots of the Legendre polynomial of degree N. Let f_i be a value of f in the node x_i: $f_i=f(x_i)$, we approximate f in the segment $[0,a]$ by interpolating function $f(x)\approx\sum_{i=1}^N f_i l_i(x)$. As an interpolation function the linear combination of Lagrange basis functions $\{l_i\}$ over the nodes $\{x_i\}$ is selected. Thus, expanding the function l_i in Legendre polynomials, we construct integrating matrix J_1,J_2. Let us consider the finite-dimensional operators

$$\mathcal{I}_1=(\mathcal{I}_{1w}^{(1)},\mathcal{I}_{1w}^{(2)}\mathcal{I}_{1u}^{(1)},\mathcal{I}_{1u}^{(2)},\mathcal{I}_{1q}^{(1)}):H_h^{\times 5}\times R^{\times 11}\to H_h^{\times 5},\ \mathcal{I}_2=(\mathcal{I}_{2w}^{(1)},\mathcal{I}_{2w}^{(2)}\mathcal{I}_{2u}^{(1)},\mathcal{I}_{2u}^{(2)},\Theta^{\times N}):H_h^{\times 5}\times R^{\times 11}\to H_h^{\times 5},$$

$$\Gamma=(\Gamma_\tau,\Gamma_{wa}^{(1)},\Gamma_{wa}^{(2)}):H_h^{\times 5}\times R^{\times 11}\to R^{\times 3},$$

$$\mathcal{Q}=(\mathcal{Q}_{11}^{(1)},\mathcal{Q}_{11}^{(2)},\mathcal{Q}_{13}^{(1)},\mathcal{Q}_{13}^{(2)},\mathcal{L}_{11}^{(1)},\mathcal{L}_{11}^{(2)}):H_h^{\times 5}\times R^{\times 11}\to R^{\times 6},$$

$$\mathcal{T}_\zeta=(\mathcal{M}_\zeta,\mathcal{T}_\zeta):H_h^{\times 5}\times R^{\times 11}\to R^{\times 2}$$

by the formulas

$$\mathcal{I}_{1w}^{(k)}X(x)=\mathcal{L}_{11}^{(k)}-D_{11}^{(k)}w_{,11}^{(k)}+H_{(k)}J_2J_1q_{1,1}+\delta_{(k)}J_2J_1\frac{E_3}{2h}\left[w_a^{(2)}-w_a^{(1)}-J_2J_1\left(w_{,11}^{(2)}-w_{,11}^{(1)}\right)\right],$$

$$\mathcal{I}_{2w}^{(k)}X(x)=J_2T_{11}^{(k)}J_1w_{,11}^{(k)},$$

$$\mathcal{I}_{1u}^{(k)}X(x)=\mathcal{Q}_{11}^{(k)}-B_{11}^{(k)}u_{1,1}^{(k)}+\delta_{(k)}J_2J_1q_{1,1},$$

$$\mathcal{I}_{2u}^{(k)} X(x) = -\frac{B_{11}^{(k)}}{2}\left(J_1 w_{,1}^{(k)}\right)^2,$$

$$\mathcal{I}_{1q} X(x) = \left(\frac{2h^3}{3E_3} + \frac{2h}{G_{13}} J_2 J_1\right) q_{1,1} + J_2 J_1\left(u_{1,1}^{(1)} - u_{1,1}^{(2)} - H_{(1)} w_{,11}^{(1)} - H_{(2)} w_{,11}^{(2)}\right);$$

$$\Gamma_\tau X(x) = J_3 q_{1,1} - \tau_1;$$

$$\Gamma_{wa}^{(k)} X(x) = \frac{\delta_{(k)}}{2h} J_3\left\{E_3\left[w_a^{(2)} - w_a^{(1)} - J_2 J_1\left(w_{,11}^{(2)} - w_{,11}^{(1)}\right)\right]\right\} + Q_{13}^{(k)} + h\tau_1;$$

$$Q_{11}^{(k)} X(x) = J_3 u_{1,1}^{(k)} - U_A + \left(z_A + \delta_{(k)} H_{(k)}\right)\varphi;$$

$$Q_{13}^{(k)} X(x) = w_a^{(k)} - \left(\zeta_A + B_2^{(+)}\right)\varphi;$$

$$\mathcal{L}_{11}^{(k)} X(x) = J_3 w_{,11}^{(k)} + \varphi;$$

$$\mathcal{M}_s X(x) = \left[\mathcal{L}_{11}^{(1)} + \mathcal{L}_{11}^{(2)} - Q_{11}^{(1)} H_{(1)} + Q_{11}^{(2)} H_{(2)} + B_2^{(+)}\left(Q_{13}^{(1)} + Q_{13}^{(2)} + 2h\tau_1\right)\right];$$

$\mathcal{T}_\zeta X(x) = Q_{11}^{(1)} + Q_{11}^{(2)}$; then formulated geometrically nonlinear problem can be represented in a finite-dimensional operator form

$$\mathcal{A}_1 X + \mathcal{A}_2(X) = F, \tag{22}$$

where $\mathcal{A}_1 = (\mathcal{I}_1, \Gamma, Q, \mathcal{T}) : H_h^{\times 5} \times R^{\times 11} \to H_h^{\times 5} \times R^{\times 11}$ is a linear operator, $\mathcal{A}_2 = (\mathcal{I}_2, \Theta^{\times 11}) : H_h^{\times 5} \times R^{\times 11} \to H_h^{\times 5} \times R^{\times 11}$ is a non-linear operator; in general, the right side is $F(T_\zeta^+, T_z^+, m_s^+, X_3^{(k)}) \in H^{\times 5} \times R^{\times 11}$. For the problem of longitudinal-transverse bending of the plate for the respective component of the external loop load and the linear moment applied to the rod and the transverse distribution of the load on the plate we set $T_z^+ = m_s^+ = X_3^{(k)} = 0$, then $F = (\Theta^{\times(5N+10)}, T_\zeta^+)$.

Thus for the functions included in the system of differential equations (1)-(3) with the highest derivative of order $2n$, the finite-dimensional scheme (22) is constructed relatively the n-th derivative of the solution of a boundary value problem. After solving the finite-dimensional problem (22) solution of the original boundary value problem is reduced by numerical integration using the previously obtained integrating matrices.

4. Iterative method and numerical experiments

To solve the projection scheme (22), we use the following two-layer iterative process with the lowering of non-linearity in the lower layer [8, 18–24] with a preconditioner, which is a linear part of the operator of the difference scheme:

$$A_1 \frac{U^{(n+1)} - U^{(n)}}{\tau} + (A_1 + A_2) U^{(n+1)} = F, \tag{23}$$

where $X^{(0)}$ is the given initial approximation, $\tau > 0$ is an iteration parameter. For the numerical realization of the iterative method (23) for the solving the problem (22) of the longitudinal and transverse bending of the plate, we have developed a software package. Numerical calculations were performed for the following geometric and stiffness characteristics of sandwich plate: plate length $2a = 20$ cm, $2h_1 = 2h_2 = 0.1$ cm, $h = 1$ см, $G_{13} = 25$ МПа, $E_3 = 50$ МПа, $E^{(k)} = 133 \cdot 10^3$ МПа, $\nu_{12}^{(k)} = \nu_{21}^{(k)} = 0.3$, $k = 1,2$, the geometric characteristics of the rod: $z_A = \zeta_A = B_2^{(+)} = h + h_1$ (Fig. 1), a value of loop load on the reinforcing rod, which provides transmission of the load to the carrier layers $T_\zeta^+ = -100 kN/m$. The number of grid points $N = 512$. The calculations according to (22) were

carried out as long as the residual norm $\left\| F-(A_1+A_2)X^{(n)} \right\|$ remained more than a given accuracy $\varepsilon = 5\cdot 10^{-8}$. As a norm of the vector $g=(g_1,g_2,\ldots,g_m)$ the value $\|g\| = \max\{|g_1|,|g_2|,\ldots,|g_m|\}$ is chosen. Iterative parameter was selected empirically.
Figures 3-8 shows the results of numerical experiments aimed at determining the parameters of the stress-strain state of the plate for a given value of loop load $T_\zeta^+ = -100 kN/m$. Under the action of this load in the end cross section of the plate and the reinforcing rod are formed by the parameters of the stress-strain state, with values
$Q_{11}^{(1)} = 0.062019 kN/m$, $Q_{11}^{(2)} = -100.062 kN/m$, $Q_{13}^{(1)} = 0.11091\ kN/m$, $Q_{13}^{(2)} = -0.23413\ kN/m$, $L_{11}^{(1)} = 0.63506 N$, $L_{11}^{(2)} = 0.066734 N$, $\tau_1 = 0.006161 MPa$, $U_A = -0.0069075\ cm$, $\varphi = -0.0032962$

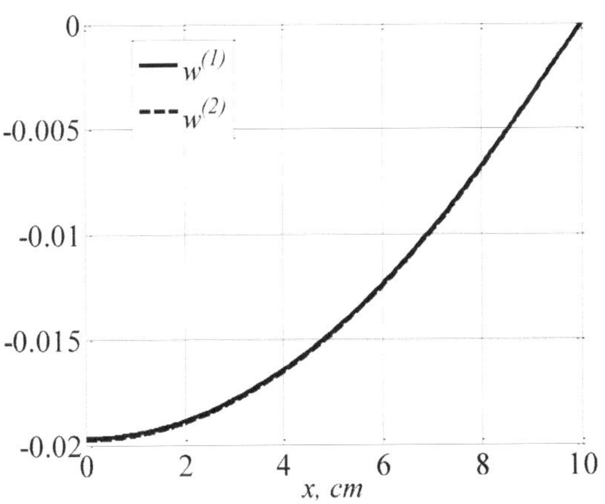

Figure 3. Deflections of the medial surfaces of the carrier layers ($w^{(k)}$, cm).

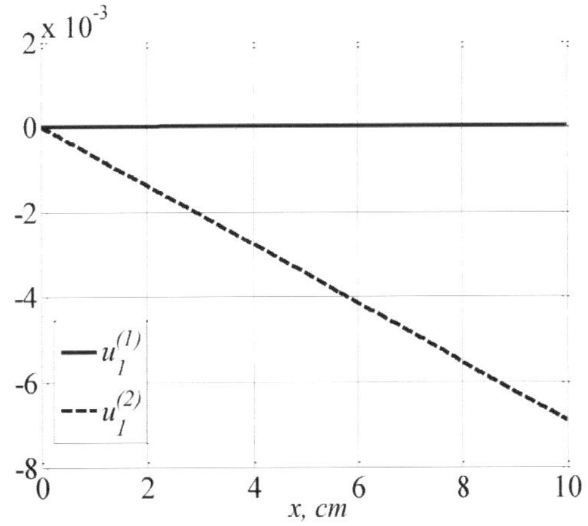

Figure 4. Axial displacements of the medial surfaces of the carrier layers ($u_1^{(k)}$, cm).

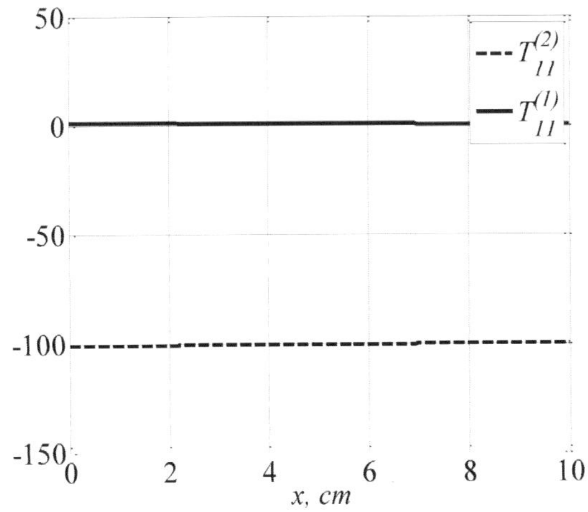

Figure 5. Membrane forces of the carrier layers ($T_{11}^{(k)}$, kN/m).

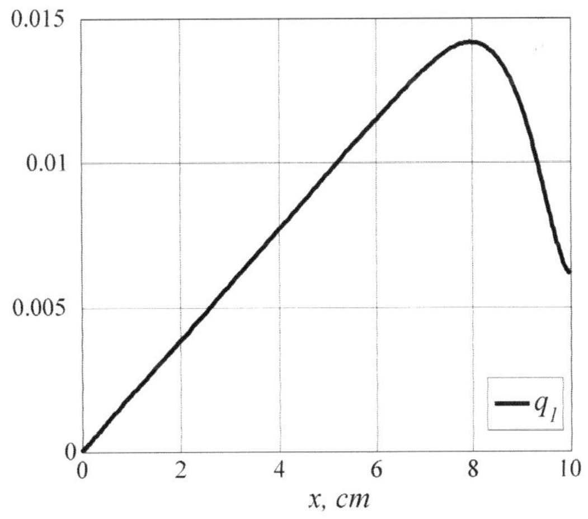

Figure 6. Transverse tangential stresses in the core (q_1, MPa).

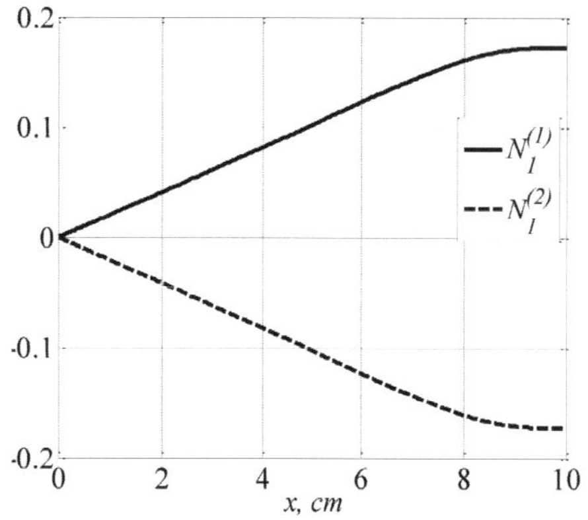

Figure 7. Generalized shear forces ($N_1^{(k)}$, kN/m).

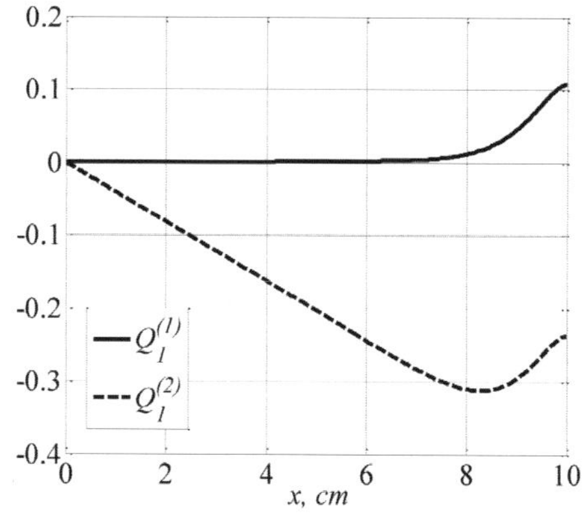

Figure 8. Shear forces of the carrier layers ($Q_1^{(k)}$, kN/m).

Figure 3 shows the dependence of the functions of deflections points of medial surfaces of the plate carrier layers on the length. It is evident that they are practically the same in view of the smallness of deformations of the transverse compression of the core in the subcritical state. Figure 4 shows the axial displacements function of points of the medial surfaces of carrier layers. Note that at the lower layer they are substantially equal to zero, and on the upper layer vary along the along the length of the plate almost linearly. From Figure 5 it follows that the plate is in the moment state due to the action on the plate eccentrically compressive load, wherein the torsional load formed in the first layer is practically zero, and the force $T_{11}^{(2)}$ along the length of the plate is not changed substantially. Figure 6 shows the variation of the tangential stresses in the core along the length of the plate. It is seen that by virtue of the acting of the eccentrically compressive force on the rod they are not equal to zero over the entire length of the plate (at the connection points of a core with the reinforcing rod, they should take the zero value in the absence of the adhesive layer) and reaches the maximum value in the neighborhood of the end cross section. Figures 7 and 8 show the functions of the generalized shear forces and shear forces in the carrier layers. By the nature of these curves we can be judged on the inclusion in the work the core and its contribution to the perception of the transverse tangential stresses.

Acknowledgments
The work was supported by the Russian Science Foundation (project 16-11-10299). The work is performed according to the Russian Government Program of Competitive Growth of Kazan Federal University.

References
[1] Grigolyuk E I and Chulkov P P 1973 *Stability and Vibrations of Sandwich Shells* (Moscow, Mashinostroenie)
[2] Bolotin V V and Novichkov Yu N 1980 Mechanics of Multilayered Structures (Moscow, Mashinostroenie)
[3] Badriev I B, Makarov M V and Paimushin V N 2015 On the interaction of composite plate having a vibration-absorbing covering with incident acoustic wave *Russian Mathematics* **59 (3)** 66-71 DOI: 10.3103/S1066369X1503007X
[4] Noor A K, Burton W S and Bert Ch W 1996 Computational models for sandwich panels and shells *Applied Mechanics Reviews* 1996 **49 (13)** 155-99.
[5] Paimushin V N 1987 Nonlinear theory of the central bending of three-layer shells with defects in the form of sections of bonding failure *Soviet Applied Mechanics* **23 (11)** 1038-43
[6] Ivanov V A and Paimushin V N 1994 Refined theory of stability of three-layer constructions

(Nonlinear equations of subcritical equilibrium of shells with transversal-soft aggregate) *Russian Mathematics* **38 (11)** 26–39

[7] Paimushin V.N and Bobrov S N 2000 Refined geometric nonlinear theory of sandwich shells with a transversely soft core of medium thickness for investigation of mixed buckling forms *Mechanics of Composite Materials* **36 (1)** 59-66

[8] Badriev I B, Banderov V V, Makarov M V and Paimushin V N 2015 Determination of stress-strain state of geometrically nonlinear sandwich plate *Applied Mathematical Sciences* **9 (78)** 3887–95. DOI: 10.12988/ams.2015.54354

[9] Badriev I B, Banderov V V, Garipova G Z, Makarov M V and Shagidullin R R 2015 On the solvability of geometrically nonlinear problem of sandwich plate theory *Applied Mathematical Sciences* **9 (82)**. 4095-102. DOI: 10.12988/ams.2015.54358

[10] Badriev I B, Makarov M V and Paimushin V N 2015 Solvability of physically and geometrically nonlinear problem of the theory of sandwich plates with transversally-soft core 2015 *Russian Mathematics* **59 (10)** 57-60. DOI: 10.3103/S1066369X15100072

[11] Badriev I B, Garipova G Z, Makarov M V, Paimushin V N and Khabibullin R F 2015 Solving physically nonlinear equilibrium problems for sandwich plates with a transversally soft core *Lobachevskii Journal of Mathematics* **36 (4)** 474-81. DOI: 10.1134/S1995080215040216

[12] Badriev I B, Garipova G Z, Makarov M V and Paimushin V N 2015 Numerical solution of the issue about geometrically nonlinear behavior of sandwich plate with transversal soft filler *Research Journal of Applied Sciences* 2015 **10 (8)** 428-35

[13] Badriev I B, Makarov M V and Paimushin V N 2016 Numerical Investigation of Physically Nonlinear Problem of Sandwich Plate Bending *Procedia Engineering* **150** 1050-5 DOI: 10.1016/j.proeng.2016.07.213

[14] Paimushin V N 2014 Contact formulation of non-linear problems in the mechanics of shells with their end sections connected by a plane curvilinear rod *Journal of Applied Mathematics and Mechanics* **78 (1)** 84-98.

[15] Lukankin S A, Paimushin V N and Kholmogorov S A 2014 Non-classical forms of loss stability of cylindrical shells joined by a stiffening ring for certain forms of loading *Journal of Applied Mathematics and Mechanics* **78 (4)** 395-408.

[16] Paimushin V N 1983 Variational methods for solving non-linear spatial problems of the joining of deformable bodies *Doklady Akad Nauk SSSR* 1983 **273 (5)** 1083-6

[17] Dautov R Z and Paimushin V N 1996 On the method of integrating matrices for the solution of boundary value problems for fourth-order ordinary equations *Russian Mathematics* **40 (10)** 11–23

[18] Badriev I B and Karchevskii M M 1994 Convergence of an iterative process in a Banach space *Journal of Mathematical Sciences* **71 (6)** 2727-35. DOI: 10.1007/bf02110578 15

[19] Badriev I B 2013 On the Solving of Variational Inequalities of Stationary Problems of Two-Phase Flow in Porous Media *Applied Mechanics and Materials* **392** 183-7 DOI: 10.4028/www.scientific.net/AMM.392.183

[20] Badriev I B and Nechaeva L A 2013 Mathematical simulation of steady filtration with multivalued law *PNRPU Mechanics Bulletin* **(3)** 37–65

[21] Badriev I B 1983 Difference schemes for nonlinear problems of filtration theory with a discontinuous law *Soviet Mathematics* **27 (5)** 1–11

[22] Badriev I and Banderov V 2014 Numerical method for solving variation problems in mathematical physics *Applied Mechanics and Materials* **668-669** 1094-97. DOI: 10.4028/www.scientific.net/AMM.668-669.1094

[23] Badriev I B and Karchevskii M M 1989 Convergence of the iterative Uzawa method for the solution of the stationary problem of seepage theory with a limit gradient *Journal of Soviet Mathematics* **45 (4)** 1302-9 DOI: 10.1007/BF01097083

[24] Badriev I B, Zadvornov O A and Lyashko A D 2004 A study of variable step iterative methods for variational inequalities of the second kind *Differential Equations* **40 (7)** 971-83

On the Finite Element Approximations of Mixed Variational Inequalities of Filtration Theory

I B Badriev, V V Banderov, E E Lavrentyeva and O V Pankratova

Kazan Federal University, 18 Kremlyovskaya Street, 420008, Kazan, Russia

E-mail: ildar.badriev1@mail.ru

Abstract. We construct the finite-element approximations for mixed variational inequalities with pseudomonotone operators and convex non-differentiable functionals in Sobolev spaces. Such variational inequalities arise in the mathematical description of the processes of an established filtration. The convergence of these approximations are investigated.

1. Introduction

Mixed variational inequalities with the monotone type operators [1, 2] arise in the mathematical description of the processes in the different areas of the science and technique, in particular, in the non linear filtration theory [3-7], in the problems of determining the equilibrium of soft shells [8-18], in the study of low-temperature plasma [19-24] etc. In this paper as a development of the research described in the paper [25] we construct and study the convergence of finite-element approximations for mixed variational inequalities of an established filtration with multivalued law. The existence of a subsequence of solutions of finite-element problems converging weakly to the solution of the original variational inequality is proved.

2. Problem statement

Let Ω be a bounded domain in R^m, $m \geq 1$, with a Lipschitz continuous boundary Γ and let $\xi \to g(\xi^2)\xi$ be the function that determines the filtration law. We assume that $g(\xi^2)\xi = g_0(\xi^2)\xi + g_1(\xi^2)\xi$, and

$$g_0(\xi^2) = 0, \quad \xi \geq \beta \tag{1}$$

($\beta \geq 0$ is a limiting gradient),

$$\text{function } \xi \to g_0(\xi^2)\xi \text{ is continuous,} \tag{2}$$

$$\text{function } \xi \to g_0(\xi^2)\xi \text{ strictly increase for } \xi \geq \beta \tag{3}$$

there exist $c_1 > 0$, $c_2 > 0$, $p > 1$, such that

$$c_1(\xi - \beta)^{p-1} \leq g_0(\xi^2)\xi \leq c_2(\xi - \beta)^{p-1} \text{ for } \xi \geq \beta, \tag{4}$$

$$g_1(\xi^2)\xi = \begin{cases} \vartheta, & \xi \geq \beta; \\ 0, & \xi < \beta. \end{cases}$$

Suppose that $\Gamma=\Gamma_0\cup\Gamma_1$, $\operatorname{mes}\Gamma_1>0$, $V=\{\eta\in W_p^{(1)}(\Omega):\eta(x)=0,\, x\in\Gamma_0\}$, $M=\{\eta\in V:\eta(x)\geq 0,\, x\in\Gamma_1\}$ (Γ_1 is the semi-permeable part of the boundary), $A_0:V\to V^*=W_{p^*}^{(-1)}(\Omega)$, $p^*=p/(p-1)$, is the operator generated by the form

$$\langle A_0 u,\eta\rangle=\int_\Omega g_0(|\nabla u|^2)(\nabla u,\nabla\eta)\,dx,\quad u,\eta\in V, \tag{5}$$

where $(\cdot,\cdot)$ and $|\cdot|$ are the inner product and norm in R^m.

We also define the functional $F_1:V\to R^1$ by the formula

$$F_1(\eta)=\int_\Omega\int_0^{|\nabla\eta|} g_1(\xi^2)\xi\,d\xi\,dx=\vartheta\int_\Omega(|\nabla\eta|-\beta)^+dx,\quad a^+=(a+|a|)/2.$$

By the solution to the filtration problem of an incompressible fluid with a multi valued filtration law with limiting gradient we mean the function $u\in M$, which is the solution of the variational inequality (see [26-28])

$$\langle A_0 u,\eta-u\rangle+F_1(\eta)-F_1(u)\geq\langle f,\eta-u\rangle\qquad\forall\eta\in M, \tag{6}$$

It is easy to verify that the set M is convex and closed. The functional F_1 is convex and Lipschitz continuous [29, 30]. If the conditions (1)–(4) hold then the operator A_0 is bounded, monotone, hemicontinuous [31] (and hence, pseudomonotone [31]) and coercive [31]. When the above conditions on the operator A_0 and the functional F_1 are imposed, problem (6) has at least one solution (see, e.g., [1, 31]).

3. Construction and investigation of finite element schemes for the filtration problems

In [25] have been considered the common case, when V is a reflexive Banach space with a uniformly convex dual space V^*, $\langle\cdot,\cdot\rangle$ is a duality relation between V and V^*, M is a closed convex set in V, the operator $A_0:V\to V^*$ is pseudomonotone and coercive on M, i.e., for some $\hat\eta\in M$ the inequality $\langle A_0\eta,\eta-\hat\eta\rangle\geq\rho(\|\eta\|_V)\|\eta\|_V$; $\lim_{\xi\to+\infty}\rho(\xi)=+\infty$, is satisfied. Recall that the operator $A_0:V\to V^*$ is called pseudomonotone [31] if it is bounded and the weak convergence of the sequence $\{u_k\}_{k=1}^{+\infty}$ in V to u^* and the inequality $\limsup_{k\to+\infty}\langle A_0u_k,u_k-u^*\rangle\leq 0$ implies that the following relation holds $\liminf_{k\to+\infty}\langle A_0u_k,u_k-\eta\rangle\geq\langle A_0u^*,u^*-\eta\rangle$ for all η. Functional $F_1:V\to R^1$ is a convex, continuous (but, in general, non-differentiable). Let $\{V_h\}_h$ be a family of spaces, where the parameter h tends to zero, such that $V_h\subset V$ for every h. Suppose that there given the linear operators $r_h:V\to V_h$ (restriction operators from V to V_h). We assume that the family $\{V_h\}_h$ approximates V, that is,

$$\lim_{h\to 0}\|r_h\eta-\eta\|_V=0\qquad\forall\eta\in V. \tag{7}$$

For each h let us consider a convex closed set $M_h\subset V_h$, approximating M, i.e., firstly, for any $\eta\in M$ an element $\eta_h\in M_h$ can be found, such that

$$\lim_{h\to 0}\|\eta_h-\eta\|_V=0, \tag{8}$$

and, secondly,

$$\text{if } \eta_h\in M_h,\ \eta_h\to\eta \text{ weakly in } V \text{ for } h\to 0, \text{ then } \eta\in M. \tag{9}$$

Note that condition (9) is satisfied if $M_h\subset M$ for each h. Indeed, in this case, the family $\{\eta_h\}_h$ belongs to the weakly closed set M, hence, $\eta\in M$.

Let us associate problem (6) with a family of approximating problems of finding the elements $u_h \in M_h$, such that

$$\langle A_0 u_h, \eta_h - u_h \rangle + F_1(\eta_h) - F_1(u_h) \geq \langle f, \eta_h - u_h \rangle \quad \forall \eta_h \in M_h. \tag{10}$$

The following theorem holds [25].

Theorem 1. Let u_h be the solution of problem (10). Then there is a subsequence $\{h_k\}_{k=1}^{+\infty}$, $h_k \to 0$ for $k \to +\infty$, such that u_{h_k} converges weakly in V to a some solution u of problem (6) for $k \to +\infty$.

Moreover, any weak limit point u^* of the family $\{u_h\}_h$ is a solution of (6), and if $\{u_{h_k}\}_{k=1}^{+\infty}$ is subsequence converging weakly in V to u^* for $k \to +\infty$, then

$$\lim_{k \to +\infty} \langle A_0 u_{h_k} - A_0 u^*, u_{h_k} - u^* \rangle = 0. \tag{11}$$

In the construction of finite element schemes we will assume for simplicity that the boundary Γ consists of s-dimensional faces ($s \leq m-1$).

Let $\Im_h$ be a family (triangulation) of m-dimensional simplexes T, union of which coincides with Ω, and the intersection of two simplexes from $\Im_h$ may be only a s-dimensional face ($s \leq m-1$), $h = \sup_{T \in \Im_h} \rho_T$, $\sigma(h) = \sup_{T \in \Im_h} \rho_T / \rho'_T$ where ρ'_T is the diameter of the largest ball contained in T and ρ_T is the diameter of the smallest ball containing T. We assume that the triangulation $\Im_h$ is regular (see [32, 33]), i. e., $\sigma(h) \leq \hat{c}_1$, where the positive constant $\hat{c}_1$ is independent of h. We now define the finite-dimensional space V_h approximating V and associated with the triangulation $\Im_h$, as the space of continuous functions vanishing at Γ_0 and linear on each simplex $T \in \Im_h$. It is clear that $V_h \subset V$.

We denote by U_h the set of all vertices x_h of the simplexes $T \in \Im_h$, and by $\overset{\circ}{U}_h$ the set of all points of U_h, which belong to $\operatorname{int}\Omega$. It is obvious that $U_h \setminus \overset{\circ}{U}_h \subset \Gamma$. As a basis in V_h we will choose the functions $x \to \omega_h(x, x_h)$ defined on Ω for any $x_h \in \overset{\circ}{U}_h$, such that $\omega_h(\hat{x}_h, x_h) = \begin{cases} 1, & \hat{x}_h = x_h; \\ 0, & \hat{x}_h \neq x_h. \end{cases}$ Any function $u_h \in V_h$ in this case can be represented as a linear combination $u_h(x) = \sum_{x_h \in \overset{\circ}{U}_h} u_h(x_h)\omega_h(x, x_h)$.

Finally, we set $M_h = \{\eta_h \in V_h : \eta_h(x) = 0, x \in \Gamma_0\}$.

We associate problem (6) with an approximating problem, which consists in finding an element $u_h \in M_h$ which is the solution of the variational inequality (10) holds.

Let $(r_h\eta)(x) = \sum_{x_h \in \overset{\circ}{U}_h} \eta(x_h)\omega_h(x, x_h)$, $x \in \Omega$. The following results (from which it follows that the constructed mesh schemes satisfy the conditions (7)–(9) hold.

Lemma 1. Let $\eta \in V$, then $r_h \eta \in V_h$, $\lim_{h \to 0} \| r_h\eta - \eta \|_V = 0$.

Lemma 2. Let $\eta \in M$, then $r_h \eta \in M_h$, $\lim_{h \to 0} \| r_h\eta - \eta \|_V = 0$, and the property {9} is satisfied.

The validity of Lemma 1 follows directly from the results of [34], and the validity of Lemma 2 follows from the definition of the sets M, M_h and Lemma 1.

We also need the following result [35].

Lemma 3. Let the conditions (1)–(4) hold, the sequence $\{u^{(n)}\}_{n=1}^{+\infty} \subset V$ converges weakly in V for $n \to +\infty$ to u and the operator $A_0 : V \to V^*$ generated by the (5) satisfies the relation

$\lim\limits_{n\to+\infty}\left\langle A_0 u^{(n)} - A_0 u, u^{(n)} - u^*\right\rangle = 0.$ Then, $\lim\limits_{n\to+\infty}\int\limits_{\Omega_{+u}} |\nabla(u^{(n)} - u)|^p\, dx = 0$, where $\Omega_{+u} = \{x\in\Omega : |\nabla u(x)| > \beta\}$ is the flow domain.

We have the following

Theorem 2. Problem (10) has at least one solution. The family of the solutions to (10) is uniformly bounded in h. Any weak limit point u of the family $\{u_h\}_h$ is a solution of {6}, and if $\{u_{h_k}\}_{k=1}^{+\infty}$ is a subsequence which converges weakly for $k\to+\infty$ in V to u, then

$$\lim_{k\to+\infty}\left\| u_{h_k} - u\right\|_{L_p(\Omega)} = 0, \qquad (12)$$

$$\lim_{k\to+\infty}\int\limits_{\Omega_{+u}} |\nabla(u_{h_k} - u)|^p\, dx = 0. \qquad (13)$$

Proof. The existence of the solutions to (10) and the fact that any weak limit point u of the family $\{u_h\}_h$ is a solution of (10) follow from Theorem 1. Relation (12) follows from the compactness of the embedding V in $L_p(\Omega)$ and relation (13) follows from (11) and Lemma 3.

Acknowledgments
The work was supported by the Russian Foundation for Basic Research (projects nos. 15-01-05686, 15-41-02315, 15-38-21099). The work is performed according to the Russian Government Program of Competitive Growth of Kazan Federal University.

References

[1] Ekeland I and Temam R 1976 *Convex Analysis and Variational Problems* (Amsterdam: North-Holland)

[2] Glowinski R Lions J L and Tremolieres R 1976 *Analyse numérique des inéquations variationnelles* (Paris: Dunod)

[3] Badriev I B and Karchevskii M M 1994 Convergence of an iterative process in a Banach space *Journal of Mathematical Sciences* **71 (6)** 2727-35. DOI: 10.1007/bf02110578 15

[4] Badriev I B and Karchevskii M M 1989 Convergence of the iterative Uzawa method for the solution of the stationary problem of seepage theory with a limit gradient *Journal of Soviet Mathematics* **45 (4)** 1302-9 DOI: 10.1007/BF01097083

[5] Badriev I B and Nechaeva L A 2013 Mathematical simulation of steady filtration with multivalued law *PNRPU Mechanics Bulletin* (3) 37–65

[6] Lapin A V 2010 Preconditioned Uzawa-Type Methods for Finite-Dimensional Constrained Saddle Point problems *Lobachevskii Journal of Mathematics* **31 (4)** 309-322. DOI 10.1134/s1995080210040013

[7] Shangaraeva A I and Shevchenko D V 2015 Speed up of the oil saturation numerical algorithm for the plane-parallel filtration *Applied Mathematical Sciences* **9 (150)** 7467-74 DOI 10.12988/ams.2015.510683

[8] Badriev I and Banderov V 2014 Numerical method for solving variation problems in mathematical physics *Applied Mechanics and Materials* **668-669** 1094-97. DOI: 10.4028/www.scientific.net/AMM.668-669.1094

[9] Abdrakhmanova A I and Sultanov L U 2016 Numerical modelling of deformation of hyperelastic incompressible solids *Materials Physics and Mechanics* **26 (1)** 30-2

[10] Badriev I B, Banderov V V, Garipova G Z, Makarov M V and Shagidullin R R 2015 On the solvability of geometrically nonlinear problem of sandwich plate theory *Applied Mathematical Sciences* **9 (82)** 4095-102. DOI: 10.12988/ams.2015.54358

[11] Badriev I B, Banderov V V, Makarov M V and Paimushin V N 2015 Determination of stress-strain state of geometrically nonlinear sandwich plate *Applied Mathematical Sciences* **9 (78)**

3887–95. DOI: 10.12988/ams.2015.54354

[12] Badriev I B, Banderov V V and Zadvornov O A 2013 On the solving of equilibrium problem for the soft network shell with a load concentrated at the point *PNRPU Mechanics Bulletin* **(3)** 17-35

[13] Badriev I B and Shagidullin R R 1995 A study of the convergence of a recursive process for solving a stationary problem of the theory of soft shells *Journal of Mathematical Sciences* **73 (5)** 519-25 DOI :10.1007/bf02367668

[14] Berezhnoi D V, Paimushin V N and Shalashilin V I 2010 Studies of quality of geometrically nonlinear elasticity theory for small strains and arbitrary displacements *Mechanics of Solids* **44 (6)** 837-51. DOI 10.3103/S0025654409060041

[15] Davydov R L and Sultanov L U 2014 Mathematical modeling of large elastic-plastic deformations *Applied Mathematical Sciences* **60 (8)** 2991-6 DOI 10.12988/ams.2014.44272

[16] Sagdatullin M and Berezhnoi D 2014 Statement of the problem of numerical modelling of finite deformations *Applied Mathematical Sciences* **8 (33-36)** 1731-8 DOI 10.12988/ams.2014.4283

[17] Sultanov L U 2014 Mathematical modeling of deformations of hyperelastic solids *Applied Mathematical Sciences* **8 (143)** 7117-24 DOI 10.12988/ams.2014.49704

[18] Sultanov L U and Davydov R L 2014 Mathematical modeling of large elastic-plastic deformations *Applied Mathematical Sciences* **8** (60) 2991-6 DOI 10.12988/ams.2014.44272

[19] Chebakova V Ju, Zheltukhin V S and Dubrovin V T 2016 Problem of radio-frequency discharge at atmospheric pressure in local statement *Applied Mathematical Sciences* **10 (21)** 1013-22

[20] Zheltukhin V S, Chebakova V J, Abaidullin R N and Kadyrov R F 2015 On solving of some nonlinear problems of plazma physics *Applied Mathematical Sciences* **9** (88) 4351-9

[21] ZheltuhinV S, Solovyev P S and Chebakova V Y 2015 Boundary conditions for electron balance equation in the stationary High-Frequency Induction discharges *Research Journal of Applied Sciences* **10 (10)** 658-62

[22] Dubrovin V T, Gabbasov F G and Chebakova V J 2016 Multidimensional central limit theorem for sums of functions of the trajectories of endomorphisms *Lobachevskii Journal of Mathematics* **37 (4)** 409-17

[23] Dubrovin V T, Chebakova V Ju and Zheltukhin V S 2016 Radio-Frequency Discharge at Low Pressure: a Non-Local Problem Statement Approach *Procedia Engineering* **150** 1041–5

[24] Zheltukhin V S, Solov'ev S I, Solov'ev P S and Chebakova V Yu 2014 Computation of the minimum eigenvalue for a nonlinear Sturm–Liouville problem *Lobachevskii J. Math.***35 (4)** 416–26

[25] Badriev I B, Banderov V V, Gnedenkova V L, Kalacheva N V, Korablev A I and Tagirov R R 2015 On the finite dimensional approximations of some mixed variational inequalities *Applied Mathematical Sciences* **9 (114)** 5697-705 DOI. 10.12988/ ams.2015.57480

[26] Badriev I B 2013 On the Solving of Variational Inequalities of Stationary Problems of Two-Phase Flow in Porous Media *Applied Mechanics and Materials* **392** 183-7 DOI: 10.4028/www.scientific.net/AMM.392.183

[27] Badriev I B 1983 Difference schemes for nonlinear problems of filtration theory with a discontinuous law *Soviet Mathematics* **27** (5) 1–11

[28] Badriev I B and Fanyuk B Y 2012 Iterative methods for solving seepage problems in multilayer beds in the presence of a point source *Lobachevskii Journal of Mathematics* **33 (4)** 386-99 DOI 10.1134/S1995080212040026

[29] Badriev I B, Zadvornov O A and Lyashko A D 2004 A study of variable step iterative methods for variational inequalities of the second kind *Differential Equations* **40 (7)** 971-83

[30] Badriev I B, Zadvornov O A and Saddek A M 2001 Convergence Analysis of Iterative Methods for Some Variational Inequalities with Pseudomonotone Operators *Differential Equations* **37 (7),** 934-42 DOI 10.1023/a:1011901503460

[31] Lions J L 1969 *Quelque problèmes méthodes de résolution des problèmes aux limites*

nonlinéaires (Paris: Dunod)
[32] Ciarlet P 1978 *The Finite Element Method For Elliptic Problems* (Amsterdam: North-Holland)
[33] Temam R 1977 *Navier-Stokes Equations* (Amsterdam: North-Holland)
[34] Ladyzhenskaya O A 1985 *The Boundary Value Problems of Mathematical Physics* (New York: Springer-Verlag)
[35] Badriev I B and Zadvornov O A 1966 Convergence of an iterative process for equations with degenerate operators *Differential Equations* **32 (7)** 902-5

Mathematical simulation of a steady process of anisotropic filtration

I B Badriev, V V Banderov, O V Pankratova and A I Shangaraeva

Kazan Federal University, 18 Kremlyovskaya Street, 420008, Kazan, Russia

E-mail: ildar.badriev1@mail.ru

Abstract. This article discusses the methods of approximate solution of mixed variational inequalities with operators of monotone type. The functional, which is included in this variational inequality, is separable, in other words, it is the sum of a number of non-differentiable functionals. These variational inequalities appear, in particular, in the description of steady incompressible filtration processes of highly viscous fluids in anisotropic medium.

1. Introduction

We consider a steady process of underground filtration of an incompressible fluid in an anisotropic porous media [1-5]. The generalized formulation of these problems is formulated in a form of mixed variational inequalities with inversely strongly monotone operator and separable, generally, non-differentiable, functional in Hilbert space. We establish the properties of an operator which is included in the variational inequality (inversely strongly monotone, coercivity) and functionals properties (Lipschitz continuity and convexity). This allows the opportunity to apply for prove of the existence theorem the known results of the theory of monotone operators [6]. To solving a variational inequality splitting iterative methods are proposed, which do not require to find the inverse operator to the original operator. Investigation of the convergence of the iterative process based on the method of successive approximations for finding a fixed point of some operator (transition operator). We prove that the transition operator is asymptotically regular [7]. This allow to prove the weakly convergence of successive approximations. At the same time for the filtration problem, each step of the iterative process actually reduces to solving boundary value problems for the Laplace operator or strongly elliptic operator. We note that the proposed methods allow to find the approximate values of not only the solution, but its characteristics (for filtration problems it is the approximate values of the gradient of solution, as well as the approximate value of flow velocities on the sets corresponding the points of multivalence in the filtration law), which is very useful from a practical points of view [8–12].

2. Statement and investigation of nonlinear stationary anisotropic filtration problems

We consider the steady-state filtration process of an incompressible fluid in a porous medium, which occupies a bounded domain $\Omega \in R^m$, $m \geq 2$ with Lipschitz continuous boundary $\Gamma = \Gamma_1 \bigcup \Gamma_2$, $\Gamma_1 \bigcap \Gamma_2 = \varnothing$, mes $\Gamma_1 > 0$, the pressure is equal to zero on the Γ_1, Γ_2 is impermeable part of the boundary. It is necessary to find a stationary fields of pressure u and fluid velocity v, that satisfy the continuity equation

$$\operatorname{div} v(x) = \tilde{f}(x), \quad x \in \Omega, \tag{1}$$

and the boundary conditions

$$u(x)=0,\quad x\in\Gamma_1,\quad (v(x),\mathrm{n})=0,\quad x\in\Gamma_2, \tag{2}$$

where **n** is the external normal to Γ_2, and the assumption that the fluid follows to the filtration multivalued anisotropic law

$$v_l(u)\in\frac{1}{m}\sum_{k=1}^{m}\left[\sum_{i=1}^{m}\alpha_{kl}^{(i)}g_i(D_i^2u(x))\right]\frac{\partial u(x)}{\partial x_k},\quad x\in\Omega \tag{3}$$

where $D_i^2u=(\mathrm{Y}_i\nabla u,\nabla u)$, $\xi\to g_i(\xi^2)\xi$, $i=1,\dots,m$, are the multivalued functions that determine the filtration law, and $\tilde{f}$ is the function characterizing the density of the external sources.

We assume that the functions $\xi\to g_i(\xi^2)\xi$ can be represented as $g_i(\xi^2)\xi=g_{0i}(\xi^2)\xi+g_{1i}(\xi^2)\xi$, where $\xi\to g_{0i}(\xi^2)\xi$ are the single-valued functions satisfying conditions

$$g_{0i}(\xi^2)\xi=0 \text{ when } \xi\le\beta_i \ (\beta_i \text{ are the limiting gradients}), \tag{4}$$

$$c_{1i}(\xi-\beta_i)\le g_{0i}(\xi^2)\xi\le c_{2i}\xi,\quad \xi\ge\beta_i,\ c_{1i},c_{2i}>0, \tag{5}$$

$$g_{0i}(\xi^2)\xi-g_{0i}(\zeta^2)\zeta\ge0 \text{ when } \xi\ge\zeta, \tag{6}$$

$$|g_{0i}(\xi^2)\xi-g_{0i}(\zeta^2)\zeta|\le c_{3i}|\xi-\zeta|,\ c_{3i}>0 \text{ when } \xi\ge\zeta, \tag{7}$$

$\xi\to g_{1i}(\xi^2)\xi$ are the multi-valued functions, which have the form $g_{1i}(\xi^2)\xi=\vartheta_i h(\xi-\beta_i)$,

$$h(\xi)=\begin{cases}0, & \xi<0,\\ [0,1], & \xi=0,\\ 1, & \xi>0,\end{cases}$$

It is obvious that the set of $g_{1i}(\xi^2)\xi$ is the subdifferential of the function μ_i in the point ξ,

$$\mu_i(\zeta)=\vartheta_i\mu(\zeta-\beta_i),\quad \mu(\zeta)=\begin{cases}0, & \zeta<0,\\ \zeta, & \zeta\ge0,\end{cases}$$

i.e.,

$$\mu_i(\zeta)-\mu_i(\xi)\ge\xi^*(\zeta-\xi)\quad \forall\xi^*\in g_{1i}(\xi^2)\xi,\ \forall\zeta,\ \xi\in R^1. \tag{8}$$

Relatively of the coefficients $\left\{\alpha_{kl}^{(i)}\right\}_{k,l=1}^{m}$ of the matrix Y_i we assume, that $\alpha_{kl}^{(i)}=\alpha_{lk}^{(i)}$, $\sum_{k,l=1}^{m}\alpha_{kl}^{(i)}\xi_k\xi_l\ge c_{4i}\sum_{k,l=1}^{m}\xi_k^2$, $c_{4i}>0$, i.e., Y_i is the symmetric, positive definite matrix. Let's denote $(\xi,\zeta)_i=\sum_{k,l=1}^{m}\alpha_{kl}^{(i)}\xi_k\zeta_l=(\mathrm{Y}_i\xi,\zeta)$.

Let us formulate the variational formulation of the problem (1)–(3). Suppose that u and v are the solution of this problem. We denote as $C_{\Gamma_1}^{\infty}(\Omega)$ the set of infinitely differentiable functions in $\bar{\Omega}$, which are equal the zero on Γ_1. Using (8), we find that the function u satisfies the variational inequality

$$\int_\Omega\tilde{f}(x)\eta(x)dx\le\frac{1}{m}\sum_{i=1}^{m}\int_\Omega g_{i0}(D_i^2u(x))(\nabla u(x),\nabla\eta(x))_i\,dx+$$
$$+\frac{1}{m}\sum_{i=1}^{m}\vartheta_i\int_\Omega\mu(D_i(u(x)+\eta(x))-\beta_i)dx-\frac{1}{m}\sum_{i=1}^{m}\vartheta_i\int_\Omega\mu(D_i(u(x))-\beta_i)dx\quad \forall\eta\in C_{\Gamma_1}^{\infty}(\Omega) \tag{9}$$

Let $V=\{\eta \in W_2^{(1)}(\Omega):\eta(x)=0, x\in\Gamma_1\}$ be the Sobolev space with the inner product $(u,\eta)_V=\int\limits_\Omega (\nabla u(x),\nabla\eta(x))dx$ and the norm $\|u\|_V=(u,u)_V^{1/2}$. It is easy to verify, that under the conditions (4), (5) the forms $a_i(u,\eta)=\int\limits_\Omega g_{i0}(D_i^2u(x))(\nabla u(x),\nabla\eta(x))_i\,dx$ are linear and bounded to the second argument, and thus they generate the operators $A_i:V\to V$ by the formula $a_i(u,\eta)=(A_iu,\eta)_V$. Let's set $A=\frac{1}{m}\sum_{i=1}^m A_i$. Further, we consider the functionals $F_i:V\to R^1$ by the formulas

$$F_i(u)=\frac{1}{m}\int\limits_\Omega \mu_i(D_i(u(x))-\beta_i)dx=\frac{1}{m}\int\limits_\Omega\int\limits_0^{D_i^2(u)} g_{i1}(\xi^2)\xi\,d\xi\,dx. \quad (12)$$

We assume that the function $\tilde{f}$ generates a continuous linear functional f on V: $\int\limits_\Omega \tilde{f}(x)\eta(x)dx=(f,\eta)_V$. Then, according to (9), under the decision of the stationary problem of filtration of an incompressible fluid, satisfying the nonlinear multi-valued anisotropic filtration law with limiting gradient, we mean the function $u\in V$, which is the solution of the variational inequality (cf. [13])

$$(Au-f,\eta-u)_V+\frac{1}{m}\sum_{i=1}^m F_i(\eta)-\frac{1}{m}\sum_{i=1}^m F_i(u)\geq 0 \quad \forall\eta\in V. \quad (12)$$

In practice the important factor for solving problems of nonlinear filtration with limiting gradient is to find the boundaries of dead zones, i.e. border regions, where there is no fluid flow.

Theorem 1. *Let the conditions (4)-(7) hold. Then the variational inequality (12) has a non-empty, convex, closed set of solutions. If $u\in V$ is the solution of (12), there is exists the function $v\in H=[L_2(\Omega)]^m$, that almost everywhere on Ω satisfied the inclusions (3), and there is the continuity equation $\int\limits_\Omega (v(x),\nabla\eta(x))dx=\int\limits_\Omega \tilde{f}(x)\eta(x)dx$ for any function $\eta\in C_{\Gamma_1}^\infty(\Omega)$. hold.*

Indeed, under the conditions (4)–(7) the operator A is inversely strongly monotone, i.e., $(Au-A\eta,\eta-u)_V\geq\sigma\|Au-A\eta\|_V^2, \sigma>0$, for all $u,\eta\in V$, in particular, it is the monotone and Lipschitz continuous with the constant $1/\sigma$, and the coercive [13]. The functionals F_i, defined by formulas (2), and therefore their sum are convex and Lipschitz continuous, therefore weakly lower semicontinuous. So, the existence of solutions of variational inequalities (12), closed convex set its decisions follows from the general theory of monotone operators [13]. If $u\in V$ is the solution of the (12), then according to the subdifferential definition

$$f-Au\in\partial(\sum_{i=1}^m F_i(u)). \quad (13)$$

Since the F_i are convex and continuous, then [13] $\partial(\sum_{i=1}^m F_i(u))=\sum_{i=1}^m \partial F_i(u)$. By analogy with the [14-16], it is easy to get in the point u the subdifferential of the functional F_i is the set of the linear continuous on V functionals l_i:

$(l_i,\eta)_V=\frac{1}{m}\int\limits_\Omega \frac{\chi_{iu}(x)}{D_i(u(x))}(\nabla u,\nabla\eta)_i\,dx,\ \chi_{iu}\in L_\infty(\Omega),\ \chi_{iu}(x)\in\vartheta_i\,h(D_iu(x))-\beta_i).$ So, the relation (13) means that the continuity equation is satisfied.

In this case, the question about the construction of the specified field of the filtration velocity v on the sets of corresponding points of multivalence in the filtration law (when $D_i u(x)) = \beta_i$) is open.

3. The iterative method for solving nonlinear stationary anisotropic filtration problems

In order to solve the variational inequalities of the form (12) iterative splitting methods are proposed [17-25]. The main difficulty in this case is the solving of arising problems of minimization in each iteration. In the case of an isotropic filtration law this problem was solved in explicit form. Due to this, it is possible to effectively calculate the subdifferential of the conjugate functional to the minimized one. In the case of anisotropic filtration law, when the minimizing functional is the sum of several functionals, the calculation of the conjugate functional is a difficult problem. We consider the splitting algorithm, which solves indicated above problem.

The variational inequalitiy (12), which characterizes the filtration problem can be written as follows:

$$(Au - f, \eta - u)_V + \frac{1}{m}\sum_{i=1}^{m} G_i(B_i\eta) - \frac{1}{m}\sum_{i=1}^{m} G_i(B_i u) \geq 0 \quad \forall \eta \in V, \tag{14}$$

if the functionals F_i have the form $F_i = G_i \circ B_i$, where the functionals G_i defined on H, according to the formula $G_i(z) = \dfrac{1}{2m}\displaystyle\int_\Omega \int_0^{|z|^2} g_{i1}(\xi)\,d\xi\,dx$, they are convex and Lipschitz continuous, $B_i : V \to H$ – linear continuous operators, $B_i = \Upsilon^{1/2}\nabla$.

To solve the variational inequality (14) we consider the following iterative process. Let's define $\tau > 0$, $r > 0$, and set $u^{(0)} \in V$, $y_i^{(0)}, \lambda_i^{(0)} \in H$, $i = 1, 2, \ldots, m$, are the arbitrary elements. For $k = 1, 2, \ldots$ we know $y_i^{(k)}, \lambda_i^{(k)}$, let's define

$$u^{(k+1)} = u^{(k)} - \tau(Au^{(k)} - f + r\sum_{i=1}^{m} B_i B_i^* u^{(k)} + \sum_{i=1}^{m} B_i^*(\lambda_i^{(k)} - r y_i^{(k)})). \tag{15}$$

Then we find $y_i^{(k+1)}$ and solve the problem

$$\begin{gathered} r(y_i^{(k+1)}, z_i - y_i^{(k+1)})_H + G_i(z_i) - G_i(y_i^{(k+1)}) \geq \\ \geq (rB_i u^{(k+1)} + \lambda_i^{(k)}, z_i - y_i^{(k+1)})_H \quad \forall z_i \in H, \quad i = 1, 2, \ldots, m. \end{gathered} \tag{16}$$

Finally, we find

$$\lambda_i^{(k+1)} = \lambda_i^{(k)} + r(B_i u^{(k+1)} - y_i^{(k+1)}), \quad i = 1, 2, \ldots, m. \tag{17}$$

In the equation (15) the operator $B_i^* : H \to V$ is the conjugate operator to the operator B_i :

$$(B_i^* y, \eta)_V = (y, B_i\eta)_H \quad \forall y \in H, \eta \in V.$$

In order to study the convergence of the described iterative process we write out the explicit form of the transition operator. Let us first recall the definition and properties of the proximal mapping (see., e.g., [13]). Suppose that V is a Hilbert space, $\xi \in V$ is the any element, $\Phi : V \to R^1$ is the convex, propex, lower semicontinuous functional. Then the variational inequality

$$(u - \xi, \eta - u)_V + \Phi(\eta) - \Phi(u) \geq 0 \quad \forall \eta \in V \tag{18}$$

has an unique solution u. Thus, the operator $\mathrm{Prox}_{\{\Phi\}} : V \to V$, $u = \mathrm{Prox}_{\{\Phi\}}(\xi)$ is defined. From the definition of the proximal mapping using the variational inequality (18), it is implied that the operator is confirmly non-extracted [13, 26], i.e.,

$$\left\| \mathrm{Prox}_{\{\Phi\}}(u) - \mathrm{Prox}_{\{\Phi\}}(\eta) \right\|_V^2 \le (\mathrm{Prox}_{\{\Phi\}}(u) - \mathrm{Prox}_{\{\Phi\}}(\eta), u-\eta)_V \quad \forall u,\eta \in V. \tag{19}$$

We denote by H^m as the direct m product of the H spaces and introduce the operator $T: Q = V \times H^m \times H^m \to Q$, mapping a $q = (q_0, q_1, \ldots, q_{2m}) = (u, y_1, \ldots, y_m, \lambda_1, \ldots, \lambda_m)$ to the element

$$T_0 q = q_0 - \tau(Aq_0 - f + r\sum_{i=1}^{m} B_i B_i^* q_0 + \sum_{i=1}^{m} B_i^*(q_{m+i} - r q_i)), \tag{20}$$

$$T_i q = \mathrm{Prox}_{\{G_i/r\}}(B_i T_0 q + \frac{1}{r} q_{m+i}),\ T_{m+i} q = q_{m+i} + r(B_i T_0 q - T_i q),\ i = 1,2,\ldots,m, \tag{21}$$

where the operators $\mathrm{Prox}_{\{G_i/r\}}$ are proximal (the functionals G_i / r are the convex, propex, lower semicontinuous functionals, because $r > 0$).

We set $q^{(k)} = (u^{(k)}, y_1^{(k)}, \ldots, y_m^{(k)}, \lambda_m^{(k)}, \ldots, \lambda_m^{(k)})$, then the iteration process (15) –(17) can be written as:

$$q^{(k+1)} = Tq^{(k)}\ , \ k = 1,2,\ldots, \tag{22}$$

$q^{(0)} \in Q$ is an arbitrary element, i.e., T is the transition operator of an iterative process.

The relation between the solution of the problem (14) and the set of fixed points of the operator T is defined by the following theorem.

Theorem 2. *The point $q = (u, y_1, \ldots, y_m, \lambda_1, \ldots, \lambda_m)$ is a fixed point of the operator T if and only if the following conditions are satisfied:*

$$y_i = B_i u\,,\ \lambda_i \in \partial G_i(B_i u)\,,\ i = 1,2,\ldots,m,\ \sum_{i=1}^{m} B_i^* \lambda_i = Au - f\,.$$

In this case, the first component of any fixed point of the T operator is a solution of problem (12).

Really, if $q = (u, y_1, \ldots, y_m, \lambda_1, \ldots, \lambda_m)$ is the fixed point of the T operator, then by virtue of (20), (21) we have the following:

$$u = u - \tau(Au - f + r\sum_{i=1}^{m} B_i^* B_i u + \sum_{i=1}^{m} B_i^*(\lambda_i - r y_i)),\ y_i = \mathrm{Prox}_{\{G_i/r\}}(B_i u + \frac{1}{r}\lambda_i),$$
$$\lambda_i = \lambda_i + r(B_i u - y_i),\ i = 1,2,\ldots,m.$$

Thus, using (18) and the subdifferential definition we get the desired result.

Let us introduce on Q the bilinear form by the formula:

$$(p,q)_Q = \frac{1 - m\tau r}{\tau}(p_0, q_0)_V + r\sum_{i=1}^{m}(p_i, q_i)_H + \frac{1}{\tau}\sum_{i=1}^{m}(p_{m+i}, q_{m+i})_H\ . \tag{23}$$

If the condition $m\tau r < 1$ then the form (23) satisfies the axioms of a scalar product. We have

Theorem 3. *Suppose that A is the inversely strongly monotone operator with the constant $\sigma > 0$, $0 < \tau < 2\sigma/(2m\sigma r + 1)$, and the problem (12) has at least one solution. Then operator T is asymptotically regular, the iterative sequence $\{q^{(k)}\}_{k=0}^{\infty}$, constructed by, converges weakly in Q for $k \to \infty$. Its limit is a fixed point of the operator T, $\left\| y_i^{(k)} - B_i u^{(k)} \right\|_H \to 0$ for $k \to \infty$.*

Indeed, by virtue of the definition of operator T, and using (19), we can establish that T is the contractive operator, moreover, for any $p, q \in Q$ the inequality

$$\|Tq-Tp\|_Q^2+\delta(Aq_0-Ap_0,q_0-p_0)_V+$$
$$+\frac{1}{1-m\tau r}\|(1-\tau r)(q_0-T_0q-(p_0-T_0p))-\tau(Aq_0-Ap_0)\|_V^2+$$
$$+r\sum_{i=1}^{m}\|q_i-B_iT_0q-(p_i-B_iT_0p)\|_H^2\le\|q-p\|_Q^2,\quad \delta=2-\tau/(\sigma(1-m\tau r)).$$

hold. This inequality implies the asymptotic regularity of the operator T, since the iterative process (22) weakly converge, and also in H the sequences $\{y_i^{(k)}-B_iu^{(k)}\}_{k=0}^{\infty}$ strongly converge to zero $i=1,2,\ldots,m$.

It follows from theorem 3 that the sequence $\{u^{(k)}\}_{k=0}^{\infty}$, $\{y_i^{(k)}\}_{k=0}^{\infty}$, constructed by (15)-(17), converge weakly to u and to B_iu, $i=1,2,\ldots,m$, respectively, where u is the solution of variational inequality (12).

Let us consider the features of the application of the iterative method (15) - (17) for the solutions of the anisotropic filtration problems.

In order to determine $u^{(k+1)}$ it is necessary firstly to solve the boundary value problem

$$\begin{cases} Rs=f-Au^{(k)}+\sum_{i=1}^{m}B_i^*(\lambda_i^{(k)}-ry_i^{(k)})+mrRu^{(k)}, & x\in\Omega,\\ s(x)=0,\quad x\in\Gamma_1,\quad (\nabla s(x),\mathrm{n})=0, & x\in\Gamma_2, \end{cases}\tag{24}$$

where $R=-\mathrm{div}(\sum_{\mathrm{i}=1}^{\mathrm{m}}\mathrm{Y}_i\nabla)$, and then put $u^{(k+1)}=u^{(k)}+\tau s$. We note that in our case $B_i^*=-\mathrm{div}\,\mathrm{Y}_i^{1/2}$.

In the numerical implementation of the iterative method the main difficulty is to solve the minimization problems (16). We write them in the form:

$$(rB_iu^{(k+1)}+\lambda_i^{(k)},z-ry_i^{(k+1)})_H\le G_i(z)+\frac{r}{2}\|z\|_H^2-G_i(y_i^{(k+1)})+\frac{r}{2}\|y_i^{(k+1)}\|_H^2\ \forall z\in H,$$

or,

$$(rB_iu^{(k+1)}+\lambda_i^{(k)},z-ry_i^{(k+1)})_H\le G_{ir}(z)-G_{ir}(y_i^{(k+1)})\quad \forall z\in H,\tag{25}$$

where $G_{ir}(z)=G_i(z)+\frac{r}{2}\|z\|_H^2$. Using the definition of the subdifferential ∂G_{ir}, we write (25) in the form of inclusion $rB_iu^{(k+1)}+\lambda_i^{(k)}\in G_{ir}(y_i^{(k+1)})$, which is equivalent [13] to the inclusion $y_i^{(k+1)}\in G_{ir}^*(rB_iu^{(k+1)}+\lambda_i^{(k)})$, where G_{ir}^* is a conjugate functional [13] to G_{ir}. The proposed scheme for solving the minimization problem (16) allows significantly alleviate the numerical implementation of the splitting method, when we can effectively calculate the subdifferential ∂G_{ir}^*. In this case the problem of finding the element $y_i^{(k+1)}$ is reduced to calculations by explicit formulas. We use this way for solving the considered anisotropic filtration problems. We have $G_{ir}^*(z)=\int_\Omega\int_0^{|z|}g_{ir}^*(\xi^2)\xi\,d\xi\,dx$, where

$$g_{ir}^{*}(\xi^2)\xi = \begin{cases} \xi / r, & \xi \le r\beta_i, \\ \beta_i, & r\beta_i < \xi \le r\beta_i + \vartheta_i / m, \\ (\xi - \vartheta_i / m) / r, & \xi > r\beta_i + \vartheta_i / m. \end{cases}$$

The functional G_{ir}^{*} is convex and differentiable with respect to the Gateaux, therefore [13] the subdifferential ∂G_{ir}^{*} consists of a single element, that coincides with its gradient, which is determined by the shape $((G_{ir}^{*})' z, y)_H = \int\limits_{\Omega} g_{ir}^{*}(|z|^2)(z, y)\,dx$. Thus, the problem (16) consists of computing $y_i^{(k+1)}$ using the formula $y_i^{(k+1)} = g_{ir}^{*}(|z|^2)z$, where $z = rB_i u^{(k+1)} + \lambda_i^{(k)}$. Therefore, each step of the considered iterative method is actually reduced to solving the boundary problem (24) with the linear strongly elliptic operator. At the same time, it follows from Theorem 2 that the approximate value of filtration velocity $v^{(k)}$ can be calculated by the formula $v_l^{(k)} = \dfrac{1}{m} \sum\limits_{k=1}^{m} \sum\limits_{i=1}^{m} \left[\alpha_{kl}^{(i)} g_{i0}(|y_i^{(k)}|^2) + \dfrac{\lambda_i^{(k)}}{|y_i^{(k)}|} \right] y_i^{(k)}$.

Acknowledgments
The work was supported by the Russian Foundation for Basic Research (projects nos. 15-01-05686, 15-41-02315, 15-38-21099). The work is performed according to the Russian Government Program of Competitive Growth of Kazan Federal University.

References

[1] Badriev I B and Karchevskii M M 1989 Convergence of the iterative Uzawa method for the solution of the stationary problem of seepage theory with a limit gradient *Journal of Soviet Mathematics* **45 (4)** 1302-9 DOI: 10.1007/BF01097083
[2] Badriev I B and Nechaeva L A 2013 Mathematical simulation of steady filtration with multivalued law *PNRPU Mechanics Bulletin* **(3)** 37–65
[3] Lapin A V 2010 Preconditioned Uzawa-Type Methods for Finite-Dimensional Constrained Saddle Point problems *Lobachevskii Journal of Mathematics* **31 (4)** 309-322. DOI 10.1134/s1995080210040013
[4] Shangaraeva A I and Shevchenko D V 2015 Speed up of the oil saturation numerical algorithm for the plane-parallel filtration *Applied Mathematical Sciences* **9 (150)** 7467-74 DOI 10.12988/ams.2015.510683
[5] Badriev I and Banderov V 2014 Numerical method for solving variation problems in mathematical physics *Applied Mechanics and Materials* **668-669** 1094-97. DOI: 10.4028/www.scientific.net/AMM.668-669.1094
[6] Lions J L 1969 *Quelque problèmes méthodes de résolution des problèmes aux limites nonlinéaires* (Paris: Dunod)
[7] Opial Z 1967 Weak convergence of the sequence of successive approximations for nonexpansive mappings *Bulletin of the American Mathematical Society* **73** (4) 591–7
[8] Badriev I B 2013 On the Solving of Variational Inequalities of Stationary Problems of Two-Phase Flow in Porous Media *Applied Mechanics and Materials* **392** 183-7 DOI: 10.4028/www.scientific.net/AMM.392.183
[9] Badriev I B 1983 Difference schemes for nonlinear problems of filtration theory with a discontinuous law *Soviet Mathematics* **27 (5)** 1–11
[10] Badriev I B and Fanyuk B Y 2012 Iterative methods for solving seepage problems in multilayer beds in the presence of a point source *Lobachevskii Journal of Mathematics* **33** (4) 386-99 DOI 10.1134/S1995080212040026

[11] Badriev I B 1983 Difference schemes for nonlinear problems of filtration theory with a discontinuous law *Soviet Mathematics* **27 (5)** 1–11

[12] Badriyev I B, Zadvornov O A, Ismagilov L N and Skvortsov E V 2009 Solution of plane seepage problems for a multivalued seepage law when there is a point source *Journal of Applied Mathematics and Mechanics* **73 (4)** 434-42 DOI 10.1016/j.jappmathmech.2009.08.007

[13] Ekeland I and Temam R 1976 *Convex Analysis and Variational Problems* (Amsterdam: North-Holland)

[14] Badriev I B and Shagidullin R R 1995 A study of the convergence of a recursive process for solving a stationary problem of the theory of soft shells *Journal of Mathematical Sciences* **73 (5)** 519-25 DOI :10.1007/bf02367668

[15] Badriev I B, Banderov V V, Gnedenkova V L, Kalacheva N V, Korablev A I and Tagirov R R 2015 On the finite dimensional approximations of some mixed variational inequalities *Applied Mathematical Sciences* **9 (114)** 5697-705 DOI. 10.12988/ ams.2015.57480

[16] Abdrakhmanova A I and Sultanov L U 2016 Numerical modelling of deformation of hyperelastic incompressible solids *Materials Physics and Mechanics* **26 (1)** 30-2

[17] Badriev I B and Karchevskii M M 1994 Convergence of an iterative process in a Banach space *Journal of Mathematical Sciences* **71** (6) 2727-35. DOI: 10.1007/bf02110578 15

[18] Badriev I B, Banderov V V, Garipova G Z, Makarov M V and Shagidullin R R 2015 On the solvability of geometrically nonlinear problem of sandwich plate theory *Applied Mathematical Sciences* **9 (82)** 4095-102. DOI: 10.12988/ams.2015.54358

[19] Badriev I B, Banderov V V, Makarov M V and Paimushin V N 2015 Determination of stress-strain state of geometrically nonlinear sandwich plate *Applied Mathematical Sciences* **9 (78)** 3887–95. DOI: 10.12988/ams.2015.54354

[20] Badriev I B, Banderov V V and Zadvornov O A 2013 On the solving of equilibrium problem for the soft network shell with a load concentrated at the point *PNRPU Mechanics Bulletin* (3) 17-35

[21] Berezhnoi D V, Paimushin V N and Shalashilin V I 2010 Studies of quality of geometrically nonlinear elasticity theory for small strains and arbitrary displacements *Mechanics of Solids* **44** (6) 837-51. DOI 10.3103/S0025654409060041

[22] Davydov R L and Sultanov L U 2014 Mathematical modeling of large elastic-plastic deformations *Applied Mathematical Sciences* **60** (8) 2991-6 DOI 10.12988/ams.2014.44272

[23] Sultanov L U 2014 Mathematical modeling of deformations of hyperelastic solids *Applied Mathematical Sciences* **8** (143) 7117-24 DOI 10.12988/ams.2014.49704

[24] Sultanov L U and Davydov R L 2014 Mathematical modeling of large elastic-plastic deformations *Applied Mathematical Sciences* **8 (60)** 2991-6 DOI 10.12988/ams.2014.44272

[25] ZheltuhinV S, Solovyev P S and Chebakova V Y 2015 Boundary conditions for electron balance equation in the stationary High-Frequency Induction discharges *Research Journal of Applied Sciences* **10 (10)** 658-62

[26] Lions P L and Merscier B 1979 Splitting algorithms for the sum of two nonlinear operators *SIAM Journal on Numerical Analysis* **16 (6)** 964–79

Numerical study of the three-degreed parametrically excited gyroscopic system

L Bakhtieva[1] **and V Bogolyubov**[2]

[1] Kazan Federal University, Kazan
[2] National Research Technical University named after A.N. Tupolev, Kazan

E-mail: lbakhtie@yandex.ru

Abstract. A mathematical model of the rotor vibratory gyroscope was constructed. The operating principle of the device based on the use of the tuned resonance of three-degreed oscillating system. Parametric excitation of a gyroscope is achieved by modulating the angular velocity of rotation of the rotor. The results of numerical calculations allow not only to illustrate the process of operation of the gyroscope, but also to find the values of the parameters under which it can be used as a device of orientation of moving objects.

1. Introduction

The development of inertial navigation tools relates with enhancing their functionality and with improving the accuracy of measurement of angular velocity (0.001-0.01 deg/hour) in conditions of exposure to significant congestion, and with minimizing their dimensions, weight characteristics and energy consumption. Such requirements are fully satisfied in the rotary vibratory gyroscope (RVG). Below we consider the case of parametrically excited three-degreed RVG, kinematic scheme of which [1] is shown in Fig. 1.

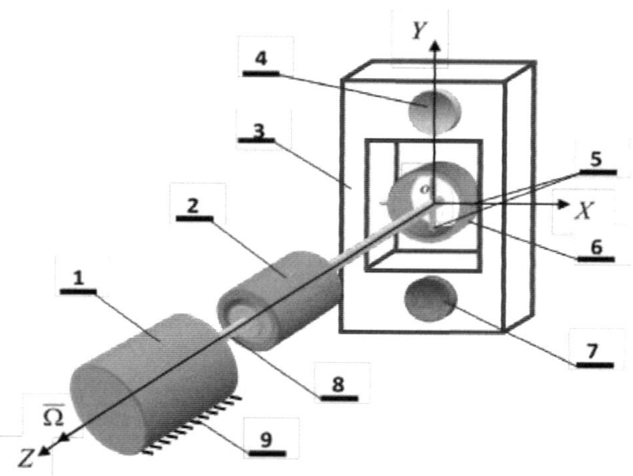

Figure 1. Kinematic scheme of the three-degreed RVG with asymmetrical rotor.

IOP Conf. Series: Materials Science and Engineering **158** (2016) 012014 doi:10.1088/1757-899X/158/1/012014

Circuit includes: a sensing element (unbalanced rotor) 3 fixed to the shaft 8 of the drive motor 1 by means of two pairs of torsions 5 and the gimbal 6. The vibration amplitude of the rotor RVG is converted to an electrical signal by the angle sensor 4, the creation of control moments applied to the rotor is effected by the torque sensor 7, controlled by signals carried from the housing 9, by using sine-cosine rotary transformer 2.

Innovative use of the gyroscope, which related to its parametric excitation, imposes special requirements to the dynamics of operation of the device. This, in turn, leads to the necessity of numerical analysis the corresponding gyro system based on a strict mathematical model.

2. The mathematical model of the parametrically excited RVG

Specificity of the modulation of the rotor vibratory gyroscope allows through the use of circuit design, without changing the mechanical design of the circuit, to raise the sensitivity of the instrument to the measured angular velocity by parametric excitation [2]. Parametric excitation of RVG as the oscillatory system is carried out by modulation of the dynamic stiffness of the torsions and gyroscopic moments by modulating the angular velocity of rotation of the rotor. Modulation is implemented by varying within a small range of frequency of the electric actuator, according to the law:

$$\Omega_m(t) = \Omega + \Delta\Omega\cos(\omega_m t) \ , \qquad (1)$$

where Ω – the nominal frequency of the motor shaft corresponding to a configured mode of operation of the gyro, $\Delta\Omega$ – frequency deviation, ω_m – frequency modulation.

The current value of the phase of the oscillations of the shaft of the drive motor relative to the nominal value defined by the expression:

$$\phi_m(t) = \int \Omega_m(t)dt = \Omega t + m\sin(\omega_m t)\,, \qquad (2)$$

where $m = \Delta\Omega/\omega_m$ – the index of the frequency modulation.

Given the properties of the Bessel functions of the first kind for values of the modulation index $m \leq 0.2$, and the expression (2), periodic functions can be written as:

$$\sin(\phi_m(t)) = \sin(\Omega t) + m(\sin(\Omega+\omega_m)t + \sin(\Omega-\omega_m)t)/2\,,$$
$$\cos(\phi_m(t)) = \cos(\Omega t) - m(\cos(\Omega-\omega_m)t - \cos(\Omega+\omega_m)t)/2\,. \qquad (3)$$

To describe the motion of RVG mounted on the moving base, we introduce a coordinate system $0\xi^*\eta^*\zeta^*$ with the origin at the center of mass of the gyro rotor.

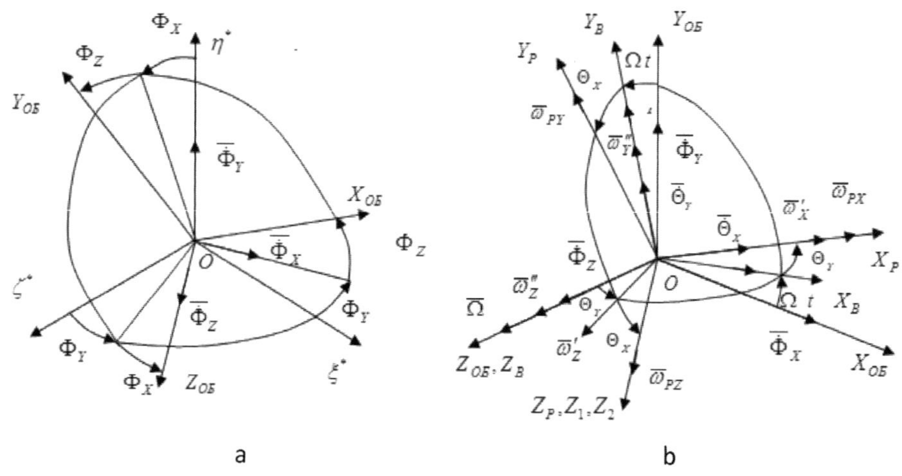

Figure 2. Coordinate systems

With a movable base we can tie the coordinate system $OX_{06}Y_{06}Z_{06}$, the axis Z_{06} of which coincides with the axis of rotation of the rotor RVG (Fig. 2a). The motion of the base of the device will considered as known, i.e. in each instant of time is known the orientation of the coordinate system $OX_{06}Y_{06}Z_{06}$ relative to the inertial $0\xi^*\eta^*\zeta^*$, and the projection $\dot{\Phi}_x, \dot{\Phi}_y, \dot{\Phi}_z$ of the vector of absolute angular velocity on the base axis system $0\xi^*\eta^*\zeta^*$ are given functions of time. In addition to the above systems are required two systems of axes $OX_BY_BZ_B$, $OX_pY_pZ_p$ associated respectively with the shaft of a drive motor and with the principal axes of inertia of the rotor (Fig. 2b).

The point of origin of the system $OX_BY_BZ_B$, $OX_pY_pZ_p$ lies in the center of mass of the EGR and its position relative to the system $OX_{06}Y_{06}Z_{06}$ is successive turns in the positive direction on appropriate angles Ωt for $OX_BY_BZ_B$ and θ_x, θ_y for $OX_pY_pZ_p$.

For deriving the equations of motion of a gyroscopic system will use the variational principle of Ostrogradsky – Hamilton (see for example [3]), from which follow equations

$$\frac{d}{dt}\frac{\partial T}{\partial \dot{q}_i} - \frac{\partial T}{\partial q_i} = Q_i\,, \tag{4}$$

where T is the kinetic energy of the considered gyroscopic system; q_i – generalized coordinates defining the position of all points of the system in space, Q_i – generalized forces acting on the system; i – the number of generalized coordinates describing the degrees of freedom of the considered electromechanical system.

Choose as generalized coordinates the angles of rotation of the rotor $\theta_x = \beta(t)$ and $\theta_y = \alpha(t)$, uniquely determining its position in the coordinate system of the shaft, and as generalized forces – damping moments of moments of elasticity of the torsions.

The differential equations of motion for the case of constant angular velocity of rotation of the base, after the procedure of factorization [4], subsequent linearization using the Jacobian matrix will be of the form

$$A\ddot{\beta} + \mu\dot{\beta} + \big(k + (C - B)\Omega(\Omega + 2m\omega_m\cos(\omega_m t))\big)\beta - (A + B - C)(\Omega + m\omega_m\cos(\omega_m t))\dot{\alpha} =$$
$$= (A + C - B)\Omega\big(\dot{\Phi}_x\sin(\Omega t) - \dot{\Phi}_y\cos(\Omega t)\big),$$

$$B\ddot{\alpha} + \mu\dot{\alpha} + \big(k + (C - A)\Omega(\Omega + 2m\omega_m\cos(\omega_m t))\big)\alpha + (A + B - C)(\Omega + m\omega_m\cos(\omega_m t))\dot{\beta} =$$
$$= (C + B - A)\Omega\big(\dot{\Phi}_x\cos(\Omega t) + \dot{\Phi}_y\sin(\Omega t)\big), \tag{5}$$

where A, B, C are respectively the equatorial and polar moments of inertia of the rotor, μ is the coefficient of viscous friction of the material of torsions, k – torsional stiffness of the torsions, $(C-A)\Omega^2$ and $(C-B)\Omega^2$ is the dynamic stiffness of the respective axes of suspension.

3. Numerical study

Feature of the system of differential equations (5) is the presence of elements associated with periodic change of the gyroscopic and the positional moments:

$$(A+B-C)m\omega_m\cos(\omega_m t)\dot{\alpha}A^{-1},\quad (A+B-C)m\omega_m\cos(\omega_m t)\dot{\beta}B^{-1},$$

$$2(C-A)m\omega_m\Omega\cos(\omega_m t)\alpha B^{-1},\qquad 2(C-B)m\omega_m\Omega\cos(\omega_m t)\beta A^{-1}.$$

The presence of such periodically varying energy parameters RVG as gyroscopic moments and dynamic stiffness included in the respective nonhomogeneous differential equations, provides favorable conditions for the parametric excitation of the mechanical contour of the considered gyro with frequency $\omega_m = \Omega$ [5].

In Fig.3 shows the results of numerical calculations carried out using the mathematical package Maple 9 when $\omega_m = \Omega$ (coherent mode) and when $\dot{\Phi}_x = 0.001\ c^{-1}$, $\dot{\Phi}_y = 0$.

Analysis of the obtained results shows that the solution of equations (5) essentially depends on values of the modulation index m. If at small values of m the solution is stable (Fig. 3a, 3b), when m = 0.0185 observed erratic behavior of the gyro (Fig. 3c). The critical value of the modulation index can be found through numerical experiments, and analytically [6].

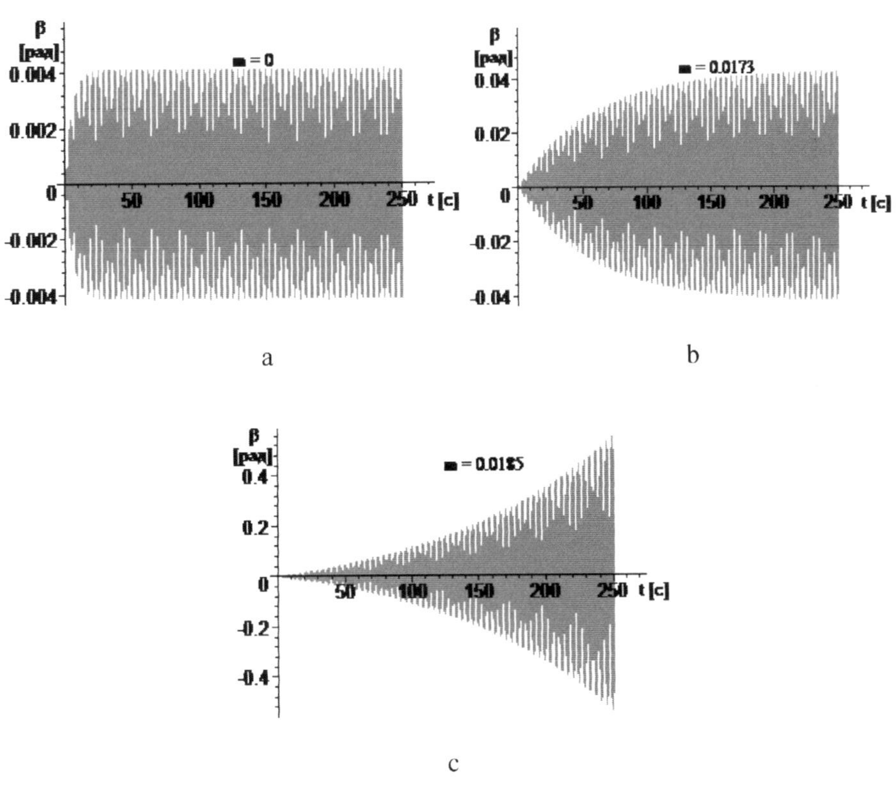

Figure 3. The graphs of the variation of rotor vibration amplitude in time at various values of modulation index m.

From the graph in Fig. 3b it follows that the modulation of the angular velocity of the gyro rotor enhances its sensitivity in the steady state (differentiating mode gyro) approximately one order of magnitude compared to the device in which parametric excitation of no (Fig. 3a). This is due to the fact, that the modulation the angular velocity of rotation of the rotor significantly reduces the coefficient of friction μ. This increases the duration of the ramp-up of amplitude, which corresponds to the operation of the device in the mode of measuring the angle of rotation of foundation (integrated mode).

It should be noted, that in the coherent mode, the sensitivity of RVG depends on to the value of the phase shift between the periodically varying dynamic stiffness of the torsions, the gyroscopic moments, which is characteristic of three-degreed gyroscope, and the external gyroscopic moment, creating by the angular velocity of rotation of the base of device.

Studies have shown that when using RVG as the gyrocompass it can be used as the coherent mode and incoherent mode of operation of the device, which is implemented when the condition $\omega_m \approx \Omega$. In the incoherent case, the amplitude of oscillation of the gyro rotor is also increasing, but the waveform is not stored and instead of harmonic oscillations with one spectral component is obtained a complex oscillation consisting of two harmonic component. It may be considered as quasi-harmonic oscillation with periodically variable amplitude.

The results obtained agree well with the experimental data given in [7].

Thus, the mathematical model describing the dynamics of motion parametrically excited by a three-degreed RVG, allows not only to illustrate the process of operation of the gyroscope, but also to find the values of the parameters under which it can be used as a device of orientation of moving objects.

References

[1] Maximov G, Jazikov B, Panshin V 1990 Way to show the sensor of the moments relative to the axes of sensitivity modulation asymmetric *DNG Gyroscopic devices for the ballistic missiles and space objects* **3** 8–12

[2] Golovan A, Mahorov G, Belugin V 1968 Gyroscope with parametric amplification of the useful signal on the auxiliary sub-harmonic motions of the sensitive element *Questions of applied mechanics* **1** (Moscow: GONTI) p 34

[3] Bakhtieva L and Tazyukov F 2014 On the stability of shells under impulsive loading *Scientific notes of Kazan University. Series of physical-mathematical science* **156 (1)** 5–11

[4] Solnicev R 1977 *Computers in the ship gyroscopy* (Leningrad: Sudostroenie)

[5] Bogolyubov V 1993 Investigation of the effect of modulation of dynamic stiffness RVG on it's accuracy *Sat. abstracts and communications VIII scientific and technical conference (Kazan)* 60

[6] Bakhtieva L and Bogolyubov V 2016 Dynamics of parametrically excited three-degree-offreedom gyro *New technologies, materials and equipment of the Russian aerospace industry: all-Russian scientific-practical conference with international participation: Collection of papers* **2** *(Kazan, Izd-vo AN RT)* 371–375

[7] Bogolyubov V and Feklina O 2014 Gyrocompass, on the basis of parametrically excited by modulation of the rotor vibratory gyroscopes *Proceedings of the XXIX conference of the memory of the outstanding designer of gyroscopic devices N. Ostryakov (Saint-Petersburg, SSC RF JSC "CONCERN" CRI" ELECTROPRIBOR")* 442

Numerical simulation of large elastoplastic deformation of shells of revolution in the "spacesuit" under the influence of pulse overload

V Bazhenov, A Demareva, M Baranova and A Kibets

Research Institute of Mechanics at the Lobachevsky State University of Nizhni Novgorod, 603950, Nizhni Novgorod, Russia

E-mail: bazhenov@mech.unn.ru

Abstract. An axisymmetric problem of high strains in a spherical lead shell enclosed into an aluminum "spacesuit" under the influence of pulse overload is considered. The shell straining is described with the use of equations of mechanics of elastoviscoplastic media in Lagrangian variables. Kinematic relations are determined in the current state metrics. Equations of state are taken in the form of equations of the flow theory with isotropic hardening. Contact interaction of a shell and a "spacesuit" is modeled by conditions of non-penetration with friction. Numerical solution under the given boundary and initial conditions is based on the moment schema of the finite element method and the explicit time integration scheme of the "cross". For sampling the space variables 4-node isoparametric finite elements with multilinear forms features are used. As it was shown by the results of calculations, spherical shell suffers significant local forming, characterized by large displacements and rotation angles of finite elements as a rigid body in the process of loading. The calculation results of the residual form are in good agreement with the experimental data.

1. Introduction

Thin-walled spherical shells are widely used in modern technology. Experimental and theoretical techniques of studies of their stability are presented in [1-5]. The analysis of computer modeling of deformation processes and the loss of stability of spherical shells under the quasi-static and impact loading, taking into account the effects of geometrical and physical nonlinearities are published in [6-15]. The results of numerical analysis of large elastic-plastic deformation of lead spherical shell in an aluminum "spacesuit" under explosive loading and contact with friction are shown below.

2. Constitutive system of equations and problem solution method

Deformation of the shell is analyzed in axisymmetric formulation and is described in Lagrangian variables using equations of mechanics of continuous media in a cylindrical coordinate system X_1, X_2 (X_1 – axis of rotation, X_2 – radius). The equations of motion are derived from the balance of virtual capacity [17, 18]:

$$\iint_{\Omega}\left(\sigma_{11}\delta\dot{\varepsilon}_{11}+\sigma_{22}\delta\dot{\varepsilon}_{22}+\sigma_{33}\delta\dot{\varepsilon}_{33}+2\sigma_{12}\delta\dot{\varepsilon}_{12}\right)X_2 d\Omega+\iint_{\Omega}\left(\rho\ddot{U}_1\delta\dot{U}_1+\rho\ddot{U}_2\delta\dot{U}_2\right)X_2 d\Omega$$
$$-\int_{G_p}\left(p_1\delta\dot{U}_1+p_2\delta\dot{U}_2\right)X_2 dG-\int_{G_q}\left(q_1\delta\dot{U}_r+q_2\delta\dot{U}u_z\right)X_2 dG=0 \tag{1}$$

where σ_{ij}, $\dot{\varepsilon}_{ij}$ are the components of the Cauchy stress tensor and the strain rate tensor (symmetric part of the displacement velocity gradient), U_i are displacement in the coordinate system X, ρ is density, p_α, q_α are the components of the surface and contact loads $(\alpha=1,2)$, Ω is the domain occupied by the construction, G_p - is the scope of the external pressure, G_q - is the part of the surface subjected to contact pressures determined in the course of solving the problem. Full stop above the symbol indicates the partial derivative over time t. The summation is over repeated indices.

Strain rates are determined in the current state metrics as:

$$\dot{\varepsilon}_{11}=\dot{U}_{1,1},\quad \dot{\varepsilon}_{22}=\dot{U}_{2,2}\quad ,\dot{\varepsilon}_{33}=X_2^{-1}\dot{U}_2\ ,\quad \dot{\varepsilon}_{12}=\tfrac{1}{2}\left(\dot{U}_{1,2}+\dot{U}_{2,1}\right),$$
$$\dot{\omega}=\tfrac{1}{2}\left(\dot{U}_{1,2}-\dot{U}_{2,1}\right),\ \dot{U}_{i,j}=\partial\dot{U}_i/\partial X_j,\quad X_j=X_j\big|_{t=0}+\int_0^t\dot{U}_j dt \tag{2}$$

Equations of state are set separately for spherical $\dot{\varepsilon}^V,\sigma^V$ and deviatoric components of the velocities of deformations and stresses $\dot{\varepsilon}'_{ij},\sigma'_{ij}$. The dependence of the spherical components of the strain rate and stress are assumed to be linear. Deviatoric strain rate components $\dot{\varepsilon}'_{ij}$ are decomposed on plastic $\dot{\varepsilon}'^p_{ij}$ and elastic components $\dot{\varepsilon}'^e_{ij}$. Deviatoric stress tensor components are calculated using the flow theory of relations with isotropic hardening [19]:

$$D_j\sigma'_{ij}=2G\dot{\varepsilon}'^e_{ij},\ \sigma'_{ij}\sigma'_{ij}\le 2/3\sigma^2_{yield}(æ,I_{2\dot{\varepsilon}})\ I_{2\dot{\varepsilon}}=\sqrt{2/3}\sqrt{\dot{\varepsilon}'^p_{ij}\dot{\varepsilon}'^p_{ij}}\ ,\ æ=\int_0^t I_{2\dot{\varepsilon}}dt,$$
$$\dot{\varepsilon}'^p_{ij}=\dot{\lambda}\sigma'_{ij}; \tag{3}$$

Here D_j is the Jaumann derivative [20] ($D_J\sigma'_{ij}=\dot{\sigma}'_{ij}-\dot{\omega}_{ik}\sigma'_{kj}-\dot{\omega}_{jk}\sigma'_{ik}$), G –shear modulus. σ_{yield} is the yield limit, æ is the Odquist parameter, $I_{2\dot{\varepsilon}}$ is the strain rate intensity, $\dot{\lambda}$ is a parameter, which is identically equal to zero in the case of elastic straining and $\dot{\lambda}>0$ determined from the condition of landing on the surface of the flow under elastoplastic deformation

The components of the contact force q_α $(\alpha=1,2)$ are calculated in the course of solving the problem in a local coordinate basis (s, ξ are the tangential and normal directions to the contact surface). The contact pressure along the normal is determined from the non-penetration condition:

$$\begin{cases}\dot{u}'_\xi=\dot{u}''_\xi\\ q'_\xi=-q''_\xi\end{cases},\ q_\xi=q'_\xi=\begin{cases}0, & q_\xi\ge 0\\ q_\xi, & q_\xi<0\end{cases} \tag{4}$$

The tangential pressure is first determined from the condition of rigid bonding and then, if the static friction force is exceeded - by the Coulomb law:

$$\begin{cases} \dot{u}'_s = \dot{u}''_s \\ q'_s = -q''_s \end{cases} \quad q_s = q'_s = \begin{cases} q_s, & |q_s| \le k_\xi |q_\xi| \\ k_\xi |q_\xi| sign(q_\xi), & |q_s| > k_\xi |q_\xi| \end{cases}, \tag{5}$$

where k_ξ is the friction coefficient.

The decision of the constitutive system of equations (1) - (5) is based on the finite element method and finite difference explicit scheme of integration over time, such as "cross" [16,21,22]. Contact nodal forces are determined based on the algorithm [22]. To reduce high-frequency nonphysical oscillations of the numerical solution, the procedure of conservative smoothing is applied [23]. The described finite element method for solving two-dimensional (axisymmetric) problems of elastic-plastic deformation of structural elements is implemented as separate software modules in the computer system "Dynamics-3" [24].

3. Results of studying dynamic buckling of the lead shell in a "spacesuit" under explosive loading

The results of the experimental investigation of this problem were reported in [25].

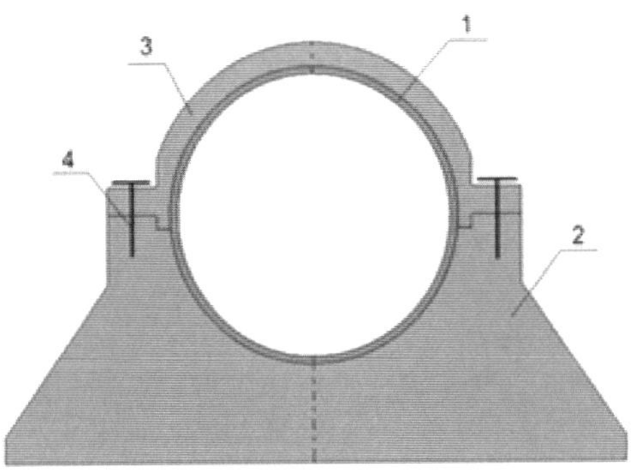

Figure 1. Experimental assembly

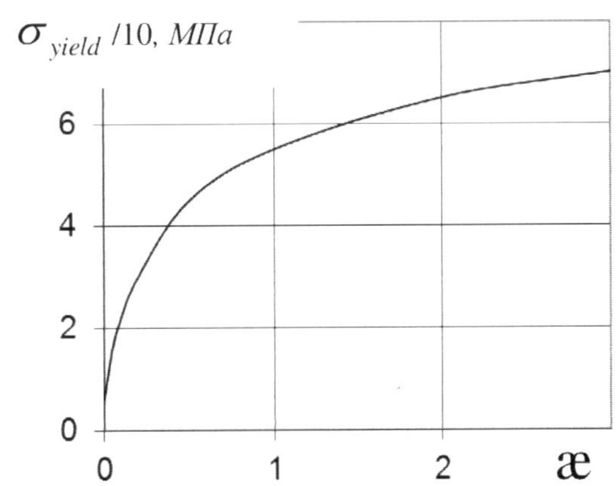

Figure 2. The dependence of the yield stress of the Odquist parameter

The experimental assembly consisting of the lead shell (1), the monolithic half-casing of the "spacesuit"(3) and the cover attached to the half-casing by screws(4) is schematically shown in figure 1. The ratio of the spherical shell thickness to its radius is h/R=0,054. The shell mass is 2,95 kg, and the total mass of the assembly is 21 kg. The shell material is S1 lead (E=1,8·10^4 MPa, ν=0,45, ρ=11,34 g/sm^3). According to [26] the stress-deformation diagram of lead (figure 2) obtained from the experimental data [27] is used in the calculation. The "spacesuit" material is the D16 aluminum alloy (E = 7,1·104 MPa, ν=0,3, ρ= 2,78 g/sm^3).). The friction coefficient k_ξ between the lead shell and the "spacesuit" is set equal to 0,2 [28].

The "spacesuit" is loaded from below by an explosive device [29] consisting of the high explosive (HE), flyer plate, and damper converting the action of the explosion on the "spacesuit" into a trapezoidal loading pulse n (figure 3).

For calculations the shell was divided into finite elements (6 finite elements over the shell thickness and 200 finite elements over the generatrix). As it was shown by preliminary calculations, the deformation of the "spacesuit" can be neglected. For this reason the "spacesuit" is modeled on the equivalent size and weight elastic prototype.

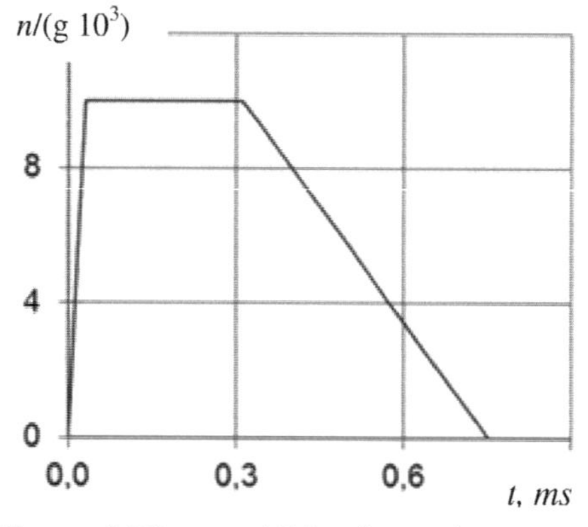

Figure 3."Spacesuit" loading pulse versus time

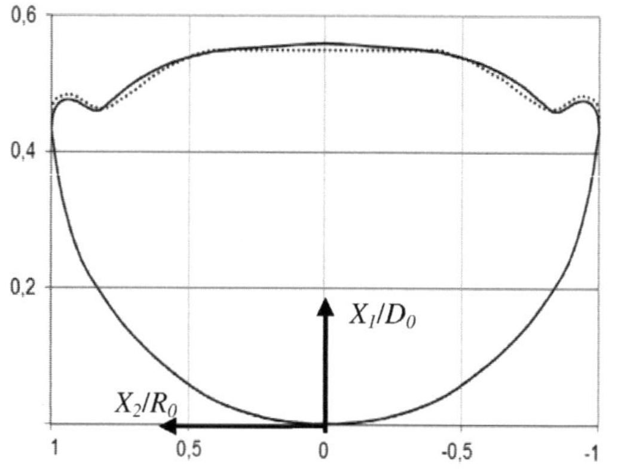

Figure 4. Residual form of a spherical shell: the solid line - experiment, dashed line - calculation

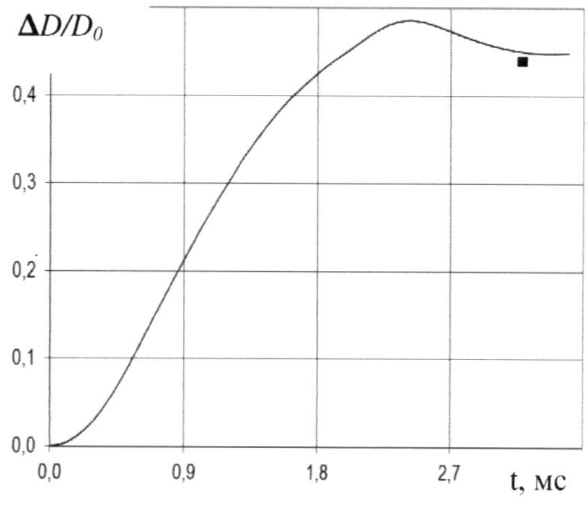

Figure 5. Convergence shell poles $\Delta D/D_0$, solid line - calculation, square - experiment

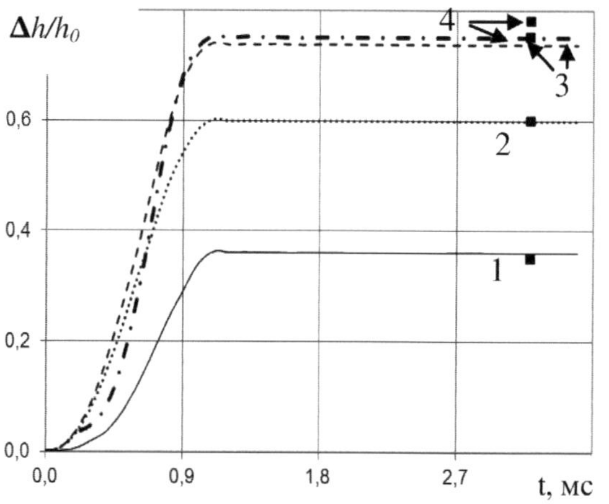

Figure 6. Changing the thickness $\Delta h/h_0$ shell along generatrix: 1 -90 degrees from the upper pole 2 - 120 degrees, 3 - 150 degrees, 4 - 180 degrees, the lines - calculation, squares - experiment.

The calculation results are shown in figures 4-6. In figure 4 the shapes of the shell in a residual state derived from natural experiment [25], and the numerical solution using the finite element method [16], are compared. The graphs of time dependence of upsetting of the shell (changes its diameter in the vertical direction) and the thickness at several locations of the lower half of the shell, spaced 30 degrees from the equator to the poles are shown in figures 5,6.

The elastoplastic shell buckling is continued throughout 2,5ms under the influence of inertial load and contact with "spacesuit". Subsequently, the velocities of the shell and the "spacesuit" are aligned. After unloading, the change of shape of the shell is substantially completed.

As a result of intensive loading, the shape of the shell and the thickness are being changed (figure 6). Increase of the thickness of the shell is irregular along its generatrix. Before buckling maximum deformations are developed in the lower hemisphere of the shell. The shell thickness is increased by 76% on rotation axis. With the removal from the lower pole shell thickness is reduced to 36% at the equator. Estimated values of the change in thickness along the generatrix are in good agreement with the experimental data in figure 5. Change in thickness of the shell stops after the formation of buckle for t> 1 ms.

Variation of friction coefficient leads to a change in the character of the distribution of shell thickening by generatrix during deformation. The change is barely noticeable, when $k_{\xi} \leq 0{,}35$. With further increase of the friction coefficient the highest thickening shell is moved toward the equator, that contradicts the experimental data.

The shape of the buckles and its location data of experiment and calculation are in good agreement. The difference between the residual vertical shell size (upsetting) based on the experimental value is 3%.

According to the calculated data, strain rate of the shell is changed from a minimum of 60 s^{-1} in the upper shell pole to the maximum value of 1200 s^{-1} in the buckle. Application of static stress-deformation diagram [27] does not allow to obtain quantitatively and qualitatively correct results at high strain rates. For example, the upsetting shell in residual state is greater than the experimental value of more than 2 times in this case.

Moment stress-strain state is realized in the zone of the ring buckles. The longitudinal strains are 70%, shear strain are 5%, the maximum rotation of the finite element as a rigid body in the area in the ring buckles is approximately 120 degrees. In these conditions, the application of Jaumann derivative provides acceptable accuracy of the numerical solution of the problem.

Computational experiments showed that the numerical schemes, based on the strain and stress velocity definition in the local basis [8,19,22], corresponding to the model shell theory of Timoshenko type based compression, allow to solve our problem on the grid with 1 layer of finite elements in thickness, which reduces computational resources on an order of magnitude or more, without loss of accuracy.

4. Conclusion

Elastoplastic straining, buckling, and further deformation of a spherical shell in a "spacesuit" under pulsed loading are numerically analyzed. There are large deformation angles of rotation of the shell elements as a rigid body and the change in thickness in the process of buckling of the lead shell. The dependence of viscosity characteristics of lead on the strain rate and the value of the friction coefficient have significant influence on forming and residual shell thickness. It is shown that the use of Jaumann derivative for small steps of integration over time provides sufficient accuracy of numerical solution of the axisymmetric problem for large deformations and rotation angles.

Acknowledgments

The reported study was funded by RFBR according to the research project № 15-48-02126.

References

[1] Volmir A S 1967 *The stability of deformable systems* (Moscow: Nauka) 984

[2] Konoplev Y G and Sachenkov A V 1967 *Investigation by theory of plates and shells* (Kazan: Kazan State University Press) **5** 161-88

[3] Vorovich I I and Minakova N I 1973 *Results of science and technology. Mechanics of solid deformable bodies* vol 7 (Moscow:VINITI) 5-86

[4] Gudramovich V S and Diskovsky I A 1977 *Doklady Physics* **232 (6)** 1283-85

[5] Grigolyuk E I and Mamai V I 1983 *Mechanics of Spherical Shell Deformation* (Moscow: Moscow State University Press) 114

[6] Yakushev V L 1997 Buckling hemispherical shells under plastic deformation //*Proceeding 18th International Conference on the Theory of Shells and Plates* vol 2 (Saratov:

Saratov State University Press) 136-41

[7] Ryabov A A, Romanov V I and Zefirov S V 1999 *Applied problems of strength and plasticity* (Nizhni Novgorod: Nizhni Novgorod State University Press) **60** 125-8

[8] Bazhenov V G, Artem'eva A A, Gonik E G, Kibets A I, Shoshin D V and Fedorova T G 2012 *Problems of Strength and Plasticity* (Nizhny Novgorod: Nizhny Novgorod State University Press) **74** 84-91

[9] Bazhenov V G, Gonik E G, Kibets A I and Shoshin D V 2014 *J. Appl. Mech. Tech. Phys.* **55 (1)** 13-8

[10] Kinkead A N, Jennings A, Newell J and Leinster C 1994 *J. of Strain Analysis* **29 (1)** 17-41

[11] Gupta N K and Venkatesh K S 2004 *Intern. J. of Impact Engng* **30 (8-9)** 1225-40

[12] Hudramovych V S 2009 *International Applied Mechanics* **45 (7)** 708-29

[13] Liang C C, Liao C C and Ma Y C 1991 *Computer and Structure* **39 (6)** 653-61

[14] Dadras A 2011 *Australian Journal of Basic and Applied Sciences* **5 (11)** 2052-8

[15] Joniak S 2004 *Journal of Theoretical and Applied Mechanics* **42 (2)** 349-56

[16] Artemyeva A A, Baranova M S, Kibetz A I, Romanov V I, Ryabov A A and Shoshin D V 2011 *Bulletin of UNN. Series Mechanics.* (Nizhni Novgorod: Nizhni Novgorod State University Press) **3 (1)** 158-62

[17] Pozdeev A A, Trusov P V and Nyashin Y I 1986 *The large elastic-plastic deformation theory, algorithms, applications* (Moscow: Nauka) 232

[18] Belytschko T., Liu W K and Moran B 2000 *Nonlinear finite elements for continua and structures* (New York: John Wiley & Sons) 600

[19] Artemyeva A A, Bazhenov V G, Kazakov D A, Kibec A I and Nagornykh E V 2015 *Journal of Applied Mathematics and Mechanics* **79 (4)** 558-69

[20] Annin B D and Korobeinikov S N 1998 *Sib. Zh. Industr. Mat.* **1 (1)** 21-34

[21] Golovanov A I, Tuleneva O N and Shigabutdinov A F 2006 *The finite element method in statics and dynamics of thin-walled structures.* (Moscow: FIZMATLIT) 391

[22] Bazhenov V G, Kibets A I and Tsvetkova I N 1995 *Problemy Mashinostraeniya i Nadezhnos'ti Mashin* **2** 20-6

[23] Bazhenov V G and Zefirov S V 2001 *Bulletin of UNN. Series Mechanics.* (Nizhni Novgorod: Nizhni Novgorod State University Press) **1** 166-73

[24] Computing system "Dynamics-3." Scientific and Technical Center for Nuclear and Radiation Safety. software tool Attestation passport. Registration certification passport number PS 325 from 04.18.2013.

[25] Ivanov I G and Novikov S A 2003 Deformation of Spherical Lead Shells under the Action of Intense Mechanical Loading *Strength of Materials under Dynamic Loading (collected scientific papers)* (Sarov: Inst. Exper. Phys. Press) 216–221

[26] *The dynamics impact* 1985 edited by S S Grigoryan (Moscow: Mir) 296

[27] Bazhenov V G, Baranova M S, Zhegalov D V, Lavrinenko V Yu and Pavlenkova E V 2013 *Russian Engineering Research* **33 (5)** 249-52

[28] Kragel'skii I V and Vinogradov I V 1955 *Friction Coefficients: Reference Book* (Moscow: Publishing House of the Machine Literature)

[29] Novikov S A, Sinitsyn V A and Pogorelov A P 1980 *Fiz. Goreniya Vzryva* **16 (6)** 111-3

Application the particle method in problems of mechanics deformable media

D V Berezhnoi, N F Gabsalikova and V V Miheev

Kazan Federal University, 18 Kremlyovskaya Street, Kazan, 420008, Russian Federation

E-mail: miheevvova@mail.ru

Abstract. The work implemented method of deformation of ground-based particle method, which is a collection of mineral grains, which are linked to some system of forces on the contact areas between the mineral particles. Two-parameter potential Lennard-Jones and it is modified version were selected for describing the behavior of ground. Some model problems of straining layer of ground in the gravity field was decided. The calculations were performed on a heterogeneous computing cluster, on each of the seven components that were installed on three GPU AMD Radeon HD 7970.

1. Introduction

Currently, most of the computational methods based on sampling the area by finite element and finite difference methods [1–10]. This approach has proven itself in the solution the problems of deformation complex heterogeneous environments of different nature. However, there is extensive class of problems, in which the loading of the medium is accompanied by the destruction of multiple fragments and slippage, intensive mass transfer, including mass mixing effects, etc. Using the mesh approach meets with considerable difficulties in solving such problems. A promising class of numerical methods of solid mechanics, adapted for modelling the destruction is a particle method. It should be noted, that in the modern sense of the term "particle method" is a collective and includes very diverse numerical methods, as belonging to the classical representatives of the discrete approach to mechanics and meshless numerical algorithms for solving equations of continuum mechanics (for example particle method in cells [11], smoothed particles method [12], SPAM [13], etc.). Moreover, in the present some current implementations conventional numerical methods, such as finite dotted element method (particle finite element method [18]) often carry to the particle methods. It should be noted, that the particle method formed by methods well-known method of molecular dynamics is used for studying the response of medium at the atomic level [14]. At the same time, the possibilities of atomic description for modelling the behavior of rigid body in interest from the point of view of engineering applications of spatial and temporal scales are extremely limited. Now, the interatomic interaction potentials for the majority of materials are known cannot be said about the potentials to describe the behavior of deformable continuous media, especially shear. One of the important advantages of the particle method is that we can do a small amount of information about the properties of the material. Complex mechanical processes can be modelling by using simple Lennard-Jones potential.

2. Formulation of the problem

Consider the aggregate from N of material particles interacting with each other in a potential field. Particle motion is described by the definitely equation of motion in this method

$$m\ddot{\mathbf{r}}_k = \sum_{n=1}^{N} \Phi(r_{kn})\mathbf{r}_{kn} + \sum_{n=1}^{N} \Psi(r_{kn}, v_{kn})\mathbf{r}_{kn} + \boldsymbol{\varphi}(\mathbf{r}_{kn}) + \boldsymbol{\psi}(\mathbf{r}_{kn}, \mathbf{v}_{kn})$$

where $\mathbf{r}_k$ and $\mathbf{v}_k$ – position vectors and velocity of k-particle,

$$\mathbf{r}_{kn} \overset{def}{=} \mathbf{r}_k - \mathbf{r}_n,\ \mathbf{v}_{kn} \overset{def}{=} \mathbf{v}_k - \mathbf{v}_n,$$

m – mass of particle, $\Phi(r)$, $\Psi(r,v)$ describe conservative and non-conservative component of the interaction between the nearby particles, $\boldsymbol{\varphi}(\mathbf{r})$ and $\boldsymbol{\psi}(\mathbf{r},\mathbf{v})$ describe external conservative and non-conservative force field. Conservative component of interaction $\Phi(r)$ we can find, according to the formula:

$$\Phi(r) \overset{def}{=} \frac{1}{r} f(r),\ f(r) \overset{def}{=} -\dot{\Pi}(r),$$

where $f(r)$ – scalar potential force, $\Pi(r)$ – interaction potential. Non-conservative force $\Psi(r,v)$ allows to describe dissipation of energy. We consider the pair potential $\Pi(r)$, interaction force $f(r)$ appropriating to this potential is defined as

$$f(r) \overset{def}{=} -\dot{\Pi}(r).$$

The most promising technology to simulate particle method in environments with different rheology and to solve problems, where the set of particles is large enough are heterogeneous computing clusters using graphics accelerators (GPU). One GPU device is equivalent about 50 Xeon CPUs for solving problems such as: calculation P-waves, S-waves propagation [15]. However, this technique has certain limitations on the amount of memory available, since the internal GPU memory devices much smaller than traditional RAM cluster nodes. For calculations on graphic accelerators used a library with open source VexCL, because this library simplifies the development of applications and strives to reduce amount of boilerplate code needed to develop GPU applications [16-18]. The calculations were implemented on a heterogeneous computing cluster. On each of the seven nodes in the cluster set of three GPU AMD Radeon HD 7970.

For comparison the proposed approaches to the solution problems of deformation and destruction of the soil, two different solutions have been implemented algorithm [19-21]. In the first case we selected Lennard-Jones potential, and the power options for interaction which takes the form of

$$D = 4*10^5,\ a = 0.4r,\ f(r) \equiv \frac{12D}{r}\left[(a/r)^{12} - (a/r)^{6}\right].$$

In the second case used a "modified" the potential, parameters and the interaction force for which takes the form of

$$D = 10^5,\ a = 0.4*1.28r,\ f(r) \equiv \frac{D}{r}\left[12(a/r)^{12} - 36(a/r)^{9} + 14.4(a/r)^{8} + 6(a/r)^{6}\right].$$

3. Numerical example

For illustration the possibilities of the proposed "modified" potential (as compared to the Lennard-Jones potential), two model problems have been solve. The first task of the soil layer is initially formed on a perfectly rigid obstacle, the central part of which at some time t=20 (it should be noted, that the time is a dummy variable and does not correspond to the real, and the decision itself is a quality) instantly disappears. The layer of soil under its own weight flexes, causing its partial destruction. Figure 1 shows a configuration that acquires the compactor mass during bending and partial destruction, depending on the time for the Lennard-Jones and "modified" potential, respectively.

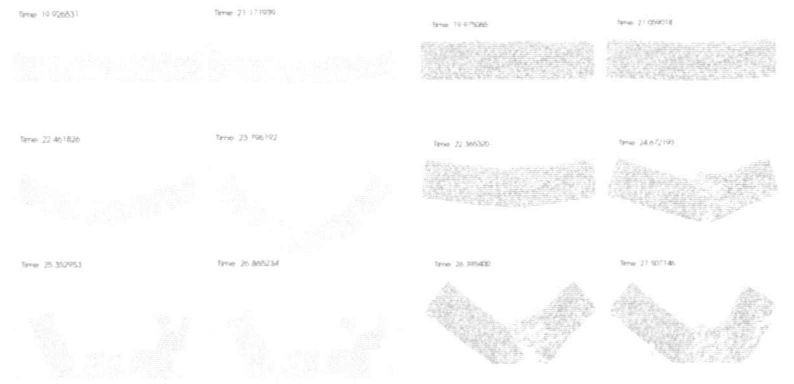

Figure 1. The configuration of the compactor mass during bending and partial destruction, depending on the time for the Lennard-Jones and "modified" potential.

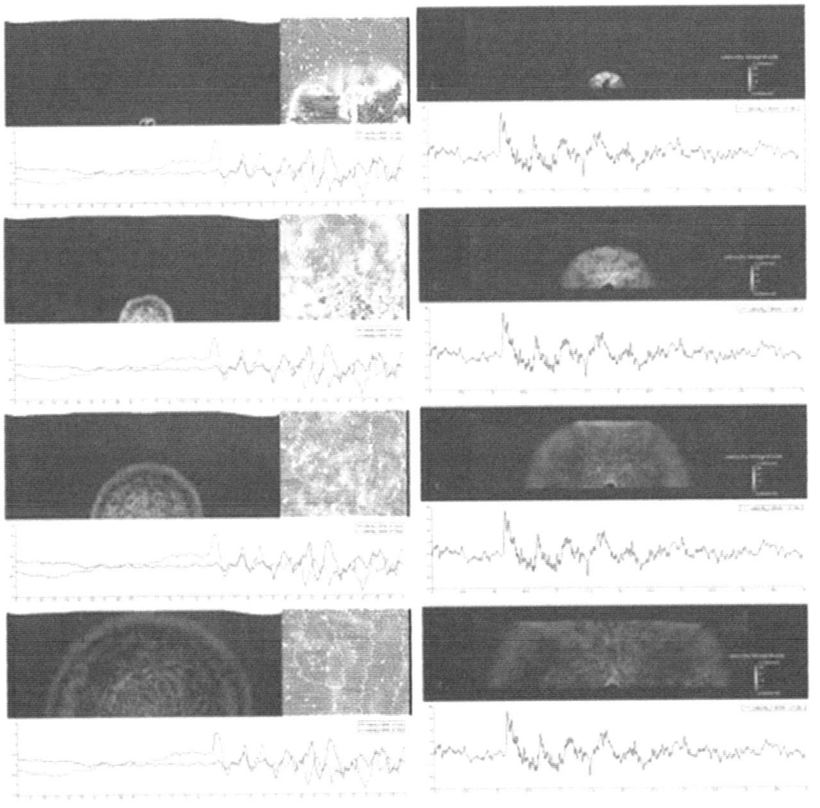

Figure 2. The distribution of per-frame velocity amplitude in the ground, modeling by "modified" potential and Lennard-Jones potential.

The second problem in the formed layer of the ground to a particle material situated in the middle of the lower border of soil is applied some initial vertical offset, which leads to the occurrence and propagation of waves in the computational domain. Figures 2 show the distribution of per-frame velocity amplitude in the ground, modeling by "modified" potential and Lennard-Jones potential, respectively. It may be noted, that using a "modified" potential can observe not only the P-wave in the ground, but and S-wave, while using Lennard-Jones S-wave is not observed.

4. Conclusion

In our work was carried particle method, which give us opportunity to investigate the interaction of a system of material points that are in a potential field. For describing the motion of the material points was used the system equations of motion. The integration is performed with using the algorithm Verlet. The problem of deformation and partial destruction of the soil mass, located in the gravity field, as well as the problem of wave propagation in a ground massif, caused some local perturbation of the computational domain were solved on the basis of the Lennard-Jones potential and the introduction of a new "modified" potential. Analysis of the results shows that the using the "modified" potential allows to realize the process of brittle disruption in the ground, while the Lennard-Jones potential allows you to reproduce only ductile disruption. In addition, in modeling the behavior of soil with Lennard-Jones potential wave propagation occur on the hydromechanical model (without shear wave), whereas the using of "modified" potential allows you to model tension-compression and shear waves.

Acknowledgments

The reported study was supported by Government of the Republic of Tatarstan research projects No. 15-07-05380, 15-41-02555.

References

[1] Badriev I B, Makarov M V and Paimushin V N 2016 *Procedia Engineering* 150 1050–1055
[2] Badriev I B, Makarov M V and Paimushin V N 2016 *Procedia Engineering* 150 1056–1062
[3] Shigapova F A, Mustakimova R F, Saleeva G T and Sachenkov O A *International Journal of Applied Engineering Research* **10 (24)** 44711–44714
[4] Sachenkov O A, Mitryaikin V I, Zaitseva T A and Konoplev Yu G 2014 *Applied Mathematical Sciences* **8 (159)** 7889–7897
[5] Sultanov L U 2015 *Procedia Earth and Planetary Science* **15** 119–124
[6] Golovanov A I and Sultanov L U 2005 *International Applied Mechanics* **41 (6)** 614–620
[7] Sultanov L U 2014 *Applied Mathematical Sciences* **8 (143)** 7117–7124
[8] Zienkiewicz O 1977 *The finite element method* (London Mc Graw-Hill) 787
[9] Wilkins M L 1967 *Vychislitelnye metody v gidrodinamice* (Mir) 212-263
[10] Evans M W and Harlow F H 1957 *Los Alamos Scientific Laboratory Report No. LA-2139* (Los Alamos)
[11] Monaghan J J 2005 *Reports on Progress in Physics* **68 (8)** 1703-1759
[12] Hoover W G and Hoover C G 2008 *American Journal of Physics* **76 (4-5)** 481-492
[13] Onate E, Idelson S R, Del Pin F and Aubry R 2004 *International Journal of Computational Methods* **1 (2)** 267-307
[14] Rieth M 2003 *Nano-engineering in Science and Technology* (Singapore: World Scientific) 151
[15] Galimov M R and Birjaltsev E V 2010 *Numerical methods and programming* **11** 77-93
[16] Demidov D *"VexCL: a C++ vector expression template library for OpenCL/CUDA"* https://github.com/ddemidov/vexcl
[17] Demidov D, Karsten A, Rupp K and Gottschling P 2013 *SIAM Journal on Scientific Computing* **35 (5)** 453-472
[18] Karsten A, Demidov D and Mulansky M 2014 *Numerical Computations with GPUs* (Springer International Publishing) 125-157
[19] Berezhnoi D V *and* Gabsalikova N F *2015* XIX International Conference on Computational Mechanics and Modern Applied Software Systems (CMMASS'2015) 220-222
[20] Berezhnoi D V, Gabsalikova N F and Kamilov M R2015 *XX International Symposium "Dynamic and Technological Problems of a Mechanics of Constructions and Continuous Mediums" (TRP)* **1** 24-26
[21] Berezhnoy D V, Gabsalikova N F and Balafendieva I S 2015 *International Journal of Applied Engineering Research* **10 (24)** 44677-44681

Stresses in the superflywheel

D V Berezhnoi and L R Gajnulina

Kazan Federal University, 18 Kremlyovskaya Street, Kazan, 420008, Russian Federation

E-mail: leysan5684@yandex.ru

Abstract. We obtain exact solutions for the problem of superflywheel that is rotating around its own axis. Superflywheel is made by the nozzle at each other with tension and subsequent gluing of the family of concentric cylinders. The effect of different methods of cylinder heads on each other to the maximum stresses in the superflywheel are analyzed, estimations are given for the possible increase of its specific energy consumption. Increasing in the specific energy consumption is marked by the decreasing in relative radius of superflywheel's the inner hole and with an increase in the Poisson ratio in material of superflywheel.

1. Introduction

In connection with the development of modern technologies the great importance is the problem of energy storage. The creation of new materials help not only improve the existing traditional electrochemical batteries, but to look for new methods of accumulation and storage of energy, including mechanical [1-3]. One is a relatively simple technique by means of energy storage is a flywheel which stores the transmitted energy as rotational kinetic energy. However, a significant limitation of such energy conservation technology is a high speed of rotation, which can lead to a sharp increase in the stress level in the flywheel and its subsequent destruction. An important factor determining the energy density of the kinetic energy storage is the ratio of the limiting rupture stress to the density of the material, which, combined with the security requirements, dictate the usage of super flywheels (flywheels made by winding filaments or tapes).

2. Kinematics

Belt super flywheels with an acceptable degree of accuracy can be represented as a system of n concentric cylinders (Fig. 1), clad on each other with a certain tension δ. In this case, before the start of loading in superflywheels, there is the some stress-strain state, which can be determined by calculating of so-called composite cylinder [4].

If the structure consists of n cylinders, clad on each other with interference δ_n, for the n-th cylinder radial σ_{rr}^n and circular $\sigma_{\varphi\varphi}^n$ stresses will be determined by the formulas

$$\sigma_{rr}^n = \frac{\delta_{n-1}E}{2}\frac{\left(r_{n-1}^2 - r_0^2\right)r_{n-1}}{\left(r_{n-1}^2 + r_0^2\right)\left(r_n^2 - r_{n-1}^2\right) + \left(r_{n-1}^2 - r_0^2\right)\left(r_n^2 + r_{n-1}^2\right)}\frac{r^2 - r_n^2}{r^2} -$$

$$-\sum_{i=n}^{N-1}\frac{\delta_i E}{2}\frac{\left(r_{i+1}^2-r_i^2\right)r_i}{\left(r_i^2+r_0^2\right)\left(r_{i+1}^2-r_i^2\right)+\left(r_i^2-r_0^2\right)\left(r_{i+1}^2+r_i^2\right)}\frac{r^2-r_0^2}{r^2},$$

$$\sigma_{\varphi\varphi}^n=\frac{\delta_{n-1}E}{2}\frac{\left(r_{n-1}^2-r_0^2\right)r_{n-1}}{\left(r_{n-1}^2+r_0^2\right)\left(r_n^2-r_{n-1}^2\right)+\left(r_{n-1}^2-r_0^2\right)\left(r_n^2+r_{n-1}^2\right)}\frac{r^2+r_n^2}{r^2}-$$

$$-\sum_{i=n}^{N-1}\frac{\delta_i E}{2}\frac{\left(r_{i+1}^2-r_i^2\right)r_i}{\left(r_i^2+r_0^2\right)\left(r_{i+1}^2-r_i^2\right)+\left(r_i^2-r_0^2\right)\left(r_{i+1}^2+r_i^2\right)}\frac{r^2+r_0^2}{r^2},$$

where E - Young's modulus of the material of the cylinder, r - the actual radius.

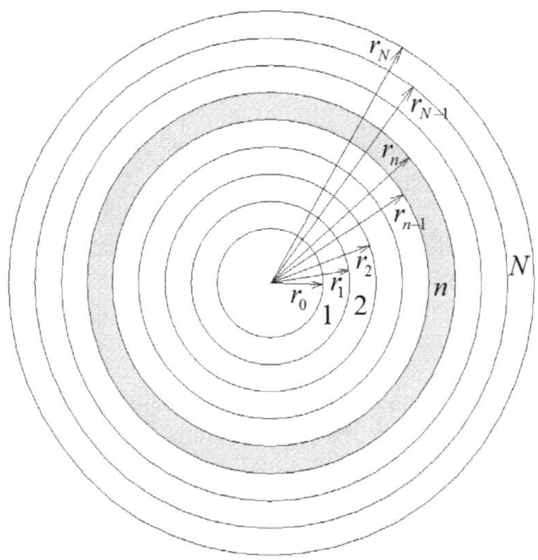

Figure 1. The system of concentric cylinders.

If the tightness of all the same, the formula for the radial and circumferential stress to the n-th cylinder will take the form

$$\sigma_{rr}^n=\frac{\delta E}{2}\left(\frac{\left(r_{n-1}^2-r_0^2\right)r_{n-1}}{\left(r_{n-1}^2+r_0^2\right)\left(r_n^2-r_{n-1}^2\right)+\left(r_{n-1}^2-r_0^2\right)\left(r_n^2+r_{n-1}^2\right)}\frac{r^2-r_n^2}{r^2}-\right.$$

$$\left.-\sum_{i=n}^{N-1}\frac{\left(r_{i+1}^2-r_i^2\right)r_i}{\left(r_i^2+r_0^2\right)\left(r_{i+1}^2-r_i^2\right)+\left(r_i^2-r_0^2\right)\left(r_{i+1}^2+r_i^2\right)}\frac{r^2-r_0^2}{r^2}\right),$$

$$\sigma_{\varphi\varphi}^n=\frac{\delta E}{2}\left(\frac{\left(r_{n-1}^2-r_0^2\right)r_{n-1}}{\left(r_{n-1}^2+r_0^2\right)\left(r_n^2-r_{n-1}^2\right)+\left(r_{n-1}^2-r_0^2\right)\left(r_n^2+r_{n-1}^2\right)}\frac{r^2+r_n^2}{r^2}-\right.$$

$$\left.-\sum_{i=n}^{N-1}\frac{\left(r_{i+1}^2-r_i^2\right)r_i}{\left(r_i^2+r_0^2\right)\left(r_{i+1}^2-r_i^2\right)+\left(r_i^2-r_0^2\right)\left(r_{i+1}^2+r_i^2\right)}\frac{r^2+r_0^2}{r^2}\right).$$

If we assume that the contact pressure between all the cylinders are the same and are equal to p_c, the ratio of stress take the form of

$$\sigma_{rr}^n=p_c\left[\frac{r^2-r_n^2}{r^2}\frac{r_{n-1}^2}{\left(r_n^2-r_{n-1}^2\right)}-\frac{r^2-r_0^2}{r^2}\sum_{i=n}^{N-1}\frac{r_i^2}{\left(r_i^2-r_0^2\right)}\right],$$

$$\sigma_{\varphi\varphi}^{n} = p_c\left[\frac{r^2+r_n^2}{r^2}\frac{r_{n-1}^2}{\left(r_n^2-r_{n-1}^2\right)} - \frac{r^2+r_0^2}{r^2}\sum_{i=n}^{N-1}\frac{r_i^2}{\left(r_i^2-r_0^2\right)}\right].$$

If we take the ratio

$$\delta_i = \frac{2p}{Er_i}\frac{\left(r_i^2+r_0^2\right)\left(r_{i+1}^2-r_i^2\right)+\left(r_i^2-r_0^2\right)\left(r_{i+1}^2+r_i^2\right)}{\left(r_{i+1}^2-r_i^2\right)},$$

the ratio of stress take the form

$$\sigma_{rr}^{n} = p\left[\frac{\left(r_{n-1}^2-r_0^2\right)}{\left(r_n^2-r_{n-1}^2\right)}\frac{r^2-r_n^2}{r^2} - \frac{\left(r^2-r_0^2\right)(N-n)}{r^2}\right],$$

$$\sigma_{\varphi\varphi}^{n} = p\left[\frac{\left(r_{n-1}^2-r_0^2\right)}{\left(r_n^2-r_{n-1}^2\right)}\frac{r^2+r_n^2}{r^2} - \frac{\left(r^2+r_0^2\right)(N-n)}{r^2}\right].$$

Consider the problem of determining the voltage (stress) in rotating with a constant angular velocity ω uniform disk density ρ, Young's modulus E and Poisson's ratio μ [5]. For a cylinder with an outer radius r_N and the central opening of radius r_0 expressions for stresses take the form ($k = r_0/r_N$)

$$\sigma_{rr}(r) = \frac{3+\mu}{8}\rho\omega^2 r_N^2\left(1+k^2\left(1-r_N^2 r^{-2}\right)-r^2 r_N^{-2}\right),$$

$$\sigma_{\varphi\varphi}(r) = \frac{3+\mu}{8}\rho\omega^2 r_N^2\left(1+k^2\left(1+r_N^2 r^{-2}\right)-\frac{1+3\mu}{3+\mu}r^2 r_N^{-2}\right).$$

3. Numerical example

Three series of calculations were performed for super flywheel with the following geometrical and mechanical properties: the Young's modulus $E = 200000\,MPa$, Poisson's ratio $\mu = 0.3$, internal radius $r_0 = 1\,m$, external radius $r_N = 2\,m\,(k = 2)$, tensile strength $\sigma_{ts} = 1100\,MPa$, material density $\rho = 7860\,kg/m^3$. Solid flywheel with these parameters permits angular velocity $\omega = 200\,rad/s$, while the maximum stress in the flywheel (hoop stresses on the inner surface of the flywheel) are equal to $\sigma_{\varphi\varphi}^{\max} = 1092.54\,MPa$.

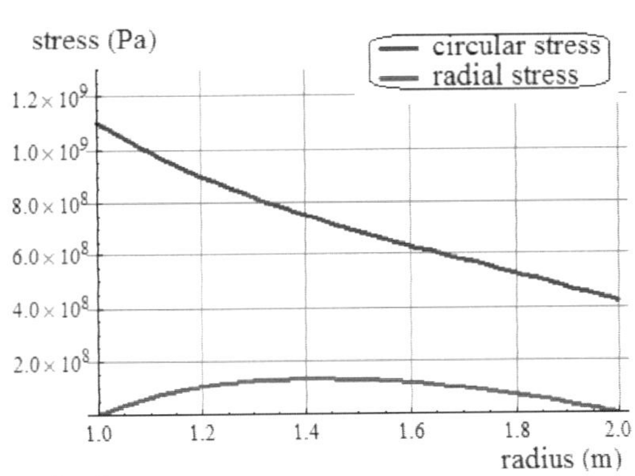

Figure 2. Curves of radial and circular stresses in the thickness of the flywheel wall.

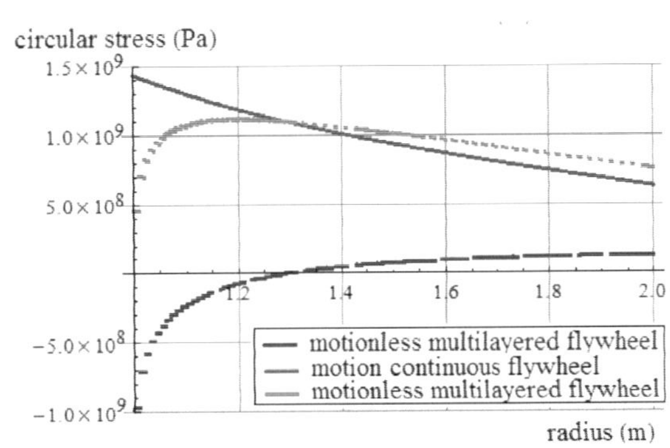

Figure 3. Circular stresses for the case of constant interference.

Figure 2 shows curves of radial and circular stresses in the thickness of the flywheel wall. Using super flywheel (in the form of composite cylinders) can increase the rotational angular velocity. Here are curves of circular stresses in a solid flywheel and super flywheels (before and after its promotion) for each variant of the corresponding angular velocity for the 100 layers, respectively, for the case of interference (Figure 3), the case of equal contact pressures (Figure 4) and the case of a monotonically changing interference (Figures 5).

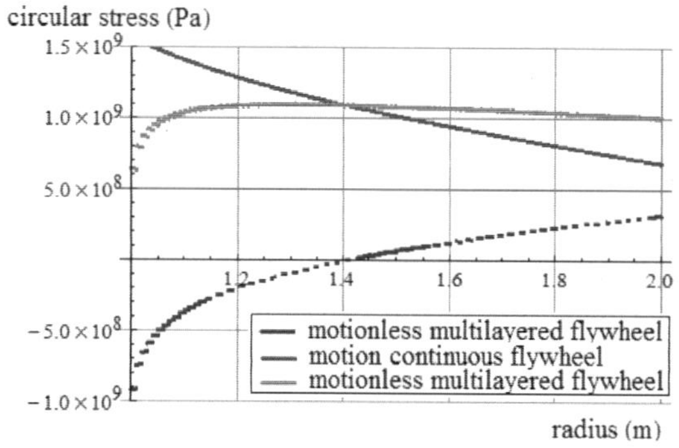

Figure 4. Circular stresses for the case of equal contact.

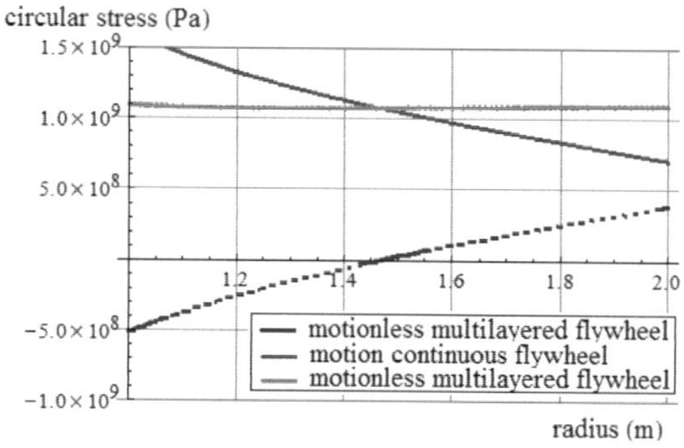

Figure 5. Circular stresses for case of a monotonically changing interference.

4. Conclusion

The exact solutions of distribution of stresses in super flywheels for different modeling tips are obtained. The effect of different modeling tips is explored for increasing the rotation speed of super flywheels. The dependence of the hoop stress is obtained in the rings of super flywheels (interference voltage) when it is winding. Dependence is obtained of super flywheels efficiency factor for a large number of layers on the Poisson ratio of the material of the flywheel and the ratio of the outer radius to the inner super flywheels.

Acknowledgments

The reported study was supported by Government of the Republic of Tatarstan research projects No. 15-01-05686, 15-41-02555.

References

[1] Drum T O 2011 *Energy storage-flywheel* http://www.resilience.org/stories/2011-10-05/energy-storage-flywheel
[2] Gulia N V 1976 *Machovichnye dvigately* (Mashinostroenie) 163
[3] Gulia N V 1980 *Nakopitely Energii* (Nauka) 152
[4] Matveev V V, Pisarenko G V and Yakovlev A P 1988 Spravochnik po soprotivleniyu materialov on resistance of materials (Naukova Dumka) 736
[5] Rabotnov Y N 1962 *Soprotivlenye materialov* (Fizmatgiz) 456
[6] Berezhnoi D V, Chickrin D E and Galimov A F 2014 *Applied Mathematical Sciences* 2014 **8 (124)** 6181–6190
[7] Berezhnoi D V, Chickrin D E, Kurchatov E Yu. and Galimov A F 2014 *Applied Mathematical Sciences* **8 (163)** 8125–8135
[8] Berezhnoi D V, Chickrin D E, Gajnulina L R, Kurchatov E Yu, Kokunin P A, Sozutov I G and Shigapov M I 2015 *Contemporary Engineering Sciences* **8 (36)** 1703–1712
[9] Berezhnoy D V, Gajnulina L R and Sekaeva L R 2015 *International Journal of Applied Engineering Research* **10 (24)** 44672–44676

Modelling of deformation of underground tunnel lining, interacting with water-saturated soil

D V Berezhnoi, I S Balafendieva, A A Sachenkov and L R Sekaeva

Kazan Federal University, 18 Kremlyovskaya Street, Kazan, 420008, Russian Federation

E-mail: berezhnoi.dmitri@mail.ru

Abstract. Built finite element method of calculating the deformation of underground tunnel lining, interacting with dry and water-saturated soils. To simulate the interaction between the lining and soils environments, including physical and non-linear, a special "contact" finite element, which allows to consider all cases of interaction between the contacting surfaces. It solved a number of problems of deformation with the ground subway tunnel lining rings.

1. Introduction

When designing and building modern transport facilities often practiced the construction of building structures in the soil, in difficult physical and geological conditions. Therefore, the establishment of methods for calculating three-dimensional substructures arranged in arrays of soil, now is especially important [1-7]. Development of mechanics in the field of modern foundation engineering and underground construction is largely determined by the achieved results so far in the field of mathematical and mechanical modeling of various physical phenomena or processes, in particular the processes of deformation and destruction of modern designs and facilities. The research problem of the stress-strain state of the soil under the influence of external forces and its own weight is a major in soil mechanics, and its solution for a variety of load cases has a direct application in the practice of construction.

Traditionally, solid mechanics to solve the geometrically nonlinear problems has spread Lagrangian description of the environment in which well formulated boundary value problem in differential or variation forms, the solution of which is possible to use different numerical algorithms [8-16]. Within modern numerical methods have been developed incremental loading techniques in accordance with the process of deformation which appears as a sequence of states of equilibrium, and the transition from the current state to a subsequent increment determined load, or a change in the boundary conditions and the computational domain, etc.

2. Modelling of deformation saturated soils and mechanical contact

When calculating saturated groundwater environments adopted quasi-two phase soil model, for which it is considered that the gas is either completely dissolved in the liquid, or is partially in bubbles, moving at a speed of filtered fluid. Taking the principle of fair Terzaghi stress and taking into account the law of filtration in the form of Darcy-Gersevanov, the basic system of variational equations describing the deformation of saturated soil, we obtain a

$$\int_V \sigma_{ij}^{ef} \delta \varepsilon_{ij}^s dV - \int_V P\delta \varepsilon_V^s dV = \int_{S_\sigma} \sigma_i^* \delta u_i^s dS + \int_V g_i \rho \, \delta u_i^s dV ,$$

$$\int_V \left\{ \frac{k}{\mu} \left(\frac{\partial P}{\partial x_i} - \rho^f g_i \right) \right\} \frac{\partial (\delta P)}{\partial x_i} dV - \int_{S_H} \frac{k}{\mu} H_n^* \delta P dS = 0 ,$$

which are unknown projection movements of material particles porous matrix u_i^s and porous pressure P. Here, k - the absolute permeability of the porous matrix, μ - fluid viscosity, ρ^f - density of the fluid, $\vec{g}$ - vector of the gravitational acceleration, ρ - the density of the two-phase ground, ε_V^s - volume deformation of the porous matrix, σ_{ij}^{ef} - effective stresses, H_n^* - fluid pressure gradient.

For modeling mechanical contact using a special contact member [17-19], which allows to consider the possibility of interaction between the various cases is contacting surfaces, in particular the separation, with sliding friction, etc. Total substructures allowing deformation equation taking into account the mechanical contact in variation form is

$$\sum_m \iiint_{\Omega_m} \{\sigma\}^T \{\delta\varepsilon\} d\Omega + \sum_k \iiint_{\Omega_k} \{\sigma_H\}^T \{\delta\varepsilon_H\} d\Omega = \sum_m \left(\iiint_{\Omega_m} \rho \{g\}^T \{\delta v\} d\Omega + \iint_{S_m^\sigma} \{p\}^T \{\delta v\} dS \right),$$

where Ω_m, Ω_k - respectively, the volume of construction elements and the contact zones; $\{\sigma\}$, $\{\varepsilon\}$ and $\{v\}$ - stress, deformation and displacement amounts of structural elements; $\{\sigma_H\}$ and $\{\varepsilon_H\}$ - the stress and strain in the contact zones; $\{g\}$ - free fall acceleration vector; $\{p\}$ - boundary load. To solve the nonlinear problem we use an iterative method, which is a variant of the method of initial stresses.

3. Numerical example

The problem of deformation of underground tunnel lining, interacting with water-saturated soil.

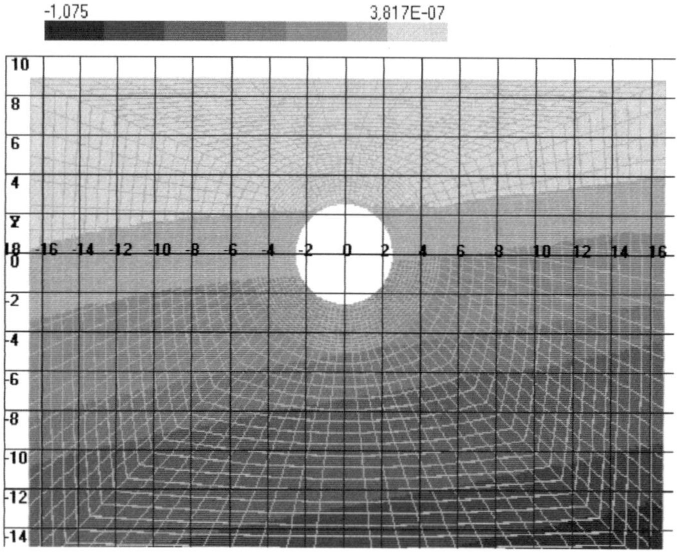

Figure 1. The distribution of pressure in the fluid phase.

It is assumed that the ground is homogenous (water-saturated fine sand) with the following characteristics: an elastic modulus $E = 33\,МПа$, Poisson's ratio $\mu = 0.3$, specific gravity $\gamma = 2040\,кг/м^3$, coefficient of adhesion $c = 2.6\,КПа$, internal friction angle $\varphi = 33^\circ$, porosity $m = 0.38$. The following options are accepted for concrete elastic modulus $E = 35000\,МПа$, Poisson's ratio

$\mu = 0.3$, specific gravity $\gamma = 2500\ кг/м^3$. Figure 1 shows the distribution of pressure in the fluid phase and in the groundwater level - $2\ m$ from the free surface. Note that the observed asymmetry of the stress in the ring lining subway tunnel with respect to the vertical plane. To illustrate the algorithm simulation of contact interaction in Figure 2 shows the horizontal displacements in the area of metro tunnel lining rings.

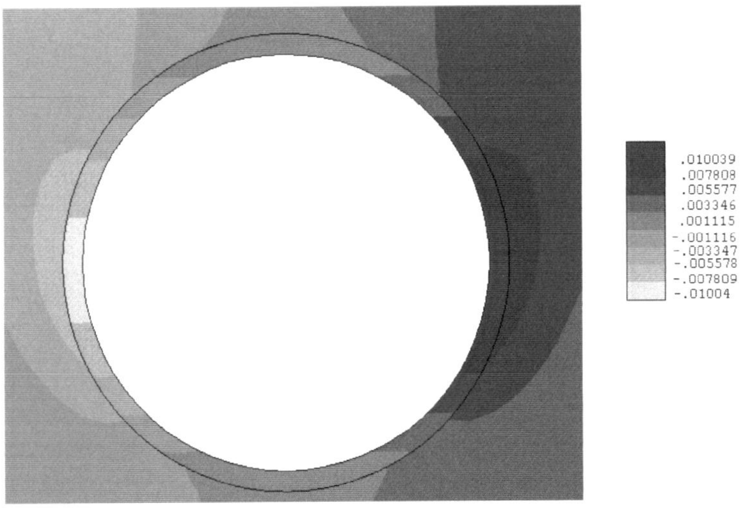

Figure 2. The horizontal displacements in the area of metro tunnel lining rings (m).

In the case of split ring underground tunnel lining (consisting of 8-th segmental blocks connected to each other by rubber gaskets), Figure 3 shows the distribution of circular stress in blocks (for the case of the lower arrangement of the locking unit).

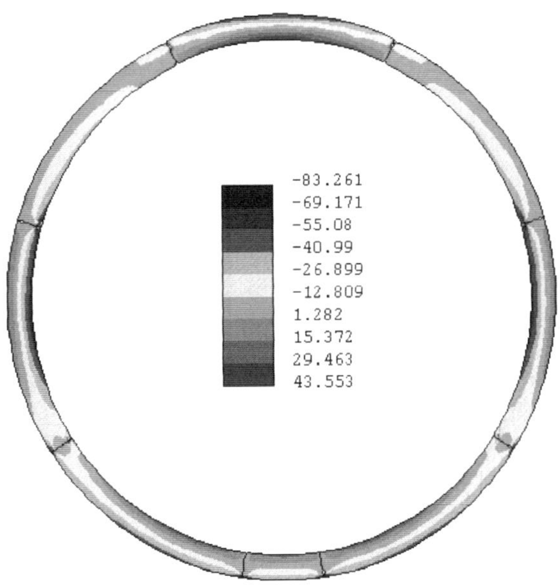

Figure 3. The circular stress in blocks ($0.1 * MPa$).

It may be noted the following features of deformation of split ring: lower location of the locking block decreases the value of the maximum and minimum circular stress. The maximum circular stress always occurs in the block opposite the locking block.

4. Conclusion

The proposed method for solving mechanical problems with specific applications refers to the modern technology of scientific support, design and construction of complex objects. Its use allows us to trace the changes in the stress-strain state and the displacement field of structurally changing the computational domain from the beginning to the end of construction. This allows a more accurate and technically correct design decisions for the various phases of construction works

Acknowledgments

The reported study was supported by Government of the Republic of Tatarstan research projects No. 15-41-02555 and RFBR research projects No. 15-31-20602.

References

[1] Badriev I B, Banderov V V, Garipova G Z, Makarov M V and Shagidullin R R 2015 *Applied Mathematical Sciences* **9 (82)** 4095–4102
[2] Badriev I B, Banderov V V, Makarov M V and Paimushin V N 2015 Applied Mathematical Sciences 9 (77-80) 3887–3895
[3] Badriev I B, Banderov V V and Zadvornov O A 2013 Applied Mechanics and Materials 392 188–190
[4] Badriev I B, Garipova G Z, Makarov M V and Paimushin V N 2015 Research Journal of Applied Sciences 10 (8) 428–435
[5] Sachenkov O, Kharislamova L, Shamsutdinova N, Kirillova E and Konoplev Yu 2015 *IOP Conference Series: Materials Science and Engineering* **98**
[6] Shigapova F A, Mustakimova R F, Saleeva G T and Sachenkov O A *International Journal of Applied Engineering Research* **10 (24)** 44711–44714
[7] Sachenkov O A, Mitryaikin V I, Zaitseva T A and Konoplev Yu G 2014 *Applied Mathematical Sciences* **8 (159)** 7889–7897
[8] Davydov R and Sultanov L 2013 *Sixth International Conference on Nonlinear Mechanics (ICNM-VI)* 64–67
[9] Davydov R L and Sultanov L U 2015 *Journal of Engineering Physics and Thermophysics* **88 (5)** 1280–1288
[10] Sultanov L U 2016 Lobachevskii Journal of Mathematics 37 (6) 787–793
[11] Abdrakhmanova A I, Gariffulin I R, Davydov R L, Sultanov L U and Fakhrutdinov L R 2015 Applied Mathematical Sciences 9 (118) 5907–5914
[12] Abdrakhmanova A I, Sultanov L U and Fakhrutdinov L R 2015 Global Journal of Pure and Applied Mathematics 11 (6) 5153–5162
[13] Sagdatullin M and Berezhnoi D 2014 *Applied Mathematical Sciences* **8 (35)** 1731–1738
[14] Sagdatullin M and Berezhnoi D 2014 *Applied Mathematical Sciences* **8 (60)** 2965–2972
[15] Berezhnoi D V, Paimushin V N and Shalashilin V I 2009 Mechanics of Solids 44 (6) 837–851
[16] Berezhnoi D V and Paimushin V N 2011 Journal of Applied Mathematics and Mechanics 75 (4) 447–462
[17] Berezhnoi D V, Sachenkov A A and Sagdatullin M K 2014 *Applied Mathematical Sciences* **8 (127)** 6341–6348
[18] Berezhnoi D V, Sachenkov A A and Sagdatullin M K 2014 *Applied Mathematical Sciences* **8 (143)** 7107–7115
[19] Berezhnoi D V and Sagdatullin M K 2015 Contemporary Engineering Sciences 8 (23) 1091–1098

The third boundary value problem for loaded differential Sobolev type equation and grid methods of their numerical implementation

M KH Beshtokov

Kabardino-Balkarian State University, Institute of physics and mathematics, Russia, Kabardino-Balkar Republic, Nalchik, 36000, 173 Chernyshevskogo St.

E-mail: beshtokov_murat@rambler.ru

Abstract. In this paper, we consider the third boundary value problem for loaded differential Sobolev type equation and grid methods of their numerical implementation. To solve the problem, a priori estimates in differential and difference settings are obtained. These a priori estimates imply uniqueness and stability of the solution with respect to the initial data and the right-hand side on the layer, as well as the convergence of the solution of each of the difference problem to the solution of the corresponding differential problem.

1. Introduction.

It is known that the solution of many practical problems arising while studying the liquid filtration process in fractured porous media [1], groundwater flow with a free surface in multi-media [2], moisture transfer [3], [4], the heat [5] and salts transfer [6] in porous media is all connected with the need to study boundary value problems for an equation of the third order

$$u_t = (k(x,t)u_x)_x + (\eta(x,t)u_x)_{xt} + r(x,t)u_x - q(x,t)u + f(x,t). \qquad (*)$$

Equation (*) is not a S.V.Kovalevskaya type equation and according to terminology [7] it is called a pseudo-parabolic equation. Equations of this type are also called Sobolev type equations or equations unsolved with respect to the derivative [8].

Interest in the loaded differential equations in partial derivatives is conditioned by the fact that linear loaded differential equations serve as mathematical models that describe the processes of long-term forecasting and regulation of groundwater and soil moisture levels, heat and mass transfer problems with a finite velocity, movement of a little compressible fluid, surrounded by a porous medium, the optimal management of agro-ecosystems.

2. Statement of the problem and a priori estimate in the differential form.

In the closed cylinder $\overline{Q}_T = \{(x,t): 0 \le x \le l,\ 0 \le t \le T\}$ we consider the third boundary value problem for a loaded equation of Sobolev type

$$u_t = (k(x,t)u_x)_x + (\eta(x,t)u_x)_{xt} + r(x,t)u_x(x_0,t) - q(x,t)u + f(x,t), \quad 0<x<l, 0<t\le T, \qquad (1)$$

$$\Pi(0,t) = \beta_1(t)u(0,t) - \mu_1(t), \ 0 \le t \le T, \qquad (2)$$

$$-\Pi(l,t) = \beta_2(t)u(l,t) - \mu_2(t), \ 0 \le t \le T, \qquad (3)$$

$$u(x,0)=u_0(x),\quad 0\le x\le l, \tag{4}$$

where

$$0<c_0\le \eta(x,t), k(x,t)\le c_1,\ |\eta_t,\eta_{xt},\eta_x,r,q,\beta_1,\beta_2|\le c_2,\ c_0,c_1,c_2-const>0, \tag{5}$$
$$u(x,t)\in C^{4,3}(\overline{Q}_T),\ \eta(x,t)\in C^{3,3}(\overline{Q}_T),\ k(x,t)\in C^{3,2}(\overline{Q}_T),\ r,q,f\in C^{2,2}(\overline{Q}_T),$$
$$u_0(x)\in C^2[0,l],\ \Pi(x,t)=ku_x+(\eta u_x)_t,\ Q_T=\{(x,t):0<x<l,0<t\le T\},$$
$$\beta_1(t),\beta_2(t),\mu_1(t),\mu_2(t)\in C[0,T].$$

Throughout the paper, we will use positive constants $M_i, i=1,2,....$, which depend only on the input data of problem (1) - (4).

Note that while constructing the difference scheme we require a higher smoothness of the solution and the coefficients of the equation to ensure the desired order of approximation of difference scheme [9].

Theorem 1. Let conditions (5), be satisfied, then for the solution of the differential problem (1) - (4) the following a priori estimate is valid

$$\|u_x\|_0^2+\|u_{xx}\|_0^2+\|u_{xx}\|_{2,Q_t}^2\le M\left(\int_0^t\left(\|f\|_0^2+\mu_1^2(t)+\mu_2^2(t)\right)d\tau+\|u_0(x)\|_0^2+\|u_0'(x)\|_0^2\right),$$

wherein M - depends only on the input data of problem (1) - (4).

3. Stability and convergence of the difference scheme.

On the grid $\overline{\omega}_{h\tau}$ to the differential problem (1) - (4) we assign a difference scheme with the order of approximation $O(h^2+\tau^{m_\sigma})$:

$$y_{t,i}=\left(ay_{\bar{x}}^{(\sigma)}\right)_x+\left(\gamma y_{\bar{x}}\right)_{xt,i}+r_i\left(y_{x,i_0}^{(\sigma)}\frac{x_{i_0+1}-x_0}{h}+y_{x,i_0+1}^{(\sigma)}\frac{x_0-x_{i_0}}{h}\right)-d_iy_i^{(\sigma)}+\varphi_i,\quad (x,t)\in\omega_{h\tau}, \tag{6}$$

$$a_1y_{x,0}^{(\sigma)}+\left(\gamma_1y_{x,0}\right)_t+0.5hr_0\left(y_{x,i_0}^{(\sigma)}\frac{x_{i_0+1}-x_0}{h}+y_{x,i_0+1}^{(\sigma)}\frac{x_0-x_{i_0}}{h}\right)=\tilde{\beta}_1y_0^{(\sigma)}+\frac{h}{2}y_{t,0}-\tilde{\mu}_1,\ t\in\overline{\omega}_\tau, \tag{7}$$

$$-\left(a_Ny_{\bar{x},N}^{(\sigma)}+\left(\gamma_Ny_{\bar{x},N}\right)_t\right)+0.5hr_N\left(y_{x,i_0}^{(\sigma)}\frac{x_{i_0+1}-x_0}{h}+y_{x,i_0+1}^{(\sigma)}\frac{x_0-x_{i_0}}{h}\right)=\tilde{\beta}_2y_N^{(\sigma)}+\frac{h}{2}y_{t,N}-\tilde{\mu}_2,\ t\in\overline{\omega}_\tau, \tag{8}$$

$$y(x,0)=u_0(x),\quad x\in\overline{\omega}_h, \tag{9}$$

where $y^{(\sigma)}=\sigma\hat{y}+(1-\sigma)y$, $a_i=k(x_{i-0.5},\bar{t})$, $\gamma_i=\eta(x_{i-0.5},\bar{t})$, $d_i=q(x_i,\bar{t})$, $\varphi_i=f(x_i,\bar{t})$,

$\tilde{\beta}_1^j=\beta_1^j+0.5hd_0$, $\tilde{\beta}_2^j=\beta_2^j+0.5hd_N$, $\tilde{\mu}_1^j=\mu_1^j+0.5h\varphi_0$, $\tilde{\mu}_2^j=\mu_2^j+0.5h\varphi_N$,

$\bar{t}=t_{j+0.5}=t_j+0{,}5\tau$, $x_{i-0.5}=x_i-0{,}5h$, h,τ - steps of the grid.

In order to get the a priori estimate we use the method of energy inequalities. Then problem (6) - (9) can be rewritten in another form, assuming $\sigma=0.5$ and denoting $Y=\hat{y}+y$:

$$y_t=0.5\tilde{\Lambda}^*(\bar{t})Y+\bar{\delta}y+\Phi, \tag{10}$$
$$y(x,0)=u_0(x), \tag{11}$$

where

$$
\tilde{\Lambda}^* Y = \begin{cases}
\tilde{\Lambda} Y = (aY_{\bar{x}})_{x,i} + r\left(Y_{0\,x,i_0} \dfrac{x_{i_0+1}-x_0}{h} + Y_{0\,x,i_0+1} \dfrac{x_0 - x_{i_0}}{h} \right) - d_i Y_i, & x \in \omega_h, \\
\Lambda^- Y = \dfrac{a_1 Y_{x,0} - \tilde{\beta}_1 Y_0 + 0.5 h r_0 \left(Y_{0\,x,i_0} \dfrac{x_{i_0+1}-x_0}{h} + Y_{0\,x,i_0+1} \dfrac{x_0 - x_{i_0}}{h} \right)}{0.5h}, & x = 0, \\
\Lambda^+ Y = \dfrac{- a_N Y_{\bar{x},N} - \tilde{\beta}_2 Y_N + 0.5 h r_N \left(Y_{0\,x,i_0} \dfrac{x_{i_0+1}-x_0}{h} + Y_{0\,x,i_0+1} \dfrac{x_0 - x_{i_0}}{h} \right)}{0.5h}, & x = l,
\end{cases}
$$

$$
\bar{\delta} y = \begin{cases}
\delta y = (\gamma y_{\bar{x}})_{xt,i}, & x \in \omega_h, \\
\delta^- y = \dfrac{(\gamma_1 y_{x,0})_t}{0.5h}, & x = 0, \\
\delta^+ y = -\dfrac{(\gamma\, y_{\bar{x}})_{t,N}}{0.5h}, & x = l,
\end{cases}
\qquad
\Phi = \begin{cases}
\varphi = \varphi_i, & x \in \omega_h, \\
\varphi^- = \varphi_0 + \dfrac{\mu_1}{0.5h}, & x = 0, \\
\varphi^+ = \varphi_N + \dfrac{\mu_2}{0.5h}, & x = l.
\end{cases}
$$

We introduce the scalar product

$$
(u,v) = \sum_{i=1}^{N-1} u_i v_i h, \quad [u,v] = \sum_{i=0}^{N} u_i v_i \hbar, \text{ где } \hbar = \begin{cases} \dfrac{h}{2}, & i = 0, i = N \\ h, & i = \overline{1, N-1} \end{cases},
$$

and the norm

$$
\| u \|_0^2 = (u,v), \ |[u]|_0^2 = [u,v], \ \| u]|_0^2 = \sum_{i=1}^{N} u_i^2 \hbar = (u,u].
$$

Theorem 2. Let conditions (5), be fulfilled, then there exist such $\tau_0, \tau_1, \tau_2, \tau_3, \tau_4$, that if $\tau \le \min\{\tau_0, \tau_1, \tau_2, \tau_3, \tau_4\}$, then with $\sigma = 0.5$ for the solution of the difference problem (10), (11) a priori estimate is valid.

$$
\| y_{\bar{x}}^{j+1} \|_0^2 + \| y_{\bar{x}x}^{j+1} \|_0^2 + \sum_{j'=0}^{j} \| Y_{\bar{x}x} \| \tau \le M \left(\sum_{j'=0}^{j} \left(|[\varphi]|^2 + \mu_1^2 + \mu_2^2 \right) \tau + |[y^0]|_0^2 + \| y_{\bar{x}}^0]|_0^2 + \| y_{\bar{x}x}^0 \|_0^2 \right), \quad (12)
$$

where M - a positive constant, not depending on h and τ.

The resulting a priori estimate (12) implies the uniqueness and stability of the solution of the difference problem (6) - (9) with respect to the initial data and the right-hand side in the grid norm $\| y_{\bar{x}}^{j+1} \|_0^2 + \| y_{\bar{x}x}^{j+1} \|_0^2 + \sum_{j'=0}^{j} \| Y_{\bar{x}x} \| \tau$ on a layer.

Let $u(x,t)$ - be the solution of problem (1) - (4), y_i^j - be the solution of the difference problem (6) - (9), then we denote by $z = y - u$ the error. Substituting $y = z + u$ into (6) - (9) and assuming that $u(x,t)$ is a given function, we obtain a problem for z:

$$
z_{t,i} = \left(a z_{\bar{x}}^{(\sigma)} \right)_x + \left(\gamma z_{\bar{x}} \right)_{xt,i} + r_i \left(z_{0\,x,i_0}^{(\sigma)} \frac{x_{i_0+1} - x_0}{h} + z_{0\,x,i_0+1}^{(\sigma)} \frac{x_0 - x_{i_0}}{h} \right) - d_i z_i^{(\sigma)} + \psi_i, \quad (13)
$$

$$a_1 z_{x,0}^{(\sigma)} + \left(\gamma_1 z_{x,0}\right)_t + 0.5hr_0\left(z_{0\ x,i_0}^{(\sigma)} \frac{x_{i_0+1}-x_0}{h} + z_{0\ x,i_0+1}^{(\sigma)} \frac{x_0 - x_{i_0}}{h}\right) = \tilde{\beta}_1 z_0^{(\sigma)} + \frac{h}{2} z_{t,0} - \nu_1,\ t \in \overline{\omega}_\tau, \quad (14)$$

$$-\left(a_N z_{\bar{x},N}^{(\sigma)} + \left(\gamma_N z_{\bar{x},N}\right)_t\right) + 0.5hr_N\left(z_{0\ x,i_0}^{(\sigma)} \frac{x_{i_0+1}-x_0}{h} + z_{0\ x,i_0+1}^{(\sigma)} \frac{x_0 - x_{i_0}}{h}\right) = \tilde{\beta}_2 z_N^{(\sigma)} + \frac{h}{2} z_{t,N} - \nu_2,\ t \in \overline{\omega}_\tau, (15)$$

$$z(x,0) = 0,\ x \in \overline{\omega}_h, \quad (16)$$

where $\psi_i = O\left(h^2+\tau^2\right)$, $\nu_1 = O\left(h^2+\tau^2\right)$, $\nu_2 = O\left(h^2+\tau^2\right)$ - are the approximation errors on the solution of the initial problem for every fixed t with $\sigma = 0.5$.

Applying a priori estimate (12) to the problem for the error, with $\sigma = 0.5$ we obtain the estimate

$$\| z_{\bar{x}}^{j+1} \|_0^2 + \| z_{\bar{x}x}^{j+1} \|_0^2 + \sum_{j'=0}^{j} \| Z_{\bar{x}x} \| \tau \le M \sum_{j'=0}^{j} \left(| [\Psi] |^2 + \nu_1^2 + \nu_2^2 \right)\tau, \quad (17)$$

where M - a positive constant, not depending on h and τ.

The a priori estimate (17) implies the convergence of the solution of the difference problem (6) - (9) to the solution of the differential problem (1) - (4) in the sense of the norm $\| z_{\bar{x}}^{j+1} \|_0^2 + \| z_{\bar{x}x}^{j+1} \|_0^2 + \sum_{j'=0}^{j} \| Z_{\bar{x}x} \| \tau$, on each layer so that there exist such $\tau_0, \tau_1, \tau_2, \tau_3, \tau_4$, that if $\tau \le \min\{\tau_0, \tau_1, \tau_2, \tau_3, \tau_4\}$, then with $\sigma = 0.5$ the following estimate is valid

$$\| y^{j+1} - u^{j+1} \|_1^2 \le M\left(h^2 + \tau^2\right).$$

In order to solve numerically the difference schemes obtained in the course of approximation of the loaded pseudo-parabolic third-order equations, it is convenient to use the parametric sweep method (see. [12, p.131]).

It is my pleasant duty to express my deep gratitude to my teacher Mohammed Habalovich Shhanukov-Lafishev.

4. The results of the numerical experiment

Consider the following example:

$$u_t = \left(k(x,t)u_x\right)_x + \left(\eta(x,t)u_x\right)_{xt} + r(x,t)u_x(x_0,t) - q(x,t)u + f(x,t),\quad 0<x<l, 0<t\le T,$$

$$\Pi(0,t) = \beta_1(t)u(0,t) - \mu_1(t),\ 0 \le t \le T,$$

$$-\Pi(l,t) = \beta_2(t)u(l,t) - \mu_2(t),\ 0 \le t \le T,$$

$$u(x,0) = u_0(x),\quad 0 \le x \le l,$$

where $l = 1$, $T = 1$, $k(x,t) = 1+\cos(x+t)$, $\eta(x,t) = 1+\sin(x+t)$, $r(x,t) = (0.5-x)\cos(x+t)$, $q(x,t) = e^x \sin(x+t)$, $\beta_1 = \sin(t)$, $\mu_1(t) = \sin(t)e^t - e^t(1+\cos(t)) - e^t(1+\sin(t)+\cos(t))$, $u_0(x) = e^x$, $\beta_2 = e^t$, $\mu_2(t) = e^{2t+l} - e^{t+l}(1+\cos(t+l)) - e^{t+l}(1+\sin(t+l)+\cos(t+l))$, $x_0 = 0.325$, $f(x,t) = e^{x+t} - e^{x+t}(2+3\cos(x+t)-\sin(x+t)) - (0.5-x)e^{x_0+t}\cos(x+t) + e^{2x+t}(\sin(x+t))$.

The exact solution is $u(x,t) = e^{x+t}$.

Table 1 compares the value of the numerical and exact solution of the problem for n=10.

Table 2 shows that when $h = \tau$ while the grid size is decreasing, then the maximum value of the error with $\sigma = 0.5$ is decreasing in accordance with the order of approximation of the difference scheme $O\left(h^2+\tau^2\right)$. The order of convergence is defined by the following formula:

The order of convergence = $\log_{\frac{h_1}{h_2}} \frac{e_1}{e_2}$

Table 1. The difference between the numerical and exact solution with $t = 1, h = \tau = 0.1$

(x_i)	**Numerical solution**	**Exact solution**	**Error solution**
0.0000	2.7170768	2.7182818	0.0012050
0.1000	3.0026299	3.0041660	0.0015361
0.2000	3.3182353	3.3201169	0.0018816
0.3000	3.6670599	3.6692967	0.0022368
0.4000	4.0526036	4.0552000	0.0025964
0.5000	4.4787343	4.4816891	0.0029548
0.6000	4.9497270	4.9530324	0.0033055
0.7000	5.4703066	5.4739474	0.0036408
0.8000	6.0456958	6.0496475	0.0039516
0.9000	6.6816678	6.6858944	0.0042266
1.0000	7.3846043	7.3890561	0.0044518

Table 2. The error alteration with decreasing grid size at t=1, where $h = \tau$

h	The maximum error	The order of convergence
1/500	23.56	
1/1000	34.64	1.9755555
1/2000	23.76	1.9830674

References

[1] Barenblatt G I and Zheltov Iu P and Kochina I N 1960 Ob osnovnyh predstavlenijah teorii fil'tracii odnorodnyh zhidkostej v treshhinovatyh porodah (Basic concepts in the theory of seepage of homogeneous liquids in fissured rocks (strata)) *PMM* **24 (5)** 852-864

[2] Dzektser E S 1975 Uravnenija dvizhenija podzemnyh vod so svobodnoj poverhnost'ju v mnogo\-slojnyh sredah (Equations of movement of underground waters with free surface in multilayer environments) *Dokl. of Akad. Nauk.* **220 (3)** 540-543

[3] Hallaire M 1966 Effektivnyy potentsial vody pri vysykhanii pochvy (Effective potential of water in the soil dries) *In.: Thermodynamics of soil moisture* 325-360

[4] Nerpin S V and Chudnovsky A F 1975 Energo I massoobmen v sistemepochva rasteniye-vozduh (Exchange of energy and mass in the system soil-plant-air) 360

[5] Chen P J and Curtin M E 1968 On a theory of heat conduction involving two temperatures - *Jornal Angew. Math. Phys* **19** 614-627

[6] Kanchukoev V Z and Shhanukov M H 1979 Krayevyye zadachi dlya uravneniy teplomassaobmena i ikh approksimatsiya ustoychivymi raznostnymi skhemami (Boundary-value problems for equations of heat and mass interchange and their approximation by stable difference schemes) *In collected articles: Boundary Value Problems for mixed-type equations and related problems of functional analysis and Applied Mathematics* **2** 143-150

[7] Showalter R E and Ting T W 1970 Pseudoparabolic partial differential equations *SIAM J. Math. Anal.* **1 (1)** 1-26

[8] Sobolev S L 1954 Ob odnoy novoy zadache matematicheskoy fiziki (On a new problem of mathematical physics) *News of USSR Academy of Sciences* Ser. Math. **18 (1)** 3–50
[9] Samarskii A A 1983 Teoriya raznostnykh skhem (The theory of difference schemes) 616
[10] Beshtokov M H 2016 On a non-local boundary value problem for the third order pseudo-parabolic equation *Computational Mathematics and Modeling* **27 (1)** 60–79
[11] Beshtokov M H 2013 Finite-difference method for a nonlocal boundary value problem for a third-order pseudoparabolic equation. *Differential equation* **49 (9)** 1–8
[12] Voevodin A F and Shugrin S M 1981 Chislennyye metody rascheta odnomernykh system (Numerical methods for calculating one-dimensional systems) 208

On the possibility of control restoration in some inverse problems of heat and mass transfer

G G Bilchenko and N G Bilchenko

Research Laboratory #16, Department of Heat Engineering and Power Engineering Machinery, Kazan National Research Technical University (KNRTU-KAI) named after A N Tupolev, 10, K Marx St., Kazan, Tatarstan 420111, Russia

E-mail: ggbil2@gmail.com

Abstract. The hypersonic aircraft permeable surfaces effective heat protection problems are considered. The physic-chemical processes (the dissociation and the ionization) in laminar boundary layer of compressible gas are appreciated in mathematical model. The statements of direct problems of heat and mass transfer are given: according to preset given controls it is necessary to compute the boundary layer mathematical model parameters and determinate the local and total heat flows and friction forces and the power of blowing system. The A.A.Dorodnicyn's generalized integral relations method has been used as calculation basis. The optimal control – the blowing into boundary layer (for continuous functions) was constructed as the solution of direct problem in extreme statement with the use of this approach. The statement of inverse problems are given: the control laws ensuring the preset given local heat flow and local tangent friction are restored. The differences between the interpolation and the approximation statements are discussed. The possibility of unique control restoration is established and proved (in the stagnation point). The computational experiments results are presented.

1. On direct problem of heat and mass transfer

The following *direct* problem was considered in [1] - [6]. According to preset given controls: $m(x)$ - the blowing into boundary layer (BL), $\tau_w(x)$ - the temperature factor ($\tau_w(x) = T_w(x)/T_{e_0}$, where $T_w(x)$ - the wall temperature, T_{e_0} - the temperature in the flow stagnation point) or the degree of beforehand cooling of surface, $s(x) = \sigma B_0^2(x)$ - magnetic field, where $x \in [0;1]$, it is necessary to compute the boundary layer mathematical model parameters [6] and determinate the local heat flow $q(x;m,\tau_w,s)$, the local tangent friction $f(x;m,\tau_w,s)$, the total heat flow $Q(m,\tau_w,s)$, the total force of friction $F(m,\tau_w,s)$ and the total blowing system power $N(m,\tau_w,s)$, i.e.

$$(m,\tau_w,s) \to (q,f;Q,F,N) \; . \tag{1}$$

In the [1] – [4] for the preset given $\tau_w(x)$ and $s(x)$ the *optimal* control $m(x)$ (for continuous functions) was constructed as the solution of extreme problem

$$Q(m,\tau_w,s) \to \inf_m \; , \tag{2}$$

$$N(m,\tau_w,s) \le N_{\max} \; . \tag{3}$$

The extreme values Q and the corresponding controls m were constructed in computational accuracy limits. Some special types of control $m(x)$, giving rather good approximations to the Q extreme values were suggested in [3].

Note. In computational experiments the $m(x)$, $\tau_w(x)$ and $s(x)$ laws were beforehand given, $q(x)$ and $f(x)$ were computed on the following grids $X=(x_j)_{j=0,\ldots,n}$:

$$x_0=0<x_1<\ldots<x_{n-1}<x_n=1.$$

2. Blowing inverse problems

By analogy with [7], [8] let's formulate the *inverse* for control $m(x)$ problems (respectively to construction (1)). Let's to be preset given

1) continuously differentiable control law $\tau_w(x)$ and $s(x)$, where $x\in[0;1]$;

2) the grid of *control*: $X_1=(x_j^{\wedge})_{j=0,\ldots,n_1}$;

3) the grid of *observation*: $X_2=(x_j^{\vee})_{j=0,\ldots,n_2}$;

4) the restrictions $I_{j,k}=[b_{j,k};t_{j,k}]$, where $j=1,\ldots,n_1$, $k=0,\ldots,\nu_1$, $\nu_1\ge 0$;

5) the "test" (or "check") values $q^{\vee}=(q_j^{\vee})_{j=0,\ldots,n_2}$ distributed on X_2.

It is necessary to find the continuous control $m^{\sim}(x)$ given [9] by elements $m^{\sim}(x)=m_j^{\sim}(x)$ for $x=[x_{j-1}^{\wedge};x_j^{\wedge}]$, $j=1,\ldots,n_1$, satisfying the conditions

$$(m^{\sim})^{(k)}(x)\in I_{j,k} \quad \text{for} \quad x=[x_{j-1}^{\wedge};x_j^{\wedge}],\quad j=1,\ldots,n_1,\quad k=0,\ldots,\nu_1, \tag{4}$$

such, that computed according to it on X_2 values $q^{\sim}=(q_j^{\sim})_{j=0,\ldots,n_2}$, $q_j^{\sim}=q(x_j^{\vee};m^{\sim},\tau_w,s)$, must be *close* to $q^{\vee}$, i.e.

$$q^{\vee}\to m^{\sim},\quad (m^{\sim},\tau_w,s)\to(q^{\sim}\approx q^{\vee},f^{\sim};Q^{\sim},F^{\sim},N^{\sim}). \tag{5}$$

Notes. 1) Instead of the consequence of values $q^{\vee}$ may be given $f^{\vee}=(f_j^{\vee})_{j=0,\ldots,n_2}$:

$$f^{\vee}\to m^{\sim},\quad (m^{\sim},\tau_w,s)\to(q^{\sim},f^{\sim}\approx f^{\vee};Q^{\sim},F^{\sim},N^{\sim}). \tag{6}$$

2) On the interval $[x_{j-1}^{\wedge};x_j^{\wedge}]$ element $m_j^{\sim}(x)$ may be presented by polynomial of degree $\mu\ge\nu_1+1$.

3) For the intervals (the windows) $[x_{j-1}^{\wedge};x_j^{\wedge}]$ of the grid X_1 let's introduce the *additional* grids (if $\nu_2=0$ they are empty)

$$X_1^{+}=(x_{j,k}^{+})_{j=1,\ldots,n_1},\ \text{where}\ x_{j-1}^{\wedge}<x_{j,k}^{+}<x_j^{\wedge},\quad j=1,\ldots,n_1,\quad k=1,\ldots,\nu_2,\quad \nu_2\ge 0.$$

4) Let's denote for $p\in[1;+\infty)$

$$R_p(q^{\sim};q^{\vee})=\left(\sum_{j=0}^{n_2}\left|q_j^{\sim}-q_j^{\vee}\right|^p\right)^{1/p} \quad \text{and} \quad R_\infty(q^{\sim};q^{\vee})=\max_{j=0,\ldots,n_2}\left|q_j^{\sim}-q_j^{\vee}\right|.$$

1) **Case** $X_2=X_1\cup X_1^{+}$, $\nu_1=\nu_2$. The proximity $q^{\sim}$ to $q^{\vee}$ must be realized in *interpolation* sense: in nodes of X_2 the values $q^{\sim}$ in limits of preset accuracy $\varepsilon_1>0$ must coincide with $q^{\vee}$:

$$R_\infty(q^{\sim};q^{\vee})\le\varepsilon_1. \tag{7}$$

For computational experiments the value ε_1 is restricted below by value $\varepsilon_{\min}>0$ which depends on the accuracy of *floating point* values representation in computer software and the applied computational methods.

Correctness. Let both the $(m^{\sim};q^{\sim})$ and $(m^{\approx};q^{\approx})$ to be solutions of (5). It is obviously, that

$$\text{if} \quad R(q^{\sim};q^{\vee}) \leq \varepsilon_1 \quad \text{and} \quad R(q^{\approx};q^{\vee}) \leq \varepsilon_1 \ , \quad \text{then} \quad R(q^{\sim};q^{\approx}) \leq \varepsilon_2 = 2\varepsilon_1 \ , \tag{8}$$

i.e. they are indistinguishable in the accuracy limits ε_2.

Hence, *if for* $\varepsilon_1 \geq \varepsilon_{\min}$ *the solution of the problem (5), (7) satisfying the (4) being a polynomial of degree* ν_1+1 *exists, then is **unique** (in sense (8)).*

2) **Case** $X_2 = X_1 \cup X_1^+$, $\nu_1 < \nu_2$. The problem is overdefined and the solution (5), (7) for small values $\varepsilon_1 > 0$ may not exist. The proximity $q^{\sim}$ to $q^{\vee}$ must be understood in *approximation* sense: for the given $p \in [1;+\infty]$ it is necessary to find $(m^{\sim};q^{\sim})$, where $m^{\sim}$ is the approximate solution of the extreme problem

$$R_p^*(q^{\vee}) = \inf_{m^{\sim}} R_p(q^{\sim};q^{\vee}) \ . \tag{9}$$

In computational experiments on the search the value $R_p^*(q^{\vee}) \geq \varepsilon_{\min}$ depends on n_1, n_2, X_1, X_2, $q^{\vee}$, p. *The solution of the problem (5), (9) satisfying the (4) being a polynomial of degree* ν_1+1 *for given* $\varepsilon_1 \geq R_p^*(q^{\vee})$ *may be constructed by **unique** (in sense (8)) way.*

3) **Case** $X_2 = X_1 \cup X_1^+$, $\nu_1 > \nu_2$. If being interpolation on X_2 and satisfying the (4) solution $(m^{\sim};q^{\sim})$ of the problem (5), (7) exists, then it must not be unique. To reduce the solution space the additional restrictions may be imposed on m and/or an additional extreme condition may be set.

Notes. 1) If the values $q^{\vee}$ (or $f^{\vee}$) are given with errors, then even if $X_2 \subseteq X_1$ the inverse problem must be considered in approximation statement (9).

2) By analogy (4)-(9) (the controls $m(x)$ and $\tau_w(x)$ change their parts; the relation between ν_1 and ν_2 is verified) the inverse problems for the control $\tau_w(x)$ are defined.

3. On the possibility of control restoration in the stagnation point

Let's formulate being true for cylindrical and spherical surfaces some statements on monotonous dependencies of local heat flows and local tangent friction on blowing, temperature factor and magnetic field in stagnation point (the notification corresponds to [6] and [10]). Let's the values of unchangeable parameters: the Mach number $M_\infty \in [10;40]$, the flight altitude $H \in [10;30]$ $[km]$, the body radius $R \in [0{,}1;1]$ $[m]$ to be *fixed*. Let's the controlling parameters changing ranges in stagnation point are *restricted* as follows:

$$m_0 = m(0) \in [0;1] \ , \quad \tau_0 = \tau_w(0) \in [0{,}1;0{,}9] \ , \quad s_0 = \sigma B_0^2(0) \in [0;5\cdot 10^4] \ [T/Ohm\cdot m] \ .$$

Then for the functions $\bar{\theta}_0(m_0,\tau_0,s_0)$, $\bar{\theta}_1(\ldots)$, $\bar{\omega}_0(\ldots)$, $\bar{\omega}_1(\ldots)$ determining the boundary layer parameters in small area of the stagnation point and being the solutions of the algebraic systems (15) for cylinder and (31) for sphere from [6] the following properties are established ([10]).

Statement 1. *For any combination of (fixed) values* τ_0, s_0 *the functions* $\bar{\theta}_0(m_0,\tau_0,s_0)$, $\bar{\theta}_1(\ldots)$, $\bar{\omega}_0(\ldots)$, $\bar{\omega}_1(\ldots)$ *strictly monotonously increases on* m_0.

Statement 2. *For any combination of (fixed) values* m_0, s_0 *the functions* $\bar{\theta}_0(m_0,\tau_0,s_0)$, $\bar{\theta}_1(\ldots)$, $\bar{\omega}_0(\ldots)$, $\bar{\omega}_1(\ldots)$ *strictly monotonously decreases on* τ_0.

Statement 3. *For any combination of (fixed) values* m_0, τ_0 *the functions* $\bar{\theta}_0(m_0,\tau_0,s_0)$, $\bar{\theta}_1(\ldots)$, $\bar{\omega}_0(\ldots)$, $\bar{\omega}_1(\ldots)$ *strictly monotonously increases on* s_0.

From the established statements it follows that function q determining the local values of heat flow in the stagnation point area is *bijective* function of m_0, τ_0, s_0 in the conditions of Statements 1, 2, 3 correspondingly. More precisely that means the correctness of following statements.

Statement 4. *For any combination of (fixed) values* τ_0, s_0 *the function* $q(0;m_0,\tau_0,s_0)$ *strictly monotonously decreases on* m_0.

Statement 5. *For any combination of (fixed) values* m_0, s_0 *the function* $q(0;m_0,\tau_0,s_0)$ *strictly monotonously decreases on* τ_0.

Statement 6. *For any combination of (fixed) values* m_0, τ_0 *the function* $q(0;m_0,\tau_0,s_0)$ *strictly monotonously decreases on* s_0.

The dependences of local heat flow on blowing values in stagnation point for the lateral surface of circular cylinder are presented on the Figure 1 for $H=10$ $[km]$, $R=0{,}1$ $[m]$, $M_\infty=10$ (it is convenient for comparison with the results of [2] – [4]). Here the curves 1 and 2 correspond to $\tau_0=0{,}1$; the curves 3 and 4 - to $\tau_0=0{,}5$; the curves 5 and 6 - to $\tau_0=0{,}9$. The dependences on temperature factor in the same conditions are presented on the Figure 2. Here the curves 1 and 2 correspond to $m_0=0$; the curves 3 and 4 - to $m_0=0{,}5$; the curves 5 and 6 - to $m_0=1$. On both the figures the curves 2, 4, 6 correspond to the applied magnetic field $s_0=5\cdot10^4$ $[T/Ohm\cdot m]$, the curves 1, 3, 5 – without magnetic field. The dependences on magnetic field in the same conditions are presented on the Figure 3. Here the curves 1, 2, 3 correspond to $m_0=0$; the curves 4, 5, 6 - to $m_0=0{,}5$; the curves 7, 8, 9 - to $m_0=1$. The curves 1, 4, 7 correspond to $\tau_0=0{,}1$; the curves 2, 5, 8 - to $\tau_0=0{,}5$; the curves 3, 6, 9 - to $\tau_0=0{,}9$.

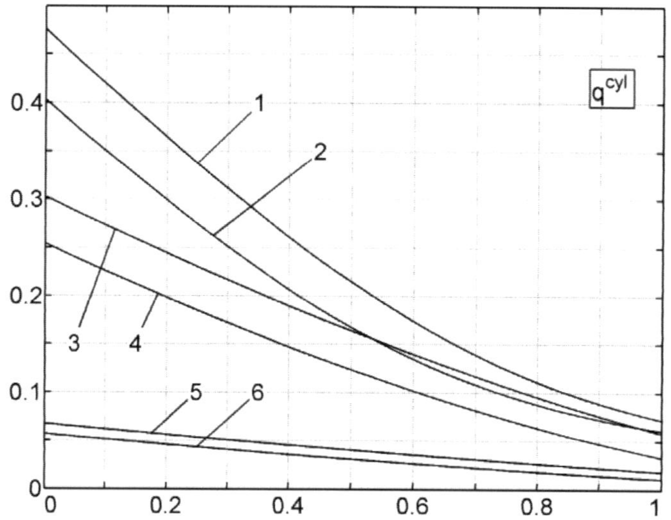

Figure 1. The dependence of local heat flow on blowing m_0

Further because the local tangent friction and parameter θ_0 of the boundary layer mathematical model [6] are dependent bijectively, so following statement are right

Statement 7. *For any combination of (fixed) values* τ_0, s_0 *the function* $\bar{f}(0;m_0,\tau_0,s_0)$ *strictly monotonously decreases on* m_0.

Statement 8. *For any combination of (fixed) values* m_0, s_0 *the function* $\bar{f}(0;m_0,\tau_0,s_0)$ *strictly monotonously increases on* τ_0.

Statement 9. *For any combination of (fixed) values* m_0, τ_0 *the function* $\bar{f}(0;m_0,\tau_0,s_0)$ *strictly monotonously decreases on* s_0.

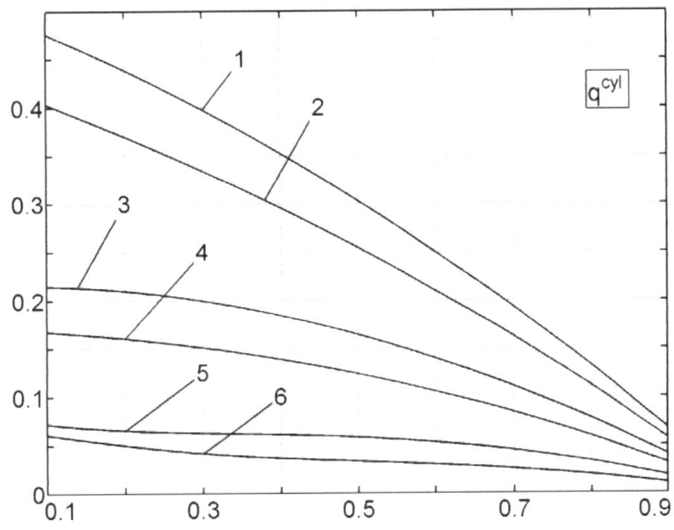

Figure 2. The dependence of local heat flow on temperature factor τ_0

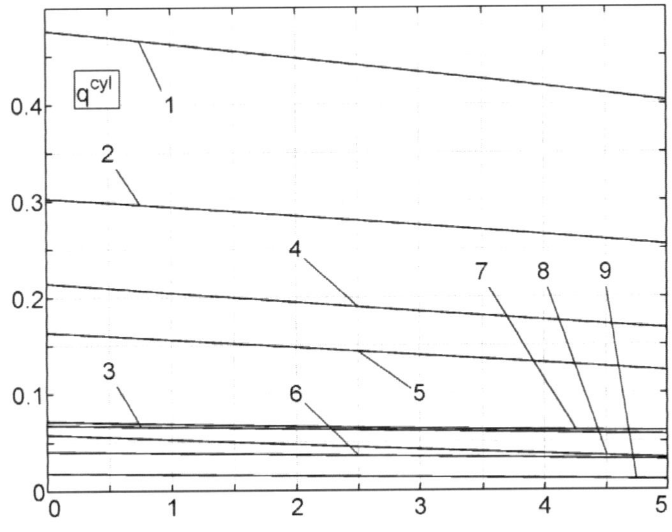

Figure 3. The dependence of local heat flow on magnetic field $s_0 \cdot 10^{-4}$

The monotonous character the algebraic systems [6] solutions dependences on the control parameters values established in the Statements 1-3 and their corollaries – the Statements 4-9 are necessary to solve the inverse problems of heat and mass transfer in *interpolation* statement – to restore the controls with the use of boundary layer local characteristics (local heat flow q and local tangent friction f).

The results of computational of the blowing restoration are discussed in continuation [11] of this paper.

Acknowledgments
This research was supported by project part of State Assignment №13.262.201.2014K.

References

[1] Garaev K G, Ovchinnikov V A and Bilchenko N G 2003 *Invariant variation problems of laminar boundary layer under various flow conditions* (Kazan: KGTU Publishers) 123 (in Russian)

[2] Bilchenko N G 2015 Computation experiments in the problems of heat and mass transfer optimal control on permeable surfaces in hypersonic flows *Proc. Voronezh State Univ. Ser. Phys. Math.* **1** 83-94 (in Russian)

[3] Bilchenko N G 2015 Computation experiments in the problems of heat and mass transfer optimal control on permeable spherical surfaces in hypersonic flows *Proc. Voronezh State Univ. Ser. System Analysis and Inform. Technologies* **1** 5-8 (in Russian)

[4] Bilchenko N G 2015 Computation experiments in the problems of heat and mass transfer optimal control on permeable surfaces in hypersonic flows: comparative analysis of "simple" blow laws use *Proc. Voronezh State Univ. Ser. Phys. Math.* **1** 95-102 (in Russian)

[5] Bilchenko N G 2015 Permeable Surfaces Hypersonic Aircraft Optimal Heat Protection Mathematical Modeling *Proc. 2015 Int. Conf. "Stability and Control Processes" in Memory of V I Zubov* ed L A Petrosyan and A P Zhabko (Russia, Saint-Petersburg, IEEE) 310-313 DOI: 10.1109/SCP.2015.7342145 URL: http://ieeexplore.ieee.org/stamp/stamp.jsp?tp=&arnumber=7342145&isnumber=7342029

[6] Bilchenko N G 2016 A A Dorodnitsyn's method in the problems of permeable surfaces in electroconductive gas laminar boundary layer heat and mass transfer optimal control *Proc. Voronezh State Univ. Ser. System Analysis and Inform. Technologies* **1** 5-14 (in Russian)

[7] Bilchenko G G and Bilchenko N G 2016 On an inverse problem of heat and mass transfer *Diff. Equations and Control Processes* **2** 2 *"Herzen Readings 2016. Some Actual Probl. of Modern Math. and Math. Educ."* ed V F Zaytsev, V D Budaev and A V Flegontov (Russia, Saint-Petersburg) 50-56 URL: http://www.math.spbu.ru/diffjournal/pdf/herzen2016.pdf (in Russian)

[8] Bilchenko G G and Bilchenko N G 2016 Inverse problem of heat and mass transfer *Proc. Int. Conf. "Stability and Oscillations of Nonlinear Control Systems" (Pyatnitskiy's Conf.)* ed V N Tkhai (Russia, Moscow, IEEE) 1-2 DOI: 10.1109/STAB.2016.7541167 URL: http://ieeexplore.ieee.org/stamp/stamp.jsp?tp=&arnumber=7541167&isnumber=7541147

[9] Bilchenko G G and Bilchenko N G 2016 The classes of solutions of the problems of boundary layer optimal control on the permeable surfaces of hypersonic aircrafts *Proc. Int. Conf. "Voronezh Winter Math. School of S G Krein"* ed V A Kostin (Russia, Voronegh, "Nauchnaya kniga" Publishers) pp 82-86 (in Russian)

[10] Bilchenko G G and Bilchenko N G 2016 On the boundary layer parameters bijective dependence on the controls in the stagnation point of hypersonic flow *Proc. Int. Conf. "Voronezh Spring Math. School "Modern Methods in Theory of Boundary Value Problems"- "Pontryagin Readings - XXVII""* (Russia, Voronegh, VSU Publishers) 44-46 (in Russian)

[11] Bilchenko G G and Bilchenko N G 2016 On computational experiments in some inverse problems of heat and mass transfer (in this book of materials)

On computational experiments in some inverse problems of heat and mass transfer

G G Bilchenko and N G Bilchenko

Research Laboratory #16, Department of Heat Engineering and Power Engineering Machinery, Kazan National Research Technical University (KNRTU-KAI) named after A N Tupolev, 10, K Marx St., Kazan, Tatarstan 420111, Russia

E-mail: ggbil2@gmail.com

Abstract. The results of mathematical modeling of effective heat and mass transfer on hypersonic aircraft permeable surfaces are considered. The physic-chemical processes (the dissociation and the ionization) in laminar boundary layer of compressible gas are appreciated. Some algorithms of control restoration are suggested for the interpolation and approximation statements of heat and mass transfer inverse problems. The differences between the methods applied for the problem solutions search for these statements are discussed. Both the algorithms are realized as programs. Many computational experiments were accomplished with the use of these programs. The parameters of boundary layer obtained by means of the A.A.Dorodnicyn's generalized integral relations method from solving the direct problems have been used to obtain the inverse problems solutions. Two types of blowing laws restoration for the inverse problem in interpolation statement are presented as the examples. The influence of the temperature factor on the blowing restoration is investigated. The different character of sensitivity of controllable parameters (the local heat flow and local tangent friction) respectively to step (discrete) changing of control (the blowing) and the switching point position is studied.

1. The inverse problems solving algorithms program realizations

This research continues [1]: the notification and the formulas numbers from the paper are saved.

The algorithm of interpolation blowing restoration is realized as the program. This algorithm for the preset given grids X_1, X_2, the consequence of "test" values $q^{\vee}$, the error $\varepsilon_1 > 0$ and restrictions $I_{j,k}$, where $j = 1,\ldots,n_1$, $k = 0,\ldots,\nu_1$, $\nu_1 \geq 0$: **either** builds the control $m^{\sim}(x)$ being *piece-polynomial* on X_1 and satisfying the conditions (4) and (7), **or** informs about impossibility of such restoration for given parameters and offers to find an approximation solution. The algorithm of search of consecutive approaches to the solution of extreme problem (9) is realized as a program. For the problem of approximation this algorithm (for the given p and $\varepsilon_1 > 0$) builds the solutions (5) satisfying the conditions (4). The search ends when the solution $(m^{\sim};q^{\sim})$ is found for the value $R_p(q^{\sim};q^{\vee}) \leq \varepsilon_1$. If $\varepsilon_1 < R_p^*(q^{\vee})$ is given, then the process ends by building of approximate solution (9) in the limits of computational accuracy.

Let's note, that the methods applied for solving the inverse problems are sufficiently different for the interpolation and approximation statements. Instead the consequence of local heat flow values $q^{\vee}$ the programs may use the consequence of local tangent friction values $f^{\vee}$. So, the programs allow to test a possibility of realization on the hypersonic aircraft surface a local heat flows $q^{\vee}$ consequence and/or local tangent frictions consequence $f^{\vee}$. The algorithm of the temperature factor τ_w restoration is realized by the same way.

Note, that before [2], [3] restoration algorithms were implemented as program for case of *piece-linear* function class of controls.

2. The influence of temperature factor on the blowing restoration

Let's present two examples of interpolation blowing restoration for local heat flows and local tangent frictions similar to considered in [4] - [6] for various constant laws of temperature factor $\tau_w \in \{0{,}10; 0{,}15;\ldots;0{,}85;0{,}90\}$ with the restriction $I_{j,0} = [0;1]$ in the case of circular cylinder lateral surface flowing round at $H = 10\ [km]$, $R = 0{,}1\ [m]$, $M_\infty = 10$.

2.1. Initial constant blowing law case

For **constant blowing** $m \equiv 0{,}3$ and $\tau_w \equiv 0{,}25$ the consequences of local heat flow and local tangent friction values were obtained with the use of the program realizing the direct problem. The procedure of blowing interpolation restoration was applied to the consequences. The control laws $m^{\sim}$ being restored for local heat flows are presented on Figure 1. The control laws $m^{\sim}$ being restored for local tangent frictions are presented on the Figure 2.

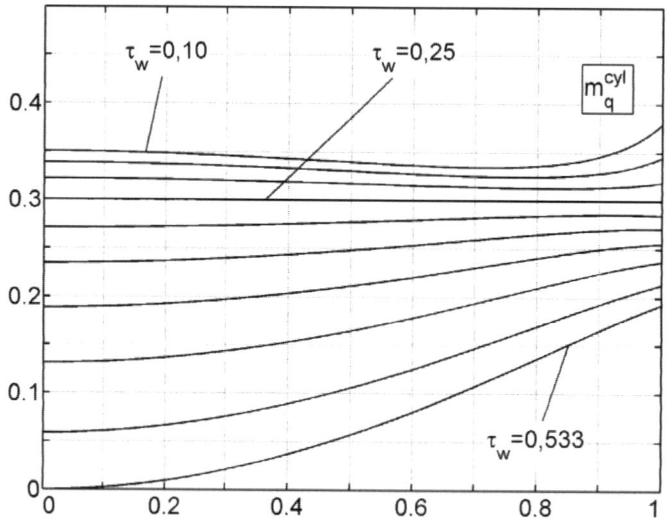

Figure 1. The blowing $m^{\sim}(x)$ restored on local heat flow for some τ_w

For the values $\tau_w \geq 0{,}534$ the restoration for local heat flows is impossible because of the restriction on blowing $b_{j,0} = 0$. Analogously, for $\tau_w \geq 0{,}755$ the restriction $t_{j,0} = 1$ spans into action for local tangent friction.

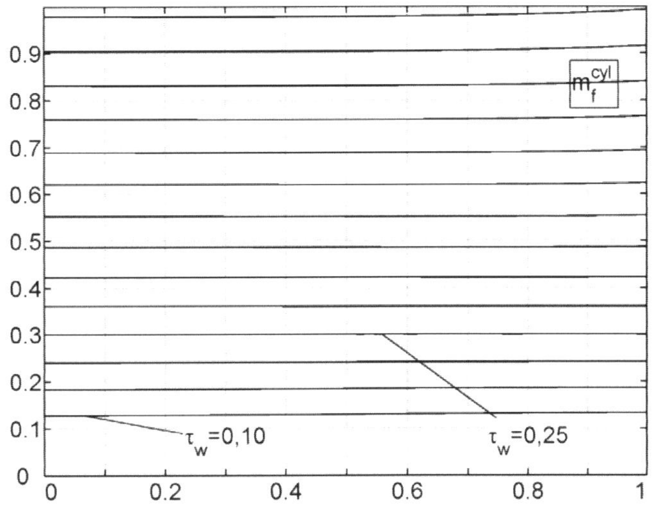

Figure 2. The blowing $m^{\sim}(x)$ restored on local tangent friction for some τ_w

2.2. Initial vertical trapezoid blowing law case

For the law "vertical trapezoid":

$$m(x;m_0,m_1)=m_0\cdot(1-x)+m_1\cdot x \quad \text{for} \quad m_0=0 \quad \text{and} \quad m_1=0{,}3996$$

(the curve 7 in the Table 6 and on the Figure 4 from [5], $\tau_w\equiv 0{,}25$) the same operations were accomplished: to the obtained local heat flows and local tangent frictions the procedure of blowing interpolation restoration was applied. The control laws constructed for local heat flows are presented on Figure 3. The control laws constructed for local tangent friction are presented on Figure 4.

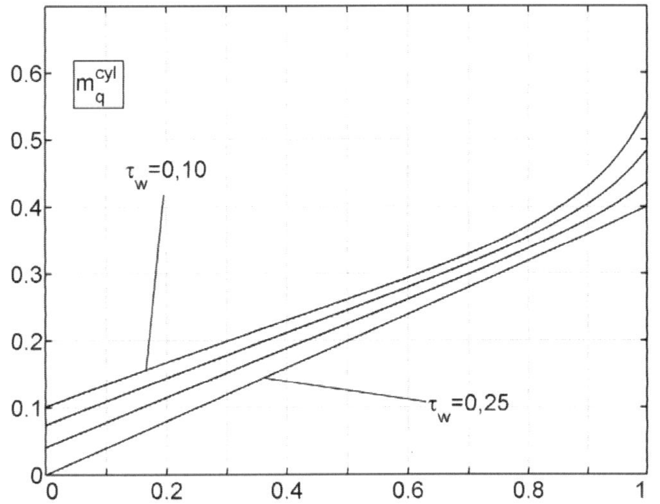

Figure 3. The blowing $m^{\sim}(x)$ restored on local heat flow for some τ_w

The blowing becomes less $b_{j,0}=0$ at $\tau_w>0{,}25$ for restoration on local heat flows, and at $\tau_w<0{,}25$ for restoration on local tangent friction. Besides, at $\tau_w\geq 0{,}67$ the restriction $t_{j,0}=1$ is obtained for the case of local tangent friction.

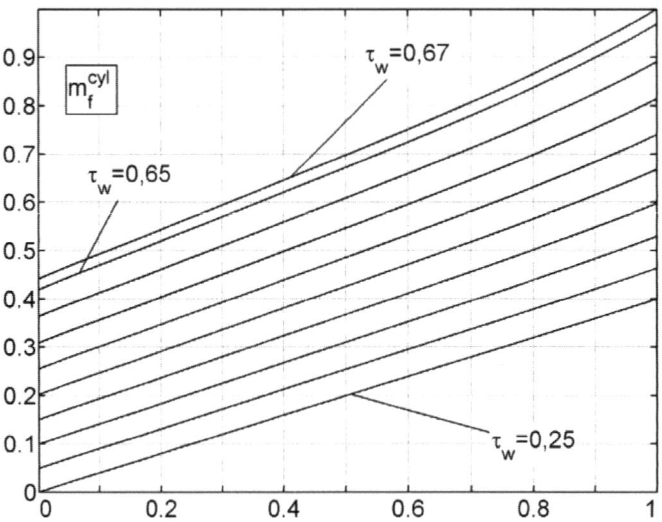

Figure 4. The blowing $m^{\sim}(x)$ restored on local tangent friction for some τ_w

3. The allowed forms of local heat flows and local tangent frictions

Let's note the different character of q and f sensitivity to step (discrete) changing of control $m(x)$. The sensitivity of q decreases sufficiently at moving of the switching point x_* to 1, it is less noticeably for f. The dependences of local heat flows and local tangent frictions on local coordinates (for the case of cylinder; the parameters are the same as in section 2) are presented on the Figure 5 and Figure 6 correspondingly. The curves correspond to the blowing laws generalizing the law "step" from [5], [6]:

$$m(x) = \begin{cases} m_1, & x \in [0; x_*); \\ m_2, & x \in (x_*; 1], \end{cases} \quad \text{where} \quad x_* \in \{0.01; 0.05; 0.10; 0.15; \ldots; 0.95\}, \quad \begin{cases} m_1 \equiv 0{,}3; \\ m_2 \in \{0; 0{,}3; 1\}. \end{cases}$$

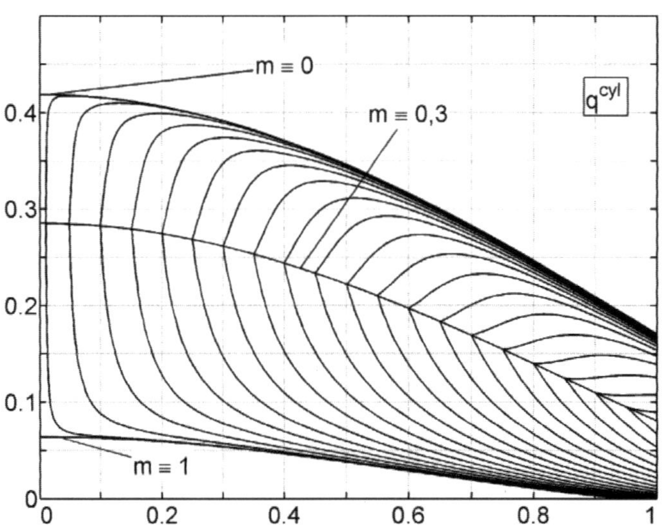

Figure 5. The reaction of local heat flows to step blowing change

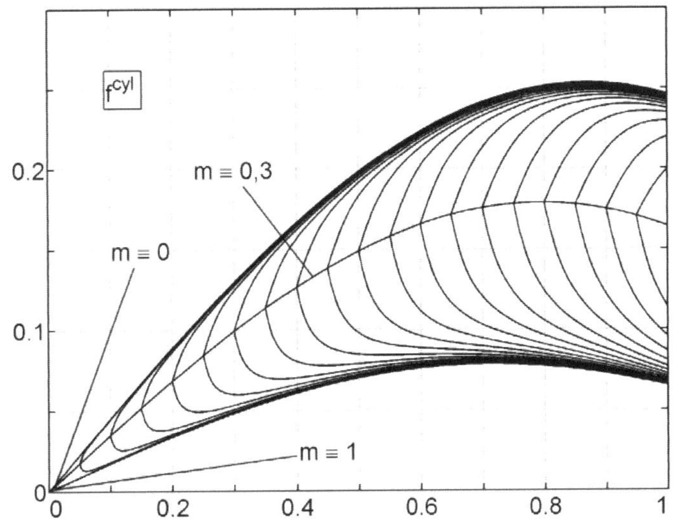

Figure 6. The reaction of local tangent frictions to step blowing change

The boundary layer characteristics using here and in [1] are obtained by numerical integration ODE system [7] derived with the use of the A A Dorodnitcyn's generalized integral relations method.

Acknowledgments
This research was supported by project part of State Assignment №13.262.201.2014K.

References

[1] Bilchenko G G and Bilchenko N G 2016 On the possibility of control restoration in some inverse problems of heat and mass transfer (in this book of materials)

[2] Bilchenko G G and Bilchenko N G 2016 On an inverse problem of heat and mass transfer *Diff. Equations and Control Processes* **2** 2 *"Herzen Readings 2016. Some Actual Probl. of Modern Math. and Math. Educ."* ed V F Zaytsev, V D Budaev and A V Flegontov (Russia, Saint-Petersburg) pp 50-56 URL: http://www.math.spbu.ru/diffjournal/pdf/herzen2016.pdf (in Russian)

[3] Bilchenko G G and Bilchenko N G 2016 Inverse problem of heat and mass transfer *Proc. Int. Conf. "Stability and Oscillations of Nonlinear Control Systems" (Pyatnitskiy's Conf.)* ed V N Tkhai (Russia, Moscow, IEEE) pp 1-2 DOI: 10.1109/STAB.2016.7541167 URL: http://ieeexplore.ieee.org/stamp/stamp.jsp?tp=&arnumber=7541167&isnumber=7541147

[4] Bilchenko N G 2015 Computation experiments in the problems of heat and mass transfer optimal control on permeable surfaces in hypersonic flows *Proc. Voronezh State Univ. Ser. Phys. Math.* **1** 83-94 (in Russian)

[5] Bilchenko N G 2015 Computation experiments in the problems of heat and mass transfer optimal control on permeable surfaces in hypersonic flows: comparative analysis of "simple" blow laws use *Proc. Voronezh State Univ. Ser. Phys. Math.* **1** 95-102 (in Russian)

[6] Bilchenko N G 2015 Permeable Surfaces Hypersonic Aircraft Optimal Heat Protection Mathematical Modeling *Proc. 2015 Int. Conf. "Stability and Control Processes" in Memory of V I Zubov* ed L A Petrosyan and A P Zhabko (Russia, Saint-Petersburg, IEEE) pp 310-313 DOI: 10.1109/SCP.2015.7342145 URL: http://ieeexplore.ieee.org/stamp/stamp.jsp?tp=&arnumber=7342145&isnumber=7342029

[7] Bilchenko N G 2016 A A Dorodnitsyn's method in the problems of permeable surfaces in electroconductive gas laminar boundary layer heat and mass transfer optimal control *Proc. Voronezh State Univ. Ser. System Analysis and Inform. Technologies* **1** 5-14 (in Russian)

IOP Conf. Series: Materials Science and Engineering **158** (2016) 012022 doi:10.1088/1757-899X/158/1/012022

Numerical research of parameters of interaction of the gas flow with rotary valve of the gas pipeline

A V Boldyrev, D L Karelin, V L Muljukin

Naberezhnye Chelny institute (branch) of Kazan Federal University, 68/19 (1/18), Naberezhnye Chelny, 423812, Russia

E-mail: alexeyboldyrev@mail.ru

Abstract. Conducted numerical research of static characteristics of the rotary gate valve at different angles of its deviation. for this purpose were set different values of pressure differential on the valve depending on which, was determined the mass flow and torque on valve axes. The mathematical model is provided by continuity equations, average on Reynolds, Navier-Stokes and energy, the equation of the perfect gas, the equations of two-layer k-ε of model of turbulence. When calculating the current near walls are used Wolfstein's model and the hybrid wall functions of Reichardt for the speed and temperature. The task is solved in three-dimensional statement with use of conditions of symmetry. The structure of the current is analyzed: zones of acceleration and flow separation, whirlwinds, etc. Noted growth of hydraulic resistance of the valve with reduction of slope angle of the valve and with the increase in mass flow. Established increase of torque with reduction of the deviation angle of the valve and with increase in the mass expense.

1. Introduction
Application of slide gate valves with the rotary valve and electromagnetic management was widely adopted in power plant engineering, and the large number of works [1, 2, 3] is devoted to research and optimization of their parameters.

In this connection, the purpose of work is numerical research of the current via the valve installed in the gas pipeline with the rotary valve, and also obtaining static characteristics of this device.

2. Materials and methods
Figure 1 shows a diagram of the computational domain, representing a portion of the tubing disposed therein valve. The course was considered to be symmetrical relative to the plane OZY. The characteristic dimensions were: internal diameter of the pipe 490 mm, the inner diameter of the valve body 330 mm, the wall thickness of the valve is 5 mm, the length of the calculated pipe section 1000 mm, valve length 346 mm.

Below are the equations of the mathematical model, written in tensor form (summation over repeated index, according to the method proposed by Einstein [4]).

Quasi-stationary three-dimensional turbulent flow of a viscous compressible gas (methane CH_4) is described:

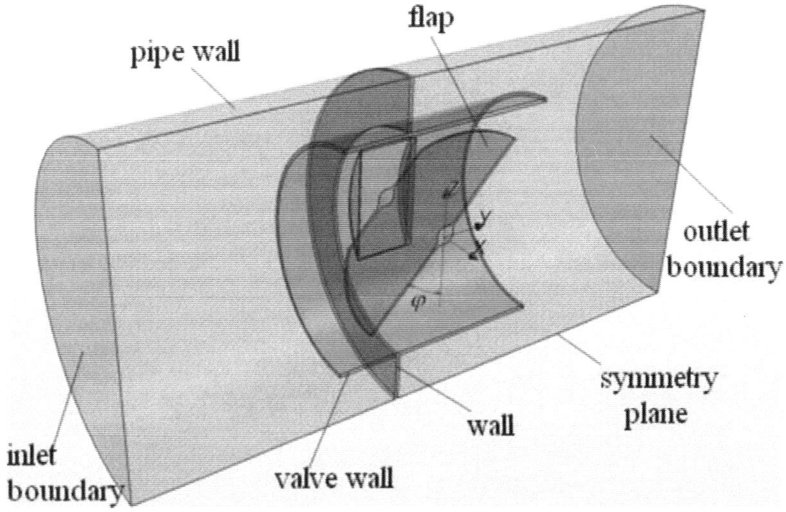

Figure 1. Scheme of computational domain. Boundary conditions.

1) Reynolds averaged continuity equation [4-6]

$$\frac{\partial(\rho \cdot V_j)}{\partial x_j} = 0, \tag{1}$$

Navier-Stokes equations [4-6]

$$\rho \cdot V_j \cdot \frac{\partial V_i}{\partial x_j} = \frac{\partial P_{ji}}{\partial x_j} \tag{2}$$

and energy equation (according to the first law of thermodynamics) [4-7]

$$\rho \cdot V_j \cdot \frac{\partial E}{\partial x_j} = -\frac{\partial q_j}{\partial x_j} + \frac{\partial}{\partial x_j}(P_{ji} \cdot V_i); \tag{3}$$

2) equation of the ideal gas (Clapeyron's equation) [4-7]

$$\rho = \frac{p_{\text{atm}} + p}{R \cdot T}; \tag{4}$$

3) equations of two-layer Realizable of $k-\varepsilon$ model of turbulence [5, 6, 8, 9]

$$\rho \cdot V_j \cdot \frac{\partial k}{\partial x_j} = \frac{\partial}{\partial x_j}\left(\left(\mu + \frac{\mu_{\text{t}}}{\sigma_{\text{k}}}\right)\frac{\partial k}{\partial x_j}\right) + G_{\text{k}} - \rho \cdot \varepsilon - \rho \cdot \Psi_{\text{M}}, \tag{5}$$

$$\rho \cdot V_j \cdot \frac{\partial \varepsilon}{\partial x_j} = \frac{\partial}{\partial x_j}\left(\left(\mu + \frac{\mu_{\text{t}}}{\sigma_{\varepsilon}}\right)\frac{\partial \varepsilon}{\partial x_j}\right) + C_{\varepsilon 1} \cdot \rho \cdot s \cdot \varepsilon - \frac{\varepsilon}{k + \sqrt{\nu \cdot \varepsilon}} \cdot C_{\varepsilon 2} \cdot \rho \cdot \varepsilon, \tag{6}$$

where i, j – the indexes accepting values 1, 2, 3; ρ – gas density, kg/m^3; $\boldsymbol{V}$ – velocity vector of fluid movement, m/s; x – coordinate, m; P – stress tensor, Pa; E – total energy of gas, J/kg; $\boldsymbol{q}$ – vector of heat flux density, W/m^2; p_{atm} – atmospheric pressure, 101325 Pas; p – excessive pressure, Pa; R – gas constant, 518 J / (kg K); T – absolute temperature of gas, K; k – kinetic turbulent energy, J/kg; μ – molecular dynamic viscosity of gas, $1.12 \cdot 10^{-5}$ Pa·s; μ_{t} – turbulent dynamic viscosity of gas, Pa·s; σ_{k} – turbulent number of Schmidt for the equation (5), 1; σ_{ε} – turbulent number of Schmidt for the equation (6), 1.2; G_{k} – the generative component of the equation of transfer of energy k (5), kg/(m·s^3); ε – speed of dissipation of kinetic turbulent energy k, W/kg; Ψ_{M}

– the component of the equation of transfer of energy k (5), k considering influence of compressibility of gas on pulsations of speed, W/kg; C_{ε} – coefficient at the generative component of the equation of transfer (6); ν – molecular kinematic viscosity of gas (μ/ρ), m^2/s; $C_{\varepsilon 2}$ – coefficient at the dissipative component of the equation of transfer (6), 1.9; s – the parameter connected with the second invariant of the tensor of speeds of deformations $\dot{S}$, s^{-1}

$$s = \sqrt{2 \cdot \dot{S}_{ij} \cdot \dot{S}_{ij}}\,, \tag{7}$$

$$\dot{S}_{ij} = \frac{1}{2} \cdot \left(\frac{\partial V_i}{\partial x_j} + \frac{\partial V_j}{\partial x_i} \right). \tag{8}$$

According to Reynolds's (RANS) approach, components of the stress tensor are defined as follows [4-6] (here the 1st index designates number of the axis perpendicular to which the site, and the 2nd index - number of the axis on which stress is projected is oriented)

$$P_{ji} = P_{\mathrm{lam}\, ji} + P_{\mathrm{turb}\, ji}\,, \tag{9}$$

where P_{lam} – viscous (laminar) stress tensor, Pa;

P_{turb} – turbulent stress tensor, Pa.

On the generalized Newton's laws for viscous compressible medium [4-6]

$$P_{\mathrm{lam}\, ji} = 2 \cdot \mu \cdot \dot{S}_{ij} - \left(p + \frac{2}{3} \cdot \mu \cdot \mathrm{div}\, \boldsymbol{V} \right) \cdot \delta_{ij}\,, \tag{10}$$

where δ – the unit tensor (the component is equal 1 at $i = j$ and it is equal 0 at $i \neq j$).

Components of the tensor of turbulent stress are calculated, according to the hypothesis of Boussinesq [4-6]

$$\mathrm{P}_{\mathrm{turb}\, ji} = 2 \cdot \mu_{\mathrm{t}} \cdot \dot{S}_{ij} - \frac{2}{3} \cdot \left(\rho \cdot k + \mu_{\mathrm{t}} \cdot \mathrm{div}\, \boldsymbol{V} \right) \cdot \delta_{ij}\,. \tag{11}$$

The total energy E represents the sum of internal and kinetic energy of gas [4-6]

$$E = C_{\mathrm{v}} \cdot T + \frac{|\boldsymbol{V}|^2}{2}\,, \tag{12}$$

where C_{v} – heat capacity of gas at the constant volume, J / (kg K).

According to Mayer [10] formula, we have

$$C_{\mathrm{v}} = C_{\mathrm{p}} - R\,, \tag{13}$$

where C_{p} – heat capacity of gas with the constant pressure, 2240 J / (kg K).

Following Reynolds's approach, the vector of heat flux density [4-6]

$$\boldsymbol{q} = \boldsymbol{q}_{\mathrm{lam}} + \boldsymbol{q}_{\mathrm{turb}}\,, \tag{14}$$

where $\boldsymbol{q}_{\mathrm{lam}}$ – viscous (laminar) component of heat flux density, W/m^2;

$\boldsymbol{q}_{\mathrm{turb}}$ – turbulent component of heat flux density, W/m^2 which are defined as [4-7]

$$\boldsymbol{q}_{\mathrm{lam}} = -\lambda \cdot \mathrm{grad}\, T\,, \tag{15}$$

$$\boldsymbol{q}_{\mathrm{turb}} = -\frac{\mu_{\mathrm{t}} \cdot C_{\mathrm{p}}}{\sigma_{\mathrm{t}}} \cdot \mathrm{grad}\, T\,, \tag{16}$$

where λ – molecular heat conductivity of gas, 0.0348 W / (m·K);

σ_{t} – turbulent number of Prandtl, 0.9.

Generation of kinetic turbulent energy depends on the turbulent stress and speeds of deformations [4-6, 8]

$$G_{\mathrm{k}} = \mu_{\mathrm{t}} \cdot s^2 - \frac{2}{3} \cdot \rho \cdot k \cdot \operatorname{div} \mathbf{V} - \frac{2}{3} \cdot \mu_{\mathrm{t}} \cdot (\operatorname{div} \mathbf{V})^2 . \tag{17}$$

Taking note of compressibility on turbulent pulsations of speed is made by introduction of the component [5, 8, 11]

$$\Psi_{\mathrm{M}} = \frac{C_{\mathrm{M}} \cdot k \cdot \varepsilon}{a^2}, \tag{18}$$

where C_{M} – Sarkar's constant, 2.0 [11];
a – sonic speed in gas depending on its properties and temperature [4, 5], m/s.

Unlike standard $k-\varepsilon$ model of turbulence the coefficient at the generative component of the equation of transfer of speed of dissipation is not the constant [5, 6, 8]

$$C_{\varepsilon 1} = \max\left(0.43; \frac{\eta}{5+\eta}\right), \tag{19}$$

where η – the coefficient

$$\eta = s \cdot \frac{k}{\varepsilon}. \tag{20}$$

In this work is used the two-layer approach for the first time offered Rodi [5, 9], allowing to calculate the flow in the viscous underlayer and which is alternative to Low-Reynolds models of turbulence.

The flow in this case is separated into two layers. In the inside layer near walls turbulent dynamic viscosity μ_{t} and speed of dissipation ε are determined from the algebraic equations as functions of distance by the normal to the next wall y and the turbulent Reynolds number

$$Re_{\mathrm{y}} = \frac{y \cdot \sqrt{k}}{\nu}. \tag{21}$$

In particular, in the inside layer the speed of dissipation ε is defined as

$$\varepsilon = \frac{\sqrt{k^3}}{l_{\varepsilon}}. \tag{22}$$

where l_{ε} – distance scale, m.

In the external layer is solved input differential equation for transfer of speed of dissipation ε (6).

Differential equation for transfer of energy (5) k (5) in this approach is applied to all area of the flow.

Smooth transition between the decisions received in internal and external layers of the flow will be reached use hybrid wall functions of Jongen [12]

$$\Lambda = \frac{1}{2} \cdot \left(1 + \operatorname{th}\left(\frac{Re_{\mathrm{y}} - Re_{\mathrm{y}}^*}{A}\right)\right), \tag{23}$$

where Re_{y}^* – the limit value of the turbulent Reynolds number (boundary between two layers of the current), 60;
A – the constant determining width of the transitional zone between internal and external layers

$$A = \frac{|\Delta Re_{\mathrm{y}}|}{\operatorname{arth}(0.98)}, \tag{24}$$

where ΔRe_{y} – the constant, 10.

In the real article to calculation of turbulent dynamic viscosity μ_{t} and speed of dissipation in the inside layer applied Wolfstein's model [5, 13].

Then the distance scale is defined as [5, 13]

$$l_{\varepsilon} = C_1 \cdot y \cdot \left(1 - e^{-\frac{Re_y}{2 \cdot C_1}}\right), \tag{25}$$

where C_1 – the coefficient defined from the following expression

$$C_1 = \kappa \cdot C_{\mu 2}^{-0.75}, \tag{26}$$

where κ – the von Kármán constant, 0.42; $C_{\mu 2}$ – the constant, 0.09.

Turbulent dynamic viscosity μ_t depends on hybrid wall function of Jongen [5, 12, 13]

$$\mu_t = \Lambda \cdot \mu_{t1} + (1 - \Lambda) \cdot \mu_{t2}, \tag{27}$$

where μ_{t1} – turbulent dynamic viscosity in the external layer of the current, Pa·s;
μ_{t2} – turbulent dynamic viscosity in the inside layer of the current, Pa·s.

$$\mu_{t1} = C_{\mu 1} \cdot \rho \cdot \frac{k^2}{\varepsilon}, \tag{28}$$

where $C_{\mu 1}$ – the coefficient defined as [5, 8]

$$C_{\mu 1} = \frac{1}{A_0 + A_S \cdot U^* \cdot \frac{k}{\varepsilon}}, \tag{29}$$

where A_0 – the constant, 4.0 [5, 8];
A_S – the coefficient calculated on the following formulas (i, j, n – the indexes accepting values 1, 2, 3)

$$A_S = \sqrt{6} \cdot \cos\left(\frac{1}{3} \cdot \arccos\left(\sqrt{6} \cdot w\right)\right), \tag{30}$$

$$w = 2\sqrt{2} \cdot \frac{\dot{S}_{ij} \cdot \dot{S}_{jn} \cdot \dot{S}_{ni}}{s^3}; \tag{31}$$

U^* – the coefficient, s^{-1} [5, 8].

$$U^* = \sqrt{\dot{S}_{ij} \cdot \dot{S}_{ij} - \dot{W}_{ij} \cdot \dot{W}_{ij}}, \tag{32}$$

where $\dot{W}$ – rotation tensor, s^{-1} [5, 8]

$$\dot{W}_{ij} = \frac{1}{2} \cdot \left(\frac{\partial V_i}{\partial x_j} - \frac{\partial V_j}{\partial x_i}\right). \tag{33}$$

Turbulent dynamic viscosity in the inside layer of the current [5, 13]

$$\mu_{t2} = \mu \cdot \mathrm{Re}_y \cdot C_{\mu 2}^{0.25} \cdot \kappa \cdot \left(1 - e^{-\frac{\mathrm{Re}_y}{A_\mu}}\right), \tag{34}$$

where A_μ – the constant, 70.

To link the parameters of the flow on the wall and in the inner layer of the flow in this work applied the hybrid wall functions of Reichardt [5, 14, 17] recorded using:

1) dimensionless distance on the normal to the next wall [4]

$$y^+ = \frac{u^* \cdot y}{\nu}, \tag{35}$$

2) the dimensionless speed directed parallel to the wall [4]

$$u^+ = \frac{u}{u^*}, \tag{36}$$

3) dimensionless temperature [5]

$$T^+ = (T - T_W) \cdot \frac{\rho \cdot C_p \cdot u^*}{q_W}, \tag{37}$$

where u – the speed directed parallel to the wall, m/s; T_W – wall temperature, K; q_W – heat flux density, passing through the wall, W/m^2; u^* – dynamic speed, m/s [4, 5, 14].

$$u^* = \sqrt{\frac{g^* \cdot u \cdot \nu}{y} + (1 - g^*) \cdot \sqrt{C_{\mu 2}} \cdot k}, \tag{38}$$

where g^* – weight function of the turbulent Reynolds number [5, 14]

$$g^* = e^{-\frac{Re_y}{11}}. \tag{39}$$

Then the generative component of the equation of transfer of kinetic turbulent energy in parietal cells is calculated as:

$$G_{k\ \text{wall}} = g^* \cdot \mu_t \cdot s^2 + (1 - g^*) \cdot \frac{1}{\mu} \cdot \left(\rho \cdot u^* \cdot \frac{u}{u^+} \right)^2 \cdot \frac{\partial u^+}{\partial y^+}, \tag{40}$$

and the rate of dissipation in near-wall cells $\varepsilon_{\text{wall}}$ is on the formula (22).

The hybrid wall function for dimensionless velocity and temperature have the form [5, 14, 15, 17]

$$u^+ = \frac{1}{\kappa} \cdot \ln(1 + \kappa \cdot y^+) + \frac{1}{\kappa} \cdot \ln\left(\frac{E^*}{\kappa}\right) \cdot \left(1 - e^{-\frac{y^+}{y_m^+}} - \frac{y^+}{y_m^+} \cdot e^{-b \cdot y^+} \right), \tag{41}$$

$$T^+ = e^{-\Gamma} \cdot (T_{\text{lam}}^+ - q_{\text{lam}}^+) + e^{-\frac{1}{\Gamma}} \cdot (T_{\text{turb}}^+ - q_{\text{turb}}^+), \tag{42}$$

where y_m^+ – coordinate of conditional boundary between viscous laminar and logarithmic turbulent underlayers (is defined during calculations), m; E^* – the constant, 9.0;
b – the coefficient

$$b = \frac{1}{2} \cdot \left(\frac{y_m^+ \cdot \kappa}{\frac{1}{\kappa} \cdot \ln\left(\frac{E^*}{\kappa}\right)} + \frac{1}{y_m^+} \right); \tag{43}$$

Γ – the coefficient

$$\Gamma = \frac{0.01 \cdot (Pr \cdot y^+)^4}{1 + 5 \cdot Pr^3 \cdot y^+}; \tag{44}$$

Pr – molecular number of Prandtl

$$Pr = \frac{\mu \cdot C_p}{\lambda}; \tag{45}$$

T_{lam}^+, T_{turb}^+ – dimensionless functions of temperature for viscous and logarithmic underlayers respectively;

q_{lam}^+, q_{turb}^+ – dimensionless functions of viscous dissipation for viscous and logarithmic underlayers respectively.

$$T^{+}_{\mathrm{lam}} = Pr \cdot y^{+}, \tag{46}$$

$$T^{+}_{\mathrm{turb}} = \sigma_{\mathrm{t}} \cdot \left(\frac{1}{\kappa} \cdot \ln\left(E^{*} \cdot y^{+}\right) + P^{+} \right), \tag{47}$$

$$q^{+}_{\mathrm{lam}} = \frac{\rho \cdot u^{*} \cdot u^{2}}{2 \cdot q_{\mathrm{W}}}, \tag{48}$$

$$q^{+}_{\mathrm{turb}} = \frac{\rho \cdot u^{*}}{2 \cdot q_{\mathrm{W}}} \cdot \left(u^{2} - \left(\frac{Pr}{\sigma_{\mathrm{t}}} - 1 \right) \cdot \left(\frac{P^{+}}{Pr - \sigma_{\mathrm{t}}} \cdot u^{*} \right)^{2} \right), \tag{49}$$

where P^{+} – function of molecular and turbulent numbers of Prandtl [16]

$$P^{+} = 9.24 \cdot \left(\left(\frac{Pr}{\sigma_{\mathrm{t}}} \right)^{0.75} - 1 \right) \cdot \left(1 + 0{,}28 \cdot \mathrm{e}^{-\frac{0.007 \cdot Pr}{\sigma_{\mathrm{t}}}} \right). \tag{50}$$

In the task are set the following boundary conditions (figure 1) [4-6]:

1) for walls – "impermeability" and "not slipping" ($\boldsymbol{V} = 0$), lack of heat exchange through the wall ($\boldsymbol{q}_{\mathrm{W}} = 0$), and for kinetic turbulent energy

$$\left. \frac{\partial k}{\partial y} \right|_{\mathrm{wall}} = 0; \tag{51}$$

2) for permeable boundaries:

2.1) on the entrance are set full excess pressure p_{1}^{*} (for obtaining static characteristics of the valve several calculations with different value are carried out p_{1}^{*} from 0.05 to 0.6 MPa), full gas temperature $T_{1}^{*} = 283$ K and turbulence parameters - intensity of turbulence $J_{\mathrm{t1}} = 0{,}01$ and viscosity turbulent dynamic ratio $(\mu_{\mathrm{t}}/\mu)_{1} = 10$.

Excessive static pressure, as well as in the case with firm walls, was in result of calculations, and the module of the vector of local speed of gas flow $|\boldsymbol{V}_{1}|$ and static temperature of gas T_{1} were calculated on formulas

$$|\boldsymbol{V}_{1}| = \sqrt{2 \cdot C_{\mathrm{p}} \cdot \left(T_{1}^{*} - T_{1}\right)}, \tag{52}$$

$$T_{1} = T_{1}^{*} \cdot \left(\frac{p_{1}^{*}}{p_{1}} \right)^{-\frac{R}{C_{\mathrm{p}}}}. \tag{53}$$

2.2) at the exit are set excessive static pressure $p_{2} = 0$ MPa, and also for inverse flows on this boundary (were used automatically only at emergence of similar flows [5]) – static temperature of gas $T_{2} = 283$ K, intensity of turbulence $J_{\mathrm{t2}} = 0.01$ and viscosity turbulent dynamic ratio $(\mu_{\mathrm{t}}/\mu)_{2} = 10$.

During calculations on output boundary were defined the local speed of gas flow $|\boldsymbol{V}_{2}|$ and static temperature T_{2}.

On the plane of symmetry are equal to zero shearing stresses, heat flux density and gradients of all parameters of the current on the normal to this boundary.

Kinetic turbulent energy and speed of its dissipation are connected with intensity of turbulence J_{t} and viscosity turbulent dynamic ratio (μ_{t}/μ) following dependences [5]

$$k = \frac{3}{2} \cdot \left(J_{\mathrm{t}} \cdot |\boldsymbol{V}|\right)^{2}, \tag{54}$$

$$\varepsilon = \frac{\rho \cdot C_{\mu 2} \cdot k^2}{\mu \cdot (\mu_t / \mu)}. \tag{55}$$

3. Results and discussion

As a result of the executed series of calculations for each corner φ at different values of the total pressure on the entrance to the valve p_1^* are found: the torsion torque on the valve axis M, resulting from influence of pressure forces and viscous friction, and also - the mass gas rate via the valve Q_m, excessive static pressure on the entrance p_1, total pressure at the exit p_2^*, losses of the total pressure Δp^* when passing gas via the valve

$$\Delta p^* = p_1^* - p_2^*. \tag{56}$$

The static characteristics of the valve received in calculations are given in figure 2. Growth of hydraulic resistance of the valve with reduction of slope angle of the valve and with increase in the mass expense, and also nonlinear dependence of losses of the total pressure on the mass expense is represented natural. However it is remarkable that square nature of this dependence is broken in the range of corners $\varphi = 60^\circ - 90^\circ$, what is especially well visible at expenses 15-30 kg/s.

As appears from schedules in figure 2, the torsion torque on the axis of the valve increases with increase in the mass expense and with reduction of flow section of the valve at reduction of the corner φ, what will well be coordinated with distribution of pressure in the vicinity of the valve (Figure 2). At the provision of the valve, parallel in relation to the pipeline axis, value of the torsion torque fluctuates about zero. It is interesting that in the range of expenses of 15-30 kg/s, as well as on dependence for full losses of pressure, there is the excess of schedules.

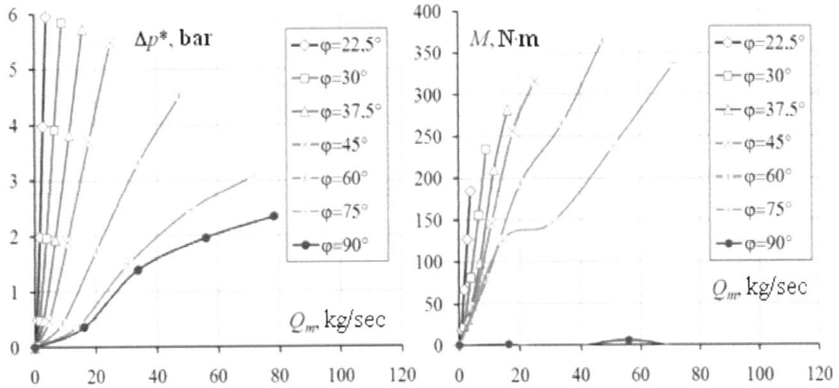

Figure 2. Dependence of losses of the total pressure (at the left) and the torsion torque (on the right) on the valve axis from the mass gas rate via the valve at different corners φ.

4. Conclusion

The analysis of the provided results of numerical modeling confirms need of specification of static characteristics of the considered valve by continuation of numerical research of the current in his flowing cavity at other intermediate values of slope angles of the valve φ, including, in the full three-dimensional problem definition without application of the plane of symmetry.

Besides, nature of fluctuations of integral characteristics of the flow during the repetitive process, testifies to possible not stationary of the current at some slope angles of the valve φ, and, so demands to consider this circumstance in mathematical model and calculations.

References

[1] Edelman A I 1970 *Fuel valves of liquid rocket engines* (Moscow: Mechanical engineering) p 244

[2] Telenkov A A 2011 *Development and deployment of the technique and engineering program of creation of optimum designs of electromagnetic valves of liquid rocket engines*. Thesis (Khimki) p 152

[3] Nizhnikov S A *Dynamics of the individual electromagnetic drive of the valve of the internal combustion engine*. Thesis (Kursk) p 102

[4] Loitsiansky L G 2003 *Fluid Mechanics* (Moscow: Drofa Publ.) p 840

[5] 2013 *User Guide STAR-CCM+ version 8.02.011* (CD–adapco)

[6] Belov I A and Isaev S A 2001 *The modeling of turbulent flows* (St. Petersburg: Baltic GTU Publ.)

[7] Cebeci T and Bradshaw P 1988 *Physical and Computational Aspects of Convective Heat Transfer* (New York: Springer)

[8] Shih T - H, Liou W W, Shabbir A, Yang Z and Zhu J 1994 *A New $k-\varepsilon$ Eddy Viscosity Model for High Reynolds Number Turbulent Flows – Model Development and Validation* NASA TM **106721**

[9] Rodi W 1991 *Experience with Two-Layer Models Combining the $k-\varepsilon$ Model with a One-Equation Model Near the Wall* 29th Aerospace Sciences Meeting January 7–10 Reno NV AIAA **91–0216**

[10] Bolgarsky A V, Mukhatchyov G A and Shchukin V K 1975 *Thermodynamics and Heat Transfer* (Moscow: Vysshaya shkola (Higher school))

[11] Sarkar S and Balakrishnan L 1990 *Application of a Reynolds-stress turbulence model to the compressible shear layer* ICASE Report **90–18** NASA CR **182002**

[12] Jongen T 1998 *Simulation and Modeling of Turbulent Incompressible Flows* Ph.D. Thesis Lausanne EPFL

[13] Wolfstein M 1969 The velocity and temperature distribution in one-dimensional flow with turbulence augmentation and pressure gradient *Int. J. Heat Mass Transfer* **12** pp 301–318

[14] Reichardt H 1951 Vollstaendige Darstellung der turbulenten Geschwindigkeitsverteilung in glatten Leitungen *Z. Angew. Math. Mech.* **31(7)** pp 208–219

[15] Kader B A 1981 Temperature and Concentration Profiles in Fully Turbulent Boundary Layers *Int. J. Heat Mass Transfer* **24** pp 1541–1544

[16] Jayatilleke C L 1969 The influence of Prandtl number and surface roughness on the resistance of the laminar sub-layer to momentum and heat transfer *Progress in Heat and Mass Transfer* **1** pp 193–330

[17] Launder B E and Spalding D B 1974 The Numerical Computation of Turbulent Flows *Computer Methods in Applied Mechanics and Engineering* **3** pp 269–289

On the solving of one type of problems of mathematical physics

V J Chebakova[1], **A V Gerasimov**[2] **and A P Kirpichnikov**[2]

[1]Kazan Federal University 420008, Kremlevskay str., 18 Kazan, Russia
[2]Kazan National Research Technological University 420015, Karl Marx str., 68 Kazan, Russia

E-mail: gerasimov@kstu.ru

Abstract. A relationship between generalized hypergeometric functions of a special type and modified Bessel functions has been established. Using this relationship the solution of inhomogeneous differential equations of Bessel type containing even degrees of an independent variable in the right-hand part can be expressed in a form convenient for engineering and technical applications.

1. Introduction

Within the frames of a relatively large class of problems [1-5] on restoration of the structure of quasi-stationary electromagnetic and temperature fields of a high-frequency induction discharge from measured values of one of its components, a necessity arises to obtain exact analytical solutions of differential equations of the form

$$y''(x)+\frac{y'(x)}{x}-Cy(x)=x^{2n} \tag{1}$$

(here $C=b^2=const$ is a nonnegative constant depending on boundary conditions), to which Maxwell's equations for electromagnetic field are reduced in the general case. In the case of C=0 (corresponding to one-dimensional statement of the problem from physical point of view), a particular solution of Eq. (1), apparently, reduces to a parabolic solution

$$y(x)=\frac{1}{4(n+1)^2}x^{2n+2}. \tag{2}$$

Let us introduce new variables $\tilde{x}=\sqrt{C}\,x=b\,x$, in which the initial equation will be rewritten as

$$y''(\tilde{x})+\frac{y'(\tilde{x})}{\tilde{x}}-y(\tilde{x})=\frac{\tilde{x}^{2n}}{b^{2(n+1)}}. \tag{3}$$

It can be easily seen that written in such a form equation (3) is one of the variants of inhomogeneous Bessel equations for unknown function $\psi(\tilde{x})$, the solution of which is given, for instance, in Ref. [6].

$$\psi(\tilde{x})=\frac{(-1)^{n+1}}{b^{2(n+1)}}\,s_{2n+1,0}\left(\tilde{x}e^{i\frac{\pi}{2}}\right). \tag{4}$$

where $s_{\mu,\nu}(x)$ is a so-called Lommel function usually defined as [7]

$$s_{\mu,\nu}(x) = \frac{x^{\mu+1}}{(\mu-\nu+1)(\mu+\nu+1)} {}_1F_2\left(1; \frac{\mu-\nu+3}{2}, \frac{\mu+\nu+3}{2}; -\frac{x^2}{4}\right) \quad (5)$$

where ${}_1F_2$ is a generalized hypergeometric function [8]. So taking account for Eq. (6), Eq. (4) can be rewritten as follows

$$\psi(\tilde{x}) = \frac{1}{\left[2(n+1)b^{n+1}\right]^2} \tilde{x}^{2(n+1)} {}_1F_2\left(1; n+2, n+2; \frac{\tilde{x}^2}{4}\right). \quad (6)$$

Solutions in the form of Eqs. (4) and (6) are inconvenient for technical and engineering applications since Lommel functions as well as hypergeometric functions, are not included into a standard tool kit of special functions used in reference tables and routine libraries. However, these expressions can be reduced to the form, which is more familiar for experts in the field of mathematical and engineering physics.

2. Method of solution

Let us examine expansion of generalized hypergeometric functions ${}_pF_q$ into their Maclaurin series. In particular, for ${}_1F_2$ function the series reads as

$${}_1F_2(a; b_1, b_2; z) = \sum_{k=0}^{\infty} \frac{(a)_k}{(b_1)_k (b_2)_k} \frac{z^k}{k!},$$

where $(a)_k$, $(b_{1,2})_k$ are the corresponding Pohgammer's symbols

$$(a)_k = a(a+1)...(a+k-1) = \frac{\Gamma(a+k)}{\Gamma(a)};$$

and

$$(b_{1,2})_k = b_{1,2}(b_{1,2}+1)...(b_{1,2}+k-1) = \frac{\Gamma(b_{1,2}+k)}{\Gamma(b_{1,2})}.$$

For natural values of a and $b_{1,2}$, we obtain

$$(n)_k = n(n+1)...(n+k-1) = \frac{(n+k-1)!}{(n-1)!}$$

and then

$${}_1F_2\left(1; n+2, n+2; \frac{\tilde{x}^2}{4}\right) = \left[(n+1)!\right]^2 \sum_{k=0}^{\infty} \frac{(1)_k}{(n+2)_k (n+2)_k} \frac{\tilde{x}^{2k}}{4^k k!} =$$

$$= \left[(n+1)!\right]^2 \sum_{k=0}^{\infty} \frac{k!}{(n+k+1)!(n+k+1)!} \frac{\tilde{x}^{2k}}{4^k k!} =$$

$$= \left[(n+1)!\right]^2 \sum_{k=0}^{\infty} \frac{\tilde{x}^{2k}}{4^k \left[(n+k+1)!\right]^2} = \left[(n+1)!\right]^2 \sum_{k=n+1}^{\infty} \frac{\tilde{x}^{2(k-n-1)}}{4^{k-n-1}(k!)^2} = .$$

$$= \left[(n+1)!\right]^2 \frac{4^{n+1}}{\tilde{x}^{2(n+1)}} \sum_{k=n+1}^{\infty} \frac{\tilde{x}^{2k}}{4^k (k!)^2} = \left[(n+1)!\right]^2 \frac{4^{n+1}}{\tilde{x}^{2(n+1)}} \sum_{k=n+1}^{\infty} \left(\frac{\tilde{x}^k}{2^k k!}\right)^2.$$

The last expression with an accuracy to the factor $\left[(n+1)!\right]^2 \frac{4^{n+1}}{\tilde{x}^{2(n+1)}}$, preceding the sum, coincides with a well-known expansion [9, 10] of a modified Bessel function of the zeroth order I_0 into its Maclaurin series and, therefore,

$${}_1F_2\left(1; n+2, n+2; \frac{\tilde{x}^2}{4}\right) = \left[(n+1)!\right]^2 \frac{4^{n+1}}{\tilde{x}^{2(n+1)}} \sum_{k=n+1}^{\infty} \left(\frac{\tilde{x}^k}{2^k k!}\right)^2 =$$

$$=\left[(n+1)!\right]^2\frac{4^{n+1}}{\tilde{x}^{2(n+1)}}\left[\sum_{k=0}^{\infty}\left(\frac{\tilde{x}^k}{2^k k!}\right)^2-1-\frac{\tilde{x}^2}{4}-\ldots-\frac{\tilde{x}^{2n}}{2^n n!}\right]=$$

$$=\left[(n+1)!\right]^2\frac{4^{n+1}}{\tilde{x}^{2(n+1)}}\left[\sum_{k=0}^{\infty}\left(\frac{\tilde{x}^k}{2^k k!}\right)^2-\sum_{k=0}^{n}\left(\frac{\tilde{x}^k}{2^k k!}\right)^2\right]=$$

$$=\left[(n+1)!\right]^2\frac{4^{n+1}}{\tilde{x}^{2(n+1)}}\left[I_0(\tilde{x})-\sum_{k=0}^{n}\left(\frac{\tilde{x}^k}{2^k k!}\right)^2\right]$$

(the relationship which first establishes a link between generalized hypergeometric functions of the special kind and modified Bessel functions) and

$$s_{2n+1,0}\left(\tilde{x}e^{i\frac{\pi}{2}}\right)=\frac{\tilde{x}^{2(n+1)}}{4(n+1)^2}\,{}_1F_2\left(1;n+2,n+2;\frac{\tilde{x}^2}{4}\right)=4^n\left(n!\right)^2\left[I_0(\tilde{x})-\sum_{k=0}^{n}\left(\frac{\tilde{x}^k}{2^k k!}\right)^2\right] \quad (7)$$

Finally we end up with

$$\psi(\tilde{x})=\left(\frac{2^n\, n!}{b^{n+1}}\right)^2\left[I_0(\tilde{x})-\sum_{p=0}^{n}\left(\frac{\tilde{x}^p}{2^p p!}\right)^2\right]$$

or

$$y(x)=\left(\frac{2^n\, n!}{b^{n+1}}\right)^2\left[I_0(bx)-\sum_{p=0}^{n}\left(\frac{b^p x^p}{2^p p!}\right)^2\right]=\frac{\left(2^n\, n!\right)^2}{C^{n+1}}\left[I_0(\sqrt{C}\,x)-\sum_{p=0}^{n}C^p\left(\frac{x^p}{2^p p!}\right)^2\right]. \quad (8)$$

3. Results and conclusion

The obtained relationship is a final formula for unknown function $y(x)$, written in a rather simple and physically transparent form. In particular, the expression in square brackets is nothing but a difference of the modified Bessel functions and the sums of the first $(n+1)$ terms of the function expansion into its Taylor's series near the point x=0. It is easy to verify that in the limit $C\to 0$, Eq. (8) gives us exactly Eq. (2), and at $b=1$, we obtain an especially simple solution

$$y(x)=\left(2^n\, n!\right)^2\left[I_0(x)-\sum_{p=0}^{n}\left(\frac{x^p}{2^p p!}\right)^2\right].$$

Let us note that the results obtained in this work can be useful both for experts in the field of mathematical simulation of various plasma devices based on the principle of inductive gas heating and, perhaps, in several adjacent regions of engineering and mathematical physics.

Acknowledgments

The work is performed according to the Russian Government Program of Competitive Growth of Kazan Federal University.

References

[1] Kirpichnikov A P 1994 On the structure of quasi-stationary electromagnetic field of a high-frequency induction discharge under atmospheric pressure *Izvestia vuzov. Fizika* **2** 77-81.

[2] Kirpichnikov A P 1995 The structure of quasi-stationary electromagnetic field of a high-frequency induction discharge in the vicinity of the axis of a plasmoid *High Temperature* **33** (1) 139-40.

[3] Kirpichnikov A P and Gerasimov A V 1995 The structure of a high-frequency induction discharge in the vicinity of plasmoid axis in the case of a work coil of finite dimensions *"Plasma technology -95". Proceedings. Zaporozhye* 28-9.

[4] Gerasimov A V and Kirpichnikov A P 1998 On the structure of temperature field of a high-frequency induction discharge of atmospheric pressure in the vicinity of the axis of a plasmoid *High Temperature* **36** (2) 342-3

[5] Gerasimov A V and Kirpichnikov A.P. 2000 Two-temperature model of the energy balance for plasma of a high-frequency induction discharge in the vicinity of the axis of a plasmoid *High Temperature* **38** (5) 710-5

[6] Erde'lyi A 1953 *Higher Transcendental Functions, Vol. II* (New York: McGraw-Hill)

[7] Prudnikov A P, Brychkov Yu A and Marichev O I 1983 *Integrals and series. Special functions*. (Moscow: Nauka) (in Russian)

[8] Prudnikov A P, Brychkov Yu A and Marichev O I 1983 *Integrals and series. Supplementary chapters* (Moscow: Nauka) (in Russian)

[9] Gradstein I S and Ryzhik I M 1962 *Tables of Integrals, Sums, Series, and Products* (Moscow: GIFML) (in Russian)

[10] Dwight G B 1957 *Tables of Integrals and other mathematical data* (New York: The Macmillan Company)

Solution of the problem of interaction between capacitive coupled radio-frequency discharge and a sample

V Ju Chebakova[1], **A F Gaisin**[2] **and V S Zheltukhin**[3]

[1]Kazan Federal University, 18 Kremlevskaya Street, Kazan, 420008, Russia
[2]Tupolev's Kazan National Research Technical University, 10, Karl Marx St., Kazan, 420111, Russia
[3]Kazan State Technological University, 68 Karl Marx Street, Kazan, 420015, Russia

E-mail: vchebakova@mail.ru

Abstract. The numerical study of interaction between the capacitive coupled radio frequency (CCRF) discharge and materials is performed. A nonlinear problem, which includes initial-boundary value problems for electron, ion, neutral atom, metastable atom, gas temperature and Poisson's equation is solved. A harmonic voltage on the loaded electrodes and Ohm's law for the sample is assumed. A results of calculations of the model problem at pressure p=760 Torr, frequency of generator f=13.76 MHz in local approximation are presented.

1. Introduction

Low-temperature plasma is applied for material treatment aimed to increase service life and reliability of machine-building products, to create light-weight and service-strong polymeric composite materials, polyethylene-plastic materials, nanodispersed metal powders and compounds, etc. Various types of discharges are used to create plasma and important role in this process is assigned to RF-discharges, and particularly, to CCRF-discharges [1-4]. Mathematical models, which allow calculating main parameters of RF-discharges is created and solved numerically in order to improve operational modes of the equipment. Models in 0D, 1D, 2D and 3D approximation can be used to simulate discharges [5-8], however, 1D models is allowed describing the physics of the processes in a discharge at low cost providing that they describe the discharge adequately. Plasma parameters change when treated specimen is inserted into the discharg. Sometimes changes of parameters of the sample by plasma is studied without studying variation of the discharge characteristics (see [9, 10], for example). Statistical simulation is also used for plasma processes researches [11-15]. This work was aimed to study influence of the sample located on an electrode on the parameters of plasma of a CCRF (13.56 MHz) discharge in argon at atmospheric pressure

2. Setting of the problem

Herein we compare results obtained for two models. Ones describes capacitive coupled radio-frequency discharge between two parallel plate electrodes, one of which is grounded and the other is connected to a CCRF-generator (Model N1). The second model describes situation, when the sample located on the grounded electrode (Model N2).

Both models are described in frame of the following physical hypothesis. Electrical field is close to the potential one and the discharge is uniform along the electrodes that allow us to use 1D model [16].

The electron energy dissipation at atmospheric pressure is significantly less than the field period, while plasma parameters change in time with the field frequency. The length of electron energy relaxation at atmospheric pressure is significantly less than the dimensions of the computational domain, therefore we can use local approximation for plasma simulation. In this approximation parameters of electronic component of the plasma depend on local value of a reduced electrical field E/N [17].

A simplified diagram of argon atom wherein the 4 lowest nearby electronically excited states is described as the single level excited atoms of density nm. Such diagram is reasoned by efficiently mixing of these levels at the electron impact [18,19].

It is assumed that the ideal gas is occurred in a discharge chamber, therefore where P is pressure, is Boltzmann's constant, is atom temperature, *N* is density of argon atoms in the ground state.

Let us introduce the *Ox* axis at the normal direction to the surface of electrodes, in this case coordinates x=0, x = b and x=a corresponds to the grounded electrode, loaded electrode and to the sample surface, respectively. Here b is distance between electrodes, a is thickness of the sample. Domain 0<x< a in the second model is occupied with the sample. Therefore boundary conditions are formulated at the points x= $a(r$-1) and x=b, where r is the model number hereinafter.

Processes in CCRF discharge are described by the following problems.

2.1. Convection– diffusion equation for electrons

$$\frac{\mathrm{d}n_e}{\mathrm{d}t}+\frac{\mathrm{d}}{\mathrm{d}x}\left(-D_e\frac{\mathrm{d}n_e}{\mathrm{d}x}-\mu_e n_e E\right)=R_1 n_e N+R_2 n_m^2+R_3 n_m n_e-R_4 n_e n_+ -$$

$$-R_5 n_e^2 n_+,\quad a(r-1)<x<b,\quad t>0$$

The boundary conditions depend from the electric field directions:

$$\begin{cases} G_e=-\gamma G_+, & \text{if the field is directed to the electord or the sample} \\ & (E\le 0 \text{ at } x=a(r-1),\quad E>0 \text{ at } x=b) \\ \dfrac{\mathrm{d}G_e}{\mathrm{d}x}=0, & \text{if the field is directed from the electrod or the sample} \\ & (E>0 \text{ at } x=a(r-1),\quad E\le 0 \text{ at } x=b) \end{cases}$$

Here n_e, n_+, n_m are densities of electrons, ions and metastable, respectively, $G_e=-D_e\,\mathrm{d}n_e/\mathrm{d}t-\mu_e n_e E$, $G_+=-D_+\,\mathrm{d}n_+/\mathrm{d}t+\mu_+ n_+ E$ are densities of electron and ion flows, μ_e μ_+ are mobility of electrons and ions, D_e, D_+ are electron and ion diffusion coefficients, γ is secondary electron emission factor E is electric field intensity, $(E=-\partial\phi/\partial x)$, φ is electric field potential. Here in after R_i,i=1...9, are factors of plasma chemical reaction velocities (table 1) (Ar+e→Ar^++2e; Ar^*+Ar^*→Ar+Ar^++e; Ar^*+e→Ar^++2e; Ar^++e→Ar+h_v; Ar^++2e→Ar+e; Ar+e→Ar^*+e; Ar^*→Ar+h_v; Ar^*+Ar→2Ar; Ar^*+e→ Ar+e). [20-26].

2.2.Convection–diffusion equation for ions

$$\frac{\mathrm{d}n_+}{\mathrm{d}t}+\frac{\mathrm{d}}{\mathrm{d}x}\left(-D_+\frac{\mathrm{d}n_+}{\mathrm{d}x}+\mu_+ n_+ E\right)=R_1 n_e N+R_2 n_m^2+R_3 n_m n_e-R_4 n_e n_+ -$$

$$-R_5 n_e^2 n_+,\quad a(r-1)<x<b,\quad t>0$$

The boundary conditions are anologues

$$\begin{cases} \dfrac{\mathrm{d}G_+}{\mathrm{d}x}=0, & \text{if the field is directed to the electord or the sample} \\ & (E\le 0 \text{ at } x=a(r-1),\quad E>0 \text{ at } x=b) \\ G_+=0, & \text{if the field is directed from the electrod or the sample} \\ & (E>0 \text{ at } x=a(r-1),\quad E\le 0 \text{ at } x=b) \end{cases}$$

2.3. Poisson's equation for the electric field potential φ

$$-\frac{d^2\phi}{dx^2}=\frac{q_e}{\varepsilon_0}(n_+-n_e),\quad a(r-1)<x<b,\quad t>0$$

where q_e istheelectron charge, ε_0 is the electrical constant, $\omega$is the angular frequency of electric field. with boundary conditions for the model No.1

$$\begin{cases}\phi=V_a\sin(\omega t),\quad \text{on the loaded electrod } (x=b)\\ \phi=0,\ \text{on the grounded electrod } (x=0)\end{cases}$$

And for model No.2

$$\begin{cases}\phi=V_a\sin(\omega t),\quad \text{on the loaded electrod } (x=b)\\ i=-\sigma\ d\phi/dx,\ \text{on the sample} (x=a)\end{cases}$$

where V_a – is the amplitude of the driven potential, σ – specific conductivity of the sample, $i=q_e(G_+-G_e)+\varepsilon_0\, dE/dt$ – total current density

2.4.Balance equation of metastable atom

$$\frac{dn_m}{dt}-\frac{d}{dx}\left(D_m\frac{dn_m}{dx}\right)=R_6Nn_e-R_2n_m^2-R_3n_mn_e-R_7n_m-R_8Nn_m-$$
$$-R_9n_mn_e,\ a(r-1)<x<b,\quad t>0$$

where D_m is full heat-transfer coefficient of the sample, T_w is electrode cooling water temperature.

3. Results of calculations

To solve non-linear systems of boundary value problems and initial-boundary value problems we used the approximation method based on preliminary finite dimensional approximation with use of the difference scheme and furtherly applying iteration process for its implementation. Difference approximation for the convection equations of charged particle diffusion is derived with use of integro-interpolation method. Drift terms are approximated by using upwind differences as in [27-29]. The research papers [30-35] are dedicated to the methods to solve non-linear problems of continuum mechanics, which describe two-layer iteration methods including methods implying nonlinearity lowering to lower layer.

Calculation was performed for the following values: γ=0.01; σ=0.01 cm; P=760 Torr, V_a=100 B, k_{full} =0.5 W/(m^2·K). Simulation is showed that placing of the sample with finite thermal conductivity on an electrode bring in asymmetry the discharge and shift the gas temperature maximum to the sample (see fig.1, fig. 2) that results in changes and asymmetry of other processes.

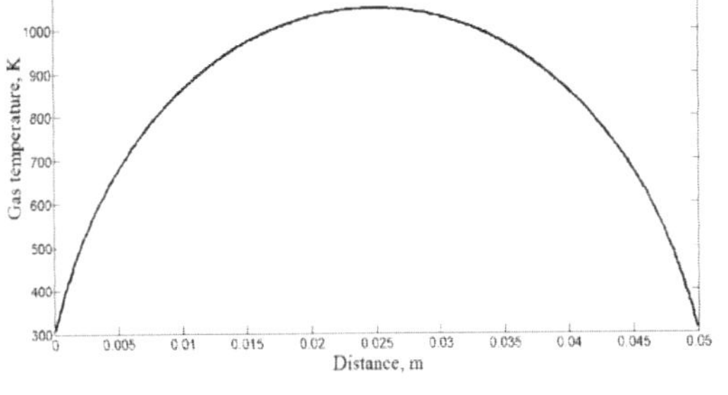

Figure1.distribution of gas temperature between electrodes for the first model.

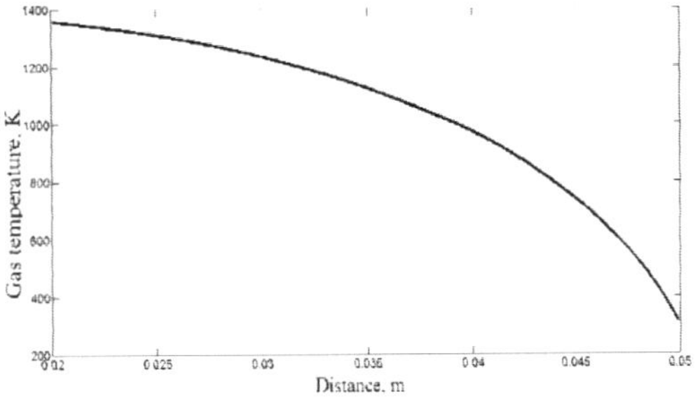

Figure 2. distribution of gas temperature between the sample and the electrode for the second model.

Due to strong dependence of molecular and atomic ions densities on the value of gas temperature and the significant difference the last ones, the kinetic scheme (table 1) shall complicate by processes with dimers and molecular ions [22].

4. Conclusion

The article describes the model of interaction between CCRF discharge and a sample, which allows calculating the characteristics of CCRF-in the interval between the sample and the powered electrode. The model includes drift–diffusion equations for electron and ions, Poisson's equation for the potential of the electrical field, balance equations for metastables and grounded state atoms as well as a steady state equation for atomic-ion gas conductivity. A numerical method is proposed for model solution. The solution for the model with non-ideal conductor is obtained. It is established that placing the sample with finite conductivity on an electrode result in asymmetry in discharge characteristics.

Acknowledgments

The work is performed according to the Russian Government Program of Competitive Growth of Kazan Federal University and work was supported by the Russian Foundation for Basic Research projects nos. 16-31-00378 and the Russian Ministry of Education and Science, Contract No. 2196 d. Feb.01,2014.

References

[1] Fridman A 2008 *Plasma chemistry* (Cambridge: Cambridge Univ. Press)

[2] Chebert Pascal, Braithwaite Nicholast 2011 *Physics of radio-frequency plasmas*(Cambridge: Cambridge Univ. Press)

[3] Lebedev Yu A and Tatarinov A V 2004 Electrodynamics of microwaves in a coaxial non-regular waveguide partly filled with plasma *Plasma Sources Science and Technology* **13** (1) 1-7.

[4] Lebedev Yu A and Mokeev M V 2001Characteristics of the electrode plasma of an electrode microwave discharge in hydrogen *Plasma Physics Reports* **27** (5) 418-23.

[5] Dautov G, Dautov I, Fayrushin I and Kashapov N 2014 Calculation of distribution of potential near the surface of metal particle in the dust-electron thermal plasma *journal of Physics:Conference Series* **56** **7**012005

[6] Epstein I L, Gavrilović M, Jovircević S, Konjević N, Lebedev Yu A and Tatarinov A V 2014 The study of a homogeneous column of argon plasma at a pressure of 0.5 torr, generated by means of the Beenakker's cavity *Eur. Phys. J. D* **68**: 334

[7] Lebedev Y A, Tatarinov A V, Epstein I L and Averin K A 2015 Mathematical modeling of the gas bubbles in the microwave discharge of boiling n-heptane *Contemporary Engineering Sciences* **8** (21) 1057-65.

[8] Zheltuhin V S, Solovyev P S and Chebakova V Y 2015 Boundary conditions for electron balance equation in the stationary High-Frequency Induction discharges *Research Journal of Applied Sciences*, **10** (10) 658-62.

[9] Memnonov V P, Ulianov P G 2011 Experimental evaluation of the distribution parameters of the surface roughness in the nanosize channels *Technical Physics Magazine* **81**(2) 104-9.[in Russian]

[10] Gorbachev Yu E, Zatevakhin M A and Kagonovich I D 1996 Growth simulation of films of hydrogenated amorphous silicon fromHF-discharged plasma*Technical Physics Magazine*. **66** (12) 89-110.[in Russian]

[11] Mokrov M S and Raiser Yu P 2008 Application of Monte-Carlo method aimed to determine ionization factors, secondary emission and current-voltage characteristics of townsend discharge on hydrogen*Technical Physics Magazine* **78**(4)47-54.[in Russian]

[12] Demkin V P, Mel'nichuk S V, Borisov A V and Bardin S S 2013 Monte-Carlo calculation of the electron energy distribution function of a He–N_2 plasma of a highvoltage pulsed

discharge, *Russian Physics Journal***56** (4) 486-8[in Russian]

[13] Rau Hans 2000 Monte Carlo simulation of a microwave plasma in hydrogen *J. Phys. D: Appl. Phys* .**33** 3214–22.

[14] Yousfi M, Hennad A and Alkaa A 1994 Monte Carlo simulation of electron swarms at low reduced electric fields *Physical review E***49**(4) 3264-73

[15] Dubrovin V T, Gabbasov F G and Chebakova V J 2016 Multidimensional central limit theorem for sums of functions of the trajectories of endomorphisms *Lobachevskii Journal of Mathematics* **37** (4) 409-17

[16] Riser Yu P, Shneider M N and Yatsenko N A 1995 *Righ-frequency capacitive discharge. Physics. Experimental procedures. Attachments*(Moscow: Publishing house affiliated with MPhTI) [in Russian]

[17] Kudryavtsev A A, Smirnov A S and Tsendin L D 2010 *Physics of glow-discharge* (Saint Petersburg: Publishing house "Lan'") [in Russian]

[18] Ferreira C M, Loureiro J and Ricard A 1985 Populations in the metastable and the resonance levels of argon and stepwise ionization effects in a low-pressure argon positive column, *Journal of Applied Physics* **57** 82–90.

[19] Diatko N A, Ionokh Yu Z, Meshanov A B and Napartovich A P 2005 Study of dark study of development of positive column of glow discharge in argon *Plasma Physics Reports***31** (10) 939-53

[20] Lymberopoulos D P, Economou D J 1993 Fluid simulations of glow discharge & Effect of metastable atoms in argon *J. Appl. Phys.* **73**(8) 3668–79

[21] Xi-Ming Zhu and Yi-Kang Pu 2010 A simple collisional–radiative model for low-temperature argon discharges with pressure ranging from 1 Pa to atmospheric pressure: kinetics of Paschen 1s and 2p levels *J. Phys. D: Appl. Phys.* **43** 015204

[22] Martinez E Castonos, Kabouzi Y, Makasheva K and Moisan M 2004 Modeling of microwave-sustained plasmas at atmospheric pressure with application to discharge contraction *Physical review E70* 066405

[23] Moravej M, Yang X, Hicks R F, Penelon J and Babayan S E 2006 A radio-frequency nonequilibrium atmospheric pressure plasma operating with argon and oxygen*J. Appl. Phys.* **99** 093305

[24] Bora B, Bhuyan H, Favre M, Wyndham E and Chuaqui H 2011 Diagnostic of Capacitively Coupled Low Pressure Radio Frequency Plasma: An Approach through Electrical Discharge Characteristic *International Journal of Applied Physics and Mathematics* **1**(2) 124–8.

[25] Balcon N P, Aanesland A, Hagelaar G J M, Boswell R and Boeuf J P 2007 Atmospheric pressure RF discharge in argon: optical diagnostic, fluid model and applications *28th ICPIG* 957–60

[26] Hagelaar G J M, Pitchford L C "Solving the Boltzmann equation to obtain electron transport coefficients and rate coefficients for fluid models" *Plasma Sources Sci. Techn.* **14** (2005) 722-733

[27] Chebakova V Ju, Zheltukhin V S and Dubrovin V T 2016 Problem of radio-frequency discharge at atmospheric pressure in local statement *Applied Mathematical Sciences* **10** (21) 1013-22.

[28] Zheltukhin V S, Chebakova V J, Abaidullin R N and Kadyrov R F 2015 On solving of some nonlinear problems of plazma physics *Applied Mathematical Sciences* **9** (88) 4351-9.

[29] Dubrovin V T, Chebakova V Ju and Zheltukhin V S 2016 Radio-Frequency Discharge at Low Pressure: a Non-Local Problem Statement Approach *Procedia Engineering* **150** 1041–5

[30] Badriev I B and Zadvornov O A 2003 A decomposition method for variational inequalities of the second kind with strongly inverse-monotone operators *Differential equations* **39** (7) 936-44

[31] Badriev I B, Zadvornov O A and Lyashko A D 2004A study of variable step iterative methods for variational inequalities of the second kind *Differential equations* **40** (7) 971-83

[32] Badriev I B and Banderov V V 2014 Numerical method for solving variation problems in

mathematical physics*Mechanical components and control engineering III*. **668-669** 1094-97

[33] Badriev I B and Zadvornov O A 2002 Construction and convergence analysis of iterative methods for variational problems with a nondifferentiable functional *Differential equations* **38** (7) 985-991

[34] Badriev I B 1983 Difference-schemes for linear-problems of the filtration theory with discontinuous law *Izvestiya Vysshikh Uchebnykh Zavedenii Matematika* **5** 3-12

[35] Zheltukhin V S , Solov'ev S I, Solov'ev P S and Chebakova V Y 2014 Computation of the minimum eigenvalue for a nonlinear Sturm-Liouville problem **35** (4) 416-26

Rare meshes FEM scheme for quasi-stationary electromagnetic fields determination 3D problems

D T Chekmarev, A V Kalinin, V V Sadovsky and A A Tiukhtina

Lobachevsky State University of Nizhnii Novgorod, 23 Gagarin av., 603950 Nizhnii Novgorod, Russia

Email: 4ekm@mm.unn.ru

Abstract. The initial-boundary value problem for the quasi-stationary magnetic approximation of the Maxwell equations in inhomogeneous media is studied. The considered problem is reduced to the variational problem of determining the vector magnetic potential. The special gauge for vector magnetic and scalar electrical potentials is used. The well-posedness of the problems is established under general conditions on the coefficients and the applicability of the projection methods for these problems is validated. For the numerical solution of this problem provides to use the effective rare mesh FEM scheme for 3D problems. This scheme is well-proven in 3D elasticity and plasticity problems solving.

1. Introduction

The solution of many actual technological problems leads to theoretical and numerical study of problems for quasi-stationary electromagnetic fields in physically heterogeneous media [1]-[3].

In the construction of numerical algorithms and in theoretical studies of the quasi-stationary problems vector magnetic and scalar electric potentials are traditionally used to describe the electromagnetic fields [4]-[10]. At the considering problems in terms of the potentials different gauges are employed, which ensure the uniqueness of the solution. In particular, in stationary and quasistationary problems the classical Coulomb gauge $\operatorname{div} \boldsymbol{A} = 0$ and Lorenz gauge $\operatorname{div} \boldsymbol{A} + \mu\sigma\varphi = 0$ are widely used [4], [6]-[8].

Problem statements are of practical interest, which allow to determine a magnetic potential independently of the electrical potential. In heterogeneous media the application of the classical Coulomb and Lorenz gauges leads to the study of the coupled system of equations for he potentials [7], [8]. In [8], [9] the modified Lorenz gauge is discussed for time-harmonic Maxwell equations, that decomposes the problem of finding the magnetic and electric potentials.

In the present paper the initial-boundary value problem for the time-dependent quasi-stationary magnetic approximation of the Maxwell equations is investigated. This problem is formulated in terms of potentials with special gauge, which generalizes the Lorenz gauge for the case of inhomogeneous media [11]. We prove the well-posedness of the problem and investigate the possibility of using the projection methods for its solution. The corresponding topics for the stationary problems are considered in [12].

2. Initial - boundary value problem

The quasi-stationary magnetic approximation of the Maxwell equations in the can be presented in the form [13]

$$\operatorname{curl} \boldsymbol{H}(x,t) = \boldsymbol{J}(x,t), \tag{1}$$

$$\operatorname{div} \boldsymbol{B}(x,t) = 0, \tag{2}$$

$$\operatorname{curl} \boldsymbol{E}(x,t) = -\frac{\partial}{\partial t}\boldsymbol{B}(x,t), \tag{3}$$

$$\operatorname{div} \boldsymbol{D}(x,t) = \rho(x,t), \tag{4}$$

where $(x,t) \in Q = \Omega \times (0,T)$, $\Omega \subset \mathbf{R}^3$, $T > 0$; $\boldsymbol{H}$, $\boldsymbol{B}$, $\boldsymbol{E}$, $\boldsymbol{D}$, $\boldsymbol{J} : Q \to \mathbf{R}^3$ and $\rho : Q \to \mathbf{R}^1$ – unknown functions.

In linear media the following constitutive relations are valid:

$$\boldsymbol{B} = \mu \boldsymbol{H}\ ,\ \boldsymbol{D} = \varepsilon \boldsymbol{E}\ ,\ \boldsymbol{J} = \sigma \boldsymbol{E} + \boldsymbol{J}^{ext}\ , \tag{5}$$

where μ is a magnetic permeability, ε is a permittivity, σ is a electrical conductivity, $\boldsymbol{J}^{ext}$ is an exterior current density.

It is assumed in this paper, that Ω is an open bounded domain, homeomorphic to a ball, with a Lipschitz boundary $\partial\Omega$. Let $\boldsymbol{v}(x)$ is unit normal vector in $x \in \partial\Omega$. For function $\boldsymbol{u} : \overline{\Omega} \to \mathbf{R}^3$ we denote by $\boldsymbol{u}_\nu$, $\boldsymbol{u}_\tau$ the normal and tangent components of $\boldsymbol{u}$ on $\partial\Omega$.

The system (1)-(5) is considered with the boundary condition

$$\boldsymbol{H}_\tau(x,t) = 0,\ x \in \partial\Omega,\ t \in (0,T), \tag{6}$$

and the initial condition

$$\boldsymbol{H}(x,0) = \boldsymbol{h}(x),\ x \in \Omega. \tag{7}$$

We assume, that $\sigma = \sigma(x)$, $\mu = \mu(x)$ and $\varepsilon = \varepsilon(x)$ are symmetric 3×3 matrices of measurable functions on Ω, satisfying for almost all $x \in \Omega$ and for all $\xi \in \mathbf{R}^3$ the conditions

$$\sigma_1|\xi|^2 \le (\sigma(x)\xi,\xi) \le \sigma_2|\xi|^2,\ \mu_1|\xi|^2 \le (\mu(x)\xi,\xi) \le \mu_2|\xi|^2,\ \varepsilon_1|\xi|^2 \le (\varepsilon(x)\xi,\xi) \le \varepsilon_2|\xi|^2,$$

where ε_i, μ_i, σ_i, $(i = 1,2)$ are given positive numbers, $\boldsymbol{J}^{ext} : Q \to \mathbf{R}^3$, $\boldsymbol{h} : \Omega \to \mathbf{R}^3$ are square integrable functions.

The generalized solutions of the problems are considered, that is all equalities have to be satisfied in the sense of distributions on $(0,T)$ with the values in the corresponding Hilbert spaces [14], and boundary conditions have to be satisfied in the sense of the trace theory [15].

The following Hilbert spaces with the respective scalar products are defined [15]:

$$H(\operatorname{div};\Omega) = \{\boldsymbol{u} \in \{L_2(\Omega)\}^3 : \operatorname{div}\boldsymbol{u} \in L_2(\Omega)\},\ K(\operatorname{div};\Omega) = \{\boldsymbol{u} \in \{L_2(\Omega)\}^3 : \operatorname{div}\boldsymbol{u} = 0\},$$

$$(\boldsymbol{u},\boldsymbol{v})_{div,\Omega} = \int_\Omega (\boldsymbol{u}\cdot\boldsymbol{v})\mathrm{d}x + \int_\Omega \operatorname{div}\boldsymbol{u}\operatorname{div}\boldsymbol{v}\mathrm{d}x,$$

$$H(\operatorname{curl};\Omega) = \{\boldsymbol{u} \in \{L_2(\Omega)\}^3 : \operatorname{curl}\boldsymbol{u} \in \{L_2(\Omega)\}^3\},\ K(\operatorname{curl};\Omega) = \{\boldsymbol{u} \in \{L_2(\Omega)\}^3 : \operatorname{curl}\boldsymbol{u} = 0\},$$

$$(\boldsymbol{u},\boldsymbol{v})_{curl,\Omega} = \int_\Omega (\boldsymbol{u}\cdot\boldsymbol{v})\mathrm{d}x + \int_\Omega (\operatorname{curl}\boldsymbol{u}\cdot\operatorname{curl}\boldsymbol{v})\mathrm{d}x.$$

$H_0(\operatorname{div};\Omega)$, $H_0(\operatorname{curl};\Omega)$ denote the closures of the set of test vector-functions in $H(\operatorname{div};\Omega)$ and $H(\operatorname{curl};\Omega)$ respectively, $K_0(\operatorname{div};\Omega) = K(\operatorname{div};\Omega) \cap H_0(\operatorname{div};\Omega)$.

The solution of the problem (1)-(7) is the set of functions $\boldsymbol{H} \in L_2(0,T,H_0(\mathrm{curl};\Omega))$, $\boldsymbol{B} \in L_2(0,T,K(\mathrm{div};\Omega))$, $\boldsymbol{J} \in L_2(0,T,K_0(\mathrm{div};\Omega))$, $\boldsymbol{E} \in L_2\left(0,T,\{L_2(\Omega)\}^3\right)$, $\boldsymbol{D} \in L_2\left(0,T,\{L_2(\Omega)\}^3\right)$, $\rho \in L_2\left(0,T,H^{-1}(\Omega)\right)$, satisfying (1), (3)–(5), (7).

The equality (3) for $\boldsymbol{B} \in L_2(0,T,K(\mathrm{div};\Omega))$, $\boldsymbol{E} \in L_2\left(0,T,\{L_2(\Omega)\}^3\right)$ implies that $\boldsymbol{H} = \mu^{-1}\boldsymbol{B} \in C\left(0,T,\{L_2(\Omega)\}^3\right)$ [14], that is the initial condition (7) makes sense.

3. Problem for vector potential

Relations (2), (3) allow to introduce the vector magnetic potential $\boldsymbol{A}$ and the scalar electric potential φ as new unknown variables by formula [13]

$$\boldsymbol{B} = \mathrm{curl}\,\boldsymbol{A},\ \boldsymbol{E} = -\mathrm{grad}\,\varphi - \frac{\partial}{\partial t}\boldsymbol{A}. \tag{9}$$

In this case, the system (1)-(5) is reduced to one equation

$$\frac{\partial}{\partial t}\sigma\boldsymbol{A} + \mathrm{curl}\,\mu^{-1}\,\mathrm{curl}\,\boldsymbol{A} = -\sigma\,\mathrm{grad}\,\varphi + \boldsymbol{J}^{ext}. \tag{10}$$

Equation (10) is provided by the boundary condition corresponding to (6)

$$\left(\mu^{-1}\,\mathrm{curl}\,\boldsymbol{A}\right)_\tau(x,t) = 0,\ x \in \partial\Omega,\ t \in (0,T), \tag{11}$$

and the initial condition

$$\boldsymbol{A}(x,0) = \boldsymbol{a}(x),\ x \in \Omega, \tag{12}$$

where $\boldsymbol{a} \in \{L_2(\Omega)\}^3$ – given function such that $\mathrm{curl}\,\boldsymbol{a} = \mu\boldsymbol{h}$.

The solution of the problem (10)–(12) is the functions $\boldsymbol{A} \in L_2(0,T,H(\mathrm{curl};\Omega))$, $\varphi \in L_2\left(0,T,H^1(\Omega)\right)$, satisfying (10), (12) in the sense of the distribution on $(0,T)$ with the values in the Hilbert spaces and (11) in the sense of the trace theory, that is $\mu^{-1}\,\mathrm{curl}\,\boldsymbol{A}(t) \in H_0(\mathrm{curl};\Omega)$ for almost all $t \in (0,T)$.

Suppose that $\boldsymbol{A}$, φ is the solution of (10) –(12) and $\boldsymbol{A} \in C^1(0,T,H(\mathrm{curl};\Omega))$. Then from (10) we obtain for all $\boldsymbol{v} \in H(\mathrm{curl};\Omega)$

$$\frac{\mathrm{d}}{\mathrm{d}t}\int_\Omega(\sigma\boldsymbol{A}\cdot\boldsymbol{v})\mathrm{d}x + \int_\Omega\left(\mu^{-1}\,\mathrm{curl}\,\boldsymbol{A}\cdot\mathrm{curl}\,\boldsymbol{v}\right)\mathrm{d}x = -\int_\Omega(\sigma\,\mathrm{grad}\,\varphi\cdot\boldsymbol{v})\mathrm{d}x + \int_\Omega\left(\boldsymbol{J}^{ext}\cdot\boldsymbol{v}\right)\mathrm{d}x. \tag{13}$$

The solution of the problem (10) –(12) is obviously not unique. The following gauge is discusses:

$$\varphi(x,t) = -\kappa\,\mathrm{div}\,\sigma\boldsymbol{A}(x,t),\ (x,t) \in Q,\ (\sigma\boldsymbol{A})_\nu(x,t) = 0,\ x \in \partial\Omega,\ t \in (0,T), \tag{14}$$

where $\kappa > 0$ is arbitrary constant.

We introduce the scalar product for $\{L_2(\Omega)\}^3$ as $(\boldsymbol{u},\boldsymbol{v})_\sigma = \int_\Omega(\sigma\boldsymbol{u}\cdot\boldsymbol{v})\mathrm{d}x$. The norm, generated by this scalar product, is equivalent to the usually norm of $\{L_2(\Omega)\}^3$, the resulting Hilbert space is denoted by $\{L_2(\sigma;\Omega)\}^3$. The following Gilbert space is also defined:

$$W(\sigma,\Omega) = \{\boldsymbol{u} \in H(\mathrm{curl};\Omega) : \sigma\boldsymbol{u} \in H_0(\mathrm{div};\Omega)\},$$

$$(\boldsymbol{u},\boldsymbol{v})_W=\int_\Omega(\boldsymbol{u}\cdot\boldsymbol{v})\mathrm{d}x+\int_\Omega(\operatorname{curl}\boldsymbol{u}\cdot\operatorname{curl}\boldsymbol{v})\mathrm{d}x+\int_\Omega(\operatorname{div}\sigma\boldsymbol{u}\cdot\operatorname{div}\sigma\boldsymbol{v})\mathrm{d}x.$$

The space $W(\sigma;\Omega)$ is dense and continuous injected in $\{L_2(\sigma;\Omega)\}^3$, the space $\left(\{L_2(\sigma;\Omega)\}^3\right)^*$ can be identified with dense subspace of $W^*(\sigma;\Omega)$. By identifying $\{L_2(\sigma;\Omega)\}^3$ and $\left(\{L_2(\sigma;\Omega)\}^3\right)^*$, we arrive at the inclusions

$$W(\sigma;\Omega)\subset\{L_2(\sigma;\Omega)\}^3\equiv\left(\{L_2(\sigma;\Omega)\}^3\right)^*\subset W^*(\sigma;\Omega).$$

The scalar product in $\{L_2(\sigma;\Omega)\}^3$ of $\boldsymbol{u}\in\{L_2(\sigma;\Omega)\}^3$ and $\boldsymbol{v}\in W(\sigma;\Omega)$ is the same as the value of functional $\boldsymbol{u}$ at the element $\boldsymbol{v}$ in the sense of the duality between $W^*(\sigma;\Omega)$ and $W(\sigma;\Omega)$.

Using (13) we obtain that the problem (10)–(12), (14) is reduced to the following problem: to find a function $\boldsymbol{A}\in L_2(0,T,W(\sigma,\Omega))$, satisfying (12), such that for all $\boldsymbol{v}\in W(\sigma,\Omega)$

$$\frac{\mathrm{d}}{\mathrm{d}t}\int_\Omega(\sigma\boldsymbol{A}\cdot\boldsymbol{v})\mathrm{d}x+\int_\Omega(\mu^{-1}\operatorname{curl}\boldsymbol{A}\cdot\operatorname{curl}\boldsymbol{v})\mathrm{d}x+\kappa\int_\Omega(\operatorname{div}\sigma\boldsymbol{A}\cdot\operatorname{div}\sigma\boldsymbol{v})\mathrm{d}x=\int_\Omega(\boldsymbol{J}^{ext}\cdot\boldsymbol{v})\mathrm{d}x. \tag{15}$$

Theorem 1. *For any* $\boldsymbol{a}\in\{L_2(\Omega)\}^3$, $\boldsymbol{J}^{ext}\in\{L_2(Q)\}^3$ *there exists a unique solution* $\boldsymbol{A}\in L_2(0,T,W(\sigma,\Omega))$ *of the problem (12,) (15). Moreover,* $\boldsymbol{A}\in C\left(0,T,\{L_2(\sigma;\Omega)\}^3\right)$ *and*

$$\sup_{t\in(0,T)}\|A(t)\|_\sigma+\|\boldsymbol{A}\|_{L_2(0,T,W)}\le C_T\left(\left\|\boldsymbol{J}^{ext}\right\|_{2,Q}+\|\boldsymbol{a}\|_\sigma\right).$$

If $\boldsymbol{a}\in W(\sigma;\Omega)$*, then* $\boldsymbol{A}\in L_2\left(0,T,\{L_2(\sigma;\Omega)\}^3\right)$,

$$\|\boldsymbol{A}'\|_{L_2\left(0,T,\{L_2(\sigma;\Omega)\}^3\right)}\le C_{T,1}\left(\left\|\boldsymbol{J}^{ext}\right\|_{2,Q}+\|\boldsymbol{a}\|_\sigma\right).$$

The constants $C_T>0$, $C_{T,1}>0$ *not depend on* $\boldsymbol{a}$, $\boldsymbol{J}^{ext}$.

Let the operator $G:W(\sigma;\Omega)\to W^*(\sigma;\Omega)$ associates with $\boldsymbol{u}\in W(\sigma;\Omega)$ the element of $W^*(\sigma;\Omega)$ such that for all $\boldsymbol{v}\in W(\sigma;\Omega)$

$$\langle G\boldsymbol{u},\boldsymbol{v}\rangle=\int_\Omega(\mu^{-1}\operatorname{curl}\cdot\operatorname{curl}\boldsymbol{v})\mathrm{d}x+\kappa\int_\Omega(\operatorname{div}\sigma\cdot\operatorname{div}\sigma\boldsymbol{v})\mathrm{d}x.$$

Then $G\boldsymbol{u}\in L_2\left(0,T,W^*(\sigma;\Omega)\right)$ for $\boldsymbol{u}\in L_2(0,T,W(\sigma;\Omega))$. If $\boldsymbol{A}\in L_2(0,T,W(\sigma;\Omega))$ is a solution of the problem (12), (15), then (15) for all $\boldsymbol{v}\in L_2(0,T,W(\sigma;\Omega))$ can be written as

$$\frac{\mathrm{d}}{\mathrm{d}t}\langle\boldsymbol{A},\boldsymbol{v}\rangle+\langle G\boldsymbol{A},\boldsymbol{v}\rangle=\langle\sigma^{-1}\boldsymbol{J}^{ext},\boldsymbol{v}\rangle.$$

Hence $\boldsymbol{A}'\in L_2\left(0,T,W^*(\sigma;\Omega)\right)$ and, therefore, $\boldsymbol{A}\in C\left(0,T,\{L_2(\sigma;\Omega)\}^3\right)$ [14].

The proof of the existence of the solution of the problem is based on the Faedo - Galerkin method [16], the feasibility of which follows from the following inequalities which generalize the obtained in [17] estimates for scalar products of vector fields in a star-shape domain.

Theorem 2. *Let* $\Omega\subset\mathbf{R}^3$ *is open bounded Lipchitz domain, homeomorphic to a ball. There exists a positive constant* $C(\Omega)$*, which depends only on* Ω*, that the inequality*

$$\left|(\boldsymbol{u},\boldsymbol{v})_{2,\Omega}\right| \le C(\Omega)\left(\|\boldsymbol{u}\|_{2,\Omega}\|\operatorname{div}\boldsymbol{v}\|_{2,\Omega} + \|\operatorname{curl}\boldsymbol{u}\|_{2,\Omega}\|\boldsymbol{v}\|_{2,\Omega} + \|\operatorname{curl}\boldsymbol{u}\|_{2,\Omega}\|\operatorname{div}\boldsymbol{v}\|_{2,\Omega}\right)$$

is true for any $\boldsymbol{u}\in H_0(\mathrm{curl};\Omega)$, $\boldsymbol{v}\in H(\mathrm{div};\Omega)$ *and for any* $\boldsymbol{u}\in H(\mathrm{curl};\Omega)$, $\boldsymbol{v}\in H_0(\mathrm{div};\Omega)$.

We will establish the equivalence the problem in terms of potentials and origin initial-boundary value problem. The following statements are valid.

Theorem 3. *Let* $\boldsymbol{A}\in L_2(0,T,W(\sigma;\Omega))$ *is the solution of the problem (12), (15), where* $\boldsymbol{a}\in W(\sigma;\Omega)$, *and* $\varphi=-\kappa\operatorname{div}\sigma\boldsymbol{A}$. *Then* $\mu^{-1}\operatorname{curl}\boldsymbol{A}\in L_2(0,T,H_0(\mathrm{curl};\Omega))$, $\varphi\in L_2(0,T,H^1(\Omega))$ *and (14) is fulfilled, that is* $\boldsymbol{A}$, φ *is the solution of the problem (10) –(12), (14).*

From theorem 3 the next theorem follows.

Theorem 6. Let $\boldsymbol{a}\in W(\sigma;\Omega)$, $\mu^{-1}\operatorname{curl}\boldsymbol{a}=\boldsymbol{h}$. *Then the problem (1)-(7) has unique solutions. Moreover the relations (9) are valid, where* $\boldsymbol{A}$, φ *is the solution of the problem (10 –(12), (14).*

4. Numerical scheme

Rare mesh scheme [18] proved the high quality of the numerical solution for 3D problems of elasticity and plasticity theory [19-21]. Scheme constructing technology is exactly the same as described in [22]. The scheme is based on the traditional scheme of the linear 4-node finite element in the form of a tetrahedron with a linear approximation of functions in the element. Integration in the element uses one point of integration that corresponds to well known reduced integration technique by O. Zienkiewich [23]. Finite element mesh is composed of a hexahedron, and the elements are located in the center of each hexahedron so that the edges of the tetrahedron are the hexahedron faces diagonals. As a result, each base hexahedron contains only one calculation element. Ullage hexahedron and it's contains extensive parameters are appended to the central tetrahedral element. To solve the problems of the quasi-stationary magnetic fields definition rare mesh scheme is applied to approximation of variational equality (15). For the time integration of the system used Crank-Nicholson numerical scheme. The explicit form of the numerical scheme for the stationary problem in the case of the uniform mesh is given in [12].

Acknowledgments

The research was supported by the Ministry of Education and Science of the Russian Federation in the framework of contract no. 2014/134 for state work in the field of scientific activity in the framework of the basic part of the state contract (project no. 2664) and in the framework of the project part of the state contract (project no. 1727) for 2014–2016 and was supported by a grant in the framework of contract no. 02.V.49.21.0003 (August 27, 2013) between the Ministry of Education and Science of the Russian Federation and Nizhnii Novgorod State University.

References

[1] Alonso Rodriguez A and Valli A 2010 *Eddy current approximation of Maxwell equations* (Milan: Springer-Verlag Italia)
[2] Galanin M P and Popov Y P 1995 *Kvazistationarnye elektromagnitnye polya v neodnorodnykh sredah* (Moscow: Nauka)
[3] Kalinin A V, Sumin M I and Tyukhtina A A 2016 *Differential Equations* **52** (5) 1–17
[4] Coulomb J-L and Sabonnadiere J-C 1985 *CAO en electrotechnique* (Paris: Hermes Publ.)
[5] Fernandes P 1995 *IEE Proc.-Sci. Meas. Technol.* **142** 299–306
[6] Jackson J D 2002 Am.J.Phys **70** (9) 917–928.
[7] Biro O and Valli A 2007 *Comput. Meth. Appl. Mech. Engrg.* 2007 **196** 890–1904
[8] Fernandes P and Valli A 2008 *Math. Meth. Appl. Sci.* **31** 71–98
[9] Bossavit A 1999 *COMPEL* **18** (3) 323–336
[10] Kang T and Kim K I 2009 *Computing* **85** 339–362
[11] Kalinin A V and Kalinkina A A 2003 *Vestnik Nizhegorod. Gos. Univ. Mat. Model. Optim.*

Upravl. **26** (1) 21–38

[12] Kalinin A V and Chekmarev D T 2014 *Mesh Methods for Boundary-Value Problems and Applications. Proc. 10th Int. Conf.* (Kazan: Kazan University) pp 356–360

[13] Landau L D and Lifshitz E M 1984 Electrodynamics of Continuous media *Course of Theoretical Phisics* vol 8 (Oxford: Pergamon press)

[14] Gajewskii H, Gröger K and Zacharias K 1974 *Nichtlineare Operatorgleichungen und Operatordifferetialgleichungen* (Berlin: Akademik-Verlag)

[15] Girault V and Raviart P 1986 *Finite element methods for Navier--Stokes equations* (N.Y.: Springler-Verlag)

[16] Temam R 1977 *Navier-Stokes Equations. Theory and Numerical Analysis* (Amsterdam: North-Holland)

[17] Kalinin A V and Kalinkina A A 2004 *Russian Mathematics (Izvestiya VUZ. Matematika)* **48 (3)** 23–31

[18] Chekmarev D T 2009 Finite element schemes on rare meshes *Probl. At. Sci. Techn Ser. Math. Model. phys. proc.* **2** 49–54

[19] Zhidkov A V, Zefirov S V, Kastalkaya K A, Spirin S V and Chekmarev D T 2011 Rare mesh scheme for numerical solution of three-dimensional dynamic problems of elasticity and plasticity *Bull. Nizhny Novgorod Univ.* **4**(4) 1480–1485

[20] Chekmarev D T 2015 Some Results of FEM Schemes Analysis by Finite Difference Method *Finite Difference Methods, Theory and Applications. Volume 9045 of the series Lecture Notes in Computer Science* 153-160

[21] Spirin S V, Chekmarev D T and Zhidkov A V 2015 Solving the 3D Elasticity Problems by Rare Mesh FEM Scheme *Finite Difference Methods, Theory and Applications. Volume 9045 of the series Lecture Notes in Computer Science* 379-384

[22] Bazhenov V G and Chekmarev D T 2000 *Solving the problems of plates and shells dynamics by variational- difference method* (Nizhny Novgorod) 118

[23] Zienkiewich O C, Too J and Taylor R L 1971 Reduced integration technique in general analysis of plates and shells *Int.J. Num.Meth.Engng.* **3 (2)** 275-290

IOP Conf. Series: Materials Science and Engineering **158** (2016) 012026 doi:10.1088/1757-899X/158/1/012026

Simulation of a liquid drop on a vibrating hydrophobic surface

A A Chernova, S P Kopysov and L E Tonkov

Institute of Mechanics UB RAS, T. Baramzinoy st., 34, 426067, Izhevsk, Russia

E-mail: alicaaa@gmail.com

Abstract. Mathematical simulation is used to study processes describing a liquid droplet oscillation on a solid surface. The pattern of generated internal flows is characterized by complex interaction between capillary and gravity waves, free surface and contact angle. Interaction process factors are analysis. The given results are compared with the experimental ones.

Introduction

Problem of a solid heaving with Faraday wave's generation in a liquid is an example of complex dynamic interaction in solid – liquid system [1]. When liquid layer with free surface oscillates in heave we can observe surface capillary waves – Faraday wave's. Its dynamics depends on the system control values such as: viscosity, surface tension and liquid density as well as external action characteristics. Such waves are initiated after they reach some threshold amplitude of oscillation.

Let's consider preliminary stage of the phenomenon with small amplitudes that endure liquid heave oscillations on its surface to initiate capillary – gravity waves. The present thesis is concerned with a research on internal flows in a small amount of liquid - a droplet and its influence on generation and initiation threshold of waves.

Most articles in experimental hydrodynamics on studying a parametrical resonance in a liquid research have the same model in common, i.e. as a rule an oscillating vessel is considered. [2, 3]. Frequency range of gravity waves initiation on a free surface in a liquid droplet is experimentally studied in scientific works [4, 5, 6, 7]. Thus, the work [4], is concerned with problems of detection of various oscillation modes for free surface of a droplet placed on a vibrating water proof surface with small hysteresis of a contact angle, as well as with contact angle and hysteresis impact on liquid oscillation modes research. Works [5, 6, 7] study oscillations of a liquid of extra small volume (5 μl) laying on a solid surface vibrating at a low frequencies (less than 800Hz) and with small amplitudes (up to 22 μm). Data on droplet free surface depending on the vibrating surface location are provided, instant droplet profiles and internal capillary flows obtained using PIV method are also considered. Moreover, articles [5, 6, 7] represent obtained droplet shapes corresponding to the various surface vibration modes.

There are only few studies [3, 8, 9, 10] that are dedicated to the solid surface vibration with a liquid of various volume simulation problems. At the same time problems of mathematical simulation of a single droplet placed on vertically vibrating at low frequencies and small amplitudes (solid) surface are concerned only in article [10] that study oscillation and spraying of a single liquid droplet of 30μl on a solid rod. In this study the physical thesis statement has several rather strong assumptions: the initial droplet configuration was determined as half sphere, in addition to this the possible movement of a contact line was not considered and the state on the border of a solid surface was assumed as sticking; in hydrodynamic equations cycling acceleration of a body is added to the gravity acceleration (system parameter) considered in equation factors or boundary conditions. Navier-Stokes equations in axially symmetrical statement are solved by marker and cell projection method. The equation system is approximated by space with the finite volume method on a structured grid. Flow reconstruction on the cell edges is performed using central difference schemes and the convective terms sampling was performed using hybrid Nichols difference scheme. The solution of given algebraic systems was achieved under Cholesky conjugate gradient scheme.

Experimental researches prove that internal flows in a droplet are characterized by a clear three-dimensionality, and wetting angle is one of the key parameter in a system. Solution methods of such tasks concerning free liquid surface require high resolution of thin layers (gas-liquid-solid contact point) and developed free surfaces implementation.

The given study is set in a following way. First chapter provides the problem description and general assumptions. Second chapter observes the features of calculating the free surface using the control volume method and VOF. Third chapter provides both results of selected schemes and algorithms tests with the task [6] having experimental description as an example, and analysis of obtained instant capillary flows patterns in an oscillating droplet, it also concerns typical topological features of flows.

1. Problem statement

We consider the problem of liquid droplet motion caused by vertical movements of a solid surface. Let the area $\Omega \in \mathrm{R}^3$(fig. 1) be filled with biphasic medium and $\Omega = \Omega_1 \bigcup \Omega_2$. We consider subdomain Ω_2 being a gas, and Ω_1 — being a liquid; Γ_0 is a phase boundary, Γ_1 — atmospheric boundary of computational region. At the initial instant $t = 0$ the surface starts to vibrate at a given frequency and amplitude. A liquid drop placed on the surface also starts to make vertical oscillations.

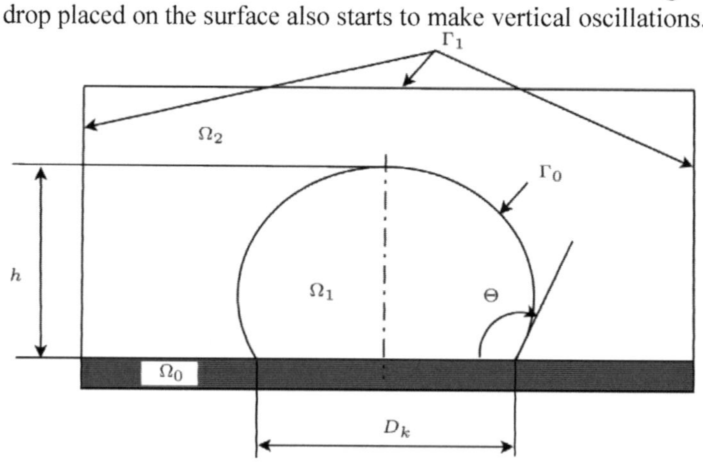

Figure 1. Computational region: droplet crosses section initial profile.

The experimental apparatus is described in article [6], the surface was carefully cleaned and covered by hydrophobic coating. De-ionised water droplet of 5μl volume was placed in a center of a surface (on a center line of a rigid surface, fig. 1) using a dropper, thus the wetting angles were equal to $\Theta = 115 \pm 1$, and the contact diameter and droplet height — 2,02 mm and 1.52 mm, respectively. Needless to say, vertical oscillations of a solid surface obtained in the study [6] are not harmonic but due to insignificance of the differences they were considered as sinusoidal oscillations under the numerical experiment.

2. Methods and algorithms

The given problem can be considered as system of two immiscible incompressible viscous fluids the motion of each is described by Navier-Stoks equations and continuity equation

$$\frac{\partial(\rho\mathbf{U})_i}{\partial t}+\nabla\cdot(\rho\mathbf{U}\mathbf{U})_i=-\nabla p_i+\nabla\cdot\boldsymbol{\tau}_i+\rho_i\mathbf{g}, \qquad (1)$$

$$\nabla\cdot\rho_i\mathbf{U}_i=0, \qquad (2)$$

where $\mathbf{e}$ — unit normal, external to the Ω_0, K — surface curvature Γ_0, σ — surface tension coefficient, Θ - wetting angle. In this study the wetting angle value will be determined basing on empirical model of dynamic contact angle [11]

$$\Theta=\Theta_0+(\Theta_A-\Theta_R)tanh\left(\frac{u_W}{u_\Theta}\right),$$

where Θ_0 — static contact angle, Θ_A — maximum permitted value of the contact angle (for the given problem $\Theta_A=116$), Θ_R — maximum value of the contact angle (for the given problem $\Theta_A=114$), u_W — solid interface velocity, u_Θ — a constant, well-behaved as scale of contact angle change rate, for the given problem $u_\Theta=1$, while the problems of impact of dynamic angle changing model on simulation results are not considered.

Geometrical characteristics of phase interface Γ_0 are unambiguously determined as

$$\mathbf{e}=\frac{\nabla\alpha}{|\nabla\alpha|},\quad K=\nabla\frac{\nabla\alpha}{|\nabla\alpha|},$$

where α – is scalar function.

$$\alpha(t,\mathbf{x})=\begin{cases}1, & \mathbf{x}\in\Omega_1,\\ 0, & \mathbf{x}\in\Omega_2\cup\Gamma_0,\end{cases}$$

in this case having the meaning of volume concentration of a liquid. By doing so we determine density, pressure and viscosity functions as $\rho=\rho_2+(\rho_1-\rho_2)\alpha$, $p=p_2+(p_1-p_2)\alpha$ and $\mu=\mu_2+(\mu_1-\mu_2)\alpha$ respectively, and rewrite the system (1), (2) considering (4), (3) as following

$$\frac{\partial\rho\mathbf{U}}{\partial t}+\nabla\cdot(\rho\mathbf{U}\mathbf{U})=-\nabla p+\nabla\cdot\boldsymbol{\tau}-\sigma K\nabla\alpha\delta(\alpha)(1-cos\Theta)+\rho\mathbf{g}, \qquad (5)$$

$$\nabla\cdot\mathbf{U}=0. \qquad (6)$$

System (5), (6) is added by transfer equation for α

$$\frac{\partial\alpha}{\partial t}=\mathbf{U}\cdot\nabla\alpha. \qquad (7)$$

We pass on to the modified pressure

$$p_d=p-\rho gh,\quad \nabla p_d=\nabla p-\rho g, \qquad (8)$$

Then the equation (5) is written as

$$\frac{\partial\rho\mathbf{U}}{\partial t}+\nabla\cdot(\rho\mathbf{U}\mathbf{U})=-\nabla p_d+\nabla\cdot\boldsymbol{\tau}-\sigma K\nabla\alpha\delta(\alpha)(1-cos\Theta). \qquad (9)$$

Surface tension force acting in a thin transient layer, that in the limit is thin infinite and corresponded to the Γ_0, can be converted to volume one [12]

$$\mathbf{G}=-\sigma K\nabla\alpha\delta(\alpha)(1-cos\Theta).$$

The system (6), (9), (7) was numerically solved using control volume method for the sampling of the original equations. In order to simplify the further stage we pass in the conventional manner on to the dimensionless variables, in this case only written form of the equation (9) will be changed

$$\frac{\partial u}{\partial t}+\nabla\cdot uu=-\nabla p+\nabla\cdot\frac{1}{\rho}\nabla u+\mathbf{G}. \tag{10}$$

Lets write an approximation of the equations (6), (10) in semi-discrete form

$$\sum_{f\in N_M} S_f\cdot u_f^{n+1}=0, \tag{11}$$

$$\frac{3u_M^{n+1}-4u_M^n+u_M^{n-1}}{2\Delta t}V_M+\sum_{f\in N_M}\mathbf{F}_f u_f^{n+1}-\sum_{f\in N_M}\frac{1}{\rho}S_f\cdot(\nabla u^{n+1})_f=-(\nabla p)_M V_M, \tag{12}$$

where u_M^n — velocity in the centre of a cell numbered M on n time interval; V_M — cell volume; $\mathbf{F}_f$ — flow through the edge with number f, N_M — set of neighboring cells numbers; S_f — outward normal vector to the edge with number f, in absolute value equal to the area of this edge.

Numerical solution of incompressible fluid dynamics equations of type (9), (6) require application of special methods to obtain velocity and pressure fields meeting the conditions of conservation and continuity by time on each step. In this study pressure implicit procedure with splitting of operator offered by Issa [13] called PISO is used.

Discretization of convective terms by area is performed using second order Van Leer's scheme. Reconstruction of flows on the edges of cells is performed using TVD scheme where approximation is based on central differences, and Van Leer's limiter [14] is applied as required limiter function. The obtained algebraic equation system is solved numerically using conjugate gradient method with incomplete Cholesky diagonal preconditioner.

The equation (7) is solved on each time step after system (9), (6) solution. In order to do so it is more convenient to rewrite (7) in conservative form

$$\frac{\partial\alpha}{\partial t}+\nabla\cdot(u\alpha)+\nabla\cdot\big(\alpha(1-\alpha)u_r\big)=0, \tag{13}$$

where $u_r=\mathbf{u}_2-\mathbf{u}_1$ — is velocity of the relative phase motion on the interface Γ_0. In initial model Γ_0 is infinitely thin and $u_r=0$, however due to the numerical diffusion in calculations transition zone inevitably appears. Following the method of control volume we write an approximation (13)

$$\frac{3\alpha_M^{n+1}-4\alpha_M^n+\alpha_M^{n-1}}{2\Delta t}+\sum_{f\in N_M}\mathbf{F}_f(\alpha_M^{n+1})+\sum_{f\in N_M}(1-\alpha)_M^n\mathbf{F}_f^*(\alpha_M^n)=0, \tag{14}$$

where

$$\mathbf{F}_f^*(\alpha)=c_\alpha\max_{\Omega}\frac{|\mathbf{e}\mathbf{u}_r\mathbf{S}_f|}{|\mathbf{S}_f|}\mathbf{e},$$

When solving the equation (14) a convective term approximation is performed is performed to that described above. And scalar coefficient c_α serves to control the artificial compression in the area of dispersion to compensate the numerical diffusion effect, in this study the coefficient $c_\alpha=1$.

3. Numerical simulation

This study investigates low-frequency (≈ 85 Hz) oscillation of a water drop 5 μl in volume on a rigid surface, that is experimentally described [6], including the analysis of droplet profile and data on time-to time variation of free surface height change.

Mathematical simulation of liquid motion under oscillations was performed using volume-of-fluid method in a cell in two stages: first – the solution of liquid droplet stabilization on a horizontal rigid surface problem; second – the vibration of the surface with a stabilized droplet on it. Droplet was set into equilibrium considering contact angle change, stabilization of the droplet from a cylindrical column of liquid of given volume under gravitation and surface tension was investigated. The droplet configuration was compared to the experimental by profile height, 3 phase contact diameter and wetting angle real value.

A Low-frequency oscillation of a liquid drop on a rigid surface simulation was performed on a grid consisting in 364140 hexagons. Let's concider methods of accounting of the contact angle in a 3 phase gas-liquid-surface contact line. Wetting angle Θ is determined not only by physical and chemical properties of a liquid but by surface material properties and is not constant even in case of liquid and surface stability. As the dynamic component of the contact angle during surface oscillations is not zero it is possible to estimate its effect both on oscillating droplet profile and on droplet local height. Fig. 2 shows diagram of the droplet height change in time resulting from vibration process simulation considering the dynamic change of wetting angle (curve 1, fig. 2) and using only static contact angle model (curve 2, fig. 2).

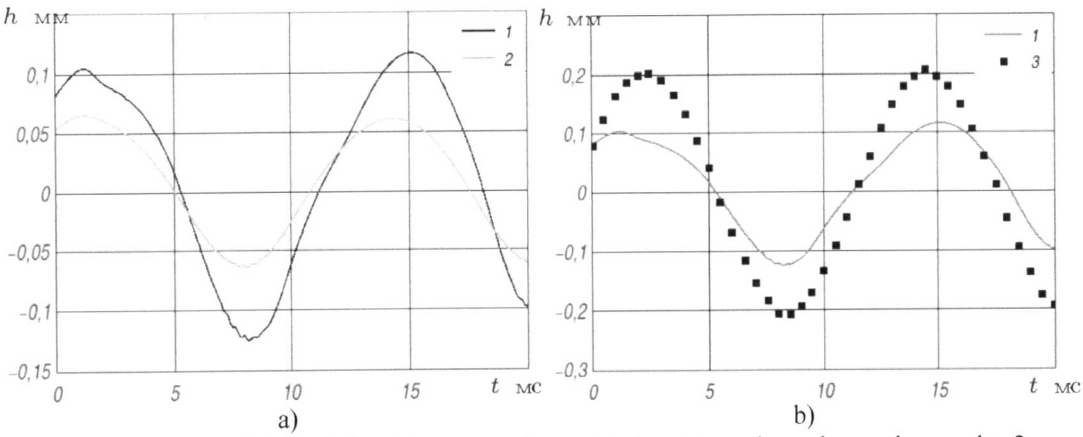

Figure 2. Droplet height variation diagrams: a) concidering: 1 — dynamic wetting angle; 2 — static contact angle b) for mode 2: 1 — experiment [6], 2 — calculation.

Given curves have similar quantities and qualitative values but the amplitude difference on peaks reach up to 15%. Thus, static wetting angle consideration alone leads to droplet height amplitude lowering. Needless to say, oscillation frequency increase leads to contact angle determination method impact increase. Free droplet surface oscillation forms on a vibrating at frequency of 85 Hz surface (fig.3) were determined, i.e. one circle nodal line on a free surface corresponding to the oscillation second mode [6] was discovered.

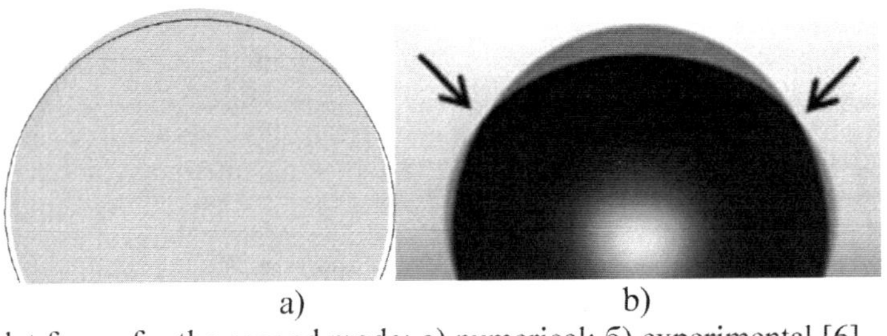

Figure 3. Droplet forms for the second mode: a) numerical; б) experimental [6].

As shown on fig. 3, the simulation with set up parameters provides oscillation modes equal to the experimental ones, but local values of calculated phase interface motions (fig. 3 a) are less that experimental ones (fig. 3 b) that results from discovered differences of equilibrium profiles and can be solved using additional algorithms of free surface reconstruction. Let's consider diagrams of droplet height change on a vibrating surface (fig. 2 b). Time-dependencies of droplet height change obtained during simulation match in qualitative values with experimental values. However, droplet oscillation amplitude values are significantly (up to 41%) less than experimental values due to small qualities of free surface motion (fig. 3).

To study the internal flow peculiarities we generated instant patterns of droplet velocity vectors at various moments of time (fig. 4).

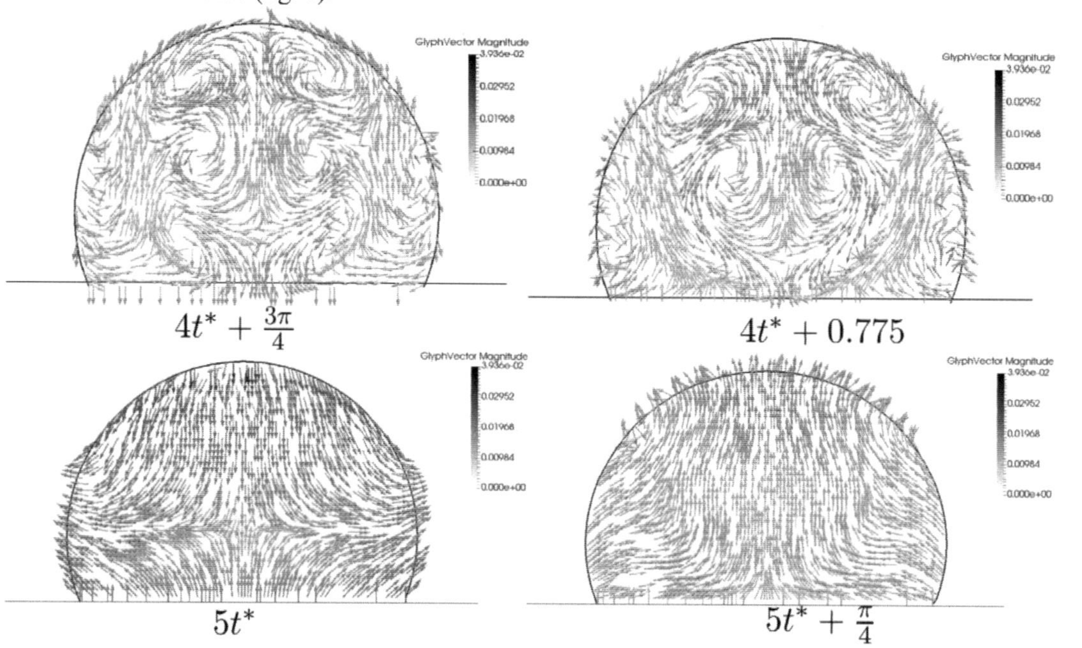

Figure 4. Droplet internal flows, $t^* = t/T$ [6].

As shown on fig. 4 the flows in a droplet are mainly determined by surface motion direction. Thus at initial moment of surface lifting on a 4th period ($4t^* + \frac{3\pi}{4}$ - lowest position of the surface and $4t^* + 0.775$ - beginning of a surface lifting) we observe velocity vectors relocation and flow direction change. Paired symmetrical whirls are clearly visualized. In this case, when at the initial moment of flowerets surface location 5 pairs of symmetrical whirl structures (three on various heights close to the droplet center line and two paired whirls at the periphery close to a free surface) and one special point of saddle type are detected the following surface lifting ($4t^* + 0.775$) leads to integration of lower whirls into large symmetrical paired whirl with two clearly defined drip lines, and integration of previously detected side whirl structures is observed. Thus internal liquid flows have delayed response to the surface motion direction change and start to change their direction from lower layers directly interacting with the surface. Interaction of differently directed internal flows results in circulation zones formation and interaction of turning flows results in drip lines and saddle point formation. Furthermore, at the moment of time related to the fifth period ($5t^*$) beginning and highest ($5t^* + \frac{\pi}{4}$) position of the surface irrotational flows are observed in the drop, and at the moment of time related to the fifth oscillation period beginning, we detect final relocation of velocity vectors, two drip lines directed along with the spreading pressure.

4. Conclusion

The studied problem of liquid droplet oscillation on a solid surface under heave low-frequency oscillations is one of the few studies that have experimental description that allows numerical schemes and algorithms testing. Comparison of the numerical simulation results and experimental data shows that applied VOF method provides appropriate description of surface-to-droplet energy transfer processes. In addition, detected significant differences of calculated drop oscillation amplitude values from experimental ones indicate a need for more detailed account of both surface forces and dynamic contact angle range. Thus algorithms for surface forces and dynamic contact angle change determination in air-liquid-surface triple point require additional improvement and testing.

Acknowledgments

This work was partially supported by RFBR according to the research project №16-38-00127-mol, №16-41-180276-r_a.

References

[1] Sorokin V I 1957 *Acoustic Journal* **3(3)** 262
[2] Kalinichenko V A, Nesterov S V and So A N 2015 *Fluid Dynamics* **50(4)** 535
[3] Ehrhorn J and Semke W 2013 *Conf. Proc. Society for Experimental Mechanics Series* **5** 243
[4] Noblin X, Buguin A and Brochard-Wyart F 2004 *Eur. Phys. Journal* **14(4)** 395
[5] Shin Y S and Lim H Ch 2014 *Eur. Phys. Journal* **37 (74)** 1
[6] Kim H and Lim H Ch 2015 *Journal Phys. Chem.* **119** 6740
[7] Park Ch S, Kim H and Lim H Ch 2016 *Experimental Thermal and Fluid Science* **78** 112
[8] Li Y and Umemura A 2014 *International Journal of Multiphase Flow* **60** 64
[9] Foote G B 1973 *Journal of Computational Physics* **11** 507
[10] James A J , Smith M and Glezer A 2003 *Journal Fluid Mech.* **476** 29
[11] Jiang T S, Oh S G and Slattery J C 1979 *Journal of Colloid and Interface Science* 69(1) 74
[12] Kopysov S P, Tonkov L E, Chernova A A and Sarmakeeva A S 2015 *Bulletin of Udmurt University Mathematics Mechanics Computer science* **25(3)** 405
[13] Issa R.I. 1986 *Journal of Computational Physics* **62** 40
[14] Van Leer B *Journal of Computational Physics* **23(3)** 263

Discontinuous Galerkin schemes for nonstationary convection-diffusion problems

R Z Dautov and E M Fedotov

Kazan Federal University, 18 Kremlevskaia Street, 420008 Kazan, Russia

E-mail: rdautov@kpfu.ru

Abstract. In this paper we analyse an approximation of linear nonstationary convection-diffusion problem based on combination of discontinues Galerkin method for time discretisation in conjunction with hybridized discontinues Galerkin for spatial approximation. Such discrete schemes can be used for the solution of equations degenerating in the leading part and are formulated in term of solution of the problem, its gradient, the diffusional flux, and the restriction of the solution to the boundaries of elements. Conditions responsible for the solvability, stability and accuracy of the schemes are presented.

1. Introduction

Discontinuous Gakerkin (DG) and hybridized DG (HDG) schemes are successfully used for an approximation of numerical solution of boundary value problems, in particular, for the diffusion equations with dominant convection. Construction and investigation of such schemes has been widely discussed in literature (see, for example, [1–7]).

Based on the mixed statement of the problem, HDG schemes are formulated in terms of solution of the problem (u), its gradient (σ), diffusion flux (q), and the restriction of the solution to boundaries of the elements (λ). Along with high accuracy, such schemes are locally and globally conservative.

In this paper we examine implicit DG schemes in conjunction with HDG spatial approximation of the source problem. Implicit DG schemes combined with standard finite element method and DG FEM were considered in [7, 8].

2. Boundary value problem

Let Ω be a polygon in R^d with boundary Γ, $d=2,3$; $T>0$ $Q_T=\Omega\times(0,T)$. Consider the convection-diffusion equation

$$\partial u/\partial t-\nabla\cdot(\varepsilon A\nabla u-bu)+au=f(x,t), \quad (x,t)\in Q_T, \tag{1}$$

$$u(x,t)=0, \quad x\in\Gamma, t\in(0,T), \tag{2}$$

$$u(x,0)=u_0(x), \quad x\in\Omega. \tag{3}$$

We assume that the original data satisfy the conditions

$$A\in[L_\infty(Q_T)]^{d\times d}, b\in[L_\infty(Q_T)]^d, \nabla\cdot b, a\in L_\infty(Q_T), f\in L_2(Q_T), u_0\in L_2(\Omega), \tag{4}$$

$$m_A |\xi|^2 \leq A(x,t)\xi\cdot\xi \leq M_A |\xi|^2 \quad \forall \xi \in R_d, (x,t) \in Q_T. \tag{5}$$

Here ε is a positive parameter, $m_A > 0$.

To construct discrete schemes, we rewrite Eq. (1) in the form of the system of first-order equations

$$\sigma = \nabla u, \quad q = \varepsilon A \sigma, \quad \partial u/\partial t - \nabla\cdot(q - bu) + au = f.$$

3. Spaces of discontinuous finite elements

3.1. Triangulation of the domain

Let h be a positive parameter; let $\hat{K}$ be a polyhedron in R^d that is a basic finite element (a unit triangle or square for $d = 2$; a tetrahedron, cube, or prism for $d = 3$); let T_h be a triangulation of Ω into finite elements of the maximum diameter h which are images of $\hat{K}$ under an affine transformation.

We assume that T_h is a conformal, regular and coordinated with a boundary partition. In what follows, K stands for an arbitrary element of T_h, and the faces of K are treated as its $d-1$-dimensional faces. The set of all faces is denoted by F_h; by F_h^D we denote the sets of faces lying on Γ.

Let $\bar{\omega}_\tau = \{t_j, j = 0,\dots, j_0\}$ be a set of grid points on $[0,T]$, $\omega_\tau = \bar{\omega}_\tau \setminus \{t_{j_0}\}$. We assume that the partition $T_\tau = \{\Delta_j\}_{j=0}^{j_0-1}$, $\Delta_j = (t_j, t_{j+1})$, $\tau_j = \tau(t_j) = t_{j+1} - t_j$ is regular, i.e. $\tau_j \sim \tau$.

3.2. Function spaces

We use the standard notation $H^s(D)$ for the scalar Sobolev space of order s. The norm and seminorm on it are denoted by $\|\cdot\|_{s,D}$ and $|\cdot|_{s,D}$. For the space of vector functions $v = (v_1,\dots,v_d)$ with components in $H^s(D)$, we use the notation $H^s(D)^d$. The norm and seminorm on it are denoted by analogy with the scalar case. The notation $(\cdot,\cdot)_D$ is used for the inner product both in the scalar case of $L_2(D)$ and in the vector case of $[L_2(D)]^d$.

Both for scalar functions and for vector functions, we set

$$(w,\eta) = \sum_{K\in T_h} (w,\eta)_K, \quad |w|_{s,T_h} = \left(\sum_{K\in T_h} |w|_{s,K}^2\right)^{1/2}, \quad \|w\|_{s,T} = \left(\sum_{K\in T_h} \|w\|_{s,K}^2\right)^{1/2}.$$

On the basis of the triangulation T_h, we define the space $H^s(T_h)$ as the subspace of functions in $L_2(\Omega)$ whose restrictions to each element $K \in T_h$ belong to the $H^s(K)$ with the norm $\|\cdot\|_{s,T_h}$. In addition, we introduce the notation

$$(w\cdot n, \eta)_{\partial T_h} = \sum_{K\in T_h} (w\cdot n_K, \eta)_{\partial K},$$

where n_K is the field of unit outward normal on K.

We also define the spaces $L_p(a,b;[H^s(D)]^d)$, $H^p(a,b;[H^s(D)]^d)$ of measurable functions~ $u(t) \in \{(a,b) \to [H^s(D)]^d\}$ with norms

$$\|u\|_{L_p(a,b;[H^s(D)]^d)} = \left(\int_a^b \|u(t)\|_{s,D}^p dt\right)^{1/p}, 1 \leq p \leq \infty;$$

$$\|u\|_{L_\infty(a,b;[H^s(D)]^d)} = \operatorname*{vrai\,max}_{t\in(a,b)} \|u(t)\|_{s,D}, \quad \|u\|_{H^p(a,b;[H^s(D)]^d)} = \left\|\|u(t)\|_{s,D}\right\|_{p,(a,b)}.$$

As before, we introduce the spaces based on triangulation $T_{h,\tau} = T_h \times T_\tau$. For brevity, the norms in the spaces $L_2(T_\tau;[H^s(T_h)]^d)$ and $H^p(T_\tau;[H^s(T_h)]^d)$ will be referred as $\|u\|_{s,0,T_{h,\tau}}$, $\|u\|_{s,p,T_{h,\tau}}$.

3.3. Spaces of finite elements

Let $P_m(D)$ be a space of polynomials of degree $m \geq 0$ defined on D. On $\hat{K}$, we introduce the finite-dimensional spaces $V(\hat{K})$ and $W(\hat{K})$ such that

$$P_0(\hat{K}) \subseteq V(\hat{K}) \subset H^1(\hat{K}), \quad [P_0(\hat{K})]^d \subseteq W(\hat{K}) \subset [H^1(\hat{K})]^d.$$

To each face $\hat{F} \subset \partial\hat{K}$ we assign the finite-dimensional space $\Lambda(\hat{F})$,

$$P_0(\hat{F}) \subseteq \Lambda(\hat{F}) \subset L_2(\hat{F}).$$

By using these spaces, we introduce the spaces $V(K)$, $W(K)$, and $\Lambda_K(F)$, $F \subset \partial K$. Let $x = F_K(\hat{x}) = B_K\hat{x} + c_K$ be an invertible affine transformation. Set

$$V(K) = \{v = \hat{v} \circ F_K^{-1}, \hat{v} \in V(\hat{K})\}, \quad \Lambda_K(F) = \{\lambda = \hat{\lambda} \circ F_K^{-1}, \hat{\lambda} \in \Lambda(\hat{K})\},$$
$$W(K) = \{\mu = |B_K|^{-1} B_K(\hat{\mu} \circ F_K^{-1}), \hat{\mu} \in W(\hat{K})\},$$

where $|B_K| = \det(B_K)$. To each face $F = F_h$ we assign a finite-dimensional space $\Lambda(F)$, by assuming that if F is a common face of two elements K and L, then

$$\Lambda(F) = \Lambda_K(F) = \Lambda_L(F),$$

and $\Lambda(F) = \Lambda_K(F)$, if K is a boundary element and its face is $F \subset \Gamma$. Let $1_D \in P_0(D)$ be the function equals to the identical unity, and let

$$\Lambda(\partial K) = \{\lambda \in L_2(\partial K) : \lambda|_F \in \Lambda(F), F \subset \partial K\},$$
$$\mathbf{V}(K) = V(K) \times \Lambda(\partial K), \quad \mathbf{V}_0(K) = \{(v,\lambda) \in \mathbf{V}(K) : (v,\lambda) = c(1_K, 1_{\partial K}), c \in R\}.$$

The main constraints for the above-introduced spaces are given by the following conditions.

(H_1) For $(v,\lambda) \in \mathbf{V}(K)$ the relation

$$-(v, \nabla\cdot\tau)_K + (\lambda, \tau\cdot n_K)_{\partial K} = 0, \quad \tau \in W(K),$$

implies that $(v,\lambda) \in \mathbf{V}_0(K)$.

(H_2) $\nabla\cdot W(K) = \{\nabla\cdot w : w \in W(K)\} \subseteq V(K)$.

(H_3) $\Lambda(F) = \{w\cdot n_K|_F : w \in W(K)\}$ for every face $F \subset \partial K$.

We introduce the spaces

$$V_h = \{v \in L_2(\Omega) : v|_K \in V(K), K \in T_h\}, \quad W_h = \{v \in [L_2(\Omega)]^d : w|_K \in W(K), K \in T_h\},$$
$$\Lambda_h = \{\lambda \in L_2(F_h) : \lambda|_F \in \Lambda(F), F \in F_h; \ \ \lambda|_{F_h^D} = 0\},$$

and also consider the spaces $V_{h\tau}$, $W_{h\tau}$, $\Lambda_{h\tau}$ of polynomials of degree $r \geq 0$ along the variable $t \in \Delta \in T_\tau$ with the values in the V_h, W_h, Λ_h, respectively.

In what follows, we use the space $V_{h\tau}$ for the approximation to the solution u, $W_{h\tau}$ — for σ and q, and $\Lambda_{h\tau}$ — for the approximation to the trace of u on element faces.

3.4. Discrete gradient

We introduce the operator $\nabla_K : \mathbf{V}(K) \to W(K)$, which is a local discrete gradient [9, 10] acting by the rule

$$(\nabla_K \mathbf{v}, w)_K = -(v, \nabla \cdot w)_K + (\mu, w \cdot n_K)_{\partial K} \quad \forall w \in W(K),$$

where $\mathbf{v} = (v, \mu)$. Note that condition (H_1) implies that the kernel ∇_K is formed only by "constant" functions; more precisely, $\ker \nabla_K = \mathbf{V}_0(K)$. We define the discrete gradient $\nabla_h : \mathbf{V}_h \to W_h$ by the relations $(\nabla_h \mathbf{v})|_K = \nabla_K (\mathbf{v}|_K)$, i.e.,

$$(\nabla_h \mathbf{v}, w)_{T_h} = -(v, \nabla \cdot w)_{T_h} + (\mu, w \cdot n)_{\partial T_h} \quad \forall w \in W_h.$$

4. Family of HDG-schemes

Consider the following discrete problem: find $(\sigma_h,\ q_h,\ u_h,\ \lambda_h) \in W_{h\tau} \times W_{h\tau} \times V_{h\tau} \times \Lambda_{h\tau}$ that satisfy the following equations

$$(\sigma_h, w)_{K \times T_\tau} + (u_h, \nabla \cdot w)_{K \times T_\tau} - (\lambda_h, w \cdot n_K)_{\partial K \times T_\tau} = 0 \quad \forall w \in W_{h\tau}, \tag{6}$$

$$(q_h - \varepsilon A \sigma_h, w)_{K \times T_\tau} = 0 \quad \forall w \in W_{h\tau}, \tag{7}$$

$$\begin{aligned} -(u_h, \partial v / \partial t)_{K \times T_\tau} &+ \sum_{t \in \omega_\tau \setminus \{0\}} (\breve{u}_h, \breve{v} - v)_K + (\breve{u}_h, \breve{v})_K |_{t=T} - (u_0, v|_{t=0})_K + \\ &+ (\mathfrak{A}_K((u_h, q_h, \lambda_h), v), 1)_{T_\tau} = (f, v)_{K \times T_\tau} \quad \forall v \in V_{h\tau},\ K \in T_h, \end{aligned} \tag{8}$$

$$(q_h \cdot n, \mu)_{\partial T_{h\tau}} - (b_\alpha(\mathbf{u}_h) \cdot n, \mu)_{\partial T_{h\tau}} = 0 \quad \forall \mu \in \Lambda_{h\tau}. \tag{9}$$

Here $v = v(t+0)$, $\breve{v} = v(t-0)$ — the limit values of the function v at the points $t \in \omega_\tau$, $\mathbf{u}_h = (u_h, \lambda_h)$,

$$\mathfrak{A}_K((u, q, \lambda), v) = (-\nabla \cdot q + au, v)_K - (bu, \nabla v)_K + (b_\alpha(\mathbf{u}_h) \cdot n_K, \mu)_{\partial K} \quad \forall v \in V_h,$$

$$b_\alpha(\mathbf{u}_h)|_{\partial K} = 0.5(u_h + \lambda_h)|_{\partial K} + \alpha(u_h - \lambda_h)|_{\partial K}\, n_K, \tag{10}$$

where $\alpha = \alpha(x)$ is some nonnegative stabilization parameter that is a scalar function on F_h.

One can readily see that if we set $\alpha|_{\partial K} = 0.5|b \cdot n_K|$, then we obtain

$$b_\alpha(\mathbf{u}_h) \cdot n_K |_{\partial K} = ((b \cdot n_K)^+ u_h - (b \cdot n_K)^- \lambda_h)|_{\partial K},$$

where $v^\pm = \max\{0, \pm v\}$ are the positive and negative parts of v. In this case, relation (8) corresponds an upwind approximation to the convection-diffusion equation [5].

Obviously, constructed scheme is purely implicit Euler scheme when $r = 0$. It can be interpreted as $r+2$ — layer implicit scheme, when $r > 0$.

In what follows, we assume that

$$\alpha_0 |b \cdot n_K| \le \alpha(x) \le C \quad \forall x \in \partial K,\ K \in T_h, \tag{11}$$

where α_0 and C are positive constants independent of h and ε.

4.1. Basic statement of the scheme

Let us present statement of the scheme (6) — (9) in terms of the unknown $\mathbf{u}_h$. It can be produced by elimination of unknowns σ_h and q_h. Basic problem for the scheme has the following form:

$$\mathbf{u}_h \in \mathbf{V}_{h\tau} = V_{h\tau} \times \Lambda_{h\tau} : D_t(\mathbf{u}_h, \mathbf{v}_h) + \mathbf{A}_h(\mathbf{u}_h, \mathbf{v}_h) = \mathbf{f}_h(\mathbf{v}_h) + (u_0, v|_{t=0})_{T_h} \ \forall \mathbf{v}_h \in \mathbf{V}_{h\tau}, \tag{12}$$

where $\mathbf{A}_h(\mathbf{u}_h,\mathbf{v}_h)=(\mathbf{a}_h(\mathbf{u}_h,\mathbf{v}_h)+\mathbf{b}_h(\mathbf{u}_h,\mathbf{v}_h),1)_{T_\tau}$, $\mathbf{f}_h(\mathbf{v}_h)=(f,v_h)_{T_{h\tau}}$,

$$D_t(\mathbf{u}_h,\mathbf{v}_h)=-(u_h,\partial v_h/\partial t)_{K\times T_\tau}+\sum_{t\in\omega_\tau\setminus\{0\}}(\breve{u}_h,\breve{v}_h-v_h)_K(t)+(\breve{u}_h,\breve{v}_h)_K|_{t=T},$$

$$\mathbf{a}_h(\mathbf{u}_h,\mathbf{v}_h)=(\varepsilon A\nabla_h\mathbf{u}_h,\nabla_h\mathbf{v}_h)_{T_h}+(au_h,v_h)_{T_h},$$

$$\mathbf{b}(\mathbf{u}_h,\mathbf{v}_h)=-(u_h,\nabla v_h)_{T_h}+(b_\alpha(\mathbf{u}_h)\cdot n,v_h-\mu_h)_{\partial T_h}.$$

After solving the Eq. (12), we determine the unknowns σ_h and q_h in accordance with relations (P_W is the L_2 projection of $[L_2(0,T;[L_2(\Omega)]^d)$ onto $W_{h\tau}$)

$$\sigma_h=\nabla_h\mathbf{u}_h,\quad q_h=P_W(\varepsilon A\sigma_h).$$

4.2. Stability of the schemes

On the space $\mathbf{V}_h$, we introduce the seminorm

$$|\mathbf{v}|_{0,h}^2=\|v\|_{0,T_h}^2+(\alpha,(\mu-v)^2)_{\partial T_h},\quad \|v\|_{1,T_h}^2=|\mathbf{v}|_{0,h}^2+\varepsilon\|\nabla_h\mathbf{v}\|_{0,T_h}^2.$$

Lemma 1. *Let conditions* (4), (5) are satisfied, and $c=\min\{m_A,1\}$. *Then*

$$\mathbf{A}_h(\mathbf{v}_h,\mathbf{v}_h)\geq c\|\mathbf{v}_h\|_{1,h}^2-\bar{c}\|v_h\|_{0,h}^2\quad\forall\mathbf{v}_h\in\mathbf{V}_h.$$

As consequence, the scheme (6) — (9) *is uniquely solvable at* $r=0$ *and* $r=1$.

Note that when $r>1$, the solvability of the scheme requires an additional conditions on the coefficients of the Eq. (1), providing non-negativity of bilinear form $\mathbf{A}_h$. In what follows, we assume $r=0$ or $r=1$.

Lemma 2. *Let conditions of Lemma 1 be satisfied and* $r\in\{0,1)$. *Then the scheme is stable and we have*

$$\max_{t\in\bar{\omega}_\tau\setminus\{0\}}\|\breve{u}_h(t)\|_{0,T_h}+\sum_{t=t_1}^{t_{j0-1}}\|u_h-\breve{u}_h\|_{0,T_h}(t)+c\|\mathbf{u}_h(t)\|_{1,0,T_{h\tau}}\leq M\left[\|\mathbf{f}_h\|_{-1,(0,T)}+\|u_0\|_{0,T_h}\right],$$

where the constant $M>0$ *not depends on* h, τ, u_0, f *and*

$$\|\mathbf{u}_h(t)\|_{1,0,T_{h\tau}}=\left(\sum_{t\in\bar{\omega}_\tau}\tau\|\mathbf{u}_h(t)\|_{1,h}^2\right)^{1/2},\quad \|\mathbf{f}_h\|_{-1,(0,T)}=\left(\int_0^T\|f_h\|_{-1,h}^2dt\right)^{1/2},$$

$$\|f_h\|_{-1,h}=\sup_{\mathbf{v}_h\in\mathbf{V}_h}\left(|\mathbf{f}_h(\mathbf{v}_h)|/\|\mathbf{v}_h\|_{1,h}\right).$$

5. Estimate for the schemes accuracy

We estimate the accuracy of the scheme (6) — (9) under extra constraints:

(H_4) $\nabla V(K)=\{\nabla v:v\in V(K)\}\subset[V(K)]^d$.

Note that if $V(K)$ is a space of polynomials, then this condition is always satisfied. We estimate the accuracy of schemes under the following conditions additional to (4) and (5):

$$b\in[W_\infty^1(T_{h\tau})]^d,\quad q\in[H^1(T_{h\tau})]^d. \tag{13}$$

Theorem 1. *Let conditions* (4), (5), ($H_{1\text{-}4}$) *be satisfied,* $r\in\{0,1\}$, and (q_h, σ_h, u_h, λ_h) *be a solution of the* (6) — (9). *Then*

$$\max_{t\in\bar{\omega}_\tau} \| u-\breve{u}_h \|_{0,\Omega} \le cE_h(u),$$

$$\| u-u_h \|_{0,0,T_{h\tau}} + \varepsilon^{1/2} \| \sigma-\sigma_h \|_{0,0,T_{h\tau}} \le cE_h(u),$$

$$\varepsilon^{-1/2} \| q-q_h \|_{0,0,T_{h\tau}} \le c(E_h(u) + \varepsilon^{-1/2} \| q-P_W q \|_{0,0,T_{h\tau}}),$$

$$\| u-\lambda_h \|_{0,0,\partial T_{h\tau}} \le c(h^{-1/2} + \varepsilon^{-1/2} h^{1/2}) E_h(u).$$

where $E_h(u) = E_{\partial T_h}(u) + E_{T_h}(u)$,

$$E_{\partial T_h}(u) = \| \pi_\tau u - P_V \pi_\tau u \|_{0,0,\partial T_{h\tau}} + \| \pi_\tau u - P_\Lambda \pi_\tau u \|_{0,0,\partial T_{h\tau}} + \| \pi_\tau u - u \|_{0,0,\partial T_{h\tau}},$$

$$E_{T_h}(u) = [\| \pi_\tau u - P_V \pi_\tau u \|_{0,0,T_{h\tau}} + \varepsilon^{-1/2} \| \pi_\tau u - u \|_{0,0,T_{h\tau}}] +$$
$$+ \varepsilon^{1/2} [\| \pi_\tau(\nabla u) - P_V \pi_\tau(\nabla u) \|_{0,0,T_{h\tau}} + \| \nabla u - \pi_\tau(\nabla u) \|_{0,0,T_{h\tau}}] +$$
$$+ \varepsilon^{-1/2} [\| \pi_\tau q - P_W \pi_\tau q \|_{0,0,T_{h\tau}} + \| \pi_\tau q - q \|_{0,0,T_{h\tau}}],$$

P_V, P_W, P_Λ *are the* L_2*-projectors of* $L_2(\Omega)$, $[L_2(\Omega)]^d$, $L_2(F)$ *onto* V_h, W_h, Λ_h, *respectively, and* $\pi_\tau = \pi_{\tau,r}$ *— special orthogonal projector* $L_\infty(T_\tau;V)$ *onto* $V_{h,\tau}$, *proposed in* [7].

6. Estimation of the accuracy of a particular family of scheme

For the practical use of the studied schemes, one should define finite-dimensional spaces $V(K)$, $W(K)$, $\Lambda(F)$ satisfying conditions (H_{1-4}) and parameter r. It was shown in [9, 10] that the spaces widely used in mixed FEM satisfy these conditions. They can be used for the definition of particular schemes whose accuracy estimate was mentioned above. As an example, we study only one family of similar schemes.

Let K be a d-dimensional simplex, and

$$V(K) = P_k(K), \quad W(K) = [P_k(K)]^d + xP_k(K), \quad \Lambda(K) = P_k(K), \quad k \ge 0. \tag{14}$$

Here $W(K)$ is the space of Raviart — Thomas polynomials [11], which is used for the definition of a conformal approximation $H(\mathrm{div},\Omega)$.

Theorem 2. *Let* (q_h, σ_h, u_h, λ_h) *be a solution of the* (6) — (9) *on the basis of the finite elements* (14) *and* $r \in \{0,1\}$. *Then for* $1 \le s \le k+1$

$$\max_{t\in\bar{\omega}_\tau\setminus\{0\}} \| u-\breve{u}_h \|_{0,\Omega} + \| u-u_h \|_{0,0,T_{h\tau}} + \varepsilon^{1/2} \| \sigma-\sigma_h \|_{0,0,T_{h\tau}} + \varepsilon^{-1/2} \| q-q_h \|_{0,0,T_{h\tau}} = O\left(\tau^{r+1}(1+\varepsilon^{-1/2}) + h^s\right).$$

Acknowledgments
This work was supported by the Russian Foundation for Basic Research, project no. 16-01- 00408, 15-01-05686

References

[1] Arnold D N, Brezzi F, Cockburn B and Marini L D 2002 Unified analysis of discontinuous Galerkin methods for elliptic problems *SIAM J. Numer. Anal.* **39(5)** pp 1749-1779

[2] Cockburn B 1999 Discontinuous Galerkin methods for convection-dominated problems *(Lecture Notes in Computational Science and Engineering vol 9)* ed T Barth and H Deconink (Berlin: Springer Verlag) pp 69-224

[3] Cockburn B, Gopalakrishnan J and Lazarov R 2009 Unified hybridization of discontinuous Galerkin, mixed and continuous Galerkin methods for second order elliptic problems *SIAM J. Numer. Anal.* **47(2)** pp 1319-1365

[4] Cockburn B, Dong B, Guzmán J, Restelli M and Sacco R 2009 A hybridizable discontinuous Galerkin method for steady-state convection-diffusion-reaction problems *SIAM J. Sci. Comput.* **31** pp 3827-3846
[5] Egger H and Schöberl J 2009 A hybrid mixed discontinuous Galerkin finite element method for convectiondiffusion problems *IMA J. Numer. Anal.* **30** pp 1206-1234
[6] Oikawa I 2014 A hybrid mixed discontinuous Galerkin finite-element method for convection-diffusion problems *Japan J. Indust. Appl. Math.* **31** pp 335-354
[7] Meidner D, Vexler B 2011 A priori error estimates for space-time finite element discretization of parabolic optimal control problems. Part I: Problems without control constraints *RICAM-Report 2007-2011* Austrian Academy of Scinces, URL: www.ricam.oeaw.ac.at
[8] Thomée V 1997 Galerkin Finite Element Methods for Parabolic Problems *Spinger Ser. Comput. Math.* **25** , Springer, Berlin chapter 12 pp 213-231
[9] Dautov R Z and Fedotov E M 2014 Abstract theory of hybridizable discontinuous Galerkin methods for second-order quasilinear elliptic problems *Computational Mathematics and Mathematical Physics* **54 (3)** pp 463-480
[10] Dautov R Z 2014 On inf-sup conditions and projections in the theory of mixed finite-element methods Differential equations *Computational Mathematics and Mathematical Physics* **50 (7)** pp 909-922
[11] Raviart P and Thomas J 1977 A mixed finite element method for 2-nd order elliptic problems *Lecture Notes in Math.* **606** pp 292-315

HDG schemes for stationary convection-diffusion problems

R Z Dautov and E M Fedotov

Kazan Federal University, 18 Kremlevskaia street, 420008 Kazan, Russia

E-mail: rdautov@kpfu.ru

Abstract. For stationary linear convection-diffusion problems, we construct and study a hybridized scheme of the discontinuous Galerkin method on the basis of an extended mixed statement of the problem. Discrete schemes can be used for the solution of equations degenerating in the leading part and are stated via approximations to the solution of the problem, its gradient, the flow, and the restriction of the solution to the boundaries of elements. For the spaces of finite elements, we represent minimal conditions responsible for the solvability, stability and accuracy of the schemes.

1. Introduction

Recently, schemes of the discontinuous Galerkin method (DG or DGFEM-schemes) have been comprehensively studied and widely used for the solution of various boundary value problems for partial differential equations. Numerous papers deal with them, which explains a variety of approaches to their construction. Of these papers, we note [1-3].

DG-schemes are based on spaces of discontinuous finite elements (of arbitrary approximation order) and are locally and globally conservative. Just as in the finite-element method (FEM), there are two basic approaches to their construction. The main statement of the boundary value problem is used in the first of them, and a mixed statement is used in the other. In the first case, the scheme is stated via approximations to the solution of the problem (u), and in the other case, the scheme is stated via approximations to the solution of the problem (u) and to the flow vector (q). For linear problems, in the numerical implementation of schemes of the second type, the unknowns corresponding to the vector q can be eliminated element by element from the system of equations.

In comparison with the standard scheme of the FEM of the same accuracy order, a system of algebraic equations of the DG-scheme (for finding u) has a much larger dimension, which is a disadvantage of the method. To eliminate it from the scheme, we introduce the new unknown corresponding to the restriction of u to the boundaries of elements. This leads to hybridized schemes of the discontinuous Galerkin method (HDG-schemes; see [4-7]).

HDG-schemes constructed on the basis of the mixed statement of the problem are quite similar to hybridized mixed schemes of the FEM [8] and are also stated via approximations to u, q, and, say, λ, where λ is the restriction of u to the boundaries of elements (see [9]). These schemes admit efficient implementation. For example, for the solution of linear problems by such methods, the unknowns corresponding to u and q are eliminated element by element from the scheme, and one obtains a system of algebraic equations for the unknown corresponding to λ alone. After finding λ, the remaining unknowns are reconstructed element by element as well.

HDG-schemes for elliptic BVP of the second order were suggested in [10] on the basis of an extended mixed statement of the original problem and were studied from abstract viewpoints. Along with approximations to the unknowns u, q, and λ, they permit straightforwardly determining the approximation to the solution gradient ($\sigma = \nabla u$). The introduction of a new unknown vector variable into the scheme distinguishes these schemes from the known HDG-schemes. The introduction of a new unknown in a scheme insignificantly increases the amount of computations, because this unknown, like u and q, can be eliminated element by element from the scheme.

In the present paper, we study two modifications of HDG-schemes in [10]. The first of them contains a stabilizing parameter and is designed for problems with dominating convection. The second scheme does not contain a stabilizing parameter and can be used for problems in which convection is not dominating. In the study of schemes, we obtain conditions on the spaces of finite elements ensuring the solvability, stability and accuracy of the schemes. The accuracy of schemes is estimated via the errors of L_2-orthoprojections of the solution onto the corresponding FE spaces.

2. Boundary value problem

Let Ω be a polygon in R^d with boundary Γ, $d = 2,3$. Consider the convection-diffusion equation

$$-\nabla\cdot(\varepsilon A\nabla u - bu) + au = f, \quad x\in\Omega. \tag{1}$$

We assume that the original data satisfy the conditions

$$A\in[L_\infty(\Omega)]^{d\times d}, \quad b\in[L_\infty(\Omega)]^d, \; \nabla\cdot b, a\in L_\infty(\Omega), \; f\in L_2(\Omega), \tag{2}$$

$$m_A|\xi|^2 \le A(x)\xi\cdot\xi \le M_A|\xi|^2 \;\; \forall\xi\in R^d, \quad a(x)+0.5\nabla\cdot b(x)\ge m_a \;\; \forall x\in\Omega. \tag{3}$$

Here ε is a positive parameter, $m_A > 0$, and $m_a \ge 0$. In what follows, to study the scheme stability and accuracy, we impose additional conditions on the problem data. We supplement Eq. (1) with the boundary conditions

$$u = g_D, \quad x\in\Gamma_D,$$
$$(\varepsilon A\nabla u - bu)\cdot n = g_N, \quad x\in\Gamma_R,$$
$$\varepsilon A\nabla u\cdot n = g_N, \quad x\in\Gamma_N.$$

Here n is the field of unit outward normals on Γ, g_D is the trace of some function $u_D\in H^1(\Omega)$, and g_N belongs to $L_2(\Gamma_R\cup\Gamma_N)$; the sets Γ_D, Γ_R, and Γ_N define a partition of the boundary Γ into disjoint parts; moreover, $\mathrm{meas}(\Gamma_D) > 0$, and

$$\Gamma_R\subseteq\Gamma^{in} = \{x\in\Gamma : b(x)\cdot n(x) < 0\}, \quad \Gamma_N\subseteq\Gamma^{out} = \{x\in\Gamma : b(x)\cdot n(x)\ge 0\}.$$

Note that either Γ_R or Γ_N can be the empty set. To construct discrete schemes, we rewrite Eq. (1) in the form of the system of first-order equations

$$\sigma = \nabla u, \quad q = \varepsilon A\sigma, \quad -\nabla\cdot q + \nabla\cdot(bu) + au = f.$$

3. Spaces of discontinuous finite elements

3.1. Triangulation of the domain

Let h be a positive parameter; let $\hat{K}$ be a polyhedron in R^d that is a basic finite element (a unit triangle or square for $d = 2$; a tetrahedron, cube, or prism for $d = 3$); let T_h be a triangulation of Ω into finite elements of the maximum diameter h, which are images of $\hat{K}$ under an affine transformation.

We assume that T_h is a conformal, regular and coordinated with a boundary partition. In what follows, K stands for an arbitrary element of T_h, and the faces of K are treated as its $(d-1)$-dimensional faces. The set of all faces is denoted by F_h; by F_h^D, F_h^R, and F_h^N we denote the sets of faces lying on Γ_D, Γ_R, and Γ_N, respectively, and $F_h^{RN} = F_h^R \cup F_h^N$.

3.2. Function spaces

We use the standard notation $H^s(D)$ for the scalar Sobolev space of order s. The norm and seminorm on it are denoted by $\|\cdot\|_{s,D}$ and $|\cdot|_{k,D}$. For the space of vector functions $\tau=(\tau_1,...,\tau_d)$ with components in $H^s(D)$, we use the notation $[H^s(D)]^d$. The norm and seminorm on it are denoted by analogy with the scalar case. The notation $(\cdot,\cdot)_D$ is used for the inner product both in the scalar case of $L_2(D)$ and in the vector case of $[L_2(D)]^d$.

Both for scalar functions and for vector functions, we set

$$(w,\eta)_{T_h} = \sum_{K\in T_h}(w,\eta)_K, \quad |w|_{s,T_h} = (\sum_{K\in T_h}|w|_{s,K}^2)^{1/2}, \quad \|w\|_{s,T_h} = (\sum_{K\in T_h}\|w\|_{s,K}^2)^{1/2}.$$

On the basis of the triangulation T_h we define the space $H^s(T_h)$ as the subspace of functions in $L_2(\Omega)$ whose restrictions to each element $K\in T_h$ belong to the same $H^s(K)$ with the norm $\|\cdot\|_{s,T_h}$. In addition, we introduce the notation

$$(w\cdot n,\eta)_{\partial T_h} = \sum_{K\in T_h}(w\cdot n_K,\eta)_{\partial K}, \quad (w,\eta)_{F_h^S} = \sum_{F\in F_h^S}(w,\eta)_F,$$

where n_K is the field of unit outward normal's on ∂K, and $S\in\{R,N,RN\}$.

3.3. Spaces of finite elements

Let $P_m(D)$ stand for the set of polynomials of degree $\le m$ defined on D. On $\hat{K}$, we introduce the finite-dimensional spaces $V(\hat{K})$ and $W(\hat{K})$ such that

$$P_0(\hat{K}) \subseteq V(\hat{K}) \subset H^1(\hat{K}), \quad [P_0(\hat{K})]^d \subseteq W(\hat{K}) \subset [H^1(\hat{K})]^d.$$

To each face $\hat{F}\subset\partial\hat{K}$ we assign the finite-dimensional space $\Lambda(\hat{F})$,

$$P_0(\hat{F}) \subseteq \Lambda(\hat{F}) \subset L_2(\hat{F}).$$

By using these spaces, we introduce the spaces $V(K)$, $W(K)$, and $\Lambda_K(F)$, $F\subset\partial K$. Let $x=F_K(\hat{x})=B_K\hat{x}+c_K$ be an invertible affine transformation $\hat{K}\to K$. Set

$$V(K)=\{v=\hat{v}\circ F_K^{-1}, \hat{v}\in V(\hat{K})\}, \quad \Lambda_K(F)=\{\lambda=\hat{\lambda}\circ F_K^{-1}, \hat{\lambda}\in\Lambda(\hat{F})\},$$

$$W(K)=\{\tau=|B_K|^{-1}B_K(\hat{\tau}\circ F_K^{-1}), \hat{\tau}\in W(\hat{K})\},$$

where $|B_K|=\det(B_K)$. To each face $F\in F_h$ we assign a finite-dimensional space $\Lambda(F)$, by assuming that if F is a common face of two elements K and L, then

$$\Lambda(F)=\Lambda_K(F)=\Lambda_L(F),$$

and $\Lambda(F)=\Lambda_K(F)$, if K is a boundary element and its face is $F\subset\Gamma$. Let $1_D\in P_0(D)$ be the function equal to the identical unity, and let

$$\Lambda(\partial K)=\{\lambda\in L_2(\partial K):\lambda|_F\in\Lambda(F), F\subset\partial K\},$$

$$\mathbf{V}(K)=V(K)\times\Lambda(\partial K), \quad \mathbf{V}_0(K)=\{(v,\lambda)\in\mathbf{V}(K):(v,\lambda)=c(1_K,1_{\partial K}), c\in R\}.$$

The main constraints for the above-introduced spaces is given by the following conditions.
(H_1) For $(v,\lambda)\in V(K)\times\Lambda(\partial K)$, the relation

$$-(v,\nabla\cdot\tau)_K+(\lambda,\tau\cdot n_K)_{\partial K}=0,\quad \tau\in W(K),$$

implies that $(v,\lambda)\in\mathbf{V}_0(K)$.
(H_2) $\nabla\cdot W(K)=\{\nabla\cdot w: w\in W(K)\}\subseteq V(K)$.
(H_3) $\Lambda(F)=\{w\cdot n_K|_F, w\in W(K)\}$ for every face $F\subset\partial K$.
We introduce the spaces of finite elements

$$V_h=\{v\in L_2(\Omega): v|_K\in V(K), K\in T_h\},$$
$$W_h=\{w\in[L_2(\Omega)]^d: w|_K\in W(K), K\in T_h\},$$
$$\Lambda_h^0=\{\lambda\in L_2(F_h):\lambda|_F\in\Lambda(F), F\in F_h;\lambda|_{F_h^D}=0\},$$

and the set of functions

$$\Lambda_h=\{\lambda\in L_2(F_h):\lambda|_F\in\Lambda(F), F\in F_h;\lambda|_{F\in F_h^D}=P_\Lambda u_D|_{F\in F_h^D}\}.$$

Here P_Λ is the L_2-projection $L_2(F_h)\to\Lambda_h$. Set $\mathbf{V}_h=V_h\times\Lambda_h$ and $\mathbf{V}_h^0=V_h\times\Lambda_h^0$.

In what follows, we use the space V_h for the approximation to the problem solution u, W_h for σ and q, and Λ_h for the approximation to the trace of u on element faces.

3.4. Discrete gradient
Let us introduce the operator $\nabla_K:\mathbf{V}(K)\to W(K)$ (a local discrete gradient [15, 16]) acting by the rule

$$(\nabla_K\mathbf{v},w)_K=-(v,\nabla\cdot w)_K+(\mu,w\cdot n_K)_{\partial K}\quad\forall w\in W(K),$$

where $\mathbf{v}=(v,\mu)$. Note that condition (H_1) implies that the kernel ∇_K is formed only by "constant" functions; more precisely, $\ker(\nabla_K)=\mathbf{V}_0(K)$. We define the discrete gradient $\nabla_h:\mathbf{V}_h\to W_h$ by the relations $(\nabla_h\mathbf{v})|_K=\nabla_K(\mathbf{v}|_K)$, i.e., $(\nabla_h\mathbf{v},w)_{T_h}=-(v,\nabla\cdot w)_{T_h}+(\mu,w\cdot n)_{\partial T_h}\quad\forall w\in W_h$.

4. Families of HDG-schemes

Let us consider the following discrete problem referred to as scheme P_h: find $(\sigma_h,\ q_h,\ u_h,\ \lambda_h)\in W_h\times W_h\times V_h\times\Lambda_h$ such that the identities

$$(\sigma_h,w)_K+(u_h,\nabla\cdot w)_K-(\lambda_h,w\cdot n_K)_{\partial K}=0,$$
$$(q_h-\varepsilon A\sigma_h,w)_K=0,$$
$$(-\nabla\cdot q_h+au_h,v)_{T_h}-(bu_h,\nabla v)_{T_h}+(b\cdot n\lambda_h,v)_{\partial T_h}=(f,v)_{T_h},\tag{4}$$
$$(q_h\cdot n,\mu)_{\partial T_h}-(b\cdot n\lambda_h,\mu)_{\partial T_h}+(b\cdot n\lambda_h,\mu)_{F_h^N}=(g_N,\mu)_{F_h^{RN}},\tag{5}$$

hold for arbitrary $w\in W_h$, $v\in V_h$, $\mu\in\Lambda_h^0$ and $K\in T_h$.
As is shown below, this scheme is stable if the convective term (or h) is sufficiently small. By introducing stabilizing terms into Eqs. (4) and (5), we obtain a scheme stable for arbitrary ε and h. We refer to it as scheme P_h^s. It differs from P_h only by two terms and has the following form: find $(\sigma_h,\ q_h,\ u_h,\ \lambda_h)\in W_h\times W_h\times V_h\times\Lambda_h$ such that

$$(\sigma_h,w)_K+(u_h,\nabla\cdot w)_K-(\lambda_h,w\cdot n_K)_{\partial K}=0,$$
$$(q_h-\varepsilon A\sigma_h,w)_K=0,$$

$$(-\nabla\cdot q_h + au_h, v)_{T_h} - (bu_h, \nabla v)_{T_h} + (b_\alpha(\mathbf{u}_h)\cdot n, v)_{\partial T_h} = (f, v)_{T_h}, \tag{6}$$
$$(q_h\cdot n, \mu)_{\partial T_h} - (b_\alpha(\mathbf{u}_h)\cdot n, \mu)_{\partial T_h} + (\lambda, \mu)_{F_h^N} = (g_N, \mu)_{F_h^{RN}}$$

for arbitrary $w\in W_h$, $v\in V_h$, $\mu\in\Lambda_h^0$ and $K\in T_h$. Here $\mathbf{u}_h=(u_h,\lambda_h)$, and

$$b_\alpha(\mathbf{u}_h)|_{\partial K} = 0.5b(u_h+\lambda_h)|_{\partial K} + \alpha(u_h-\lambda_h)|_{\partial K}\, n_K, \tag{7}$$

where $\alpha=\alpha(x)$ is some nonnegative stabilization parameter that is a scalar function on F_h.

One can readily see that the scheme P_h is obtained from P_h^s for $\alpha|_{\partial K}=-0.5b\cdot n_K$ in relation (7). If we set $\alpha|_{\partial K}=0.5|b\cdot n_K|$, then we obtain

$$b_\alpha(\mathbf{u}_h)\cdot n_K|_{\partial K} = ((b\cdot n_K)^+ u_h - (b\cdot n_K)^- \lambda_h)|_{\partial K},$$

where $v^\pm=\max\{0,\pm v\}$ are the positive and negative parts of v. In this case, relation (6) is an upwind approximation to the convection-diffusion equation [6].

In what follows, we assume that

$$\alpha_0|b\cdot n_K| \le \alpha(x) \le C \quad \forall x\in\partial K, \quad K\in T_h, \tag{8}$$

where α_0 and C are positive constants independent of h and ε.

For the numerical implementation of the scheme, as the stabilizing function α one can choose a piecewise constant function on F_h satisfying condition (8). In this case, the function $b_\alpha(\mathbf{u}_h)\cdot n_K|_F$ has the same smoothness as $b\cdot n_K|_F$ (unlike the upwind scheme) if $\lambda_h|_F$ Ë $u_h|_F$ are smooth functions. This is important from the viewpoint of an approximate computation of the corresponding integrals for the implementation of the scheme.

4.1. Basic statement of the schemes

Let us present other statements of the above-constructed schemes, which are referred below to as basic statements and which are convenient both for the study and for the numerical implementation. They are obtained by the elimination of the unknowns σ_h and q_h elementwise from the schemes and are stated for the pair of unknowns $\mathbf{u}_h=(u_h,\lambda_h)$. Basic problem for scheme P_h has the following form:

$$\mathbf{u}_h\in\mathbf{V}_h:\quad \mathbf{A}_h(\mathbf{u}_h,\mathbf{v}_h)=\mathbf{f}_h(\mathbf{v}_h)\quad \forall\mathbf{v}_h\in\mathbf{V}_h, \tag{9}$$

where

$$\mathbf{A}_h(\mathbf{u}_h,\mathbf{v}_h)=\mathbf{a}_h(\mathbf{u}_h,\mathbf{v}_h)+\mathbf{b}_h(\mathbf{u}_h,\mathbf{v}_h),\quad \mathbf{f}_h(\mathbf{v}_h)=(f,v_h)_{T_h}+(g_N,\mu_h)_{F_h^{RN}},$$
$$\mathbf{a}_h(\mathbf{u}_h,\mathbf{v}_h)=(\varepsilon A\nabla_h\mathbf{u}_h,\nabla_h\mathbf{v}_h)_{T_h}+(au_h,v_h)_{T_h},$$
$$\mathbf{b}_h(\mathbf{u}_h,\mathbf{v}_h)=-(bu_h,\nabla v_h)_{T_h}+(b\cdot n\lambda_h, v_h-\mu_h)_{\partial T_h}+(b\cdot n\lambda_h,\mu_h)_{F_h^N}.$$

After the solution of Eq. (9), the unknowns σ_h and q_h are determined elementwise in accordance with relations (P_W is the L_2 projection of $[L_2(\Omega)]^d$ onto W_h)

$$\sigma_h=\nabla_h\mathbf{u}_h,\quad q_h=P_W(\varepsilon A\sigma_h). \tag{10}$$

We can prove that the scheme P_h can be reduced to system (9), (10). By reversing the argument, one can readily show that the solution of system (9), (10) can be reduced to the solution of the original scheme P_h. The basic statement of the scheme P_h^s is obtained by analogy. It has the form

$$\mathbf{u}_h\in\mathbf{V}_h:\quad \mathbf{A}_h^s(\mathbf{u}_h,\mathbf{v}_h)=\mathbf{a}_h(\mathbf{u}_h,\mathbf{v}_h)+\mathbf{b}_h^s(\mathbf{u}_h,\mathbf{v}_h)=\mathbf{f}_h(\mathbf{v}_h)\quad\forall\mathbf{v}_h\in\mathbf{V}_h, \tag{11}$$
$$\mathbf{b}_h^s(\mathbf{u}_h,\mathbf{v}_h)=-(bu_h,\nabla v_h)_{T_h}+(b_\alpha(\mathbf{u}_h)\cdot n, v_h-\mu_h)_{\partial T_h}+(b\cdot n\lambda_h,\mu_h)_{F_h^N},$$

where $\mathbf{a}_h$ and $\mathbf{f}_h$ are the same forms as above. After the solution of Eq. (11), the other unknowns of the scheme are found elementwise in accordance with (10).

4.2. Stability of the schemes
On the space $\mathbf{V}_h$, we introduce the seminorm $|\cdot|_{0,h}$ and the norm $\|\cdot\|_{1,h}$ by the relations

$$|\mathbf{v}|_{0,h}^2 = m_a\|v\|_{0,T_h}^2 + (\alpha,(\mu-v)^2)_{\partial T_h} + 0.5(|b\cdot n|,\mu^2)_{F_h^{RN}}, \quad \|\mathbf{v}\|_{1,h}^2 = |\mathbf{v}|_{0,h}^2 + \varepsilon\|\nabla_h \mathbf{v}\|_{0,\Omega}^2.$$

Lemma 1. *Let conditions (2), (3) be satisfied, and let $c = \min\{m_A, 1\}$. Then*

$$\mathbf{A}_h^s(\mathbf{v}_h,\mathbf{v}_h) \ge c\|\mathbf{v}_h\|_{1,h}^2 \qquad \forall \mathbf{v}_h \in \mathbf{V}_h^0.$$

As consequence, the scheme P_h^s is uniquely solvable.

Let C_b be defined such that $\alpha + 0.5|b| \le C_b$. In the study of the scheme P_h, we assume that

$$\varepsilon = 1, \quad m_a \ge 0, \quad m_A - C_\Lambda^2 C_b h\|b\|_{L_\infty(\Omega)} \ge \tilde{m}_A > 0. \tag{12}$$

Lemma 2. *Let conditions (2), (3) and (12) be satisfied. Then*

$$\mathbf{A}_h(\mathbf{v}_h,\mathbf{v}_h) \ge \tilde{m}_A\|\nabla_h \mathbf{v}_h\|_{0,\Omega}^2.$$

As consequence, the scheme P_h is uniquely solvable.

5. Estimate for the schemes accuracy

We estimate the accuracy of the schemes P_h^s and P_h under extra constraints:

(H_4) $\nabla V(K) = \{\nabla v : v \in V(K)\} \subset [V(K)]^d$.

We estimate the accuracy of schemes under the following conditions additional to (2) and (3):

$$b \in [W_\infty^1(T_h)]^d, \quad q \in [H^1(T_h)]^d. \tag{13}$$

Theorem 1. *Let conditions (2), (3) with $m_a > 0$, (12), (13), and $(H_{1\text{-}4})$ be satisfied, and $(q_h, \sigma_h, u_h, \lambda_h)$ be a solution of the scheme P_h^s. Then*

$$\|u-u_h\|_{0,\Omega} + \varepsilon^{1/2}\|\sigma-\sigma_h\|_{0,\Omega} \le c\,E_h^s(u), \quad \|q-q_h\|_{0,\Omega} \le c\,\varepsilon^{1/2}E_h^s(u) + \|q - P_W q\|_{0,\Omega},$$

$$\|u-\lambda_h\|_{0,\partial T_h} \le c(h^{-1/2} + \varepsilon^{-1/2}h^{1/2})E_h^s(u),$$

where

$$E_h^s(u) = \|u - P_V u\|_{0,\partial T_h} + \|u - P_\Lambda u\|_{0,\partial T_h} + \|u - P_V u\|_{0,T_h} + \varepsilon^{1/2}\|\nabla u - P_W(\nabla u)\|_{0,T_h} + \varepsilon^{-1/2}\|q - P_W q\|_{0,\Omega}.$$

Theorem 2. *Let conditions (2), (3), (13) and $(H_{1\text{-}4})$ be satisfied, and $(q_h, \sigma_h, u_h, \lambda_h)$ be a solution of the scheme P_h. Then*

$$\|u-u_h\|_{0,T_h} + \|\sigma-\sigma_h\|_{0,\Omega} \le c\,E_h(u), \quad \|q-q_h\|_{0,\Omega} \le c\,E_h(u) + \|q - P_W q\|_{0,\Omega},$$

$$\|u-\lambda_h\|_{0,\partial T_h} \le c\,h^{-1/2}E_h(u),$$

where $E_h(u) = h^{1/2}\|u - P_\Lambda u\|_{0,\partial T_h} + \|u - P_V u\|_{0,T_h} + \|\nabla u - P_W(\nabla u)\|_{0,T_h} + \|q - P_W q\|_{0,\Omega}$.

6. Estimation of the accuracy of a particular family of schemes

For the practical use of the studied schemes P_h^s and P_h, one should define finite-dimensional spaces $V(K)$, $W(K)$, and $\Lambda(F)$ satisfying conditions $(H_{1\text{-}4})$. It was shown in [10, 11] that the spaces widely used in mixed FEM satisfy these conditions. They can be used for the definition of particular schemes P_h^s and P_h whose accuracy estimate was mentioned above. As an example, we study only one family of similar schemes. Let K be a d-dimensional simplex, and let

$$W(K) = [P_k(K)]^d + xP_k(K), \quad V(K) = P_k(K), \quad \Lambda(F) = P_k(F), \; k \ge 0. \tag{14}$$

Here $W(K)$ is the space of Raviart-Thomas polynomials [12].

Theorem 3. *Let* $u \in H^{s+1}(T_h)$, $q = \varepsilon A\nabla u \in H^s(T_h)$, *and* $(q_h, \sigma_h, u_h, \lambda_h)$ *be a solution of the scheme* P_h^s *on the basis of the finite elements (14). Then for* $1 \le s \le k+1$

$$\| u - u_h \|_{0,\Omega} + \varepsilon^{1/2} \| \sigma - \sigma_h \|_{0,\Omega} + \varepsilon^{-1/2} \| q - q_h \|_{0,\Omega} \le c\, E_h^s(u),$$

$$\| u - \lambda_h \|_{0,\partial T_h} \le c(h^{-1/2} + \varepsilon^{-1/2} h^{1/2}) E_h^s(u),$$

under conditions (2), (3) and (13. Here $E_h^s(u) = h^{s-1}(h^{1/2} | u |_{s,T_h} + \varepsilon^{1/2} h(| u |_{s+1,T_h} + | A\nabla u |_{s,T_h}))$.

Theorem 4. *Let* $u \in H^{s+1}(T_h)$, $q = \varepsilon A\nabla u \in H^s(T_h)$, *and* $(q_h, \sigma_h, u_h, \lambda_h)$ *be a solution of the scheme* P_h *on the basis of the finite elements (3). Then for* $1 \le s \le k+1$ *the estimate*

$$\| u - u_h \|_{0,\Omega} + \| \sigma - \sigma_h \|_{0,\Omega} + \| q - q_h \|_{0,\Omega} + h^{1/2} \| u - \lambda_h \|_{0,\partial T_h} \le c\, h^s [u]_{s+1,T_h}$$

hold under conditions (2), (3), (12) and (13). Here $[u]_{s+1,T_h} = | u |_{s,T_h} + | u |_{s+1,T_h} + | A\nabla u |_{s,T_h}$.

Acknowledgments
This work was supported by the Russian Foundation for Basic Research, project no. 16-01- 00408, 15-01-05686

References

[1] Arnold D N, Brezzi F, Cockburn B and Marini L D 2002 Unified analysis of discontinuous Galerkin methods for elliptic problems *SIAM J. Numer. Anal.* **(5)** 1749-79

[2] Arnold D N 1982 An interior penalty finite element method with discontinuous elements *SIAM J. Numer. Anal.* **(4)** 742-60.

[3] Cockburn B 1999 *Discontinuous Galerkin methods for convection-dominated problems* (Lecture Notes in Computational Science and Engineering vol 9) ed T Barth and H Deconink (Berlin: Springer Verlag) pp 69-224

[4] Cockburn B, Gopalakrishnan J and Lazarov R 2009 Unified hybridization of discontinuous Galerkin, mixed and continuous Galerkin methods for second order elliptic problems *SIAM J. Numer. Anal.* **(2)** 1319-65

[5] Cockburn B, Dong B, Guzmán J, Restelli M and Sacco R 2009 A hybridizable discontinuous Galerkin method for steady-state convection-diffusion-reaction problems *SIAM J. Sci. Comput.* 3827-46

[6] Egger H and Schöberl J 2009 A hybrid mixed discontinuous Galerkin finite element method for convection-diffusion problems *IMA J. Numer. Anal.* 1206-34

[7] Oikawa I 2014 A hybrid mixed discontinuous Galerkin finite-element method for convection-diffusion problems *Japan J. Indust. Appl. Math.* 335-54

[8] Arnold D N and Brezzi F 1985 Mixed and non-conforming finite element methods: implementation, post-processing and error estimates *Math. Modell. Numer. Anal.* **(1)** 7-32

[9] Cockburn B, Gopalakrishnan J and Sayas F 2010 A projection-based error analysis of HDG methods *Math. Comp.* 1351-67

[10] Dautov R Z and Fedotov E M 2014 Abstract theory of hybridizable discontinuous Galerkin methods for second-order quasilinear elliptic problems *Computational Mathematics and Mathematical Physics* **(3)** 463-80

[11] Dautov R Z 2014 On inf-sup conditions and projections in the theory of mixed finite-element methods *Differential equations* **(7)** 909-22

[12] Raviart P and Thomas J 1977 A mixed finite element method for 2-ond order elliptic problems *Lecture Notes in Math.* 292-315

IOP Conf. Series: Materials Science and Engineering **158** (2016) 012029 doi:10.1088/1757-899X/158/1/012029

Numerical modeling of photonic crystal fibers using the finite element method

R Z Dautov and E M Karchevskii

Kazan Federal University, 18 Kremlevskaya street, 420008 Kazan, Russia

E-mail: rdautov@kpfu.ru

Abstract. Propagation constants and amplitudes of eigenwaves of photonic crystal fibers are calculated numerically using an algorithm based on combination of the exact nonlocal boundary conditions method and the finite-element method. The design of fibers has a central large core filled with nematic liquid crystal. We investigate the influence of radii of the cladding air holes and their number as well as radius of the central liquid crystal on the spectral characteristics of fibers. Our results strongly suggest that radius of the crystal in contrast to the size and the number of capillaries has a significant influence on eigenwaves and propagation constants. Varying this radius we control the number of solutions of the problem for a fixed wavelength.

1. Introduction

Spectral problems of the theory of photonic crystal fibers (PCF) attract a lot of attention (see, for example, [1], [2]). PCF having a central large core filled with nematic liquid crystal (NLC) is a modern component of micro-devices used in photonics and laser technology [3]. The development of efficient numerical methods for accurate and stable computations of spectral characteristics of NLC-PCF is essential for designing and optimizing of such devices. Mesh methods, namely, various modifications of the finite-element method (FEM, see, for example, [3], [4]) and the finite-difference method (see, for example, [5], [6]) are used extensively to solve these important applied problems. Often the authors concentrate on the algorithm's features and physical interpretation of the numerical results rather than on fundamental mathematical aspects, including correctness of used models.

The original problems for eigenwaves of open dielectric waveguides, particularly, NLC-PCF, are formulated on the whole plane. From the mathematical point of view, the main difficulty in applying the mesh methods to solve such problems is the transfer of radiation conditions from infinity to the boundary of the finite mesh domain. This obstacle was overcome in two different manners by the method of exact nonlocal boundary conditions in [7] and [8]. The original problems were reduced to problems on a bounded domain (on a circle) through the use of the nonlocal boundary operators defined by the Fourier series. This approach allows us to give a new correct formulation of the problem for eigenwaves of NLC-PCF as a generalized spectral problem with a nonlinear dependence on the spectral parameter, which is applicable for the numerical solution. A more convenient for the numerical solution formulation of the spectral problem for open dielectric waveguides was proposed in [9]. It is also a problem in a bounded domain and is formulated as a generalized eigenvalue problem for self-adjoint operators in a Hilbert space, but the spectral parameter enters into it linearly. The latter property significantly simplifies the numerical solution of this class of problems and allows us to develop efficient numerical methods. An algorithm for the numerical solution of such problems based on the finite-element approximation was proposed in [9]. The convergence of this method was proved in [10].

In the presented paper, using mentioned above algorithm, we calculate numerically propagation constants and amplitudes of eigenwaves of NLC-PCF and investigate the influence of radii of the cladding air holes and their number as well as radius of the central liquid crystal on the spectral characteristics of the fibers. Our results strongly suggest that radius of the crystal in contrast to the size and the number of capillaries has a significant influence on eigenwaves and propagation constants. Varying this radius we control the number of solutions of the problem for a fixed wavelength.

2. Problem statement and the exact nonlocal boundary conditions method

In this section, following [3], we state the problem for the TE-eigenwaves of NLC-PCF. We use Cartesian coordinates and assume that the generating line of the fiber is parallel to Ox_3 axis. Figure 1 shows a schematic diagram of the investigated NLC-PCF. All the cladding air holes have the same radius r and are arranged with a hole pitch L. The big central hole has a radius r_0 and is infiltrated with NLC material. We assume that the refractive index n of this material is equal to n_0 = 1:5024. The structure of all capillaries Ω_i is surrounded by silica with the refractive index n_s = 1:45. As usual, the refractive index of air is one and the magnetic permeability of all dielectric materials is equal to the magnetic constant μ_0. Note that our consideration is true for a much more general case, namely, if the refractive index is a sufficiently smooth function of the transverse coordinates and Ω_i is a bounded domain with a piecewise smooth boundary γ.

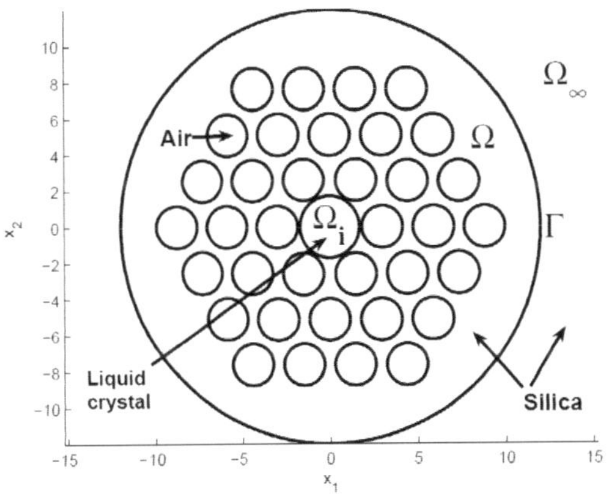

Figure 1. The structure of the suggested NLC-PCF and the decomposition of the coordinate plane R^2.

The original problem is formulated as follows: *find pairs of the numbers* $(\beta,k)\in\Lambda$ *and nonzero real-valued vanished at infinity functions u satisfying the equation*

$$-\Delta u+\beta^2 u=k^2n^2u,\quad x\in R^2\backslash\gamma, \tag{1}$$

and the boundary conditions

$$u^+=u^-,\ \frac{\partial u^+}{\partial \nu}=\frac{\partial u^-}{\partial \nu},\quad x\in\gamma. \tag{2}$$

Here $x=(x_1,x_2)$, ν is the unit outward normal vector for the curve γ, $k=(\varepsilon_0\mu_0)^{1/2}\omega$ is the longitudinal wavenumber, ε_0 is the vacuum permittivity, ω is the angular frequency, β is the propagation constant, u is the amplitude of the longitudinal component of the electric field, $\Lambda=\{(\beta,k):\ \beta/n_0<k<\beta/n_s,\ \beta>0\}$. If $(\beta,k)\in\Lambda$, then the transverse wavenumber

$$p = (\beta^2 - k^2 n_s^2)^{1/2} \tag{3}$$

is real and positive. Equation (1) has the form

$$-\Delta u + p^2 u = 0 \tag{4}$$

on the domain $\Omega_\infty = R^2 \setminus \bar{\Omega}_i$, and p defines the rate of decay of u at infinity, namely (see [11], for example),

$$u = \exp(-p|x|)O\left(1/\sqrt{|x|}\right) \; as\, |x| \to \infty.$$

We can solve the original problem as the parametric eigenvalue problem, where the parameter is either β or k, but let us introduce the new pair of the unknown parameters (β, p) and the set

$$K = \{(\beta, p) : \beta > 0,\; 0 < p < \sqrt{(n_0^2 - n_s^2)/n_0^2}\; \beta\}.$$

It is easy to see that formula (3) defines the one-to-one correspondence between the sets Λ and K. Let us consider the new problem: *find pairs of the numbers* $(\beta, p) \in K$ *and nonzero real-valued vanished at infinity functions u satisfying the following equation on* $R^2 \setminus \gamma$:

$$-\Delta u + p^2 \sigma u = \beta^2 (\sigma - 1) u, \quad \sigma = n^2 / n_s^2, \tag{5}$$

and transparency conditions (2). Using (3), we see that equation (5) transforms to equation (1). The converse is also true. Therefore the two problems are equivalent in the following sense: (β, k, u) is a solution of (1) if and only if (β, p, u) is a solution of (5). Now we reduce problem (5) to a problem posed on a circle.

We assume that the origin belongs to Ω_i. Let $B_{R_0} = \{x : |x| < R_0\}$ be the circle of the minimal radius R_0 such that $\Omega_i \subseteq B_{R_0}$. We choose a number R that more or equal to R_0 and put $\Omega = B_R$, $\Gamma = \partial\Omega$, and $\Omega_\infty = R^2 \setminus \bar{\Omega}$ (see Fig. 1). Each solution of problem (5) satisfies equation (4) on $R^2 \setminus \bar{\Omega}_i$, hence it is smooth on this domain. By u_p (u) we denote its restriction to the domain $\bar{\Omega}_\infty$ $(\bar{\Omega})$. Then the function u_p satisfies equation (4), and for $x \in \Gamma$ the following equalities hold:

$$u_p = u, \quad u_{p\nu} = u_\nu.$$

Here ν is the unit outward normal vector for the curve Γ, u_ν is the derivative of u in the direction of ν. The function u_p is the solution of the exterior boundary value problem

$$-\Delta u_p + p^2 u_p = 0, \; x \in \Omega_\infty, \quad u_p = u, \; x \in \Gamma. \tag{6}$$

Using separation of variables, we easily get

$$u_p = \sum_{n=-\infty}^{\infty} \frac{K_n(pr)}{K_n(pR)} a_n(u) e^{in\varphi}, \quad a_n(u) = \frac{1}{2\pi} \int_0^{2\pi} u\Big|_{r=R} e^{-in\varphi}\, d\varphi. \tag{7}$$

Here (r, φ) are the polar coordinates of x in the system where the pole is the center of B_R, $K_n(r)$ is the modified Bessel function of the second kind of order n. Thus, we see that (β, p, u) satisfies the following equation on the circle with the nonlocal boundary condition:

$$-\Delta u + p^2 \sigma u = \beta^2 (\sigma - 1) u, \; x \in \Omega, \quad u_\nu + S_\Gamma(p) u = 0, \; x \in \Gamma. \tag{8}$$

IOP Conf. Series: Materials Science and Engineering **158** (2016) 012029 doi:10.1088/1757-899X/158/1/012029

Here,

$$S_\Gamma(p)u = -u_{p\nu} = \frac{1}{R}\sum_{n=-\infty}^{\infty} \mathbf{K}_n(Rp)a_n(u)e^{in\varphi}, \quad \mathbf{K}_n(r) = -r\frac{K_n'(r)}{K_n(r)}. \tag{9}$$

The operator $S_\Gamma(p)$ is the mentioned above nonlocal boundary operator. Problem (8) is the desired problem on the bounded domain. In [12] we investigated the generalized solvability of problem (8) and proved its equivalence to the original problem, in [9], [10] we proposed and theoretically investigated a numerical method for problem (8) based on FEM. We solve problem (8) as the parametric eigenvalue problem for β^2, where $p > 0$ is the parameter. In the next section we describe the numerical results obtained using this method.

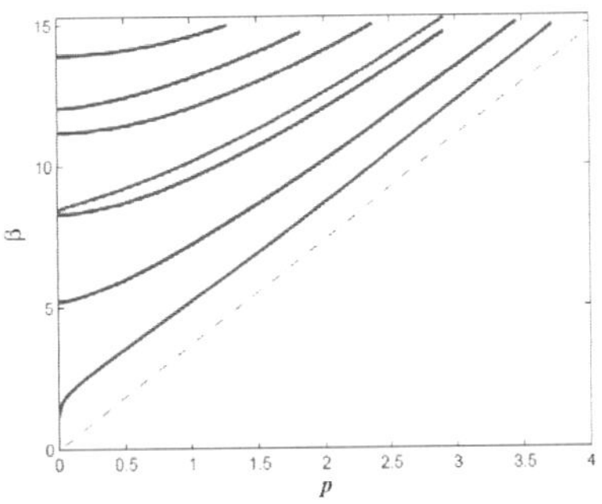

Figure 2. Variation of the propagation constant β with the transverse wavenumber p for the fiber without cladding air holes.

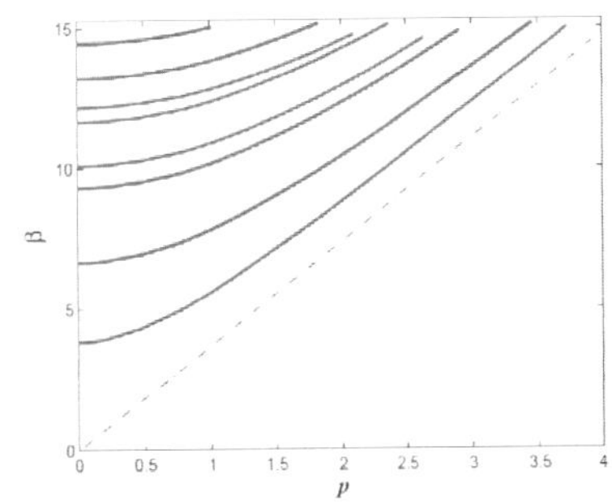

Figure 3. Variation of the propagation constant β with the transverse wavenumber p for the fiber with 36 cladding air holes.

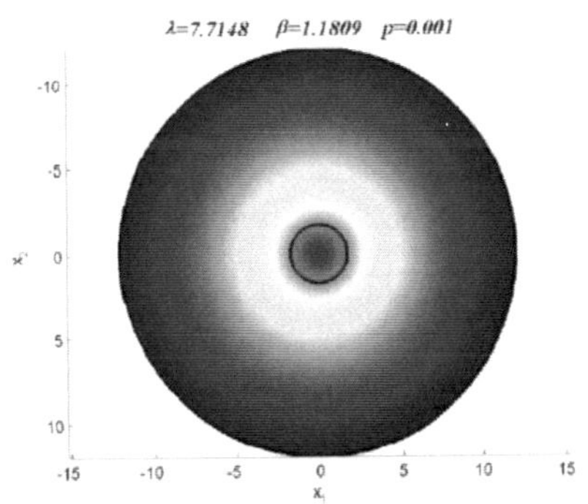

Figure 4. The fundamental mode of the fiber without cladding air holes.

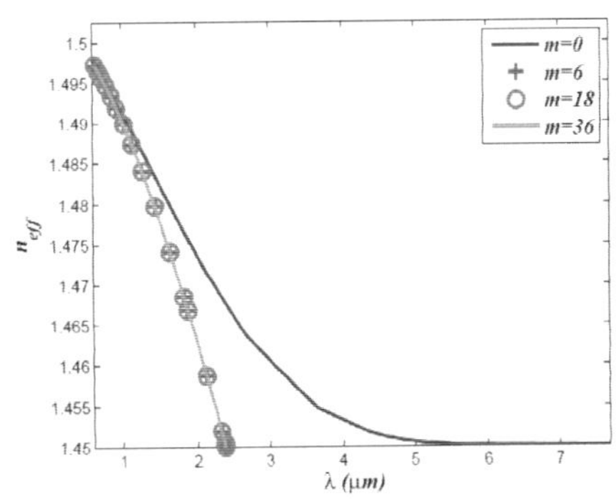

Figure 5. Variation of the effective index n_{eff} of the fundamental mode with the wavelength λ for a rank of number m of cladding air holes.

3. Numerical results and discussion

Figure 2 shows the variation of the propagation constant β with the transverse wavenumber p for the fiber without cladding air holes and core radius $r_0 = 1.7\mu m$. This structure satisfies the classical optical fiber. It is well known that the fundamental modes of such fibers have no cut-off values, i.e. the fundamental (propagating) mode can propagate for any positive wavelength λ (see [12], for example). The bottom curve starting at (0,0) is the dispersion curve of the fundamental mode, its diagram for $\lambda = 7.7148 \mu m$ is presented in Fig. 4. The variation of the effective index $n_{eff} = \beta / k$ of the fundamental mode with λ, for this fiber is presented in Fig. 5 by the blue line.

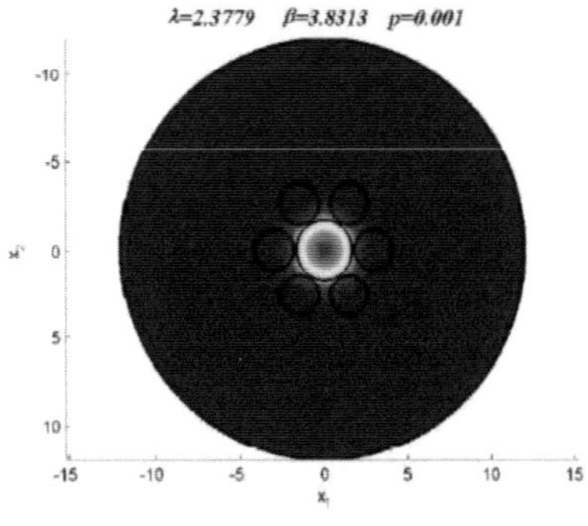

Figure 6. The fundamental mode of the fiber having 6 cladding air holes.

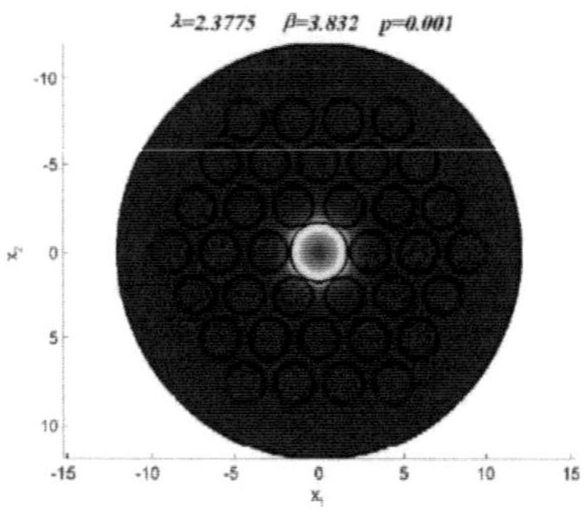

Figure 7. The fundamental mode of the fiber having 36 cladding air holes.

If the core is surrounded by the cladding air holes, then the fundamental mode has a cut-off value. Figure 3 shows the variation of the propagation constant β with the transverse wavenumber *p* for the NLC-PCF having 36 cladding air holes with radius $r = 0.4L$ and the hole pitch $L = 2.9 \mu m$, where $r_0 = 1.7 \mu m$. The bottom curve of the fundamental mode starts from the cut-off value. Variation of the effective index n_{eff} of the fundamental mode with the wavelength λ is presented in Fig. 5 by the azure line. We see that it has the cut-off value. There are not any solutions of the problem with λ more than $2.378 \mu m$. The diagram of the fundamental mode for $\lambda = 2.3775 \mu m$ is presented in Fig. 7.

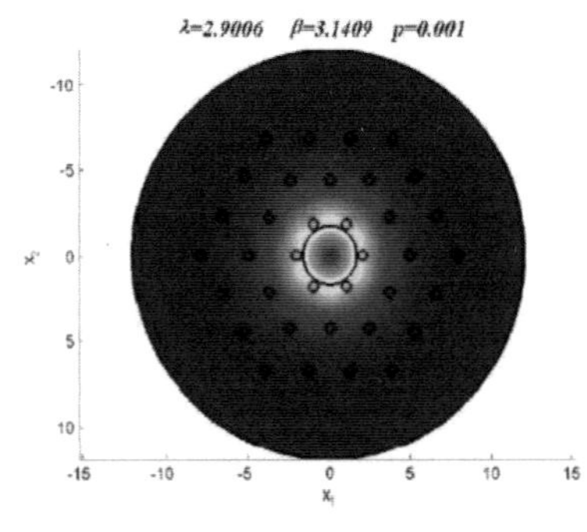

Figure 8. The fundamental mode of the NLC-PCF with 36 small cladding air holes of radius $r = 0.1L$.

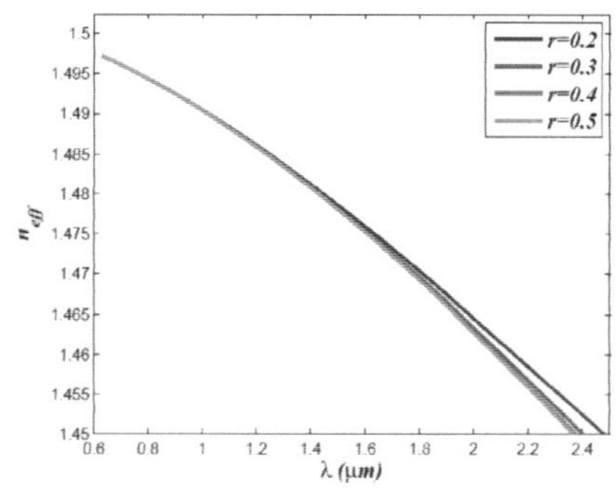

Figure 9. Variation of the effective index n_{eff} of the fundamental mode with the wavelength λ for a rank of radius of the cladding air holes *r*.

We investigated the influence of radii of the cladding air holes and their number as well as radius of the central liquid crystal on the fundamental mode. It is interesting that the size and the number of capillaries have no significant influence on the mode diagrams and the propagation constants (see Figs. 5-9).

We investigated the influence of radius of the central liquid crystal on the spectral characteristics of the fibers. Our results strongly suggest that radius of the crystal in contrast to the size and the number of capillaries has a significant influence on eigenwaves and propagation constants (see Figs. 10, 11).

Varying this radius we can control the number of solutions of the problem for a fixed wavelength (see Fig. 11).

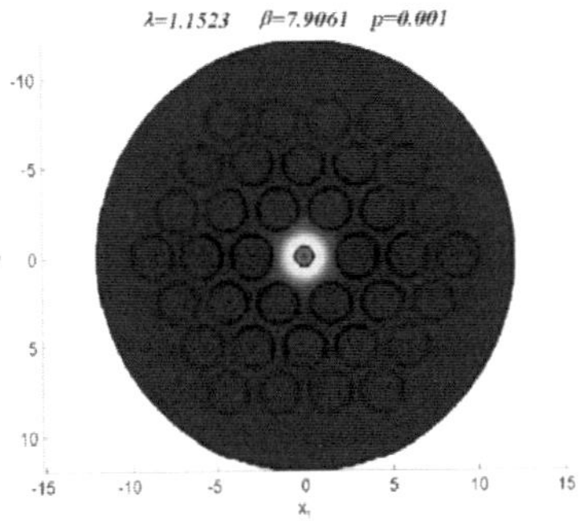

Figure 10. The fundamental mode of the fiber with 36 cladding air holes and a small central core with $r_0 = 0.5 \mu m$.

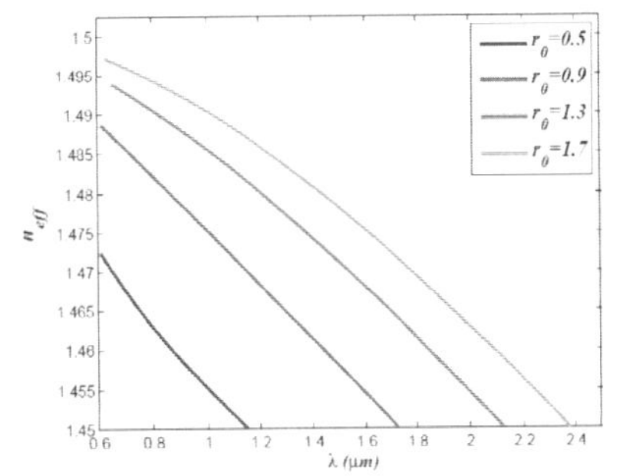

Figure 11. Variation of the effective index n_{eff} of the fundamental mode with the wavelength λ for a rank of radius of the central core r_0.

4. Conclusions

The exact nonlocal boundary conditions method together with the finite-element method gives the reliable tool for numerical modeling of PCF. The future development of these methods is urgent to calculate leaky modes of PCF.

Acknowledgments
This work was supported by the Russian Foundation for Basic Research, project no. 16-01-00408.

References

[1] Obayya S 2011 *Computational Photonics* (John Wiley and Sons)
[2] Coldren L, Corzine S, and Mashanovitch M 2012 *Diode Lasers and Photonic Integrated Circuits* (John Wiley and Sons)
[3] Hussein R A et al 2015 *Opt. Quant. Electron.* **47** 2993-3007
[4] Pintus P 2014 *Optics Express* **22** 15737-15756
[5] Kowalczyk P et al 2011 *Optics Express* **19** 19354-19364
[6] Monfared Y et al 2013 *Optik* **124** 7049-7052
[7] Joly P and Poirier C 1999 *Math. Meth. Appl. Sci.* **22** 389-447
[8] Dautov R Z and Karchevskii E M 2002 *Computational Mathematics and Mathematical Physics* **42** 1012- 1027
[9] Dautov R Z and Karchevskii E M 2005 *Computational Mathematics and Mathematical Physics* **45** 2119- 2134
[10] Dautov R Z and Karchevskii E M 2016 *Lobachevskii Journal of Mathematics* **37(5)** 610–625
[11] Kartchevski E M, Nosich A I and Hanson G W 2005 *SIAM Journal on Applied Mathematics* **65(6)** 2033-2048
[12] Dautov R Z and Karchevskii E M 2009 *Intrgral Equation Method and Exact Nonlocal Boundary Conditions in the Theory of Dielectric Waveguides* (Kazan University Press)

Additive formulation elastoplasticity at finite strain

R L Davydov and L U Sultanov

Kazan Federal University, 18 Kremlyovskaya Street, Kazan, 420008, Russian Federation

E-mail: Ruslan.Davydov@mail.ru

Abstract. In the article a method of investigation of elastic-plastic deformation for solids under complex loading is presents. Resolving equation based on the principle of virtual power. For simulation of plastic deformation taken the associated flow law with updating the stressed state. Numerical implementation is based on using 8-node finite element discretization. Numerical calculations show the suitability of the proposed methodology.

1. Introduction

There are many publications, where solutions of nonlinear problems of a solid mechanic are discussed, for example [10-13].In this paper a numerical algorithm of the investigation of stress-strain state of the elastic–plastic solids with large deformations is described.

2. Kinematics

The deformation gradient tensor $\mathbf{F}$, the left Cauchy–Green tensor $\mathbf{B}=\mathbf{F}\cdot\mathbf{F}^{\mathrm{T}}$, the corresponding spatial gradient of velocity $\mathbf{h}=\dot{\mathbf{F}}\cdot\mathbf{F}^{-1}$ and the rate of deformation $\mathbf{d}=\mathrm{sym}(\mathbf{h})$ are used for describing of kinematics of a continuum. The basic relationships are described in [1, 6-9].

3. Constitutive equations

The constitutive equations are obtained using the free energy function ψ that plays the role of an elastic potential and yield function. The Cauchy stress tensor is defined as [1, 5]

$$\mathbf{\Sigma}=\frac{2}{J}\mathbf{B}\cdot\frac{\partial\psi}{\partial\mathbf{B}}, \qquad (1)$$

where $J=\det(\mathbf{F})$ is a changing of volume. For isotropic material the free energy ψ is defined as $\psi=\psi\left(I_{1\mathbf{B}},I_{2\mathbf{B}},I_{3\mathbf{B}}\right)$, where $I_{i\mathbf{B}}$ is the corresponding invariants of tensor $\mathbf{B}$.

After linearization (1) the rate of Cauchy stress is defined as

$$\dot{\mathbf{\Sigma}}=2\left\{\frac{1}{J}\dot{\mathbf{B}}\cdot\frac{\partial\psi}{\partial\mathbf{B}}+\frac{1}{J}\left[\mathbf{B}\cdot\frac{\partial^2\psi}{\partial\mathbf{B}^2}\right]\cdot\cdot\dot{\mathbf{B}}-\frac{1}{J}\mathbf{B}\cdot\frac{\partial\psi}{\partial\mathbf{B}}I_{1\mathrm{d}}\right\}=\mathbf{\Lambda}_{\Sigma}\cdot\cdot\mathbf{d}+\mathbf{h}\cdot\mathbf{\Sigma}+\mathbf{\Sigma}\cdot\mathbf{h}^{\mathrm{T}}-\mathbf{\Sigma}I_{1\mathrm{d}},$$

where $\mathbf{\Lambda}_{\Sigma}=\frac{4}{J}\mathbf{B}\cdot\frac{\partial^2 W}{\partial\mathbf{B}\partial\mathbf{B}}\cdot\mathbf{B}$ – fourth order constant tensor.

Or

$$\mathbf{\Sigma}^{Tr}=\mathbf{\Lambda}_{\Sigma}\cdot\cdot\mathbf{d}, \qquad (2)$$

where $\mathbf{\Sigma}^{Tr} = \dot{\mathbf{\Sigma}} + \mathbf{h}\cdot\mathbf{\Sigma} + \mathbf{\Sigma}\cdot\mathbf{h}^{\mathrm{T}} - I_{1\mathbf{d}}\mathbf{\Sigma}$ is the Truesdell stress rate. Therefore, constitutive model can be defined as linear relation between an objective derivative of Cauchy stress tensor and rate of deformation.

The theory of flow is used for describing plastic deformation [2-5]. The total deformation rate is represented as a sum of elastic and plastic parts: $\mathbf{d} = \mathbf{d}^e + \mathbf{d}^p$ [2-5]. For the plastic deformation rate must hold association flow rule:

$$\mathbf{d}^p = \dot{\gamma}\frac{\partial \Phi}{\partial \mathbf{\Sigma}}, \tag{3}$$

where $\dot{\gamma}$ is the consistency parameter, Φ is a yield function.

4. Integration algorithm of the flow rules

For the solution problem of a plastic flow the general return method is used [5, 14-16]. The solution at time ${}^{k}t$ is known. The stress at time ${}^{k+1}t = {}^{k}t + \Delta^{k+1}t$ is defined from the initial conditions at time ${}^{k}t$. The trial stress is defined as

$${}^{k+1}\tilde{\mathbf{\Sigma}} = {}^{k}\mathbf{\Sigma} + {}^{k+1}\Delta\mathbf{\Sigma}. \tag{4}$$

The plastic flow $\Delta^{k+1}\gamma$ can be computed from equation

$${}^{k+1}\mathbf{\Sigma} + \Delta^{k+1}\gamma\frac{\partial^{k+1}\Phi}{\partial^{k+1}\Sigma}\cdot\cdot{}^{k+1}\mathbf{\Sigma} = {}^{k+1}\tilde{\mathbf{\Sigma}}. \tag{5}$$

The solution is obtained by solving a nonlinear system of equations (5) with linearized Newton scheme using (2)–(4).

5. Variation formulation

The research algorithm is based on an Update Lagrange formulation. The principle of virtual work in terms of the virtual velocity is used [1-5]:

$$\int_{\Omega}\mathbf{\Sigma}:\delta\mathbf{d}d\Omega = \int_{\Omega}\mathbf{f}\cdot\delta\mathbf{v}d\Omega + \int_{S^{\sigma}}\mathbf{p}\cdot\delta\mathbf{v}dS,$$

where Ω is the current volume, S^{σ} is the surface on which the force $\mathbf{p}$ is applied, $\mathbf{f}$ is the body force vector, $\mathbf{v}$ is a velocity vector. After linearization the system of linear equations is obtained, where the unknown is the increment of displacement in the current state $\mathbf{\Delta}^{k+1}\mathbf{u}$. For solving general system of equations the arc-length method is applied [5]. The current state is defined as ${}^{k+1}\mathbf{R} = {}^{k}\mathbf{R} + \Delta^{k+1}\mathbf{u}$. Then the trial stress is calculated by (4). And if $\Phi({}^{\mathbf{k+1}}\tilde{\Sigma}) \le 0$ then the Cauchy stress ${}^{k+1}\mathbf{\Sigma} = {}^{k+1}\tilde{\mathbf{\Sigma}}$, else the radial return method with an iterative refinement of the current mode of deformation is applied [5].

6. Numerical example

As an example the potential of elastic deformation is considered:

$$\psi = \frac{\lambda + 2\mu}{8}\left(I_{1\mathbf{B}} - 3\right)^2 + \mu\left(I_{1\mathbf{B}} - 3\right) - \frac{\mu}{2}\left(I_{2\mathbf{B}} - 3\right),$$

here λ, μ are Lame parameters. The von Mises yield criterion with isotropic hardening is used: $\Phi = \sigma_i - \xi(\chi) \le 0$, where $\sigma_i = \sqrt{1.5\, dev\mathbf{\Sigma}: dev\mathbf{\Sigma}}$, $\xi(\chi) = \sigma_T + h\chi + (\sigma_\infty - \sigma_T)(1 - e^{-\delta\chi})$ is the hardening function. The material data for isotropic elasticity and the von Mises yield condition are given as follows: E=206.9 GPa, $\nu = 0.29$, $\sigma_\infty = 0.715$ GPa, $\sigma_T = 0.450$ GPa, $h = 0.129$, $\delta = 16.93$.

The numerical implementation is based on the finite element method. An 8-node brick element is used [2-5, 9].

6.1. Necking of a circular bar. The necking of a circular bar is an example widely investigated in the literature; see e.g. [14] or [15]. To initialize the necking process radius in the center is reduced by 1.8 %. Figure 1 displays the final deformed structure and the equivalent plastic strain, which concentrates in the necking zone. The results are in very good agreement with the computational reference solutions of [14] and [15].

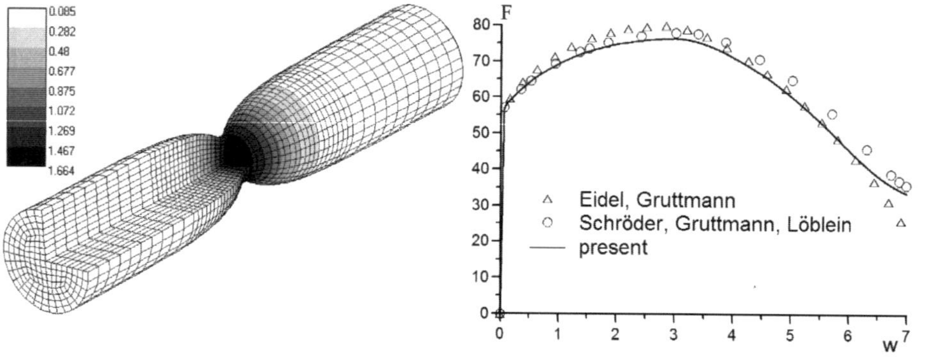

Figure 1. Left: Equivalent plastic strain at final structure.
Right: Computational results of applied force F [kN] versus axial elongation w [mm].

6.2. Conical shell. In the second example of isotropic elastoplasticity a conical shell subjecting to a constant ring load is considered [9–10]. Figure 2 displays equivalent plastic strain for the several deformed structure.

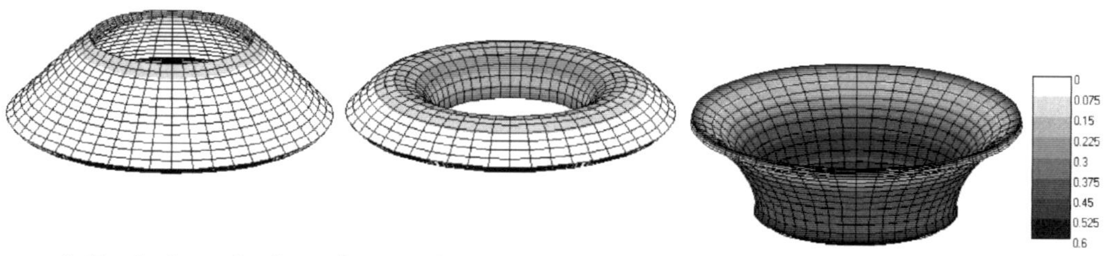

Figure 2. Equivalent plastic strain on deformed structures at different stages of the punching process.

6.3. Drawing of a Circular Blank. Two different materials are considered for orthotropic yielding. For material A the shear stresses dominate in the yield criterion, for material B the normal stresses are predominant in yielding. As expected the plastic strains concentrate for material A at a 45° angle in the (x, y)-plane and for material B along the x-and y-axes, see Figure 3.

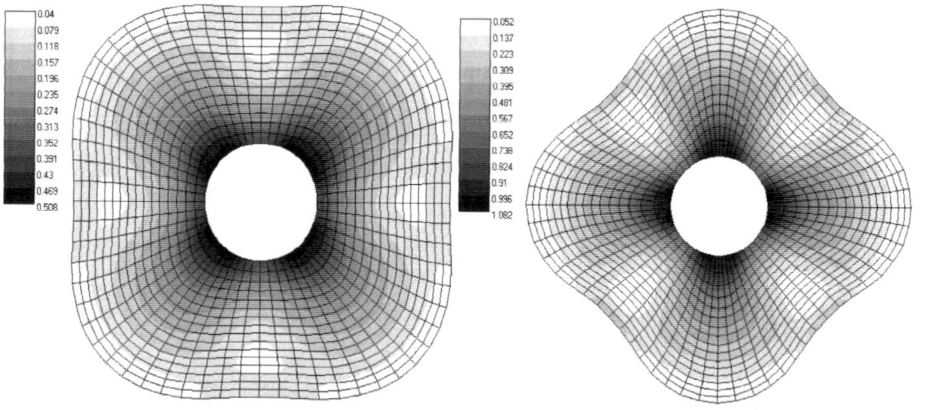

Figure 3. Left: Equivalent plastic strain on deformed structures (material A).
Right: Equivalent plastic strain on deformed structures (material B).

7. Conclusions

In the paper an additive formulation of the elastic-plastic deformation presents and finite element implementation is considered. The physical conditions obtained using the free energy function. The linearized equations derived from the principle of virtual work for the current state. Mises yield criterion with isotropic hardening is used. The effectiveness of present research algorithm of large elastic-plastic deformation demonstrate resolved problems, results are compared with the results of other authors.

Acknowledgements

The reported study was supported by Government of the Republic of Tatarstan research projects No. 15-41-02557 and RFBR research projects No. 15-01-08733.

References

[1] Paimushin V N 2008 *Journal of Applied Mathematics and Mechanics* **72 (5)** 597–610
[2] Simo J S 1988 *Computer Methods in Applied Mechanics and Engineering* **66 (2)** 199–219
[3] Basar Y and Itskov M 1999 *Computational mechanics* **23 (5)** 466–481
[4] Meyers A, Schievbe P and Bruhns O T 2000 *Acta Mech.* **139 (1)** 91–103
[5] Sagdatullin M and Berezhnoi D 2014 *Applied Mathematical Sciences* **8 (35)** 1731–1738
[6] Sagdatullin M and Berezhnoi D 2014 *Applied Mathematical Sciences* **8 (60)** 2965–2972
[7] Badriev I B, Banderov V V, Garipova G Z, Makarov M V and Shagidullin R R 2015 *Applied Mathematical Sciences* **9 (82)** 4095–4102
[8] Badriev I B and Shagidullin R R 1995 *Journal of Mathematical Sciences* **73 (5)** 519–525
[9] Sachenkov O, Kharislamova L, Shamsutdinova N, Kirillova E and Konoplev Yu 2015 *IOP Conference Series: Materials Science and Engineering* **98**
[10] Shigapova F A, Mustakimova R F, Saleeva G T and Sachenkov O A *International Journal of Applied Engineering Research* **10 (24)** 44711–44714
[11] Golovanov A I, Konoplev Yu G and Sultanov L U 2008 *Uchenye Zapiski Kazanskogo Universiteta. Seriya Fiziko-Matematicheskie Nauki* **150 (1)** 29–37
[12] Golovanov A I, Konoplev Yu G and Sultanov L U 2008 *Uchenye Zapiski Kazanskogo Universiteta. Seriya Fiziko-Matematicheskie Nauki* **150 (3)** 122–132
[13] Golovanov A I, Konoplev Yu G and Sultanov L U 2010 *Uchenye Zapiski Kazanskogo Universiteta. Seriya Fiziko-Matematicheskie Nauki* **151 (3)** 108–120
[14] Golovanov A I, Konoplev Yu G and Sultanov L U 2010 *Uchenye Zapiski Kazanskogo Universiteta. Seriya Fiziko-Matematicheskie Nauki* **152 (4)** 115–126
[15] Bonet J and Wood R D 1997 *Nonlinear continuum mechanics for finite element analysis* (Cambridge University Press) 283
[16] Davydov R L and Sultanov L U 2013 *PNRPU Mechanics Bulletin* **1** 81–93
[17] Golovanov A I and Sultanov L U 2005 *International Applied Mechanics* **41 (6)** 614–620
[18] Sultanov L U 2014 *Applied Mathematical Sciences* **8 (143)** 7117–7124
[19] Sultanov L U and Davydov R L and 2014 *Applied Mathematical Sciences* **8 (60)** 2991–2996
[20] Sultanov L U 2015 *Procedia Earth and Planetary Science* **15** 119–124
[21] Abdrakhmanova A I and Sultanov L U 2016 *Materials Physics and Mechanics* **26 (1)** 30–32
[22] Davydov R and Sultanov L 2013 *Sixth International Conference on Nonlinear Mechanics (ICNM-VI)* 64–67
[23] Davydov R L and Sultanov L U 2015 *Journal of Engineering Physics and Thermophysics* **88 (5)** 1280–1288
[24] Sultanov L U and Fakhrutdinov L R 2013 *Magazine of Civil Engineering* **44(9)** 69–74
[25] Sultanov L U and Davydov R L 2013 *Magazine of Civil Engineering* **44(9)** 64–68
[26] Sultanov L U 2016 *Lobachevskii Journal of Mathematics* **37 (6)** 787–793
[27] Eidel B and Gruttmann F 2003 *Computational Materials Science* **28 (3-4)** 732–742
[28] Schröder J, Gruttmann F and Löblein J 2002 *Computational Mechanics* 48–4
[29] Eidel B and Gruttmann F 2003 *Springer Lecture Notes in Mechanics* **10** 51–78

Numerical investigation of large strains of hyperelastic solids

L R Fakhrutdinov and L U Sultanov

Kazan Federal University, 18 Kremlyovskaya str., Kazan, Russia

E-mail: bishchumbek@gmail.com

Abstract. The work is concerned with numerical investigation of large deformations of hyperelastic solids. The kinematics of continua is described in terms of the deformation gradient tensor and left Cauchy–Green tensor. The stress state is described by the Cauchy stress tensor. An incremental method is used to solve the nonlinear problems. The linearized physical equations for the Cauchy stresses for Mooney–Rivlin material are given.

1. Introduction

Different rubber-like materials, elastic polymers, foams and even biological tissues [8, 9] can be described by nonlinear rheological models. According the mechanics of solids, hyperelastic materials are nonlinear. So while studying their deformations we must consider geometrical nonlinearity in terms of large deformations. There are plenty of different approaches described in literature [1–7].

2. Kinematics. Constitutive equations

Basic tensor of the kinematics of continua is deformation gradient tensor (F). The following tensors are used to describe strain and strain rate [10, 15–25]:

$(B)=(F)\cdot(F)^T$ – the left Cauchy–Green tensor (Finger deformation tensor),

$(h)=(\dot{F})\cdot(F^{-1})$ – the velocity gradient tensor,

$(d)=\frac{1}{2}\left[(h)+(h)^T\right]$ – the rate of deformation tensor.

Stress state is described by Cauchy tensor of true stresses (Σ) defined in the actual state. As arguments of strain energy function we take Finger tensor [11, 15–25]:

$$W=W(B). \tag{1}$$

To divide volumetric and shearing strains non-volumetric strain measures are considered:

$$(\hat{B})=J^{-2/3}(B),$$

here $J=dV/dV_0$ is the ratio of the deformed elastic volume over the reference (undeformed) volume of materials.

Then (1) will become:

$$W=W_0(J)+W'(I_{1\hat{B}},I_{2\hat{B}}).$$

Cauchy stress tensor in this case expresses as:

$$(\Sigma)=\frac{2}{J}(F)\cdot(F)^T\cdot\left(\frac{\partial W}{\partial B}\right)=\frac{2}{J}(B)\cdot\left(\frac{\partial W}{\partial B}\right).$$

Cauchy stress rate:

$$(\dot{\Sigma})=(\Lambda_\Sigma)\cdot\cdot(d)+(h)\cdot(\Sigma)+(\Sigma)\cdot(h)^T-(\Sigma)I_{1d},$$

where $(\Lambda_\Sigma)=\frac{4}{J}(B)\cdot\left(\frac{\partial^2 W}{\partial B \partial B}\right)\cdot(B)$ [12, 15–25].

3. Calculation algorithm

To obtain numerical solution of the task Upgrade Lagrange formulation is used. As resolving expression principle of virtual work, written for the step $(k+1)$ is taken [12, 17, 23]:

$$\iiint\limits_{V_{k+1}}\left({}^{k+1}\Sigma\right)\cdot\cdot\left(\delta\,{}^{k+1}d\right)dV=\iiint\limits_{V_{k+1}}{}^{k+1}\vec{f}\cdot\delta\vec{\upsilon}dV+\iint\limits_{S^{\sigma}_{k+1}}{}^{k+1}\vec{t}_n\cdot\delta\vec{\upsilon}dS, \tag{2}$$

here V_{k+1} – actual volume, S^{σ}_{k+1} – loaded surface, $\vec{f},\vec{t}_n$ – body and surface force vectors.

After linearization and rewriting Cauchy stress tensor in increments:

$$\left({}^{k+1}\Sigma\right)=\left({}^{k}\Sigma\right)+\left(\Delta^k\Sigma\right) \tag{3}$$

in expression (2) displacement vector for current configuration $\Delta^k\vec{u}=\Delta^k x_i\vec{e}_i$ could be obtained. This vector defines next step configuration:

$${}^{k+1}\vec{R}={}^{k}\vec{R}+\Delta^k\vec{u}. \tag{4}$$

And stress tensor in this case will be:

$$\left({}^{k+1}\Sigma\right)=\frac{2}{J}\left({}^{k+1}B\right)\cdot\left(\frac{\partial W}{\partial\,{}^{k+1}B}\right). \tag{5}$$

4. Numerical example

As an example obtaining physical equations for material described by the Mooney–Rivlin elastic strains potential function [14]:

$$W=U_1(I_{1B}-3)+U_2(I_{2B}-3)+\frac{K}{2}(J-1)^2,$$

here $W_0(J)=\frac{K}{2}(J-1)^2$, $W'(I_{1B},I_{2B})=U_1(I_{1B}-3)+U_2(I_{2B}-3)$, U_1,U_2 – material constants.

For this material Cauchy stress tensor expresses as:

$$(\Sigma)=(\Sigma_0)+(\Sigma'),$$

where $(\Sigma')=\frac{2}{J}(B)\cdot\left(\frac{\partial W'}{\partial B}\right)$ – non-volumetric part, $(\Sigma_0)=\frac{2}{J}(B)\cdot\left(\frac{\partial W_0}{\partial B}\right)$ – volumetric part.

So:

$$(\Sigma)=2U_1J^{-\frac{5}{3}}\left[(B)-\frac{1}{3}I_{1B}(I)\right]+2U_2J^{-\frac{7}{3}}\left[I_{1B}(B)-\frac{1}{3}I^2_{1B}(I)-\frac{2}{3}(B^2)\right]+K(J-1)(I). \tag{6}$$

Linearizing (6), expression for Cauchy stress rate could be obtained:

$$(\Delta\Sigma)=(\Lambda_\Sigma)\cdot\cdot(d)+(h)\cdot(\Sigma)+(\Sigma)\cdot(h)^T-(\Sigma)I_{1d},$$

where $(\Lambda_\Sigma) = (\Lambda_{\Sigma'}) + (\Lambda_{\sigma 0})$, $\Lambda_{\Sigma'} = \frac{4}{J}(B) \cdot \left(\frac{\partial^2 W'}{\partial B \partial B} \right)$ – non-volumetric part, $\Lambda_{\sigma 0} = \frac{4}{J}(B) \cdot \left(\frac{\partial^2 W_0}{\partial B \partial B} \right)$ – volumetric part.

It gives

$$\Lambda_\Sigma = \frac{4}{3} J^{-\frac{5}{3}} U_1 \left\{ \frac{1}{3} I_{1B} (I)(I) - (I)(B) - (B)(I) + I_{1B} (C_{II}) \right\} +$$
$$+ \frac{4}{3} J^{-\frac{7}{3}} U_2 \left\{ \frac{2}{3} I_{1B}^2 (I)(I) - 2 I_{1B} (I)(B) + \frac{4}{3} (I)(B^2) + 3(B)(B) - 2 I_{1B} (B)(I) - \right.$$
$$\left. - I_{1B}^2 (C_{II}) - 2(B) \cdot (C_{II}) \cdot (B) \right\} + K \left\{ (J-1)(I)(I) + J(I)(I) + 2(J-1)(C_{II}) \right\}$$

So system (2)–(5) can be implemented to define stress state of hyperelastic materials which can be discretized by finite element method to obtain numerical solutions.

To prove efficiency of described algorithm it was implemented to some test problems. Here is one of them:

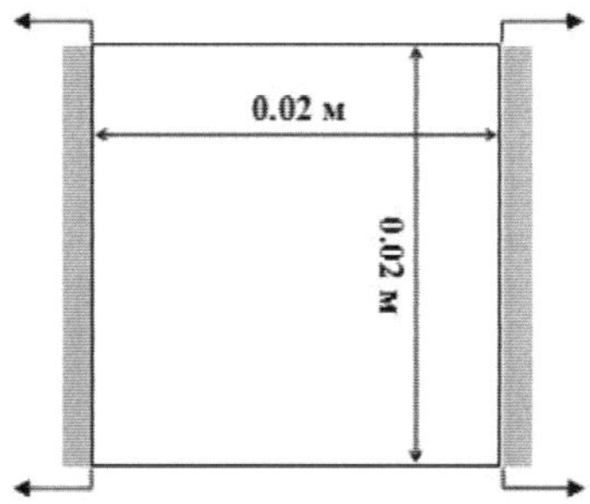

Figure 1. Plain deformation of square strip.

Plain deformation of square strip [13, 14, 17, 23] $0.02 \times 0.02 \text{ m}^2$, material properties set as $U_1 = 0.36 MPa$, $U_2 = 0{,}22 MPa$, $K = 2000 \text{MPa}$; displacements defined on the vertical sides of the strip as shown on figure 1 taking into account symmetry of the problem only 0.25 of strip was considered. On figure 2 vertical displacements field of the 0.25 part of the strip is shown.

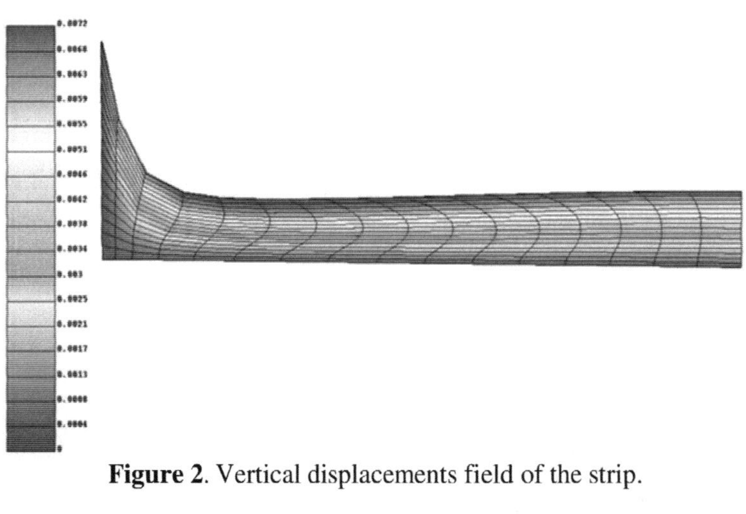

Figure 2. Vertical displacements field of the strip.

5. Conclusion

Numerical algorithm of investigation of hyperelastic solids using Finger strain measure is described. Physical equations are defined using strain potential function. For Mooney–Rivlin potential linearized constitutive relations and resolving expression are obtained. Numerical implementation is based on finite element method using 8-node brick element.

Acknowledgments

The reported study was supported by Government of the Republic of Tatarstan research projects No. 15-41-02557 and RFBR research projects No. 15-01-08733.

References

[1] Paimushin V N 2008 *Journal of Applied Mathematics and Mechanics* **72 (5)** 597–610
[2] Simo J S 1988 *Computer Methods in Applied Mechanics and Engineering* **66 (2)** 199–219
[3] Basar Y and Itskov M 1999 *Computational mechanics* **23 (5)** 466–481
[4] Sagdatullin M and Berezhnoi D 2014 *Applied Mathematical Sciences* **8 (35)** 1731–1738
[5] Sagdatullin M and Berezhnoi D 2014 *Applied Mathematical Sciences* **8 (60)** 2965–2972
[6] Badriev I B, Banderov V V, Garipova G Z, Makarov M V and Shagidullin R R 2015 *Applied Mathematical Sciences* **9 (82)** 4095–4102
[7] Badriev I B and Shagidullin R R 1995 *Journal of Mathematical Sciences* **73 (5)** 519–525
[8] Sachenkov O, Kharislamova L, Shamsutdinova N, Kirillova E and Konoplev Yu 2015 *IOP Conference Series: Materials Science and Engineering* **98**
[9] Shigapova F A, Mustakimova R F, Saleeva G T and Sachenkov O A *International Journal of Applied Engineering Research* **10 (24)** 44711–44714
[10] Golovanov A I, Konoplev Yu G and Sultanov L U 2008 *Uchenye Zapiski Kazanskogo Universiteta. Seriya Fiziko-Matematicheskie Nauki* **150 (1)** 29–37
[11] Golovanov A I, Konoplev Yu G and Sultanov L U 2008 *Uchenye Zapiski Kazanskogo Universiteta. Seriya Fiziko-Matematicheskie Nauki* **150 (3)** 122–132
[12] Golovanov A I, Konoplev Yu G and Sultanov L U 2010 *Uchenye Zapiski Kazanskogo Universiteta. Seriya Fiziko-Matematicheskie Nauki* **151 (3)** 108–120
[13] Golovanov A I, Konoplev Yu G and Sultanov L U 2010 *Uchenye Zapiski Kazanskogo Universiteta. Seriya Fiziko-Matematicheskie Nauki* **152 (4)** 115–126
[14] Bonet J and Wood R D 1997 *Nonlinear continuum mechanics for finite element analysis* (Cambridge University Press) 283
[15] Davydov R L and Sultanov L U 2013 *PNRPU Mechanics Bulletin* **1** 81–93
[16] Golovanov A I and Sultanov L U 2005 *International Applied Mechanics* **41 (6)** 614–620
[17] Sultanov L U 2014 *Applied Mathematical Sciences* **8 (143)** 7117–7124
[18] Sultanov L U and Davydov R L and 2014 *Applied Mathematical Sciences* **8 (60)** 2991–2996
[19] Sultanov L U 2015 *Procedia Earth and Planetary Science* **15** 119–124
[20] Abdrakhmanova A I and Sultanov L U 2016 *Materials Physics and Mechanics* **26 (1)** 30–32
[21] Davydov R and Sultanov L 2013 *Sixth International Conference on Nonlinear Mechanics (ICNM-VI)* 64–67
[22] Davydov R L and Sultanov L U 2015 *Journal of Engineering Physics and Thermophysics* **88 (5)** 1280–1288
[23] Sultanov L U and Fakhrutdinov L R 2013 *Magazine of Civil Engineering* **44(9)** 69–74
[24] Sultanov L U and Davydov R L 2013 *Magazine of Civil Engineering* **44(9)** 64–68
[25] Sultanov L U 2016 *Lobachevskii Journal of Mathematics* **37 (6)** 787–793

IOP Conf. Series: Materials Science and Engineering **158** (2016) 012032 doi:10.1088/1757-899X/158/1/012032

Vector random fields in mathematical modelling of electron motion

F G Gabbasov[1], V T Dubrovin[2] and V J Chebakova[2]

[1]Kazan state university of architecture and engineering, 1 Zelenaya Str., Kazan, 420043, Russia
[2]Kazan Federal University, 18 Kremlevskaya Street, Kazan, 420008, Russia

E-mail: vchebakova@mail.ru

Abstract. In order to use electron motion for mathematical modelling we assessed convergence rate in central limit theorem for homogeneous vector random fields, which fulfil condition of strong mixing.

1. Introduction

Capacitive coupled radio-frequency (CCRF) discharges are widely used for treatment of organic materials because they have low gas temperature. When making numerical calculation of CCRF-discharge characteristics in approximation of continuous medium, we need apply the values of the factors of the initial-boundary value problems, which are the part of the mathematical model of the discharge (factors of diffusion, mobility of charged factors, velocity of plasma chemical reaction) [1-3]. We have very little experimental information, that is why the factors for electron gas are determined by solving Boltzmann equation, and Maxwellian distribution function is used for ion gas as a rule. However, these methods are not accurate under high values of the electrical field observed in electrode sheath of glow discharge and CCRF discharge; in order to account impact of the field, they apply statistical calculation methods for velocity distribution function of discharged particles [4-7]. Here we use statistical analysis of random fields. It is known that limit theorems are significantly important for statistical methods. Many authors use central limit theorem for random fields under some low dependency limits [7-10]. This research paper is dedicated to application of asymptotic decomposition of the characteristic functions and assessment of sum semi-invariants of random vectors in order to obtain values of convergence rate in limit theorem for weakly dependent vector random fields.

2. Statement of a theorem

Let us assume that $\xi(t_1,...,t_n)$ – homogenous, in restricted sense, vector random field, assumed on integer lattice Z^n, in other words the random function reflecting Z^n in m-dimensional Euclidean space R^m.

We shall set $S_{T_1},...,_{T_n} = \sum_{t_1=1}^{T_1}...\sum_{t_n=1}^{T_n}\xi(t_1,...,t_n)$, $d(E_1,E_2) = \inf_{t\in E_1, s\in E_2}\sqrt{\sum_{i=1}^{n}(t_i - s_i)^2}$, $t=(t_1,...,t_n)$, $s=(s_1,...,s_n)$.

We shall introduce the function $\alpha(E_1, E_2) = \sup|P(AB) - P(A)P(B)|$ for any invariant sets E_1, E_2 from Z^n, where the upper boundary is taken at all events A, B belonging to minimal σ-algebras, resulted by $\xi(t_1,...,t_n)$, when the vector $(t_1,...,t_n)$ belongs to E_1, E_2, correspondingly.

Theorem. Let us assume that the following conditions are applied:

1) Random vectors $\xi(t_1,...,t_n)$ have zero vector of mathematical mean value;

2) $|\xi(t_1,...,t_n)| \le B$, B- positive constant;

3) random field $\xi(t_1,...,t_n)$ corresponds to the condition of strong mixing with the factor $\alpha(E_1, E_2) \le Ad(E_1, E_2)^{-\omega}$, where $E_1 \cap E_2 = \varnothing$, ω - significantly large fixed positive number, A - positive constant;

4) covariance matrix of the sum $S_{T_1,...,T_n} / \sqrt{T_1 T_2 ... T_n}$ is not confluent and under $T_1,...,T_n \to \infty$ is approaching to the unit matrix.

So, $\sup\limits_M \left|P(S_{T_1,...,T_n} / \sqrt{T_1...T_n} \in M) - \Phi(M)\right| = O((T_1...T_n)^{-\frac{1}{2n}+\varepsilon})$, under $T_1,...,T_n \to \infty$, where the upper boundary is taken at all convex measurable sets $M \subset R^m$, $\varepsilon = 3n / ((n-1)\sqrt[8]{\omega})$, Φ - normal distribution with unit matrix of co-variations and zero mean vector

3. Theorem proof.

The theorem is proved by using "step-by-step approximations" method developed by V.T. Dubrovin, D.A. Moskvin to study rates of convergence in boundary value theorems for the sum functions from slightly dependent values; development of this method was continued by F.G. Gabbasov for multidimensional boundary value theorems for slightly dependent vectors [11-15].

Let us provide the values used to prove the theorem. We shall introduce a random value, where $x = (x_1,...,x_m)$, $|x| = \sqrt{\sum_{i=1}^{m} x_i^2}$. We shall specify $\chi_\nu(Q_1,...,Q_n)$ a semi-invariant of ν - order of the sum $\sum_{k_1=1}^{Q_1}...\sum_{k_n=1}^{Q_n} \tau_{k_1,...,k_n}$.

Lemma 1. The following estimate is fair

$$|\chi_\nu(Q_1,...,Q_n)| = O(Q_1 Q_2 ... Q_n).$$

The lemma is proved with the same way as in [12].

Lemma 2. Under fixed ν, $1 \le \nu < \omega / n$, the estimate is fair

$$E\left|\sum_{k_1=1}^{h_1}...\sum_{k_n=1}^{h_n} \tau_{k_1...k_n}\right|^{2\nu} = O((h_1 h_2 ... h_n)^{\nu+2n\nu^2/\omega}$$

The lemma shall be proved as lemma 2 from [12].

Let us assume that P_Q-distribution of the sum $\sum_{k_1=1}^{Q_1}...\sum_{k_n=1}^{Q_n} \xi(k_1,...,k_n)$, and $f_Q(x)$-its characteristic function.

Lemma 3. If under some $0 < \alpha \le 1/2$, and some $\gamma, \gamma > 1$ asymptotic equality takes place

$$|P_Q(M) - \Phi(M)| = O(\frac{1}{1+\beta^\gamma(M)} \frac{1}{(Q_1...Q_n)^\alpha}) \tag{1}$$

under $Q_1,...,Q_n \to \infty$, for every $\delta > 0$ there are such $\Delta, \varepsilon > 0$, the inequation $\max\limits_{\delta < |x| < \varepsilon(Q_1...Q_n)^\alpha} |f_Q(x)| \le 1 - \Delta$ holds, $\beta(M) = \inf\limits_{x \in \delta(M)} |x|$, where $\delta(M)$ is boundary of convex set M.

The lemma shall be proved with the similar way as lemma 2 from the research paper [13]

Let us introduce the following symbols. Let us assume that N and Q_i are random, growing together with $T_1,...,T_n$, natural numbers. Let us assume that

$$E_{k_1,...,k_n}=\{(t_1,...,t_n)\in Z^n:(k_i-1)(Q_i+N)<t_i\le k_i(Q_i+N),i=\overline{1,n}\},$$

$$E'_{k_1,...k_n}=\{(t_1,...,t_n)\in Z^n:(k_i-1)(Q_i+N)<t_i\le(k_i-1)(Q_i+N)+Q_i,i=\overline{1,n}\}$$

$$E^0_{p_1+1,...,p_n+1}=\{(t_1,...,t_n)\in Z^n:p_i(Q_i+N)<t_i\le T_i,i=\overline{1,n}\},$$

$$E^0_{k_1,...,k_n}=E_{k_1,...,k_n}/E'_{k_1,...,k_n},$$

where $1\le k_i\le p_i, p_i=[T_i/(Q_i+N)]$, [ ] – the sign of an integral part. Moreover, let us assume, that

$$|T_i-p_i(Q_i+N)|\le p_i,i=\overline{1,n}. \quad (2)$$

Let us indicate $Q=Q_1Q_2...Q_n$, $p=p_1p_2...p_n$.
Further, let us compose the sums
$\eta_{k_1,...,k_n}=S_{E'_{k_1...k_n}}/\sqrt{Q}$, $\eta^0_{k_1,...,k_n}=S_{E^0_{k_1...k_n}}/\sqrt{Q}$, $\eta^0_{p_1+1,...,p_n+1}=S_{E^0_{p_1+1...p_n+1}}/\sqrt{Q}$ where summing $\xi(t_1,...,t_n)$ is made according to the stated subsets of integral lattice Z^n. We shall compose the sums

$$\varsigma_p=\sum_{k_1=1}^{p_1}...\sum_{k_n=1}^{p_n}\eta_{k_1,...,k_n},\quad \varsigma^0_p=\sum_{k_1=1}^{p_1+1}...\sum_{k_n=1}^{p_n+1}\eta^0_{k_1,...,k_n}.$$

Now the sum $S_{T_1...T_n}=\varsigma_p\sqrt{Q}+\varsigma^0_p\sqrt{Q}$. Here, due to the condition of strong mixing 3) theorems, random values in the sum ς_p are almost independent and we can apply the central limit theorem for the sums of independent random vectors including evaluation of convergence rate. The sum ς^0_p does not affect general limit distribution and its contribution of distribution of the sum $S_{T_1...T_n}$ passes to a residual member of the theorem.

Let us demonstrate it. We shall indicate $\hat{\eta}_{k_1,...,k_n}$ random vectors, which are distributed by the same way as $\eta_{1,...,1}$ and $\hat{\varsigma}_p=\sum_{k_1=1}^{p_1}...\sum_{k_n=1}^{p_n}\hat{\eta}_{k_1,...,k_n}$.

Let us assume that Λ is the matrix of vector covariate $\eta_{1,...,1}$. Taking into account condition 4) of the theorem, we can show that the elements Λ differ from the elements of unit matrix at the value $O(1/Q)$. Let us indicate A such matrix, that $A'A=\Lambda^{-1}$, where A' - is transposed matrix A. It is obviously that the vector $A\hat{\varsigma}_p/\sqrt{p}$ has unit matrix of covariate.

Let us assume that G_p - is distribution of $\varsigma_p/\sqrt{p}$; G_p^A - is distribution of $A\varsigma_p/\sqrt{p}$; $f_p(x)$, $\hat{f}_p(x)$ - are characteristic functions of $A\varsigma_p/\sqrt{p}$, $A\hat{\varsigma}_p/\sqrt{p}$, respectively.

By use of condition of strong mixing 3) of the theorem, we can get (see.[13], balance(6))

$$\left|f_p(x)-\hat{f}_p(x)\right|=O(p/N^{\sqrt[4]{\sigma r^3}}).$$

More over, under $|x|\le\sqrt{p}/(8(E|A\eta_{1,...,1}|^{\nu+1})^{1/(\nu+1)})=T_{\nu p}$ is right (see [13], ratios (7) and (8)) asymptotic decomposition

$$\hat{f}_p(x)=\exp(-|x|^2/2)(1+\sum_{r=1}^{\nu}P_r(ix)(1/p^{r/2}))+O(3^{\nu+2}|x|^{\nu+3}\exp(-|x|^2/4)/T_{\nu p}^{\nu+1}),$$

where $P_r(ix)$ are dependent on semi-invariants of the value $(A\eta_{1,1,...,1},ix)$ of the order not higher than r+2 (see [13] of ratios (7) and (8)). Lemma 1 implies that $|P_r(ix)|=O(r^{2r}|x|^{r+2}/Q^{r/2})$.
In further we use S.M. Sadikova's inequation – linking difference between the characteristic functions with the difference between corresponding distributions. The methods of its use for this specific case

is described in [13], where we shall choose $T = O(Q^{\alpha}\ T_{\nu p})$ and consider that the conditions of lemma 3 are fulfilled.
Thus we get,

$$\sup_M \left| G_p^A(M) - \Phi(M) \right| = O(\ln^{m/4}(T)(pT^m / N^{\sqrt[4]{\omega}} + 1/\sqrt{pQ} + 1/T_{\nu p}^{\nu+1} + Q^{m\alpha} p^{m/2} \exp(-cp) + \ln T / T))$$

In further, as well as in [13] we can show that evaluation of $\sup_M \left| G_p(M) - \Phi(M) \right|$ differs from preliminary evaluation on $O(1/Q)$. Than after evaluation of distribution $G_p(M) = P(\varsigma_p / \sqrt{p} \in M)$ we start evaluating distribution $P((\varsigma_p + \varsigma_p^0)/\sqrt{p} \in M) = P(1/\sqrt{pQ}\sum_{k_1=1}^{T_1} \ldots \sum_{k_n=1}^{T_n} \xi(k_1,\ldots,k_n) \in M)$ as well as in [13], by using lemma 2. Thus, we get order difference $O((N+1)/Q^{1/(2n)})$. The reason of such rough evaluation is in large amount of summands in ς_p^0 and in the applied method, which can be improved in future. Similarly, we start evaluating distribution $P(1/\sqrt{n}\sum_{j=1}^{n} \xi_j \in M)$ and get order difference $O((T_1 \ldots T_n)^{\nu+2n\nu^2/\omega} / (Q^{2\nu} p^{\nu}))$. Than due to choice of $\nu = \sqrt[3]{\omega}, N = (T_1 \ldots T_n)^{1/\sqrt[4]{\omega}},\ \ p_i = [T_i^{(1-2n\alpha)/(n+1-2n\alpha)}]$, and Q_i from the condition (2) we get

$$\left| P(S_{T_1,\ldots,T_n} / \sqrt{T_1 \ldots T_n} \in M) - \Phi(M) \right| = O((T_1 \ldots T_n)^{\frac{1}{2(n+1-2n\alpha)} + \frac{1}{\sqrt[4]{\omega}}})$$

Thus, the evaluation (see [13], lemma 5) follows:

$$\left| P(S_{T_1,\ldots,T_n} / \sqrt{T_1 \ldots T_n} \in M) - \Phi(M) \right| = O(1 / ((1+\beta^{\sqrt[12]{\omega}})(T_1 \ldots T_n)^{\frac{1}{2(n+1-2n\alpha)} - \frac{2}{\sqrt[4]{\omega}}}))$$

This evaluation is obtained in presumption that condition (1) of lemma 3 is fulfilled. If we assume that α =0, so we get

$$\left| P(S_{T_1,\ldots,T_n} / \sqrt{T_1 \ldots T_n} \in M) - \Phi(M) \right| = O(1 / ((1+\beta^{\sqrt[12]{\omega}})(T_1 \ldots T_n)^{\frac{1}{2(n+1)} - \frac{2}{\sqrt[4]{\omega}}})) \qquad (3)$$

without using lemma 3. This result can be considered as the initial approximation for our method. Due to use of condition (1) of lemma 3 under (3) is fulfilled, we get more precise residual member in the theorem.

As a result of step-by-step approximation, we get proof for our theorem as well as in [12]. We shall notice that condition 4) of unicity of covariance matrix does not limit the community as the distance between distribution of random vectors is invariant in relation to non-generate linear transformation of these vectors.

4. Conclusion

Thus, as a result of use of step-by-step approximation method we managed to distribute evaluation of convergence rate within boundary value theorem for uniform random fields one dimensional case to multidimensional case. In further there is possibility to improve this evaluation in order to reach the level of the boundary value theorems for independent random fields.

Acknowledgments

The work is performed according to the Russian Government Program of Competitive Growth of Kazan Federal University and work was supported by the Russian Foundation for Basic Research projects nos. 16-31-00378 and 16-01-00301.

References

[1] Chebakova V Ju, Zheltukhin V S and Dubrovin V T 2016 Problem of radio-frequency discharge at atmospheric pressure in local statement *Applied Mathematical Sciences* **10** (21) 1013-22

[2] Zheltuhin V S, Solovyev P S and Chebakova V Y 2015 Boundary conditions for electron balance equation in the stationary High-Frequency Induction discharges *Research Journal of Applied Sciences* **10** (10) 658-62

[3] Dubrovin V T, Chebakova V Ju and Zheltukhin V S. 2016 Radio-Frequency Discharge at Low Pressure: a Non-Local Problem Statement Approach *Procedia Engineering*. **150** 1041–5

[4] Demkin V P, Mel'nichuk S V, Borisov A V and Bardin S S 2013 Monte-Carlo calculation of the electron energy distribution function of a He–N2 plasma of a highvoltage pulsed discharge *Russian Physics Journal* **56** (4) 486-8

[5] Hans Rau 2000 Monte Carlo simulation of a microwave plasma in hydrogen *J. Phys. D: Appl. Phys*. **33** 3214–22

[6] Yousfi M, Hennad A and Alkaa A 1994 Monte Carlo simulation of electron swarms at low reduced electric fields *Physical review E* **49**(4) 3264-73

[7] Bulinskiy A V 1979 Central limit theorem for the random fields with strong and slight dependence. *The Paper of Academy of Science of USSR* **248** (1) 17-9 [in Russian]

[8] Bulinski A V and Shashkin A P 2004 Rates in the central limit theorem for the sums of dependent multiindexed random vectors *Journal of Math. Sciences* **122** (4) 3343-58

[9] Bulinskiy A V and Shashkin A P 2008 *Boundary value theorems for associated random fields and related systems* (Moscow: PHYSMATHLIT). [in Russian]

[10] Zuev N M 2000 *Boundary value theorems for slightly dependent random values* (Mn: BGU) [in Russian]

[11] Dubrovin V T, Gabbasov F G, Chebakova V J 2016 Multidimensional central limit theorem for sums of functions of the trajectories of endomorphisms *Lobachevskii Journal of Mathematics*, **37**(4) 409-17

[12] Aminev F A and Dubrovin V T 1989 A rate of convergence bound in the central limit theorem for weakly dependent homogeneous random fields *Journal of Soviet Mathematics* **44** (5) 559-67

[13] Gabbasov F G 1977 A multidimensional central limit theorem for sums of functions of sequences with mixing *Lithuanian Mathematical Journal* **17** (4) 494-505

[14] Gabbasov F G and Dubrovin V T 1988 Probability that the normed sum of weakly dependent vectors falls in the exterior of a convex set *Journal of Soviet Mathematics* **43** (1) 2187-93.

[15] Dubrovin V T 2014 Convergence rate in limit theorems for weakly dependent random values *Lobachevskii Journal of Mathematics* **35**(4) 390-6

Solving of variational inequalities by reducing to the linear complementarity problem

Z R Gabidullina

Kazan Federal University, 18 Kremlejovkaja street, Kazan, Russia, 420008

E-mail: zgabid@mail.ru

Abstract. We study the variational inequalities closely connected with the linear separation problem of the convex polyhedrain the Euclidean space. For solving of these inequalities, we apply the reduction to the linear complementarity problem. Such reduction allows one to solve the variational inequalities with the help of the Matlab software package.

1. Reduction to the projection problem

In this section, we demonstrate that the variational inequalities of our interest are reducible to the problem of projecting the origin of the Euclidean space onto the convex polyhedron.

In [1–4], we establish and study the close connection between the solutions of some variational inequalities and linear separation problem for two convex and closed sets A and B from the Euclidean space R^n. These inequalities consist in finding a vector $c \in R^n$ such that

$$\langle c, x-y \rangle \geq 0, \quad x \in A, \; y \in B, \; \Delta > 0; \tag{1}$$

$$\langle c, x-y \rangle > 0, \quad x \in A, \; y \in B; \tag{2}$$

$$\langle c, x-y-c \rangle \geq 0, \quad x \in A, \; y \in B; \tag{3}$$

$$\langle c, x-y \rangle \geq 0, \quad x \in A, \; y \in B. \tag{4}$$

In regard the inequalities (3)-(4), of course, we are interested in nontrivial solutions.

The solution to the problem of the best strong linear separation of two sets A and B is also the solution for all above-enumerated variational inequalities. The best strong linear separation consists in finding the best (in the sense of separation thickness) support vector for the hyperplane strongly separating the sets A and B. This problem can be reduced to the problem of projecting the origin of R^n onto the Minkowski difference of these sets A-B. Let us note that the formulas for determining of the Minkowski difference for the different kinds of sets were proposed, for instance, in [2], [4]. In [2], there was proved that it holds

$$C := A - B = conv\{a_i - b_j\}, i = 1, \cdots, m_1; j = 1, \cdots, m_2,$$

$$where \; A = conv\{a_i\}, i = 1, \cdots, m_1; B = conv\{b_j\}, j = 1, \cdots, m_2,$$

i.e. A-B is a convex polyhedron given as the convex hull of $m_1 \bullet m_2$ vectors $a_i - b_j$.

If for the projection of the origin onto C, it holds $P_C(\odot) = \odot$, then the inequalities (1)–(3) have no solutions. In this case, we can not state that (4) has no solutions too (there can exist the separator with null thickness). In this event, to solve (4) it is necessary to use, for example, maxi-min optimization problem presented in [3]. In [1], there was originally established the connection between the problem of projecting [5] and the quadratic programming ones of a special type, associated with the cone of the strong support vectors for sets, onto which the projection is carried out.

Note that the terminology relating to the different kinds of the support vectors and the types of linear separation of sets, is studied in detail in [3]. In this paper, using the Kuhn-Tucker theorem, the quadratic programming problem (associated with the projection problem) is reduced to the classical linear complementarity problem. The solution of the linear complementarity problem can be obtained by using the function LCPSolve in MATLAB [5].

2. Connection of the variational inequalities with the linear complementarity problem

In the previous section, we noted that solving of (1)–(4) can be reduced to the projection problem. In this section, we study the reduction of the projection problem to the linear complementarity problem. As a result, we establish the connection of (1)–(4) with the linear complementarity problem.

Let us consider the following setting of the projection problem connected with the so-called inner representation of the convex polyhedron $C = conv\{z_i\}, i=1,\cdots,m$ [6].

$$min\ \varphi(\alpha) = \|\sum_{i=1}^{m}\alpha_i z_i\|^2, \tag{5}$$

$$\sum_{i=1}^{m}\alpha_i = 1, \tag{6}$$

$$\alpha_i \geq 0,\ i=1,\dots,m. \tag{7}$$

Further, the problem (5)–(7) can be easily rewritten in a vector-matrix form as follows

$$min\ \varphi(\alpha) = \langle\alpha, D\alpha\rangle, \tag{8}$$

$$\langle e,\alpha\rangle = 1, \tag{9}$$

$$\alpha \geq 0, \tag{10}$$

Where D is m×m-dimensional matrix with elements

$$d_{ij} = z_i, z_j,\ i,j=1,\dots,m;\ e=\begin{pmatrix}1\\ \vdots\\ 1\end{pmatrix},\ e\in\boldsymbol{R}^m,\ \alpha=\begin{pmatrix}\alpha_1\\ \alpha_2\\ \vdots\\ \alpha_m\end{pmatrix}.$$

Let α^* be an optimizer for the problem (8)-(10), then

$$P_C(\odot) = \sum_{i=1}^{m}\alpha_i^* z_i. \tag{11}$$

In [5], we prove that the matrix B is nonnegative definite. Consequently, the problem (8)-(10) is characterized as the convex quadratic programming one. We can rewrite the constraint (9) with the help of the two inequality constraints as follows.

$$min\ \varphi(\alpha) = \langle\alpha, D\alpha\rangle,$$

$$\langle e,\alpha\rangle \geq 1,$$

$$\langle -e,\alpha\rangle \geq -1,$$

$$\alpha \geq 0.$$

Let us further construct the Lagrange function for the previous optimization problem:

$$L(\alpha, f) = \langle\alpha, D\alpha\rangle + \langle f, b - F\alpha\rangle, f \geq 0, f \in R^2.$$

Where f is the vector of the Lagrange multipliers, F is the $2\times m$ matrix

$$F = \begin{pmatrix}1 & \dots & 1\\ -1 & \dots & -1\end{pmatrix}, b = \begin{pmatrix}1\\ -1\end{pmatrix}.$$

It is not hard to check that it is fulfilled

$$L'_\alpha(\alpha,f)=2D\alpha-F^T f,$$
$$L'_f(\alpha,f)=b-F\alpha.$$

It is well known that in the case of minimizing the convex function subject to the linear constraints, the optimality conditions for the problem (8)-(10) are determined as the saddle point conditions for the Lagrange function. Due to the Kuhn-Tucker theorem, it is necessary and sufficient to determine the vectors α, f such that

$$L'_\alpha(\alpha,f)\ge 0,$$
$$L'_f(\alpha,f)\ge 0,$$
$$\langle \alpha,\ L'_\alpha(\alpha,f)\rangle=0,$$
$$\langle f,\ L'_f(\alpha,f)\rangle=0,$$
$$f,\alpha\ge 0.$$

This system can be expressed as a linear complementarity problem [7] which consists in determining the vectors w and v such that

$$w=Mv+q,$$
$$w\ge 0,\ v\ge 0,$$
$$\langle w,\ v\rangle=0,$$

where $M=\begin{pmatrix} 2D & | & -F^T \\ F & | & \theta \end{pmatrix}$ is the matrix with block structure from $R^{l\times l}$, $l=m+2$;

w, v, q are m+2-dimensional vectors; θ is a null matrix with dimension 2×2,

$$w=\begin{pmatrix} u \\ r \end{pmatrix},\ v=\begin{pmatrix} \alpha \\ f \end{pmatrix},\ q=\begin{pmatrix} 0 \\ \vdots \\ 0 \\ -b \end{pmatrix},\quad u=2D\alpha-F^T f,$$

$$r=F\alpha-b;\ \ \alpha,f,u,r\ge 0;\ \langle u,\alpha\rangle=0,\ \ \langle r,f\rangle=0.$$

We further recall briefly the following cones defined in [3]:

$$W_C=\left\{\mathrm{p}\in \boldsymbol{R}^n:\ \inf_{x\in C}\langle \mathrm{p},\ \mathrm{x}\rangle\ge 0\right\},$$

$$V_C=\left\{\mathrm{h}\in \boldsymbol{R}^n:\ \inf_{x\in C}\langle \mathrm{h},\ \mathrm{x}\rangle>0\right\},\qquad Q_C=\left\{\mathrm{s}\in \boldsymbol{R}^n:\langle \mathrm{s},\ \mathrm{x}\rangle>0\ \ \forall x\in C\right\},$$

$$\Omega_C=\left\{g\in \boldsymbol{R}^n:\ \langle \mathrm{g},\ \mathrm{x}\rangle\ \ge\ \|g\|^2\ \ \forall x\in C\right\},$$

$$E(\Omega_C)=\left\{x\in \boldsymbol{R}^n:\ \ \mathrm{x}=\lambda\,\mathrm{g},\ \ \lambda\ge 0,\ \ g\in\Omega_C\right\},$$

where C is a nonempty subset of R^n, $W_C\setminus\{\odot\}$ is a cone of generalized support vectors of the set C. The notations V_C, Q_C correspond to the cones of generalized strong and strict support vectors of the set C, respectively. The main properties of the above-mentioned cones and their relationship with well-known other ones have been learned in [3]. The necessary and sufficient conditions for emptiness of the cones of generalized support vectors have been explored in [4]. We refer the reader to [3]-[4] for definitions, terminology, and details. Let us recall the important relationship of above-enumerated cones:

$$V_C\subseteq Q_C\subseteq W_C, \tag{12}$$
$$V_C=E(\Omega_C)\setminus\{\odot\}. \tag{13}$$

It is obvious that the variational inequalities (1),(2),(3), and (4) are associated with the cones V_C, Q_C, Ω_C, and W_C, respectively. In [3], for the case of the convex and compact set C it was proved that $V_C = Q_C$. If C is a convex and closed subset of R^n, then the operation of projection onto C is well defined. Due to Theorem 3.3 from [3], it holds $P_C(\odot) \in Bd(\Omega_C)$, where $Bd(\Omega_C)$ stands for the boundary of Ω_C. Due to Theorem 2.7 from [3], if the set C is convex and compact, then Ω_C is also a convex compact. Consequently, it is fulfilled $P_C(\odot) \in \Omega_C$, since $Bd(\Omega_C) \subseteq \Omega_C$. If we find the projection of the origin of R^n onto C and observe that $P_C(\odot) = \odot$. This means that $\odot \in C$. Owing to Lemma 3.6, we have $\Omega_C = \{\odot\}$. This implies that (3) is without non-zero solutions. Then from (13), by virtue of the fact $\Omega_C = \{\odot\}$, it follows that $V_C = \varnothing$, i.e. C is not strongly separable from the origin of R^n. Due to General theorem on strong separability from [3], the variational inequality (1) has no solutions as well. As for (4), we can not unambiguously asserts (under condition of $\odot \in C$) (4) has no non-zero solutions, too. For further analysis, we need to know whether or not the origin of R^n is the interior point of C. If it is, then (4) is without nontrivial solutions. In opposite case, i.e. when the origin belongs to the boundary of the set C [4], there can be found the non-zero solution for (4) using other tools (see [3]). If we obtain $P_C(\odot) \neq \odot$, then from above, due to (12)–(13), it holds $P_C(\odot) \in V_C, \Omega_C, W_C$. This means that $P_C(\odot)$ satisfies to (1)–(4).

References

[1] Gabidullina Z R 2006 A theorem on separability of a convex polyhedron from zero point of the space and its applications in optimization *Izvestiya VUZ. Matematika* **50 (12)** 18–23

[2] Gabidullina Z R 2011A theorem on strict separability of convex polyhedra and its applications in optimization *J.Optim.Theory Appl.* **148 (3)** 550–570

[3] Gabidullina Z R 2013 A linear separability criterion for sets of euclidean space *J.Optim.Theory Appl.* **158 (1)** 145–171

[4] Gabidullina Z R 2015 Necessary and sufficient conditions for emptiness of the cones of generalized support vectors *Optim Lett.* **9 (4)** 693–729

[5] Gabidullina Z R 2016 The problem of projecting the origin of euclidean space onto the convex polyhedron http://arxiv.org/abs/1605.05351

[6] Wolfe P 1976 Finding the nearest point in a polytope *Math Program.* **11** 128–149

[7] Reklaitis G V, Ravindran A and Ragsdell K M 2006 *Engineering Optimization: Methods and applications, 2nd Edition* (John Wiley&Sons,Inc)

Numerical analysis of nonlinear straining of non-shallow spherical shell subjected to thermal and force loadings

M S Ganeeva, V E Moiseeva and Z V Skvortsova

Institute of Mechanics and Engineering, Kazan Science Center, Russian Academy of Sciences, 2/31 Lobachevsky str., Kazan 420111, Russia

E-mail: ganeeva@kfti.knc.ru

Abstract. The influence of low and high temperatures on the process of nonlinear straining and the critical load buckling of the elastic-plastic deep spherical shells under external axisymmetric and non-axisymmetric loading has been investigated in this paper. We have studied the cases where reached temperature is substantially different from the initial one. For the analyzed thermal and force loading it has been found that the highest critical loads for the deep spherical shell are observed at cryogenic temperatures, and they considerably decrease at higher temperatures.

1. Statement of the problem and the solution method

Solution of the non-linear thermo-force problem of the shell theory is a challenging task with the considerable effort devoted to it [1-4]. In this paper, we numerically investigate nonlinear straining of the deep spherical segments under the action of axisymmetric and non-axisymmetric thermo-force loading.

A shell of revolution (Figure 1) that is closed in the circumferential direction [4] is considered. As coordinate lines we take the meridians s, parallels φ, and external normal z to the middle surface. In this case, $s_1 \le s \le s_N$, $0 \le \varphi \le 2\pi$, $-0{,}5h \le z \le 0{,}5h$, h is the shell thickness, $\theta \in [\theta_1, \theta_N]$ is the angle between the rotation axis of the shell x and the normal to the middle surface z.

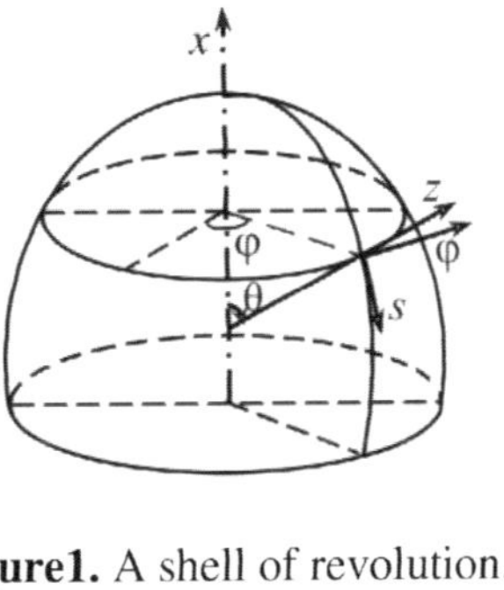

Figure1. A shell of revolution.

The loads $X_i^{\text{H}}(s,\varphi), i=\overline{1,3}$ and uniform temperature T act upon the shell. The problem is solved on the basis of the Kirchhoff-Love theory equations in the geometrically nonlinear statement with moderate turns. It is assumed that the material operates beyond the elastic limit, and the stress-strain dependence is established by the equations of small elastic-plastic strains. In the case where there is a considerable difference between the reached temperature of the shell and its initial temperature T_0 it will be necessary to take into account the dependence of material properties on current temperature T. Further, we use relationships and notation taken from [4].

Let the loads acting upon the shell will be presented as trigonometric series relative to the circumferential coordinate φ :

$$X_i^{\text{H}}(s,\varphi)=\sum_{k=0}^{n} X_{i,k}^{\text{H}}(s)\cos k\varphi,\ \ i=1,3;\ \ X_2^{\text{H}}(s,\varphi)=\sum_{k=1}^{n} X_{2,k}^{\text{H}}(s)\sin k\varphi.$$

To solve the problem, we use method of expanding the sought functions

$$\boldsymbol{Y}=(T_{11}^{\text{H}},T_{12}^{\text{H}},Q_1^{\text{H}}+\frac{1}{r}\frac{\partial M_{12}}{\partial\varphi},M_{11},v_1,v_2,w,\vartheta_1)'$$

into the trigonometric series relative to the circumferential coordinate [1,2]:

$$w=\sum_{k=0}^{L} w_k(s)\cos k\varphi,\ \ v_2=\sum_{k=1}^{L} v_{2,k}(s)\sin k\varphi,\ldots \tag{1.1}$$

Since the problem is nonlinear $L>n$, and the value L is specified by the numerical experiment.

We introduce a vector of decision functions for the amplitudes of series (1.1):

$$\mathbf{Z}_k=(T_{11,k},T_{12,k},Q_{1,k},M_{11,k},v_{1,k},v_{2,k},w_k,\vartheta_{1,k})',\quad k=\overline{0,L}. \tag{1.2}$$

The problem is reduced to integration of a number of systems for the nonlinear ordinary differential equations:

$$\frac{d\boldsymbol{Z}_k}{ds}=\boldsymbol{H}_k(s)\boldsymbol{Z}_k+\boldsymbol{g}_k(s,X_{i,k}^{\text{H}},T-T_0)+\boldsymbol{G}_k(s,\boldsymbol{Y})+\boldsymbol{P}_k(s,\boldsymbol{Y}),\ k=\overline{0,L} \tag{1.3}$$

With the corresponding boundary conditions:

$$\begin{aligned}&\boldsymbol{A}_k\boldsymbol{Z}_k=\boldsymbol{a}_k \quad \text{at} \quad s=s_1;\\ &\boldsymbol{B}_k\boldsymbol{Z}_k=\boldsymbol{b}_k \quad \text{at} \quad s=s_N,\quad k=\overline{0,L}.\end{aligned} \tag{1.4}$$

Here $\boldsymbol{H}_k(s)$ is the (8×8) matrix of coefficients; $\boldsymbol{g}_k$ is the (8×1) vector of the amplitudes for loading and temperature terms; $\boldsymbol{A}_k,\boldsymbol{B}_k$ are the (4×8) matrices; $\boldsymbol{a}_k,\boldsymbol{b}_k$ are (4×1) vectors. The Fourier coefficients $\boldsymbol{G}_k(s,\boldsymbol{Y}),\boldsymbol{P}_k(s,\boldsymbol{Y})$ of the expansions of the geometrically and physically nonlinear terms $\boldsymbol{G}(s,\boldsymbol{Y})$, $\boldsymbol{P}(s,\boldsymbol{Y})$ in the series of the form (1.1) are calculated numerically using Bessel's formulas, wherein the nonlinear terms bind systems (1.3) with each other.

Nonlinear boundary problem (1.3), (1.4) is solved using the method of successive approximations in combination with the method of orthogonal sweep running. The process of successive approximations for the given temperature range is organized for a number of leading parameter values. A good choice of the leading parameter value will ensure the process convergence, obtaining a full picture of the distinctive curve of the load parameters and deflection, and finding the limit points. The basis of the technique is the numerical solution continuation algorithm using some integral parameter. This parameter is determined according to the half-wave epure on the amplitude of one of the main deflection harmonics in the circumferential coordinate trigonometric series (1.1) [5].

2. Example and Results Analysis

In this work we have numerically studied the straining of the deep spherical shell kept under the normal pressure X_3^H and uniform temperature T. The following notations are used below: h -

thickness, a – foundation radius, $H = a/2$ – pole height above the foundation, wherein $\theta_1 = 0, \theta_N = 0{,}9273$. The spherical shell is made from the stainless steel grade 12X18H10T, which characteristics are listed in Table 1, at $T_0 = 20^o C$. The shell material behaves in accordance with the linear hardening law, where $\lambda = 0{,}9$ is the hardening factor; $\nu = 0{,}3$ – Poisson's factor; σ_S – yield point stress, σ_S – break point stress; E – elastic modulus; α – thermal linear expansion coefficient; $\tilde{\alpha}$ – average thermal linear expansion coefficient for the temperature range $[T_0,T]$.

Material characteristics dependence on the temperatures T_0,T was taken into account in the problem equations:

$$E = E(T),\ \sigma_S = \sigma_S(T),\ \tilde{\alpha} = \tilde{\alpha}(T_0,T). \quad (2.1)$$

Shell foundation is rigidly fixed: $v_1 = 0,\ v_2 = 0,\ w = 0,\ \vartheta_1 = 0.$

The following cases of external normal pressure X_3^H were studied:
axisymmetrical loading -

$$X_3^H = Q; \quad (2.2)$$

non-axisymmetrical loading -

$$X_3^H = \begin{cases} Q, & \theta \in [0,\theta_0],\quad \theta_0 = 0{,}048; \\ Q[1+\sin(\theta-\theta_0)(1{,}2976\cos\varphi + 0{,}6488\cos 2\varphi)], & \theta \in [\theta_0,\theta_N]. \end{cases} \quad (2.3)$$

Table 1. Steel characteristics.

$T, {}^oC$	−253	−193	−103	−70	20	300	500
$E \cdot 10^{-5}$, MPa	2.23	2.12		2.09	2.03	1.74	1.53
σ_S, MPa	520	448		380	250		137
σ_B, MPa	1420	1000			660		392
$\alpha(T) \cdot 10^5, 1/{}^oC$	0.08	0.66	1.42		1.60	1.80	2.07
$\tilde{\alpha}(T_0,T) \cdot 10^5, 1/{}^oC$	1.12	1.32	1.49		1.58	1.72	1.79

The numerical results of the spherical segment with the parameters $a/h = 200$, $H/a = 0{,}5$ dependency on the thermo-force loading are presented on the Figures 2,3. On Figure 2 the following dependencies on temperature T are shown: the dependency of the loading parameter Q_H of the non-

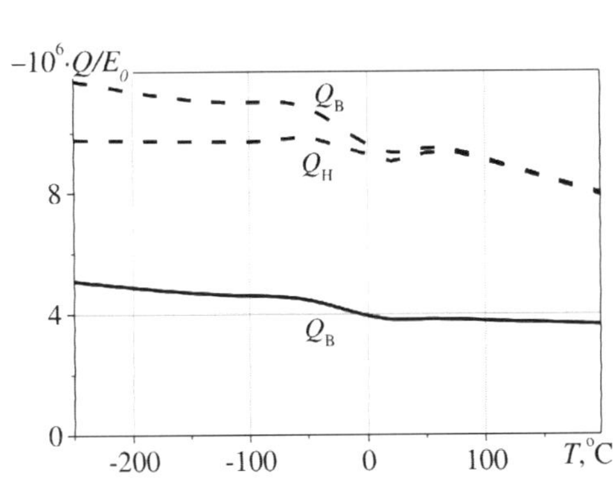

Figure 2. Dependences $Q_H(T)$ under the loading (2.2) and Q_B (T) under the loadings (2.2) and (2.3).

axisymmetrical wave formation along the parallel with the number of waves $l \in [15,17]$, when under pressure load (2.2) (depicted in a dashed line); the dependency of the upper critical loading parameter Q_B of axysymmetrical stability loss , when under pressure load (2.2) (dashed line); the dependency of the upper critical loading parameter Q_B under non-axisymmetrical loading (2.3) (solid line).

One can notice a significant dependency of the stress-stain state segment on the loading type and temperature $T - T_0$. The calculations have shown that in case of the joint impact of temperature T and loading (2.3) the solution in the vicinity Q_B is found when the number of components $L=8$ (in the series of sought functions) (1.1). It has been seen that for the deep spherical shell (Figure 2) under the loading (2.2) the wave formation loadings Q_H changes in the examined temperature range not more than 22%. The maximum values of the loading parameter Q_B for the loadings (2.2), (2.3) are observed under cryogenic temperatures, and their values decrease under elevated temperatures.

Figure 3 shows the dependency of the loading parameter Q/E_0 on maximum bending parameter $\max\limits_{s,\varphi}(w/h)$, dependency of the maximum stress intensity $\max\limits_{s,\varphi,z}(\sigma_i/E_0)$ on loading parameter for the cases of the axisymmetric loading (2.2) (dashed lines) and non-axisymmetric loading (2.3) (solid lines), both under the temperature $T = -100^o, 20^o, 100^o C$; points where the functions took the values Q_B and Q_H have been marked by the icons •, × accordingly. For the studied temperature range dependencies $Q(w)$, $\sigma_i(Q)$ for the loadings (2.2) and (2.3) differ significantly starting from the initial phase of straining (see Figure 3).

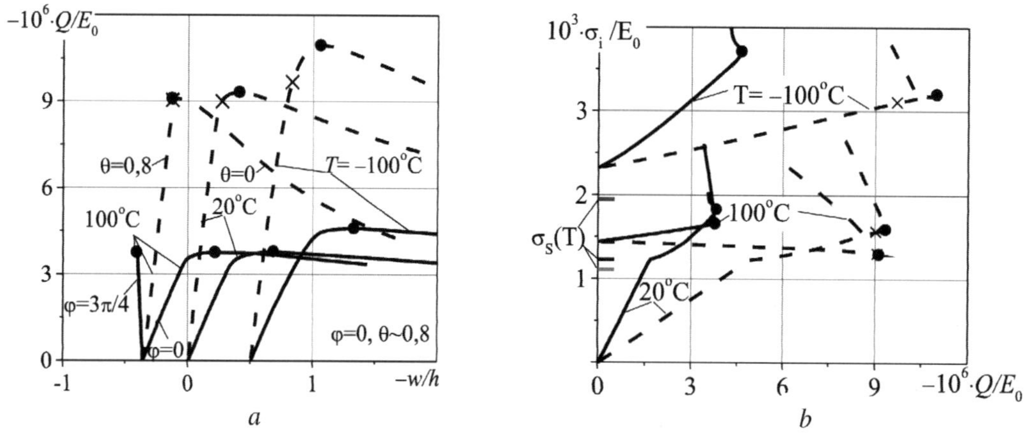

Figure 3. *a* – dependencies $Q(w)$; *b* - $\sigma_i(Q)$ under $T = -100^o, 20^o, 100^o C$ for (2.2),(2.3).

References

[1] Grigolyuk E I and Kabanov V V 1978 Ustoi'chivoct' obolochek (Moscow: Nauka)
[2] Thornton E A 1993 *Appl. Mech. Rev.* **46 (10)** pp 485-506
[3] Merzlyakov V A and Shevchenko Y N 1999 *Int. Appl. Mech.* **35 (5)** pp 431-461
[4] Ganeeva M S 1985 Termosilovaya zadacha v geometricheski i fizicheski nelineinoi teorii netonkikh i tonkikh obolochek (Kazan: VINITI 4459-85)
[5] Ganeeva M S, Moiseeva V E and Skvortsova Z V 2009 *Izv. Vuz. Problemyi Energetiki* **9–10** pp 89–98

The algorithm for investigation a structure of composite materials

I R Garifullin and L U Sultanov

Kazan Federal University, 18 Kremlyovskaya str., 420008, Kazan, Russia

E-mail: islam-garifullin@mail.ru

Abstract. This paper presents a model of deformation of hyperelastic reinforced composite. The constitutive equations are derived using the free strain energy. The numerical algorithm to solve such problems is compiled. A computational algorithm is based on a finite element method.

1. Introduction

In different industries to increase structural strength often use composite materials. Besides, in biomechanics tissues can be described as composite materials with specific properties [13, 14]. Also modelling of mechanical behavior of composite materials is important in contact mechanics [7]. There are many different methods of calculating the structure of these materials [6, 27–30]. It is a promising improvement and development of new techniques. In this paper, an algorithm for calculating a composite material structure is considered with hyperelastic properties. Many papers are devoted to research of hyperelastic isotropic continuum [1–5, 8–12, 25, 26]. They lay down on the basis of this paper. In the first part is considered the free strain energy function for hyperelastic reinforced composite. The basic physical relations are constructed. A computational algorithm is described. The second part is devoted to finite element discretization.

2. Constitutive relations

Hyperelastic reinforced composite material is characterized by a free energy function strain in the form [20–22]:

$$\mathbf{\Psi}(\mathbf{C},\mathbf{A})=\frac{\lambda}{2}J_1^2+\mu_t J_2+\alpha J_3 J_1+2(\mu_l-\mu_t)J_4+\frac{\beta}{2}J_3^2.$$

There $\lambda, \mu_l, \mu_t, \alpha, \beta$ – the mechanical characteristics of the material determined from the experimental [23, 24]. Where invariants are defined as follows:

$$J_1=\frac{1}{2}tr\left[\mathbf{C}-\mathbf{I}\right],\ J_2=\frac{1}{4}tr\left[(\mathbf{C}-\mathbf{I})^2\right],\ J_3=\frac{1}{2}tr\left[\mathbf{A}\cdot(\mathbf{C}-\mathbf{I})\right],\ J_4=\frac{1}{4}tr\left[\mathbf{A}\cdot(\mathbf{C}-\mathbf{I})^2\right].$$

They depended on the measure of the left Cauchy–Green deformation tensor and the structural tensor of $\mathbf{A}$, which describes the position of the fibers in the material. At the initial time:

$$\mathbf{A}_0 = \vec{a}_0 \otimes \vec{a}_0,$$

$\vec{a}_0$ – the preferential fiber orientation vector in the reference configuration. After deformation, the fibers change their direction, and the tensor structure will

$$\mathbf{A} = \vec{a} \otimes \vec{a},$$

where $\vec{a}$ – the preferential fiber orientation vector in the deformation configuration.

To create a physical relationship is used equation of the form [21, 22, 27]:

$$\delta\mathbf{\Psi} = \frac{1}{2}\mathbf{S} \cdot\cdot \delta\mathbf{C}.$$

This implies that the second Piola–Kirchhoff stress tensor is defined as follows:

$$\mathbf{S} = 2\frac{\partial \mathbf{\Psi}}{\partial \mathbf{C}} = \frac{1}{2}\Big[\lambda tr(\mathbf{C}-\mathbf{I})\cdot\mathbf{I} + 2\mu_t(\mathbf{C}-\mathbf{I}) + \alpha\big(tr(\mathbf{A}\cdot(\mathbf{C}-\mathbf{I})) + tr((\mathbf{C}-\mathbf{B})\cdot\mathbf{A})\cdot\mathbf{I}\big) +$$
$$+2(\mu_l - \mu_t)\big((\mathbf{C}-\mathbf{I})\cdot\mathbf{A} + \mathbf{A}\cdot(\mathbf{C}-\mathbf{I})\big) + \beta tr\big(\mathbf{A}\cdot(\mathbf{C}-\mathbf{I})\cdot\mathbf{A}\big)\Big].$$

To use the incremental method of calculation is necessary to define the stress increments, which are calculated as a material derivative of the stress tensor by time. In the general case we can write [1, 3]:

$$\dot{\mathbf{S}} = 2\frac{\partial^2 \mathbf{\Psi}}{\partial \mathbf{C}^2} \cdot\cdot \dot{\mathbf{C}}.$$

After transformations the following expression:

$$\frac{\partial^2 \mathbf{\Psi}}{\partial \mathbf{C}^2} = \mathbf{\Lambda}^{'} = \frac{1}{2}\big(\lambda \mathbf{I} \otimes \mathbf{I} + 2\mu_t \mathbf{\Lambda} + \alpha(\mathbf{I} \otimes \mathbf{A} + \mathbf{A} \otimes \mathbf{I}) + 2(\mu_l - \mu_t)\mathbf{\Lambda}_A + \beta \mathbf{A} \otimes \mathbf{A}\big),$$

where $\Lambda_{ijkl} = \frac{1}{2}(\delta_{ik}\delta_{jl} + \delta_{il}\delta_{jk})$ – components of the 4th order tensor $\mathbf{\Lambda}$, $\Lambda_{A_{ijkl}} = A_{im}\Lambda_{jmkl} + A_{jm}\Lambda_{mikl}$ – components 4th order tensor $\mathbf{\Lambda}_A$, which components depend on the fiber direction.

The equation of principle of virtual power in the initial configuration is written as

$$\int_{V_0} \frac{1}{2}(\mathbf{S} \cdot\cdot \delta\dot{\mathbf{C}})dV_0 = \int_{V_0} \vec{f}_0 \cdot \delta\vec{\upsilon}dV_0 + \int_{S_0^\sigma} \vec{t}_{0n}^{\,*}\delta\vec{\upsilon}dS_0$$

where $\vec{\upsilon}$ – velocity vector of a material point; S_0^σ – part of the surface on which are defined forces; $\vec{t}_{0n}^{\,*}$ – vector of surface forces, $\vec{f}_0$ – vector of body forces.

A total Lagrangian formulation is used to solve this problem [27, 28]:

$$\int_{V_0} \frac{1}{2}(\dot{\mathbf{S}} \cdot\cdot \delta\dot{\mathbf{C}} + \mathbf{S} \cdot\cdot \delta\ddot{\mathbf{C}})dV_0 = \int_{V_0} \dot{\vec{f}}_0 \cdot \delta\vec{\upsilon}dV_0 + \int_{S_0^\sigma} \dot{\vec{t}}_{0n}^{\,*}\delta\vec{\upsilon}dS_0. \tag{1}$$

3. Finite element discretization

An 8 node isoparametric finite element is used for computer implementation. We introduce approximation geometry and velocities:

$${}^k y^i(\xi^j) = \sum_{t=1}^{8} {}^k y_t^i N_t(\xi^j),\; {}^k\upsilon^i(\xi^j) = \sum_{t=1}^{8} {}^k\upsilon_t^i N_t(\xi^j),$$

where $N_t\left(\xi^i\right)=\frac{1}{8}\left(1+\xi_t^1\xi^1\right)\left(1+\xi_t^2\xi^2\right)\left(1+\xi_t^3\xi^3\right)$ – interpolation function, $-1\le\xi^1,\xi^2,\xi^3\le 1$, ${}^k y_t^i$ – coordinates of nodes, $\xi_t^i=\pm 1$ the coordinates of the respective nodes in the local coordinate system, ${}^k\upsilon_t^i$ – velocity of nodes.

Define the following components of the tensors

$$C^{ij}=\sum_{t,s=1}^{8} y_t^m y_s^m N_{t,i}N_{s,j},\quad F^{ij}=\sum_{t=1}^{8} y_t^i N_{t,j},\quad \dot F^{ij}=\sum_{t=1}^{8}\upsilon_t^i N_{t,j}.$$

For the equation (1) we can write:

$$\begin{aligned}
&\dot{\mathbf{S}}\cdot\cdot\delta\dot{\mathbf{C}}=\dot{\mathbf{S}}\cdot\cdot(\mathbf{F}^T\cdot\delta\dot{\mathbf{F}}+\delta\dot{\mathbf{F}}^T\mathbf{F})=\mathbf{\Lambda}^{\circ}\cdot\cdot\dot{\mathbf{C}}\cdot\cdot(\mathbf{F}^T\cdot\delta\dot{\mathbf{F}}+\delta\dot{\mathbf{F}}^T\mathbf{F})=\\
&=\mathbf{\Lambda}^{\circ}\cdot\cdot(\dot{\mathbf{F}}^T\cdot\mathbf{F}+\mathbf{F}^T\cdot\dot{\mathbf{F}})\cdot\cdot(\mathbf{F}^T\cdot\delta\dot{\mathbf{F}}+\delta\dot{\mathbf{F}}^T\cdot\mathbf{F})=\\
&=\Lambda_{ijkl}\upsilon_t^m(N_{t,l}F_{mk}F_{ri}N_{s,j}+N_{t,l}F_{mk}F_{rj}N_{s,i}+N_{t,k}F_{ml}F_{ri}N_{s,j}+N_{t,k}F_{ml}F_{rj}N_{s,i})\delta\upsilon_s^r,
\end{aligned}$$

$$\mathbf{S}\cdot\cdot\left(\dot{\mathbf{F}}^T\cdot\delta\dot{\mathbf{F}}+\delta\dot{\mathbf{F}}^T\cdot\dot{\mathbf{F}}\right)=2S_{ij}\upsilon_t^m N_{t,k}\delta\upsilon_s^m N_{s,r}=2\upsilon_t^m S_{ij}\upsilon_t^m N_{t,j}N_{s,i}\delta_r^m\delta\upsilon_s^r$$

since the second Piola–Kirchhoff tensor is a symmetric tensor.

$$\int_{V_0}\dot{\vec f}_0\cdot\delta\vec\upsilon dV_0+\int_{S_0^\sigma}\dot{\vec t}_{0n}^{\,*}\cdot\delta\vec\upsilon dS_0=\left\{\mathbf{P}\right\}^T\cdot\left\{\delta\vec\upsilon\right\},$$

$$-\frac{1}{\Delta t}\left\{\int_{V_0}(\mathbf{S}\cdot\cdot\delta\mathbf{C})dV_0-\int_{V_0}\vec f_0^{\,*}\cdot\delta\vec\upsilon\, dV_0-\int_{S_0^\sigma}\vec t_{0n}^{\,*}\cdot\delta\vec\upsilon dS\right\}=\frac{1}{\Delta t}\left\{\mathbf{H}\right\}^T\cdot\left\{\delta\vec\upsilon\right\}.$$

After integration (1) we obtain a system of linear algebraic equations:

$$\left[{}^k\mathbf{K}\right]\left\{\Delta^k\vec u\right\}=\left\{\Delta^k\mathbf{P}\right\}+\frac{1}{\Delta t}\left\{{}^k\mathbf{H}\right\}. \tag{2}$$

Solving equation (2), give the increment of the displacements $\Delta^k\vec u$, ${}^{k+1}y^i={}^k y^i+\Delta^k u^i$ and stresses ${}^{k+1}\mathbf{S}=2\dfrac{\partial\mathbf{\Psi}}{\partial\,{}^{k+1}\mathbf{C}}$.

4. Conclusion

The paper is constructed procedure of investigation of hyperelastic composites. The constitutive relationships are obtained for composite materials. The computational algorithm is created. A total Lagrangian formulation is used. The basic equations are obtained for computational algorithm. The numerical implementation is based on the finite element method.

Acknowledgments

The reported study was supported by Government of the Republic of Tatarstan research projects No. 15-41-02557 and 15-31-20602.

References

[1] Golovanov A I, Konoplev Yu G, Sultanov L U 2008 *Uchenye Zapiski Kazanskogo Universiteta. Seriya Fiziko-Matematicheskie Nauki* **150 (3)** 122–132

[2] Golovanov A I, Konoplev Yu G, Sultanov L U 2010 *Uchenye Zapiski Kazanskogo Universiteta. Seriya Fiziko-Matematicheskie Nauki* **151 (3)** 108–120

[3] Bonet J and Wood R D 1997 *Nonlinear continuum mechanics for finite element analysis* (Cambridge University Press) 283
[4] Davydov R L and Sultanov L U 2013 *PNRPU Mechanics Bulletin* **1** 81–93
[5] Golovanov A I and Sultanov L U 2005 *International Applied Mechanics* **41 (6)** 614–620
[6] Paimushin V N 2008 *Journal of Applied Mathematics and Mechanics* **72 (5)** 597–610
[7] Sachenkov O A, Mitryaikin V I, Zaitseva T A and Konoplev Yu G 2014 *Applied Mathematical Sciences* **8 (159)** 7889–7897
[8] Sagdatullin M and Berezhnoi D 2014 *Applied Mathematical Sciences* **8 (35)** 1731–1738
[9] Sagdatullin M and Berezhnoi D 2014 *Applied Mathematical Sciences* **8 (60)** 2965–2972
[10] Sultanov L U 2014 *Applied Mathematical Sciences* **8 (143)** 7117–7124
[11] Sultanov L U and Davydov R L and 2014 *Applied Mathematical Sciences* **8 (60)** 2991–2996
[12] Sultanov L U 2015 *Procedia Earth and Planetary Science* **15** 119–124
[13] Sachenkov O, Kharislamova L, Shamsutdinova N, Kirillova E and Konoplev Yu 2015 *IOP Conference Series: Materials Science and Engineering* **98**
[14] Shigapova F A, Mustakimova R F, Saleeva G T and Sachenkov O A *International Journal of Applied Engineering Research* **10 (24)** 44711–44714
[15] Abdrakhmanova A I and Sultanov L U 2016 *Materials Physics and Mechanics* **26(1)** 30–32
[16] Davydov R L and Sultanov L U 2013 *Sixth International Conference on Nonlinear Mechanics (ICNM-VI)* 64–67
[17] Davydov R L and Sultanov L U 2015 *Journal of Engineering Physics and Thermophysics* **88(5)** 1280–1288
[18] Sultanov L U and Fakhrutdinov L R 2013 *Magazine of Civil Engineering* **44 (9)** 69–74
[19] Sultanov L U and Davydov R L 2013 *Magazine of Civil Engineering* **44 (9)** 64–68
[20] Spencer A J M 1984 *Continuum theory of the mechanics of fibre-reinforced composites* (Springer Verlag GMBH) 284
[21] Behrens B A, Rolfes R, Vucetic M, Reinoso J, Vogler M and Grbic N 2014 *Procedia CIRP* **18** 250–255
[22] Advani S G and Talreja R 1993 *Mechanics of Composite Materia*ls **29 (2)** 171–183
[23] Mortazavian S, Fatemi A 2015 *Composites:Part B* **72** 116–129
[24] Vogler M, Rolfes R, Camanho P P 2013 *Mechanics of Materials* **59** 50–64
[25] Golovanov A I and Sultanov L U 2008 *Russian Aeronautics* **51 (4)** 2008 362–368
[26] Sultanov L U 2016 *Lobachevskii Journal of Mathematics* **37 (6)** 787–793
[27] Badriev I B, Banderov V V, Garipova G Z, Makarov M V and Shagidullin R R 2015 *Applied Mathematical Sciences* **9 (82)** 4095–4102
[28] Badriev I B, Banderov V V, Makarov M V, Paimushin V N 2015 *Applied Mathematical Sciences* **9(82)** 3887-3895
[29] Badriev I B, Makarov M V and Paimushin V N 2016 *Procedia Engineering* **150(3)** 1050–1055
[30] Badriev I B, Garipova G Z, Makarov M V, Paymushin V N 2015 *Research Journal of Applied Sciences* **10(8)** 428–435

Evaluation of the stress-strain state of a one-dimensional heterogeneous porous structure

O Gerasimov, F Shigapova, Yu Konoplev and O Sachenkov

Kazan Federal University, Institute of Mathematics and Mechanics, Russia, Tatarstan, Kazan, 18 Kremlevskaya St.

E-mail: 4works@bk.ru

Abstract. The paper deals with the problem of determining the stress-strain state of the distal part of the pelvic girdle bones. The area was modeled using a rod loaded by a compressive force and was described by physical relations linking the stress-strain tensor through the elastic constants, the fabric tensor, and the solid volume fraction of the material. Taking into account the law of porosity variation, we considered the problem of evaluating the stress-strain state depending on the nature of the porous structure, and the relationship of the structure with mechanical macroparameters. In this work, we present the results of calculations for a single load, construct the diagrams for the components of the strain tensor, and carry out an assessment of deformations for various system parameters. To evaluate the macroparameters, we built the dependence of the Poisson ratio of the material on the rotation angle α and the pore ellipticity parameter λ. The sensitivity of the deformations to the elastic constants was also estimated.

1. Introduction

When solving the problems of determining the stress-strain state (SSS) of porous structures, it is necessary to take into account structural peculiarities of the material [1–5]. Today, when solving various problems of bone biomechanics [6–8], structures are commonly described by means of the fabric tensor. In this case, it is assumed that the fabric tensor is a quadratic form that describes the shape of a pore. Some physical relations have currently been built which connect the stress-strain tensor through the elastic constants, the fabric tensor, and the solid volume fraction in the material [9]. Analysis of the SSS of the distal part of the pelvic girdle bones is one of the urgent problems in bone biomechanics. In this work, the distal area is modeled using a rod loaded by a compressive force and described by the relations mentioned above. It is important to evaluate SSS taking into account the nature of the porous structure and the relationship of this structure with easily measurable mechanical macroparameters.

2. Materials and Methods

Let us consider a block diagram loaded by the compressive force P (see figure 1). Assume that it is uniformly filled with pores having the porosity v_0. The pores have an elliptical shape with the fixed semi-radii and form an angle $\pi/2-\alpha$ with the line of action of the applied force. Let us formulate the SSS problem for this rod. Obviously, the constant tension is retained in each section:

$$\sigma_i = \frac{P}{A}, \tag{1}$$

where A is the cross-section area.

For description of the porous structure, we will use the formalism associated with the fabric tensor. Then we will supplement the system by introducing physical relations in the form [9-12]:

$$\begin{aligned}\tilde{\sigma} &= (g_1 + g_2 e) tr\tilde{\varepsilon}\cdot E + (g_3 + g_4 e)\tilde{\varepsilon} + g_5(\tilde{\varepsilon}\tilde{K} + \tilde{K}\tilde{\varepsilon}) + \\ &+ g_6\left(tr(\tilde{K}\tilde{\varepsilon})\cdot E + tr\tilde{\varepsilon}\cdot\tilde{K}\right), \vec{x}\in\bar{S}, t\geq 0,\end{aligned} \tag{2}$$

where e is the variation of the solid bone volume fraction, $\tilde{K}$ is the fabric tensor deviator, g_i are the elastic constants.

The fabric tensor is by definition normalized in such a way that

$$Tr\tilde{H} = 1,$$

its deviator is defined by the formula

$$\tilde{K} = \tilde{H} - \frac{1}{3}\tilde{E}.$$

As a result in this case we obtain

$$Tr\tilde{K} = 0.$$

The law of porosity variation was used in the following form:

$$\nu = \frac{\nu_0 + \sigma_i\left({}^{1}\!/_{3K_{sp}} - {}^{1}\!/_{3K_s}\right)}{1 + {}^{\sigma_i}\!/_{3K_{sp}}} \tag{3}$$

where K_{sp} is the bulk modulus of elasticity of the skeleton with pores, K_s is the bulk modulus of elasticity of the skeleton, ν and ν_0 are the actual and initial porosities respectively, and σ_i is the compressive stress.

The following parameters were introduced during the research:

$$n = \frac{K_{sp}}{K_s}; \lambda = \frac{r_1}{r_2},$$

where n characterizes the volume liquid loss of the material, and λ is the pore ellipticity.

In this case the fabric tensor can be written in the following form:

$$\tilde{H} = \begin{pmatrix} \lambda\cdot h & 0 \\ 0 & h \end{pmatrix}$$

Rotation of the tensor H is implemented by means of multiplication by a rotation matrix, wherein the rotation angle α is a system parameter.

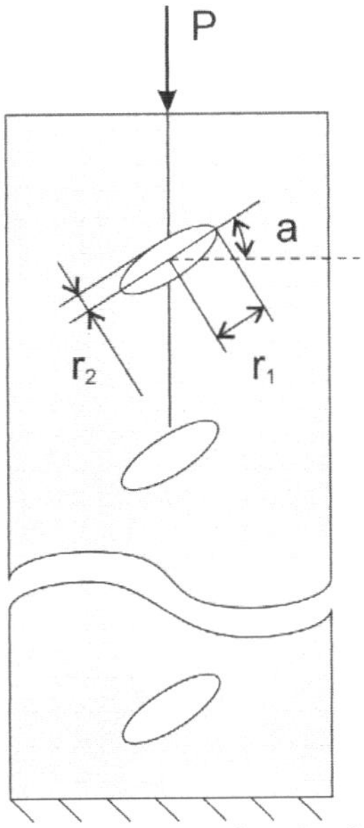

Figure 1. Computational scheme.

The aim of the study is to evaluate deformations for various system parameters. We used the following constant values in the numerical simulation [Cowin, Kichenko]: g_1 = 154.9 GPa, g_2 = 1147 GPa, g_3 = 612.9 GPa, g_4 = 4536 GPa, g_5 = 2384 GPa, g_6 = 510.8 GPa. We also evaluated the solution sensitivity to the deviation from these parameters in case of their variation within 20%.

3. Results and Discussion

We carried out calculations for a single load and constructed diagrams for the strain tensor components (see figure 2) on the λ-α axes. In extreme cases when λ=1 (round pores), the solution coincides with the known one and does not depend on the rotation angle; there are no shear deformations. After analyzing longitudinal deformations, we can note that they reach the largest values when the pore with the bigger radius is directed along the load action line (see figure 2a). The distribution of the transverse deformations is symmetrical relative to the line $\alpha = \pi / 4$, and their absolute value increases when approaching $\alpha = \pi / 4$ (see figure 2b). As far as shear deformations are concerned, they reach the maximum in the case of $\alpha = \pi / 6$ and $\lambda \to 0$ (see figure 2c).

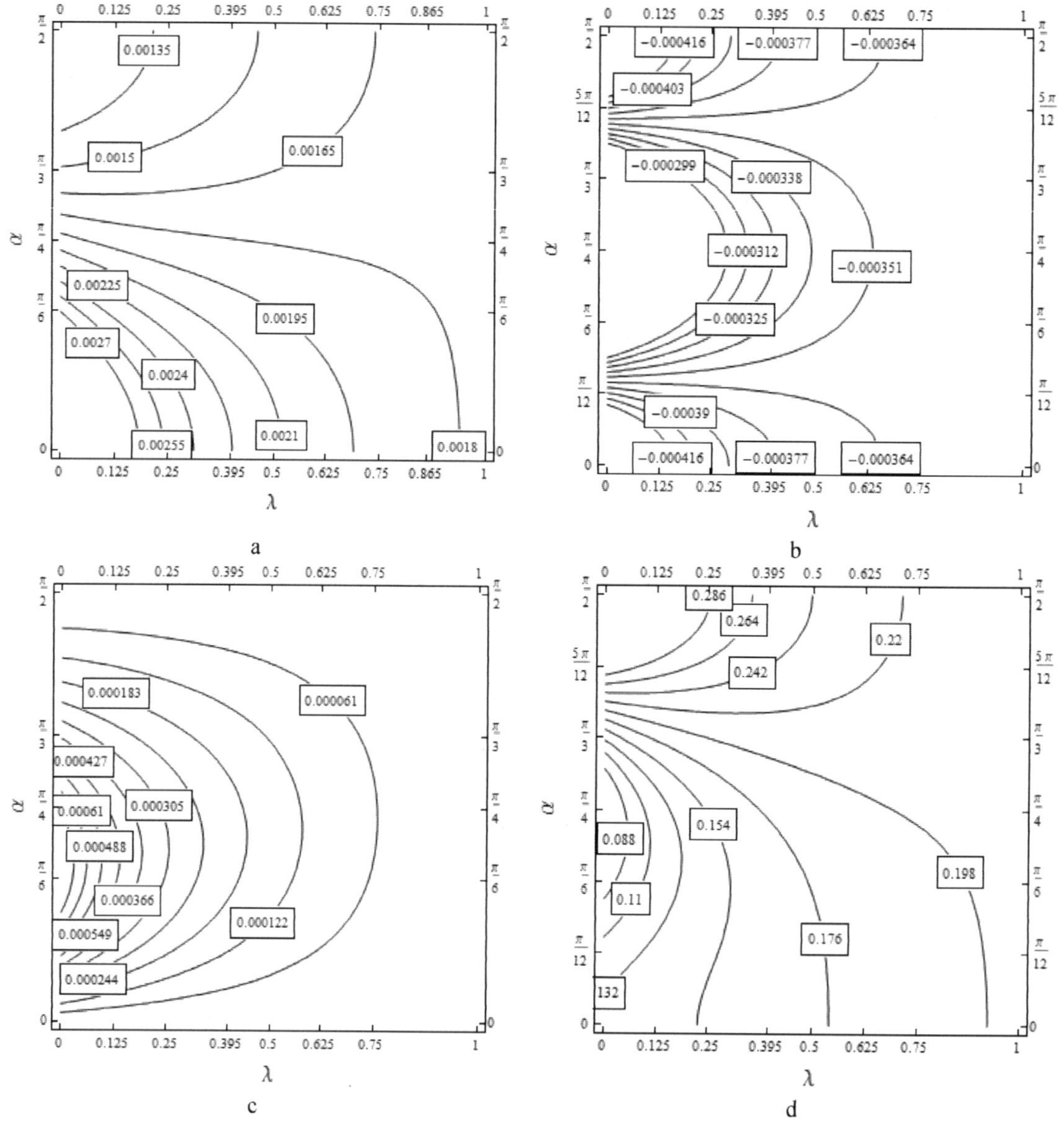

Figure 2. Diagrams for the components of the strain tensor.

For evaluating the macroparameters, we built a dependence of the Poisson ratio of the material on the λ-α axes (see figure 2d). It was noted that in the case of circular pores, the obtained dependence does not depend on the parameters n, v_0, K_s (but obviously depends on the elastic constants g_i) and tend to 0.2. The largest value of the Poisson ratio is achieved when $\alpha \to \pi/2$ and $\lambda \to 0$ is within 0.3–0.35, depending on the initial porosity and elastic constants. Evaluation of sensitivity to the elastic constants showed their underestimation to the value of up to 20% results in the deformation increase by up to 30%. With an increase of the elastic constants by 20%, the deformations decrease by 20%.

4. Conclusion

This study examines the problem of finding the stress-strain state of the distal area in the pelvic girdle bones. The area was modeled by a rod loaded with a compressive force and was described by the physical relations linking the stress-strain tensor through the elastic constants, the fabric tensor, and the solid volume fraction of the material. For the stated law of porosity variation, we considered the problem of evaluating the stress-strain state, depending on the nature of the porous structure and the relationship of the structure with the mechanical macroparameters. The calculation results are given for a single load. We constructed the diagrams for the strain tensor components and made an assessment of deformations for various system parameters. To evaluate the macroparameters, we built the dependence of the Poisson ratio of the material on the rotation angle α and the pore ellipticity parameter λ. The sensitivity of the deformations to the elastic constants was found to be 20–30%.

Acknowledgements

The work was partly supported by the Russian Foundation for Basic Research within the scientific project No. 16-04-00772 and by the subsidy allocated to Kazan Federal University for the state assignment in the sphere of scientific activities.

References

[1] Berezhnoi D V, Sachenkov A A and Sagdatullin M K 2014 Geometrically nonlinear deformation elastoplastic soil *Applied Mathematical Sciences* **8** 6341-6348 http://dx.doi.org/10.12988/ams.2014.48672

[2] Berezhnoi D V, Sachenkov A A and Sagdatullin M K 2014 Research of interaction of the deformable designs located in the soil *Applied Mathematical Sciences* **8** 7107-7115 http://dx.doi.org/10.12988/ams.2014.49706

[3] Berezhnoi D V and Sagdatullin M K 2015 Calculation of interaction of deformable designs taking into account friction in the contact zone by finite element method *Contemporary Engineering Sciences* **8** 1091-1098 http://dx.doi.org/10.12988/ces.2015.58237

[4] Sultanov L U 2015 Numerical Modelling of Deformation of Soil Foundations *Procedia Earth and Planetary Science* **15** 119–124 doi:10.1016/j.proeps.2015.08.029

[5] Davydov R L, Sultanov L U and Kharzhavina V S 2015 Elastoplastic Model of Deformation of Three- Dimensional Bodies in Terms of Large Strains *Global Journal of Pure and Applied Mathematics* **11** 5099-5108

[6] Sachenkov O, Kharislamova L, Shamsutdinova N, Kirillova Y and Konoplev Yu 2015 Evaluation of the bone tissue mechanical parameters after induced alimentary Cu-deficiency followed by supplementary injection of Cu nanoparticles in rats *IOP Conference Series: Materials Science and Engineering* **98** 012015 doi:10.1088/1757-899X/98/1/012015

[7] Shigapova F A, Mustakimova M F, Saleeva G T and Sachenkov O A 2015 Definition of the intense and deformable jaw state under the masseters hyper tone *International Journal of Applied Engineering Research* **10** 44711-44714

[8] Galiullin R R, Sachenkov O A, Khasanov R F and Andreev P S 2015 Evaluation of external fixation device stiffness for rotary osteotomy *International Journal of Applied Engineering Research* **10** 44855-44860

[9] Kichenko A A, Tverier V M, Nyashin Y I, Osipenko M A and Lokhov V A 2012 On application of the theory of trabecular bone tissue remodeling *Russian J. of Biomechanics* **16** 46–64

[10] Cowin S C and Benalla M 2011 Graphical illustrations for the nur-byerlee-carroll proof of the formula for the biot effective stress coefficient in poroelasticity *Journal of Elasticity* **104** 133-141 DOI 10.1007/s10659-011-9324-7

[11] Thompson M and Willis J R 1991 A reformation of the equations of anisotropic poroelasticity *Journal of Applied Mechanics* **58** 612-616 doi:10.1115/1.2897239

[12] Rho J Y, Roy M E, Tsui T Y and Pharr G M 1999 Elastic properties of microstructural components of human bone tissue as measured by indentation *J. Biomed. Mater. Res.* **45** 45–48

The evolution of the bone in the half-plane under the influence of external pressure

O Gerasimov, F Shigapova, Yu Konoplev and O Sachenkov

Kazan Federal University, Institute of Mathematics and Mechanics, Russia, Tatarstan, Kazan, 18 Kremlevskaya St.

E-mail: 4works@bk.ru

Abstract. The paper deals with the problem of restructuring the trabecular bone tissue in the half-plane under the influence of a distributed load. We used evolutionary relations in Cowin's formulation, describing changes in pore orientation in terms of the fabric tensor and the fraction of the solid bone volume in the region under consideration. We modeled the process of the half-plane loading by the distributed pressure. The lazy zone theory and physical relations were used to obtain the initial value of the change in the solid bone volume fraction. As a result, we found the pore distribution field and represented it for clarity as a field of the first eigenvectors of the fabric tensor, which corresponds to the direction of an ellipse's larger radius. This study reveals variation of pore geometry in time and with the distance from the load action line as well as describes the behavior of the solid bone volume fraction with the distance from the pressure line.

1. Introduction

Description of porous structures is currently a topical problem in and solids mechanics [1-3] and bone biomechanics [4–6]. Recently, much attention has been paid to the types of pores in the bone tissue and the methods of describing their direction. Two types of pores are commonly distinguished: channel pores (vascular porosity) and pores formed by the rebuilding of trabeculae (lacunar-canalicular porosity). The use of the fabric tensor in description of such structures has lately become widespread [7–9]. It is known that the trabecular bone tissue is an inhomogeneous anisotropic porous structure, and its mechanical properties are largely determined by its internal structure. Many problems of the locomotive system require the stress-strain state of the bone tissue to be described taking into account the formation of its structure in time when the external load changes. For example, when setting problems of mechanical interaction in joints, it is necessary to consider peculiarities of the tissue under the articular surfaces, as well as to be able to evaluate structural changes during orthopedic interventions into the joint, since these changes can have a significant impact on the strength and stiffness properties of the bone [10, 11].

In this work, we consider the problem of evolution of the trabecular bone tissue and the problem of a half-plane loaded with distributed pressure. The initial state of the tissue is assumed to be uniformly filled by circular pores. The paper also examines the evolutionary model in Cowin's formulation [7–9].

2. Materials and Methods

Let us consider a half-plane (see figure 1) subjected to the distributed load q at the origin of coordinates. We will formulate the problem of the stress-strain state for the half-plane and the adaptive processes in it. Let us designate the region in question as S. Then the complete problem must contain the equilibrium equations:

$$\vec{\nabla}\tilde{\sigma}=0,\ \vec{x}\in S,\ t\geq 0 \tag{1}$$

physical relations [7–9]:

$$\begin{aligned}\tilde{\sigma} &= (g_1+g_2 e)tr\tilde{\varepsilon}\cdot E + (g_3+g_4 e)\tilde{\varepsilon} + g_5(\tilde{\varepsilon}\tilde{K} + \tilde{K}\tilde{\varepsilon}) + \\ &+g_6\left(tr(\tilde{K}\tilde{\varepsilon})\cdot E + tr\tilde{\varepsilon}\cdot\tilde{K}\right),\ \vec{x}\in\bar{S}, t\geq 0,\end{aligned} \tag{2}$$

where e is the variation of the solid bone volume, $\tilde{K}$ is the fabric tensor deviator, g_i are the elastic constants.

The evolutionary relation describing variation of the pore orientation in the region under consideration has the form:

$$\begin{aligned}\frac{d\tilde{K}}{dt} &= h_1\left(\tilde{\varepsilon} - tr\tilde{\varepsilon}\cdot\frac{E}{3} - \left(\tilde{\varepsilon}_0 - tr\tilde{\varepsilon}_0\cdot\frac{E}{3}\right)\right) + \\ &+h_2\left(tr\left(\tilde{K}(\tilde{\varepsilon} - \tilde{\varepsilon}_0)\right)E - \frac{3}{2}\left(\tilde{K}(\tilde{\varepsilon} - \tilde{\varepsilon}_0) + (\tilde{\varepsilon} - \tilde{\varepsilon}_0)\tilde{K}\right)\right),\ \vec{x}\in\bar{S},\ t\geq 0.\end{aligned} \tag{3}$$

The evolutionary relation describing variation in the fraction of the solid bone volume in the region under consideration has the form:

$$\frac{de}{dt}=(f_1 + f_2 e)(tr\tilde{\varepsilon} - tr\tilde{\varepsilon}_0),\ \vec{x}\in\bar{S},\ t\geq 0\ . \tag{4}$$

Cauchy's geometrical relations:

$$\tilde{\varepsilon}=\frac{1}{2}\left(\vec{\nabla}\vec{u}+\vec{u}\vec{\nabla}\right),\ \vec{x}\in\bar{S},\ t\geq 0 \tag{5}$$

Boundary conditions:

$$\begin{aligned}&\vec{n}\cdot\tilde{\sigma}=\vec{q},\ x\in S_q,\ t\geq 0\\ &\vec{n}\cdot\tilde{\sigma}=0,\ x\notin S_q,\ t\geq 0\end{aligned} \tag{6}$$

Initial conditions:

$$\tilde{K}=\tilde{K}_0, e=e_0,\ x\in\bar{S}, t=0 \tag{7}$$

The stress field for the case of loading the half-plane is known and can be represented by the form:

$$\sigma_x(x,y)=\frac{q}{2\pi}\left(2(\theta_2-\theta_1)+Sin2\theta_2-Sin2\theta_1\right) \tag{8.1}$$

$$\sigma_y(x,y)=\frac{q}{2\pi}\left(2(\theta_2-\theta_1)-Sin2\theta_2+Sin2\theta_1\right) \tag{8.2}$$

$$\tau_{xy}(x,y)=\frac{q}{\pi}\left(Sin^2\theta_2-Sin^2\theta_1\right) \tag{8.3}$$

where $\theta_1(x,y)=Arctg\left(\frac{q_s-x}{y}\right)$, $\theta_2(x,y)=Arctg\left(\frac{q_e-x}{y}\right)$.

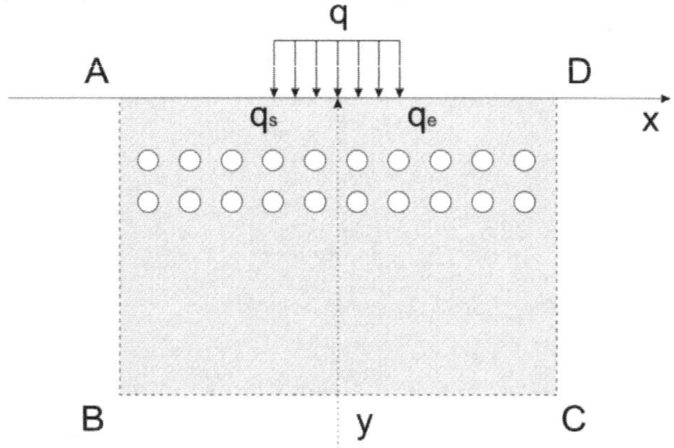

Figure 1. Computational scheme.

During the simulation, we assumed that at the initial time moment the bone is composed of uniform circular pores, and, therefore, the fabric tensor deviator is $\tilde{K}_0=0$. The following formal approach was used to determine the initial variation in the solid volume fraction. It is known that the reconstruction of the bone tissue does not occur when the strains are equal to $\tilde{\varepsilon}_{lz}$ ($\varepsilon_{ij}=-15\cdot10^{-4}\delta_{ij}$), i.e. in the case of the so-called lazy zone. Then, substituting the strain tensor of the lazy zone and the stress tensor in Equation (2), it is possible to determine the initial value e_0 of the solid bone volume fraction.

Since Equations (2)–(4) represent a system of nonlinear algebraic differential equations, the solution of the problem was implemented numerically. In this case, we considered the region ABCD (see figure 1), which was broken up into a regular grid where the equations were then solved. Numerical simulation was performed using the following values for the constants [7-9, 12]: g_1 = 154.9 GPa, g_2 = 1147 GPa, g_3 = 612.9 GPa, g_4 = 4536 GPa, g_5 = 2384 GPa, g_6 = 510.8 GPa, h_1 = 0.01 1/day, h_2 = 0.02 1/day, f_1 = –2.5 1/day, f_2 = 5 1/day. When making calculations, a single load was applied. Let us consider the properties of the fabric tensor, which is of interest in this work. The fabric tensor is by definition normalized in such a way that

$$Tr\tilde{H}=1,$$

Its deviator can be defined by the formula

$$\tilde{K}=\tilde{H}-\frac{1}{3}\tilde{E}.$$

As a result we obtain in this case

$$Tr\tilde{K}=0.$$

Since the problem under consideration is a plane one, we assumed that the pores do not change their directions relative to the chart's plane; i.e. the fabric tensor deviator in its actual state can be represented as follows:

$$\tilde{K} = \begin{pmatrix} K_{11} & K_{12} & 0 \\ K_{12} & K_{22} & 0 \\ 0 & 0 & -K_{11}-K_{22} \end{pmatrix}.$$

3. Results

We made calculations for regions with a load which makes up 8–15% of their length (relation q_sq_e/AD in figure 1). Considering the symmetry relative to the y axis, we show only the results for the positive abscissa axis. Let us present the results of calculation of adaptive processes for the fabric tensor. For clarity, we determined eigenvectors for the fabric tensor and built a field for the first eigenvector (see figure 2 a, b). The second eigenvector is directed orthogonally to the first. The geometrical meaning of these vectors is the direction of the semi-radii of elliptically shaped pores in the half-plane. To evaluate ellipticity, we built a distribution of the eigenvalues ratio for the fabric tensor (see figure 2 c, d).

At the beginning of the adaptation process, the pores begin to flatten under the load action line (see figure 2 a).When approaching the free surface, they begin to line up at an angle of about 30°, but at a distance they behave chaotically. The pores in this case still retain their original shape – the semi-radii ratio is 0.99 (see figure 2 c). With the increase of adaptation time, the direction of the pores changes: it remains the same under the load action line and is lined up along the hyperbolic lines at a distance (see figure 2 b). The pores get flattened by 15% in the region of load application (see figure 2 d); at a distance they remain close to round shape, the ratio of the semi0radii is equal to 0.93.

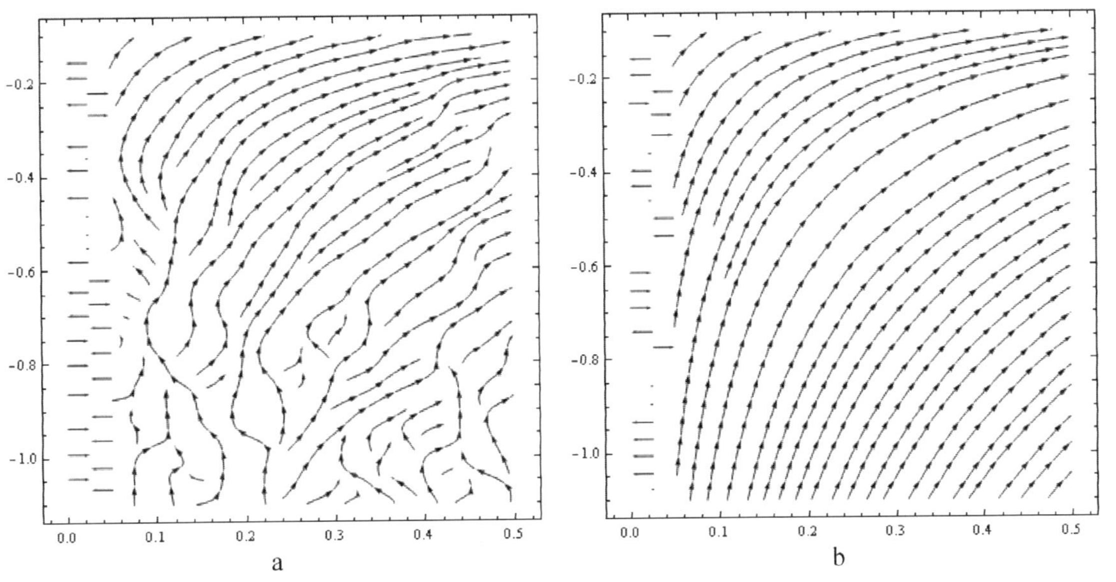

IOP Conf. Series: Materials Science and Engineering **158** (2016) 012037 doi:10.1088/1757-899X/158/1/012037

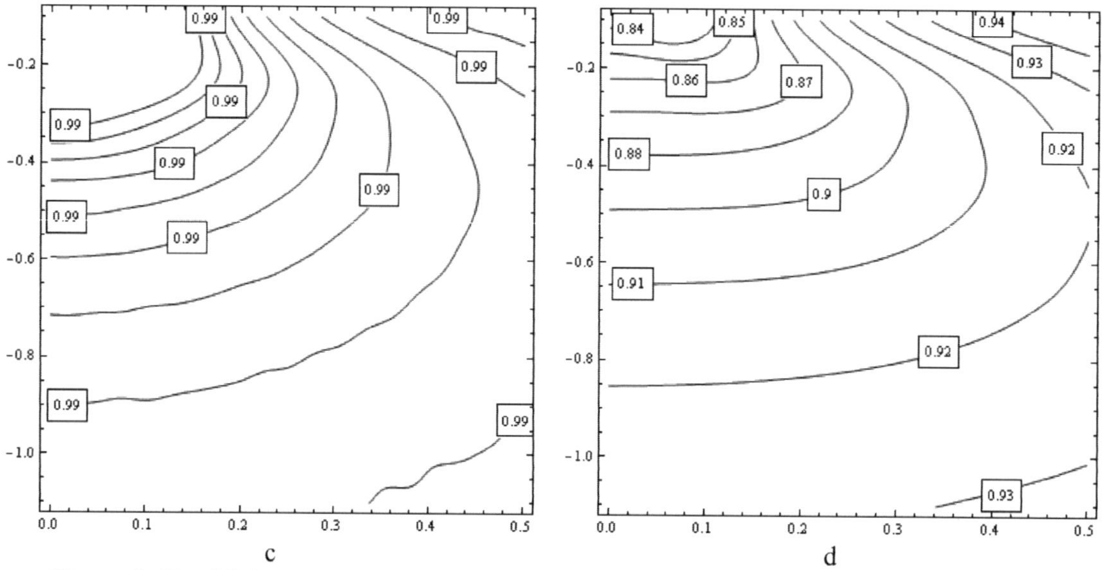

c d

Figure 2. The fabric tensor calculation results: a – the directions of the first eigenvector at the beginning of the adaptation process; b – the directions of the first eigenvector at the end of the adaptation process; c – the ratio of the eigenvalues at the beginning of the adaptation process; d – the ratio of the eigenvalues at the end of the adaptation process.

Let us consider the results of the adaptation process. The variation in the solid bone volume fraction decreases dramatically down to certain minimum value, whereupon the solid volume fraction begins to increase slightly, reaching the asymptote. Figure 3a shows the distribution of changes in the fraction of the solid bone volume at the above-mentioned extreme moment. It is worth noting that all values are less than zero, which indicates that the porosity increases. If the process of adaptation continues, the values become asymptotic (see figure 3b), but remain negative. In this case the largest in magnitude values of variation in the solid bone volume fraction are at a distance from the applied load.

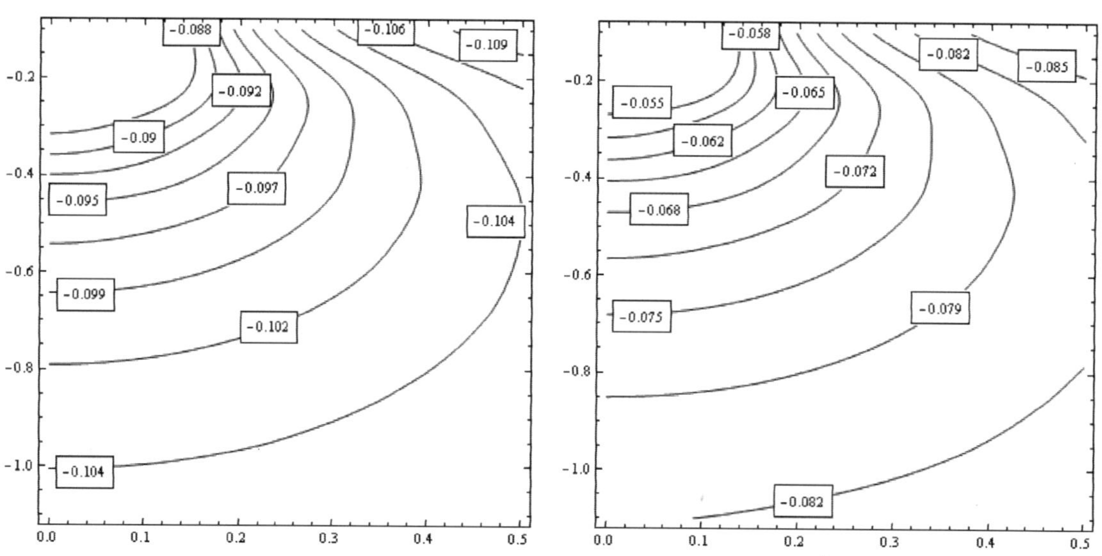

a b

Figure 3. Calculation results for the variation in the solid bone volume fraction: a – the peak change in the solid bone volume fraction during the adaptation process, b – the variation of the solid bone volume fraction after the end of the adaptation process.

4. Discussion

As a result of simulation of adaptive processes for the case of a half-plane loaded by a distributed load, it was revealed that the pores become elongated under the load action line in the orthogonal direction; when getting remote from the load action line, the bigger semi-radii are directed along the hyperbolic curves, their asymptotes being the free surface and the load action line. The highest degree of ellipticity (semi-radii ratio) is achieved in the region of load application and decreases with the distance in the radial direction from zero. The variation of the solid bone volume fraction is negative, which shows the increase of porosity. We also found that the porosity increases sharply until a certain time, after which there is a smooth decrease of the porosity down to reaching the asymptote. The lowest value of porosity is observed in the under-load region and increases with the distance from zero in the radial direction.

5. Conclusion

This paper analyses the problem of restructuring the trabecular bone in a half-plane under the action of a distributed load, using evolutionary relations in Cowin's formulation, which describe the change in orientation of the pores in terms of the fabric tensor and the solid bone volume fraction in the region under consideration. We modeled the process of the half-plane loading by distributed pressure. Using the lazy zone theory and physical relations, we obtained the initial value of variation in the solid bone volume fraction. The field of pore distribution was determined as a result. For clarity it is represented as a field of the first eigenvectors of the fabric tensor, which corresponds to the direction of the bigger semi-radius of an ellipse. We reveal the nature of changes in the pore geometry with time and with the distance from the action line of the applied load and describe variation in the fraction of the solid bone volume with the distance from the pressure action line.

Acknowledgements

The work was partly supported by the Russian Foundation for Basic Research within the scientific project No. 16-04-00772.

References

[1] Berezhnoi D V, Sachenkov A A and Sagdatullin M K 2014 Geometrically nonlinear deformation elastoplastic soil *Applied Mathematical Sciences* **8** 6341-6348 http://dx.doi.org/10.12988/ams.2014.48672

[2] Berezhnoi D V, Sachenkov A A and Sagdatullin M K 2014 Research of interaction of the deformable designs located in the soil *Applied Mathematical Sciences* **8** 7107-7115 http://dx.doi.org/10.12988/ams.2014.49706

[3] Berezhnoi D V and Sagdatullin M K 2015 Calculation of interaction of deformable designs taking into account friction in the contact zone by finite element method *Contemporary Engineering Sciences* **8** 1091-1098 http://dx.doi.org/10.12988/ces.2015.58237

[4] Shigapova F A, Mustakimova M F, Saleeva G T and Sachenkov O A 2015 Definition of the intense and deformable jaw state under the masseters hyper tone International *Journal of Applied Engineering Research* **10** 44711-44714

[5] Galiullin R R, Sachenkov O A, Khasanov R F and Andreev P S 2015 Evaluation of external fixation device stiffness for rotary osteotomy *International Journal of Applied Engineering Research* **10**, 44855-44860

[6] Ambrosi D, Ateshian G A, Arruda E M, Cowin S C, Dumais J, Goriely A, Holzapfel G A, Humphrey J D, Kemkemer R, Kuhl E, Olberding J E, Taber L A and Garikipati K 2011 Perspectives on biological growth and remodeling *Journal of the Mechanics and Physics of Solids* **59** 863-883 doi:10.1016/j.jmps.2010.12.011

[7] Cowin S C 1992 An evolution Wolff's law for trabecular architecture *J. Biomech. Engng* **114** 129–136

IOP Conf. Series: Materials Science and Engineering **158** (2016) 012037 doi:10.1088/1757-899X/158/1/012037

[8] Kichenko A A, Tverier V M, Nyashin Y I, Osipenko M A and Lokhov V A 2012 On application of the theory of trabecular bone tissue remodeling *Russian J. of Biomechanics* **16** 46–64

[9] Kichenko A A, Tverier V M, Nyashin Y I, Osipenko M A and Lokhov V A 2012 Statement of initial boundary value problem on the trabecular bone tissuer remodelling *Russian J. of Biomechanics* **16** 30–45

[10] Rho J Y, Roy M E, Tsui T Y and Pharr G M 1999 Elastic properties of microstructural components of human bone tissue as measured by indentation *J. Biomed. Mater. Res.* **45** 45–48

[11] Sachenkov O, Kharislamova L, Shamsutdinova N, Kirillova E and Konoplev Yu 2015 Evaluation of the bone tissue mechanical parameters after induced alimentary Cu-deficiency followed by supplementary injection of Cu nanoparticles in rats IOP Conference Series: Materials Science and Engineering **98** 012015 doi:10.1088/1757-899X/98/1/012015

[12] Cowin S C and Benalla M 2011 Graphical illustrations for the nur-byerlee-carroll proof of the formula for the biot effective stress coefficient in poroelasticity *Journal of Elasticity* **104** 133-141 DOI 10.1007/s10659-011-9324-7

Mathematical modelling of the mass transfer process between two coaxial cylinders in the problem of thermal creep

O V Germider and V N Popov

Northern (Arctic) Federal University named after M.V. Lomonosov, 17, Severnaya Dvina emb., Arhangelsk, 163002, Russia

E-mail: o.germider@narfu.ru

Abstract. An analytic solution to the Williams equation in the thermal creep flow problem in the channel, formed by two coaxial cylinders, has been constructed using the kinetic approach. For the boundary condition on the channel walls, the diffusion model is used. Taking into account the constructed distribution function for various values of the ratio of radii of cylinders and Knudsen number, the value of the mass flow rate in channel is calculated. The analysis of the obtained results during the transition to free molecular and hydrodynamic regimes is done.

1. Introduction

From the point of view of applications, one of the most important problems in the dynamics of a rarefied gas is the study of gas flow in a channel [1]. In the wake of the development of micro and nanotechnologies, considerable attention has been paid recently to analysis of gas flows in channels with arbitrary cross sections. For example, the flow of a rarefied gas in a channel, formed by two coaxial cylinders, in a channel with constant rectangular and elliptical cross sections was considered in [2-4], respectively. In [5], the problem of mass and heat transfer was viewed in a cylindrical channel with constant pressure and temperature gradients by using Shakhov's model of Boltzmann kinetic equation for cylindrical channels. In [6] the profiles of the mass flow velocity of the gas and the heat flux vector in the channel in free molecular regime have been constructed. A more comprehensive analysis of the current state of the issue on the study of rarefied gas flows in channels one can found in [7]. The aim of this work is to construct an analytic solution to the problem of the thermal creep flow in the channel, formed by two coaxial cylinders under the action of a small thermal pressure drop by using the Williams equation. In this work, for the boundary conditions on the channel walls, we used diffuse reflection model. As a supplement, we constructed a profile of the gas mass velocity in the channel for various values of the ratio of radii of cylinders and Knudsen number.

2. Statement of the problem and mathematical model

Let us consider a channel, formed by two coaxial cylinders with radii R_1' and R_2' ($R_2' > R_1'$). We assume that a constant temperature gradient is maintained in the channel along the Oz' axis. We use the diffuse reflection model for the boundary condition at the channel walls. We denote by $\mathbf{r}'$ the radius vector of gas molecules, by ρ', φ, z' their cylindrical coordinates in configuration space, and by $\mathrm{v}_\rho, \mathrm{v}_\varphi, \mathrm{v}_z$ - projection of velocity vector $\mathbf{v}$ on the coordinate axes.

Mass flux through the channel cross section we define according to [7] as

$$J'_M = 2\pi mn \int_{R'_1}^{R'_2} u'_z(\rho')\rho' d\rho' .$$

Hear m and n are the mass and concentration of the gas molecules, $u'_z(\rho')$ – z'-component of the mass velocity of gas.

The mass velocity of gas we obtained proceeding from the statistical meaning of the distribution function $f(\mathbf{r}', \mathbf{v})$ [8]

$$u_z(\rho') = \frac{1}{n(z')} \int \mathrm{v}_z f(\mathbf{r}', \mathbf{v}) d^3\mathbf{v} .$$

For the sake of convenience in further computations, we pass in previous expressions to the dimensionless variable, by considering $\rho = \rho'/R'_2$, $z = z'/R'_2$ as the coordinates of the dimensionless radius vector; $\mathbf{C} = \beta^{1/2}\mathbf{v}$ as the dimensionless velocity of gas molecules, there $\beta = m/(2k_B T_0)$, k_B is the Boltzmann constant, T_0 is the gas temperature at a point chosen as the origin of coordinates. As a result, we obtain the following expressions for dimensionless mass velocity and a mass flux:

$$U_z(\rho) = \pi^{-3/2} \int C_z \exp(-C^2) f(\mathbf{r}, \mathbf{C}) d^3\mathbf{C} , \tag{1}$$

$$J_M = \frac{4}{(1 - R_1^2)} \int_{R_1}^{1} U_z(\rho)\rho \, d\rho \tag{2}$$

The distribution function of gas molecules $f(\mathbf{r}', \mathbf{v})$ we obtain from Williams equation, which in the chosen coordinate system can be written in the form [9]

$$\mathrm{v}_\rho \frac{\partial f}{\partial \rho'} + \frac{\mathrm{v}_\varphi}{\rho'} \frac{\partial f}{\partial \varphi} + \mathrm{v}_z \frac{\partial f}{\partial z'} + \frac{\mathrm{v}_\varphi^2}{\rho'} \frac{\partial f}{\partial \mathrm{v}_\rho} - \frac{\mathrm{v}_\rho \mathrm{v}_\varphi}{\rho'} \frac{\partial f}{\partial \mathrm{v}_\varphi} = \frac{\omega}{\gamma l_g}(f_* - f) , \tag{3}$$

$$\omega = |\mathbf{v} - \mathbf{u}(\mathbf{r}')| , \quad f_* = n_* \left(\frac{m}{2\pi k_B T_*} \right)^{3/2} \exp\left(-\frac{m}{2k_B T_*} (\mathbf{v} - \mathbf{u}_*)^2 \right), \quad \gamma = 5\sqrt{\pi}/4 .$$

Here quantities n_*, $\mathbf{u}_*$ and T_*, entering into f_*, are not the local density, temperature, and velocity, but are certain parameters that are chosen from the condition that the model collision integral satisfies the laws of conservation of the number of particles, momentum, and energy [9]. For the diffuse reflection model for the boundary condition at the channel walls we can write

$$f^+(\mathbf{r}'_s, \mathbf{v}) = f_s(\mathbf{r}', \mathbf{v}), \quad \mathbf{n}_i \mathbf{v} > 0 , \ i = 1, 2 . \tag{4}$$

Here $f^+(\mathbf{r}'_s, \mathbf{v})$ – distribution function of gas molecules, reflected from the surface of the channel, $f_s(\mathbf{r}', \mathbf{v})$ – local equilibrium distribution function with the parameters on the streamlined gas cylindrical surfaces, $\mathbf{n}_1$ and $\mathbf{n}_2$ are the normal vectors to the surfaces of the cylinders, which are directed into the gas,

$$f_s(\mathbf{r}', \mathbf{v}) = n(z') \left(\frac{m}{2\pi k_B T(z')} \right)^{3/2} \exp\left(-\frac{m}{2k_B T(z')} \mathbf{v}^2 \right) . \tag{5}$$

We assume that a relative temperature drop over the mean free path of gas molecules is small. Then, the problem allows linearization, and the coordinate and velocity distribution function of gas molecules can be written in the form

$$f(\mathbf{r}', \mathbf{v}) = f_0(C)(1 + h(\mathbf{r}, \mathbf{C}) + C_z Z_0(\rho, \mathbf{C})) , \tag{6}$$

$$h(\mathbf{r},\mathbf{C}) = 2C_z U_0 + G_T\left[\left(z - \gamma Kn\frac{C_z}{C}\right)\left(C^2 - \frac{5}{2}\right) - \frac{2\gamma Kn}{3\sqrt{\pi}}C_z\right].$$

Here $f_0(C) = n_0(\beta/\pi)^{3/2}\exp(-C^2)$ is absolute Maxwellian, G_T is the dimensionless temperature gradient, and $Z_0(\rho,\mathbf{C})$ is the linear correction to the local equilibrium distribution function, which takes into account the influence of the walls, $Kn = l_g / R_2'$ is Knudsen number, $l_g = \eta_g\, \beta^{-1/2}/p$ is the mean free path of the gas molecules.
Substituting (6) into (3), we arrive at the equation for finding function $Z_0(\rho,\mathbf{C})$:

$$\left(\cos\psi\frac{\partial Z_0}{\partial\rho} - \frac{\sin\psi}{\rho}\frac{\partial Z_0}{\partial\psi}\right)\gamma Kn\sin\theta + Z_0(\rho,\mathbf{C}) = \frac{3}{4\pi}\int C'\exp(-C'^2)C_z'^2 Z_0(\rho,\mathbf{C}')d^3\mathbf{C}'. \quad (7)$$

For the sake of convenience in further computations, we pass in (7) to the spherical coordinate system in the velocity space, assuming that $C_\rho = C\cos\psi\sin\theta$, $C_\varphi = C\sin\psi\sin\theta$, $C_z = C\cos\theta$, where the angles ψ and θ are measured from the positive directions of the axes C_ρ and C_z. The boundary conditions by taking into account (4), (5) we rewrite as follows

$$Z_0(\rho,\mathbf{C})\big|_s = -2U_0 + \gamma KnG_T\left(C - \frac{5}{2C}\right) + \frac{2\gamma KnG_T}{3\sqrt{\pi}},\ \mathbf{n}_1\mathbf{C} > 0,\ \mathbf{n}_2\mathbf{C} > 0. \quad (8)$$

We seek a solution of (7) in the form of decomposition in two orthogonal functions $e_1(C) = 1$ and $e_2(C) = C - 5/(2C)$:

$$Z_0(\rho,\varphi,\mathbf{C}) = Z_1(\rho,\psi,\sin\theta) + \left(C - \frac{5}{2C}\right)\gamma KnG_T[Z_2(\rho,\psi,\sin\theta) + 1]. \quad (9)$$

The orthogonally here is understood as the equality to zero of the integral $\int_0^{+\infty} e_1(C)e_2(C)C^5\exp(-C^2)dC$. Substituting (9) into (7), we arrive at the following system of equations

$$\gamma Kn\left(\cos\psi\frac{\partial Z_1}{\partial\rho} - \frac{\sin\psi}{\rho}\frac{\partial Z_1}{\partial\psi}\right)\sin\theta + Z_1(\rho,\psi,\sin\theta) =$$

$$= \frac{3}{4\pi}\int_0^{\pi}\cos^2\theta'\sin\theta' d\theta'\int_0^{2\pi} Z_1(\rho,\psi',\sin\theta')d\psi', \quad (10)$$

$$\gamma Kn\left(\cos\psi\frac{\partial Z_2}{\partial\rho} - \frac{\sin\psi}{\rho}\frac{\partial Z_2}{\partial\psi}\right)\sin\theta + Z_2(\rho,\psi,\sin\theta) + 1 = 0, \quad (11)$$

with boundary conditions

$$Z_1(\rho,\psi,\sin\theta)\big|_s = -2U_0 + \frac{2\gamma KnG_T}{3\sqrt{\pi}},\quad \mathbf{n}_1\mathbf{C} > 0,\ \mathbf{n}_2\mathbf{C} > 0, \quad (12)$$

$$Z_2(\rho,\psi,\sin\theta)\big|_s = 0,\quad \mathbf{n}_1\mathbf{C} > 0,\quad \mathbf{n}_2\mathbf{C} > 0. \quad (13)$$

It is easy to see from direct substitution that the solution (10) with boundary conditions (12) has the form:

$$Z_1(\rho,\psi,\sin\theta) = -2U_0 + \frac{2\gamma KnG_T}{3\sqrt{\pi}}. \tag{14}$$

From (13) we can see, that the boundary conditions for $Z_2(R_1,\psi,\sin\theta)$ are homogeneous:

$$Z_2(R_1,\psi,\sin\theta) = 0,\ \cos\psi < 0, \tag{15}$$

$$Z_2(1,\psi,\sin\theta) = 0,\ \cos\psi > 0. \tag{16}$$

Here $R_1 = R'_1 / R'_2$.

The solution of equation (11) with boundary conditions (15) and (16) is sought by the method of characteristics [10]. The system of equations for characteristic of equation (11) has the form:

$$\frac{d\rho}{\gamma Kn\sin\theta\cos\psi} = -\frac{\rho d\psi}{\gamma Kn\sin\theta\sin\psi} = -\frac{dZ_2}{Z_2(\rho,\psi,\sin\theta)+1} = dt \tag{17}$$

Integrating this system of equations, we obtain the two first independent integrals:

$$\rho|\sin\psi| = C_1, \qquad (Z_2(\rho,\psi,\sin\theta)+1)\exp\left(\frac{\rho\cos\psi}{\gamma Kn\sin\theta}\right) = C_2. \tag{18}$$

The integration constants C_1 and C_2 in (18) we exclude using boundary conditions (15) and (16). Thus we obtain

$$Z_2(\rho,\psi,\sin\theta) = \exp(-t_k)-1,\ k = 1,2.$$

The value of the variable t we determine from the condition that the reflection of molecules from the surface of the inner cylinder occurs at $-\alpha \le \psi \le \alpha$:

$$t_1 = \frac{\rho\cos\psi - \sqrt{R_1^2 - \rho^2\sin^2\psi}}{\gamma Kn\sin\theta},\ \alpha = \arccos\frac{\sqrt{\rho^2 - R_1^2}}{\rho};$$

and at $\alpha \le \psi \le 2\pi - \alpha$ reflection occurs from the surface of the outer cylinder:

$$t_2 = \frac{\rho\cos\psi + \sqrt{1-\rho\sin\psi^2}}{\gamma Kn\sin\theta}.$$

Substituting the distribution function (6) in (1) and taking into account the obtained results, we find dimensionless mass velocity of gas in the channel

$$U_z(\rho) = \frac{G_T\gamma Kn}{3\sqrt{\pi}}\left(1 - \frac{3}{2\pi}\int_0^{\pi}\cos^2\theta\sin\theta\, d\theta\left(\int_0^{\alpha}\exp(-t_1)d\psi + \int_{\alpha}^{\pi}\exp(-t_2)d\psi\right)\right). \tag{19}$$

The values of J_M that we found according to (2) using the computer algebra system Maple 17 under different Knudsen number and relationship of the radii of the cylinders R_1' and R_2' are given in table 1.

Table 1. Values of J_M for various ratios R_1'/R_2'.

Kn	$R_1 = R_1'/R_2'$					
	0	0 [5]	0.01	0.1	0.5	0.9
0.0001	0.0001	-	0.0001	0.0001	0.0001	0.0001
0.0010	0.0008	-	0.0008	0.0008	0.0008	0.0008

0.0100	0.0083	0.0116	0.0083	0.0083	0.0082	0.0076
0.1000	0.0764	0.1020	0.0764	0.0757	0.0698	0.0414
0.5000	0.2705	0.3027	0.2695	0.2601	0.2014	0.0761
1.0000	0.3881	0.3968	0.3862	0.3684	0.2660	0.0878
2.0000	0.4977	0.4784	0.4948	0.4677	0.3208	0.0963
5.0000	0.6080	-	0.6040	0.5666	0.3724	0.1034
10.000	0.6632	0.6280	0.6586	0.6158	0.3970	0.1065
100.00	0.7376	0.7210	0.7321	0.6816	0.4290	0.1103
1000.0	0.7502	0.7493	0.7446	0.6927	0.4342	0.1108
10000	0.7520	-	0.7463	0.6943	0.4349	0.1109

3. Results and discussion

Profiles of the z-components of the mass velocity in the channel, calculated according to (19), are given on figures 1 and 2. The case $R_1 = 0$ is described the processes of mass transfer through the cross section of round pipe without internal cylinder.

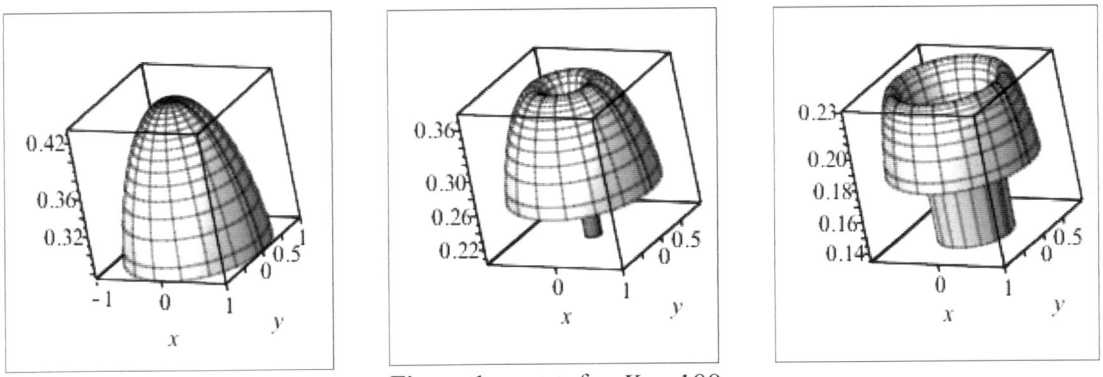

Figure 1. $U_z(\rho)$ for $Kn = 100$
$R_1 = 0$, 0.1 and 0.5 (from left to right)..

From the presented figures one can see that at $Kn >> 0$ the distribution of mass velocity has a maximum, which shifts toward the inner cylinder with the decrease of its radius. When $R_1 << 1$ and $Kn >> 1$ the distribution of the velocity profile close to a paraboloid of rotation.

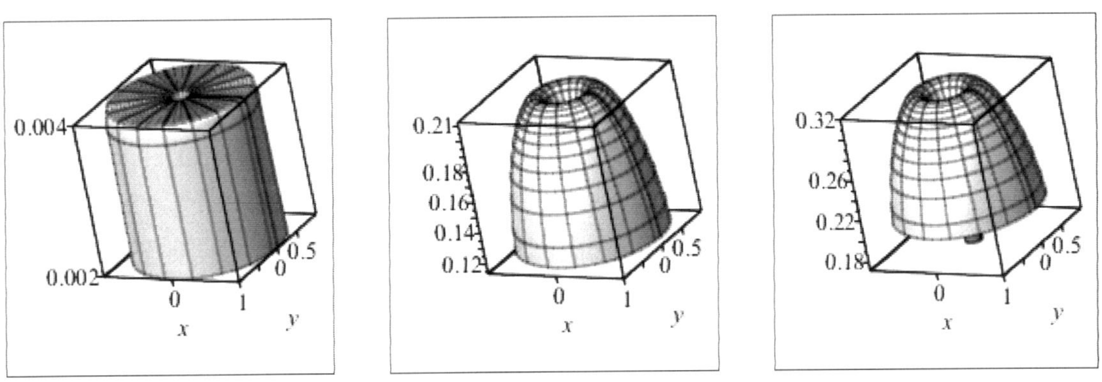

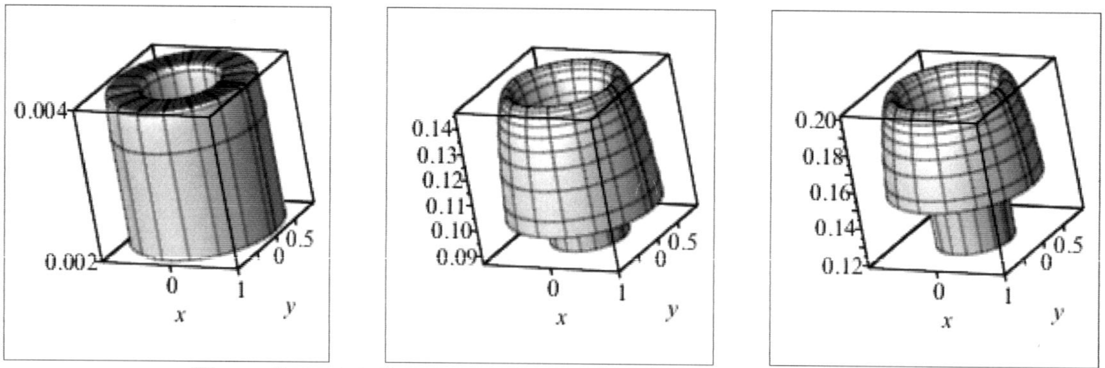

Figure 2. $U_z(\rho)$ for $R_1 = 0.1$ (top row) and $R_1 = 0.5$ (bottom row) for $Kn = 0.01$, 1 and 5 (from left to right)..

For the close to the free molecular flow regime expression (19) for the mass velocity we can to decompose in a series in the small parameter $1/Kn$. In this case, by limited to the linear terms of the expansion, according to (2), we find

$$J_M = \frac{G_T}{\pi^{1/2}(1-R_1^2)} \int_{R_1}^{1} \left(\int_{\alpha}^{\pi} \sqrt{1-\rho^2 \sin^2 \psi}\, d\psi - \int_{0}^{\alpha} \sqrt{R_1^2 - \rho^2 \sin^2 \psi}\, d\psi \right) \rho d\rho . \quad (20)$$

The expression (20) defines a specific mass flow through the channel with cross-section formed by two coaxial cylinders, in a free molecular flow regime, under the effect of the longitudinal temperature gradient. For the case $R_1 = 0$ integrals in (20) may be calculated analytically. In this case $J_M / G_T = 4/(3\sqrt{\pi})$, that coincides with the result of the cylindrical channel [7]. For flow regimes close to the hydrodynamic regime, analysis of the expression (19) gives the following result

$$J_M = \frac{5 G_T Kn}{6}. \quad (21)$$

Thus, for the regime close to the hydrodynamic, specific mass flux does not depend on the radii of the cylinders. The last statement is confirmed by the results presented in table 1 for $Kn \le 0.001$. From (21) one can obtained, that the thermal slip coefficient is equal $5/6$ and it is considering with the corresponding result, obtained in [11].
The results of calculations for the specific mass flow for $0.0001 \le Kn \le 10000$ and $R_1 = 0.00$, 0.01, 0.10, 0.50 and 0.90 are presented in table 1. For comparison, let us consider the values obtained in [5] using Shakhov's model of Boltzmann kinetic equation for cylindrical channels. The difference between the values in table 1 we can explained by the fact, that the macro parameters of gas in the problem on thermal creep strongly depend on the choice of model of collision integral [7].

4. Conclusions

In this work, we have calculated the gas mass flux in a channel, formed by two coaxial cylinders. The profiles of the mass flow velocity of the gas in the channel have been constructed. For different values of Knudsen number and various ratios of cylinder radii the specific gas mass flux through the channel cross section have been calculated. The analysis of the obtained expressions is done. It is shown that in extreme cases then $Kn \ll 1$ and $Kn \gg 1$, the results obtained in this work are moving in similar results for the hydrodynamic and free-regimes, respectively.

Acknowledgments
This work was supported by the state project "Development of Computational Infrastructure for Solving Science-intensive Applied Problems" (project No. 3628).

References

[1] Koshmarov Yu A, Ryzhov Yu A 1977 *Applied Dynamics of Rarified Gas* (Moscow: Mashinostroenie)
[2] Shakhov E M 2003 *Zh. Vychisl. Mat. Mat. Fiz.* **43** 7
[3] Titarev V A, Shakhov E M 2010 *Zh. Vychisl. Mat. Mat. Fiz.* **50** 7
[4] Graur I, Sharipov F 2008 *Eur. J. Mech. B/Fluid* **27** 335
[5] Siewert C E, Valougeorgis D 2002 J. of Quantitative Spectroscopy & Radiative Transfer **72**
[6] Germider O V, Popov V N, Yushkanov A A 2016 Technical Physics **61** 6
[7] Sharipov F M, Seleznev V D 2008, *Motion of Rarefied Gases in Channels and Microchannels* (Yekaterinburg: UrO RAN)
[8] Kogan M N 1967 *Rarefied Gas Dynamics. Kinetic Theory* (Moscow: Nauka)
[9] Cercignani C 1969 *Mathematical Methods in Kinetic Theory* (New York: Plenum Press)
[10] Courant P 1964 *Partial Differential Equations* (Moscow: Mir)
[11] Latyshev A V, Yushkanov A A 2004 Analytic solution of boundary problems for kinetic equations (Moscow: MGOU)

Numerical study of nanoparticle formation in a free turbulent jet

A K Gilfanov[1], W Koch[2], S K Zaripov[1] and O D Rybdylova[3]

[1] Department of Ecological System Modelling, Kazan Federal University, 18 Kremlyovskaya St., 420087 Kazan, Russia
[2] Department of Aerosol Technology, Fraunhofer Institute of Toxicology and Experimental Medicine, Nikolai-Fuchs-Strasse 1, 30625 Hannover, Germany
[3] School of Computing, Engineering and Mathematics, University of Brighton, Cockcroft Building, Lewes Road, Brighton, BN2 4GJ

E-mail: artur.gilfanov@kpfu.ru

Abstract. Di-ethyl-hexyl-sebacate (DEHS) aerosol nanoparticle formation in a free turbulent jet as a result of nucleation, condensation and coagulation is studied using fluid flow simulation and the method of moments under the assumption of lognormal particle size distribution. The case of high nucleation rates and the coagulation-controlled growth of particles is considered. The formed aerosol performance is jet is numerically investigated for the various nozzle diameters and two approximations of the saturation pressure dependence on the temperature. It is demonstrated that a higher polydispersity of the aerosol is obtained for smaller nozzle diameters.

1. Introduction

Nanoparticle formation in a turbulent jet occurs in nature and industry for various practical cases. Aerosol nanoparticles are formed in engine exhaust plumes and propagated in the environment. The turbulent gas-vapour jets in aerosol reactors are used to generate the particles of submicron size for calibration systems. As a step of the creation of a reference submicron particle generator, the aerosol formation and growth in the continuously stirred tank reactor was theoretically and experimentally investigated by Koch et. al. [1]. The reactor provides nucleation of nanosize droplets and their growth by condensation and coagulation in the hot gas-vapour free turbulent jet. The influence of the saturation temperature and the flow rate on the particle size distribution (PSD) of the feed aerosol in the reactor was studied theoretically by Gilfanov et. al. [2]. The next important parameter that influences on the properties of aerosol is the jet nozzle diameter. The objective of this contribution is the study of the influence of the variation of the jet nozzle diameter on the PSD of the generated aerosol on the base of the early developed mathematical model.

2. Problem formulation

A hot gas-vapour turbulent jet that is a feed aerosol source in a continuously stirred tank reactor is considered (figure 1) [1, 2]. Di-ethyl-hexyl-sebacate (DEHS) vapour, saturated at temperature T_0, and nitrogen mixture streams with the velocity U_0 through a small nozzle of diameter d into the cubical mixing reactor with the side length of 60 cm. The gas temperature T_{amb} in the camera is assumed to be much lower than the vapour saturation temperature. Due to mixing of hot jet with the surrounding cold gas the high supersaturation zone occurs at a distance of few diameters adjacent to the nozzle. As a result at high nucleation rates the formation of nanosize droplets takes place. Further growth of particles is performed due to condensation and coagulation in the jet mixing zone.

The gas phase is assumed to be a viscous compressible fluid. The steady turbulent fluid flow is described by the Reynolds-averaged Navier-Stokes equations. The realizable *k-ε* model is used to model turbulence. The temperature field is governed by the energy equation in the form

$$\nabla\cdot(\bar{u}(\rho E+p))=\nabla\cdot((k+\mu_t/\mathrm{Pr}_t)\nabla T), \tag{1}$$

where $\bar{u}$ is the averaged-gas velocity, ρ is the gas phase density, p is the pressure, $E=h-p/\rho+u^2/2$ is the specific energy, h is the sensible enthalpy, k is the thermal conductivity, μ_t is the turbulent viscosity, Pr_t is the turbulent Prandtl number. In the assumption of the axial symmetry the gas phase flow equations are solved in a cylindrical coordinate system. The ideal gas state law is used to close the system of fluid flow equations. The mathematical model of aerosol particle dynamics is based on the general dynamic equation (GDE) [3]

$$\begin{aligned}&\frac{\partial\rho n}{\partial t}+\nabla\cdot(\rho\bar{u}n)+\frac{\partial(G\rho n)}{\partial v}=\nabla\cdot(D\nabla n)+I(v^*)\delta(v-v^*)+\\&+\frac{1}{2}\int_0^v\beta(v-v',v')\rho n(v-v',t)\rho n(v',t)dv'-\rho n(v,t)\int_0^\infty\beta(v,v')\rho n(v',t)dv',\end{aligned} \tag{2}$$

where $n(v, t)$ is the PSD, $G(v)$ is the rate of condensational growth in the particle volume interval [v, $v+dv$], D is the diffusivity of particles, I is the nucleation rate of new particles of critical volume v^*. The birth and loss of particles by coagulation in the size interval [v, $v+dv$] are described by the last two terms in the RHS of Eq. (2), respectively. To solve Eq. (2) the method of moments is applied in the assumption of a lognormal shape of the PSD. This method was developed in previous works of Pratsinis [4], Lee [5, 6], Koch [7], Settumba and Garrick [8]. The transport equations for the moments M_k of the PSD obtained by multiplying Eq. (2) by v^k and integrating over the entire volume range are written in the form

$$\frac{\partial\rho M_k}{\partial t}+\nabla\cdot(\rho\bar{u}M_k)=\nabla\cdot(D\nabla M_k)+S_{M_k}, \tag{3}$$

where S_{Mk} is the source term accounting for internal processes: nucleation, condensation and coagulation. The total diffusivity is $D = D_m + D_t$, where D_m is the molecular diffusion coefficient and D_t is the turbulent diffusion coefficient ($D_t = \mu_t / \mathrm{Sc}_t$, μ_t is the turbulent viscosity, Sc_t is the turbulent Schmidt number). Finally the problem is reduced to the system of equations for the moments M_0, M_1 and M_2. The introduced moments M_0 and M_1 give the total number concentration of particles and the volume fraction of particles, respectively. To take into account gas phase-to-particle conversion the equation for the vapour concentration is solved together with Eq.(3).

The detailed description of the mathematical model of particle nucleation, condensation and coagulation for the considered problem was given in our previous paper [2]. The vapour saturation ratio, which is a ratio of the actual vapour pressure to the saturation pressure P_s, strongly affects the particle nucleation and condensation intensity. The dependence of saturation pressure on temperature can be found from experimental studies. The following two formulas were used to approximate $P_s(T)$ for the considered DEHS vapour

$$\log_{10}P_s=15.3487-7230.347/(T+51.865) \tag{4}$$

$$\log_{10}P_s=10.371-3276.8155/(T-91.4724) \tag{5}$$

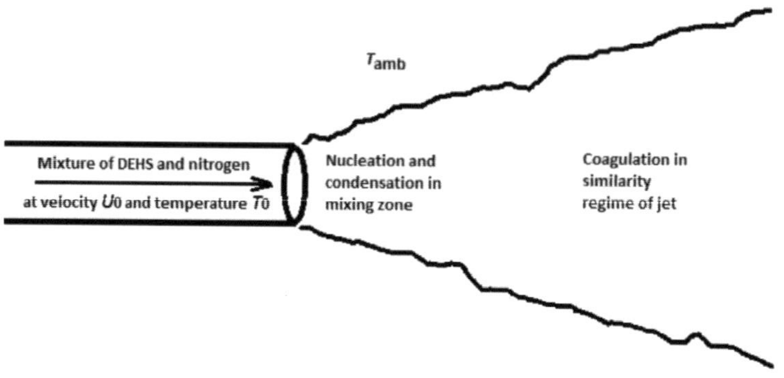

Figure 1. Flow diagram.

3. Results and discussion

To solve the fluid flow and aerosol dynamics equations the CFD package ANSYS Fluent was applied. For the accepted axial symmetry case the computational domain was a trapezoidal area with the length of 600*d* and heights of 80*d* and 120*d*. The structured grid composed by 601 and 401 points in axial and radial directions respectively is used for numerical calculations. The boundary conditions were the same as the conditions used in works [1, 2]: the exit velocity is $U_0 = 45$ m/s, the saturation temperature is $T_0 = 453$K, the ambient temperature is $T_{amb} = 295$ K. For the considered inlet diameters $d = 0.5$, 1 and 2 mm the Reynolds numbers are Re ≈ 700, 1400 and 2800, respectively. The turbulent Prandtl and Schmidt numbers were set to 0.85. The equations (3) for the moments were solved using User-Defined Scalars (UDS). The source terms in (3) were implemented by User-Defined Functions (UDF). The numerical scheme MUSCLE of third order accuracy was applied.

The lognormal PSD is determined by the geometric mean diameter d_g and the geometric standard deviation σ_g

$$n(v,t) = \frac{M_0}{\sqrt{2\pi} d_p \ln \sigma_g} \exp\left(-\frac{\ln^2 (d/d_g)}{2\ln^2 \sigma_g} \right), \tag{6}$$

where d_p is the particle diameter. The particle number current $\dot{N}$ is calculated as

$$\dot{N}(x) = 2\pi \int_0^\infty r u_x(x,r) \rho M_0(x,r) dr, \tag{7}$$

where x and r are the axial and radial coordinates, respectively.

Using the described mathematical model the effect of various nozzles diameters was investigated numerically. The calculated axial gas phase velocity profiles are shown in figure 2a. The free jet with the smaller nozzle diameter is quenched faster that influences on the particle growth. The particle formation depends on the particle residence time τ_{res} that is defined as time for a particle to reach the current position on the axis. The residence time against the axial coordinate x for the various nozzle diameters is shown in figure 2b. The ratio of τ_{res} for the considered diameters is almost constant along the jet axis. The ratio of τ_{res} for $d = 0.5$ mm and 1 mm is about 2, for $d = 0.5$ mm and 2 mm it is 3.8, for $d = 1$ mm and 2 mm the ratio of τ_{res} is 1.9. This means that, for instance, for $d = 0.5$ mm the particles move to the final point 3.8 times slower than for $d = 2$ mm and internal physical processes occur 3.8 times longer.

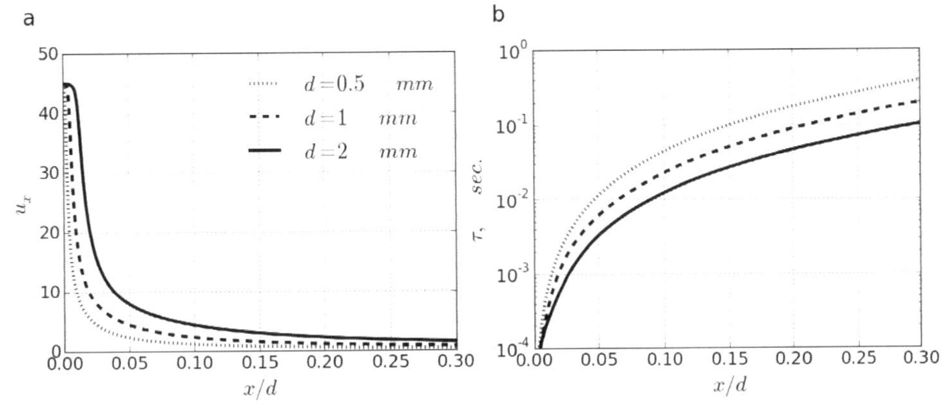

Figure 2. Profiles of axial velocity (a) and residence time of the particle (b) on the jet axis for different nozzle diameters.

The distributions of the particle number current $\dot{N}$, the geometric mean diameter d_g and the geometric standard deviation σ_g along the jet axis is shown in figure 3. The curves without and with markers are obtained using Eq. (4) and (5) respectively. Note that at T_0 = 453K the aerosol volume fractions of $1.3 \cdot 10^{-6}$ and $2.5 \cdot 10^{-6}$ are obtained for formulas (4) and (5), respectively. This leads to more notable production and growth effects of particles for the second approximation. As was mentioned the increase of nozzle diameter leads to the corresponding increase of flow rate. For a fixed value of the exit velocity (U_0 = 45 m/s), a twofold increase in the nozzle diameter leads to the same increase of Reynolds number, and a fourfold increase in flow rate. This results in the increase of the particle number current $\dot{N}$. However, the maximal values of $\dot{N}$ are sufficiently close to each other: $2.9 \cdot 10^{12}$-$4.1 \cdot 10^{12}$, $4.7 \cdot 10^{12}$-$6.4 \cdot 10^{12}$ and $7.2 \cdot 10^{12}$-$1.1 \cdot 10^{13}$ s^{-1}. The two numbers in the given ranges here and further correspond to the limits found using Eq. (4) and Eq. (5). The initial growth of particles by condensation has the larger contribution for larger diameters. For example at a distance of 20*d* from the nozzle the values d_g are 19-22, 26-30 and 36-40 nm for d = 0.5, 1 and 2 mm, respectively. It is seen from figure 3b that the growth of particles by coagulation is more considerable for smaller nozzle diameters. The number current decreases 4.9-6.6 times (to $5.9 \cdot 10^{11}$-$6.2 \cdot 10^{11}$ s^{-1}), 2.5-3 times (to $1.9 \cdot 10^{12}$-$2.1 \cdot 10^{12}$ s^{-1}) and 1.5-1.8 times (to $4.7 \cdot 10^{12}$-$6.0 \cdot 10^{12}$ s^{-1}), the geometric mean diameter increases 1.57-1.68 times (to 30-37 nm), 1.27-1.33 times (to 33-40 nm) and 1.1-1.15 times (to 40-46 nm) for d = 0.5, 1 and 2 mm, respectively. As it can be expected all the effects (nucleation, condensation and coagulation) are more essential for Eq. (5) as in this case the mass fraction of aerosol is two times larger. The geometric standard deviation σ_g increases quickly to 1.3 for the smallest diameter (especially for Eq. (5)) and sufficiently slow to 1.23 for largest one. The values of $\sigma_g > 1.25$ and $\sigma_g < 1.25$ for d = 0.5 mm and d = 2 allow us to conclude that we have polydisperse and monodisperse aerosol formation in these two cases. The effect can be explained by the fact that for the smaller diameter locally the higher particle number concentration is produced by nucleation. This is shown in figure 4a where axial profiles of the first moment M_0 are plotted. The maximum of the number concentration is about 2 and 4 times larger for d = 0.5 m than for d = 1 mm and d = 2 mm respectively. The collision time τ_{coll} is a characteristic time of coagulation process and is inversely related with particle concentration M_0. We can see in figure 4b that the collision time for the smallest diameter is lower at the initial stage of the free jet and coagulation is much more intensive. However, at the late stages of the free jet the pattern is reversed. The higher particle concentration with the higher residence time makes the particle formation for smaller d more controlled by coagulation. It was shown in [2] that the coagulation-controlled particle growth regime is more stable relatively to the change of external conditions. So in the context of using free turbulent jet as a feeding system for the reference particle size distribution generator smaller nozzle diameters can be recommended. Another effect in behavior of collision time for different saturation pressure curves can be seen in figure 4b.

The collision time obtained using Eq. (5) is less than one obtained using Eq. (4) at the beginning of the jet as much more particles are produced in this case (figure 4a). However particles are spent to coagulation more intensively for Eq. (5) and the collision time grows faster along jet axis.

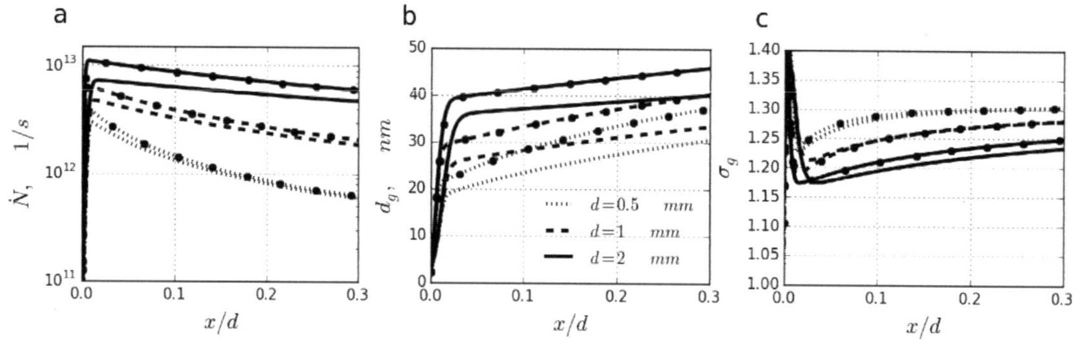

Figure 3. Profiles of particle number current (a), geometric mean diameter (b) and standard geometric deviation (c) on the jet axis for different nozzle diameters.

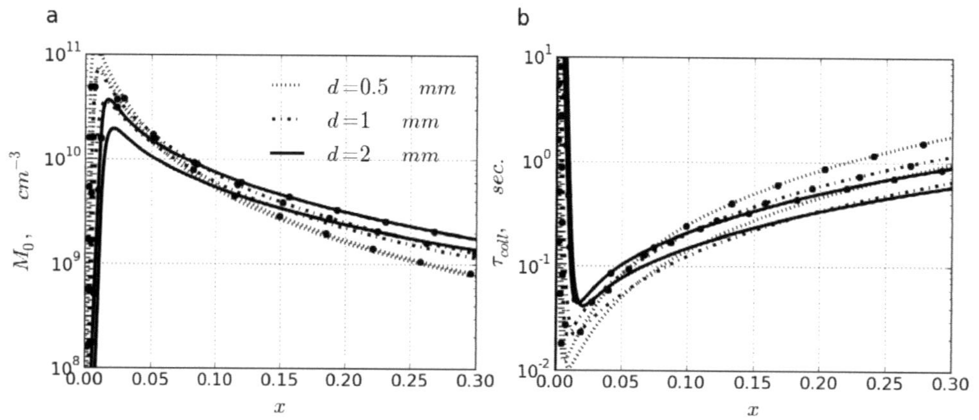

Figure 4. Profiles of particle concentration (a) and collision time (b) on the jet axis for different nozzle diameters.

4. Conclusions

The mathematical model of aerosol particle formation in a hot turbulent gas-vapour jet is used to simulate the nanoparticles generation as a result of nucleation, condensation and coagulation processes. The model includes the Reynolds-averaged Navier-Stokes equations for the fluid flow and general aerosol dynamics equation. The moment method for the lognormal particle size distribution is applied. The influence of the variation of the nozzle diameter on the generated aerosol characteristics is investigated. As a result of the parametrical calculations it is found that for the larger gas-vapour nozzle diameters an aerosol with higher particle number current and mean particle size is obtained. The use of the smaller nozzle diameters leads to higher particle concentration and aerosol growth in the coagulation-controlled mode that allows obtaining the aerosol with higher polydispersity. It means that the smaller nozzle diameters can be recommended for the design of the reference particle size distribution generator to realize more stable aerosol formation.

Acknowledgments

This work was supported by the grant of the President of the Russian Federation for the support of young Russian scientists (MK-6235.2015.1), the Russian Foundation for Basic Research (grant No. 16-51-10024).

References

[1] Koch W, Lödding H and Pohlmann G. 2012 *J. Aerosol Sci* **49** 1-8.
[2] Gilfanov A K, Koch W and Zaripov S K 2016 *J. Aerosol Sci* **96** 124-139.
[3] Friedlander S K 2000 *Smoke, dust, and haze. Fundamentals of aerosol dynamics* (New York: Oxford University Press).
[4] Pratsinis S E 1988 *J Colloid Interf Sci* **124** 416-427.
[5] Lee K W and Chen H 1984 *Aerosol Sci Tech* **3** 327-334.
[6] Lee K W, Chen H and Gieseke J A 1984 *J. Aerosol Sci* **3** 53-62.
[7] Koch W, Windt H and Karfich N 1993 *J. Aerosol Sci* **28** 909-918.
[8] Settumba N and Garrick S C 2003 *J. Aerosol Sci* **34** 149-167.

The finite element method scheme for a solution of an evolution variational inequality with a nonlocal space operator

O V Glazyrina and M F Pavlova

Kazan Federal University, 18 Kremlyovskaya str., Kazan, 420008 Russia

E-mail: glazyrina-olga@ya.ru

Abstract. We consider the parabolic inequality with monotone with respect to a gradient space operator, which is depended on integral with respect to space variables solution characteristic. We construct a two-layer differential scheme for this problem with use of penalty method, semidiscretization with respect to time variable method and the finite element method (FEM) with respect to space variables. We proved a convergence of constructed mothod.

1. Statement of the problem

In this paper we construct and investigate the convergence of an approximate solution method for the following problem: find a function $u \in K$ such that

$$\frac{\partial u}{\partial t} \in L_{p'}(0,T;W_{p'}^{-1}(\Omega)), \tag{1}$$

$$u(x,0) = u_0(x) \text{ almost everywhere (a.e) in } \Omega, \tag{2}$$

and the following inequality holds:

$$\int_0^T \langle \frac{\partial u}{\partial t}, v-u\rangle dt + \int_0^T \langle Lu, v-u\rangle dt \geq \int_0^T \langle f, v-u\rangle dt, \qquad \forall v \in K. \tag{3}$$

Here Ω is a convex bounded domain in the space R^n, $Q_T = \Omega \times (0,T)$, K is a set of the form

$$K = \left\{ v, v \in L_p(0,T;\overset{0}{W}{}_p^1(\Omega)) \bigcap L_\infty(0,T;L_2(\Omega)), v \geq 0 \text{ a.e. in } Q_T \right\}, p > 1,$$

$p' = p/(p-1)$, $\langle w, v\rangle$ is the value of a functional $w \in W_{p'}^{-1}(\Omega)$ on an element $v \in \overset{0}{W}{}_p^1(\Omega)$, L is the operator given by the formula

$$Lu = -\sum_{i=1}^n \frac{\partial}{\partial x_i} k_i(x, u, \nabla u, Bu), \tag{4}$$

∇u is the gradient of u, k_i, u_0 are known functions, B is an operator of the form

$$(Bu)(t)=\int_{\Omega'} g(x,u(x,t))\,dt, \tag{5}$$

where Ω' is a domain that is contained in Ω or coincides with it and g is a given function.

Spatial operators with nonlocalities of the form (5) arise, for example, in the mathematical description of diffusion of bacteria population when it is assumed that the propagation rate of bacteria at a point is specified by the global state of environment (e.g., see [1]–[3]).

Throughout the following, we assume that $k_i(x,\xi_0,\xi,\nu)$, $i=1,\ldots,n$, are continuous with respect to ξ_0,ν,ξ, are measurable with respect to x and satisfy the following conditions for arbitrary values of the arguments $x\in\Omega$, $\xi_0,\nu\in R$, and $\xi^1,\xi^2,\xi\in R^n$:

$$|k_i(x,\xi_0,\xi,\nu)|\le d_0\sum_{j=1}^{n}|\xi_j|^{p-1}+d_1, \quad d_0=const>0, \quad d_1=const\ge 0, \tag{6}$$

$$\sum_{i=1}^{n}k_i(x,\xi_0,\xi,\nu)\xi_i\ge d_2\sum_{i=1}^{n}|\xi_i|^{p}-d_3, \quad d_2=const>0, \quad d_3=const\ge 0, \tag{7}$$

$$\sum_{i=1}^{n}(k_i(x,\xi_0,\xi^1,\nu)-k_i(x,\xi_0,\xi^2,\nu))(\xi_i^1-\xi_i^2)\ge 0, \tag{8}$$

$$\sum_{i=1}^{n}k_i(x,\xi_0,\xi,\nu)\,\xi_i\ge 0. \tag{9}$$

It follows from these assumptions that the operator L acting from $\overset{\circ}{W}{}_p^1(\Omega)$ to $W_{p'}^{-1}(\Omega)$ is continuous, bounded, coercive, and monotone with respect to the gradient.

The existence and uniqueness of a generalized solution were proved in papers [4], [5] for problem (1)–(3).

2. Description of the approximate method

We use the semidiscretization method with respect to the variable t and FEM with respect to space variables to construct an implicit scheme with penalty for problem (1)–(3). To this end, on [0,T] we define a uniform grid $\bar{\omega}_\tau$, where τ is the grid increment, and set $\omega_\tau=\bar{\omega}_\tau\setminus\{0\}$, $M\tau=T$. We approximate the domain Ω by the n-dimensional polyhedron $\Omega_h\subset\Omega$ so as to ensure that, for any point $x\in\Gamma$, there exists a point $\xi\in\Gamma_h$ (Γ_h is the boundary of Ω_h) lying from x at a distance not more than h. The triangulation of Ω_h is performed with the use of simplexes. The approximate solution on each layer $t\in\omega_\tau$ is taken from the set of functions $V_h\in W_p^1(\Omega_h)\cap L_2(\Omega_h)$, whose restriction to each finite element is a linear function, and by $\overset{\circ}{V}_h\subset V$ we denote functions vanishing on Γ_h.

Definition 1. A function $y_\varepsilon(t)\in\overset{\circ}{V}_h$, $t\in\omega_\tau$ as referred to as an approximate solution of problem (1)–(3) if

$$\int_{\Omega_h} y_\varepsilon(x,0)w\,dx=\int_{\Omega_h}u_{0,h}(x)w\,dx \quad \forall w\in\overset{\circ}{V} \tag{10}$$

and for all $t\in\{0,\tau,\ldots,T-\tau\}$ the following equations are completed

$$\int_{\Omega_h}\Big(\frac{\hat{y}_\varepsilon - y_\varepsilon}{\tau}w+\sum_{i=1}^{n} k_i\big(x, y_\varepsilon, \nabla\hat{y}_\varepsilon, By_\varepsilon\big)\frac{\partial w}{\partial x_i}+\frac{1}{\varepsilon}\beta\hat{y}_\varepsilon w\Big)dx=\int_{\Omega_h}\sum_{i=1}^{n} f_{i,\tau}(t)\frac{\partial w}{\partial x_i}dx \quad \forall w\in \overset{\circ}{V}_h, \tag{11}$$

where $f_{i,\tau}(x,t)=\frac{1}{\tau}\int_t^{t+\tau} f_i(x,\xi)d\xi$, $\hat{y}(t)=y(t+\tau)$, β is a penalty operator which is defined by equation $\beta w=-|w^-|^{q-2}w^-$, $q>1$, $w^-=(|w|-w)/2$, $u_{0,h}$ is an interpolant from $\overset{\circ}{V}_h$ for the function u_0.

For problem (10), (11) the unique solvability was proved when $q\le p$. We obtained uniform with respect to h,τ,ε estimates for the set $\{\Pi^- y_\varepsilon\}$ in $L_p(0,T;\overset{0}{W}{}_p^1(\Omega_h))$ and for the sets $\{\Pi^\pm y_\varepsilon\}$ in $L_\infty(0,T;L_2(\Omega_h))$. Here $\Pi^+ z, \Pi^- z$ are piecewise-constant extensions of the mesh function z which are equal to a value of $z(t^*)$ on the set $[t^*,t^*+\tau]$ and $[t^*-\tau,t^*]$ respectively. Also we obtained estimates of the form

$$\frac{1}{\varepsilon}\sum_{t=0}^{t'}\tau\|\Pi^+ y_\varepsilon^-(t)\|^q_{L_q(\Omega_h)}\le const \quad \forall t'\in\omega_\tau, \tag{12}$$

$$\frac{1}{\varepsilon^{q'}}\sum_{t=0}^{T-\tau}\tau\|\Pi^+ y_\varepsilon^-(t)\|^q_{L_q(\Omega_h)}\le const, \tag{13}$$

$$\sum_{t=0}^{T-\tau}\tau\|\Pi^+ (y_\varepsilon)_t(t)\|^{p'}_{W_{p'}^{-1}(\Omega_h)}\le \; const, \tag{14}$$

$$\frac{1}{k\tau}\sum_{i=0}^{M-k}\tau\int_{\Omega_h}\big(\Pi^+ y_\varepsilon(t_i+k\tau)-\Pi^+ y_\varepsilon(t_i)\big)^2 dx\le const \quad \forall k=1,2,...,M. \tag{15}$$

From it obviously follows that for the function $\tilde{y}_\varepsilon$, which is extension of the function y_ε on domain $\Omega\setminus\Omega_h$ by zero , similar uniform with respect to h,τ,ε estimates and inequality similar to (15) (with an integration over Ω) hold in $L_p(0,T;W_p^1(\Omega))$, $L_\infty(0,T;L_2(\Omega))$. These results and the compactness theorem (see [4], lemma 2) provide us the existence of function $u\in L_p(0,T;W_p^1(\Omega))\bigcap L_\infty(0,T;L_2(\Omega))$ and subsequence $\{\tilde{y}_{\varepsilon,k}\}_{k=0}^\infty\subset\{\tilde{y}_\varepsilon\}$ such that as $\tau,\varepsilon\to 0$ and $k\to+\infty$

$$\Pi^-\tilde{y}_{\varepsilon,k} \rightharpoonup u \quad \text{in} \quad L_p(0,T;W_p^1(\Omega)), \tag{16}$$

$$\Pi^\pm\tilde{y}_{\varepsilon,k} \overset{*}{\rightharpoonup} u \quad \text{in} \quad L_\infty(0,T;L_2(\Omega)), \tag{17}$$

$$\Pi^\pm\tilde{y}_{\varepsilon,k} \to u \quad \text{in} \quad L_{p^*}(Q_T), \tag{18}$$

$$\Pi^\pm\tilde{y}_{\varepsilon,k} \to u \quad \text{a.e. in} \quad Q_T, \tag{19}$$

where

$$\begin{aligned} &p^*=2, \ \text{ if } \ (p\ge n)\vee\big((pn/(n-p)>2)\wedge(p<n)\big),\\ &p^*\in[1,pn/(n-p)), \ \text{ if } \ (pn/(n-p)\le 2)\wedge(p<n).\end{aligned} \tag{20}$$

Furthermore by using limit relations (16),(19) and estimate (13) we prove that limit function $u\ge 0$ almost everywhere in Q_T, from it follows that $u\in K$.

The next considerations rely on the following results.

Lemma 1. (see [4]) Let Ω be a bounded domain in the space R^n, let $\{u_m\}$ be a sequence bounded in $L_{q_1}(0,T;L_{p_1}(\Omega))$ and converging to u strongly in $L_{q_0}(0,T;L_{p_0}(\Omega))$, $1<p_0<p_1<\infty, 1<q_0<q_1<\infty$. Then the sequence $\{u_m\}$ converges to u strongly in $L_q(0,T;L_p(\Omega))$ for arbitrary $p_0 \le p < p_1$, $q_0 \le q < q_1$.

Lemma 2. (see [6]) Let Ω be a bounded domain in the space R^n, let the sequence $\{u_m\}_{m=1}^{\infty}$ converges to u in the space $L_1(0,T;L_{p_1}(\Omega))$, $p_1>1$, and let the function $g(x,\xi)$, defining the operator B be measurable with respect to x for any $\xi \in R$, be continuous with respect to ξ for almost all $x \in \Omega$, and satisfy condition

$$|g(x,\xi)| \le g_0(x) + c_g |\xi|^s \quad \forall x \in \Omega, \quad \forall \xi \in R, \tag{21}$$

here $c_g > 0, s \in [0, p_1]$ are given constants, g_0 is not negative function integrable over Ω. Then Bu_m converges to Bu strongly in $L_1(0,T)$.

From a priory estimates and Lemma 1 we have that sequence $\{\Pi^{\pm}\tilde{y}_k\}$ converges strongly in $L_{p^*}(0,T;L_{\tilde{p}}(\Omega)) \in R$, also .

$$\begin{cases} \tilde{p}<2, & \text{if} \quad (n>p)\wedge(np/(n-p)\le 2), \\ \tilde{p}=np/(n-p), & \text{if} \quad (n>p)\wedge(np/(n-p)>2), \\ \tilde{p}\in[2,+\infty), & \text{if} \quad n\le p. \end{cases} \tag{22}$$

It follows that $\tilde{p} < q_{max}$, where q_{max} is a maximum value such that $L_q(\Omega) \supset \left(\overset{\circ}{W}{}_p^1(\Omega) \cap L_2(\Omega)\right)$.

From Lemma 2 and considerations above we have

$$B(\Pi^{\pm}\tilde{y}_{\varepsilon,k}) \to Bu \quad \text{in} \quad L_1(0,T), \tag{23}$$

if $s \in [0,\tilde{p})$, where $\tilde{p}$ is a parameter from (22).

Next we ascertain that the limit function u satisfies inequality (3). For this end we pick arbitrary functions $z \in C_0^{\infty}(\Omega), \eta \in C^{\infty}(0,T): \eta(T)=0$, such that $z\eta \in K$. For the function z we construct an interpolant $z_h \in \overset{\circ}{V}_h$. As $\tilde{z}_h$ we denote the extension by zero of the function z_h on domain $\Omega \setminus \Omega_h$. We rewrite equation (11) for $w = \hat{\tilde{y}}_{\varepsilon} - \tilde{z}_h \hat{\eta}_{\tau}$, with help of operators $\Pi^{\pm}$ it is extended on $[0,T]$. We integrate the result over the segment $[0,t'], t' \in [0,T]$. Then we obtain

$$\begin{gathered} \int_0^{t'}\int_{\Omega} \Pi^+(\tilde{y}_{\varepsilon})_t \, \Pi^- \tilde{y}_{\varepsilon} \, dxdt - \int_0^{t'}\int_{\Omega} \Pi^+(\tilde{y}_{\varepsilon})_t \, \tilde{z}_h \, \Pi^+ \hat{\eta}_{\tau} \, dxdt + \\ + \int_0^{t'}\int_{\Omega} \sum_{i=1}^{n} k_i\left(x, \Pi^+\tilde{y}_{\varepsilon}, \Pi^- \nabla \tilde{y}_{\varepsilon}, B(\Pi^+\tilde{y}_{\varepsilon})\right) \Pi^- \frac{\partial(\tilde{y}_{\varepsilon} - \tilde{z}_h \eta_{\tau})}{\partial x_i} dxdt = \\ = \int_0^{t'}\int_{\Omega} \sum_{i=1}^{n} \Pi^+ f_{i,\tau} \Pi^- \frac{\partial(\tilde{y}_{\varepsilon} - \tilde{z}_h \eta_{\tau})}{\partial x_i} dxdt + \frac{1}{\varepsilon} \int_0^{t'}\int_{\Omega} \sum_{i=1}^{n} -\Pi^- \beta \tilde{y}_{\varepsilon} \Pi^- (\tilde{y}_{\varepsilon} - \tilde{z}_h \eta_{\tau}) dxdt. \end{gathered} \tag{24}$$

Let's notice that

$$J=\int_0^t\int_\Omega \Pi^+(\tilde{y}_\varepsilon)_t \Pi^-\tilde{y}_\varepsilon\,dxdt \geq \frac{1}{2\tau}\int_0^t\int_\Omega (\Pi^-\tilde{y}_\varepsilon^2-\Pi^+\tilde{y}_\varepsilon^2)\,dxdt=\frac{1}{2\tau}\int_{t'-\tau}^{t'}\int_\Omega (\Pi^-\tilde{y}_\varepsilon)^2\,dxdt-\frac{1}{2}\int_\Omega u_0^2(x)dx.$$

From the convexity of the functional $\int_\Omega (\Pi^-\hat{y}_\varepsilon)^2\,dx$ follows that

$$\frac{1}{2\tau}\int_{t'-\tau}^{t'}\int_\Omega (\Pi^-\tilde{y}_\varepsilon)^2\,dxdt \geq \frac{1}{2}\| \Lambda_\tau \tilde{y}_\varepsilon(t')\|^2_{L_2(\Omega)},$$

where $\Lambda_\tau \tilde{y}_\varepsilon$ is a piecewise linear extension of the function y_ε with respect to t; therefore,

$$J\geq \frac{1}{2}\| \Lambda_\tau \tilde{y}_\varepsilon(t')\|^2_{L_2(\Omega)}-\frac{1}{2}\| u_0(x)\|^2_{L_2(\Omega)}\,dx. \tag{25}$$

Next, it follows from condition (9) that

$$\begin{aligned}&\int_0^t\int_\Omega\sum_{i=1}^n k_i\big(x,\Pi^+\tilde{y}_\varepsilon,\Pi^-\nabla\tilde{y}_\varepsilon,B(\Pi^+\tilde{y}_\varepsilon)\big)\Pi^-\frac{\partial(y_\varepsilon-\tilde{z}_h\eta_\tau)}{\partial x_i}dxdt\geq\\ &\geq\int_0^t\int_\Omega\sum_{i=1}^n k_i\big(x,\Pi^+\tilde{y}_\varepsilon,\Pi^+\nabla(\tilde{z}_h\hat{\eta}_\tau),B(\Pi^+\tilde{y}_\varepsilon)\big)\Pi^-\frac{\partial(\tilde{y}_\varepsilon-\tilde{z}_h\eta_\tau)}{\partial x_i}dxdt.\end{aligned} \tag{26}$$

Since β is a monotone operator and $\beta\tilde{z}_h\hat{\eta}_\tau=0$, we have

$$\frac{1}{\varepsilon}\int_0^t\int_\Omega\sum_{i=1}^n -\Pi^-(\beta\tilde{y}_\varepsilon)\Pi^-\frac{\partial(\tilde{y}_\varepsilon-\tilde{z}_h\eta_\tau)}{\partial x_i}dxdt\leq 0. \tag{27}$$

Let us transform relation (24) with the use of inequalities (25)–(27). We multiply the resulting inequality by $1/\lambda, \lambda>0$, and integrate with respect to t' from $T-\lambda$ to T,

$$\begin{aligned}&\frac{1}{\lambda}\int_{T-\lambda}^T\Big\{\frac{1}{2}\| \Lambda_\tau\tilde{y}_\varepsilon(t'))\|^2_{L_2(\Omega)}-\frac{1}{2}\| u_0(x)\|^2_{L_2(\Omega)}-\int_0^t\int_\Omega\Pi^+(\tilde{y}_\varepsilon)_t\,\Pi^+(\tilde{z}_h\hat{\eta}_\tau)dxdt+\\ &+\int_0^t\int_\Omega\sum_{i=1}^n k_i\big(x,\Pi^+\tilde{y}_\varepsilon,\Pi^+\nabla(\tilde{z}_h\hat{\eta}_\tau),B(\Pi^+\tilde{y}_\varepsilon)\big)\Pi^-\frac{\partial(\tilde{y}_\varepsilon-\tilde{z}_h\eta_\tau)}{\partial x_i}dxdt\Big\}dt'\leq\\ &\leq\frac{1}{\lambda}\int_{T-\lambda}^T\int_0^t\int_\Omega\sum_{i=1}^n\Pi^+f_{i,\tau}\Pi^-\frac{\partial(\tilde{y}_\varepsilon-\tilde{z}_h\eta_\tau)}{\partial x_i}dxdtdt'.\end{aligned} \tag{28}$$

The main part of the further considerations is a proof of passage to the limit in inequality (28) when $h,\tau,\varepsilon\to 0$. We will write below the least obvious parts of this proof in our opinions.

Let us show that

$$\begin{aligned}&\int_0^t\int_\Omega\sum_{i=1}^n k_i\big(x,\Pi^+\tilde{y}_\varepsilon,\Pi^+\nabla(\tilde{z}_h\hat{\eta}_\tau),B(\Pi^+\tilde{y}_\varepsilon)\big)\Pi^-\frac{\partial(\tilde{y}_\varepsilon-\tilde{z}_h\eta_\tau)}{\partial x_i}dxdt\to\\ &\to\int_0^t\int_\Omega\sum_{i=1}^n k_i(x,u,\nabla z\eta,Bu)\frac{\partial(u-z\eta)}{\partial x_i}dxdt\end{aligned} \tag{29}$$

as $h, \tau, \varepsilon \to 0$. To this end, we show that

$$k_i\big(x, \Pi^+ \tilde{y}_\varepsilon, \Pi^+ \nabla(\tilde{z}_h \hat{\eta}_\tau), B(\Pi^+ \tilde{y}_\varepsilon)\big) \to k_i(x, u, \nabla z\eta, Bu) \quad \text{in} \quad L_{p'}(Q_T). \tag{30}$$

Since the functions $k_i(x, \xi_0, \xi, v)$ are continuous with respect to v, it follows from (23) that

$$k_i\big(x, \Pi^+ \tilde{y}_\varepsilon, \Pi^+ \nabla(\tilde{z}_h \hat{\eta}_\tau), B(\Pi^+ \tilde{y}_\varepsilon)\big) \to k_i(x, u, \nabla z\eta, Bu) \text{ a.e.in } Q_T. \tag{31}$$

Let

$$J_2 = \int_{Q_T} \big|k_i\big(x, \Pi^+ \tilde{y}_\varepsilon, \Pi^+ \nabla(\tilde{z}_h \hat{\eta}_\tau), B(\Pi^+ \tilde{y}_\varepsilon)\big) - k_i(x, u, \nabla z\eta, Bu)\big|^{p'} \, dxdt. \tag{32}$$

It follows from (31) that the integrant in (32) function tends to zero as $h, \tau, \varepsilon \to 0$ almost everywhere in Q_T. Furthermore, if follows from the estimate (6) that it is bounded above by a function depending only on $z\eta$. Therefore, by the Lebesgue dominated convergence theorem, we have $J_2 \to 0$ as $h, \tau, \varepsilon \to 0$, i.e. the limit relation (30) holds. Obviously, relation (29) follows from (30) and (16).

By using the relation (29), we pass to the limit in inequality (28) as $h, \tau, \varepsilon \to 0$. Here we use the fact that, by the virtue of the weak lower semicontinuity of the norm,

$$\underline{\lim}_{\tau, \varepsilon \to 0} \int_{T-\lambda}^{T} \| \Lambda_\tau \tilde{y}_\varepsilon(t') \|^2_{L_2(\Omega)} \, dt' \geq \int_{T-\lambda}^{T} \| u(t') \|^2_{L_2(\Omega)} \, dt'. \tag{33}$$

Then we obtain

$$\begin{gathered} \frac{1}{2} \int_{T-\lambda}^{T} \| u(t') \|^2_{L_2(\Omega)} \, dt' - \lambda \frac{1}{2} \| u_0(x) \|^2_{L_2(\Omega)} \leq \int_{T-\lambda}^{T} \Big\{ \int_0^{t'} \langle f, u - z\eta \rangle \, dt + \int_0^{t'} \Big\langle \frac{\partial u}{\partial t}, z\eta \Big\rangle \, dt - \\ - \int_0^{t'} \int_\Omega \sum_{i=0}^{n} k_i(x, u, \nabla z\eta, Bu) \frac{\partial (u - z\eta)}{\partial x_i} \, dxdt \Big\} dt' \equiv \int_{T-\lambda}^{T} J_1(t') \, dt'. \end{gathered} \tag{34}$$

Obviously, inequality (34) also holds for each function v from K.

We rewrite inequality (34) in the form

$$\frac{1}{2\lambda} \int_{T-\lambda}^{T} \| u(t') \|^2_{L_2(\Omega)} \, dt' - \frac{1}{2} \| u_0 \|^2_{L_2(\Omega)} \leq \frac{1}{\lambda} \int_{T-\lambda}^{T} J_1(t') \, dt'. \tag{35}$$

Since $u \in L_2(0, T; \overset{\circ}{W}{}^1_p(\Omega))$ and $\frac{\partial u}{\partial t} \in L_2(0, T; W_p^{-1}(\Omega))$, it follows (see [7], p. 177) that $u \in C(0, T; L_2(\Omega))$. Therefore, $\| u(t) \|_{L_2(\Omega)}$ is a continuous function on $(0, T)$. By a property of the Lebesgue integral $J_1(t')$ is continuous with respect to t' as well. Therefore, in inequality (35) we admit the passage to the limit as $\lambda \to 0$; then we obtain

$$\begin{gathered} \frac{1}{2} \| u(T) \|^2_{L_2(\Omega)} - \frac{1}{2} \| u_0 \|^2_{L_2(\Omega)} = \int_0^T \int_\Omega f(u - v) \, dxdt + \int_0^T \langle \frac{\partial u}{\partial t}, v \rangle \, dt - \\ - \int_0^T \int_\Omega \sum_{i=0}^{n} k_i(x, u, \nabla v, Bu) \frac{\partial (u - v)}{\partial x_i} \, dxdt. \end{gathered} \tag{36}$$

The last relation, together with the equation (see [7], p.177)

$$\frac{1}{2}\| u(T)\|^2_{L_2(\Omega)} - \frac{1}{2}\| u_0\|^2_{L_2(\Omega)} = \int_0^T \langle \frac{\partial u}{\partial t}, u\rangle dt,$$

implies that u satisfies the inequality

$$\int_0^T \langle \frac{\partial u}{\partial t}, u-v\rangle dt + \int_0^T\int_\Omega \sum_{i=1}^n (k_i(x,u,\nabla v, Bu)\frac{\partial (u-v)}{\partial x_i} dxdt \le \int_0^T \langle f, u-v\rangle dt \quad \forall\, v\in K. \tag{37}$$

As it's known (see [4], lemma 1) inequality (37) is equivalent to inequation (3).

In conclusion, we state the main result of the present paper.

Theorem 1. Let the functions k_i satisfy conditions (4)–(9), let the function f belong to the space $L_{p'}(Q_T)$, $p>1$; let $u_0 \subset L_2(\Omega): u_0(x)\ge 0$ almost everywhere in Ω, let $g\in L_{p_1}(\Omega')$. Then for each $s\in[0,\tilde{p})$, $q\le p$, there exists a subsequence of piecewise constant, with respect to *t*, extensions of the solution of semidiscrete problem (10), (11), which is strongly converges in $L_{p^*}(0,T;L_{\tilde{p}}(\Omega))$ to a solution of the problem (1)–(3); here p^* and $\tilde{p}$ are the parameters defined by relations (20), (22).

Acknowledgments
The research was supported by the Russian Foundation for research (projects 15-01-05686, 15-41-02315)

References

[1] Chipot M and Molinet L 2001 Asymptotic behavior of some nonlocal diffusion problems *Applicable Analysis* **80** (3/4) 279 – 315

[2] Chipot M and Lovat B 2001 Existence and uniqueness results for a class of nonlocal elliptic and parabolic problems *Dyn. Contin. Discrete Impuls. Syst., Ser. A: Math. Anal.* **8** (1) 35 – 51

[3] Simon L 2004 On quasilinear parabolic functional differential equation with discontinuous terms *Annales Univ. Shi. Budapest* **47** 211 – 29

[4] Glazyrina O V and Pavlova M F 2014 On the solvability of an evolution variational inequality with a nonlocal space operator *Differential equations* **50** (7) 873 – 87

[5] Glazyrina O V and Pavlova M F 2014 The solution uniqueness theorem of evolution variational inequation with nonlocal space operator *Proc. of the 10th Int. Conf. «Mesh methods for boundary-value problems and applications»* (Kazan: Kazan University Publishing House) 205 – 08

[6] Glazyrina O V and Pavlova M F 2015 Study of the Convergence of the Finite-Element Method for solving Parabolic Equations with Nonlineal Nonlocal Space Operator *Differential equations* **51** (7) 876 – 89

[7] Gajevskii H, Gröger K and Zacharias K 1974 *Nichtlineare Operatorgleichungen und Operatordifferemtialgleichungen* (Berlin: Akademic–Verlag). Translated under the title *Nelineynye operatornye uravneniya I operatornye differencial'nye uravnenia* (Moscow: Mir, 1978)

The finite element method explicit scheme for a solution of one problem of surface and ground water combined movement

L L Glazyrina and M F Pavlova

Kazan Federal University, 18 Kremlyovskaya str., Kazan, 420008 Russia

E-mail: glazyrina-ludmila@ya.ru

Abstract. We investigate the initial-boundary value problem for the system of two nonlinear degenerate parabolic combined equations, when the one of equations is set in the domain we consider, and the second is set on the conducted section of this domain. With help of the semidiscretization methods with respect to time variable and finite element method (FEM) with respect to space variable we construct an explicit difference scheme. We obtained conditions which provide us convergence of constructed method. The result of convergence was obtained with the minimum conditions for smoothness of the initial data.

1. Statement of the problem

This paper is devoted to the convergence investigation of an approximate solution method for the following initial-boundary problem:

$$\frac{\partial\varphi_1(u)}{\partial t}(x,t)-\sum_{i=1}^{2}\frac{\partial}{\partial x_i}\big(k_i\big(x,u(x,t),\nabla u(x,t)\big)\big)=f_1(x,t),\quad x\in\Omega_\Pi=\Omega\setminus\Pi,\ t\in(0,T) \tag{1}$$

$$\frac{\partial\varphi_2(u)}{\partial t}(x,t)-\frac{\partial}{\partial s}\Big(k_\Pi\Big(x,u(x,t),\frac{\partial u}{\partial s}(x,t)\Big)\Big)+$$
$$+\Big[\sum_{i=1}^{2}k_i\big(x,u(x,t),\nabla u(x,t)\big)\cos(n,x_i)\Big]_\Pi=f_2(x,t), \tag{2}$$
$$[u]_\Pi=0,\quad x\in\Pi,\ t\in(0,T),$$

$$u(x,0)=u_0(x),\ \ x\in\Omega;\qquad u(x,t)=0,\ \ x\in\Gamma,t\in(0,T). \tag{3}$$

Here Ω is a bounded domain in space R^2, Γ is its boundary, Π is a conducted section of Ω and dividing it into two connected domains, $[\cdot]_\Pi$ is a discontinuity of a function when its passing through the section Π, n is a normal to Π.

Problem (1)–(3) describes (see [1]) the process of surface-water and groundwater joint motion, in addition the section Π is a bed river or channel, the required function u defines the height of the water face above the impenetrable base.

We assume that functions $\varphi_i(\xi)$ are absolutely continuous, strongly increasing and satisfy the following conditions for arbitrary $\xi\in R^1$

$$b_{0i}|\xi|^{\alpha_i} - b_{1i} \le \Phi_i(\xi) \equiv \int_0^{\xi} \varphi_i'(t)t\,dt \le b_{2i}|\xi|^{\alpha_i} + b_3, \quad \alpha_i > 1, \tag{4}$$

$$|\varphi_i(\xi)| \le b_{4i}|\xi|^{\alpha_i - 1} + b_5, \tag{5}$$

$$(\varphi'(\xi)\xi)' \ge 0, \tag{6}$$

here

$$b_{0i} > 0, b_{1i} \ge 0, b_{2i} > 0, b_{3i} \ge 0, b_{4i} > 0, b_{5i} \ge 0, i = 1,2.$$

We assume functions $k_i(x,\xi_0,\xi)(i=1,2), k_\Pi(x,\xi_0,\xi)$ are continuous with respect to x,ξ_0,ξ, are measurable with respect to x and satisfy the following conditions for arbitrary $x \in \Omega$, $\overline{x} \in \Pi$, $\xi_0, \zeta, \zeta^1, \zeta^2 \in R^1$, $\xi^1, \xi^2, \xi \in R^2$

$$|k_i(x,\xi_0,\xi)| \le M_{01}\sum_{j=1}^{2}|\xi_j|^{p_1-1} + M_{11}, \quad M_{01} > 0, M_{11} \ge 0, p_1 > 1, \tag{7}$$

$$\sum_{i=1}^{2} k_i(x,\xi_0,\xi)\xi_i \ge M_{21}\sum_{i=1}^{2}|\xi_i|^{p_1} - M_{31}, \quad M_{21} > 0, M_{31} \ge 0, \tag{8}$$

$$\sum_{i=1}^{2}(k_i(x,\xi_0,\xi^1) - k_i(x,\xi_0,\xi^2))(\xi_i^1 - \xi_i^2) \ge 0, \tag{9}$$

$$|k_\Pi(\overline{x},\zeta_0,\zeta)| \le M_{02}|\zeta|^{p_2-1} + M_{12}, \quad M_{02} > 0, M_{12} \ge 0, p_2 > 1, \tag{10}$$

$$k_\Pi(\overline{x},\zeta_0,\zeta)\zeta \ge M_{22}|\zeta|^{p_2} - M_{32}, \quad M_{22} > 0, M_{32} \ge 0, \tag{11}$$

$$(k_\Pi(\overline{x},\zeta_0,\zeta^1) - k_\Pi(\overline{x},\zeta_0,\zeta^2))(\zeta^1 - \zeta^2) \ge 0. \tag{12}$$

Throughout the following, to define the generalized solution for the problem (1)–(3) we use normed spaces $\overset{\circ}{V}, \overset{\circ}{V}(0,T), \overset{\circ}{W}(0,T)$, obtained by closing $C_0^\infty(\Omega)$ and $C^\infty(0,T;C_0^\infty(\Omega))$ in the following norms

$$\|u\|_{\overset{\circ}{V}} = \|u\|_{\overset{\circ}{W}{}^1_{p_1}(\Omega)} + \|u\|_{\overset{\circ}{W}{}^1_{p_2}(\Pi)},$$

$$\|u\|_{\overset{\circ}{V}(0,T)} = \|u\|_{L_{p_1}(0,T;\overset{\circ}{W}{}^1_{p_1}(\Omega))} + \|u\|_{L_{p_2}(0,T;\overset{\circ}{W}{}^1_{p_2}(\Pi))},$$

$$\|u\|_{\overset{\circ}{W}(0,T)} = \|u\|_{\overset{\circ}{V}(0,T)} + \|u\|_{L_\infty(0,T;L_{\alpha_1}(\Omega))} + \|u\|_{L_\infty(0,T;L_{\alpha_2}(\Pi))},$$

and we denote the functional J such that its value on element $v \in \overset{\circ}{V}$ can be defined as

$$\langle J(z(t)), v\rangle_* = \frac{d}{dt}\Big(\int_\Omega \varphi_1(z(t))v(x)dx + \int_\Pi \varphi_2(z(t))v(s)ds\Big), \tag{13}$$

here $z \in \overset{\circ}{V}(0,T)$, $\langle F, v\rangle_*$ is a value of functional $F \in (\overset{\circ}{V})^*$ on the element.

In papers [2], [3] we proved that space $\overset{\circ}{V}$ coincides with the set of functions from $\overset{\circ}{W}{}^1_{p_1}(\Omega)$, and their increment on Π from $\overset{\circ}{W}{}^1_{p_2}(\Pi)$. Also in papers [2], [3] the similar result was obtained for $\overset{\circ}{V}(0,T)$.

Definition 1. Function $u \in \overset{\circ}{W}(0,T)$ such that

$$u(x,0) = u_0(x) \text{ a.e. in } \Omega \text{ and } \Pi,$$

$$\int_0^T \langle J(u), \cdot \rangle_* \, dt \in (\overset{\circ}{V}(0,T))^*, \tag{14}$$

is called generalized solution of the problem (1)–(3) if the integral identity

$$\begin{gathered}\int_0^T \langle J(u), v \rangle_* \, dt + \int_0^T \int_\Omega \sum_{i=1}^2 k_i(x,u,\nabla u) \frac{\partial v}{\partial x_i} dxdt + \\ + \int_0^T \int_\Pi k_\Pi\left(x,u,\frac{\partial u}{\partial s}\right) \frac{\partial v}{\partial s} dsdt = \int_0^T \langle f_1, v \rangle dt + \int_0^T \langle f_2, v \rangle_\Pi \, dt\end{gathered} \tag{15}$$

holds for arbitrary $v \in \overset{\circ}{W}(0,T)$; here $\langle f,v \rangle (\langle f,v \rangle_\Pi)$ is the value of functional f from the space $L_{p_1'}(0,T;W_{p_1'}^{-1}(\Omega))$ (and from $L_{p_2'}(0,T;W_{p_2'}^{-1}(\Pi))$ respectively) on element v from $\overset{\circ}{W}(0,T)$.

This paper is a continuation of the investigations which have been begun in papers [4], [5], where was given the generalized statement of the considering problem, the theorems of existence and uniqueness of generalized solution have been proved.

2. Description of the approximate method. The convergence investigation.

For problem (1)–(3) we construct the approximate solution method with respect to variable t and FEM with respect to space variables. To this end, on [0,T] we define a uniform grid $\bar{\omega}_\tau = \{0, \tau, \ldots, N\tau = T\}$, where τ is the grid increment. Under the assumption that Ω is a convex domain, we define the polygon Ω_h inscribed in Ω with boundary Γ_h. This polyhedron satisfies the following conditions:

1) for any point $x \in \Gamma$, there exists a point $\xi \in \Gamma_h$ lying from it at a distance not more than h (h is the grid increment with respect to space variables);
2) intersection points of Γ and Π are the vertexes of polygon Ω_h.

The triangulation of domain Ω_h is carried out the following way. For each of the subdomains $\Omega_h^1 \subset \Omega_h$ and $\Omega_h^2 \subset \Omega_h$ on which the section Π divides the domain Ω_h, the triangulation is performed autonomously such that the sets of mesh nodes constructed on Ω_h^1 and Ω_h^2 and lying on Π coincide. Such fragmentation in neighborhood of section Π provides us triangles with one curvilinear side. Let the $\overset{\circ}{V}_{h,l}$ be a set of the functions from $\overset{\circ}{V}$ such that their restriction on each finite element of the triangulation is an image of a polynomial of degree l on the basic element (see [6], p.113, [7] p. 47).

Definition 2. A function $y(t) \in \overset{\circ}{V}_{h,l}$ is called semidiscrete problem solution if for arbitrary $z \in \overset{\circ}{V}_{h,l}$ and $t \in \{0, \tau, \ldots, T-\tau\}$ the following conditions are satisfied

$$\begin{gathered}\int_{\Omega_h} \Big(\varphi_{1t}(y(t))\, z(x) + \sum_{i=1}^2 k_i(x,y,\nabla y) \frac{\partial z}{\partial x_i}(x)\Big) dx + \\ + \int_\Pi \Big(\varphi_{2t}(y(t))\, z(s) + k_\Pi\Big(x,y,\frac{\partial y}{\partial s}\Big) \frac{\partial z}{\partial s}(s)\Big) ds =\end{gathered} \tag{16}$$

$$= \int_{\Omega_h} f_{1\tau}(x,t)z(x)\,dx + \int_{\Pi} f_{2\tau}(s,t)z(s)\,ds,$$

$$y(x,0) = u_0(x) \ \text{ a.e. in } \ \Omega_h \ \text{ and } \ \Pi. \tag{17}$$

Here $f_{i\tau}(x,t) = \frac{1}{\tau}\int_t^{t+\tau} f_i(x,\xi)\,d\xi,\ i=1,2.$

The following statements were proved.

Lemma 1. Let the conditions (4)–(12) be satisfied. In addition,

$$f_1 \in L_{p_1'}(0,T;W_{p_1'}^{-1}(\Omega)),\ f_2 \in L_{p_2'}(0,T;W_{p_2'}^{-1}(\Pi)),\ u_0 \in \overset{\circ}{V} \cap L_{\alpha_1}(\Omega),$$

increments τ and h are picked the way that the following inequality holds

$$\tau \le c h^{\rho}, \tag{18}$$

where

$$\rho = \begin{cases} \max\{\alpha_1, \alpha_2\}, & 1 < p_i < \alpha_i, \\ \max\left\{ p_1 + \dfrac{2(p_1 - \alpha_1)}{\alpha_1}, p_2 + \dfrac{p_2 - \alpha_2}{\alpha_2} \right\}, & p_i \ge \alpha_i, \\ \max\left\{ p_1 + \dfrac{2(p_1 - \alpha_1)}{\alpha_1}, \alpha_2 \right\}, & p_1 \ge \alpha_1, 1 < p_2 < \alpha_2, \\ \max\left\{ \alpha_1, p_2 + \dfrac{p_2 - \alpha_2}{\alpha_2} \right\}, & 1 < p_1 < \alpha_1,\ p_2 \ge \alpha_2. \end{cases}$$

Then the following priori estimates hold for the solution of problem (16), (17)

$$\| y(t')\|_{L_{\alpha_1}(\Omega_h)} + \| y(t')\|_{L_{\alpha_2}(\Pi)} \le c, \tag{19}$$

$$\sum_{t=0}^{t'} \tau \left[\| y(t)\|^{p_1}_{\overset{\circ}{W}{}^1_{p_1}(\Omega_h)} + \| y(t)\|^{p_2}_{\overset{\circ}{W}{}^1_{p_2}(\Pi)} \right] \le c, \tag{20}$$

$$\sum_{t=0}^{t'} \tau^{\alpha_1} \| y_t(t)\|^{\alpha_1}_{L_{\alpha_1}(\Omega_h)} \le c, \tag{21}$$

$$\sum_{t=0}^{t'} \tau^{\alpha_2} \| y_t(t)\|^{\alpha_2}_{L_{\alpha_2}(\Pi)} \le c \quad \forall t' \in \omega_\tau, \tag{22}$$

$$\sum_{t=0}^{T-k\tau} \tau \int_{\Omega_{ih}} \big(\varphi_i(y(t+k\tau)) - \varphi_i(y(t))\big)\big(y(t+k\tau) - y(t)\big)\,dx \le k\tau c, \tag{23}$$

$i = 1,2,\ 1 \le k \le N,\ \Omega_{1h} = \Omega_h, \Omega_{2h} = \Pi.$

Lemma 2. Let $\{y\}$ be a sequence of functions for which a priori estimates (19)–(23) are valid. Then there exists the function $u \in \overset{\circ}{W}(0,T)$ and increment sequences $\{\tau\}$ and $\{h\}$ that satisfy the following limit relations as $\tau \to 0,\ h \to 0$:

$$\Pi^{\pm} \tilde{y}(t) \rightharpoonup u \ \text{ in } \ \overset{\circ}{V}(0,T), \tag{24}$$

$$\Pi^{\pm}\tilde{y}(t) \rightharpoonup u \text{ *-weakly in } L_{\infty}(0,T;L_{\alpha_1}(\Omega)), \tag{25}$$

$$\Pi^{\pm}\tilde{y}(t), \rightharpoonup u \text{ *-weakly in } L_{\infty}(0,T;L_{\alpha_2}(\Pi)), \tag{26}$$

$$\Pi^{+}\tilde{y}(t) \to u \text{ a.e. in } Q_T \text{ and } \Pi_T, \tag{27}$$

where

$$\tilde{y}(x,t) = \begin{cases} y(x,t), & (x,t) \in \Omega_h \times \bar{\omega}_\tau, \\ 0, & (x,t) \in (\Omega \setminus \Omega_h) \times \bar{\omega}_\tau, \end{cases}$$

$\Pi^{+}z, \Pi^{-}z$ are the piecewise constant extensions of the mesh function z. They are equal to value $z(t^*)$ on the sets $[t^*, t^*+\tau], [t^*-\tau, t^*]$, respectively, $t^* \in \omega_\tau$.

Let's consider some proof steps of Lemma 2. A priori estimates (19)–(20) provide us boundedness of sequences $\{\Pi^{\pm}\tilde{y}(t)\}$ in $\overset{\circ}{V}(0,T)$, $L_{\infty}(0,T;L_{\alpha_1}(\Omega))$ and in $L_{\infty}(0,T;L_{\alpha_2}(\Pi))$. That is why there exist function u from $\overset{\circ}{W}(0,T)$ and subsequence of sequences $\{\tau\},\{h\}$ such that as $\tau \to 0$ and $h \to 0$ relations (24)–(26) hold. Then we take subsequences $\{\tau\},\{h\}$ such that a priori estimates (19)–(23), the compactness theorem (see [8], lemmas 1.8, 1.9) and (24)–(26) provide us the limit relation (27). And relations (28), (29) follow from the estimate (20) and inequalities (7), (10)

$$\Pi^{\pm}\ k_i(x,\tilde{y},\nabla\tilde{y}) \rightharpoonup \bar{K}_i \text{ in } L_{p_1'}(Q_T), \tag{28}$$

$$\Pi^{\pm}\ k_{\Pi}(x,\tilde{y},\frac{\partial\tilde{y}}{\partial s}) \rightharpoonup \bar{K}_{\Pi} \text{ in } L_{p_2'}(\Pi_T). \tag{29}$$

The next we proved that the limit function u satisfies (1)–(3). For that we perform the passage of the limit in the following equation

$$-\int_0^T\int_{\Omega} \Pi^{+}\varphi_1(\tilde{y}(t))v_h \Pi^{+}\eta_{\tau\bar{t}}\,dxdt - \int_{\Omega}\varphi_1(u_0)v_h\eta(0)\,dx -$$

$$-\int_0^T\int_{\Pi} \Pi^{+}\varphi_2(\tilde{y}(t))v_h \Pi^{+}\eta_{\tau\bar{t}}\,dsdt - \int_{\Pi}\varphi_2(u_0)v_h\eta(0)\,ds +$$

$$+\int_0^T\int_{\Omega}\sum_{i=1}^{2} \Pi^{+}k_i(x,\tilde{y},\nabla\tilde{y})\frac{\partial v_h}{\partial x_i}\Pi^{+}\eta_\tau\,dxdt + \int_0^T\int_{\Pi}\Pi^{+}k_{\Pi}(x,\tilde{y},\frac{\partial\tilde{y}}{\partial s})\frac{\partial v_h}{\partial s}\Pi^{+}\eta_\tau\,dsdt =$$

$$= \int_0^T \Pi^{+}\langle f_{1\tau}, v_h\eta_\tau\rangle\,dt + \int_0^T \Pi^{+}\langle f_{2\tau}, v_h\eta_\tau\rangle_{\Pi}\,dt,$$

which is obtained from (16) as $z = v_h\eta_\tau$ (v_h is an interpolant from $\overset{\circ}{V}_{h,l}$ constructed for the function $v \in C_0^{\infty}(\Omega)$, $\eta \in C^{\infty}(0,T): \eta(T)=0$) after the transformation of the first and the third summands with t help of the summation by parts formula and the use of extension with respect to variable t and integration over $[0,T]$. As a result we obtain the following equation as $\tau \to 0$, $h \to 0$

$$-\int_0^T\int_{\Omega}\varphi_1(u)v\frac{d\eta}{dt}\,dxdt - \int_{\Omega}\varphi_1(u_0)v\eta(0)\,dx + \int_0^T\int_{\Omega}\sum_{i=1}^{2}\bar{K}_i\frac{\partial v}{\partial x_i}\eta\,dxdt -$$

$$-\int_0^T\int_\Pi \varphi_2(u)v\frac{d\eta}{dt}dsdt-\int_\Pi \varphi_2(u_0)v\eta(0)ds+\int_0^T\int_\Pi \bar{K}_\Pi\frac{\partial v}{\partial s}\eta dsdt= \tag{30}$$

$$=\int_0^T\langle f_1,v\eta\rangle dt+\int_0^T\langle f_2,v\eta\rangle_\Pi dt.$$

In consideration of $\eta\in C_0^\infty(0,T)$ we rewrite (30) in the form

$$-\int_0^T\{\int_\Omega \varphi_1(u)v\,dx+\int_\Pi \varphi_2(u)v\,ds\}\frac{\partial\eta}{\partial t}dt=\int_0^T\{\langle f_1,v\rangle-$$

$$-\int_\Omega\sum_{i=1}^2\bar{K}_i\frac{\partial v}{\partial x_i}dx+\langle f_2,v\rangle_\Pi-\int_\Pi\bar{K}_\Pi\frac{\partial v}{\partial s}ds\}\eta\,dt=\int_0^T G(t)\eta\,dt. \tag{31}$$

From the equation (31) follows that function $G(t)$ is a generalized derivative from $L_1(0,T)$ of the function

$$\int_\Omega\varphi_1(u)v\,dx+\int_\Pi\varphi_2(u)v\,ds,$$

from it follows that (see (13))

$$\langle J(u(t)),v\rangle_*=\langle f_1,v\rangle-\int_\Omega\sum_{i=1}^2\bar{K}_i\frac{\partial v}{\partial x_i}dx+\langle f_2,v\rangle_\Pi-\int_\Pi\bar{K}_\Pi\frac{\partial v}{\partial s}ds \tag{32}$$

almost everywhere in [0,T].

Next we use the monotone method to determine that

$$\int_0^T\{\int_\Omega\sum_{i=1}^2\overline{K_i}\frac{\partial v}{\partial x_i}dx+\int_\Pi\overline{K_\Pi}\frac{\partial v}{\partial s}ds\}dt=$$

$$=\int_0^T\{\int_\Omega\sum_{i=1}^2 k_i(x,u,\nabla u)\frac{\partial v}{\partial x_i}dx+\int_\Pi k_\Pi(x,u,\frac{\partial u}{\partial s})\frac{\partial v}{\partial s}ds\}dt.$$

By using it we prove that

$$\int_0^{t_1}\int_\Omega\sum_{i=1}^2\Pi^+k_i(x,\tilde{y},\nabla\tilde{y})\Pi^+\frac{\partial(\hat{\tilde{y}}(t)-\tilde{y}(t))}{\partial x_i}dxdt\to 0,$$

$$\int_0^{t_1}\int_\Pi\Pi^+k_\Pi(x,\tilde{y},\nabla\tilde{y})\Pi^+\frac{\partial(\hat{\tilde{y}}(t)-\tilde{y}(t))}{\partial s}dsdt\to 0$$

as $\tau\to 0,\ h\to 0$ such that

$$\tau/h^\rho\to 0, \tag{33}$$

where ρ is from lemma 2, $\hat{\tilde{y}}(t)=\tilde{y}(t+\tau)$. The main result of this paper is

Theorem 1. Let Ω be a convex domain in the space R^2, functions φ_i, $k_i(i=1,2),k_\Pi$ satisfy conditions (19)–(23), and let

$$f_1\in L_{p_1'}(0,T;W_{p_1'}^{-1}(\Omega)),\ f_2\in L_{p_2'}(0,T;W_{p_2'}^{-1}(\Pi)),\ u_0\in\overset{\circ}{V}\cap L_{\alpha_1}(\Omega)$$

and condition (29) is satisfied. Then the piecewise constant extensions subsequence of scheme solution (16)–(17), which is satisfying relations (24)–(29), converges to generalized solution of the problem (1)–(3). In conditions of solution uniqueness (see [5]) the whole sequence has this property.

Acknowledgments
The research was supported by the Russian Foundation for research (projects 15-01-05686, 15-41-02315).

References

[1] Antontsev S N and Meyrmanov A M 1979 *Matematicheskie modeli covmestnogo dvigeniya poverhnostnyh I podzemnyh vod* (Novosibirsk: Novosibirsk State University publishing)
[2] Timerbaev M R 2004 Spaces with a graph norm and strengthened Sobolev spaces, I *Russian Mathmatics* **5** 55 – 65
[3] Timerbaev M R 2004 Spaces with a graph norm and strengthened Sobolev spaces, II *Russian Mathmatics* **9** 46 – 53
[4] Glazyrina L L and Pavlova M F 1994 On solvability of certain problem on surface and ground water joint motion *Russian Mathmatics* **9** 15 – 26
[5] Glazyrina L L and Pavlova M F 2000 A uniqueness theorem for the solution of a problem in the theory of the join motion of channel and underground *Russian Mathmatics* **11** 10 – 22
[6] Dautov R Z and Karchevskii M M 2011 *Vvedenie d teoriyu metoda konechnyh elementov* (Kazan: Kazan State University publishing)
[7] Korneev V G 1977 *Shemy metoda konechnyh elementov vysokih poryadkov tochnosti* (Leningrad: Lenigrad State University publishing)
[8] Alt H.W. and Luckhaus S. 1983 Quasilinear elliptic-parabolic differential equation *Mat.Z.* **8** 311 – 41

On norms of operators generated by shift transformations arising in signal and image processing on meshes supplied with semigroup structures

R N Gumerov

Chair of Mathematical Analysis, N.I. Lobachevskii Institute of Mathematics and Mechanics, Kazan (Volga region) Federal University, Kremlevskaya 35, Kazan,420008, Tatarstan, Russian Federation

E-mail: rn.gumeroff@gmail.com

Abstract. Shift transformations and linear operators generated by shifts have a number of applications in signal and image processing. This note is concerned with a problem which has arisen in studying properties of real-world signals and images defined on meshes. For processing we suggest to introduce in domains of signals and images different semigroup structures. Semigroup operations give us opportunities to introduce shift transformations of signals and images. We study norms of polynomial filters generated by shift operators.

Introduction

We deal with a problem of operator theory which has arisen in studying properties of real-world signals and images. A part of motivation for our work has come from processing of experimental data in pressure pulse testing. Recall that a basic element in pulse testing is a well pair consisting of a pulsing well and an adjacent responding well [1]. There is a pair of signals, respectively. In practice, studying properties of these signals is often used for estimating a number of characteristics, for example, an average permeability. As is well known, different tools from functional analysis are involved in that investigation. For example, the Fourier transform is one such tool, and a wavelet expansion is another. The most basic linear operators in analysis of signals are shift transformations and their compositions. In particular, in time-frequency analysis these are the shifts in time and in frequency. Furthermore, there are a number of important operations in signal and image processing, such as the convolution and the correlation of signals, which use shift operators. It is also worth noting that shift transformations and linear operators generated by shifts play a significant role in different areas of functional analysis, in complex analysis and in dynamical systems. Another part of motivation for our work is a desire to give a self-contained elementary proof for a special case of the Coburn theorem from the theory of operator algebras (see, e.g., [2, Theorem 3.5.18]).

This note consists of Introduction and two Sections. In Section 1 we introduce the notations and recall some definitions and results that will be used in the sequel. In Section 2 we present results about norms of linear operators generated by shift transformations on the Hilbert space of square summable sequences.

1. Preliminaries

In practice signals and images are often defined on meshes of different types. We usually handle these meshes as sets without any additional structures.

Here, we suggest to consider a domain S of a signal $f : S \to \mathbb{F}$ as a mesh supplied with an additional structure, namely, a semigroup operation. In what follows, we shall make use of the additive notation. Here $\mathbb{F}$ denotes either the field of all real numbers $\mathbb{R}$ or the field of complex numbers $\mathbb{C}$.

Let $l^2(S)$ be a Hilbert space of all square summable, in general, complex-valued signals:

$$l^2(S) = \{f : S \to \mathbb{F} : \sum_{s \in S} |f(s)|^2 < +\infty\}.$$

Recall that $l^2(S)$ is a Hilbert space under the coordinatewise linear operations and with the inner product defined by

$$< f, g >= \sum_{s \in S} f(s)\overline{g(s)}.$$

Let $\{e_s : s \in S\}$ be the canonical orthonormal basis in $l^2(S)$ given by the formula

$$e_a(b) = \begin{cases} 1, & \text{if } a = b; \\ 0, & \text{if } a \neq b. \end{cases}$$

A semigroup structure on a mesh allows us to introduce in consideration for every element $a \in S$ the shift operator defined by

$$T_a : l^2(S) \longrightarrow l^2(S) : e_b \longmapsto e_{a+b},$$

where $b \in S$.

In what follows, for simplicity, we assume that S is the semigroup of all non-negative integers $\mathbb{Z}^+ := \{0, 1, 2, \ldots\}$ or any its subsemigroup. In practice, for signal and image processing the author has also made use of the Cartesian products $\mathbb{Z}^+ \times \mathbb{Z}^+$ and $\mathbb{Z}^+ \times \mathbb{Z}^+ \times \mathbb{Z}^+$ with the natural operations and with some others in accordance with problems. Of course, those semigroups are used for one-dimensional, two-dimensional and three-dimensional signals, respectively. Moreover, the semigroups $\mathbb{Z}^+ \times \mathbb{Z}^+$ and $\mathbb{Z}^+ \times \mathbb{Z}^+ \times \mathbb{Z}^+$ and their subsemigroups have been used for matricisations and tensorisations of signals and images (see [3, Part II]).

In the case of the semigroup $S = \mathbb{Z}^+$, we have the right shift operator

$$T := T_1 : l^2(\mathbb{Z}^+) \longrightarrow l^2(\mathbb{Z}^+) : e_n \longmapsto e_{n+1}.$$

It is clear that the following infinite Toeplitz matrix M corresponds to the operator T with respect to the canonical basis $\{e_n : n \in \mathbb{Z}^+\}$:

$$M = \begin{pmatrix} 0 & 0 & 0 & 0\ldots \\ 1 & 0 & 0 & 0\ldots \\ 0 & 1 & 0 & 0\ldots \\ 0 & 0 & 1 & 0\ldots \\ 0 & 0 & 0 & 1\ldots \\ \vdots & \vdots & \vdots & \vdots\ddots \end{pmatrix}.$$

Of course, the right shift operator T is a linear bounded operator. Thus it is an element of the C^*-algebra $\mathcal{B}(l^2(\mathbb{Z}^+))$ of all bounded linear operators on the Hilbert space $l^2(\mathbb{Z}^+)$. Moreover, the operator T is a non-unitary isometry, i.e.,

$$\|Tx\| = \|x\|$$

for each $x \in l^2(\mathbb{Z}^+)$. Note that T is injective but not surjective.

It is easy to see that the adjoint T^* of T is the left shift operator defined by

$$T^*(e_n) = \begin{cases} e_{n-1}, & \text{if } n \geqslant 1; \\ 0, & \text{if } n = 0. \end{cases}$$

Obviously, we have $(T^*)^m = (T^m)^*$, $m \in \mathbb{Z}^+$. We denote this operator by T^{*m}.

One can easily see that the matrix of the left shift operator T^* with respect to the canonical basis is an infinite Toeplitz matrix which has the following form:

$$M^* = \begin{pmatrix} 0 & 1 & 0 & 0\dots \\ 0 & 0 & 1 & 0\dots \\ 0 & 0 & 0 & 1\dots \\ 0 & 0 & 0 & 0\dots \\ \vdots & \vdots & \vdots & \vdots\ddots \end{pmatrix}.$$

Note that the left shift operator T^* is surjective but not injective.

Let $I : l^2(\mathbb{Z}^+) \longrightarrow l^2(\mathbb{Z}^+)$ be the identity operator. Denote by $< e_0 >$ the one-dimensional linear subspace $\{\lambda e_0 : \lambda \in \mathbb{C}\}$ in $l^2(\mathbb{Z}^+)$. Let $Q_0 : l^2(\mathbb{Z}^+) \longrightarrow < e_0 >$ be the orthogonal projector onto $< e_0 >$:

$$Q_0(\sum_{i=0}^{\infty} \lambda_i e_i) = \lambda_0, \quad \text{where} \quad \sum_{i=0}^{\infty} \lambda_i e_i \in l^2(\mathbb{Z}^+).$$

Then we have the following relations for the composition operators:

$$T^*T = I, \tag{1}$$

$$TT^* = I - Q_0, \tag{2}$$

In other words, the composition TT^* is the orthogonal projector onto the subspace of $l^2(\mathbb{Z}^+)$ spanned by the vectors e_n, where $n = 1, 2, \dots$.

We consider the shifts T and T^* as operator generators of filters for signal and image processing on meshes with semigroup structures. Notice that some standard operations in digital signal processing have suitable analogs among those filters. That is, we have representations of operations as operators written by means of the shifts.

In practice, we have used these filters for denoising signals and in combination with matricisation and tensorisation operations of the Fourier transformation and a wavelet expansion for further signal and image processing.

2. Main results

We consider an arbitrary linear operator $p(T, T^*) \in \mathcal{B}(l^2(\mathbb{Z}^+))$ which is a polynomial in the variables T and T^* with complex coefficients. Equivalently, one can say that we take an element of the complex $*$-algebra $\mathcal{P}$ generated by T and T^*. It is worth noting that the algebra $\mathcal{P}$ is not complete. We call elements of $\mathcal{P}$ *polynomial filters*. We have the following lemma about representations of polynomial filters.

Lemma. *Every polynomial filter $p(T, T^*) \in \mathcal{P}$ can be represented as follows:*

$$p(T, T^*) = \lambda_1 T^{n_1} T^{*m_1} + \lambda_2 T^{n_2} T^{*m_2} + \dots + \lambda_d T^{n_d} T^{*m_d}, \tag{3}$$

IOP Conf. Series: Materials Science and Engineering **158** (2016) 012042 doi:10.1088/1757-899X/158/1/012042

where $\lambda_i \in \mathbb{C}$, (n_i, m_i) are distinct pairs of non-negative integers, $i = 1, \dots, d$, and $d \in \mathbb{Z}^+ \setminus \{0\}$. Furthermore, for a non-zero polynomial filter representation (3) with non-trivial coefficients is unique up to an order of summands.

The sketch of the proof. The possibility to represent a polynomial filter as linear combination (3) follows from relations (1) and (2). To prove the uniqueness of representation (3) one can show that the family of operators $\{T^n T^{*m} : (n, m) \in \mathbb{Z}^+ \times \mathbb{Z}^+\}$ is linearly independent.□

Let us fix a number $k \in \mathbb{Z}^+ \setminus \{0\}$. For $j = 0, 1, \dots, k-1$ we define the Hilbert spaces

$$H_j = \{x \in l^2(\mathbb{Z}^+) : x = \sum_{l=0}^{+\infty} \lambda_{j+kl} e_{j+kl}\},$$

which are subspaces of the Hilbert space $l^2(\mathbb{Z}^+)$. We also define the shift operators as follows:

$$T_j : H_j \longrightarrow H_j : e_{j+kl} \mapsto e_{j+k(l+1)},$$

where $l = 0, 1, 2, \dots$

Below we present the results about polynomial filters and their norms.

Proposition 1. *Let $p(T, T^*)$ be a polynomial filter and $k \in \mathbb{Z}^+ \setminus \{0\}$. Then there exists a unitary operator $U : l^2(\mathbb{Z}^+) \longrightarrow \bigoplus_{j=0}^{k-1} H_j$ such that the following diagram*

$$\begin{array}{ccc} l^2(\mathbb{Z}_+) & \xrightarrow{p(T^k, T^{*k})} & l^2(\mathbb{Z}_+) \\ {\scriptstyle U}\big\downarrow & & \big\downarrow{\scriptstyle U} \\ \bigoplus\limits_{j=0}^{k-1} H_j & \xrightarrow{\bigoplus_{j=0}^{k-1} p(T_j, T_j^*)} & \bigoplus\limits_{j=0}^{k-1} H_j, \end{array}$$

is commutative,i.e., we have the operator equality

$$p(T^k, T^{*k}) \circ U = U \circ \bigoplus_{j=0}^{k-1} p(T_j, T_j^*). \tag{4}$$

Standard properties of operators and their norms together with equality (4) imply

Corollary 1. *Let $p(T, T^*)$ be a polynomial filter and $k \in \mathbb{Z}^+ \setminus \{0\}$. Then the following equality holds:*

$$\|p(T^k, T^{*k})\| = \|p(T_0, T_0^*)\|.$$

In the similar way we have the next two statements.

Proposition 2. *Let $p(T, T^*)$ be a polynomial filter and $k \in \mathbb{Z}^+ \setminus \{0\}$. Then there exists a unitary operator $W : H_j \longrightarrow l^2(\mathbb{Z}^+)$ such that the following diagram*

$$\begin{array}{ccc} H_0 & \xrightarrow{p(T_0, T_0^*)} & H_0 \\ {\scriptstyle W}\big\downarrow & & \big\downarrow{\scriptstyle W} \\ l^2(\mathbb{Z}_+) & \xrightarrow{p(T, T^*)} & l^2(\mathbb{Z}_+), \end{array}$$

is commutative,i.e., we have the operator equality

$$p(T,T^*)\circ W = W\circ p(T_0,T_0^*).$$

Corollary 2. *Let $p(T,T^*)$ be a polynomial filter and $k\in\mathbb{Z}^+\setminus\{0\}$. Then the following equality holds:*

$$\|p(T,T^*)\| = \|p(T_0,T_0^*)\|.$$

As an immediate consequence of Corollaries 1 and 2 we obtain

Corollary 3. *Let $p(T,T^*)$ be a polynomial filter and $k\in\mathbb{Z}^+\setminus\{0\}$. Then the following equality holds:*

$$\|p(T^k,T^{*k})\| = \|p(T,T^*)\|.$$

Let $\mathcal{T}$ be the C^*-subalgebra of $\mathcal{B}(l^2(\mathbb{Z}^+))$ generated by the operator T. The C^*-algebra $\mathcal{T}$ is called *the Toeplitz algebra generated by the semigroup* , or *the reduced semigroup C^*-algebra of* $\mathbb{Z}^+$. Note that the $*$-algeba $\mathcal{P}$ is dense in the C^*-algebra $\mathcal{T}$ with respect to the norm topology of $\mathcal{B}(l^2(\mathbb{Z}^+))$. From the above statements the following special case of the Coburn theorem follows:

Proposition 3. *Let $k\in\mathbb{Z}^+\setminus\{0\}$. Then there exists a unique $*$-endomorphism $\varphi:\mathcal{T}\longrightarrow\mathcal{T}$ of the Toeplitz algebra such that $\varphi(T)=T^k$.*

The sketch of the proof. The proof consists of two steps. First, we construct the $*$-homomorphism $\varphi_0:\mathcal{P}\longrightarrow\mathcal{T}$ which is well-defined in view of Lemma by the formula

$$\varphi_0(T^nT^{*m}) = T^{kn}T^{*km}.$$

Second, using Corollary 3, we extend the $*$-homomorphism φ_0 by continuity to the $*$-endomorphism on the whole space $\mathcal{T}$. □

References

[1] Johnson C, Greenkorn R and Woods E 1966 *J. Petroleum Technology*, **18 (12)**, 1599–1604.
[2] Murphy G J 1990 *C^*-algebras and operator theory* (New York:Academic Press)
[3] Hackbusch W 2012 *Tensor Spaces and Numerical Tensor Calculus* (Series Comput. Math. 42, Berlin: Springer)

Numerical simulaion of dynamics of a gas bubble in liquid near a rigid wall during its growth and collapse

M A Ilgamov, L A Kosolapova and V G Malakhov

Institute of Mechanics and Engineering, Kazan Science Center Russian Academy of Sciences (IME KazSC RAS) 2/31, Lobachevsky str., Kazan 420111 Russia

E-mail: kosolapova@kfti.knc.ru

Abstract. A numerical technique of calculating the dynamics of a cavitation bubble near a plane rigid wall is presented. The bubble at its collapse can become toroidal. The liquid is assumed inviscid and incompressible, its flow being potential. The bubble surface movement is determined by the Euler method, the normal component of the liquid velocity on the bubble surface is found by the boundary element method. The technique also includes an algorithm for calculating the velocity and pressure fields in the liquid. The convergence of the numerical solution with refining the temporal and spatial steps is demonstrated. The results of the present technique are compared with some known numerical and experimental data by other authors, their satisfactory agreement is found. To illustrate the capabilities of the present technique the process of growth and collapse of a bubble in water near a wall is considered. The liquid pressure contours in the stage of the bubble collapse are given and the radial liquid pressure profiles on the wall and at a small distance from the wall where the liquid pressure is maximum are shown.

1. Inroduction

The vapor bubbles near a wall are known to lead to its cavitation damage. In particular, this happens to the blades of marine propulsion propellers, the walls of pipe systems, the blades of hydraulic turbines, etc. It was first believed that the cavitation damage is caused by the pressure pulses resulting from the spherical collapse of bubbles. However, in 1944, Kornfeld and Suvorov [1] suggested that the destructive action of bubbles on a solid is realized by means of jets formed at collapse of cavitation bubbles near a solid wall. This hypothesis was later confirmed by Plesset and Chapman [2] using the finite difference method. After that, the dynamics of bubbles near a wall with the formation of the jets was widely studied, for example, in [3-7]. In most papers devoted to investigating the dynamics of a bubble in an inviscid incompressible fluid the boundary element method was used. In [8] a numerical technique based on the boundary element method was first proposed, which was later applied in [3] for studying the axisymmetric deformations of an empty spheroidal cavity near a rigid wall. Much later a similar approach was utilized in [4] for studying the growth and collapse of a bubble near a wall, in [5, 6] it was adapted to the spatial problems of bubble dynamics, and in [9-13] it was modified to simulate the toroidal phase of bubble dynamics. The toroidal phase is characteristic of the liquid flow circulation around the toroidal bubble.

In this paper, a numerical technique for studying the evolution of a bubble near a wall is presented. The liquid flow rotation in the toroidal phase of bubble dynamics is taken into account by introducing an artificial plane section perpendicular to the axis of symmetry [9]. Unlike the other authors [11, 14], the numerical instability is eliminated by a smoothing cubic spline. The capabilities of the technique are illustrated by computing a problem of growth and collapse of a bubble in water at a small distance from a wall.

2. Main relations and solution method

Axisymmetric growth and collapse of a cavitation bubble in liquid at a distance from a plane motionless rigid wall is considered. The axis of symmetry is orthogonal the wall. The pressure in the bubble varies as

$$p_b = p_{b0}(V_0/V)^{\kappa} \quad (1)$$

where p_{b0} is the initial pressure in the bubble, V_0, V are the initial and current volumes of the bubble, respectively, κ is the adiabatic index. The shape of the bubble at its maximum size due to the presence of the wall is non-spherical. Upon collapse a cumulative jet can arise on the bubble surface. This jet hits the opposite side of the bubble surface so that the bubble becomes toroidal.

It is assumed that the liquid is inviscid and incompressible, its flow is potential and is governed by the equations

$$\Delta\Phi = 0\,, \quad (2)$$

$$\Phi_t + \frac{1}{2}(\nabla\Phi)^2 + \frac{p - p_\infty}{\rho_f} = 0 \quad (3)$$

where p, p_∞ are the liquid pressures at any point and at a great distance from the bubble, respectively, ρ_f is the liquid density, Φ is the liquid velocity potential, $\mathbf{v} = \nabla\Phi$ is the liquid velocity, t is the time.

On the bubble surface, the following dynamic

$$p^+ = p^- - 2\sigma H \quad (4)$$

and kinematic

$$d\mathbf{x}_c/dt = \mathbf{v}_c \quad (5)$$

boundary conditions are posed. In (4), (5), p^+, p^- are the liquid and gas pressure on the bubble surface, respectively, σ is the surface tension, H is the mean curvature at a point of the bubble surface; $\mathbf{x}_c$, $\mathbf{v}_c = (\nabla\Phi)|_c$ are respectively the position-vector and the velocity of a surface point.

On the rigid wall the condition of impermeability

$$v_z = \partial\Phi/\partial z = 0 \quad (6)$$

is posed where z is the axial coordinate, v_z is the axial component of the liquid velocity.

In the process of evolution of a bubble with the formation of the cumulative jet on its surface one can separate out two phases. The first phase corresponds to the formation of the jet. It ends at the moment of impact of the jet t_c on the opposite side of the bubble surface. The second phase corresponds to the motion of the toroidal bubble resulting from that impact. Accordingly, the numerical technique of the present paper is also divided into two stages, with the first (second) stage corresponding to the first (second) phase. At that, the bubble surface movement and deformation are determined by the Euler method from kinematic (4) and dynamic (5) boundary conditions, the liquid velocity and its potential on the bubble surface are evaluated by the boundary element method [15].

The first stage of the numerical technique used for computing the first phase of the bubble dynamics is described in detail in [7]. Its utilizing without any smoothing procedure can lead to numerical instability that manifests itself in high-frequency distortions of the bubble shape. To suppress such distortions, a smoothing procedure proposed in [7] is adopted in the present technique. That procedure is based on a cubic spline. It is used for smoothing the functions $\Phi(s)$, $r(s)$, and $z(s)$, which respectively define the liquid velocity potential, the radial and axial coordinates of the bubble surface along its contour in its axial section (s is the contour arc length).

The second phase of the bubble dynamics begins with formation of a toroidal bubble resulting from the jet impact on the opposite side of the bubble surface. In the present numerical technique the transition from singly connected to doubly connected bubble surface in its axial section is performed by the algorithm proposed in [9]. In the doubly connected flow region the velocity potential is not uniquely defined. To resolve this ambiguity, a straight infinitely-thin cut perpendicular to the symmetry axis and connecting the symmetry axis with a point of the torus section is introduced in the area of the jet. Across this cut the velocity potential is assumed discontinuous. The value of this discontinuity is constant along the cut. It is equal to the difference $\Delta\Phi$ of the axial potential values at the points of the jet end and the opposite side of the bubble surface attained at the end of the first phase of the bubble dynamics. The only difference of the second stage of the present numerical technique from the first one is that an additional term containing an integral along the cut appears in the boundary integral equation [9].

In addition to the first and second stages, the present numerical technique includes the construction of the liquid velocity and pressure fields. To this end, the potential Φ is determined by the boundary integral equation whereas the velocity components are calculated by the finite-differences of the second order of accuracy. The pressure in the liquid is found by equation (3) where $\partial\Phi/\partial t$ is also calculated by the finite-differences.

3. Numerical convergence. Comparison with known numerical and experimental results

The following dimensionless parameters are introduced: $R_0^* = R_0/R_{max}$, $\gamma = d_0/R_{max}$, $r^* = r/R_{max}$, $z^* = z/R_{max}$, $v^* = v/(p_\infty/\rho_f)^{1/2}$, $p^* = p/p_\infty$, $t^* = t/[R_{max}(\rho_f/p_\infty)^{1/2}]$ where R_0 is the initial bubble radius, R_{max} is the mean radius of the bubble at its maximum size, d_0 is the initial distance from the center of the bubble to the wall, r is the radial coordinate, $v = |\mathbf{v}|$ is the liquid velocity value.

To estimate the numerical convergence of the present technique in computing the first phase of the bubble dynamics (until the moment t_c^* at which the jet touches the opposite side of the bubble surface), a problem of initially spherical bubble growth and collapse in water is considered for $R_0^* = 0.159$, $\gamma = 0.7$, $\kappa = 1.33$, $p_{b0}^* = 37.76$, $v_0^* = 10.37$, $R_{max} = 6.3 \cdot 10^{-4}$ m, $p_\infty = 10^5$ Pa.

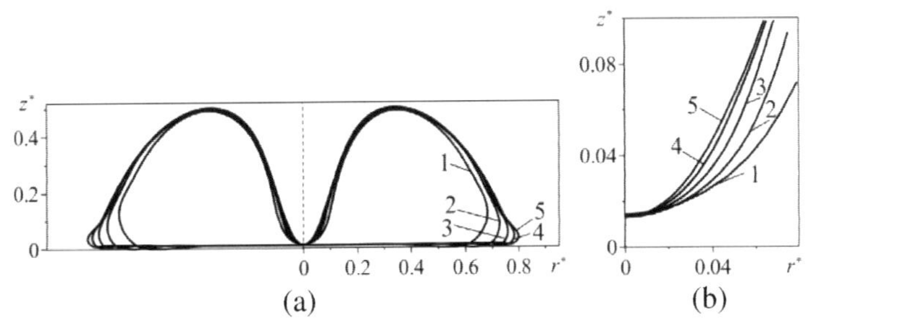

Figure 1. The contours of the bubble surface at the moment t_c^* for n = 50 (curve 1), 100 (curve 2), 200 (curve 3), 400 (curve 4), 800 (curve 5) (a) and their enlargements near the jet end for the same values of n (b).

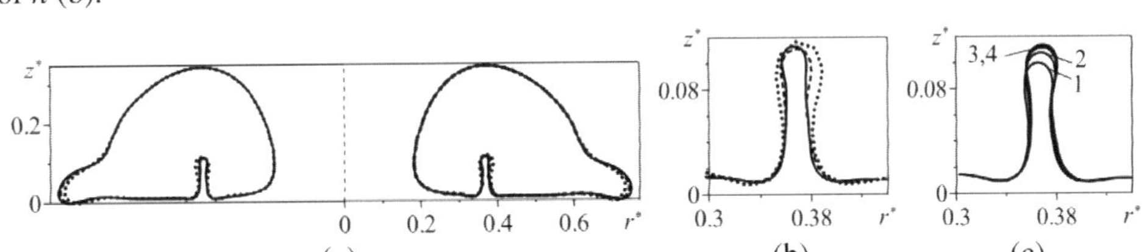

Figure 2. The contours of the bubble surface in the toroidal phase of the bubble dynamics at the moment $t^* = 0.0635$ for $\tau_0^* = 0.4 \cdot 10^{-4}$ and n = 100 (dotted curve), 200 (dashed curve) and 400 (solid curve) (a); their enlargements in the vicinity of the splash for the same values of n (b); the similar

fragments of the bubble surface contours for n = 200 and the initial time steps $\tau_0^* = 2.1 \cdot 10^{-4}$ (curve 1), $0.9 \cdot 10^{-4}$ (curve 2), $0.4 \cdot 10^{-4}$ (curve 3), $0.2 \cdot 10^{-4}$ (curve 4) (c).

Figure 1 shows the contours of the bubble surface in its axial section at the moment t_c^*, computed for consecutively increasing contour discretization defined by the contour segment number n. One can see numerical converge with increasing n. At that, the contours for $n > 100$ are only different in the areas with large curvature (at the end of the jet and in the region most distant from the symmetry axis). Moreover, the difference of the solutions for n = 400 and 800 is small.

In the toroidal phase of the bubble dynamics following the jet impact, liquid circulation through the hole in the toroidal bubble occurs. This liquid flows onto the wall and radially spreads on it. At the same time, there is an oppositely directed liquid flow from the far field. The collision of these flows can result in an upward splash on the bubble surface (Figure 2) [12, 17, 18]. The curvature of the contour within the splash is greater than outside it, which impairs the numerical convergence. Figure 2 illustrates this feature in the case of increasing n (at $t^* = 0.0635$). It should be noted that for n = 200 and 400 the shape of the bubble is almost the same. In Figure 2c, the numerical convergence of the splash is demonstrated by refining the initial time step (in the case of n=200). One can see that the solutions corresponding to $\tau_0^* = 0.4 \cdot 10^{-4}$ and $\tau_0^* = 0.2 \cdot 10^{-4}$ are graphically coincident. Similar convergence in the other significantly-smoother part of the bubble contour is derived for all the considered values of τ_0^*.

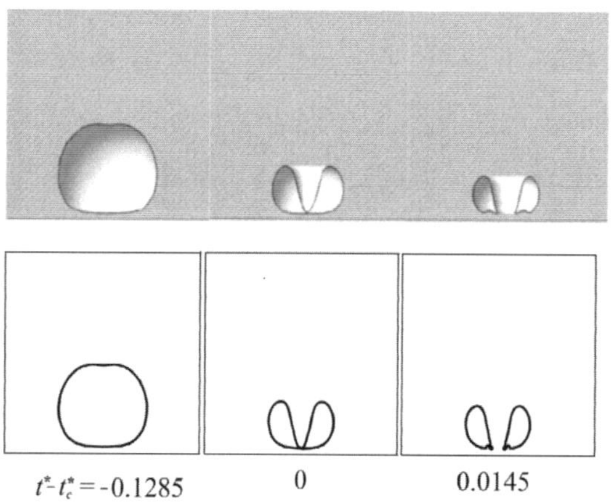

Figure 3. Comparison of the numerical results of the present work (bottom row) with those of [12] (top row).

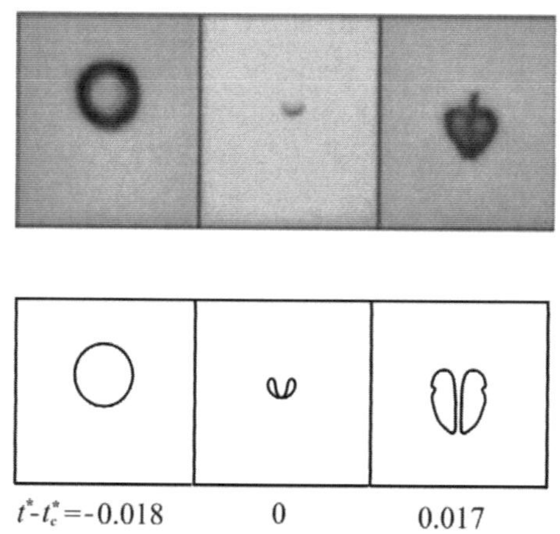

Figure 4. Comparison of the numerical results of the present work (bottom row) with the experimental ones of [18] (top row).

To assess the reliability of the technique of the present work its results are compared with the numerical results of [12] (Figure 3) and with the experimental data of [18] (Figure 4). In the both cases a problem of growth and collapse of an initially spherical bubble in water near a solid wall is considered. In the first case the simulation is carried out for $R_0^* = 0.159$, $\gamma = 1.0$, $\kappa = 1.4$, $p_{b0}^* = 100$, $p_\infty = 10^5\,\mathrm{Pa}$. In the second case it is performed for $R_0^* = 2.5 \cdot 10^{-4}$, $\gamma = 2.0$, $R_{\max} = 1.45 \cdot 10^{-3}\,\mathrm{m}$, $\rho_f = 1000\,\mathrm{kg/m^3}$, $p_\infty = 10^5\,\mathrm{Pa}$ where R_0, $R_{\max}$, and γ are taken from [18]. It follows from Figures 3 and 4 that in both cases the agreement is quite satisfactory.

4. Growth and collapse of a cavitation bubble near a rigid wall

To illustrate the capabilities of the present technique a problem of growth and collapse of a cavitation bubble in water near a rigid wall is considered with

$R_0^*=0.159$, $\gamma=0.7$, $p_{b0}^*=37.76$, $v_0^*=10.37$, $\kappa=1.33$, $p_\infty=10^5$ Pa, $\rho_f=1000$ kg/m^3, $n=400$.

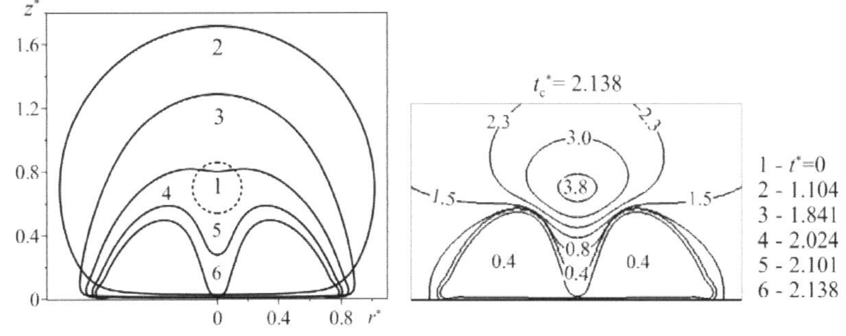

(a) (b)

Figure 5. The shape of the bubble at six time moments (a) and the pressure field in the liquid surrounding the bubble at the moment t_c^* (b).

Figure 5 (a) shows the change in the shape of the bubble from its inception (moment 1) until the transformation into a torus (moment 6). The maximum growth is reached at moment 2 ($t^*=1.104$). At moment 6 ($t_c^*=2.138$) the cumulative jet impacts on the bubble surface side nearest to the wall. One can see that at the moment of maximum bubble size the lower part of the bubble surface is flattened, which predetermines the bubble shape at the moment of the jet impact. The jet velocity at the moment t_c^* is equal to 7.22. Figure 5 (b) shows the liquid pressure field near the jet end at the moment $t_c^*=2.138$. Maximum pressure $p^*=3.8$ is observed at the bottom of the jet.

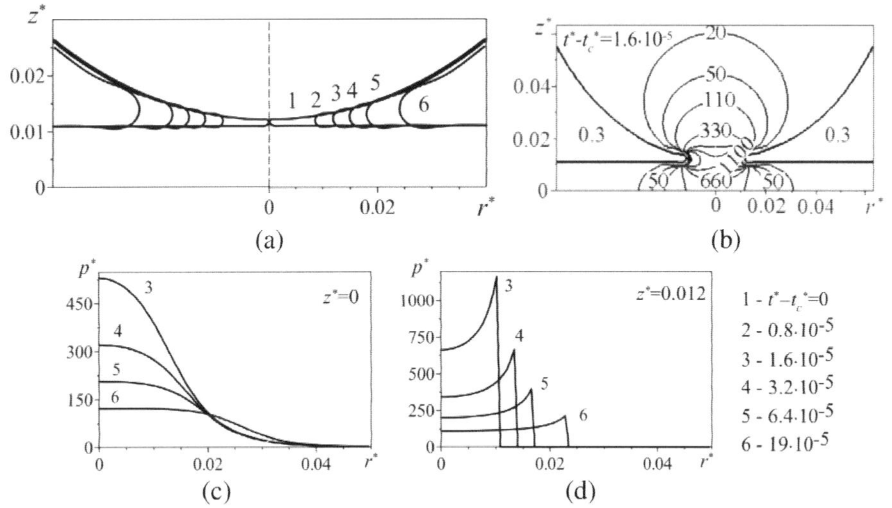

(c) (d)

Figure 6. The bubble contours at six time moments at the beginning of the toroidal phase of the bubble dynamics (a), the liquid pressure field near the jet end at $t^*-t_c^*=1.6\cdot10^{-5}$ (b), the radial pressure profiles on the wall $z^*=0$ (c) and at the level $z^*=0.012$ (d) at four time moments.

Figure 6 (a) shows the change of the bubble contour in the vicinity of the jet end at the beginning of the toroidal phase of the bubble dynamics. Figure 6 (b) presents the liquid pressure field near the jet end at a moment at the beginning of the toroidal phase. The maximum pressure in the liquid $p^*=1100$ is observed on the periphery of the circular area of contact of the jet end with the thin liquid layer between the bubble and the wall. The maximum pressure on the wall equal to 660 is located at a point of the axis of symmetry. The radial pressure profiles on the wall $z^* = 0$ and at the level $z^* = 0.012$ are shown in figures 6 (c), (d) for a number of time moments. One can see that with time the pressure on the wall as well as at the level $z^* = 0.012$ significantly decreases. The liquid layer between the bubble and the wall weakens the jet action on the wall.

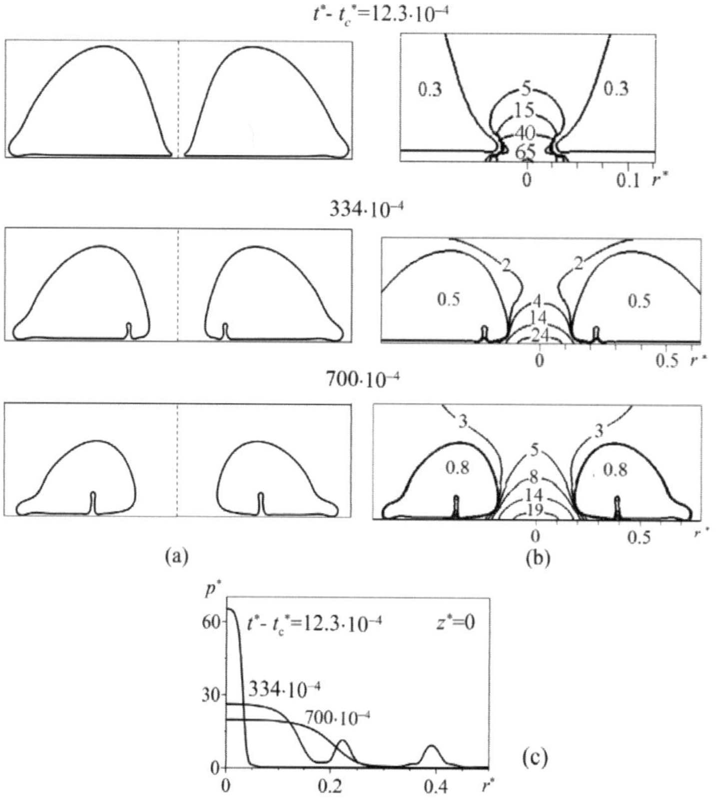

Figure 7. The shapes of the bubble (a), the pressure fields in the liquid surrounding the bubble (b) and the radial pressure profiles on the wall at three time moments in the toroidal phase of the bubble dynamics (c).

Further dynamics of the toroidal bubble is demonstrated in figure 7 where the shapes of the bubble (a), the pressure fields in the liquid surrounding the bubble (b) and the pressure profiles on the wall (c) are shown for three time moments. One can see that as the process of the bubble collapse progresses a splash arises on the bubble surface near the wall. At that, the pressure on the wall decreases everywhere except the neighborhood of the splash where it slightly grows.

5. Conclusion

A numerical technique for investigating the dynamics of a cavitation bubble near a rigid wall in an inviscid and incompressible liquid is presented. The technique is based on the boundary element method and consists of two stages. The first stage is devoted to calculating the bubble dynamics until the impact of the cumulative liquid jet arising on the surface of the bubble at its collapse on the bubble surface side closest to the wall. In the second stage the dynamics of the toroidal bubble resulting from that impact is evaluated. The numerical instability is eliminated by smoothing the functions defining the bubble surface and its liquid velocity potential by a smoothing procedure using a cubic spline. The numerical convergence of the technique with increasing the spatial and temporal discretization is shown. A comparison with numerical and experimental results of other authors shows satisfactory agreement. The problem of growth and collapse of an initially spherical bubble near a wall is computed. The corresponding pressure fields in the liquid surrounding the bubble and the radial pressure profiles on the wall are presented.

References

[1] Kornfeld M and Suvorov L 1944 *J. Appl. Phys.* **15** p 495-506
[2] Plesset M S and Chapman R B 1971 *J. Fluid Mech.* **47** No 2 p 283-290
[3] Voinov O V and Voinov V V 1976 *Soviet Physics. Doklady* **21** p 133-136
[4] Blake J R, Taib B B and Doherty G 1986 *J. Fluid Mech.* **170** p 479-497
[5] Wang Q X 1998 *Theoret. Comput. Fluid Dynamics* **12** p 29-51
[6] Afanas-ev K E and Grigor-eva IV 2006 *Vychislitel-nye tekhnologii* (instityt vychislitel-nykh tekhnologij SO RAN Novosibirsk Rossia) **11** special-nyj vypusk s 4-25
[7] Aganin A A, Ilgamov M A, Kosolapova L A and Malakhov V G 2013 *Vestnik Bashkirskogo universiteta* **18** No 1 s 15-21
[8] Voinov O V and Voinov V V 1975 *Soviet Physics. Doklady* **20** p 179-182
[9] Best J P 1994 Bubble Dynamics and Interface Phenomena (Dordrecht: Kluwer Academic Publishers) p 405-412
[10] Zhang S, Duncan J H and Chahine G L 1993 *J. Fluid Mech.* **257** p 147-181
[11] Wang Q X, Yeo K S, Khoo B C and Lam K Y 1996 *Computers and Fluids* **25** No 7 p 607-628
[12] Pearson A, Blake J R and Otto S R 2004 *J. Engineering Mathematics* **48** p 391-412.
[13] Aganin A A, Ilgamov M A, Kosolapova L A and Malakhov V G 2015 *Izvestiya ufimskogo nauchnogo tsentra RAN* **3** s 5-11
[14] Zhang Z Y and Zhang H S 2004 *Physical Review* E **70** 056 310
[15] Brebbia C A, Telles J C F and Wrobel L 1984 *Boundary element techniques: Theory and Applications in Engineering* (Berlin-Heidelberg - NewYork - Tokyo - Springer-Verlag) p 464
[16] Tong R P, Schiffers W P, Shaw S J, Blake J R and Emmony D C 1999 *J. Fluid Mech.* **380** p 339-361.
[17] Brujan E A, Keen G S, Vogel A and Blake J R 2002 *Physics of fluids* **14** No 1 p 85-92
[18] Philipp A and Lauterborn W 1998 *J. Fluid Mech.* **361** p 75-116

Correctness investigation of a boundary value problem for a parabolic equation with nonlocal space operator with respect to the solution gradient

A S Ivanova and M F Pavlova

Department of Computational Mathematics, Institute of Computational Mathematics and Information Technologies, Kazan Federal University, 18 Kremlyovskaya Street, Kazan 420008, Russia

E-mail: m.f.pavlova@mail.ru

Abstract. We consider the parabolic equation whose space operator is product of nonlinear bounded function which depends on nonlocal characteristic with respect to a solution gradient and strongly monotone, potential operator. The existence and uniqueness of the solution in the class of the vector-valued functions with values in Sobolev space are proved.

1. Introduction

The paper is devoted to investigation of questions of existence and uniqueness of a solution to initial boundary value problem for parabolic equation with a space operator

$$Bu = -a\left(\|\nabla u\|^2_{L_2(\Omega)}\right) Au,$$

where a is nonlinear, continuous function bounded in $[0,\infty)$, A is nonlinear second order differential operator in divergence form. The results of this paper summarize results of [1], [2], where the same problem was solved for a parabolic equation with a spatial operator $Bu = -a\left(\|\nabla u\|^2_{L_2(\Omega)}\right)\Delta u$, where Δu is Laplace operator (see [3], [4]). It should be noted that the parabolic equation with nonlocal space operators occur in the theory of plasticity, elasticity and in the description of bacteria diffusion processes in biology (see [3]–[8]). The presence of non-local dependence on the gradient of the solution greatly complicates the study on the existence and uniqueness of solution.

2. Formulation of the problem

The in $\Omega \subset R^n$ with boundary Γ we consider the following initial boundary value problem:

$$\frac{\partial u}{\partial t}(x,t) - a\left(\| \nabla u(t)\|^2\right)\operatorname{div}\left(k(x,\nabla u(x,t))\right) = f(x,t), \quad (x,t)\in Q_T = \Omega\times(0,T), \quad (1)$$

$$u(x,t) = 0, \quad (x,t)\in \Gamma\times(0,T], \quad u(x,0) = u_0(x), \quad x\in\Omega, \quad (2)$$

where $a(\xi)$, $k=(k_1,...,k_n)$, $k_i(x,\xi)$, $i=1,...,n$ are given functions. Here and below, the following notations will be denoted shortly by $\|\cdot\|=\|\cdot\|_{L_2(\Omega)}$, $\|\cdot\|=\|\cdot\|_{W_2^1(\Omega)}$.

It is supposed that function $a(\xi)$ is continuous on $[0,\infty)$ and

$$0<\gamma_1\le a(\xi)\le\gamma_2 \quad \forall\xi\in[0,\infty), \tag{3}$$

functions $k_i(x,\xi)$, $i=1,...,n$, are measurable in x for any $\xi\in R^n$, continuous and differentiable with respect to ξ for almost all $x\in\Omega$ and satisfy the following conditions

$$\frac{\partial k_i}{\partial\xi_j}(x,\xi)=\frac{\partial k_j}{\partial\xi_i}(x,\xi) \quad i,j=1,...,n, \tag{4}$$

$$\sum_{i=1}^{n}\left(k_i(x,\bar\xi-k_i(x,\xi)\right)(\bar\xi_i-\xi_i)\ge c_0\sum_{i=1}^{n}|\bar\xi_i-\xi_i|^2 \quad \forall\xi,\bar\xi\in R^n, x\in\Omega, \tag{5}$$

$$|k_i(x,\xi)|\le c_1\sum_{j=1}^{n}|\xi_j|+c_2 \quad \forall\xi\in R^n,\ x\in\Omega, \tag{6}$$

where $c_0>0$, $c_1>0$, $c_2\ge0$ are positive constants. Note that operator limitation follows from (6)

$$Au=-\sum_{i=1}^{n}\frac{\partial}{\partial x_i}k_i(x,\nabla u)$$

valid from $W_2^1(\Omega)$ into $W_2^{-1}(\Omega)$. Conditions (4), (5) provide potentiality and strong monotony for the operator.

Definition 1. Function $u\in L_2\left(0,T;\ H_0^1(\Omega)\cap H^2(\Omega)\right)$ such that

$$\frac{\partial u}{\partial t}\in L_2(Q_T),\quad u(x,0)=u_0(x) \text{ almost everywhere in the } \Omega, \tag{7}$$

we call generalized solution of (1)–(2) problem, if

$$\int_{Q_T}\frac{\partial u}{\partial t}v\,dxdt+\int_0^T a\left(\|\nabla u\|^2_{L_2(\Omega)}\right)\int_\Omega\sum_{i=1}^{n}k_i(x,\nabla u)\frac{\partial v}{\partial x_i}\,dxdt=\int_{Q_T}fv\,dx\,dt. \tag{8}$$

3. Existence theorem

Theorem 1. Let conditions (4)–(5) be satisfied and for the solution of homogeneous Dirichlet problem

$$\int_\Omega\sum_{i=1}^{n}k_i(x,\nabla\tilde u)\frac{\partial v}{\partial x_i}\,dxdt=\int_{Q_T}\tilde f v\,dx\,dt \qquad \forall v\in H_0^1(\Omega)$$

for any function $\tilde f\in L_2(\Omega)$ estimate

$$\|\tilde u\|_{H^2(\Omega)}\le C\|\tilde f\|, \tag{9}$$

holds true, then for any $u_0\in L_2(\Omega)$, $f\in L_2\left(0,T;\ L_2(\Omega)\right)$ the problem (1)–(2) has generalized solution from $L_2\left(0,T;\ H_0^1(\Omega)\cap H^2(\Omega)\right)$.

We describe main steps of the proof of this assertion. Function y, defined on the set $\bar{\Omega}\times\bar{\omega}_\tau, (\bar{\omega}_\tau=\{0,\tau,2\tau,...,M\tau=T\})$ is called semidiscrete approximate solution of the problem (1)–(2), if for any $t\in\omega_\tau=\bar{\omega}_\tau\setminus\{0\}$ function $y(t)$ belongs to the space $H_0^1(\Omega)$, $y(x,0)=u_0(x)$ almost everywhere in Ω, for any function $\eta\in H_0^1(\Omega)$ for all $t\in\bar{\omega}_\tau\setminus\{T\}$ equality

$$\int_\Omega \frac{\hat{y}(t)-y(t)}{\tau}\eta\,dx+a\left(\|\nabla y(t)\|^2\right)\int_\Omega\sum_{i=1}^n k_i\left(x,\nabla\hat{y}(t)\right)\frac{\partial\eta}{\partial x_i}\,dx=\int_\Omega f_\tau\eta\,dx\,, \tag{10}$$

holds true, where $\hat{y}(t)=y(t+\tau)$, $f_\tau(x,t)=1/\tau\int_t^{t+\tau}\left(f(x,\xi)\right)d\xi$.

Due to the fact that in (10) nonlocal characteristic $a\left(\|\nabla y(t)\|^2\right)$ is calculated from the lower layer, the function $\hat{y}(t)$ with known $y(t)$ is a generalized solution of elliptic equations with strongly monotone operator. Following, for example, [9], c. 237, it is easy to see the unique solvability of the problem (10).

Next we prove that for the semidiscrete problem solution following a priori estimates hold:

$$\|y(t)\|^2=\tilde{c}\left(\|u_0\|^2+1+\|f\|^2\right)\qquad\forall t\in\bar{\omega}_\tau, \tag{11}$$

$$\sum_{t\in\omega_\tau}\tau\|y(t)\|_1^2\le\tilde{c}\left(\|u_0\|^2+1+\|f\|^2\right), \tag{12}$$

$$\sum_{t\in\omega_\tau}\tau\|k_j\left(\cdot,\nabla y(t)\right)\|^2\le\tilde{c}\left(\|u_0\|^2+1+\|f\|^2\right),\qquad j=1,\dots,n, \tag{13}$$

$$\sum_{t=0}^{t'-\tau}\|y_t(t)\|_1^2\le\tilde{c}\left(\|u_0\|_1^2+1+\sum_{t=0}^{t'-\tau}\tau\|f_\tau\|^2\right), \tag{14}$$

$$\max_{t\in\omega_\tau}\|y_t(t')\|_1^2\le\tilde{c}\left(\|u_0\|_1^2+1+\sum_{t=0}^{t'-\tau}\tau\|f_\tau\|^2\right), \tag{15}$$

$$\sum_{t\in\omega_\tau}\tau\|y(t)\|_{H^2(\Omega)}^2\le\tilde{c}\left(\|u_0\|^2+1+\|f\|^2\right), \tag{16}$$

here $\tilde{c}$ is a constant independent of τ. Justification (11)–(13) are conventional: in (10) we assume that $\eta=\hat{y}$, result is summed (with respect to) t from 0 to $T-t$, and converted using (5) and (6). To obtain a priori estimates (14)–(15) we use potentiality of the operator A, wherein we select $\eta=\dfrac{\tau y_t(t)}{a\left(\|\nabla y\|^2\right)}$ in (10). Conditions (3) provide that this function belongs to the space $H_0^1(\Omega)$. Next we prove that $\hat{y}\in H^2(\Omega)$. For that in the equality (10) we select $\eta=\dfrac{\omega}{a\left(\|\nabla y\|^2\right)}$, where ω — arbitrary function from the space $H_0^1(\Omega)$. The result is written in the form

$$\int_\Omega\sum_{i=1}^n k_i\left(x,\nabla\hat{y}(t)\right)\frac{\partial\omega}{\partial x_i}\,dx=\frac{1}{a\left(\|\nabla y\|^2\right)}\int_\Omega\left(f_\tau-y_t\right)(x,t)\omega\,dx\,. \tag{17}$$

From the equation (17), the estimate (14) and condition (9) we obtain that (16) follows.

These a priori estimates provide existence of sequence $\{y_\tau\}$ and element $u \in L_2\left(0,T;\, H_0^1(\Omega) \cap H^2(\Omega)\right)$ such that following limit relations hold

$$\prod_\tau^{\pm} y_\tau \to u \quad \text{in } L_2\left(0,T;\, H_0^1(\Omega)\right), \qquad \prod_\tau^{+}(y_\tau)_t \to \frac{\partial u}{\partial t} \quad \text{in } L_2\left(Q_T\right), \tag{18}$$

$$\prod_\tau^{\pm} a\left(\|\nabla y_\tau\|^2\right) \to a\left(\|\nabla u\|^2\right) \quad \text{in } L_2(0,T), \qquad \prod_\tau^{\pm} k_i\left(x, \nabla y_\tau\right) \to k_i\left(x, \nabla u\right) \quad \text{in } L_2(0,T). \tag{19}$$

Here $\prod_\tau^{\pm} z(t')$ are operators of function z continuation, defined on $\bar{\omega}_\tau$, on interval $[0,T]$ defined by the equalities

$$\prod_\tau^{-} z(t') = z(t) \text{ for } (t-\tau) < t' \le t, \qquad \prod_\tau^{+} z(t') = z(t) \text{ for } (t) < t' \le (t+\tau).$$

Note, that in this case strong convergence $\prod_\tau^{\pm} y_\tau$ to u in space $L_2\left(0,T;\, H_0^1(\Omega)\right)$ holds, that is what allows us to obtain limit relations (19).

Passage to the limit in equality ends the proof

$$\int_{Q_T} \prod_\tau^{-} y_t(t)\eta(x)z(t)\,dx\,dt + \tag{20}$$

$$+ \int_0^T a\left(\prod_\tau^{-} \|\nabla y(t)\|^2\right) \int_\Omega \sum_{i=1}^n \prod_\tau^{-} k_i\left(x, \nabla \hat{y}(t)\right) \frac{\partial \eta}{\partial x_i} z(t)\,dx\,dt = \int_{Q_T} \prod_\tau^{-} f_\tau\, \eta(x) z(t)\,dx\,dt$$

obtained from (10) after multiplying this equation by an arbitrary function z from $L_2(0,T)$ and integration with respect to variable t. Limit relations (18), (19) are sufficient to justify the passage to the limit in (20).

4. Uniqueness theorem

Theorem 2. Let functions $k_i(x,\xi)$, $i = 1, 2, \ldots, n$, satisfy conditions (5)–(7). In addition, following condition holds

$$\left(\gamma_1 c_0 - \tilde{C} \max_{0 \le \xi \le M_\nabla} |a'(\xi)|\right) \ge 0, \tag{21}$$

where $\tilde{C} = 2M_\nabla n^{\frac{1}{2}} (\tilde{c})^2 \left(n M_\nabla^2 + 2n^{\frac{1}{2}} (\text{mes}\,\Omega)^{\frac{1}{2}} M_\nabla + \text{mes}\,\Omega\right)^{\frac{1}{2}}$, $M_\nabla = const > 0$, satisfying inequality: $\|u\|_{L_r(0,T;H_0^1(\Omega))} \le M_\nabla$. Then solution of problem (7)–(8) is uniquely determined.

(Note that estimate (15) obtained in the proof of the previous theorem can effectively determine the value of the constant M_∇.)

Indeed, let u, $\tilde{u}$ be two solutions of problem (7)–(8). Then

$$\int_\Omega \frac{\partial (u - \tilde{u})}{\partial t} v\,dx\,dt + a\left(\|\nabla u\|^2\right) \int_\Omega \sum_{i=1}^n k_i(x, \nabla u) \frac{\partial v}{\partial x_i}\,dx - \tag{22}$$

$$- a\left(\|\nabla \tilde{u}\|^2\right) \int_\Omega \sum_{i=1}^n k_i(x, \nabla \tilde{u}) \frac{\partial v}{\partial x_i}\,dx = 0, \qquad \forall\, v \in L_2\left(0,T;\, H_0^1(\Omega)\right).$$

Assuming $v = u - \tilde{u}$, write (22) in form

$$\int_\Omega \frac{\partial (u-\tilde{u})}{\partial t}(u-\tilde{u})\,dx + a(\|\nabla u\|^2)\int_\Omega \sum_{i=1}^n \left(k_i(x,\nabla u) - k_i(x,\nabla \tilde{u})\right)\frac{\partial (u-\tilde{u})}{\partial x_i}\,dx =$$

$$= -\left(a(\|\nabla u\|^2) - a(\|\nabla \tilde{u}\|^2)\right)\int_\Omega \sum_{i=1}^n k_i(x,\nabla\tilde{u})\frac{\partial(u-\tilde{u})}{\partial x_i}\,dx \equiv I.$$

From last equation, conditions (22) and strong monotonicity of A it follows that

$$\frac{\partial}{\partial t}\int_\Omega (u-\tilde{u})^2\,dx + \gamma_1 c_0\,\|u-\tilde{u}\|_1^2 \le I. \tag{23}$$

It is easy to see that

$$I_1 = a(\|\nabla \tilde{u}\|^2) - a(\|\nabla u\|^2) \le 2\max_{0\le\xi\le M_\nabla}|a'(\xi)|\,M_\nabla \|\nabla\tilde{u} - \nabla u\|,$$

$$I_2 = \left|\int_\Omega \sum_{i=1}^n k_i(x,\nabla\tilde{u})\,\frac{\partial(u-\tilde{u})}{\partial x_i}\,dx\right| \le n^{\frac{1}{2}}(\tilde{c})^2\left(nM_\nabla^2 + 2n^{\frac{1}{2}}(\text{mes}\,\Omega)^{\frac{1}{2}}M_\nabla + \text{mes}\,\Omega\right)^{\frac{1}{2}}\|\tilde{u}-u\|_1.$$

Hence,

$$|I| = |I_1||I_2| \le \tilde{C}\max_{0\le\xi\le M_\nabla}|a'(\xi)|\;\|\nabla\tilde{u}-\nabla u\|^2. \tag{24}$$

Using (24) to estimate the right-hand side of inequality (23), we have

$$\frac{\partial}{\partial t}\int_\Omega (u-\tilde{u})^2\,dx + \gamma_1 c_0\,\|u-\tilde{u}\|_1^2 \le \tilde{C}\max_{0\le\xi\le M_\nabla}|a'(\xi)|\;\|\nabla\tilde{u}-\nabla u\|^2. \tag{25}$$

From (25) and condition (21) implies that $u = \tilde{u}$. This completes the proof of the theorem.

Acknowledgments
The research was supported by the Russian Foundation for research (projects 15-01-05686, 15-41-02315)

References

[1] Zheng S and Chipot M 2005 *Asymptot. Anal.* **45** pp 301-312
[2] Chang N H and Chipot M 2003 *RACSAM – Rev. R. Acad. Cien. Serie A. Mat.* **97** pp 393-415
[3] Chipot M and Molinet L 2001 *Applicable Analysis* 80 pp 279-315
[4] Chipot M and Lovat B 2001 *Dyn. Contin. Discrete Impuls. Syst., Ser. A: Math. Anal.* 8 pp 35-51
[5] Pavlova M F 2011 Differential Equations **47 (8)** pp 1161-75
[6] Glazyrina O V and Pavlova M F 2012 *Izv. Vyssh. Uchebn. Zaved.* 3 pp 83-86
[7] Glazyrina O V and Pavlova 2014 *Differential Equations* **50 (7)** pp 873-887
[8] Glazyrina O V and Pavlova 2015 *Differential Equations* **50 (7)** pp 876-889
[9] Karchevskii M M and Pavlova M.F. 2016 *Uravneniya matematicheskoi fiziki* (Lan, Sankt-Peterburg)

Physical processes and modeling of plasma deposition and hardening of coatings-switched electrical parameters

A M Kadyrmetov[1], **S N Sharifullin**[2]

[1]Voronezh State Forestry academy, 8 Timiryazev Street., Voronezh, 394613, Russian Federation
[2]Kazan Federal University, 18 Kremlyovskaya Street, Kazan, 420008, Russian Federation

E-mail: Saidchist@mail.ru

Abstract. This paper presents the results of simulation of plasma deposition and hardening of coatings in modulating the electrical parameters. Mathematical models are based on physical models of gas-dynamic mechanisms more dynamic and thermal processes of the plasma jet. As an example the modeling of dynamic processes of heterogeneous plasma jet, modulated current pulses indirect arc plasma torch.

1. Introduction.

One of the most progressive and effective lines of creating parts with desired properties on their working surfaces is a protective coating of gas-thermal methods. In this group, one of the most effective and universal method is plasma spraying. Existing ways of improving the plasma spraying, aimed at improving the efficiency of stationary coating, almost exhausted, and limited to an insufficient level of physical-mechanical and tribological properties of coatings subjected to high dynamic alternating and shock loads performance. Some work is devoted to improving the plasma deposition methods dynamizing processes using acoustic impact on the plasma jet (SSI "Institute of Heat and Mass Transfer" NASB), and pulsed arc power bypass (SSTU, St. Petersburg, MSTU. NE Bauman). However, questions remained unstudied theoretical description of processes at the modulation parameters of the plasma spraying and methodical study of optimization.

Based on the study of general questions of plasma spraying process can be concluded that improving the efficiency of plasma spray technology is advantageously carried out by the improvement, development and universalization of both the technological operation spray coating and subsequent processing steps of hardening or combining with them [1].

Improvement of the process step of spraying may be achieved by modulating the parameters of the plasma torch, by adding additives to the plasma gas (the plasma for the air - by the addition of carbohydrate-gen-containing gas), gas dynamic processes, and by combining these methods using [2, 3].

Modulation of the electrical parameters of the plasma spraying (current arc plasma torch) is technological method to simply and effectively regulate the power and energy of the arc by increasing or decreasing the pulse arc power at a certain frequency. It allows you to: generate a plasma jet at low intensity shock waves and intense acoustic waves; increase the drag coefficient of the spray particles and heat them to two or more times and thus enhance the energy parameters of the spray particles (increase of the speed and temperature of 30 ... 50%) [2, 3]. This ensures increasing the density and strength of the coating.

The technological method of adding to the hydrocarbon-containing gas plasma air can increase the enthalpy and the thermal conductivity of the plasma jet to 2 times, create a neutral and a reducing atmosphere; increase the thermal efficiency of the plasma torch; increase the temperature of the plasma jet (from 5300 °K air to 6200 °K for propane in air) and the speed of the plasma jet and the sprayed particles by 10 ... 30% [3, 4].

For the development of technical advice in the process of plasma deposition and hardening of coatings requires knowledge of patterns and relationships between technological factors in the production of coatings in the mode of modulation parameters and gas-dynamic and thermal coatings quality criteria that determine the level of residual stresses in the coatings, their strength and tribological characteristics.

To accomplish this task have been developed mathematical models describing the technological system during application and hardening coatings mode modulation parameters.

2. Research methods.

Methodological basis of theoretical studies have made the scientific basis of mechanical engineering, the theory of gas dynamics in terms of jet streams, traveling wave velocity and entropy waves temperature, the theory of thermal welding processes, methods of computational mathematics (finite difference method and discrete components), advanced graphics and computer systems for the computer.

3. Physical application and hardening coatings

The principle of the arc current modulation action of the plasma torch is applied to the stationary current pulse arc current forward or reverse polarity, and bipolar pulses (Fig. 1). Technological use of dormant plasma deposition of hardening coatings carried modulated electrical parameters based on physical models dormant gas dynamic and thermal parameters of the mechanism that causes the conversion of electrical energy pulse current indirect or remote arc at a modulation in the thermal and gas-dynamic arc energy and space around it (Fig. 2).

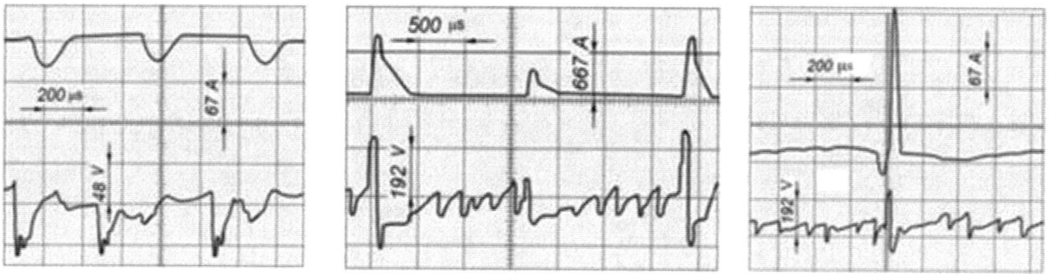

Figure 1. Examples of current waveforms (top curve) and voltage (lower curves) arc torch: a) - Reverse polarity modulation (J_{av} = 109 A; U_{av} = 87 V; G_g = 35 l /min); b) - the modulation of direct polarity (J_{av} = 210 A; U_{av} = 90 V; G_g = 40 l /min); c) - bipolar modulation (J_{av} = 118 k = 92 In Usr; G_g = 40 l / min); J_{av}; U_{av} - average current and voltage; G_g - plasma flow rate of air

Physical processes in the plasma jet in the modulation of indirect arc plasma generator with a frequency modulation v_m duration τ_m, amplitude ΔN and steep power dN/dt pulses is to generate in the plasma jet of entropy heat waves with the amplitude of the temperature ΔT_p and shock waves with Mach number M, appropriate strong or weak shock waves [5]. The result of the impact of these waves on the plasma jet is heterogeneous: increased turbulence in the plasma jet, leading to an increase in the uniformity of its cross-profile A range of a temperature L_T and enthalpy $L_{\Delta H}$; increasing the drag coefficient C_D and particle heat transfer them α by the plasma jet; increasing the speed V_p particles and their heating rate dT_p /dt. Such an increase in the energy state of the particles provides improvement of physical contact of the particles with the substrate and connected to it by increasing the pressure and the pulse pressure of the particles to the substrate upon impact, which results in a decrease of porosity P and increase strength coating performance (adhesion σ_a and cohesion σ_c).

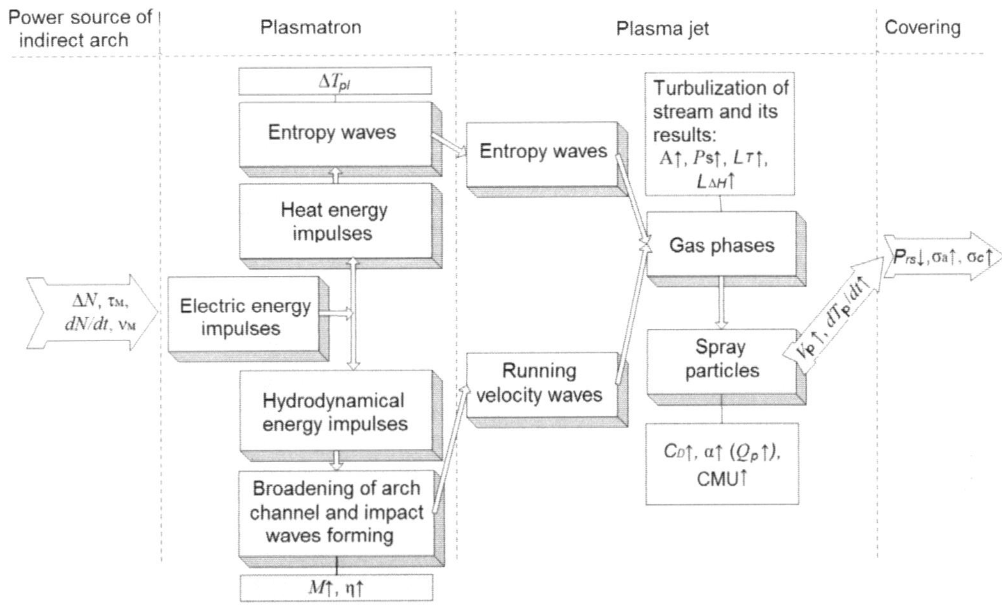

Figure 2. Schematic of a physical model of plasma deposition processes by modulating the electrical parameters of the indirect arc plasma torch: ΔN- amplitude of the arc power; τ_{imp}- pulse duration; dN/dt- slew rate output pulse; ν_m - modulation frequency; ΔT_{pl} - the amplitude of the temperature of the plasma in the arc; M - Mach number; η - the thermal efficiency of the plasma torch; A - uniformity parameter of the plasma jet cross-sections; P_s - the sound pressure, L_T and $L_{\Delta H}$ - long range plasma jet temperature and enthalpy; C_D - drag coefficient; α - heat transfer coefficient from the plasma jet to the particle; Q_p - the heat flow to the particle CMU - utilization of the powder material; V_P, T_p - speed and temperature of the particles, respectively; t - time; P_{rs} - coating porosity; σ_a, σ_c - adhesive and cohesive strength of the coating; ↑, ↓ - symbol to increase or decrease the parameter, respectively

Physical processes in the coating in the modulation of the direct arc torch with an amplitude ΔN_d and pulse time τ_m consist of fusion of the coating to the transition zone to the substrate and the formation of physical contact for the time t_c and chemical interaction covering the substrate in point local zones bind direct arc to the base or coating (Fig. 3). On the scale of the entire surface with the influence of the speed V and feed S torch formed certain density distribution of discrete zones of penetration and thermal effect on the substrate and the coating that contribute to the uniformity of the temperature field bases and improving mechanical and tribological characteristics of the coating.

Formation and quality of the coating conditions are determined by the energy impact on formed by coating on the three channels from the deposited particles (with parameters of V_p speed, temperature T_p and heat flow q_p) from the gas phase of the plasma jet {using parameters V_g velocity, temperature T_g, heat flows heat q_g and cooling q_c, direct arc power N_d and its amplitude (Fig. 3, 4)}, and thermal and electromechanical effects on the sprayed surface and coating (direct arc power parameter N_p and its amplitude, pressure roller and its amplitude ΔP, heat flows from the thermal q_t and electromechanical q_e effects, thermal conductivity λ and heat capacity c_p materials, radii shaft R_{sh} and run-roller Rr, shaft speed n_{sh} and pitch of turns spraying H).

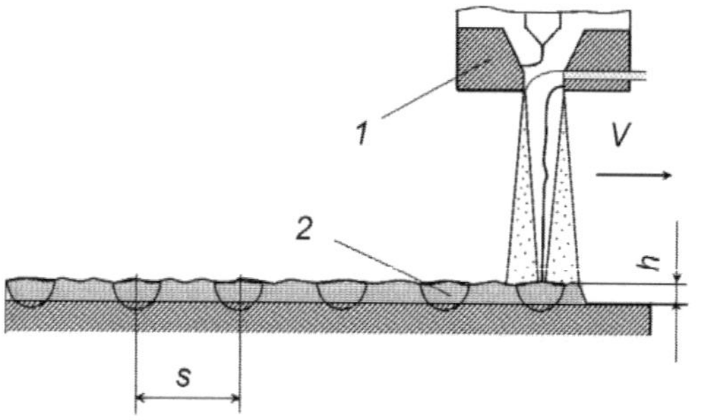

Figure 3. Scheme of the plasma spray coating process with pulse modulation of direct arc burning between the anode of the plasma torch and the substrate: 1 - plasma torch; 2 - penetration of coverage; h - thickness of the coating; V - speed of the plasma torch; S – step

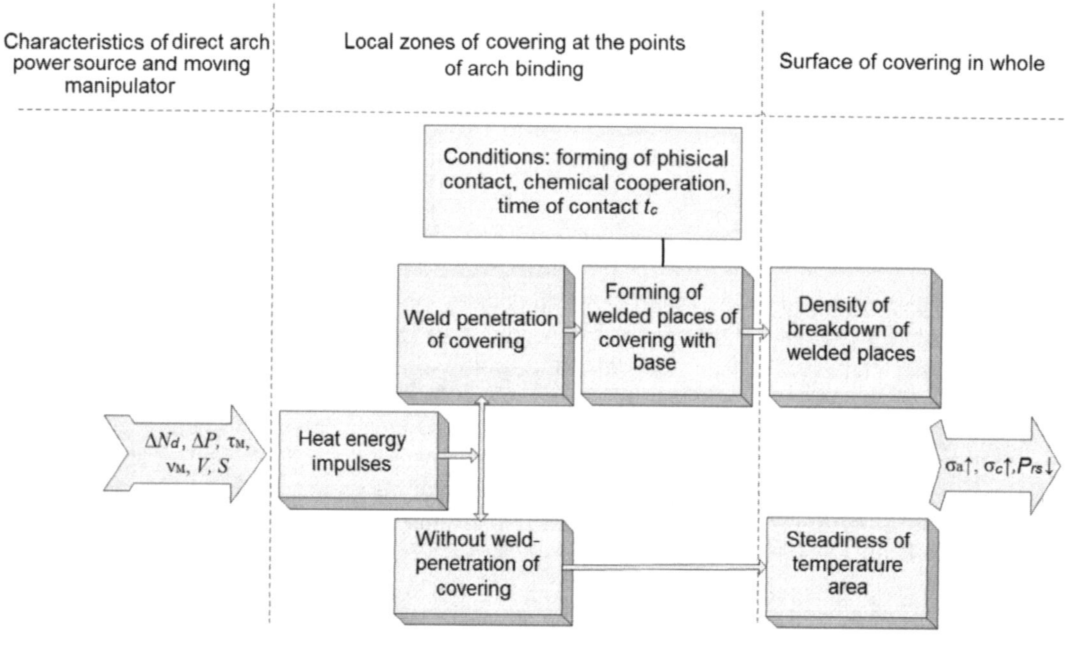

Figure 4. Schematic of a physical model of the plasma coating process by modulating the electrical parameters of the direct arc torch: ΔN_d - the amplitude of the arc power

The model shows that for the process to ensure the quality of decisive importance is the effect of two physical components of the plasma process: the energy state of the particles of the coating material and the thermodynamic state of the "coating-base." Therefore, the complexity of the physical processes described in the "indirect and direct arc - plasma jet - Floor - base" in the modulation mode, the electrical parameters of the plasma torch arcs led to the need to provide the system of partial models. Their study allows us to solve the problem
of getting better quality coatings in a complex mathematical models that include a description
of the processes with the modulation parameters: gas-dynamic, temperature field distribution of the "coating - substrate" with regard to the cooling surface deposition, thermal and electro-coating treatment.

4. Mathematical model of modulating the plasma jet

Here we consider one of the private mathematical models developed for the dynamic processes of heterogeneous plasma jet, modulated current pulses indirect arc plasma torch. It includes a stationary-phase empirical description of the plasma jet, an analytical description of the stationary distribution in the jet traveling wave velocity and temperature and entropy waves the description of the behavior of spray particles in modulating plasma jet.

On the basis of the equations of motion and continuity of the gas phase distribution of the plasma jet traveling wave velocity jet axis x stands for a solution:

$$x=\left(\frac{\chi+1}{2}\cdot V_v+V_{om}+c_{om}\right)\cdot t+f(V_v) \qquad (1)$$

where the gas velocity is $V_g = V_v + V_{om}$; V_v, V_{om} – respectively variable and fixed components of the velocity of the plasma torch to cut, m/s; c_{om} – the speed of sound in the undisturbed environment at $V_g = V_v$, m/s; t – time, s; χ - polytropic index. Function speed $f(V_g)$ determined from the boundary conditions change the speed at the nozzle exit of the plasma torch generated triangular wave power indirect arc of dependencies for the first and second half pulses, respectively:

$$V_g=V_{om}+\frac{2V_a}{\tau_{imp}}\cdot\tau \text{ at } n\cdot\tau_n<\tau<\frac{\tau_{imp}}{2}+n\cdot\tau_n,$$

$$V_g=V_{om}-\frac{2V_a}{\tau_{imp}}\cdot\tau+2V_a \text{ at } n\cdot\tau_n+\frac{\tau_{imp}}{2}<\tau<\tau_{imp}+n\cdot\tau_n \qquad (2)$$

where $n = 0, 1, 2, 3,\ldots$; V_{g0}, V_a - respectively, the gas velocity and amplitude of its fluctuations at the nozzle exit of the plasma torch, m/s; τ_{imp}, τ_n - respectively, the pulse duration and the period of its pulsations, s.

The initial conditions corresponding to the distribution of the parameters of the plasma jet along its x-axis describes the dependence:

$$V_m(x)=V_{om}\cdot\sqrt{\pi\cdot\left[\phi(x)+(1-\phi(x))\cdot\frac{T_n-300}{T_{om}-300}\right]},$$

$$T_m(x)=\phi(x)\cdot T_{om}+(1-\phi(x))\cdot T_n,$$

$$c_m(x)=\sqrt{\chi\cdot P/\rho_m(x)} \qquad (3)$$

where for the initial portion of the plasma jet $\varphi = 1$, $\pi = 1$; for the transition - $\varphi \sim 0{,}5$, $\pi \sim 0{,}6$; for the main - $\varphi \sim 0{,}1$, $\pi \sim 0{,}1$; $V_m(x)$ - velocity at the axis of the plasma jet in m/s; T_n - normal temperature, $T_n = 300$ K; T_n (x) - the temperature on the axis of the plasma jet, K; $\rho_m(x)$ - the gas density at the jet axis, kg/m3; $c_m(x)$ - the sound velocity on the axis of the plasma jet in m/s; P - pressure, Pa.

For the initial portion of the plasma jet at intervals of time corresponding to the first half of the pulse arc power, based on the expressions (1) and (2) the dependence of the variable speed of the gas phase in the form:

$$V_v(x,t)=\frac{V_a\cdot t}{\tau_{imp}}-\frac{c_{om}+V_{om}}{\gamma+1}+\sqrt{\left(\frac{V_a\cdot t}{\tau_{imp}}-\frac{c_{om}+V_{om}}{\gamma+1}\right)^2-\frac{4\cdot V_a}{\tau_{imp}\cdot(\gamma+1)}\cdot[x-(c_{om}+V_{om})\cdot t]}$$

and for times corresponding to the second half of the pulse, - in the form:

$$V_v(x,t)=\frac{V_a\cdot t}{\tau_{imp}}\cdot\frac{c_{om}+V_{om}}{\gamma+1}+V_a+\left\{\left(-\frac{V_a\cdot t}{\tau_{imp}}-\frac{c_{om}+V_{om}}{\gamma+1}+\upsilon_a\right)^2+\right.$$

$$\left.+\frac{4V_a}{\tau_{imp}\cdot(\gamma+1)}\left[\frac{4\cdot V_a}{\tau_{imp}\cdot(\gamma+1)}\cdot\left[x-(c_{om}+V_{om})\cdot(t-\tau_{imp}\right]\right]\right\}^{1/2}.$$

Entropy waves carried along with the flow, so the general solution of the temperature distribution in the longitudinal coordinate stream of x and time t is given by: $T(x,t) = T(0, t - x/V_{om})$.

Based on the boundary conditions corresponding to the change in temperature at the nozzle exit of the plasma torch in the form of a triangular wave generated by the pulse power of indirect arc:

$$T_{go} = T_{om} + \frac{2T_a}{\tau_{imp}}\tau \text{ at } n\cdot\tau_n < \tau < \frac{\tau_{imp}}{2} + n\cdot\tau_n,$$

$$T_{go} = T_{om} - \frac{2T_a}{\tau_{imp}}\tau + 2T_a \text{ at } n\cdot\tau_n + \frac{\tau_{imp}}{2} < \tau < \tau_{imp} + n\cdot\tau_n$$

$$T_{go} = T_{om} \text{ at } n\cdot\tau_n + _{imp} < \tau < (1+n)\cdot\tau_n$$

and initial conditions (3) analytical temperature versus distance x and time t, where Ta - the amplitude of the temperature, K.

In the initial section of the plasma jet for the time corresponding to the first half of the pulse power of the arc, the expression for the temperature of the gas phase is of the form:

$$T_{ini}(x,t) = T_{om} + \frac{2\cdot T_a}{\tau_{imp}}\cdot\left(t - \frac{x}{V_{om}}\right),$$

for the time corresponding to the second half of the pulse - form:

$$T_{ini}(x,t) = T_{om} + 2\cdot T_a - \frac{2\cdot T_a}{\tau_{imp}}\cdot\left(t - \frac{x}{V_{om}}\right).$$

Similar expressions variable speed and temperature of the gas phase are obtained for the transition and main portions of the plasma jet.

Changing the speed and temperature of the particles described by the equations of motion and heat balance in the quasi-stationary approximation, corresponds to what the drag coefficient and the heat transfer particles depend only on the instantaneous values of the hydrodynamic parameters. This is justified by the smallness of the time of establishment of the stationary boundary layer of particles equal to 0.1-10 microseconds, compared to the duration of the modulating pulses of the order of 100-1000 microseconds. A mathematical model of a heterogeneous plasma jet is implemented as an algorithm for calculating the Euler method using the 2nd order.

We were determined according to the speed of motion of particles from materials PG-CP4 (self-fluxing material based on Ni), PN85YU15 (85% Ni + 15% Al) and PN55T45 (55% Ni + 45% Ti) in an air plasma stream generated by the plasmatron, the parameters modulation of the electrical parameters of self-setting arc torch during spraying air plasma jet [6]. They show that in the range of changes in the duration τ^+ pulses from 20 to 300 microseconds and a pulse output amplitude ΔN + from 50 to 500 kW increase in arc current modulation frequency ν_m to 3 kHz spraying distance of 160 mm increases the velocity of the particles on average of 30%. More sensitive to modulation particles are less dense material. It is shown that the effect of modulation frequency on the deposition process increases with the amplitude and duration output pulses. Increasing the duration and power of the pulse amplitude also increases the velocity of the particles is 40 - 60%. Effect modulation plasmatron arc current temperature on particle acceleration is reflected in their heating and melting. The melting velocity and particle motion and increases in proportion.

5. Conclusions.

Pulse modulation indirect arc power leads to entropy and shock waves in the plasma jet, its turbulence. This enhances the energy exchange in the plasma torch and the plasma jet, increases the powder utilization factor and the thermal efficiency of the plasma torch, as well as speed and heating the particles. This increase in the energy state of the particle provides improved conditions of physical contact with the substrate particles and compounds with it and increases the strength characteristics of the coating.

To direct the arc electrical pulse power level and duration with modulation determine the possibility of penetration of the coating to the transition zone to the substrate and the formation of physical contact and chemical interaction of the coating to the substrate. The wide surface of modulation frequency and speed of the torch relative to the substrate determine the density distribution of digital penetration coverage, uniformity of temperature field bases and increasing the strength characteristics of the coating.

It is shown that the description of the technological processes of the system and application hardening coatings in the modulation mode settings possible on the basis of mathematical models developed on the basis of physical models of gas-dynamic mechanisms more dynamic and thermal parameters of the plasma jet.

The concrete mathematical models developed for the dynamic processes of heterogeneous plasma jets modulated pulses of current indirect arc plasma torch. A mathematical model of heterogeneous plasma jet is realized numerically Euler 2nd order. The results of these studies indicate the possibility of generating a plasma arc indirect modulation power of weak shock waves with a Mach number less than 1,2. Waves are rapidly absorbed into the plasma jet and to intensify the exchange of energy therein, thereby increasing the speed of the spray particles is 50-60% and the particle proportion to heat their velocity.

References

[1] Stanchev D I, Kadyrmetov A M, Kluchnikov V I and Yakovlev K A 2002 *Recovery of machine parts thermal spray methods: the manual* (Voronezh: VGLTA)

[2] Kadyrmetov A M and Suhochev G A 2009 Features of the air-plasma deposition and hardening of coatings *Hardening and coating technology* **4 (52)** 25-8

[3] Suhochev G A and Kadyrmetov A M 2008 Experimental studies of the parameters of control of air-plasma deposition and hardening of coatings *Hardening and coating technology* **11 (47)** 53-6

[4] Klubnikin B C, Karasev M V and Petrov G K 1990 *Plasma spraying in the active media* (Leningrad: Society "Knowledge"of the RSFSR)

[5] Kadyrmetov A M 2014 *The theoretical basis and technological support for the quality of plasma deposition and hardening of coatings modulated electrical parameters: Author. Dis. on soisk. scientists. step. Doctor. tehn.* (Voronezh: VGLTA)

[6] Kadyrmetov A M 2013 Intensification of energy exchange in a heterogeneous plasma jet by modulation of the electric parameters of the plasma-spraying process *Journal of Engineering Physics and Thermophysics* **86 (4)** 789-97

Modified Coulomb and Lorenz gauges in the modeling of low-frequency electromagnetic processes

A V Kalinin, A A Tiukhtina and S R Lavrova

Lobachevsky State University of Nizhnii Novgorod, 23 Gagarin av., 603950 Nizhnii Novgorod, Russia

E-mail: kalinmm@yandex.ru

Abstract. The boundary value problem for the quasistationary magnetic approximation of the time-harmonic Maxwell equations in inhomogeneous media is studied. The considered problem is reduced to the variational problem of determining vector magnetic and scalar electric potentials. The special gauges are discussed, that generalize the Coulomb and Lorenz gauges and allow to formulate the problems of the independent definitions of the vector magnetic potential. The correctness of the problems are established under general conditions on the coefficients. The relation between solutions of the problems with different gauges is studied. The equivalence of the problems for potentials to the original boundary value problem is proved.

1. Introduction

The solution of many actual technological problems leads to theoretical and numerical study of problems for quasi-stationary electromagnetic fields in physically heterogeneous media [1]. Quasi-stationary magnetic or low-frequency approximation of the Maxwell's equations used in the design of electromagnetic devices in metallurgical industry and power transformers, for remote sensing, in solving problems of magnetic levitation and in advanced medicine [1]–[3].

The essential stage in the development and justification of efficient numerical methods for solving applied problems is the study questions of the correctness of their various mathematical formulations. Electromagnetic processes in theoretical and numerical researches are traditionally described in terms of magnetic or electric fields [4]-[6] either in terms of potentials – vector electrical and scalar magnetic (the $\boldsymbol{T}-\psi$ formulations) [7]-[9] or vector magnetic and scalar electrical (the $\boldsymbol{A}-\varphi$ formulations) [10]-[16].

The application of potentials always presupposes the choosing of certain gauges that ensure the uniqueness of the solution of the problem. In particular, for the description of stationary and quasistationary electromagnetic fields the classical Coulomb gauge $\operatorname{div}\boldsymbol{A}=0$ and Lorenz gauge $\operatorname{div}\boldsymbol{A}+\mu\sigma\varphi=0$ are used [11]-[14], which allow to uniquely determine the vector magnetic and scalar electric potentials.

In the case of inhomogeneous media the classic Coulomb and Lorenz gauges do not lead to the decoupled problem for the electric and magnetic potentials [13], [14], however this decoupling plays an important role in the algorithms development for the numerical solution. In this regard, various modifications of the gauges are considered. In [14], [15] the possibility of applying the modified

Lorenz gauge, that allows to separate the problems of determining the magnetic and electric potentials, is discussed.

In the present paper the theoretical issues are investigated related to the use of the modified gauges, which generalize the Coulomb and Lorenz gauges for the case of inhomogeneous media and decouple the problem for the vector magnetic and scalar electric potentials. We consider the time - harmonic Maxwell equations in the low frequency approximation in physical domains filled with conductive and non-conductive material. The respective topics for conductors are considered for the stationary problems in [17], [18], for time-dependent problems – in [19].

For the considered class of problems this paper provides a full justification of the suitability of these gauges, namely, we prove the well-posedness of the problems and their equivalence to the original boundary value problem. The relation between solutions of the problems under different gauges is examined. In the case of homogeneous media, this relation is similar to the ratio between the solutions for the system of Stokes equations for incompressible and weakly compressible liquid [20].

The study of the properties of the boundary value problems is based on the inequalities which combine the scalar product of vector fields and norms of their curl and divergence. These inequalities generalize the estimates for scalar products of vector fields obtained in [17], [21] for the case of domains homeomorphic to a ball.

2. The boundary value problem

The time-harmonic Maxwell equations in the quasistationary magnetic approximation can be presented in the form [1]

$$\operatorname{curl} \boldsymbol{H}(x) = \boldsymbol{J}(x), \tag{1}$$

$$\operatorname{div} \boldsymbol{B}(x) = 0, \tag{2}$$

$$\operatorname{curl} \boldsymbol{E}(x) = -\mathrm{i}\,\omega \boldsymbol{B}(x), \tag{3}$$

$$\operatorname{div} \boldsymbol{D}(x) = \rho(x), \tag{4}$$

where $x \in \Omega \subset \mathbf{R}^3$; $\boldsymbol{H}$, $\boldsymbol{B}$, $\boldsymbol{E}$, $\boldsymbol{D}$, $\boldsymbol{J}: \Omega \to \mathbf{C}^3$ and $\rho: \Omega \to \mathbf{C}^1$ are unknown functions.

In linear media the following constitutive relations are valid:

$$\boldsymbol{B} = \mu \boldsymbol{H}, \; \boldsymbol{D} = \varepsilon \boldsymbol{E}, \; \boldsymbol{J} = \sigma \boldsymbol{E} + \boldsymbol{J}^{ext}, \tag{5}$$

where μ is a magnetic permeability, ε is a permittivity, σ is a electrical conductivity, $\boldsymbol{J}^{ext}$ is an exterior current density.

We suppose $\omega \geq 0$. In case $\omega = 0$ let $\boldsymbol{H}$, $\boldsymbol{B}$, $\boldsymbol{E}$, $\boldsymbol{D}$, $\boldsymbol{J}$, ρ are real valued functions and thus (1)-(4) is the stationary system of the Maxwell equations.

It is assumed in this paper, that Ω is an open bounded domain, homeomorphic to a ball, with a Lipschitz boundary $\partial\Omega$. Let $\boldsymbol{\nu}(x)$ is the unit normal vector in $x \in \partial\Omega$. For function $\boldsymbol{u}: \overline{\Omega} \to \mathbf{C}^3$ we denote by $\boldsymbol{u}_\nu$, $\boldsymbol{u}_\tau$ the normal and tangent components of $\boldsymbol{u}$ on $\partial\Omega$.

The system (1)–(5) is considered with the boundary condition

$$\boldsymbol{H}_\tau(x) = 0, \; x \in \partial\Omega. \tag{6}$$

We assume, the domain consists from the conductor Ω_C and the isolator $\Omega_I = \Omega \setminus \overline{\Omega}_C$. Let Ω_C is an open bounded homeomorphic to a boll domain with Lipschitz boundary Γ, $\overline{\Omega}_C \subset \Omega$, so $\partial\Omega_I = \partial\Omega \cup \Gamma$. The unit normal vectors in $x \in \Gamma$ to Ω_C and Ω_I are denoted by $\boldsymbol{\nu}_C(x)$ and $\boldsymbol{\nu}_I(x)$ respectively, $\boldsymbol{\nu}_C(x) + \boldsymbol{\nu}_I(x) = 0$. For functions $\boldsymbol{u}: \Omega \to \mathbf{C}^3$ and $u: \Omega \to \mathbf{C}^1$ we denote by $\boldsymbol{u}_C$, u_C their restrictions on Ω_C, by $\boldsymbol{u}_I$, u_I their restrictions on Ω_I.

The ε and μ are self-ajoint linear operators from $\{L_2(\Omega)\}^3$ into $\{L_2(\Omega)\}^3$, satisfying the following conditions:

$$\varepsilon_1\|\boldsymbol{u}\|^2_{2,\Omega} \le (\varepsilon\boldsymbol{u},\boldsymbol{u})_{2,\Omega} \le \varepsilon_2\|\boldsymbol{u}\|^2_{2,\Omega},\ \mu_1\|\boldsymbol{u}\|^2_{2,\Omega} \le (\mu\boldsymbol{u},\boldsymbol{u})_{2,\Omega} \le \mu_2\|\boldsymbol{u}\|^2_{2,\Omega},$$

$\sigma = \sigma(x)$ is symmetric 3×3 matrix of measurable functions on Ω, satisfying the conditions

$$\sigma_{ij}(x) = 0,\ i,j = 1,2,3, \text{ for almost all } x \in \Omega_I,$$

$$\sigma_1|\xi|^2 \le (\sigma(x)\xi,\xi) \le \sigma_2|\xi|^2 \text{ for almost all } x \in \Omega_C \text{ and for all } \xi \in \mathbf{R}^3,$$

where ε_i, μ_i, σ_i, ($i = 1,2$) are given positive numbers, by $\|\cdot\|_{2,\Omega}$ and $(\cdot,\cdot)_{2,\Omega}$ the norm and the scalar product in $\{L_2(\Omega)\}^3$ are denoted.

$\boldsymbol{J}^{ext}: \Omega \to \mathbf{C}^3$ is a square integrable function, such that

$$\operatorname{div} \boldsymbol{J}_I^{ext} = 0,\ \left(\boldsymbol{J}_I^{ext}\right)_\nu(x) = 0,\ x \in \partial\Omega.$$

The generalized solutions of the problems will be considered, that is all equalities have to be satisfied in the sense of distributions and boundary conditions have to be satisfied in the sense of the trace theory [22].

The following Hilbert spaces with the respective scalar products are defined [22]:

$$H(\mathrm{div};\Omega) = \left\{\boldsymbol{u} \in \{L_2(\Omega)\}^3 : \operatorname{div}\boldsymbol{u} \in L_2(\Omega)\right\},\ K(\mathrm{div};\Omega) = \left\{\boldsymbol{u} \in \{L_2(\Omega)\}^3 : \operatorname{div}\boldsymbol{u} = 0\right\},$$

$$(\boldsymbol{u},\boldsymbol{v})_{div,\Omega} = \int_\Omega (\boldsymbol{u}\cdot\bar{\boldsymbol{v}})dx + \int_\Omega \operatorname{div}\boldsymbol{u}\operatorname{div}\bar{\boldsymbol{v}}dx,$$

$$H(\mathrm{curl};\Omega) = \left\{\boldsymbol{u} \in \{L_2(\Omega)\}^3 : \operatorname{curl}\boldsymbol{u} \in \{L_2(\Omega)\}^3\right\},\ K(\mathrm{curl};\Omega) = \left\{\boldsymbol{u} \in \{L_2(\Omega)\}^3 : \operatorname{curl}\boldsymbol{u} = 0\right\},$$

$$(\boldsymbol{u},\boldsymbol{v})_{curl,\Omega} = \int_\Omega (\boldsymbol{u}\cdot\bar{\boldsymbol{v}})dx + \int_\Omega (\operatorname{curl}\boldsymbol{u}\operatorname{curl}\bar{\boldsymbol{v}})dx.$$

$H_0(\mathrm{div};\Omega)$, $H_0(\mathrm{curl};\Omega)$ denote the closures of the set of test vector-functions in $H(\mathrm{div};\Omega)$ and $H(\mathrm{curl};\Omega)$ respectively, $K_0(\mathrm{div};\Omega) = K(\mathrm{div};\Omega) \cap H_0(\mathrm{div};\Omega)$.

The solution of the problem (1)–(6) is the set of functions $\boldsymbol{H} \in H_0(\mathrm{curl};\Omega)$, $\boldsymbol{B} \in K(\mathrm{div};\Omega)$, $\boldsymbol{J} \in K_0(\mathrm{div};\Omega)$, $\boldsymbol{E} \in H(\mathrm{curl};\Omega)$, $\boldsymbol{D} \in \{L_2(\Omega)\}^3$, $\rho \in H^{-1}(\Omega)$, satisfying (1), (3)–(5).

To uniquely identify function $\boldsymbol{E}$ in the whole domain Ω, we suppose that unknown functions $\boldsymbol{E}$, ρ satisfy the conditions

$$\rho_I = \rho_0,\ (\varepsilon\boldsymbol{E})_\nu(x) = 0,\ x \in \partial\Omega, \tag{7}$$

where $\rho_0 \in L_2(\Omega_I)$ is a given function.

In case $\omega = 0$ the problem also may be regarded under conditions

$$\rho_I = \rho_0,\ \boldsymbol{E}_\tau(x) = 0,\ x \in \partial\Omega,\ \left\langle(\varepsilon\boldsymbol{E})_{\nu_I},1\right\rangle_\Gamma = Q, \tag{8}$$

Q is a given constant.

3. Problems for the vector potential

Relations (2), (3) allow to introduce the vector magnetic potential $\boldsymbol{A}$ and the scalar electric potential φ as new unknown variables by the formulas [23]

$$\boldsymbol{B} = \operatorname{curl} \boldsymbol{A}, \ \boldsymbol{E} = -\operatorname{grad} \varphi - \mathrm{i}\omega \boldsymbol{A} \tag{9}$$

In this case, the system (1)–(5) is reduced to one equation

$$\mathrm{i}\omega\sigma\boldsymbol{A} + \operatorname{curl} \mu^{-1} \operatorname{curl} \boldsymbol{A} = -\sigma \operatorname{grad} \varphi + \boldsymbol{J}^{ext}. \tag{10}$$

Equation (10) is provided by the boundary condition corresponding to (6):

$$\left(\mu^{-1} \operatorname{curl} \boldsymbol{A}\right)_\tau (x) = 0, \ x \in \partial\Omega. \tag{11}$$

The solution of the problem (10), (11) is the functions $\boldsymbol{A} \in H(\operatorname{curl};\Omega)$, $\varphi \in H^1(\Omega)$, satisfying (10) in the sense of the distribution and (11) in the sense of the trace theory, that is $\mu^{-1} \operatorname{curl} \boldsymbol{A} \in H_0(\operatorname{curl};\Omega)$.

Suppose that $\boldsymbol{A}$, φ is the solution of (10), (11). Then from (10) we obtain for all $\boldsymbol{v} \in H(\operatorname{curl};\Omega)$

$$\mathrm{i}\omega \int_{\Omega_C} (\sigma\boldsymbol{A} \cdot \bar{\boldsymbol{v}}) \mathrm{d}x + \int_{\Omega} \left(\mu^{-1} \operatorname{curl} \boldsymbol{A} \cdot \operatorname{curl} \bar{\boldsymbol{v}}\right) \mathrm{d}x = -\int_{\Omega_C} (\sigma \operatorname{grad} \varphi \cdot \bar{\boldsymbol{v}}) \mathrm{d}x + \int_{\Omega} \left(\boldsymbol{J}^{ext} \cdot \bar{\boldsymbol{v}}\right) \mathrm{d}x. \tag{12}$$

The solution of the problem (10), (11) is obviously not unique. The following two types of gauges are discussed:

$$\operatorname{div} \sigma\boldsymbol{A}_C = 0, \ (\sigma\boldsymbol{A})_{\nu_C}(x) = 0, \ x \in \Gamma, \ \operatorname{div} \boldsymbol{A}_I = 0, \ \boldsymbol{A}_\nu(x) = 0, \ x \in \partial\Omega, \tag{13}$$

and

$$\varphi_C = -\kappa \operatorname{div} \sigma\boldsymbol{A}_C, \ (\sigma\boldsymbol{A})_{\nu_C}(x) = 0, \ x \in \Gamma, \ \operatorname{div} \boldsymbol{A}_I = 0, \ \boldsymbol{A}_\nu(x) = 0, \ x \in \partial\Omega, \tag{14}$$

where $\kappa > 0$ is an arbitrary constant.

We define the Gilbert spaces

$$W(\sigma;\Omega_C,\Omega) = \{\boldsymbol{u} \in H(\operatorname{curl};\Omega) : \sigma\boldsymbol{u}_C \in H_0(\operatorname{div};\Omega_C), \operatorname{div} \boldsymbol{u}_I = 0, (\boldsymbol{u}_I)_\nu(x) = 0, x \in \partial\Omega\},$$

$$(\boldsymbol{u},\boldsymbol{v})_W = \int_{\Omega} (\boldsymbol{u} \cdot \bar{\boldsymbol{v}}) \mathrm{d}x + \int_{\Omega} (\operatorname{curl} \boldsymbol{u} \cdot \operatorname{curl} \bar{\boldsymbol{v}}) \mathrm{d}x + \int_{\Omega_C} (\operatorname{div} \sigma\boldsymbol{u} \cdot \operatorname{div} \sigma\bar{\boldsymbol{v}}) \mathrm{d}x,$$

$$V(\sigma;\Omega_C,\Omega) = \{\boldsymbol{u} \in W(\sigma;\Omega_C,\Omega) : \operatorname{div} \sigma\boldsymbol{u}_C = 0\}, \ (\boldsymbol{u},\boldsymbol{v})_V = (\boldsymbol{u},\boldsymbol{v})_{curl,\Omega}.$$

Using (12) we obtain that the problem (10), (11), (13) is reduced to the following problem: to find a function $\boldsymbol{A} \in V(\sigma;\Omega_C,\Omega)$ such that for all $\boldsymbol{v} \in V(\sigma;\Omega_C,\Omega)$

$$\mathrm{i}\omega \int_{\Omega_C} (\sigma\boldsymbol{A} \cdot \bar{\boldsymbol{v}}) \mathrm{d}x + \int_{\Omega} \left(\mu^{-1} \operatorname{curl} \boldsymbol{A} \cdot \operatorname{curl} \bar{\boldsymbol{v}}\right) \mathrm{d}x = \int_{\Omega} \left(\boldsymbol{J}^{ext} \cdot \bar{\boldsymbol{v}}\right) \mathrm{d}x, \tag{15}$$

the problem (10), (11), (14) is reduced to the problem of determining $\boldsymbol{A} \in W(\sigma;\Omega_C,\Omega)$ such that for all $\boldsymbol{v} \in W(\sigma;\Omega_C,\Omega)$

$$\mathrm{i}\omega \int_{\Omega_C} (\sigma\boldsymbol{A} \cdot \bar{\boldsymbol{v}}) \mathrm{d}x + \int_{\Omega} \left(\mu^{-1} \operatorname{curl} \boldsymbol{A} \cdot \operatorname{curl} \bar{\boldsymbol{v}}\right) \mathrm{d}x + \kappa \int_{\Omega_C} (\operatorname{div} \sigma\boldsymbol{A} \cdot \operatorname{div} \sigma\bar{\boldsymbol{v}}) \mathrm{d}x = \int_{\Omega} \left(\boldsymbol{J}^{ext} \cdot \bar{\boldsymbol{v}}\right) \mathrm{d}x. \tag{16}$$

Theorem 1. *The problems (15) and (16) have unique solutions.*

The next theorem establishes the relation between the solutions of the problems under different gauges.

Theorem 2. *Let* $\boldsymbol{A}$ *and* $\boldsymbol{A}_\kappa$ *are the solutions of (15) and (16) respectively. Then* $\operatorname{curl} \boldsymbol{A} = \operatorname{curl} \boldsymbol{A}_\kappa$, $\boldsymbol{A}_\kappa \to \boldsymbol{A}$ *in* $W(\sigma;\Omega_C,\Omega)$ *as* $\kappa \to \infty$ *and the following estimation is valid*

$$\|A_\kappa - A\|_W \le C\kappa^{-1}\|J^{ext}\|_{2,\Omega},$$

where the constant $C > 0$ *depends only on* Ω_C *and* Ω.

The proof of the theorems is based on the following inequalities which generalize the obtained in [18], [22] estimates for scalar products of vector fields in a star-shape domain.

Theorem 3. *Let* $\Omega \subset \mathbf{R}^3$ *is an open bounded Lipschitz domain, homeomorphic to a ball. There exists a positive constant* $C(\Omega)$, *which depends only on* Ω, *such that the inequality*

$$|(\boldsymbol{u},\boldsymbol{v})_{2,\Omega}| \le C(\Omega)\left(\|\boldsymbol{u}\|_{2,\Omega}\|\operatorname{div}\boldsymbol{v}\|_{2,\Omega} + \|\operatorname{curl}\boldsymbol{u}\|_{2,\Omega}\|\boldsymbol{v}\|_{2,\Omega} + \|\operatorname{curl}\boldsymbol{u}\|_{2,\Omega}\|\operatorname{div}\boldsymbol{v}\|_{2,\Omega}\right)$$

holds for any $\boldsymbol{u} \in H_0(\mathrm{curl};\Omega)$, $\boldsymbol{v} \in H(\mathrm{div};\Omega)$ *and for any* $\boldsymbol{u} \in H(\mathrm{curl};\Omega)$, $\boldsymbol{v} \in H_0(\mathrm{div};\Omega)$.

4. The correctness of application the gauges

The using of potentials requires that $\boldsymbol{H}$, $\boldsymbol{E}$, which defined by (9), satisfy the Maxwell equations. Thus the correctness of application the gauges implies the equivalence the problems in terms of potentials and the origin boundary value problem.

The conditions (7) in terms of potentials means

$$\operatorname{div}\varepsilon(\mathrm{i}\omega\boldsymbol{A} + \operatorname{grad}\varphi)_I = -\rho_0,\ \varepsilon(\mathrm{i}\omega\boldsymbol{A} + \operatorname{grad}\varphi)_\nu(x) = 0,\ x \in \partial\Omega. \tag{17}$$

The conditions (8) for $\omega = 0$ means

$$\operatorname{div}\varepsilon\operatorname{grad}\varphi_I = -\rho_0,\ \varphi(x) = \mathrm{const},\ x \in \partial\Omega,\ \langle(\varepsilon\operatorname{grad}\varphi)_{\nu_I}, 1\rangle_\Gamma = -Q. \tag{18}$$

The following statements are valid.

Theorem 4. *Let* $\boldsymbol{A} \in W(\sigma;\Omega_C,\Omega)$ *is the solution of the problem (15) or (16). Then there is a function* $\varphi \in H^1(\Omega)$, *unique up to an additive constant, such that* $\boldsymbol{A}$, φ *is the solution of the problem (10), (11), (17). In case* $\omega = 0$ *there is a unique function* $\varphi \in H_0^1(\Omega)$ *such that* $\boldsymbol{A}$, φ *is the solution of the problem (10), (11), (18).*

Theorem 5. *Let* $\boldsymbol{A} \in W(\sigma;\Omega_C,\Omega)$ *is the solution of the problem (16). Then there is a unique function* $\varphi \in H^1(\Omega)$, *such that* $\boldsymbol{A}$, φ *is the solution .of the problem (10), (11), (17) and* $\varphi_C = -\kappa\operatorname{div}\sigma\boldsymbol{A}_C$. *In case* $\omega = 0$ *there is a unique function* $\varphi \in H^1(\Omega)$, *such that* $\boldsymbol{A}$, φ *is the solution of the problem (10), (11), (18) and* $\varphi_C = -\kappa\operatorname{div}\sigma\boldsymbol{A}_C$.

From theorems 4, 5 the next theorem follows

Theorem 6. *The problems (1)-(7) and (1)-(6), (8) (at* $\omega = 0$*) have unique solutions. Moreover the relations (9) are valid, where* $\boldsymbol{A}$ *is the solution of (15) or (16).*

Acknowledgments

The research was supported by the Ministry of Education and Science of the Russian Federation in the framework of contract no. 2014/134 for state work in the field of scientific activity in the framework of the basic part of the state contract (project no. 2664) and in the framework of the project part of the state contract (project no. 1727) for 2014–2016 and was supported by a grant in the framework of contract no. 02.V.49.21.0003 (August 27, 2013) between the Ministry of Education and Science of the Russian Federation and Nizhnii Novgorod State University.

References

[1] Alonso Rodriguez A and Valli A 2010 *Eddy current approximation of Maxwell equations* (Milan: Springer-Verlag Italia)
[2] Galanin M P and Popov Y P 1995 *Kvazistationarnye elektromagnitnye polya v neodnorodnykh*

sredah (Moscow: Nauka)
[3] Kalinin A V, Sumin M I and Tyukhtina A A 2016 *Differential Equations* **52 (5)** 1–17
[4] Meddahi S and Selgas V 2008 *App. Num. Math.* **58** 1061–1083
[5] Prato Torres R A, Stephan E P and Leydecker F 2010 *Computing* **88** 131–154
[6] Arnold L and Harrach B 2012 *SIAM J. Appl. Math.* **72 (2)** 558–576
[7] Albanese R and Rubinacci G 1998 Finite Element Methods for the Solution of 3D Eddy Current Problems *Advances in Imaging and Electron Physics* **102** ed. P W Hawkes (San Diego: Academic Press) pp 2–86
[8] Chen T, Kang T, Lu G and Wu L 2014 *Math. Meth. Appl. Sci.* **37** 343–359
[9] Preis K, Biro O and Ticar I 2000 *IEEE Trans. Magn.* **36** 840–843
[10] Fernandes P 1995 *IEE Proc.-Sci. Meas. Technol.* **142** 299–306
[11] Coulomb J-L and Sabonnadiere J-C 1985 *CAO en electrotechnique* (Paris: Hermes Publ.)
[12] Jackson J D 2002 *Am.J.Phys* **70** (9) 917–928.
[13] Biro O and Valli A 2007 *Comput. Meth. Appl. Mech. Engrg*. 2007 **196** 890–1904
[14] Fernandes P and Valli A 2008 *Math. Meth. Appl. Sci.* **31** 71–98
[15] Bossavit A 1999 *COMPEL* **18** (3) 323–336
[16] Kang T and Kim K I 2009 *Computing* **85** 339–362
[17] Kalinin A V and Kalinkina A A 2002 *Vestnik Nizhegorod. Gos. Univ. Mat. Model. Optim. Upravl.* **25 (1)** 95–106
[18] Kalinin A V and Chekmarev D T 2014 *Mesh Methods for Boundary-Value Problems and Applications. Proc. 10th Int. Conf.* (Kazan: Kazan University) pp 356–360
[19] Kalinin A V and Kalinkina A A 2003 *Vestnik Nizhegorod. Gos. Univ. Mat. Model. Optim. Upravl.* **26 (1)** 21–38
[20] Kobelkov G M 1991 *Computational processes and systems* **8** ed. G I Marchuk (Moskow: Nauka) pp. 204-236.
[21] Kalinin A V and Kalinkina A A 2004 *Russian Mathematics (Izvestiya VUZ. Matematika)* **48 (3)** 23–31
[22] Girault V and Raviart P 1986 *Finite element methods for Navier-Stokes equations* (N.Y.: Springler-Verlag)
[23] Bossavit A 1998 *Computational electromagnetism. Variational formulations, complementarity, edge elements* (Boston: Academic Press)

IOP Conf. Series: Materials Science and Engineering **158** (2016) 012047 doi:10.1088/1757-899X/158/1/012047

Composite mesh generator for CFD problems

E I Kalinin[1], A B Mazo[1] and S A Isaev[2]

[1] Kazan (Volga region) Federal University, 18 Kremlevskaya Street, Kazan 420008, Tatarstan, Russia
[2] Saint Petersburg State University of Civil Aviation, 38 Pilotov Street, Saint Petersburg 196210, Russia

E-mail: kalininei@yandex.ru

Abstract. In present paper a brief introduction of HybMesh grid generator which uses composite approach is given. The process of complicated area meshing using HybMesh generator consists of sequential building structured prototype grids in relatively simple geometry, mapping them to a non-regular domains and superposing to assemble resulting grid. Transitional areas between two superposed low level grids are filled with triangular cells. Currently only 2D algorithms of such approach are implemented; 3D grids can only be restored as a result of extrusion or revolution of 2D objects.

1. Introduction

Composite grid generation approach is based on meshing of given arbitrary domain by geometric union of lower level grids built in more primitive domains. Advantages of such approach are the simplicity of meshing domains with complicated geometry and convenient definition of appropriate mesh refinement. Furthermore, resulting grid is partly structured and this feature can be utilized for building robust numerical solution schemes. In current paper grid generator HybMesh which is based on composite approach is presented. Building a grid using HybMesh generator includes three basic steps: constructing structured prototype grids, mapping these grids to non-regular geometry (if necessary) and final superposition of low level grids into the final one. Resulting grid could be exported into vtk, Tecplot, gmsh or ANSYS Fluent file formats for further calculations.

HybMesh is a cross platform open source application which is written in C++ and Python programming languages. Currently HybMesh provides scripting interface based on python syntax. Latest version installation distributive, source code and documentation can be downloaded from git repository [1].

2. Algorithms

HybMesh 2D grid object is a set of cells which are defined as sequences of vertices. There is no restriction on the number of vertices used in a cell. All basic geometrical operations such as translation, scaling, reflecting are naturally supported. 2D grids could be clipped by a closed contour. 3D grids could be built by extrusion or revolution of 2D grid objects.

2.1. Grid superposition

This is the basic HybMesh operation. Generally it takes two independent grids which have non-zero domain intersection and composes them into a single grid. The domain of resulting grid is exactly equal to the domain of geometrical union of parent grid domains and its cells reflect the original grids cells everywhere except for a zone around the line of parent grids contact which is triangulated providing smooth cell size transition. This zone is later referenced as a buffer zone.

Order of superposition does matter. For clarity sake we call the first of two original grids the base grid and the second one – overlaid grid. Buffer is always built within the base grid. Cells of the overlaid grid are transferred to the resulting grid mostly untouched. So, as you can see in figure 1, by swapping the grid roles we obtain different resulting grid geometry.

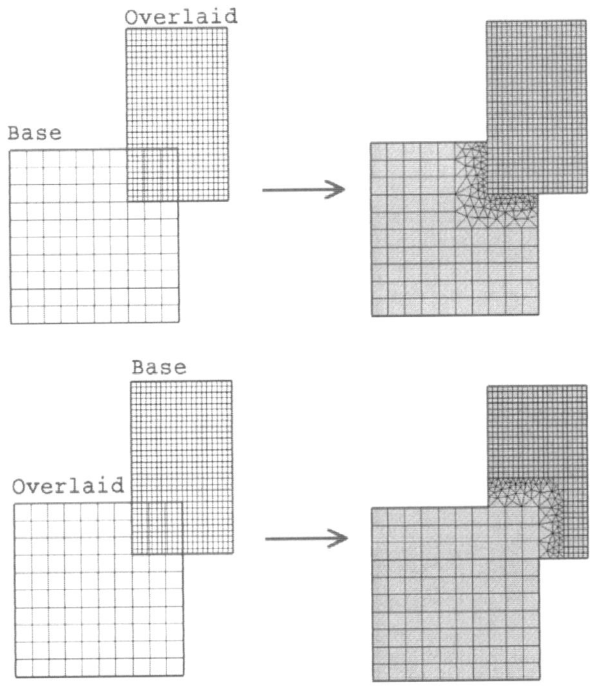

Figure 1. Basic superposition example.

Buffer zone is constructed as an area of all base grid cells which contain a vertex located no further than given buffer zone size from the contact line. Larger buffer zone provides smoother triangle grid within the buffer (see figure 2).

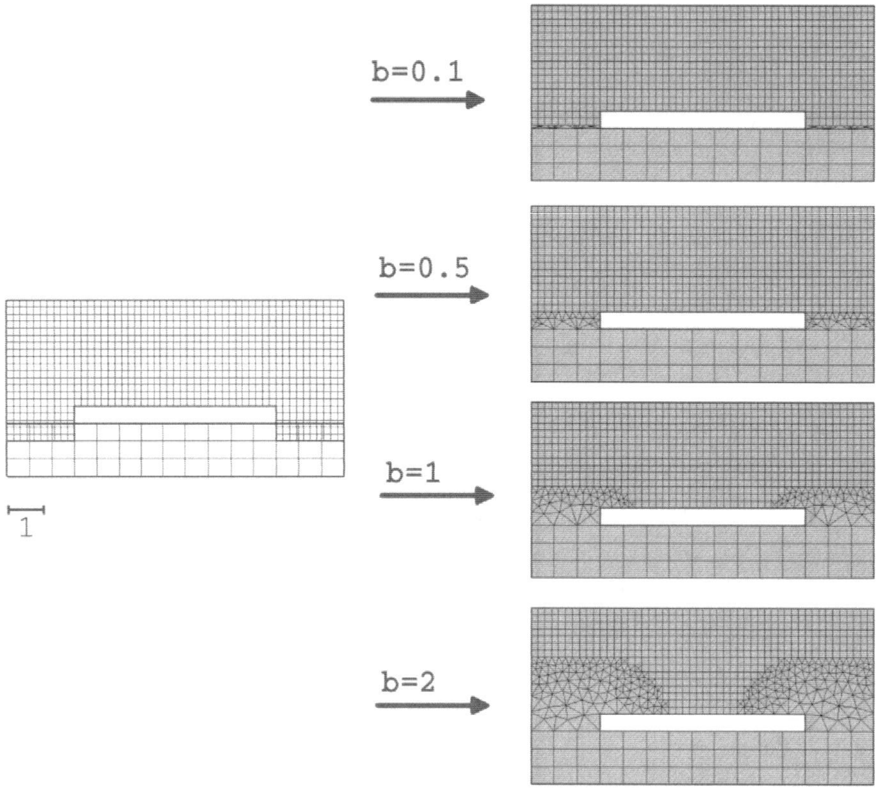

Figure 2. Superposition with different buffer sizes.

2.2. Mapping

This procedure maps the domain containing a grid (base domain) to any other domain with equal connectivity (target domain) and uses this mapping to translate the grid. Mapping is built as a result of solution of the Laplace equation with boundary conditions of the first kind [2]. Boundary values are calculated using domain's boundary mapping defined by user as a set of corresponding points at base and target boundaries. Boundary value problem is solved by a finite element method using auxiliary triangular grid built within a solution domain. There are no restrictions on the input grid and target domain except for their connectivities should be equal and boundary mapping for each of bounding contours should be provided. Example of the ring grid which was mapped to doubly connected domain could be seen in figure 3.

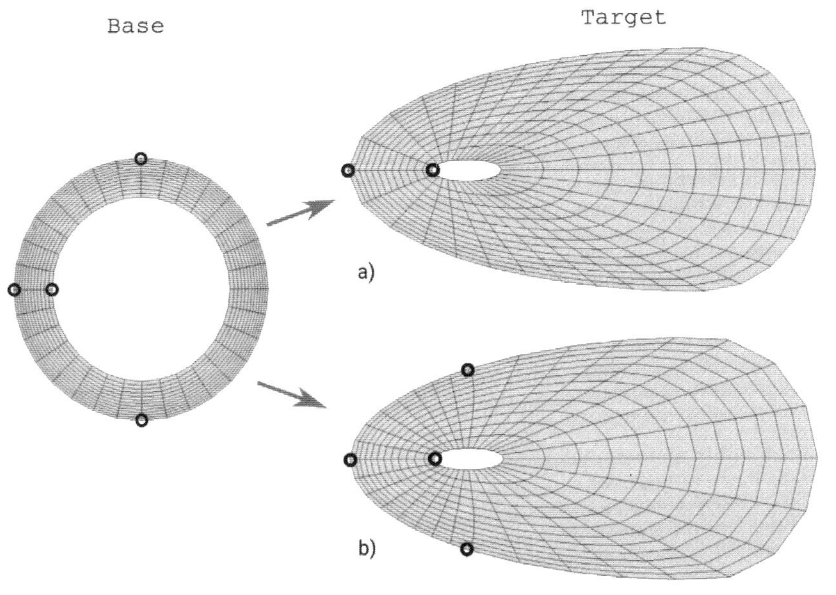

Figure 3. Mapping of ring grid using two (a) and four (b) reference points.

2.3. Grid prototypes

HybMesh provides the collection of instruments for building structured grid prototypes. This includes simple building blocks like structured quadrangular grid in rectangle, radial grid in circle/ring domains, structured grid in triangle domain, algorithms for building custom structured grid in curvilinear quadrangle or circle areas.

There is also a special algorithm for building a prototype grid around any given contour which allows user to define a mostly orthogonal boundary layer grid. Later this grid is used as any other low level grid so the connection of boundary grid with domain grid is implemented using general superposition algorithm.

3. Examples

Figure 4 illustrates a complicated shark-like multiply connected area meshed in HybMesh using three simple steps. First, grid prototypes for "body" and "fins" were built in appropriate triangular and hexagonal areas which were mapped to non-regular geometry. Then these grids were coupled using chain superposition procedure with non-zero buffer size. Finally, "gills", "eye" and "mouth" cuts were built as boundary layer grids and superposed on the basic grid.

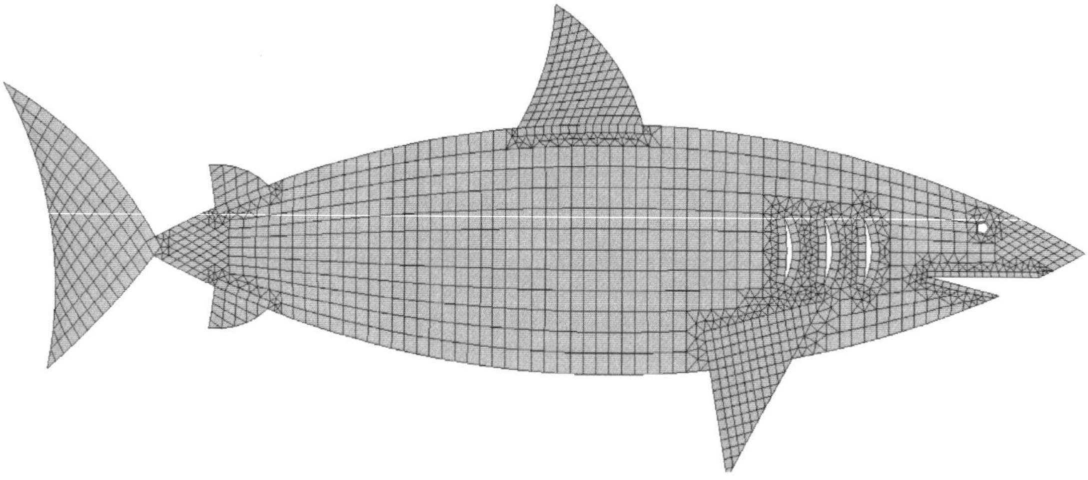

Figure 4. Result of chain superposition of multiple prototype grids.

A practical example of using HybMesh approach towards building a grid for typical CFD problem is shown in figure 5. Here external boundary layer grids were built around foil and generator contours. Then they were superposed on a structured substrate grid. Finally, a patch grid was added to the wake region of vortex generator to provide an appropriate cell size refinement.

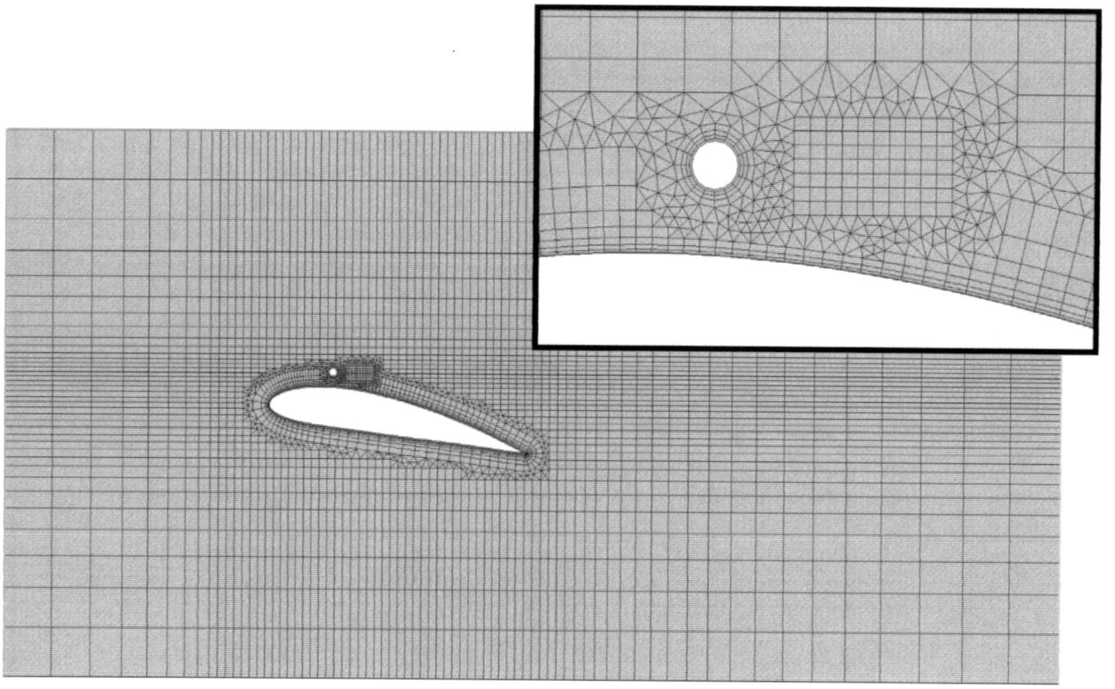

Figure 5. Airfoil with circular vortex generator.

4. Further development

Further HybMesh grid generator development involves several aspects. New grid prototypes including those widely used in CFD problems (potential grids, H-grids, C-grids etc.) will be added. 3D superposition algorithm using tetrahedral cells as buffer fillers will be implemented. It is also planned to provide HybMesh with graphical user interface and enhance the set of supported import/export mesh formats.

Acknowledgments

This work was supported by Russian Fundamental Research Foundation (RFFI) Grant No 15-01-06172 A.

References

[1] HybMesh latest release https://github.com/kalininei/HybMesh/releases/latest
[2] Liseikin V D 2009 *Grid Generation Methods (2nd ed.)* (Springer Publishing Company Incorporated) p 171

IOP Conf. Series: Materials Science and Engineering **158** (2016) 012048 doi:10.1088/1757-899X/158/1/012048

Error estimations of mixed finite element methods for nonlinear problems of shallow shell theory

M Karchevsky

Department of Computational Mathematics, Institute of Computational Mathematics and Information Technologies, Kazan Federal University, 18 Kremlyovskaya Street, Kazan 420008, Russia

E-mail: mkarchev@kpfu.ru

Abstract. The variational formulations of problems of equilibrium of a shallow shell in the framework of the geometrically and physically nonlinear theory by boundary conditions of different main types, including non-classical, are considered. Necessary and sufficient conditions for their solvability are derived. Mixed finite element methods for the approximate solutions to these problems based on the use of second derivatives of the bending as auxiliary variables are proposed. Estimations of accuracy of approximate solutions are established.

1. Introduction

In present paper we propose and investigate a new class of finite element methods for solution of variational problems that describe the equilibrium of shallow shell with non-classical boundary conditions. These methods allow to use of simple Lagrangian finite elements and have an optimal error estimations.

2. Statement of the problem

The problem of equilibrium of a shallow shell in the framework of the geometrically and physically nonlinear theory can be formulated [1] as the problem of minimization of the potential energy functional

$$F(u) = \int\limits_{\Omega} \varphi(\varepsilon, \varkappa) dx - \int\limits_{\Omega} f \cdot u dx - \int\limits_{\Gamma} g \cdot u dx \tag{1}$$

on the space of displacements $u = (u_1, u_2, u_3)$ satisfying the main (geometric) boundary conditions. Here $u = (u_1, u_2, u_3)$ are tangent displacements and u_3 is the bending, Ω is a bounded domain on R^2 identified with the middle surface of the shell, referred to the Cartesian coordinate system, Γ is the boundary of Ω, f, g are given vector-functions describing the density of the external loads. Tangential and bending deformations of the shell are determined by the following relations: $\varepsilon = \varepsilon(u) = \{\varepsilon_{ij}(u)\}_{ij=1}^{2}$, $\varkappa = \varkappa(u) = \{\varkappa_{ij}(u)\}_{ij=1}^{2}$,

$$\varepsilon_{ij}(u) = e_{ij}(u) + k_{ij}u_3 + \frac{1}{2}\frac{\partial u_3}{\partial x_i}\frac{\partial u_3}{\partial x_j}, \quad e_{ij}(u) = \frac{1}{2}\left(\frac{\partial u_i}{\partial x_j} + \frac{\partial u_j}{\partial x_i}\right), \quad i,j = 1,2, \tag{2}$$

$$\varkappa_{ij}(u) = -\frac{\partial^2 u_3}{\partial x_i \partial x_j}, \quad i,j = 1,2, \tag{3}$$

where k_{ij}, $i, j = 1, 2$ are initial curvatures of the shell. The real-valued function $\varphi(\xi)$, $\xi \in R^6$ is the density of the strain energy of the shell. It is assumed that the function φ is continuous and convex and there exist the constants c_0, c_1, $p \in (1, \infty)$ such that

$$c_0 \, |\xi|^p \leq \varphi(\xi) \leq c_1 \, |\xi|^p \quad \forall \xi \in R^6. \tag{4}$$

We introduce the Banach space $V = W_p^1(\Omega) \times W_p^1(\Omega) \times W_p^2(\Omega)$, where $W_p^l(\Omega)$, $l = 1, 2$, $p > 1$, is the Sobolev space. We denote by V_0 the subspace of the space V obtained by closure of the space of the functions $C_0^\infty(\Omega) \times C_0^\infty(\Omega) \times C_0^\infty(\Omega)$. By V_τ we denote the subspace of the space V obtained by closure of the set of functions $C^\infty(\Omega) \times C^\infty(\Omega) \times C_0^\infty(\Omega)$ satisfying the boundary condition $\bar{u}(x) \cdot \nu(x) = 0$, $x \in \Gamma$. Here ν is the unit normal field on Γ and $\bar{u} = (u_1, u_2)$ is the vector field of tangential displacements. Next, let V_ν be the subspace of the space V obtained by closure of the set of functions $C^\infty(\Omega) \times C^\infty(\Omega) \times C_0^\infty(\Omega)$ satisfying the boundary condition $\bar{u}(x) \cdot \tau(x) = 0$, $x \in \Gamma$, where τ is the unit tangential field on Γ.

We suppose that $f \in [L_q(\Omega)]^3$, $g = (g_1, g_2) \in [L_q(\Gamma)]^2$, $1/p + 1/q = 1$, $k_{ij} \in L_{p_1}(\Omega)$, $p_1 > p$.

The problem of minimization of the functional F on the space V_0 is called the Dirichlet problem (Problem I).

The problem of minimization of the functional F on the space V_τ it is natural to call the problem of hard contact of the border of the shell (Problem II).

The problem of minimization of the functional F on the space V_ν we call the problem of normal load on the boundary of the shell (Problem III).

3. Mixed finite element method

For each problem from I to III we propose mixed finite element methods of the type [2, 3] based on the use of second derivatives of functions as the auxiliary variables. Their approximate values $w_{ij}^h(y_3) \in H_l$ are defined by the following relations:

$$\int_\Omega w_{ij}^h(y_3)\eta dx = -\frac{1}{2}\int_\Omega \left(\frac{\partial y_3}{\partial x_i}\frac{\partial \eta}{\partial x_j} + \frac{\partial y_3}{\partial x_j}\frac{\partial \eta}{\partial x_i} \right) dx \ . \ \forall \eta \in H_l. \tag{5}$$

Here H_l is the finite element space, corresponding to polynomials of degree $l \geq 1$, and y_3 is the approximation of the bending. A characteristic feature of these schemes is that they are based on the simplest Lagrangian triangular or rectangular finite elements of the class C^0.

4. Investigation of mixed finite element method

The solvability of the corresponding discrete problems and convergence of their solutions can be establish under the same assumptions that are used in the justification of the solvability of the original variational problems [1].

Under additional assumptions on the function φ:

if $p \geq 2$ then

$$|\nabla\varphi(\xi) - \nabla\varphi(\eta)| \leq c_1|\xi - \xi|(|\xi| + |\eta|)^{p-2} \quad \forall \xi, \eta \in R^6,$$

$$(\nabla\varphi(\xi) - \nabla\varphi(\eta)) \cdot (\xi - \eta) \geq c_2|\xi - \eta|^p \quad \forall \xi, \eta \in R^6,$$

if $1 < p < 2$ then

$$|\nabla\varphi(\xi) - \nabla\varphi(\eta)| \leq c_3|\xi - \xi|^{p-1} \quad \forall \xi, \eta \in R^6,$$

$$(\nabla\varphi(\xi) - \nabla\varphi(\eta))(|\xi| + |\eta|)^{2-p} \cdot (\xi - \eta) \geq c_4|\xi - \eta|^2 \quad \forall \xi, \eta \in R^6,$$

c_1, c_2, c_3, c_4 are positive constants, providing the implementation of inequalities of the type of strong convexity of the energy functional and the smoothness of the solution of the original problem we can establish the error estimates of the approximate solutions. The following estimates, which generalize the estimates of [4], are typical:

$$\sum_{ij=1}^{2}\left\|\frac{\partial^2 u_3}{\partial x_i \partial x_j} - w_{ij}^h(y_3)\right\|_{L_p(\Omega)} + \sum_{i=1}^{2}\|u_i - y_i\|_{W_p^1(\Omega)} \leq ch^{(l-1)/(p-1)}, \quad p \geq 2,$$

$$\sum_{ij=1}^{2}\left\|\frac{\partial^2 u_3}{\partial x_i \partial x_j} - w_{ij}^h(y_3)\right\|_{L_p(\Omega)} + \sum_{i=1}^{2}\|u_i - y_i\|_{W_p^1(\Omega)} \leq ch^{(l-2/p)/(p-1)}, \quad 1 < p < 2.$$

Acknowledgments
This research was supported by Russian Foundation for Basic Research (projects 15-01-05686, 15-41-02315, 15-41-02569).

References
[1] Karchevsky M M 1991 *Differ. Uravn.* **27:7** 1196–1203
[2] Astrakhantsev G P 1989 *Zh.vychisl.Mat.mat.Fiz.* **29:10** 1492–1504
[3] Karchevsky M M 1992 *Izv. Vyssh. Uchebn. Zaved. Mat.* **7** 12–19
[4] Karchevsky M M, Zabotina L Sh 1995 *Math zapiski, Yakutsk, YaGU* **2:2** 121–139

Some results of mesh convergence estimation for the spectral element method of different orders in FIDESYS industrial package

V S Karpenko[1,2], A V Vershinin[1,2], V A Levin[1] and K M Zingerman[3]

[1]Lomonosov Moscow State University, Faculty of Mechanics and Mathematics, Russia, 119991, Moscow, GSP-1, 1 Leninskiye Gory, Main Building
[2]Fidesys LLC, Office 402, 1 bld. 77, MSU Science Park, Leninskie Gory, Moscow 119234
[3]Tver State University, Russia, 170100, Tver, 32 Zhelyabova

E-mail: vitkarpenko@gmail.com

Abstract. The article describes the Mesh convergence estimation for the method of spectral elements of different orders in FIDESYS industrial package. Estimation is done using the automated testing system developed for FIDESYS package. The tasks with known exact solutions were used for estimation. By changing the spectral element order the exponential decrease computational errors is shown against the degree of approximating polynomials.

1. Introduction

Using the method of spectral elements for sufficiently smooth functions one can show exponential decrease of computational error against the degree of the approximating polynomials.

In this paper, this feature is checked using the FIDESYS package with the help of automated testing system allowing to carry out a repeated calculation of the same problem while modifying some parameters (in this case, the degree of the approximating polynomials) for problems with known analytical solutions: Kirsch problem and Lame problem.

2. Two-dimensional case (Kirsch problem)

Let's examine a tension with tensile stress 1e6 N/m of square plate made of isotropic material with a Young's modulus of 2e11 Pa and a Poisson's ratio of 0.3, with a side length of 10 m and a hole with a diameter of 0.5 min the the center.

The main advantage of the method of spectral elements is rapid convergence of solutions to the exact solution with a fewer number of nodes if compared with the finite element analysis method.

In connection with this, for more clear demonstration of exponential convergence, a rather coarse mesh was built:

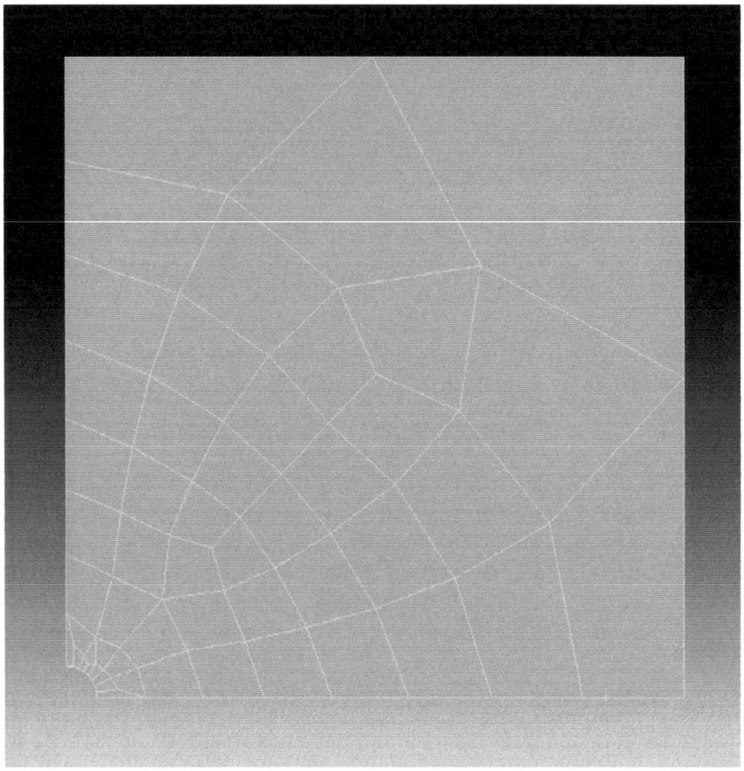

Figure 1. Mesh for Kirsch problem.

This problem is simulated using the FIDESYS package with the help of the following log file:

```
create surface rectangle width 10 zplane
create surface circle radius 0.25 zplane
subtract volume 2 from volume 1
webcut volume 1 with plane xplane preview
webcut volume 1 with plane xplane imprint nomerge
webcut volume 1 with plane yplane preview
webcut volume 1 with plane yplane imprint nomerge
delete volume 1
delete volume 3
curve 20  scheme bias fine size 0.4 factor 1.1 start vertex 18
curve 7  scheme bias fine size 0.4 factor 1.1 start vertex 8
curve 22  interval 8
surface 7  size auto factor 9
surface 7 scheme Auto
mesh surface 7
create displacement on curve 7 dof 1 fix 0

create displacement on curve 20 dof 2 fix 0
create pressure  on curve 11  magnitude -1e6
create material  elastic_modulus 2e11 poisson_ratio 0.3
set duplicate block elements off
block 1 surface 7
block 1  material 'Material 1'
```

block 1 element type QUAD8
analysis type static dim2

As it is known, in the polar co-ordinates a tension on the hole border should be three times greater than the tensile stress, i.e., should be equal to 3e6.

With the help of automated testing system the analysis of this problem was carried out with the help of finite element analysis method, and using the method of spectral elements with approximating polynomials degrees from 2 to 9.

Deliverables:

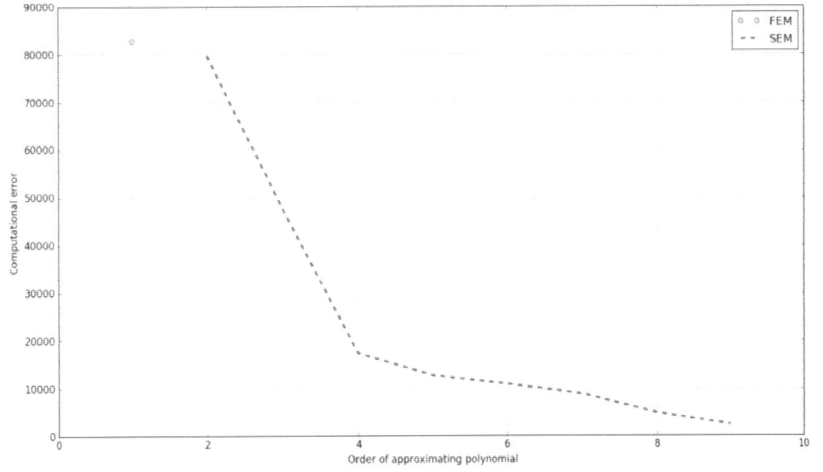

Figure 2. Computational errors against the approximating polynomials degree for Kirsch problem.

Yellow dot indicate the result for the finite element analysis method, the blue dotted line represents the results for the different degrees of the approximating polynomial for spectral elements method.

In order to verify that the computer error decreases exponentially, let's display the same results on the logarithmic axis.

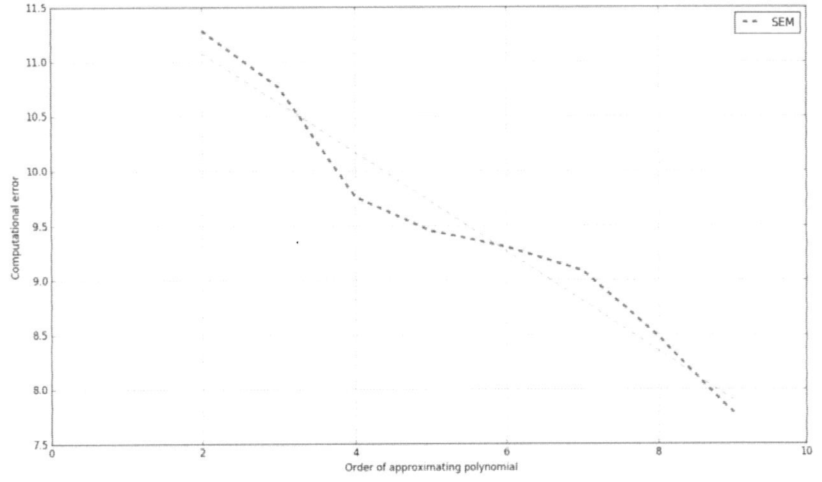

Figure 3. Dependence of computational errors on the degree of approximating polynomials for Kirsch problem; yellow dotted line represents a linear interpolation obtained by the method of least squares (logarithmic scale).

IOP Conf. Series: Materials Science and Engineering **158** (2016) 012049 doi:10.1088/1757-899X/158/1/012049

The resulting broken line, as expected, is quite close to straight line.

3. Three-dimensional case (Lame problem)

Let's examine a thick-walled cylinder with a height of 0.5 m and a radius of 2 m, a wall thickness of 1 m, made of isotropic material with a Young's modulus 2e11 Pa and a Poisson's ratio of 0.3, loaded with internal and external pressures of 1e6 and 0.5e6, respectively.

In this case, the generated mesh was even more coarse in order to demonstrate the exponential convergence.

This problem is simulated using the FIDESYS package with the help of the following log file:

```
create Cylinder height 0.5 radius 2
create Cylinder height 0.5 radius 1
subtract volume 2 from volume 1
webcut volume 1 with plane xplane offset 0 noimprint nomerge
webcut volume 1 with plane yplane offset 0 noimprint nomerge
delete Volume 1 3
volume 4  size auto factor 8
mesh volume 4
create Displacement  on surface 27  dof 2 fix
create Displacement  on surface 11  dof 1 fix
create Displacement  on surface 29 31  dof 3 fix
create pressure  on surface 30  magnitude 1e6
create pressure  on surface 28  magnitude 0.5e6
create material  elastic_modulus 200e9 poisson_ratio 0.3
set duplicate block elements off
block 1 volume 4
block 1  material 'Material 1'
block 1 element type HEX20
analysis type static dim3
```

One can check, e.g. stress σ_{zz} at the inner boundary of the wall (in cylindrical coordinates). In accordance with known analytical solution, it must be equal to $\sigma_{zz} = 2\nu\frac{p_a a^2 - p_b b^2}{b^2 - a^2} = -200{,}000$, Where a - inner radius, b - outer radius, p_a and p_b - internal and external pressures, respectively, and ν - Poisson's ratio.

In this case, the results are as follows:

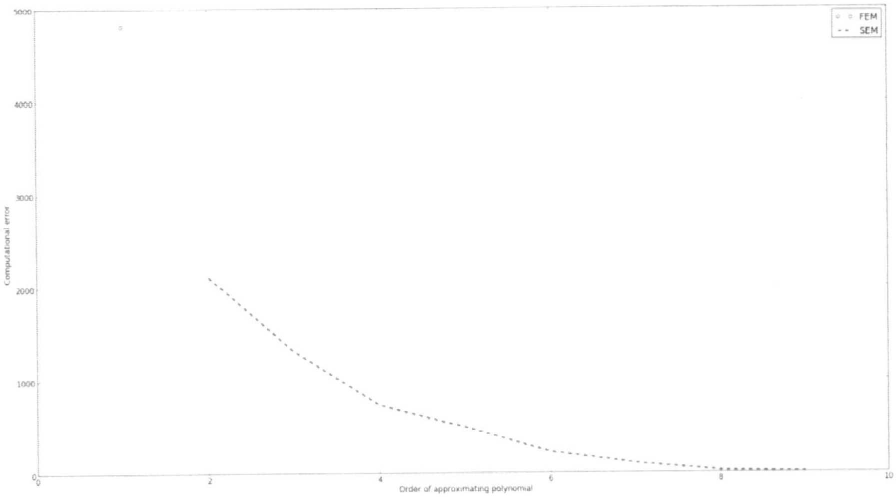

Figure 4. Computational errors against the approximating polynomials degree for Lame problem.

Now let's just use a logarithmic scale:

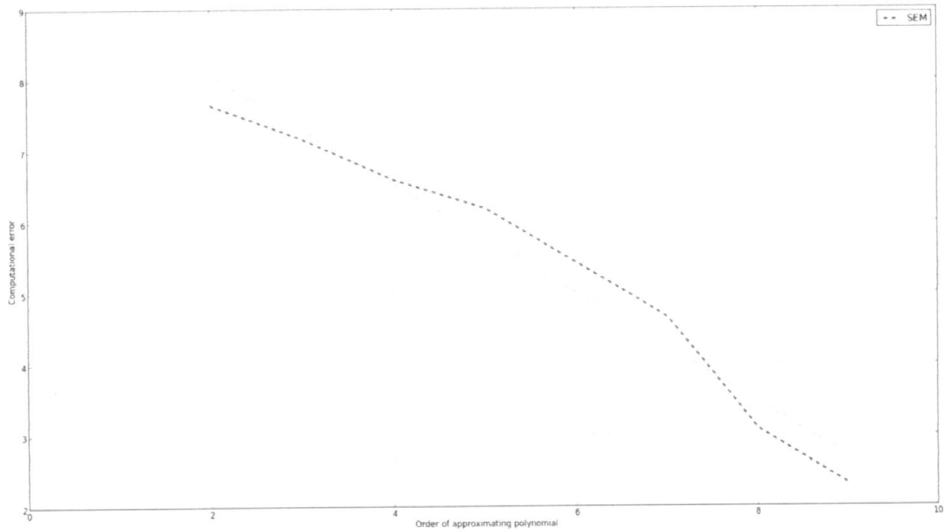

Figure 5. Dependence of computational errors on the degree of approximating polynomials for Lame problem; yellow dotted line represents a linear interpolation obtained by the method of least squares (logarithmic scale)

In this case, the proximity of the resulting broken line to straight line is even more obvious than it is for the two-dimensional case.

4. Conclusion

The article describes the checking of exponentially fast decrease of computational errors for the spectral element method depending on the degree of the approximating polynomials using the FIDESYS package for problems with known analytical solutions: the Kirsch problem and the Lame problem.

The obtained results confirm the exponential rate of convergence to the exact solution.

Acknowledgments
The research for this article was performed in FIDESYS LLC and was financially supported by Russian Ministry of Education and Science (project №14.579.21.0007, project ID RFMEFI57914X0007).

References

[1] Grinchenko V T, Ulitko A F, 1970 Numerical analysis of effective mechanical properties of rubber-cord composites under finite strains *Soviet Applied Mechanics* **6(5)** pp 455–460.
[2] Barashkov V N 2015 *Solution of axisymmetric plane problem of the theory of elasticity and plasticity for solids of revolution with elastic and elastic-plastic deformations* [in Russian] (TSUAB: Tomsk, Russia) 35-38 pp.
[3] Fidesys LLC official website. Available online: http://cae-fidesys.com

Development of film- and- fabric composite materials durability assessing methodology under time-dependent influences of temperature and solar radiation

R A Kayumov[1], I Z Muhamedova[1], A M Suleymanov[1], B F Tazyukov[2]

[1]Kazan State University of Architecture and Enginieering, 1 Zelenaya Street, Kazan, 420043, Russian Federation
[2]Kazan Federal University, 18 Kremlyovskaya Street, Kazan, 420008, Russian Federation

E-mail: muhamedova-inzilija@mail.ru

Abstract. In this paper, we present the design of stress-strain state calculation and film-and-fabric composite materials durability under stresses and solar radiation. We have constructed a two-dimensional finite-state-element computer model of the deforming process of the low-level cell of film-and-fabric-based composite material for the evaluation of its durability which takes into account non-linear viscoelasticity, temperature variations, ageing of the material, the process of upbuilding of microdamage and photodegradation. Qualitative research of operational factors influence (UV, temperature) on film-and-fabric composite materials durability was conducted.

1. Introduction

Nowadays pneumatic and canvass structures made of film-and-fabric composite materials are widely used. With the advent of new polymeric materials the diversity and application area of such structures increase. Advantages of film-and-fabric composite materials structures are low investment costs, low material consumption, mobility and quick installation.

Film-and-fabric composite materials is a composite with fabric reinforcing foundation of high-strength synthetic yarns and film coating of elastomers or thermoplastics, protecting the reinforcing foundation from weather conditions and giving material air permeability.

Analyzing the application of mobile structures it turns out that unlike traditional structures, the greatest efficiency of their use is determined by not so much maximal as economically proved optimal service life. For that matter the problem of film-and-fabric composite materials creating not only with maximal, but also with preset durability under certain conditions of operation arises. Currently, one of the most important issues is fabric composite durability evaluation considering the aging of material under weather conditions, temperature and UV irradiation.

To assess the long-term strength of the given material we need to know the stress-strain state of each composite component on a scale of fabric unit cell [2]. Obviously, because of the small size of film-and-fabric composite materials material cross section (0.8-3 mm) solving this problem with methods and means of full-scale strain measurement or other empirical methods only is not possible. Clearly, creating optimal structures, therefore optimal property performance requires development of film-and-fabric composite materials structurally-simulated model adequately imitating real material [1,4,5]. Creating such models is possible using numerical methods only, focused on modern computers with advanced visual system. The term structurally-simulated model means finite element model of composite unit cell, where physical, mechanical and geometrical parameters of structural components vary.

2. Structure defining (physical) relations for film-and-fabric composite materials component considering static, kinematic and structural parameters of the material with ultraviolet radiation

When choosing a structure defining relations we take into account the well-known experimental facts concerning polymeric materials. First, the film-and-fabric composite materials matrix is assumed to be isotropic. Defining relations for the aging viscoelastic material are taken as:

$$\varepsilon = S\sigma + \int_0^t H(t,\sigma,\omega,W,T,v,...)d\tau\,, \tag{1}$$

Here ω, W – some parameters of the deformation process, v – vector of structural parameters (specific volume of various types of modifying additives, regulating mechanical and operational characteristics of the film-and-fabric composite materials matrix – plasticisers, fillers, stabilizers, etc.) .

To determine the model of all possible kinds of parametres, consider the following. First, we use damage parameter ω which describes the accumulation of material defects like microcracks, micropores. For ω we take up the kinetic equation:

$$d\omega / dt = \Omega(\sigma,\omega,T,W,v,...),... \tag{2}$$

Further, affected by external non-bearing aggressive action, UV radiation in particular, phase transformation and polymer matrix mechanical properties changes occur which is called material degradation (UV affection causes photodestruction). As a result of the secondary reactions the process develops inside the material on some layer of height h, and it comes from the surface side exposed to radiation. Microcracks appear on the surface and increase in size, which leading to h layer height increasing. In this regard, we introduce the scalar parameter W, which we call parameter of photodegradation, considering it proportional to the exposure intensity γ. For it, we take evolutionary equation as a constitutive relation:

$$dW / dt = U(\sigma,\omega,T,W,h,...),... \tag{3}$$

Fick's law is not used due to the fact that as we stop radiation process photodegradation also stops. For simplicity we assume that the radiation surface is flat, the penetration of photodegradation inside the material will be describe by equation similar to (3):

$$dh / dt = R(\sigma,\omega,T,W,h,...),... \tag{4}$$

Material stiffness characteristics, included in matrix D=S-1, at a first approximation can be assumed as time-related t due to aging of structural parameters and material, accumulation of microdamages, temperature, photodegradation parameter:

$$D = D(\omega,W,T,v,t). \tag{5}$$

Material strength condition will be described by the following equation:

$$f(\sigma, \varepsilon, \omega, W, g, T) = 1, \tag{6}$$

here g – structural parameters including, strength limit or microcrack's typical length in particular.

In this paper, to conduct the qualitative analysis of the given composite behavior, task was formulated simpler. First, the strength of the matrix was investigated, because its destruction causes destructions of the fabric due to UV access to foundation. Secondly, it was assumed that film-and-fabric composite materials element is flat and deformed. Thirdly, strains and displacements were minor.

3. Development of finite element analysis method of representative film-and-fabric composite materials cell deformation process

Film-and-fabric material regular structure allows to allocate as a representative element for modeling material only one cell (Figure 1) formed by two adjacent pairs of warp yarns (**1**) and weft (**4** and **5**). Layers between matrix and yarns are also introduced in this model (**2**, **3**, **6**, **7**).

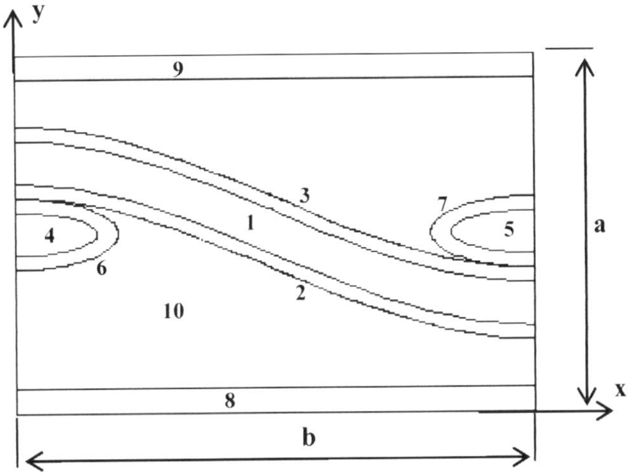

Figure 1. Geometry of common film-woven material two-dimensional cell

Geometry of the basis is totally determined by two components - the upper and lower contours changing according to cosine law. Sectional shape of the weft in presented structural cell is an ellipse. Light protecting coating which is applied directly to the matrix (**10**), also presented as separate fields (**8** and **9**) in composite cell structure.

Space wise problem discretisation is carried out by finite elements method, consisting of 6-noded triangular elements with quadratic approximation of displacement [3]. Following boundary conditions were taken for the cell. Surface of the element $y=a$ is irradiated with ultraviolet. On the left end when x=0 there are no displacements along the axis x. At the point with coordinates (0,0) there is no any displacement along the axis y. The upper and lower bounders ($y=0$ and $y=a$) are neither loaded nor fixed. Right end $x=b$ is given displacement $u_x = \Delta b$. In time, stress relaxation occurs, but as a result of irradiation and microdamages at some point strength can be broken. This value we will $t=t^*$ call durability further in the paper.

Total strain vector ε considering viscoelastic properties of phases of microdamades accumulation process is written as (further the vectors are in curly braces and matrices in square):

$$\{\varepsilon\} = \{e\} + \{\varepsilon^c\} + \{\varepsilon^\omega\}, \tag{7}$$

where $\{e\}$ is the vector of the elastic of deformation, $\{\varepsilon^C\}$ is creep strain vector, $\{\varepsilon^\omega\}$ is vector of deformations arising from scattered damage accumulation. Elastic law in numerical calculations was taken as linear:

$$\{\sigma\} = [D]\{e\}. \tag{8}$$

For creep strain $\{\varepsilon^c\}$, according to Kachanov's hypothesis the following constitutive relation was accepted:

$$\varepsilon^{\tilde{n}} = \int_0^t H[t-\tau, \sigma_i(\tau)]\sigma_i(\tau)d\tau \tag{9}$$

To provide flat strain state, further it was assumed that matrices $D^{-1}$and H are proportional:

$$H = \lambda(t-\tau,\sigma_i)D^{-1} \tag{10}$$

In numerical experiments, the function λ was taken as generalization of Abel's kernel:

$$\lambda(t-\tau,\sigma_i) = \frac{C(\sigma_i)}{(t-\tau)^{\alpha(\sigma_i)}}, \qquad C \geq 0,\ 0<\alpha<1; \tag{11}$$

Creeping parameters approximated as follows:

$$\begin{cases} C = (c_0 + c_1\sigma_i)^2 \\ \alpha = 1 - \dfrac{1}{\sqrt{1+(\alpha_0+\alpha_1\sigma_i)^2}} \end{cases} \tag{12}$$

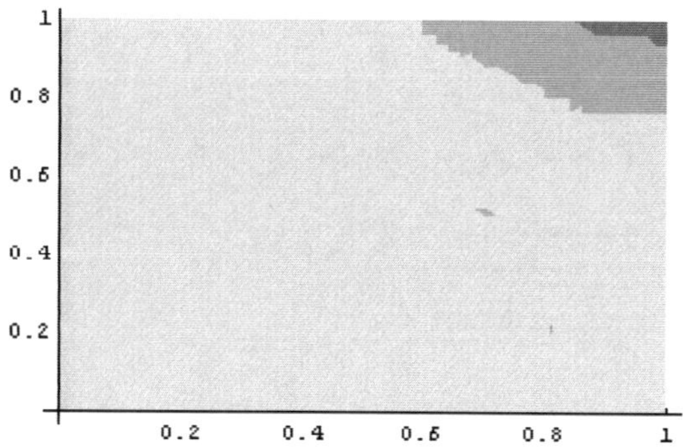

Figure 2. Damage parameter distribution ω in film-and-fabric composite materials cell at a time t*/3.

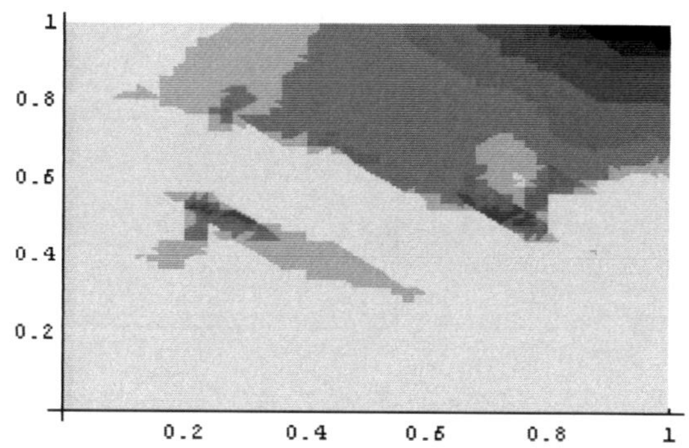

Figure 3. Damage parameter distribution ω in film-and-fabric composite materials cell before destruction.

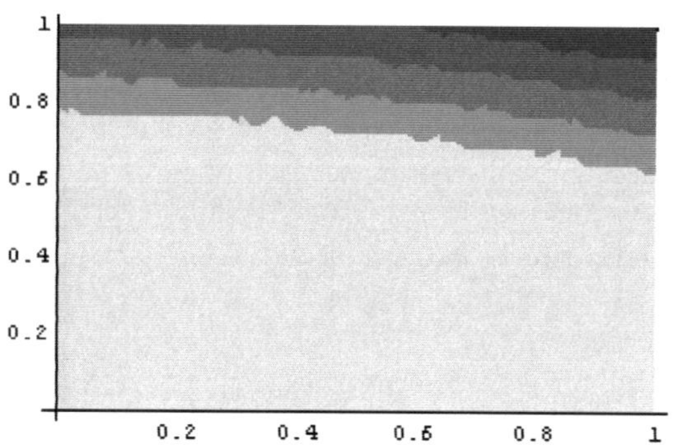

Figure 4. Photodegradation distribution level in film-and-fabric composite materials cell at a time t* / 3.

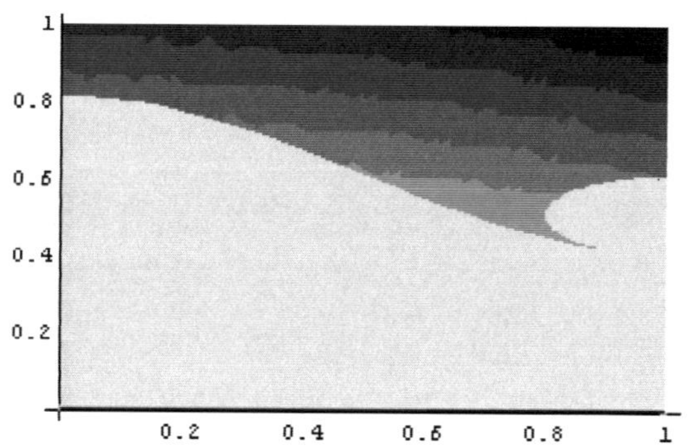

Figure 5. Photodegradation distribution level in film-and-fabric composite materials cell before fracture.

Based on PVC film coating full-scale analysis we get the following matrix creep parameters $c_0 = 0.096\ (\sqrt{c}\,)^{-1}$, $c_1 = 0.026\ (MPa*\sqrt{c})^{-1}, \alpha_0 = 9.15$, $\alpha_1 = 2.402$.

Figure 2 shows the damage parameter distribution ω in film-and-fabric composite materials cell at time $t^*/3$, and figure 3 – shows the sate before material destruction. Figure 4 shows photodegradation distribution level at a time $t^*/3$, and figure 5 – shows the state before fracture.

4. Solutions under unsteady temperature variations

This section describes the results of a qualitative research of influence patterns of temperature variations on durability of film-and-fabric materials. A year round temperature variation formula can be approximated as follows:

$$T = T^{**} + \Delta T \cdot Sin(\omega t). \tag{13}$$

Consequently (see 13), the temperature variation diagram is represented in fig 6.

Durability prediction calculations were based on the linear addition of damageability principle which states that damage occurs under the following conditions [6]:

$$\int_0^{t^*} \frac{dt}{t^*(T)} = 1. \tag{14}$$

Considering that durability t^* is determined numerically, thus, (14) was substituted for an approximate ratio:

$$\sum_i \Delta t_i / t_i^* = 1 \tag{15}$$

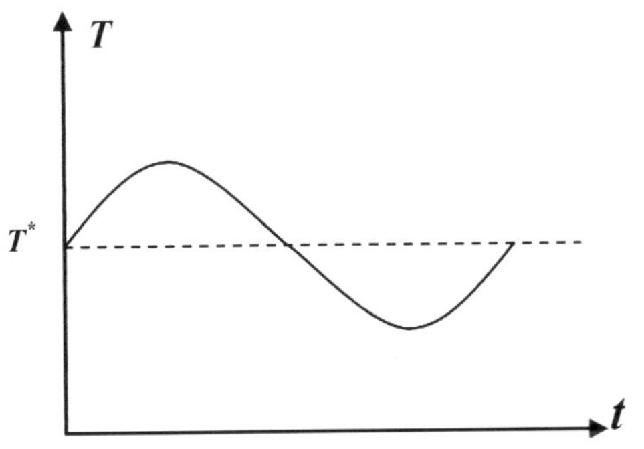

Figure 6. Time Temperature changes law.

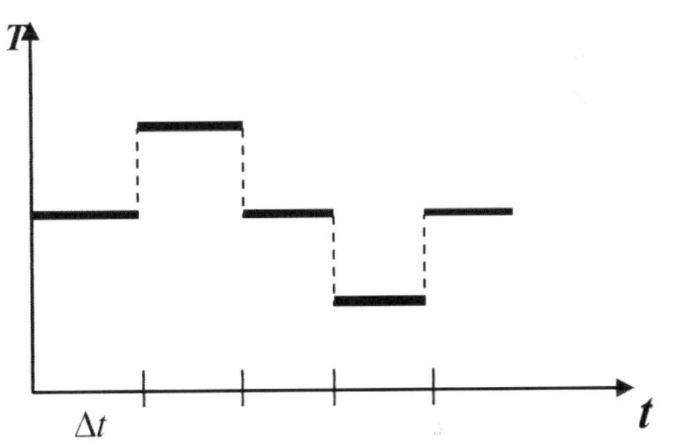

Figure 7. Jump function temperature approximation.

Suppose Δt is the time step estimated, for instance, as three months of spring, three months of summer, three months of autumn and three months of winter. Thus, temperature variations occur four times a year. Consequently, T is approximated by a step function as is shown on fig. 7.

Therefore, the problem can be written as follows (15):

$$N_{years} \cdot \left(\frac{\Delta t}{t^*_{spring}} + \frac{\Delta t}{t^*_{summer}} + \frac{\Delta t}{t^*_{autumn}} + \frac{\Delta t}{t^*_{winter}}\right) = 1, \tag{16}$$

where N_{years} is the unknown number of years which determines the durability of the composite material in question given a specific temperature variation ΔT.

The formation of defining correlations should be based on the known experimental facts referred to polymeric materials, due to the fact that mechanical properties that constitute creep flow ratio (19), will depend on the temperature T. Let us introduce as an example, the following function f_1, that determines the law of pattern change of mechanical properties of creep flow depending on the temperature:

$$f_1 = \left(f_0 \cdot \left(Arctg\left[\left(T - T^{**} + \frac{tg\left[\left[\frac{1}{f_0} - 1\right] \cdot \frac{\pi}{2}\right]}{\theta_1}\right) \cdot \theta_1\right] + \frac{\pi}{2}\right) / \frac{\pi}{2}\right)^2, \tag{17}$$

here, the constants f_0, θ_1 determine the nature of changes in creep flow parameters depending on the temperature. Mechanical properties of the creep flow, included in the ratio (12), as shown in figure 8, are equated as follows:

$$c_0 = c_{01} \cdot f_1(T), \quad c_1 = c_{11} \cdot f_1(T), \quad \alpha_0 = \alpha_{01} \cdot f_1(T), \quad \alpha_1 = \alpha_{11} \cdot f_1(T). \tag{18}$$

Here, $c_{01}, c_{11}, \alpha_{01}, \alpha_{11}$ are the mechanical parameters that correspond to the average operating temperature T^{**}. Figure 8 illustrates the variation of c_0 with T.

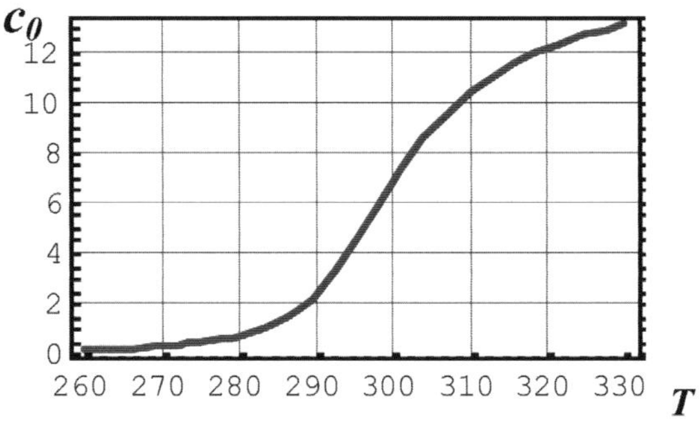

Figure 8. Mechanical Creeping Properties depending on temperature T.

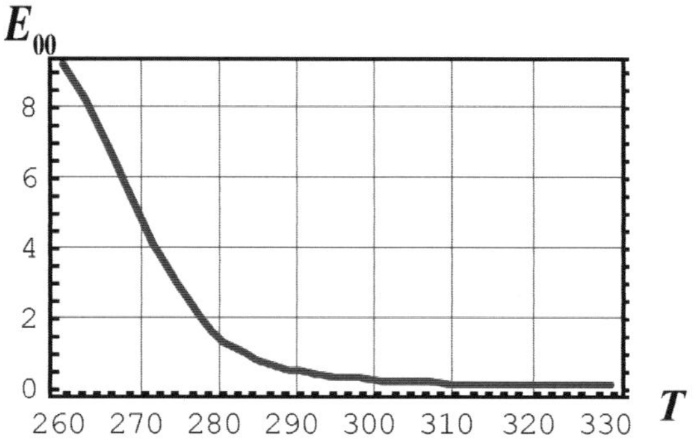

Figure 9. Modulus elasticity changes E_{00} depending on temperature T.

In much the same way as the above-said, coefficient of elasticity is also temperature-dependent. The equation f_2 illustrates its variations:

$$f_2 = (f_0 \cdot (-Arctg[(T - T^{**} - \frac{tg[[\frac{1}{f_0} - 1] \cdot \frac{\pi}{2}]}{\theta_2}) \cdot \theta_2] + \frac{\pi}{2}) / \frac{\pi}{2})^2 \tag{19}$$

Here, f_0, θ_2 are the parameters that determine the nature of elasticity coefficient variations depending on the temperature. Parameter E_{00} shall be equated as follows:

$$E_{00} = \tilde{E}_{00} \cdot f_2(T), \tag{20}$$

where $\tilde{E}_{00}$ is the Young's modulus, which corresponds to the average operating temperature T^{**}. Figure 9 illustrates the variations of E_{00} with T.

The mechanical property C_B for the damage parameter ω, shall be similarly (19) represented as follows:

$$C_B = C_{B0} \cdot f_3(T), \tag{21}$$

where C_{B0} is the parameter which corresponds to the average operating temperature T^{**}.

For the sake of intensification, UV-radiation γ shall be similarly represented as follows:

$$\gamma = \gamma_0 \cdot f_4(T). \tag{22}$$

Next, we carried out numerical experiments. For the sake of simplicity, we assumed that $f_3 = f_1$, $f_4 = f_2$. The calculations were made on the assumption of smallness of deformity in accordance with the developed model. $T^{**} = 283\ {}^0K$ was taken as the average equivalent temperature on the Kelvin temperature scale.

Figures 10 - 11 represent the diagrams of dependence of durability t^* on temperature variations ΔT with different positions of the right-side abutting end.

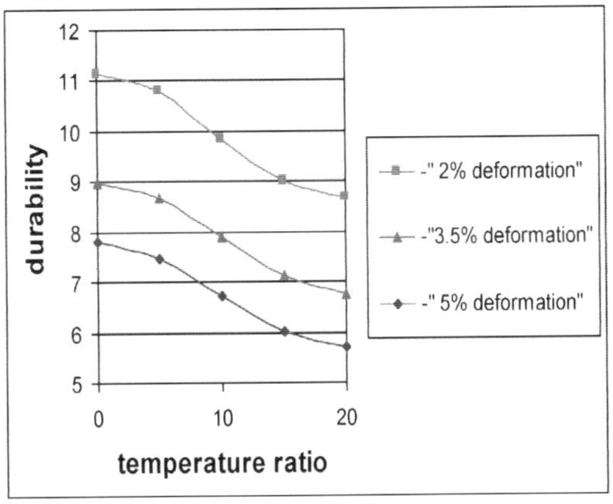

Figure 10. Durability function t^* of temperature changes ΔT for different deformations ε under $\theta_1 = 0.045$, $\theta_2 = 0.045$

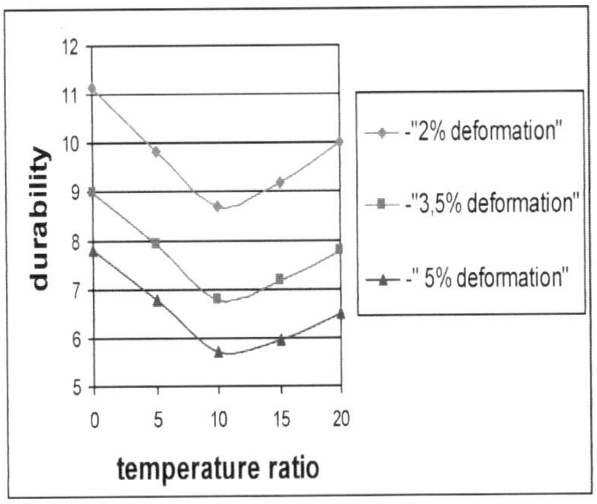

Figure 11. Durability function t^* of temperature changes ΔT for different deformations ε under $\theta_1 = 0.18$, $\theta_2 = 0.18$

5. Results and discussion

Conducting numerical experiments we studied the effect of geometrical parameters on the durability of fabric composite cells. Numerical experiments revealed some expected and unexpected results. The following principles were obtained. As it was predicted the higher the film-and-fabric composite materials thickness (under constant reinforcing foundation and weft flattening) the higher t^* durability. This can be determined by the increasing of matrix thickness above and below the weft. As step of weaving also increases the longevity increases too. The bigger the shift of the right-side abutting end, the lower the durability, exceedingly at that.

Due to modifiers mechanical properties of the matrix and light shielding layer can be changed, for example, it can influence matrix sensitivity and the light shielding layer to photodegradation. It is possible to vary the coefficients $E_{00}, c_0, c_1, \alpha_0, \alpha_1$, characterizing the viscosity and stiffness of the matrix and light shielding layer in the formula (19), and the modulus of elasticity and Poisson's ratio of the warp and weft. Therefore further durability was investigated by means of varying the mechanical characteristics of film-and-fabric composite materials structural components. Other cases of different sensitivity to photodegradation and light shielding layer matrix were studied.

Studies showed that stiffness reduction of the surface layer or matrix increases the durability. But increasing the warp or weft stiffness causes a slight decrease in durability.

We also revealed the following qualitative regularities of temperature-dependent ΔT changes in durability t^*. Temperature variations result in slight changes in elasticity properties and creep flow parameters; an increase of ΔT results in a decrease of durability t^*. However, with sufficiently high θ_1 and θ_2 durability starts to increase from a certain value of ΔT, as is shown in figure 11. This is attributable to the fact that with higher parameters of θ_1, θ_2 and larger ΔT, a material starts to soften so fast that damages in material accumulate slower. Thus, our numerical experiments led to scope for structure optimization and this type of composites durability increase.

Acknowledgments

The reported study was supported by RFBR research projects No. 15-08-06018.

References

[1] Kayumov R A, Muhamedova I Z and Kovtunova T I 2012 *News of the KSUAE* **4 (22)** 296–301

[2] Kayumov R A, Abdullin I C, Serasutdinov M N, Hammatova V V., Muhamedova I Z, Kumpan E V and Chakirova A M 2010 *Herald of Kazan Technological University* **6** 42–53

[3] Obrazcov I F, Saveljev L M and Hazanov H S 1985 *The finite element method in problems of aircraft structural mechanic* 329–64

[4] Boisse P, Gasser A, Hagege B and Billoet J L 2005 *Int. Journal of Material Science* **40** 5955-5962

[5] Heß H, Roth Y C and Himmel N 2007 *Composites Science and Technology* **67** **(6)** 1081–1095

[6] Teregulov I G 1984 *Strength of materials and fundamentals of the theory elasticity and plasticity* 472-178

Nonlinear problems of stability cylindrical panels with imperfection

R A Kayumov[1], B F Tazyukov[2] and I Z Muhamedova[1]

[1]Kazan State University of Architecture and Engineering, 1 Zelenaya Street, Kazan, 420043, Russian Federation
[2]Kazan Kazan Federal University, 18 Kremlyovskaya Street, Kazan, 420008, Russian Federation

E-mail: Bulat.Tazioukov@kpfu.ru

Abstract. The article examines a problem of large deflections and elastic jumps from one equilibrium state to another in a thin-walled structure in the form of a thin elastic cylindrical panel with imperfections in the form of a small-scale additional wave formation. An analytical solution of the problem was obtained and an analysis of the influence of geometric characteristics on the panel deformation process was made.

1. Basic relations

We investigate thin elastic shallow cylindrical panel with imperfections in the form of n half-waves, subjected to a transverse load P (Figure1, a panel without imperfections is shown by dash lines). We believe that the force P is applied with some certain eccentricity ξ_0 and that the wave formation is represented according to the law $a_1(\cos(\pi n\xi/2)-\cos(\pi n/2))$. We search for dependence of the w panel moving due to a transverse linear load P, as well as impact of the wave formation on the critical load.

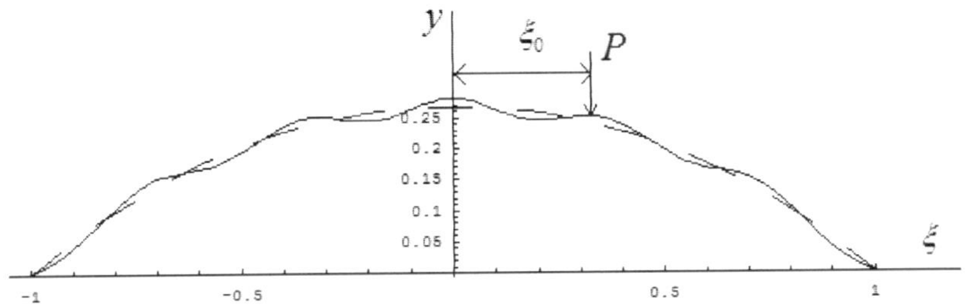

Figure 1. n=11, $a_1/a_0=0.1$, $r_0/b=2$.

The following dimensionless values are introduced,

$$\xi=\frac{x}{b};\; a=\frac{a_0}{b};\tilde{a}=\frac{a_1}{b};\quad q=\frac{Pb^2}{D};\; D=\frac{2Eh^3}{3\left(1-\nu^2\right)};\; w=\frac{W}{b};\; v=\frac{V}{b},$$

where $2h$ and $2b$ – are thickness and length of the panel; W, V – are moving along the normal and in the tangent directions regarding the panel; E – is a Young module of the material; a_0 – is an arrow of the panel rise; a_1 – is an amplitude of the wave formation.

The equation of the bent axis is represented as [1]:

$$y=\sqrt{\left(\frac{r_0}{k}\right)^2-\xi^2}-\frac{\sqrt{r_0^2-k^2}}{k}+\frac{a_1}{k}\left(\cos(\pi n\xi/2)-\cos(\pi n/2)\right),\quad k=\sqrt{r_0^2-(r_0-a_0)^2}\,, \tag{1}$$

where r_0 – is a radius of the cylindrical panel.

The arch element equilibrium equations take the form [2]:

$$T^{'}-\frac{b}{r}N=0,\ N^{'}+\frac{b}{r}T=0,\ G^{'}-bN=0, \tag{2}$$

where dash lines from now on mean a derivative with respect to dimensionless coordinate ξ, T and N – is tangential and transverse forces, G – is a moment of deflection.

Because of the flatness of the arch we can accept [2]:

$$T=const;\ \aleph\approx\frac{W^{''}}{b^2}=\frac{w^{''}}{b^2}.$$

It follows from the second and the third equations (2) that:

$$G^{''}-\mu^2\frac{D}{r}=0, \tag{3}$$

where $\mu^2=-T\dfrac{b^2}{D}$ – is a dimensionless compressive force.

The equations for each element take the form (3), but for the right and left parts of the panel their solutions will be different because of the load P.

The relation between the bending moment and deflection for the elastic material takes the form:

$$G=-D\aleph=-\frac{D}{b}\ddot{w}.$$

The relation between the curvature of the panel and its deflection:

$$\frac{1}{r}=(y+wb)^{''}_{\xi\xi}\frac{1}{b^2}=\frac{1}{r_0}+w^{''}\frac{1}{b}. \tag{4}$$

Substituting the expression for G into the equation of equilibrium (3), we obtain:

$$w^{''''}+\mu^2 w^{''}+\mu^2\left(\frac{b}{r_0}+n^2\tilde{a}\cos\frac{n\pi\xi}{2}\right)=0. \tag{5}$$

2. Solutions

The general solution of the equation (5) takes this form

$$w=B_1\cos\mu\xi+B_2\sin\mu\xi+B_3\xi+B_4+f(\xi),$$
$$f(\xi)=4(-\mu^2)\left(\frac{a\cos\frac{\pi\xi}{2}}{\pi^2-4\mu^2}+\frac{\tilde{\alpha}}{n^2\pi^2-4\mu^2}\cos\frac{n\pi\xi}{2}\right), \tag{6}$$

or the matrix form:

$$w=[B]\{X\}+f(\xi);$$
$$[B]=[B_1,\ B_2,\ B_3,\ B_4]\ ;\ \{X\}^T=\{\cos\mu\xi,\ \sin\mu\xi,\ \xi,\ 1\}. \tag{7}$$

The unknown coefficients B_i are found from the boundary conditions. Let us denote a deflection in the left and right parts from the point of force application P as follows:

$$w_l = [C]\{X\} + f(\xi)\ ;\ w_r = [B]\{X\} + f(\xi). \tag{8}$$

Next, let us write the conditions of docking at the point of force application

$$\begin{gathered} w_l(\xi_0) = w_r(\xi_0)\ ;\ w_l''(\xi_0) = w_r''(\xi_0)\ ; \\ w_l'(\xi_0) = w_r'(\xi_0)\ ;\ w_l'''(\xi_0) = w_r'''(\xi_0) - q. \end{gathered} \tag{9}$$

Let us rewrite the conditions of docking with $\xi = \xi_0$ in the matrix form

$$[H]\{C\} = [H]\{B\} + \{0,0,0,-q\}^T,\ \{C\} = [C]^T,\ \{B\} = [B]^T,$$

$$[H] = \begin{bmatrix} \cos\mu\xi_0 & \sin\mu\xi_0 & \xi_0 & 1 \\ -\mu\sin\mu\xi_0 & \mu\cos\mu\xi_0 & 1 & 0 \\ -\mu^2\cos\mu\xi_0 & \mu^2\sin\mu\xi_0 & 0 & 0 \\ \mu^3\sin\mu\xi_0 & -\mu^3\cos\mu\xi_0 & 0 & 0 \end{bmatrix}.$$

The relation between the coefficients for deflections w_l and w_r will take the form

$$\{C\} = \{B\} + [H]^{-1}\{0,\ 0,\ 0,\ -q\}^T. \tag{10}$$

As it follows from the relation (10), it is necessary to find only the last column of the matrix $[H]^{-1}$. Let us write it for $\{Z\}$. Then

$$[H]\{Z\} = \{0,\ 0,\ 0,\ 1\}^T. \tag{11}$$

The solution of the equation (11) takes the form

$$\{Z\} = \left\{\frac{\sin\mu\xi_0}{\mu^3},\ -\frac{\cos\mu\xi_0}{\mu^3},\ \frac{1}{\mu^2},\ -\frac{\xi_0}{\mu^2}\right\}^T.$$

It follows from (10) that

$$[C] = [B] + [F],\ [F] = [-\sin\mu\xi_0,\ \cos\mu\xi,\ -\mu,\ \mu\xi_0]\, q/\mu^3. \tag{12}$$

In order to get a connection between the power q and the force of compression μ^2 it is necessary to find one more equation. For this purpose we substitute the expression for ε, expressed in terms of displacement w и v, into the relation of Hook's law, connecting the force μ^2 and the tangential deformation ε. From the obtained equation we can find v for the left and right parts. The sought equation regarding q and μ will be a condition of equality of the displacement value v at the point of force application q, calculated according to the formulas for the right and left parts.

Given the Hook's law $T = K\varepsilon = 3D/h^2\varepsilon$, we obtain:

$$\varepsilon = -\mu^2\frac{h^2}{3b^2}. \tag{13}$$

The deformation ε is expressed in terms of w, v according to the formula

$$\varepsilon = \frac{dv}{d\xi} - \frac{b}{r_0}w + \frac{1}{2}(\dot{w})^2.$$

Hence, for the left and right sides, taking into account (13) we obtain the expression for v:

$$\begin{gathered} v_l = -\mu^2\frac{h^2}{3b^2}(\xi+1) + \int_{-1}^{\xi}\frac{b}{r_0}w_l d\xi - \int_{-1}^{\xi}\frac{1}{2}(\dot{w}_l)^2 d\xi\ ; \\ v_r = -\mu^2\frac{h^2}{3b^2}(\xi-1) + \int_{-1}^{\xi}\frac{b}{r_0}w_r d\xi - \int_{-1}^{\xi}\frac{1}{2}(\dot{w}_r)^2 d\xi\ . \end{gathered} \tag{14}$$

Here we have the boundary conditions $v(\pm1) = 0$ in mind. The condition of docking at the point of force application takes the form

$$v_l(\xi_0) - v_r(\xi_0) = 0. \tag{15}$$

After substitution of w_l and w_r from (8) we obtain a quadratic equation regarding q:

$$\eta\, q^2 + \chi\, q + \tau = 0, \tag{16}$$

By setting different values for μ, we obtain the appropriate values for q. According to the formulas (8), (12) we can find displacements at any point.

3. Results and discussion

Farther, let us consider a case of panel rigid packing. The boundary conditions are written through a deflection:

$$w_l(-1) = 0;\ \ w_r(1) = 0;\ \ w_l'(-1) = 0;\ \ w_r'(1) = 0. \tag{17}$$

Then the coefficients, included into the expression for the deflection (7) are derived from the system (17) and takes the form:

$$\{B\} = \frac{q}{2\mu^3}\left\{ -\frac{\left(-q + 2\frac{\mu^2}{r_0} + q\cos(\mu(1+\xi_0))\right)coseq(\mu)}{2q} + \frac{8a_1 m\mu^4 coseq(\mu)\sin\left(\frac{n\pi}{2}\right)}{\pi q(m^2\pi^2 - 4\mu^2)}, \right.$$

$$\frac{-\mu\xi_0 - \mu\cos(\mu(1+\xi_0)) + \sin(\mu(1+\xi_0))}{2\mu\cos(\mu) - 2Sin(\mu)}, \frac{\mu(\mu(1+\xi_0))\cos(\mu) - \sin(\mu) - \sin(\mu\xi_0)}{2\mu\cos(\mu) - 2\sin(\mu)},$$

$$\left(32a_1\mu^5\cos\left(\frac{n\pi}{2}\right) + \pi^2(m^2\pi^2 - 4\mu^2)\left(-\mu\left(q - \frac{\mu^2}{r_0} + q\xi_0\right) + q\sin(\mu(1+\xi_0))\right) + \right. \tag{18}$$

$$\left. \left. +\pi\cot(\mu)\left(\begin{array}{l}\pi(m^2\pi^2 - 4\mu^2)\left(-q + 2\frac{\mu^2}{r_0}\right) + \\ \pi q(m^2\pi^2 - 4\mu^2)\cos(\mu(1+\xi_0)) - 16a_1 n\mu^4\sin\left(\frac{n\pi}{2}\right)\end{array}\right)\right)\left(2\pi^2 q(m^2\pi^2 - 4\mu^2)\right)^{-1}\right\}.$$

Let us consider the case of the centrally applied force under ξ_0=0, given that the loading is rigid, providing the symmetrical deformation.

Figures 2–3 shows the dependences $q = q(\mu)$ and $q = q(w_0)$, $w_0 = w(0)$. Hereinafter, the dash lines show the dependences for the cases in which there are no imperfections. We can see that q>0 all the time, it means that after the loss of panel stability it is necessary to apply an additional load in order to keep it back.

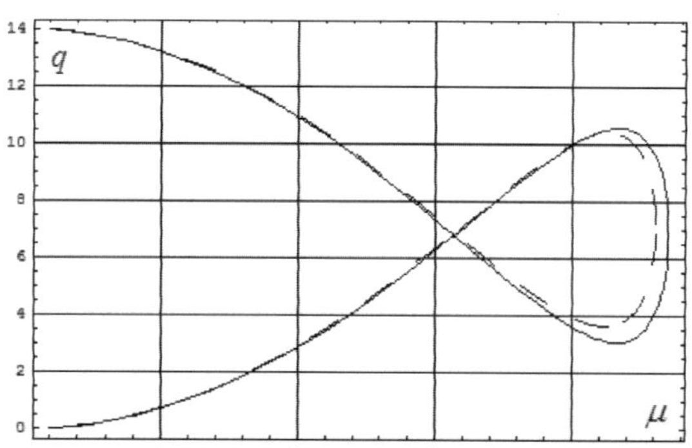

Figure 2. n=7, $a_1/a_0 = 0.1$, $r_0/b = 2$, $\xi_0 = 0$.

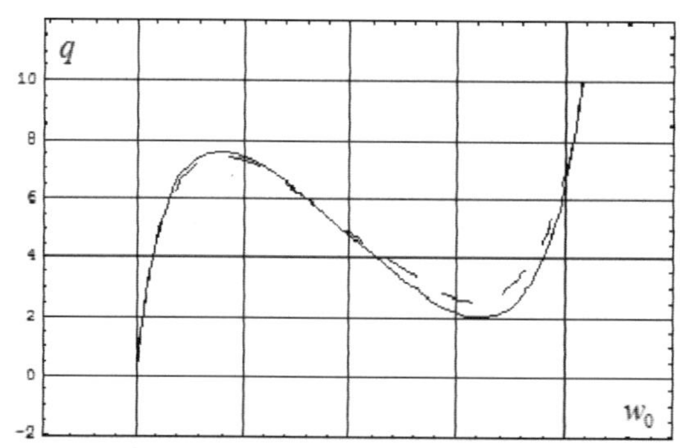

Figure 3. n=7, $a_1/a_0 = 0.1$, $r_0/b = 2$, $\xi_0 = 0$.

Figure 4 shows the dependences of the critical load q_* of the type and amplitude of wave formation. Figure 5 represents the dependences of the critical load q_* on the amount of half-waves n. The diagram shows, that under small values of n there is a noticeable change in the critical load. The critical load under sufficiently large n is close to the value for the case in which there are no imperfections.

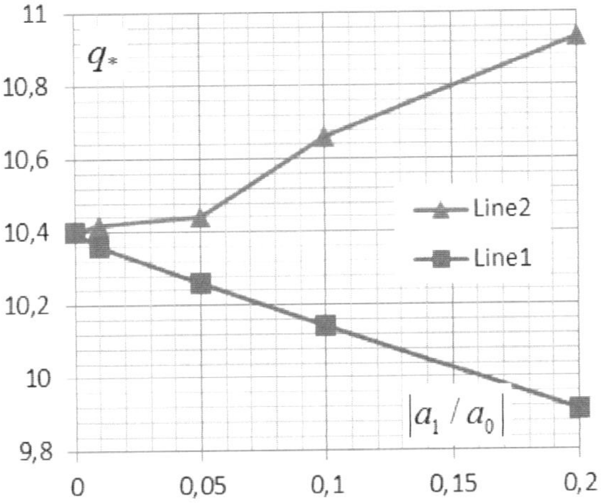

Figure 4. n=7, $r_0/b=2$, $\xi_0=0$, $a_1/a_0=-0.1$, (*Line* 1), $a_1/a_0=0.1$, (*Line* 2)

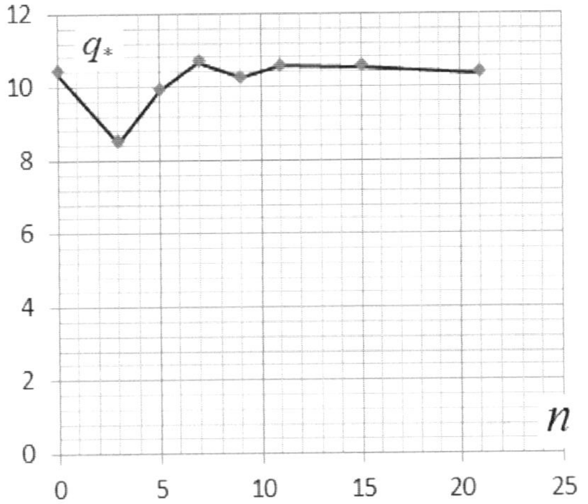

Figure 5. n=7, $r_0/b=2$, $\xi_0=0$, $a_1/a_0=-0.1$.

The dependences of the critical load on the application point is shown in Figure 6. The forms of the deformed panel in the case of a centrally applied force under the different values of q are shown in Figure 7. Because of the hard boundary conditions, only the symmetric form of loss in stability is realized.

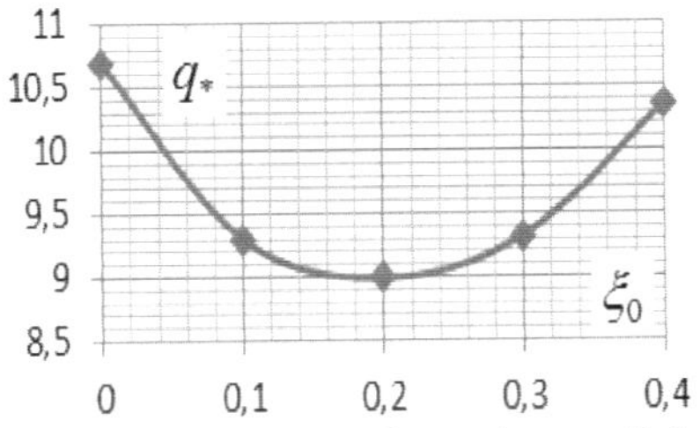

Figure 6. n=7, $r_0=2$, $a_1/a_0=-0.1$

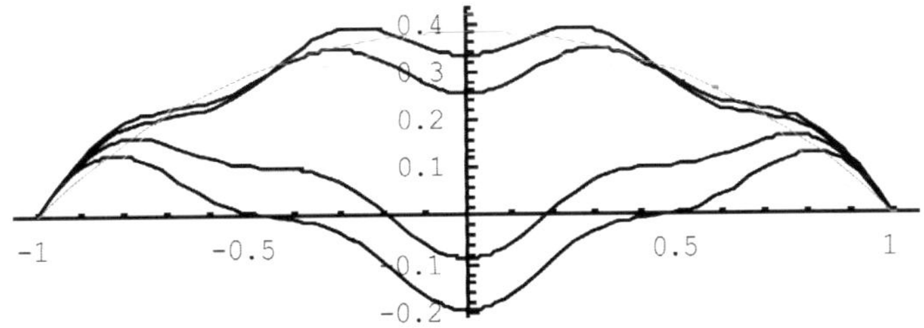

Figure 7. n=7, $r_0/b=2$, $\xi_0=0$, $a_1/a_0=-0.1$

References

[1] Kayumov R A and Tazyukov B F 2001 *Izv. VUZ. Aviatsionnaya Tekhnika.* **4** 12-15
[2] Kornishin M S and Mushtari K M 1995 *Math. USSR Academy of Sciences. Ser. Sci. and tehn. Sciences* 7 36-50
[3] Abdrakhmanova A I and Sultanov L U 2016 *Materials Physics and Mechanics* **26 (1)** 30–32
[4] Davydov R L and Sultanov L U 2015 *Journal of Engineering Physics and Thermophysics* **88 (5)** 1280–1288
[5] Stroganov V F, Kayumov R A, Akhmetshin A S and Stroganov I V 2011 *Polymer Science – Series D* **4 (4)** 334-338

Numerical investigation of strong compression of vapor inside spherical cavitation bubbles

T F Khalitova and D Yu Toporkov

Institute of Mechanics and Engineering, Kazan Science Center, Russian Academy of Sciences (IME KazSC RAS), 2/31, Lobachevsky str., Kazan 420111 Russia

E-mail: top.dmtr@gmail.com

Abstract. Numerical investigation of possibility of realizing shock waves in cavitation bubbles during their collapse in water, acetone, and tetradecane is performed. The radius of the bubble is 500 μm, the liquid pressure and temperature are in the ranges of 1-100 bar and 293-313 K, respectively. A numerical technique is used in which the movement of the interphase boundary is governed by the Rayleigh-Plesset equation. The thermodynamic parameters of the vapor are assumed uniform, the state of the vapor being described by the modified Van der Waals equation. The shock waves inside a bubble in tetradecane are found to arise in all the conditions under consideration. By contrast, inside the bubble in acetone they do not appear at relatively low pressures, while inside the bubble in water they never arise. At equal initial data the shock wave is formed much closer to the interface in the case of tetradecane.

1. Introduction

Dynamics of gas bubbles in liquids is of large scientific and practical significance. Of particular interest is energy accumulation during high-speed cavitation bubble collapse taking place near solids at both small and great distances from their surfaces. Such accumulation can lead to cavitation erosion (solid surface destruction), chemical transformations, bubble glowing, etc. For example, if a bubble at collapse remains close to spherical, very high pressures, densities, and temperatures can be achieved inside its cavity during the final stage of compression of its content.

The compression of the bubble content can be with [1] or without [2] formation of converging shock waves. Most important case is that with the shock waves since they result in the highest pressures and temperatures. Moreover, at their focusing in the central region of the bubble the reflected shock waves occur, propagating in the surrounding liquid. Such diverging shock waves can act on the nearby solid surface as localized short-run pulses of high pressure to result in the surface damages. Therefore, knowledge on the bubble collapse scenarios (with or without the shock waves) is of interest for applications.

The implementation of a shock or shockless scenario of bubble collapse depends on a number of factors. A criterion of formation of converging shock waves in a bubble during its high-speed collapse is proposed [3]. That criterion allows one to estimate the effect of some of the most important physical factors. It is shown in [3] that the shock bubble content compression scenario depends on the vapor properties (molecular weight, specific heat ratio) as well as the conditions of the bubble collapse (liquid pressure and temperature, the initial radius of the bubble). It should be noted that the criterion of [3] implies the use of some results of full hydrodynamic simulation with realistic wide-range equations of state which usually are rather complex.

In this paper, a comparison of water, acetone, and tetradecane is conducted about possibility of formation of shock waves inside cavitation bubbles collapsing in these liquids. A similar comparison of acetone and water was made in [3] at a pressure 15 bar and a temperature 293 K. It was found in [3] that for implementing shock waves in those conditions acetone is much more favorable. The present paper essentially differs from [3] in that much larger liquid pressure and temperature ranges are considered here, namely, those from 1 through 100 bar and from 293 through 313 K, respectively. Another difference is that not only acetone and water but also tetradecane are studied here. Tetradecane is taken following the recommendations in [3], according to which a liquid with the higher molecular weight and the lower specific heats ratio is more preferable for the shock waves. The formation of shock waves in a bubble is stated by a criterion based on results of applying a simple mathematical model in which the movement of the interface is governed by the Rayleigh-Plesset equation, the vapor in the bubble is assumed uniform, the vapor state is described by a modified Van der Waals equation [4].

2. Problem statement

Collapse of a single spherical cavitation bubble in liquid (water, acetone and tetradecane) due to the difference between the initial pressure in the bubble and the pressure in the surrounding liquid is considered. At the beginning of the collapse the temperature T of the liquid and vapor is T_0 everywhere, the liquid pressure far from the bubble is p_0, the vapor pressure p_b is equal to the saturation pressure $p_S(T_0)$ throughout the bubble. At the initial time moment $t = 0$ the bubble radius R is 500 μm, the velocity of the vapor in the bubble, the surrounding liquid, and their interface is zero.

The possibility of forming a shock wave in the cavitation bubble during its high-speed collapse in water, acetone and tetradecane is studied at various liquid pressure p_0 and temperature T_0 ranging from 1 through 100 bar and from 293 through 313 K, respectively.

3. Technique of estimating formation of shock waves

The appearance of a shock wave in a bubble during its collapse is estimated by using a quite simple numerical technique, in which the bubble radius R is governed by the Rayleigh-Plesset equation

$$\left(1-\frac{\dot{R}}{c_l}\right)R\ddot{R}+\frac{3}{2}\left(1-\frac{\dot{R}}{3c_l}\right)\dot{R}^2=\left(1+\frac{\dot{R}}{c_l}\right)\frac{p_b-p_0}{\rho_l}+\frac{R}{c_l}\frac{\dot{p}_b-\dot{p}_0}{\rho_l} \quad (1)$$

where c_l is the speed of sound in liquid, ρ_l is the liquid density, the overdot means time derivative.

The vapor pressure p_b in the bubble is governed by a modified Van der Waals equation of state [4]

$$p_b=\left(p_{b,0}+A\rho_{b,0}^{\alpha}\right)\left(\frac{\rho_{b,0}^{-1}-B}{\rho_b^{-1}-B}\right)^{\gamma}-A\rho_b^{\alpha} \quad (2)$$

where $\rho_{b,0}$, ρ_b are the initial and current vapor densities in the bubble, γ is the adiabatic exponent, A, B, α are constants defined by the following expressions

$$A=\frac{V_{cr}^{\alpha+1}R_gT_{cr}}{\alpha\left(V_{cr}-B\right)^2},\quad B=V_{cr}-\frac{2V_{cr}}{\alpha+1},\quad \alpha=\frac{1}{2}\left(\frac{4V_{cr}p_{cr}}{R_gT_{cr}}+\sqrt{\left(\frac{4V_{cr}p_{cr}}{R_gT_{cr}}\right)^2+4}\right),\ V_{cr}=\frac{1}{\rho_{cr}},\ R_g=\frac{R_{ug}}{M} \quad (3)$$

in which T_{cr}, p_{cr}, ρ_{cr} are the critical values of the temperature, pressure and density, R_{ug}, R_g are the universal and specific gas constants, M is the molecular weight of the vapor. The parameters used in (1)-(3) are given in Table 1.

Table 1.

	p_{cr} (bar)	T_{cr} (K)	ρ_{cr} (kg/m^3)	M (g/mol)	γ	ρ_l (kg/m^3)	c_l (m/s)
Water	221.2	374.12	317.8	18	1.325	998.2	1483
Acetone	46.6	508.2	280	58	1.125	810	1189
Tetradecane	16.2	693	240	198.38	1.0265	763	1331

Equations (1)-(3) are solved numerically by the Runge-Kutta method of high-order accuracy with a variable time step [5].

In the present paper as in [3] it is assumed that the shock wave arises inside a bubble in the course of its collapse if the following inequality is satisfied

$$\min_t \frac{\Delta R_{sh}^*}{R} < 1 \tag{4}$$

where ΔR_{sh}^* is the distance from the bubble surface (at the time moment t) to the point of possible occurrence of the shock wave.

The value of ΔR_{sh}^* is determined by the expression

$$\Delta R_{sh}^* \approx -\left(1-\frac{\dot{R}}{c}\right)\frac{c^2}{k\ddot{R}}, \quad k = 1+\frac{\rho c'(\rho)}{c} \tag{5}$$

where ρ, c are the vapor density and sound speed (the prime means the derivative with respect to density at constant entropy).

4. Validity of the shock wave formation estimate technique

The validation of the shock wave formation estimate technique based on bubble dynamics model (1)-(3) is performed by comparing its results for the water and acetone with those obtained, like in [3], by the full hydrodynamic simulation with the wide-range equations of state (but with dropping the effects of the thermal conductivity and condensation on the interface to eliminate the influence of the thin thermal layer on the boundary vapor density).

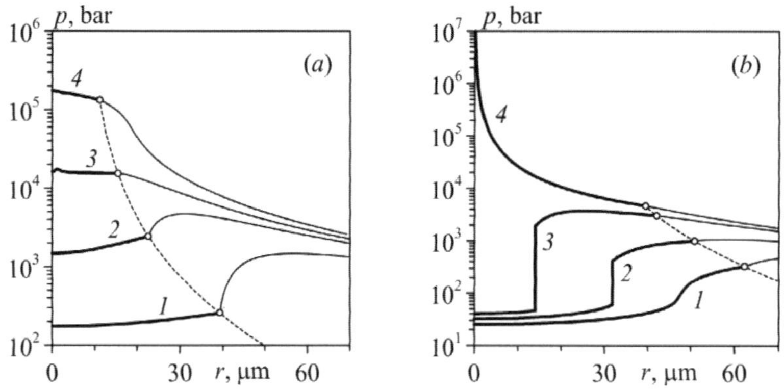

Figure 1. Radial distributions of the vapor and liquid pressure at four moments (curves *1-4*) of the final stage of bubble collapse in water (*a*) and acetone (*b*). Curves *4* correspond to the extreme compression of the vapor in the bubble. Circles show the interface values (the vapor is on their left).

Fig. 1 shows the change of the pressure inside the bubble and its surroundings in the end of the cavitation bubble collapse in water and acetone at $p_0 = 15$ bar and $T_0 = 293$ K. These results were obtained by the full hydrodynamic model of [3]. Fig. 1*a* corresponds to the bubble in water, fig. 1*b* to the bubble in acetone. In the former case the bubble content is compressed without a shock wave while in the latter one a converging shock wave is seen to arise in the bubble at some distance from its surface. The intensity of the shock wave rapidly increases due to the spherical convergence. Owing to the shock wave, the maximum values of the thermodynamic parameters in a small central area of the bubble in acetone are many times higher than those in the bubble in water.

Dependences of ratio $\Delta R_{sh}^*/R$ on the bubble radius R are presented in fig. 2 for the bubbles collapsing in water and acetone. The minimum value of such a function is used in criterion (4) to draw a conclusion about shock wave appearance in the bubble. The curves shown in fig. 2 were obtained by model (1)-(3) and by the much more complex full hydrodynamic simulation with the model of [3]. It follows from fig. 2 that the minimum values of $\Delta R_{sh}^*/R$ for these two models are very close to each other in the case of the bubble in water, and the minimum of $\Delta R_{sh}^*/R$ obtained by (1)-(3) is 2.3 times less than that obtained by model of [3] in the case of the bubble in acetone. Both models lead to the similar conclusions that the shock wave arises in the bubble in acetone and it does not appear in the bubble in water. This comparison indicates that model (1)-(3) can be used for estimating the minimum value of $\Delta R_{sh}^*/R$ during bubble collapse.

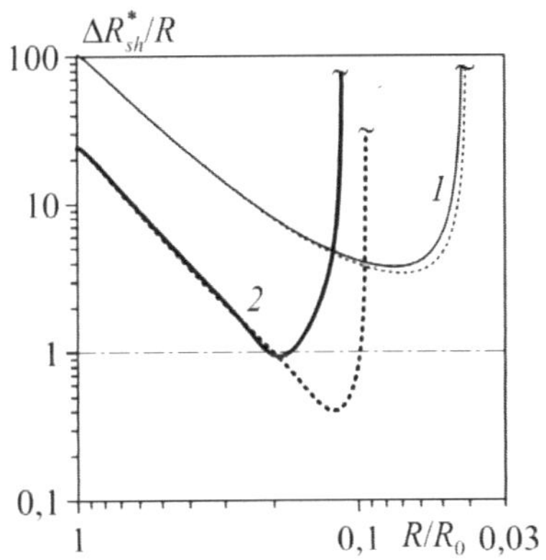

Figure 2. Evolution of ratio $\Delta R_{sh}^*/R$ during cavitation bubble collapse in water (curves *1*) and acetone (curves *2*) determined by the model of [3] (solid curves) and model (1)-(3) (dotted curves). The horizontal dash-dotted line corresponds to the critical value $\Delta R_{sh}^*/R = 1$ in criterion (4) (below this value the shock waves are assumed to arise in the bubble).

5. Results

Fig. 3 shows time-dependences of the bubble radius R during bubble collapse in water, acetone and tetradecane at the liquid pressure $p_0 = 15$ bar for the liquid temperatures $T_0 = 293$ and 313 K (the initial bubble radius is 500 μm). Under these conditions the mass of the vapor in the bubble in the case of tetradecane is considerably less than that in the cases of water (by more than 300 times) and acetone (by more than 10^4 times). As a result, the bubble in tetradecane collapses deeper (its minimum radius R_{min} is 1.89 μm against 41.48 μm in water and 15.27 μm in acetone). Increasing the initial liquid temperature leads to the growth of the mass of the vapor in a bubble resulting in an increase of R_{min}. A bubble in tetradecane with the vapor mass equal to that in the bubble in acetone at $T_0 = 293$ K can be obtained by increasing the liquid temperature by more than 100 degrees.

Fig. 4 shows the dependences of the minimum values of $\Delta R_{sh}^*/R$ achieved during the cavitation bubble collapse in water, acetone, and tetradecane on the liquid pressure p_0. Using criterion (4), it can be concluded that under the conditions considered ($1 \leq p_0 \leq 100$ bar and $293 \leq T_0 \leq 313$ K) the shock wave in the bubble in water does not occur. By contrast, the shockless collapse of the bubble in acetone takes place only if $p_0 \leq 5$ bar for $T_0 = 293$ K and $p_0 < 10$ bar for $T_0 = 313$ K, whereas in the bubble in tetradecane the shock waves always form.

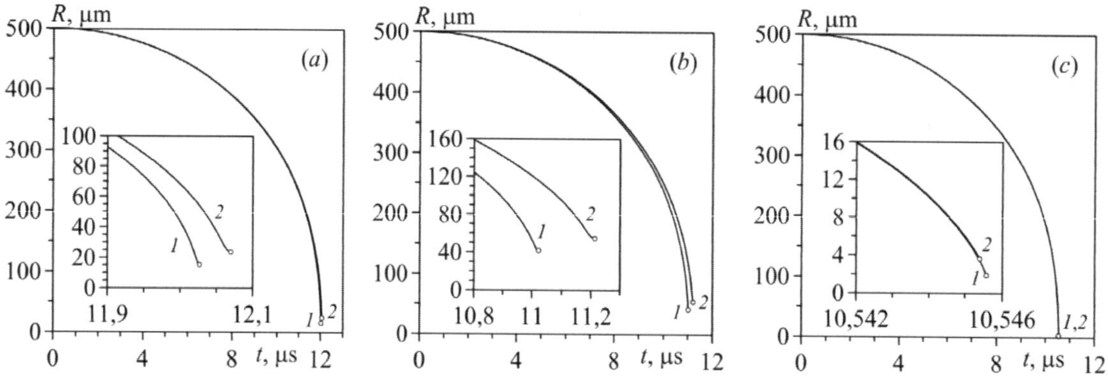

Figure 3. Evolution of the bubble radius R during bubble collapse in water (a), acetone (b), and tetradecane (c). The liquid temperature $T_0 = 293$ K (curves *1*) and 313 K (curves *2*). Circles indicate R_{min}.

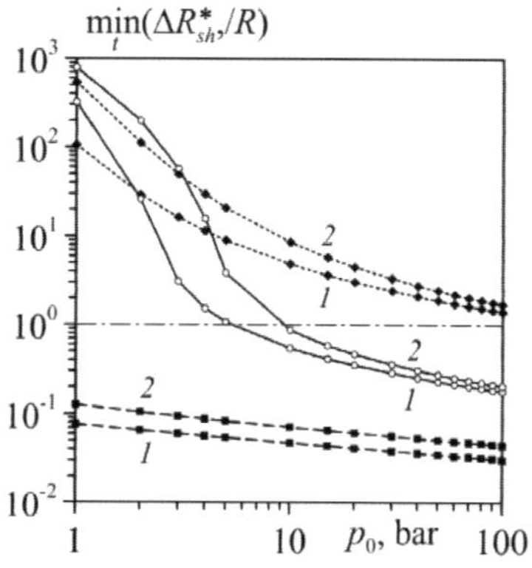

Figure 4. The minimum values of $\Delta R_{sh}^*/R$ achieved during cavitation bubble collapse in water (dotted curves), acetone (solid curves) and tetradecane (dashed curves), as functions of the liquid pressure p_0 for the liquid temperature $T_0 = 293$ (curves *1*) and 313 K (curves *2*). The horizontal dash-dotted line corresponds to the critical value $\Delta R_{sh}^*/R = 1$ in criterion (4).

6. Conclusion

Investigation of shock wave formation in a cavitation bubble during its collapse in water, acetone, and tetradecane has been performed for the bubble with the initial radius 500 μm at the liquid pressure p_0 from 1 through 100 bar and the liquid temperature T_0 from 293 through 313 K. A simple mathematical model is used, in which the movement of the interphase boundary is governed by the Rayleigh-Plesset equation, the vapor in the bubble is assumed uniform, being described by the modified Van der Waals equation of state. It has been shown that in all the conditions under consideration the shock waves do not appear inside the bubble in the case of bubble in water. In the case of bubble in acetone at $T_0 = 293$ K and 313 K the shock waves occur at $p_0 > 5$ bar and 10 bar, respectively. In the case of bubble in tetradecane the shock waves always arise.

References

[1] Nigmatulin R I, Akhatov I Sh, Topolnikov A S, Bolotnova R Kh, Vakhitova N K, Lahey R T (Jr) and Taleyarkhan R P 2005 *Phys. Fluids* **17** 107106
[2] Khabeev N S 2010 *Fluid Dyn.* **45** 208
[3] Nigmatulin R I, Aganin A A, Toporkov D Yu and Ilgamov M A 2014 *Dokl. Phys.* **59** 431
[4] Nigmatulin R I, Aganin A A, Ilgamov M A and Toporkov D Yu 2014 *J. Appl. Mech. Tech. Phys.* **55** 444
[5] Hairer E, Nersett S P and Wanner G 1987 *Solving Ordinary Differential Equations. 1. Nonstiff Problems* (Springer-Verlag, Berlin, Heidelberg)

Waterflooding simulation of reservoir containing horizontal well stimulated by multistage hydraulic fracturing

A B Mazo, M R Khamidullin and K A Potashev

Department of Aerohydromechanics, Kazan Federal University, Kazan, Russia

E-mail: marseille.ksu@mail.ru

Abstract. The article presents a three-dimensional mathematical model for two-phase fluid flow near a multistage hydraulically fractured horizontal well (MSHFHW). The flow in the reservoir and in the fractures is simulated separately, and the flow rate is governed by Darcy's law. Finite volume method is used for spatial approximation. The obtained systems of linear equations for pressure in the reservoir and in the fractures are solved simultaneously, which allows us to avoid using iterative process for solution adjustment both in the fractures and the reservoir. Saturation is calculated by the implicit adaptive scheme AIM.
Keywords. horizontal well, multistage hydraulic fracturing, two-phase flow in porous media

1. Introduction

Multistage hydraulically fractured horizontal wells (MSHFHW) are often used for development of oil fields characterized by weak reservoir properties, high heterogeneity, or low permeability. With the growing popularity of this technique, there is a need to develop mathematical models describing complex filtration processes in the "reservoir–hydraulic fractures–well" system.

Paper [1] proposes a mathematical model of two-phase flow in porous media with fractures. Sampling of equations is performed on an unstructured hybrid grid using implicit schemes for pressure and saturation. Work [2] is devoted to solving the problem of two-phase flow near a horizontal well with MSHF in a three-dimensional formulation in the Cartesian coordinate system.

A model of two-phase flow near HW with MSHF, which takes into account the variation of fractures conductivity in time, is presented in [3]. Spatial approximation of the equations was performed using finite difference method. The IMPES scheme was used for temporal sampling [4]. The pressure in this case was calculated by the implicit scheme, and saturation – by the explicit one. The results showed that consideration of saturation changes in the fracture has a significant effect on the pressure distribution near the fractures. Paper [5] proposes an iterative method for simultaneous solution of the problems for saturation and pressure, which accelerates calculations by 30% compared to the full implicit method (FIM) [6] and, unlike IMPES, is convergent at a bigger Courant number:

$$C = \frac{\tau \max|u|}{h} \ll 1,$$

where h, τ are the grid steps in space and time, u is the flow rate.

This paper presents a three-dimensional mathematical model describing a two-phase flow near multistage hydraulically fractured horizontal wells.

2. Constitutive equations

We consider the problem of two-phase flow near HW stimulated by transverse hydraulic fractures in a three-dimensional formulation. The solution region D is a part of the reservoir, having the form of a rectangular parallelepiped with the height $2H$ and rounded edges, at the center of which there is a cut of a cylindrical shape – a well with the length L and the radius r_w (figure 1). The region D is vertically bounded by the planes $z=\pm H$, modeling the top and bottom of the reservoir. The lateral surface Γ located at the distance l from the well γ models the external boundary. The Oy axis is directed along the borehole, and the hydraulic fractures are located orthogonally to it. The walls of the fracture are two rectangular planes F_i^- and F_i^+ with the normals in the directions $n^{\pm}=\pm y_i$ and having the sizes $2H\times 2h_i$, the distance between them being 2δ (fracture opening). The fractures have permeability far exceeding the absolute permeability of the reservoir. The reservoir is considered to be uniform. Capillary and gravitational forces are neglected.

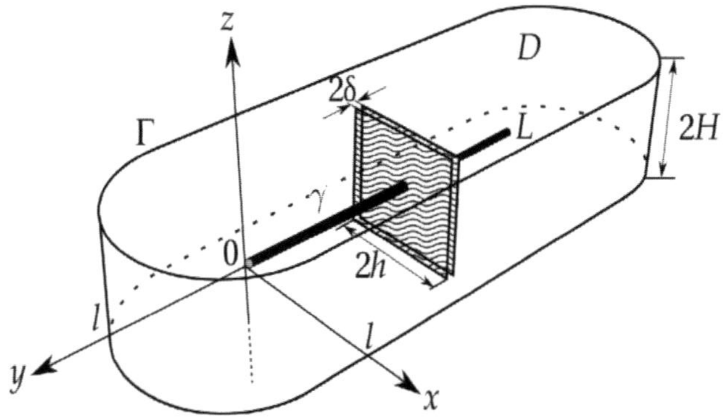

Figure 1. Scheme of HW with MSHF.

In dimensionless variables, the equations for pressure and saturation in the reservoir D have the form:

$$\beta\frac{\partial p}{\partial t}+\text{ div }\vec{u}=0,\qquad \vec{u}=-\sigma\text{ grad }p.,\tag{1}$$

$$\frac{\partial s}{\partial t}+\text{ div }\left(f(s)\vec{u}\right)=0;.\tag{2}$$

$$f(s)=\frac{k_w(s)}{\sigma},\quad \sigma=k_w(s)+\text{K}_\mu k_o(s),$$

$$k_w(s)=s^A,\quad k_o(s)=(1-s)^B.$$

Here $\beta\sim 10^{-3}$ is the compressibility; $f(s)$ is the Buckley-Leverett function; $k_w(s)$, $k_o(s)$ are the relative permeabilities (RP) of water and oil respectively; A, B are the RP coefficients; $\sigma(S)$ is the transmissibility; $\text{K}_\mu<1$ is the water and oil viscosity ratio.

The dimensionless initial conditions

$$t=0,\ (x,y,z)\in D:\quad p=1,\quad s=0..\tag{3}$$

show that the reservoir is saturated by oil under hydrostatic pressure. The top and bottom of the reservoir (spatial coordinates are normalized on H) are impermeable,

$$z=\pm 1:\ u_n=-\sigma\frac{\partial p}{\partial n}=0..\tag{4}$$

The process of waterflooding is modeled at $t>0$, when the constant pressure $p=1$ and water saturation $s=1$ are maintained at the external boundary Γ. In this case the pressure $p=p_\gamma=0$ is set on the well γ. The hydrodynamic interaction between the reservoir, the surface γ of the well and the surfaces $F_i^{\pm}$ of the fractures is expressed in the continuity of the pressure and the normal flow rate.

The equation for the dimensionless pressure p_f, averaged by the fracture opening 2δ, is written as follows:

$$\Delta_{xz} p^f + \frac{1}{2M}\sigma \frac{\partial p}{\partial y}\Bigg|_{F^-}^{F^+} = 0 \quad -h_i < x < h_i,\ -1 < z < 1, \qquad . \tag{5}$$
$$x = \pm h_i,\ z = \pm 1: \ \frac{\partial p}{\partial n} = 0, \quad (x, y, z) \in \gamma: \ \ p^f = p_\gamma = 0.$$

The equation for the water saturation s^f in the dimensionless variables has the form:

$$m^f \frac{\partial S^f}{\partial t} + \nabla\left(f\left(S^f\right)\vec{u}^f \right) + \frac{H}{2\delta}\left(f(S)u_y \right)\Bigg|_{F_i^-}^{F_i^+} = 0.. \tag{6}$$

Let us note that the Buckley-Leverett function in the last term of this equation, modeling the water inflow from the reservoir to the well, is calculated at the fracture faces F_i^+ and F_i^- from the reservoir's side, just as the total flow rates $u_y^{\pm} = -\sigma\partial p / \partial y,$. The boundary conditions for this equation are not necessary.

Thus, the mathematical model of two-phase flow towards the production horizontal well with MSHF in the dimensionless form consists of Equations (1)–(4) in the reservoir D and Equations (5), (6) in the fractures.

The method of covering the area by the condensing grid, the approximation of Equations (1) and (5), and the algorithms for solving the problem for pressure using the implicit scheme are presented in [7]. To calculate saturation, we used the adaptive implicit method AIM [8].

3. Numerical scheme

To solve the problem, we were seeking the average pressure and saturation in the finite volumes (FV) V

$$P = \frac{1}{|V|}\int_V p\,\mathrm{d}V, \quad S = \frac{1}{|V|}\int_V s\,\mathrm{d}V. \tag{7}$$

For each FV, Equation (2) integrated through the volume V_i can be written (taking into account the divergence theorem) in the form:

$$m_i \frac{\hat{S}^i - S^i}{\tau} + \sum_j f^{i,j}\tilde{u}_n^{i,j}\left|\Gamma_{V_i}^j\right| = 0, \tag{8}$$

in the fractures

$$m_f \frac{\hat{S}_f^i - S_f^i}{\tau} + \sum_j \left(\tilde{u}_f\right)_n^{i,j}\left|\Gamma_f^{i,j}\right| + \frac{1}{M^i}\left|V_f^i\right|\frac{P - P_f^i}{h_d^i} = 0, \tag{9}$$

where $\Gamma_V = \bigcup_j \Gamma_V^j$ is the surface of the finite volume V, which consists of the plane faces Γ_V^j; $\left|\Gamma_V^j\right|$ is the area of the face Γ_V^j; $\tilde{u}_n^j$ is the average projection of the rate on the external normal of this face.

The problem of finding the pressure P, P_f by Equations (1), (5) and the saturation on a new temporal layer $\hat{S}, \hat{S}_f$ by Equations (2), (6) is solved in several stages.

1) We calculate the pressures P_i, P_{fi} on the new temporal layer using the saturation from the previous layer S_i, S_{fi}.

2) We calculate the total flow rates through all the faces of the finite volumes:

$$\tilde{u}_n^{i,j} = -\sigma^{i,j} \frac{P_j - P_i}{h_d^{i,j}},$$

where $h_d^{i,j}$ is the distance from the center of the ith FV to the center of the jth one. If the centers of the FV are not located on the line, through which the vector of the external normal passes to the face separating the two FV, then we use the four-point linear approximation [9] through the average values of pressure at collocation points located at the centers of the corresponding finite volumes.

3) We calculate the saturations S_i, S_{fi} using

a) explicit scheme

$$\overset{\frown}{S}^i = S^i - \frac{\tau}{m_i} \sum_j f^{i,j}(S) \tilde{u}_n^{i,j}(S) \left| \Gamma_{V_i}^j \right| = 0;$$

б) implicit scheme

$$\overset{\frown}{S}^i + \frac{\tau}{m_i} \sum_j f^{i,j}\left(\overset{\frown}{S}\right) \tilde{u}_n^{i,j}\left(\overset{\frown}{S}\right) \left| \Gamma_{V_i}^j \right| = S^i.$$

4. Results

Figure 2 shows the pressure distribution map and the isobar field in a uniform reservoir containing a HW with a single hydraulic fracture for $h = 5$, $M = 100$. It can be seen that the fracture substantially changes the pressure distribution, and the concentric surfaces of the isobars stretch with the increase of antigradient due to the presence of the fracture.

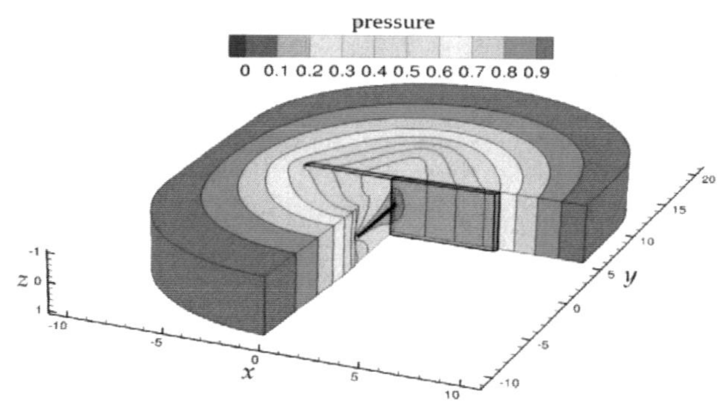

Figure 2. Pressure distribution and isobar map in computational domain D.

Figure 3 shows the saturation maps for the case of the reservoir waterflooding at different moments of time. We can see that the presence of the fracture accelerates the movement of the isosats in that part of the reservoir where there are low-pressure zones (figure 2). It is worth mentioning that the isosats move towards the well ends much more slowly than towards the main part of the borehole. This is due to the fact that the total withdrawal of the fluid in the case of spherical symmetry is carried out over a longer period of time than in the case of radial symmetry [10].

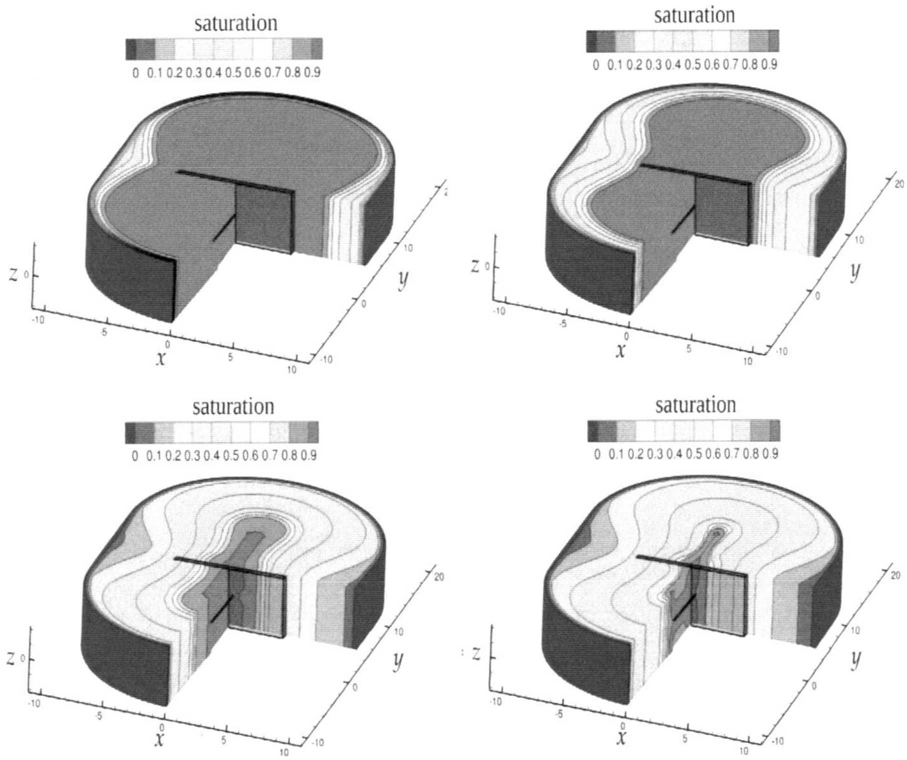

Figure 3. Saturation map for waterflooding of the reservoir containing HW with MSHF.

The pressure distribution and streamlines near the wellbore area and the high-conductivity hydraulic fracture are shown in figure 4. It is clearly seen how the fluid inflow separates from the external boundary: one part enters the fractures, the other gets directly to the well.

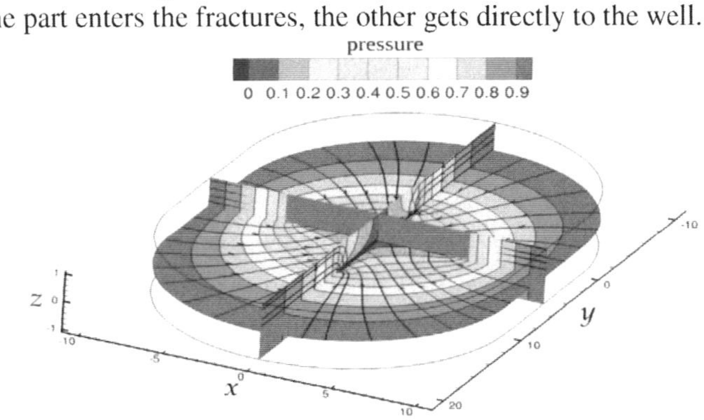

Figure 4. Pressure distribution and streamlines in cross-section areas of computational domain D

Figure 5 demonstrates the dynamics of water cut and cumulative oil production of the well for the case of: a) the absence of hydraulic fractures, b) the presence of a single fracture at $h = 5$, $M = 10^2$, intersecting the borehole in the middle, and c) the presence of two fractures with similar parameters equidistant from each other and from the ends of the well. It is evident that water content grows earlier in HW with MSHF, but at the same time the increase in the well production rate results in the growth of the cumulative oil production at a particular moment of time as compared to HW without MSHF.

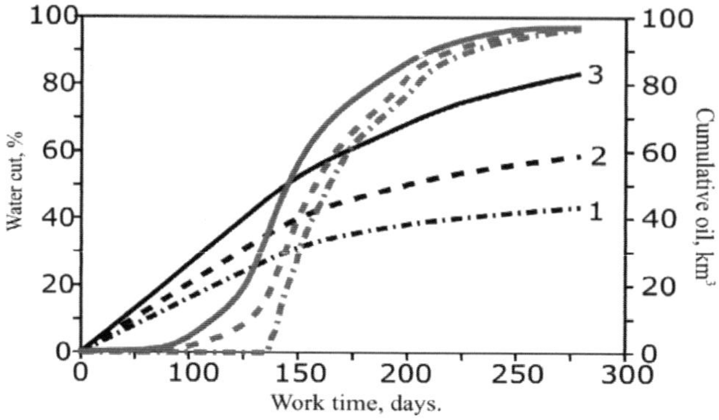

Figure 5. Dynamics of water cut and cumulative oil production of the well: 1 – HW without MSHF, 2 – HW with a single fracture, 3 – HW with two hydraulic fractures.

5. Conclusion

We presented a mathematical model of two-phase flow near a horizontal well with multistage hydraulic fracturing. Calculation of saturation was accelerated by using the adaptive implicit scheme. We analyzed the effect of hydraulic fractures on the reservoir waterflooding. It was shown that water cut grows earlier in HW with MSHF compared to the traditional HW, but the total cumulative oil production increases significantly.

Acknowledgments

The work was financially supported by the Russian Foundation for Basic Research and the Government of the Republic of Tatarstan (nos. 15-41-02698, 15-41-02699).

References

[1] Matthai S K, Mezentsev A and Belayneh M 2005 SPE J. **93341-MS.**
[2] Xie W 2016 **SPE-180995-MS**.
[3] Shen R and Gao S 2012 3rd International conference on computer and electrical engineering (Singapore: IACSIT Press).
[4] Aziz Kh and Settati A 1979 Pertoleum reservoir simulation (London: Applied Science publishers Ltd.).
[5] Lu B and Alshaalan T 2007 **SPE-110114-MS.**
[6] Appleyard J, Cheshire I, and Pollard R 1981 Proceedings of the European Symposium on Enhanced Oil Recovery (Harwell: Atomic Energy Research Establishment, Computer Science and Systems Division).
[7] Khamidullin M R 2016 *Uch. Zap. Kaz. Univ., Ser. Phys.-math. Sci.* **158** 287 [*in russian*]
[8] Thomas GW and Thurnau D H 1983 SPE J. **10120-PA** 759.
Fung L S-K et al 1998 SPE J. **16003-PA** 45.
[9] Mazo A B, Potashev K A, Kalinin E I and Bulygin D V 2013 *Mat. Model.* **25** 51 [*in russian*].
[10] Basniev K 1993 Podzemnaya gidromehanica (Subsurface hydromechanics) (Moscow: Nedra) p 416 [*in russian*].

Decomposition method for zonal resource allocation problems in telecommunication networks

I V Konnov[1] **and A Yu Kashuba**[2]

[1] Department of System Analysis and Information Technologies, Kazan Federal University, Kazan 420008, Russia
[2] LLC "AST Povolzhye", Kazan, 420029, Russia

E-mail: Igor.Konnov@kpfu.ru

Abstract. We consider problems of optimal resource allocation in telecommunication networks. We first give an optimization formulation for the case where the network manager aims to distribute some homogeneous resource (bandwidth) among users of one region with quadratic charge and fee functions and present simple and efficient solution methods. Next, we consider a more general problem for a provider of a wireless communication network divided into zones (clusters) with common capacity constraints. We obtain a convex quadratic optimization problem involving capacity and balance constraints. By using the dual Lagrangian method with respect to the capacity constraint, we suggest to reduce the initial problem to a single-dimensional optimization problem, but calculation of the cost function value leads to independent solution of zonal problems, which coincide with the above single region problem. Some results of computational experiments confirm the applicability of the new methods.

1. Introduction

Despite the existence of powerful processing and transmission devices, increasing demand of different telecommunication services and its variability lead to serious congestion effects and inefficient utilization of network resources. This situation forces one to replace the fixed allocation rules with more flexible mechanisms, which are based on proper mathematical models; see e.g. [1]–[3]. In particular, spectrum sharing is now one of the most critical issues in this field and various adaptive mechanisms have been suggested. Most papers are devoted to game-theoretic models and implementation of decentralized iterative methods for finding the Nash equilibrium points or their generalizations; see e.g. [4, 5]. At the same time, various optimization based mechanisms are also suggested; see e.g. [6, 7, 5, 3].

In this paper, we consider some problems of optimal allocation of a homogeneous resource in telecommunication networks such that the income received from users payments is maximized and the implementation costs of the network operator are minimized. We first present an optimization formulation for the case where the network manager aims to distribute some homogeneous resource (bandwidth) among users of one region with quadratic charge and fee functions. These convex quadratic optimization problems can be solved by simple and efficient solution methods. We describe some modifications for this special problem. Next, we consider a more general resource allocation problem for a provider of a wireless communication network divided into zones (clusters); which was formulated as a convex optimization problem in [8, 9]. Now, since the price functions are affine, we obtain again a convex quadratic optimization problem having capacity and zonal balance constraints. Unlike [8, 9], we now suggest to apply the dual Lagrangian method with respect to only capacity constraint. Therefore, we replace the initial problem with a single-dimensional optimization problem, however, calculation of its cost function value requires independent solution of zonal problems. Each of these problems coincides with the above single region resource allocation problem and can be solved by the simple algorithms suggested. In such a way we develop a new dual decomposition approach for solution finding, whose implementation is simpler essentially in comparison with the methods from [8, 9]. We present results of computational experiments which confirm the applicability of the new method.

2. Simple resource allocation model

Let us consider a single telecommunication network with nodes (users). The general problem of a network manager is to find an optimal allocation of a limited homogeneous resource among the users in order to maximize the total payment received from the users and to minimize the total network implementation expenses. That is, x is an unknown quantity of the resource offered by the network, with the capacity bounds $x \in [0,b]$, which yields the network expense (cost of implementation) $u(x)$. Similarly, y_i is the unknown resource offered to user $i \in I$ and $\varphi_i(y_i)$ is the fee (incentive) value paid by node i with the capacity bounds $y_i \in [0,a_i]$, where I is the index set of users. The network manager problem is formulated as follows:

$$\max_{(x,y)\in D} \to \sum_{i\in I} \varphi_i(y_i) - u(x), \tag{1}$$

where $y = (y_i)_{i\in I}$,

$$D = \left\{ (x,y) \ \sum_{i\in I} y_i = x, 0 \le y_i \le a_i, i \in I, 0 \le x \le b \right\}.$$

Suppose that the set D is non-empty, functions $u(x)$ and $-\varphi_i(y_i)$ are convex and quadratic, i.e.,

$$\varphi_i(y_i) = 0.5\alpha_i y_i^2 + \beta_i y_i, \alpha_i < 0, i \in I; u(x) = 0.5\gamma x^2 + \delta x, \gamma > 0.$$

Then (1) is a convex quadratic optimization problem, which can also be treated as a two-side auction models with one trader where all the participants have affine price functions; see [10, 11]. For these problems there exist many rather efficient solution methods; see e.g. [12] and references therein. They are mostly based on duality theory.

Following this approach, write the Lagrange function of problem (1) with the negative sign:

$$\begin{aligned} M(x,y,p) &= u(x) - \sum_{i\in I} \varphi_i(y_i) - p\left(x - \sum_{i\in I} y_i \right) \\ &= (u(x) - px) - \sum_{i\in I} (\varphi_i(y_i) - py_i) \\ &= (0.5\gamma x^2 + \delta x - px) - \sum_{i\in I} (0.5\alpha_i y_i^2 + \beta_i y_i - py_i) \end{aligned}$$

In order to find a value of the dual cost function

$$\theta(p) = \min_{x\in[0,b],\ y\in[0,a]} M(x,y,p),$$

where $a = (a_i)_{i\in I}$, we have to solve one- dimensional problems:

$$\min_{0 \le x_k \le b_k} \to (0.5\gamma x^2 + \delta x - px),$$

and

$$\min_{0\le y_i\le a_i} \to (-0.5\alpha_i y_i^2 - \beta_i y_i + p y_i),$$

for $i \in I$. Solutions of these problems denoted by $x(p)$ and $y_i(p)$, $i \in I$, respectively, are defined uniquely. Set $\tilde{x}(p) = (p-\delta)/\gamma$ and $\tilde{y}_i(p) = -(\beta_i - p)/\alpha_i$, then

$$x(p) = \begin{cases} 0 & \text{if } p \le \delta, \\ b & \text{if } p \ge \delta + \gamma b, \\ \tilde{x}(p) & \text{otherwise}; \end{cases} \quad y_i(p) = \begin{cases} 0 & \text{if } p \ge \beta_i, \\ a_i & \text{if } p \le \beta_i + \alpha_i a_i, \\ \tilde{y}_i(p) & \text{otherwise}; \end{cases} \quad \text{for } i \in I; \tag{2}$$

It follows that the function $\theta(p)$ is concave and differentiable with

$$\theta'(p) = \sum_{i\in I} y_i(p) - x(p).$$

Besides, the one- dimensional dual problem

$$\max_p \to \theta(p)$$

coincides with the simple equation

$$\theta'(p) = 0, \tag{3}$$

where $\theta'(p)$ is non-increasing. If p^* is the solution of (3), then we can find the solution of the initial problem (1) from (2) by setting $p = p^*$.

If we set $p'' = \max_{i\in I} \beta_i$ and $p' = \delta$, then the case $p'' \le p'$ gives immediately the zero solutions in accordance with (2). So we can consider only the non-trivial case where $p' < p''$. Then by (2) we must have $\theta'(p') > 0$ and $\theta'(p'') < 0$. These properties enable us to find a solution of (3) by the simple bisection algorithm, denoted as **Algorithm (BS)**. Given an accuracy $\varepsilon > 0$ and the initial segment $[p', p'']$, we take $\tilde{p} = 0.5(p' + p'')$, calculate $\theta'(\tilde{p})$. Then we set $p' = \tilde{p}$ if $\theta'(\tilde{p}) > 0$ and $p' = \tilde{p}$ otherwise, until $(p'' - p') < \varepsilon$.

We can also utilize various heuristic algorithms. For instance, we describe a simple **Algorithm (SQ)**. Define $I_a = \{i \in I \mid \beta_i > p'\}$, set $y_i^* = 0$ for $i \notin I_a$ and re-arrange the indices in I_a to have the descending order for the values of β_i. Then find two sequential indices i_l and i_{l+1} in I_a such that $\Delta_l < 0$ and $\Delta_{l+1} > 0$, where

$$\Delta_l = \sum_{s=1}^{l} y_{i_s}(\beta_{i_l}) - x(\beta_{i_l}).$$

Then find p^* such that $\theta'(p^*) = 0$ in the segment $[\beta_{i_l}, \beta_{i_{l+1}}]$.

3. Multi-zonal network problem

Let us consider a more general model where a telecommunication network is divided into several zones (clusters). The problem of a manager of the network is to find the optimal allocation of a limited homogeneous network resource among the zones in order to maximize the total profit containing the total income received from consumers' fees and negative resource implementation costs; see [8, 9].

Let us use the following notation:

- n is the number of zones;
- I_k is the index set of users (currently) located in zone k ($k = 1,\dots,n$);
- B is the total resource supply (the total bandwidth) for the system (network);
- x_k is an unknown quantity of the resource allotted to zone k with the upper bound b_k and $f_k(x_k)$ is the cost of implementation of this quantity of the resource for zone k ($k = 1,\dots,n$);
- y_i is the resource amount received by user i with the upper bound a_i and $\varphi_i(y_i)$ is the charge value paid by user i for the resource value y_i.

The network manager problem is the optimization problem involving capacity and balance constraints:

$$\max \to \sum_{k=1}^{n}\left[\sum_{i\in I_k}\varphi_i(y_i) - f_k(x_k)\right], \tag{4}$$

subject to

$$\sum_{k=1}^{n} x_k \le B; \tag{5}$$

$$\sum_{i\in I_k} y_i = x_k,\ k = 1,\dots,n; \tag{6}$$

$$0 \le y_i \le a_i, i \in I_k, 0 \le x_k \le b_k, k = 1,\dots,n. \tag{7}$$

That is, (6) provides the balance for demand and supply in each zone, (7) involves capacity constraints for users and network supply values in each zone, and (5) gives the upper bound for the total resource supply.

In what follows we assume that there exists at least one feasible point satisfying conditions (5)–(7), all the functions $f_k(x_k)$ and $-\varphi_i(y_i)$ are convex and quadratic, i.e.

$$\begin{aligned} &\varphi_i(y_i) = 0.5\alpha_i y_i^2 + \beta_i y_i, \alpha_i < 0, i \in I_k,\\ &f_k(x_k) = 0.5\gamma_k x_k^2 + \delta_k x_k, \gamma_k > 0; k = 1,\dots,n. \end{aligned} \tag{8}$$

This means that (4)–(8) is a convex quadratic optimization problem. However, due to large dimensionality and inexact data one can meet serious drawbacks in solving this problem with usual finite or penalty solution methods. In order to create an efficient method, we have to take into account its separability and apply certain decomposition approach. However, the standard duality approach using the Lagrangian function with respect to all the functional constraints leads to the multi-dimensional dual optimization problem. We will apply another approach, which was suggested in [13]. Let us define the Lagrange function of problem (4)–(7) as follows:

$$L(x,y,\lambda) = \sum_{k=1}^{n}\left[\sum_{i\in I_k}\varphi_i(y_i) - f_k(x_k)\right] - \lambda\left(\sum_{k=1}^{n} x_k - B\right).$$

We utilize the Lagrangian multiplier λ only for the total resource bound. We can now replace problem (4)–(7) with its dual:

$$\min_{\lambda\ge 0} \to \psi(\lambda), \tag{9}$$

where

$$\psi(\lambda) = \max_{(x,y)\in W} L(x,y,\lambda) = \lambda B + \max_{(x,y)\in W} \sum_{k=1}^{n} \left[\sum_{i\in I_k} \varphi_i(y_i) - f_k(x_k) - \lambda x_k \right],$$

$$W = \left\{ (x,y) \begin{array}{l} \sum_{i\in I_k} y_i = x_k, 0 \le y_i \le a_i, i \in I_k, \\ 0 \le x_k \le b_k, k = 1,\ldots,n \end{array} \right\}.$$

By duality (see e.g. [14, 15]), problems (4)–(7) and (9) have the same optimal value. But solution of (9) can be found by one of well-known single-dimensional optimization algorithms; see e.g. [15]. In order to calculate the value of $\psi(\lambda)$ we have to solve the inner problem:

$$\max \to \sum_{k=1}^{n} \left[\sum_{i\in I_k} \varphi_i(y_i) - f_k(x_k) - \lambda x_k \right],$$

subject to

$$\sum_{i\in I_k} y_i = x_k, 0 \le y_i \le a_i, i \in I_k,$$
$$0 \le x_k \le b_k, k = 1,\ldots,n.$$

Obviously, this problem decomposes into n independent zonal optimization problems

$$\max \to \left[\sum_{i\in I_k} \varphi_i(y_i) - f_k(x_k) - \lambda x_k \right], \tag{10}$$

subject to

$$\sum_{i\in I_k} y_i = x_k, 0 \le y_i \le a_i, i \in I_k,$$
$$0 \le x_k \le b_k;$$

for $k = 1,\ldots,n$. Each k -th independent zonal problem (10) clearly coincides with problem (1) where

$$\varphi_i(y_i) = 0.5\alpha_i y_i^2 + \beta_i y_i, i \in I_k,$$
$$u(x) = f_k(x_k) + \lambda x_k = 0.5\gamma_k x_k^2 + (\delta_k + \lambda)x_k.$$

Therefore, we can find its solution by the algorithms of Section 2.

4. Numerical experiments

The methods were implemented in C++ with a PC with the following facilities: Intel(R) Core(TM) i7-4500, CPU 1.80 GHz, RAM 6 Gb.

The initial intervals for choosing the dual variable λ were taken as [0,1000]. Values of b_k were chosen by trigonometric functions in [1,11], values of a_i were chosen by trigonometric functions in [1,2]. Value B were taken equal 1000. The number of zones was varied from 5 to 105, the number of users was varied from 210 to 1010. Users were distributed in zones either uniformly or according to the normal distribution.

The coefficients of the functions $f_k(x_k)$ and $\varphi_i(y_i)$ from (8) were taken as

$$\gamma_k = 2(|sin(2k+2)|+1), \delta_k = |cos(k+1)|+3,$$

and

$$\alpha_i = -6|cos(2i+1)|-6, \beta_i =|sin(i+2)|-1.$$

For all the methods of finding solution of problem (4)–(7) the accuracy of upper dual problem solution were varied from 10^{-1} to 10^{-4}. The accuracy of lower level problem solution in Algorithm (BS), was fixed and equal to 10^{-2}. For each set of the parameters made 1000 tests. Let J denote the total number of users, T_ε the total processor time in seconds. The results of computations are given in Tables 1–3.

Table 1. Results of testing with $J = 510,\ n = 70$.

ε_λ	T_ε: Algorithm (SQ)	T_ε: *Algorithm (BS)*
10^{-1}	0.0183	0.0018
10^{-2}	0.0217	0.0017
10^{-3}	0.0247	0.0020
10^{-4}	0.0289	0.0022

Table 2. Results of testing with $n = 70,\ \varepsilon = 10^{-2}$.

J	T_ε: Algorithm (SQ)	T_ε: *Algorithm (BS)*
210	0.0043	0.0009
310	0.0088	0.0009
410	0.0142	0.0012
510	0.0217	0.0017
610	0.0303	0.0022
710	0.0397	0.0026
810	0.0515	0.0032
910	0.0639	0.0039
1010	0.0791	0.0048

Table 3. Results of testing with $J = 510,\ \varepsilon = 10^{-2}$.

n	T_ε: Algorithm (SQ)	T_ε: *Algorithm (BS)*
5	0.0221	0.0004
15	0.0217	0.0005
25	0.0218	0.0006
35	0.0217	0.0006
45	0.0214	0.0017
55	0.0220	0.0010
65	0.0215	0.0011
75	0.0223	0.0010
85	0.0216	0.0012
95	0.0220	0.0012
105	0.0219	0.0015

As we can see from the results in the tables, in all the cases the suggested methods were rather effective in finding a solution. Moreover, for the same accuracy, both the methods gave the same numbers of upper iterations, so that the main difference was in the processor time which showed that utilization of Algorithm (BS) for inner optimization problems give better performance.

5. Conclusions

We considered several problems of optimal resource allocation in telecommunication networks. We presented simple and efficient solution methods for the case where the network manager aims to distribute some homogeneous resource (bandwidth) among users of one region with quadratic charge and fee functions. Next, we considered a more general problem for a provider of a wireless communication network divided into zones. By using the dual Lagrangian method with respect to the capacity constraint, we suggest to reduce the initial problem to a single-dimensional optimization problem, where calculation of the cost function value leads to independent solution of zonal problems, which coincide with the above single region problem. Some results of computational experiments confirm the applicability of the new methods.

Acknowledgments
In this work, the authors were supported by the RFBR grant, project No. 16-01-00109a. Also, the first author was supported by grant No. 297689 from Academy of Finland.

References

[1] Courcoubetis C and Weber R 2003 *Pricing Communication Networks: Economics, Technology and Modelling* (Chichester: John Wiley & Sons)

[2] Stańczak S, Wiczanowski M and Boche H 2006 *Resource Allocation in Wireless Networks. Theory and Algorithms* (Berlin: Springer)

[3] Wyglinski A M, Nekovee M and Hou Y T (eds.) 2010 *Cognitive Radio Communications and Networks: Principles and Practice* (Amsterdam: Elsevier)

[4] Leshem A and Zehavi E 2009 Game theory and the frequency selective interference channel: A practical and theoretic point of view *IEEE Signal Process* **26** pp 28–40

[5] Raoof O and Al-Raweshidy H 2010 Auction and game-based spectrum sharing in cognitive radio networks. in: *Game Theory* ed Q Huang (Rijeka: Sciyo) ch 2 pp 13–40

[6] Huang J, Berry R A and Honig M L (2006) Auction-based spectrum sharing *ACM/Springer Mobile Networks and Appl.* **11** pp 405–418

[7] Koutsopoulos I and Iosifidis G 2010 Auction mechanisms for network resource allocation. in: *Proc. of Workshop on Resource Allocation in Wireless Networks, WiOpt 2010* pp 554–563

[8] Konnov I V, Kashina O A and Laitinen E 2011 Optimisation problems for control of distributed resources *Int. J. Model., Ident. and Contr.* **14** pp 65–72

[9] Konnov I V, Kashina O A and Laitinen E 2012 Two-level decomposition method for resource allocation in telecommunication network *Int. J. Dig. Inf. Wirel. Comm.* **2** pp 150–155

[10] Konnov I V 2006 On modeling of auction type markets *Issled. Inform.* **10** pp 73–76, Available at SSRN: http://ssrn.com/abstract=2482282

[11] Konnov I V 2015 An alternative economic equilibrium model with different implementation mechanisms *Adv. Model. Optim.* **17** pp 245–265

[12] Patriksson M and Strömberg C 2015 Algorithms for the continuous nonlinear resource allocation problem: New implementations and numerical studies *Eur. J. Oper. Res.* **243** pp 703–722

[13] Konnov I V, Kashuba A Yu and Laitinen E 2015 A simple dual method for optimal allocation of total network resources *Recent Advances in Mathematics. Proceedings of the International Conference "PMAMCM 2015"* ed I J Rudas (Zakynthos) pp 19–21

[14] Polyak B T 1983 *Introduction to Optimization* (Moscow: Nauka) (Engl. transl. in Optimization Software, New York, 1987)

[15] Konnov I V 2013 *Nonlinear Optimization and Variational Inequalities* (Kazan: Kazan Univ. Press)

IOP Conf. Series: Materials Science and Engineering **158** (2016) 012055 doi:10.1088/1757-899X/158/1/012055

Scalability of parallel finite element algorithms on multi-core platforms

S P Kopysov, A K Novikov, N S Nedozhogin and V N Rychkov

Institute of Mechanics, Ural Branch of the Russian Academy of Sciences, 34 T. Baramzinoy, Izhevsk, Russia 426067

E-mail: s.kopysov@gmail.com

Abstract. The speedup of element-by-element FEM algorithms depends not only on peak processor performance but also on access time to shared mesh data. Eliminating memory boundness would significantly speed up unstructured mesh computations on hybrid multi-core architectures, where the gap between processor and memory performance continues to grow. The speedup can be achieved by ordering unknowns so that only those elements are processed in parallel which do not have common nodes. Therefore, memory conflicts are minimized. FEM assembly is performed with respect to the ordering, which defines how to compose vectors. Mesh can be partitioned into disjoint subdomains by using different layer-by-layer schemes. In this work, we evaluated several partitioning schemes (block, odd, even, and their modifications) on multi-core platforms, using Gunther's Universal Law of Computational Scalability. We performed numerical experiments with element-by-element matrix-vector multiplication on unstructured meshes on multi-core processors accelerated by MIC and GPU. With ordering, we achieved 5-times speedup on CPU, 40-times speedup on MIC, and 200-times speedup on GPU.

1. Introduction

Performance and scalability modeling of computational algorithms on modern multi-core/many-core platforms is a crucial problem. On multi-core platforms, it is important to take into account different types of delays, which can be caused by: exchange of shared rewritable data between the processor caches and between the processors and main memory (i), synchronization locks (serialization) shared data available for recording (ii), waiting for memory access to complete operations (iii), etc. Evolution of parallel speedup models derived from Amdahl's law is described in [1,2]. In particular, they refer to Hill-Marty model [3], which includes additional parameters defining a number of computing hardware resources. However, estimation of these parameters for real platforms is not trivial.

For over a decade, the Universal Scalability Law has been successfully applied to model diverse software applications on modern hardware platforms [4,5] but it has been largely ignored by the parallel simulation community. This model accounts for waiting for access to shared resources, retrograde scaling (latency due to exchange of data between caches) and resource limitations. In this work, we apply this model to FEM simulations. To the best of our knowledge, this is the first application of this model to parallel numerical algorithms.

Predictive models based on the results of real computational experiments allow for detecting and eliminating the sources of delays. Many FEM algorithms demonstrate irregular memory accesses, which significantly reduces the efficiency of parallel processing. The irregularity is caused by unstructured meshes. As a result, the global assembly phase becomes the main performance bottleneck in many FEM algorithms [6].

This paper is structured as follows. In Section 2, we show how element-by-element FEM can be parallelized on shared memory architecture. In Section 3, we present new partitioning schemes, which significantly improve data locality and performance of element-by-element FEM algorithms. In Section 4, we describe experimental platforms and their memory access features. In Section 5, we develop the Gunther's Universal Law of computational scalability in relation to proposed algorithms and multi-core architectures. Section 6 concludes the paper.

2. Element-by-element FEM on shared memory platforms

In element-by-element schemes [7], assembling is a part of solving the finite element system of equations. Assembly operator is applied not to matrices but to vectors resulting from the matrix-vector multiplication:

$$q = Kp = \sum_{e=1}^{m} C_e^T \widetilde{K}_e C_e p = \sum_{e=1}^{m} C_e^T \tilde{q}_e = \sum_{e=1}^{m} q_{e,} \tag{1}$$

where q, p, q_e are the vectors of size N; K is the global matrix $N \times N$; $\widetilde{K}_e$ is the local stiffness matrix $N_e \times N_e$ of a finite element e; C_e is the incidence matrix $N_e \times N$ that maps the local space if numbers of unknowns (degrees of freedom) $[1,2,\dots,N_e]$ into the global space $[1,2,\dots,N]$; m is the number of finite elements. This mapping can be done either by indirect indexing of mesh nodes to unknowns (i) or by multiplying by the incidence matrix, which can be efficiently implemented on GPUs (ii).

Let us consider a parallel version of (1) for shared memory. Here vectors p and q are in the shared memory of parallel processes/threads, and matrices $\widetilde{K}_e$, C_e, C_e^T and sparse vector q_e are in the local memory of processes. Then summation of the vectors qe that belong to different processes but have nonzero components with the same indices may result in conflicts and cause errors.

The product $q = Kp$ can be replaced by two operations: the element-by-element multiplication $\tilde{q}_e = \widetilde{K}_e C_e p$, $e = 1,2,\dots,m$ and assembling $q_e = \sum_{e=1}^{m} C_e^T \tilde{q}_e$. These operations have different computation and communication costs because of different data locality, especially in case of unstructured meshes. They have different memory access patterns, determined by local and global node indexing, and therefore have different potential for parallelization.

Assume that each process computes over $m_i \approx m / n$ finite elements, where i is the process number, and n is the number of processes. Then the product can be expressed in the matrix form as follows:

$$q = Kp = \sum_{i=1}^{n} C^{T(i)} \widetilde{K}^{(i)} C^{(i)} p = \sum_{i=1}^{n} C^{T(i)} \tilde{q}^{(i)} = \mathcal{A}(\tilde{q}), \tag{2}$$

where $C^{(i)}$ is the incidence matrix assembled from m_i matrices C_e; $\widetilde{K}^{(i)}$ is the block-diagonal matrix united from m_i matrices $\widetilde{K}_e$ as blocks; $\tilde{q}^{(i)}$ is the vector assembled from m_i vectors $\tilde{q}_e$; and $\mathcal{A}$ is the assembly operator for the vector q.

To make this algorithm efficient on shared memory platforms, it is necessary to minimize concurrent memory accesses. This can be achieved by splitting finite elements into sets that do not share nodes with each other. Then, these sets are assigned to different processes/threads, and the assembly operation (2) is performed on nonadjacent elements.

3. FEM algorithms with layer-by-layer partitioning

In this section, we present new partitioning schemes, which significantly improve data locality and performance of element-by-element FEM algorithms on shared memory platforms.

We assume that two subsets of mesh cells are nonadjacent at some moment of time if the cells simultaneously taken from these subsets do not contain common vertices. The moment of time is taken when the subsets are accessed for computations. This means that the cells in a subset and the subsets themselves can be connected topologically.

We partition the mesh Ω into nonoverlapping layers of cells $s_j, j = 1,2,\dots,n_s$ (figure 1(a)), then combine the layers to get the subsets of cells (subdomains) $\Omega_i, i = 1,2,\dots,n_\Omega$. We use different schemes to construct subdomains Ω_i from layers. Our target is to reorder cells so that the cells with common vertices are not accessed simultaneously from parallel processes (threads).

In the block scheme, the layers are combined consecutively, then these combined layers are divided into subdomains with approximately equal number of finite elements (figure 2(a)). In the layer index parity scheme, the layers are enumerated (figure 2(c)). The dashed line in figure 2 (b) represents the end of one parallel OpenMP section and the beginning of another.

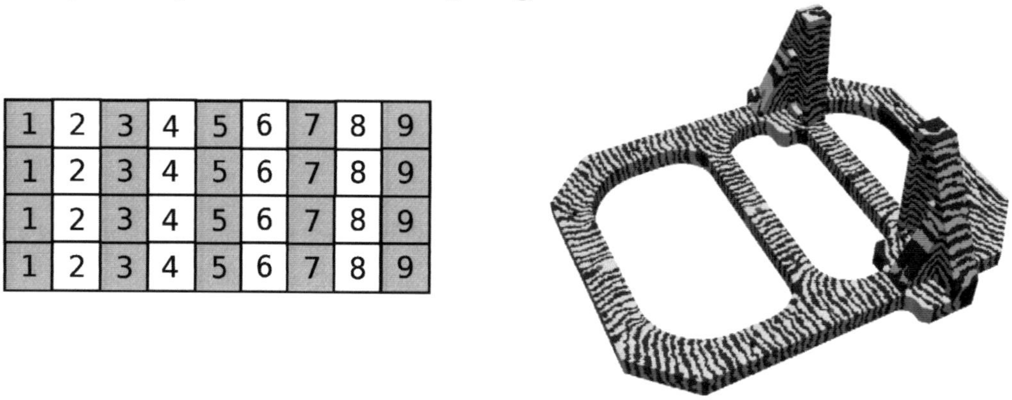

(a) (b)

Figure 1. Layer-by-layer mesh partitioning: (a) scheme; (b) result: unstructured mesh with 137 layers, 485843 tetrahedrons.

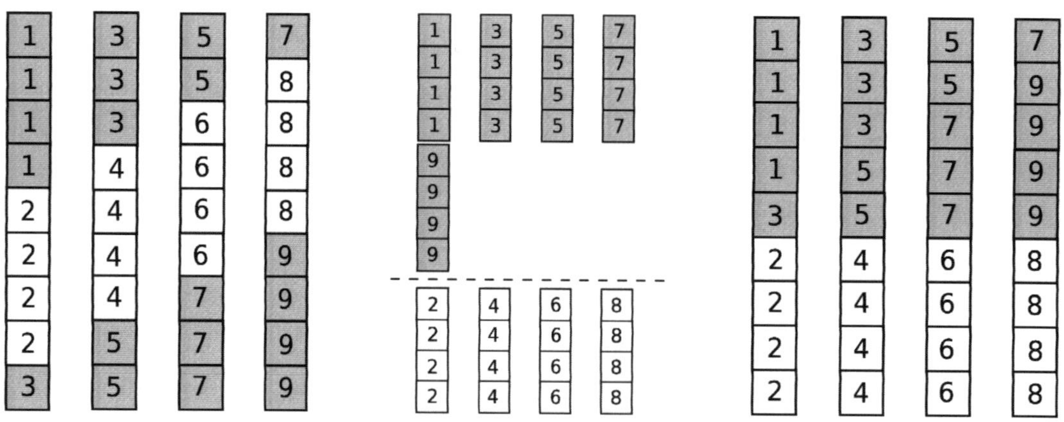

(a) (b) (c)

Figure 2. Combining layers into blocks for four processes ($n_\Omega = 4$): (a) block scheme (bl); (b) layer index parity scheme (ev); (c) balanced layer index parity scheme (ev+bal).

We compared these schemes by measuring the performance and load imbalance of the element-by-element matrix-vector product without assembling the vectors (this operation is fully parallel). We performed experiments with the unstructured mesh shown in figure 1 on eight-core processor Xeon E5-2690. Figure 3 shows the speedup obtained with four types of the mesh partitioning: block (bl), odd-parity (ev), balanced odd-parity (ev+bal), and multilevel graph partitioning by METIS). The load imbalance in percent to equal load is shown in percentage above the bars. The load imbalance of (bl) and (ev+bal) is neglected as it was essentially less than 1% in all experiments. Achieving the linear speedup is limited by load imbalance. The load imbalance caused by the proposed mesh partitioning schemes proved to be less than the one caused by the multilevel partitioning of the mesh dual graph [8].

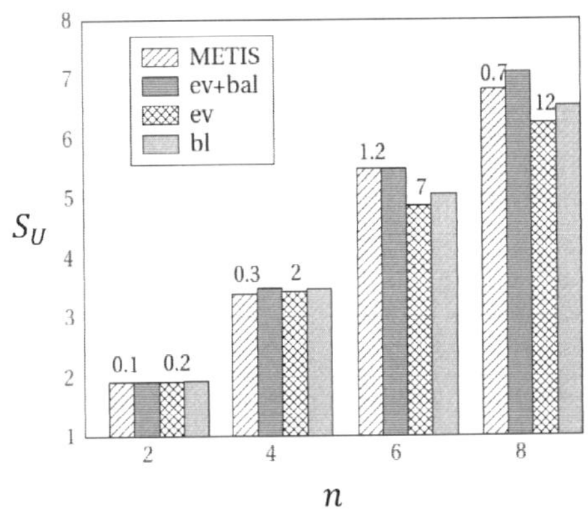

Figure 3. The speedup of the matrix-vector product on Xeon E5-2690 for different mesh partitioning schemes and numbers of processes ($n = 2, 4, 6, 8$).

The rest of the paper will be detected the sources of delays in some parallel FEM operations using layer-by-layer partitioning on multi-core/many-core platforms.

4. Experimental platforms and their memory access features

In this section, we describe experimental platforms and their memory access features.

Parallel execution of FEM algorithms on hybrid multi-core platforms significantly increases the frequency of memory access so that memory bandwidth becomes a performance bottleneck. Multiple processes/threads compete for memory, which results in memory contention. This situation is further complicated by multi-level caches, which allow for reusing frequently accessed data but require maintaining data consistency (cache coherence). Memory can be utilized more efficiently if multiple writes to the same memory location issued by different processes are executed sequentially (i), and there is a gap in time between write and read operations issued from different processes to the same memory location (ii).

We performed experiments with parallel element-by-element FEM algorithms on several platforms, which represent most popular multi-core architectures: CPU — Xeon E5-2609, Opteron 8435, Xeon E5-2690; MIC — Xeon Phi 7110X; GPU — GTX 980. Platforms based on these architectures have significant differences in memory access.

Important for the scalability analyzing of parallel algorithms are methods to solve problems of memory access organization in modern architecture. Some of them settled in the hardware or the compiler and should be somehow taken into account when describing the architecture of the computer system on which the computation will be performed and the behavior that we want to predict.

5. Universal scalability estimation

Model of computing is the next higher level of abstraction and is more formal model of the appropriate architecture model. It allows you to build a cost function reflecting the time required to perform the algorithm on the resources of a computing system, a given architecture model. This computing model should provide a simple method for the design and evaluation of parallel algorithms.

In the work of Gunther [5] (see Bibliography therein) was demonstrated the possibility of the representation and evaluation of scalability / speedup on the basis of generalization of Amdahl's law, taking into account overheads relating to the agreement of calculations between identical cores / processors and the maintenance of a coherent state of the system.

Gunther's model (USL) is considered to be universal in the sense that it is not expected to have any properties of the algorithm, software, or a specific multi-core architecture of the system, and others.

The model assumes that speedup is function that has the global maximum value and depends on the number of cores/processors, the degree of contention, and the lack of coherency.

The use of this model is to measure the acceleration of parallel computing $S(n) = t_1 / t_n$, where t_1, t_n — algorithm execution time of one or n — threads and to obtain the most accurate form for an speedup function of the specific algorithm and computing system with estimation of two model parameters.

The speedup of an arbitrary computing system (algorithmic, program and architectural) is performed a rational function.

Speedup of S_U with an increase in the number of parallel processes n in the USL model is written as

$$S_U(n) = \frac{n}{1 + \sigma(n-1) + \lambda n(n-1)}, \quad (3)$$

where parameters σ, λ defined within $0 \leq \sigma$, $0 < \lambda < 1$. At $\lambda = 0$ ratio (3) takes the form of Amdahl's law with the part of sequential computing, defined as $\sigma = (1 - f)$, where f — parallel fraction. The denominator in (3) consists of three terms, each has a definite physical interpretation: parallelism — linear scalability is possible if the individual components of the system (CPU, threads, etc.) can work without any interaction $\sigma, \lambda = 0$; scalability is limited contention — conflicts between different parts of a computer system (algorithm), caused by the effects of serialization and queuing $\sigma > 0, \lambda = 0$; scalability is limited coherence — delays caused by the maintenance of the system in a consistent state, is associated with a cache memory in a different parts of software and hardware of the computer system $\sigma, \lambda > 0$.

Thus, the parameters of the model can be represented in charge of two specific mechanisms that limit the linear acceleration of the parallel algorithm with increasing of the number of parallel processes, or cores. Note, that the second term increases linearly and the third — square with increasing of the number n of interacting processes. Then, computational algorithms, architectures can be classified and described in accordance with the settings σ, λ. The absolute values of the model parameters do not specify the detailed mechanisms and data for computing organizing and working with memory, but fairly well characterized model of parallelism and architecture of the system when considering the obtained speedup for one and the same algorithm as data partitioning on different types of multi-core architectures. However, it is possible to assess and predict the incidence of excessive costs on the agreement and synchronization of access to shared data between threads.

Based on the data obtained of parallel speedup $S = t_1 / t_n$ for the above algorithms of layer-by-layer partitioning determined scalability (3), (4) of parallel operation of vectors assembly $q = \mathcal{A}(\tilde{q})$ on the unstructured mesh for several architectures and technologies of parallel programming with increasing of the parallel processes number n (for technologies OpenMP and CUDA see table 1). Numerical experiments carried out over the assembly vectors results element-by-element matrix-vector product for adaptive unstructured mesh, shown in figure 1. The partition by parity numbers of layers with alignment load was chosen for comparison parameters of Gunther's model on different processor architectures. This optimized data movement and memory access. Thus, the calculations were performed on the same data, to separate them by threads number used the same single algorithm and software OpenMP with parallelization of cycle for finite elements by thread number. To improve the effectiveness of the assembly operations on the GPU algorithm was modified by the introduction of new levels of parallelization.

Table 1. USL parameters of the assembly operation $\boldsymbol{q} = \mathcal{A}(\tilde{\boldsymbol{q}})$ on the unstructured mesh (figure 1) for experimental platforms.

USL parameter	Xeon E5-2609	Opteron 8435	Xeon E5-2690	Xeon Phi 7110X	GeForce GTX 980
n	4	6	8	61	2048
σ	0.1	0.021	0.08	0.006	0.0018
λ	0.0001	0.02	0.0001	0.00006	0.00000044
R^2	0.99	0.982	0.9965	0.994	0.999
$S_{U_{max}} / n_{U_{max}}$	8.41 / 94	3.56 / 7	10.09 / 96	46.76 / 129	318.96 / 1511
S_{max} / n_{max}	3.1 / 4	3.48 / 5	5.19 / 8	39.43 / 60	219 / 384

The non-linear regression analysis was used to find the model parameters in (3). The results are shown in table 1. The degree of correspondence between experimental S and data computed by USL — S_U is estimated with R^2 measure. The obtained values of R^2 explain the change of speedup with dependence from n for some σ and λ.

The obtained model parameters σ, λ allowed us to estimate the number of processor cores, at which the maximum speedup of $n_{U_{max}} = \sqrt{(1-\sigma)\lambda}$ is achieved for the algorithm executed on the CPU and computing accelerators. Among the obtained results, we note a slowdown of computations at parallel vectors assembly in the shared memory model for graph partitioning with a critical section. The critical section is not required in the proposed layer-by-layer partitioning algorithms, all restrictions on access to the shared memory are allowed, and as a result, good scalability are achieved on universal computing processors and accelerators. The figure 4 shows the experimental measuring results (marks) of the speedup for assembly operations are for different architectures and universal processors and curves constructed by the model of USL on (3).

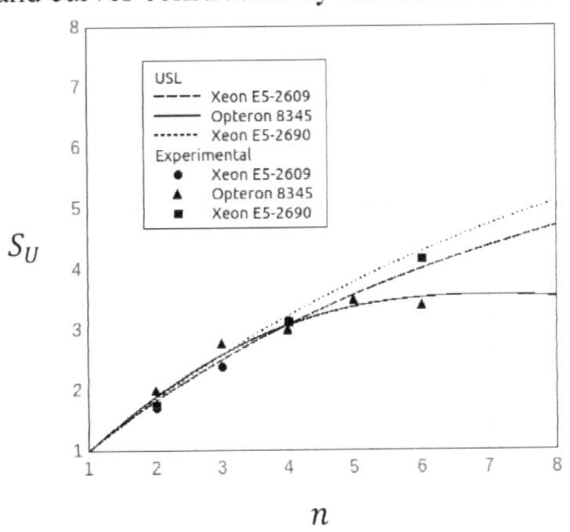

Figure 4. Scalability of assembly operation on multi-core CPUs.

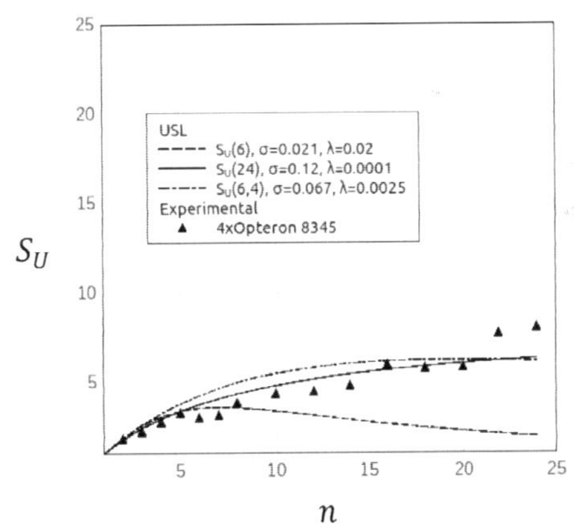

Figure 5. Scalability of assembly operation on multi-socket node.

A significant three-five-fold speedup of parallel assembly is achieved even for unstructured meshes, which is about half the maximum possible speedup while increasing the number of cores on the order for the architecture of Xeon E5 processors. In the six-core Opteron processors, with a three-level cache organization costs for agreement of data location lead to the fact that the maximum speedup occurs only five cores involved.

Scalability on computing accelerators

Element-by-element assembly vectors can efficiently be carried out on the same computing accelerators (see figure 6), which is confirmed by the experimental results and the estimation of the USL model parameters for the advanced accelerators MIC with enlarged cores number. For comparison, experiments were performed on quasi-structured meshes. The assembly operation on MIC is performed with almost the same scalability, which corresponds to the obtained values of the model parameters σ, λ, characterizing cache coherency among all used cores (speedup of about 40 times on the 61 cores). In this assembly algorithm, memory access is hidden behind the calculation rate (minimum parameters σ, λ among all architectures) at computing on the graphics card GTX 980 and the maximum acceleration could be achieved with such an architecture with 1511 cores (see figure 7). Experimentally achieved speedup is limited only by the ability at a given mesh to obtain the desired number of layers and the presence of an appropriate amount of RAM.

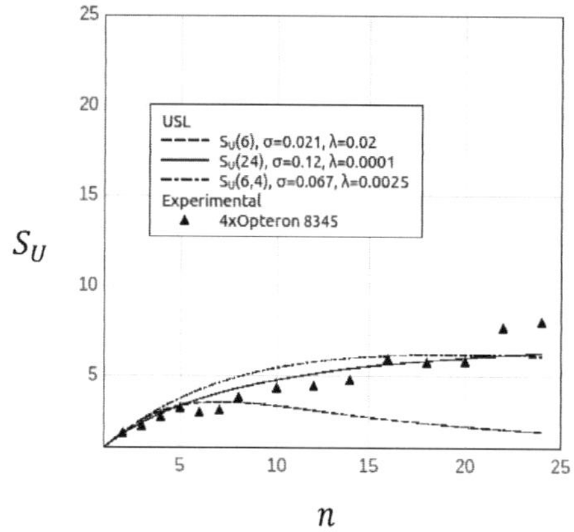

Figure 6. Scalability of assembly operation on MIC.

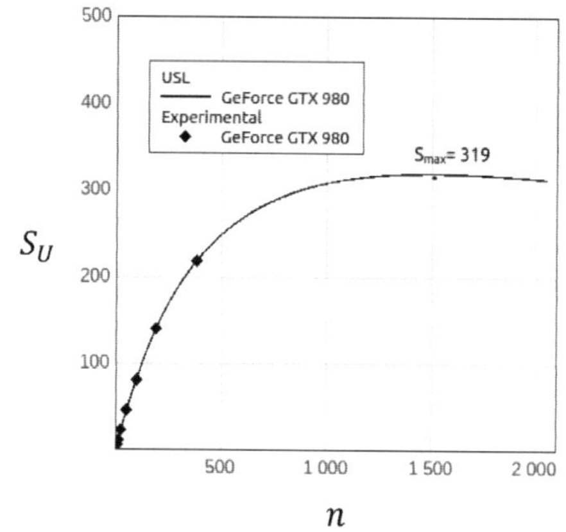

Figure 7. Scalability of assembly operation on GPU.

Scalability on multi-socket node

Since the universal model has no internal structure, it can be used for more general simulation, but only homogeneous computing system, such as multiprocessor (multi-socket) computing nodes containing the same processors and clusters consisting of computing nodes with processors of the same architecture and performance.

Define the parameters of a multiprocessor computing node as: p — the number of multicore processors in a single compute node; $n_p n / p$ — number of cores per one processor; σ_p — parameter characterizing contention between processors; λ_p — data consistency across multiple processors of a computing node.

Then we can write a generalization of ratio (3) to define a universal scalability of computational node containing p processors/sockets

$$S_U(n,p) = \frac{pS_U(n)}{1+\sigma_p(p-1)S_U(n)+\lambda_p(p-1)pS_U^2(n)}, \tag{4}$$

where $S_U(n,p)$ — speedup, and determined taking into account the overhead multi-sockets node based on the speedup of a multi-core processor $S_U(n)$. At $p = 1$, $\sigma_p = 0$ last expression is converted to (3).

To use this model to consider when evaluating the speedup of multi-sockets node $S_U(n,p)$, that all of its processors should also be uniform. In the future, the ratio of (3) can also be generalized to evaluate the speedup computing cluster $S_U(n,p,n_c)$, containing n_c — computing nodes connected by a communication network and determining the appropriate parameters σ_c, λ_c.

For the four sockets node with Opteron 8345 processors and non-uniform memory access were measured computational costs and three speedup estimations were realized (see figure 5). The first curve $S_U(6)$ constructed according to a six-core processor, and shows a significant difference from the experimental results with the model of increasing the number of cores.

The curve denoted by $S_U(24)$ is obtained in the approximation of the results for the model (3) when performing assembly operations on all available twenty four cores excluding the non-uniform memory access neighboring processors. In this case, evaluation of the acceleration reaches seven in the presence of a large number of cores. When model accounting for one processor (3) and NUMA architecture four sockets node in the model (4) obtained more reliable results predicted maximum acceleration of about six.

6. Conclusion

The results of the speedup simulation characterize fairly well used model of parallel computing for consideration obtained speedup of the same algorithm and data sharing on different types of multi-core architectures. On the basis of this model it is possible to consider the applicability of the algorithms change, architecture or software in order to minimize the values of the parameters σ, λ and reduce costs.

In this work, we used USL model for scalability analysis of element-by-element FEM operations on modern multi-core platforms. We extended this model for multi-socket, MIC, GPU architectures, and demonstrated that it adequately approximates the scalability.

Acknowledgments

This work was partially supported by RFBR according to the research projects №14–01–00055-a, №16–01–00129-a.

References

[1] Yavits L, Morad A and Ginosar R 2014 *Parallel Computing* D **40** 1
[2] Al-Babtain B M, Al-Kanderi F J, Al-Fahad M F and Ahmad I 2013 *Int. J. of New Comp. Arch. and their Appl.* **3** 30
[3] Hill M D and Marty M R 2008 *Computer* **41** 33
[4] Gunther N J 2002 *CoRR.* **cs.DC/0210017**
[5] Gunther N J, Puglia P and Tomasette K 2015 *Comm. of the ACM* **58** 46
[6] Markall G R, Slemmer A, Ham D A, Kelly P H J, Cantwell C D and J S S 2013 *Int. J. Numer. Meth. Fluids* **71** 80
[7] Kopysov S P, Kuzmin I M, Nedozhogin N S, Novikov A K, Rychkov V N, Sagdeeva Y A and Tonkov L E 2014 *Computer Research and Modeling* **6** 79
[8] Karypis G and Kumar V 1998 *SIAM Journal on Scientific Computing* **20** 359

Consistent robust a posteriori error majorants for approximate solutions of diffusion-reaction equations

V G Korneev

Sankt-Petersburg state university, 7/9 Universitetskaya Emb., 199034, Saint Petersburg, Russia

E-mail: vad.korneev2011@yandex.ru

Abstract. Efficiency of the error control of numerical solutions of partial differential equations entirely depends on the two factors: accuracy of an a posteriori error majorant and the computational cost of its evaluation for some test function/vector-function plus the cost of the latter. In the paper consistency of an a posteriori bound implies that it is the same in the order with the respective unimprovable a priori bound. Therefore, it is the basic characteristic related to the first factor. The paper is dedicated to the elliptic diffusion-reaction equations. We present a guaranteed robust a posteriori error majorant effective at any nonnegative constant reaction coefficient (r.c.). For a wide range of finite element solutions on a quasiuniform meshes the majorant is consistent. For big values of r.c. the majorant coincides with the majorant of Aubin (1972), which, as it is known, for relatively small r.c. ($< ch^{-2}$) is inconsistent and looses its sense at r.c. approaching zero. Our majorant improves also some other majorants derived for the Poisson and reaction-diffusion equations.

1. Introduction.

For the successful error control of approximate solutions to the boundary value problems, the guaranteed a posteriori error majorant must be sufficiently accurate and cheap in a sense of the computational work. The first requirement can be considered as satisfied at least in part, if the majorant is consistent in respect to the order of accuracy with the a priori convergence estimate of the numerical method. Obviously, a consistent majorant is unimprovable in the order, if the a priori convergence estimate is unimprovable in the same sense. An error majorant usually depends on the approximate solution and on some other function or functions which are termed test functions. In this paper consistency assumes that it can be approved by an easily calculated test function with the use of some procedure of the linear complexity.

The term "functional a posteriory error majorants" is usually related to a posteriory error bounds possessing significant generality and some other positive properties. However, sometimes generality is attained for the price of the lost of consistency, resulting in the necessity of attracting some majorant minimization procedures over some set of the admissible test functions [8, 9]. The computational cost of such procedures can exceed the cost of the numerical solution of the boundary value problem.

The majorant of Aubin [3] is one of the earliest. Let us illustrate it on a model problem

$$-\operatorname{div}(\mathbf{A}\operatorname{grad} u) + \sigma u = f(x), \quad x \in \Omega \subset R^m,$$
$$u\,|_{\Gamma_D} = \psi\,|_D, \qquad -\mathbf{A}\nabla u \cdot \nu\,|_D = \psi_N,$$

where Γ_D, Γ_N are not intersecting parts of the boundary $\partial\Omega=\Gamma_D\cup\Gamma_N$, $\text{mes}\Gamma_D>0$ ν is the internal unite normal to the boundary, $\mathbf{A}$ is a symmetric $m\times m$ matrix, and $0<\sigma=\text{const}$. It is assumed that the matrix $\mathbf{A}$ satisfies the inequalities

$$\mu_1\xi\cdot\xi\le\mathbf{A}\xi\cdot\xi\le\mu_2\xi\cdot\xi, \qquad 0<\mu_1,\mu_2=\text{const}$$

for any $x\in\Omega$ and $\xi\in\mathbf{R}^m$. Everywhere in the paper, the boundary $\partial\Omega$, the coefficients of the matrix $\mathbf{A}$ and the right hand part f are assumed to be sufficiently smooth, if more specific requirements to their smoothness are absent.

The error bounds in the energy norm

$$|||u|||=(\|u\|_{\mathbf{A}}^2+\sigma\|u\|_{L_2(\Omega)}^2)^{1/2} \quad \|u\|_{\mathbf{A}}^2=\int_\Omega\nabla u\cdot\mathbf{A}\nabla u \tag{2}$$

are most usable in applications. For vectors $\mathbf{y}\in\mathbf{R}^m$ we introduce also the spaces $\mathbf{L}_2(\Omega)=(L_2(\Omega))^m$, $\mathbf{H}(\Omega,\text{div}\,u)=\{\mathbf{y}\in\mathbf{L}_2(\Omega):\text{div}\,\mathbf{y}\in L_2(\Omega)$, and the norm $]|\mathbf{y}|[_{\mathbf{A}^{-1}}=(\int_\Omega\mathbf{A}^{-1}\mathbf{y}\cdot\mathbf{y})^{1/2}$.

Theorem 1. *Let* $f\in L_2(\Omega)$, $\psi\in H^1(\Omega)$, $\psi_N\in L_2(\Gamma_N)$, v *is any function from* $H^1(\Omega)$ *satisfying the boundary condition on.* Γ_D. *Then for any* $\mathbf{z}\in\mathbf{H}(\Omega,\text{div})$, *satisfying on* Γ_N *the boundary condition* $\mathbf{z}\cdot\mathbf{v}=\psi_N$, *we have*

$$|||v-u|||\le]|\mathbf{A}\nabla v+\mathbf{z}|[_{\mathbf{A}^{-1}}^2+\frac{1}{\sigma}\|f-\sigma v-\text{div}\,\mathbf{z}\|_{L_2(\Omega)}^2.$$

The bound (3) is a particular case of the results of [3], see, *e. g.*, Theorem 22 in Introduction and additionally Theorems 1.2, 1.4, 1.6 of ch. 10. It can be found also in [15, 17]. Obviously, this majorant becomes meaningless at $\sigma\to0$. Let us add that in (3) one can use $\mathbf{z}=-\mathbf{A}\nabla w$ with any $w\in H^1(\Omega,\Im)$, satisfying the boundary condition in (1) on Γ_N. Here $\Im=\text{-div}(A\,\text{grad}\,u)$ and $H^1(\Omega,\Im)=\{w:w\in H^1(\Omega),\Im w\in L_2(\Omega)\}$. If $v\in C(\Omega)\cap H^1(\Omega)$ is the finite element solution, then $w=\tilde{w}(v)$ most often is obtained from v by some recovery technique [13, 1, 4, 5].

Let $H_0^1(\Omega):=\{v\in H^1(\Omega):v|_{\partial\Omega}=0\}$, for simplicity $\Gamma_D=\partial\Omega$, $\psi_D=0$ and $\mathbf{A}=\mathbf{I}$, where $\mathbf{I}$ is the unity matrix. In [16] for the case $\sigma=0$, it was suggested the majorant

$$\|\nabla(v-u)\|_{\mathbf{L}_2(\Omega)}^2\le(1+\varepsilon)\|\nabla v+\mathbf{z}\|_{\mathbf{L}_2(\Omega)}^2+c_\Omega(1+\frac{1}{\varepsilon})\|\nabla\cdot\mathbf{z}-f\|_{L_2(\Omega)}^2 \quad \forall\varepsilon>0 \tag{4}$$

where v and $\mathbf{z}$ are any function and vector from $H_0^1(\Omega)$ and $\mathbf{H}(\Omega,\text{div})$, respectively, and c_Ω is the constant from the Friedrichs inequality

Attempts to modify the Aubin's majorant in such a way that it provided admissible accuracy for all $\sigma\ge0$ were made in the papers [17, 18]. The latter suggests the majorant for all $\sigma=\text{const}\ge0$ of the form

$$|||v-u|||^2\le(1+\varepsilon)]|\mathbf{A}\nabla v+\mathbf{z}|[_{\mathbf{A}^{-1}}^2+\frac{1}{\sigma+\varepsilon/(c_\Omega(1+\varepsilon))}\|f-\sigma v-\text{div}\,\mathbf{z}\|_{L_2(\Omega)}^2 \tag{5}$$

It was shown in [2, 11] that the correction of arbitrary vectors $\mathbf{z}\in\mathbf{H}(\Omega,\text{div})$ into the vectors $\boldsymbol{\tau}$ satisfying the balance/equilibrium equations can be done by quite a few rather simple techniques. In particular, it is true for the correction of the flux vector ∇u_{fem} into $\boldsymbol{\tau}(u_{\text{fem}})$. This allows to implement the a posteriori bound $|||v-u|||\le]|\mathbf{A}\nabla u_{\text{fem}}+\boldsymbol{\tau}(u_{\text{fem}})|[_{\mathbf{A}^{-1}}^2$ or the bound with the additional free vector-function in the right part, which we present below. For simplicity, we restrict considerations to the same homogenious Dirichlet problem for the Poisson equation in a two-dimensional convex domain. Let T_k be the projection of the domain Ω on the axis x_{3-k} and the equations of the left and lower parts of the boundary be $x_k = a_k(x_{3-k})$, $x_{3-k} \in T_k$. If β_k are arbitrary bounded functions and $\beta_1 + \beta_2 \equiv 1$, then according to [2, 11]

$$\| \nabla(v-u) \|_{\mathbf{L}_2(\Omega)} \leq \| \nabla v + \mathbf{z} \|_{\mathbf{L}_2(\Omega)} + c\sum_{k=1}^{2} \| \int_{a_k(x_{3-k})}^{x_k} \beta_k (f - \nabla \cdot \mathbf{z})(\eta_k, x_{3-k}) d\eta_k \|_{L_2(\Omega)} \tag{6}$$

In (6) on the right we have integrals from the residual and this helps to make the majorant more accurate. Besides there is an additional free function β_1 or β_2 and it's right choice (for instance, with the use of the found approximate solution v) can accelerate the process of the minimization of the right part. Nevertheless, the majorant (6), as well as majorants (4), (5), are not consistent. Since it is practically obvious, below we discuss this matter very briefly.

The inconsistency is the most clearly visible for finite element methods of a higher smoothness. Let us turn to (4) in the case when $v = u_{\text{fem}} \in C^1(\omega) \cap H^2(\Omega)$ and $f \in H^1(\Omega)$ Therefore, $u \in H^3(\Omega)$, and the unimprovable a priori convergence estimates $\| u - v \|_{H^k(\Omega)} \leq ch^{3-k} \| u \|_{H^3(\Omega)}$, $k = 0,1,2$, hold with the mesh parameter h under assumption that the finite element assemblage satisfies the conditions of the generalized quasiuniformity, see *e. g.* [10, 12]. For the latter conditions see, e. g., Section 3.2 in [12]. Now we see that the left part of (4) is estimated from above with the order h^4. One can set $\mathbf{z} = -\nabla v$ making the first term in the right part equal to zero. At the same time, the second term in the right part is estimated from above only with the order h^2. More over since the estimates of the convergence are exact there are functions $f \in H^1(\Omega)$ for which the second term is estimated with the order h^2 from below. The proofs of the inconsistency of the majorants (5), and (6) are also straightforward.

If the FEM belongs to the class $\wp$, then we can use the so called recovered flux $\mathbf{z} = \tilde{\mathbf{z}}(u_{\text{fem}})$, whose components $\tilde{z}_k(u_{\text{fem}})$ in the simplest case are defined as functions of the same finite element space, to which u_{fem} belongs. Several cheap averaging procedures were developed for defining the nodal parameters of $\tilde{z}_k$, which were desined to provide at least the same orders of accuracy for $\partial u / \partial x_k - \tilde{z}_k$ and $\partial(u - u_{\text{fem}})/\partial x_k$, see [13, 1, 4, 5]. If $f \in L_2(\Omega)$, then in this case the order of the left part is again by the multiplier h^2 higher than the order of the right part.

Let σ_* be the value from the inequality

$$\| u - v \|_{L_2(\Omega)}^2 \leq \sigma_*^{-1} \| u - v \|_{\mathbf{A}}^2 . \tag{7}$$

There is the multiplier $1/\sigma$ before the second norm in the right part of (3). In view of this, it can be shown that at $\sigma \geq \sigma_*$ the Aubin's majorant is consistent for approximate solutions by FEM, if $\sigma_*^{-1} \leq c_+ h^2$, $c_+ = \text{const}$ and some natural conditions are fulfilled, see Lemma 1 in the next section. However, at $\sigma << \sigma_*$ the consistency deteriorates and with σ tending to zero the majorant becomes meaningless.

2. Consistent error majorant for any nonnegative reaction coefficient

We start from the presentation of a guaranteed robust error majorant valid for all $\sigma \in [0, \infty)$ which at application to the FEM solutions is consistent.

Theorem 2. *Let $\Gamma_D = \partial\Omega$, the conditions of Theorem 1 be fulfilled, and σ_* satisfy the inequality (7). Then*

$$|||\, v-u \,|||^2 \leq \aleph(\sigma, f, v, \mathbf{z}) = \Theta\{|\,|\,\mathbf{A}\nabla v+\mathbf{z}\,|\,[^2_{\mathbf{A}^{-1}} + \theta \,\|\, f-\sigma v-\operatorname{div}\mathbf{z}\,\|^2_{L_2(\Omega)}\} \tag{8}$$

where

$$\Theta = \begin{cases} 2/(1+\kappa), & \forall\sigma\in[0,\sigma_*] \\ 1, & \forall\sigma>\sigma_* \end{cases}, \quad \theta = \begin{cases} 1/\sigma_*, & \forall\sigma\in[0,\sigma_*] \\ 1/\sigma, & \forall\sigma>\sigma_* \end{cases}, \quad \kappa=\sigma/\sigma_* \tag{9}$$

As was noted above, for $\sigma \geq \sigma_*$ the majorant (8), (9) coincides with the majorant of Aubin. In the contrast to Aubin's majorant, for all $\sigma \geq 0$ it is well defined and, vore over, belongs to the class of consistent majorants when applied to the solutions by the finite element method satisfying quite natural conditions. Before formulating the respective result in Lemma 1 below, we briefly discuss these conditions.

It is assumed that the finite element space $\mathbf{V}_h(\Omega)$, $\mathbf{V}_h(\Omega) \subset C(\Omega)\cap H^1(\Omega)$, is induced by the assemblage of the finite elements, in general curvilinear, which satisfy the generalized conditions of quasiuniformity with the mesh parameter h, see e.g. [10, 12], and $\mathbf{V}_h^0(\Omega) = \{v \in \mathbf{V}_h(\Omega) : v|_{\partial\Omega} = 0\}$. For simplicity, we consider the FEM of the first order of accuracy, *i. e.* with finite elements associated with the triangular linear and square bilinear reference elemens. If $f \in L_2(\Omega)$, boundary $\partial\Omega$ and the coefficients of the matrix $\mathbf{A}$ are sufficiently smooth, then the following convergence estimates can be proved:

$$\begin{aligned} &\| u-u_{\text{fem}} \|_{k,\Omega} \leq c_{k,l} h^{l-k} \| u \|_{l,\Omega}, \quad k=0,1, \quad l=1,2, \\ &\| u-u_{\text{fem}} \|_{0,\Omega} \leq \min(c_0 c_{0,2} h^2 \sigma^{-1}) \| f \|_{0,\Omega}, \quad \forall\sigma\geq 0, \end{aligned} \tag{15}$$

where $\|\cdot\|_{k,\Omega}$ are the norms in the space $L_2(\Omega)$ for $k=0$ and in the spaces $H^k(\Omega)$ for $k>0$, whereas $c_0, c_{k,l} = \text{const}$. If $\sigma=0$, they are the well known FEM convergence estimates for regular elliptic problems, see, *e. g.*, [14, 7, 10]. If $\sigma \leq c_+^{-1} h^{-2}$, the results of [6] on elliptic projections in the space L_2 together with the fact that at least $u \in H^2(\Omega)$ can be used for their proof. Indeed, it is easily shown that

$$\| u \|_{H^2(\Omega)} \leq c_0 \| f \|_{L_2(\Omega)}, \quad c_0 = \text{const} \tag{16}$$

at any $\sigma \geq 0$, if it is true (with different constant) for $\sigma = 0$ The second bound (15) takes additionally into account in the bound $\| u-u_{\text{fem}} \|_{0,\Omega} \leq \sigma^{-1} \| f \|_{0,\Omega}$.

Acknowledgments
The author expresses his sincere gratitude to professor M.R. Timerbaev and professor M.M. Karchevsky for helpful discussions and improvements of the paper and to doctor V. S. Kostylev for making-ready the text files. The author has been partially supported by the Johann Radon Institute for Computational and Applied Mathematics (RICAM) of the Austrian Academy of Sciences during his research visits at Linz.

References

[1] Ainsworth M and Oden J T 2000 *A posteriori estimation in finite element analysis* (New York: John Wiley & Sons Inc.)
[2] Anufriev I E, Korneev V G and Kostylev V S 2006 Exactly equilibrated fields, can they be efficiently used for a posteriori error estimation? *Uchenye zapiski Kazanskogo gos. Universiteta Serija Fiziko-matematicheskie nauki* **148** (4) 94-143 (in Russian)
[3] Aubin J-P 1972 *Approximation of elliptic boundary-value problems* (Wiley-Interscience)

[4] Babuska I and Strouboulis T 2001 *Finite element method and its reliability* (New York: Oxford Univ. Press)

[5] Babuska I, Witeman J R and Strouboulis T 2011 *Finite elements An introduction to the method and error estimation* (Oxford Univ. Press)

[6] Bramble J H and Xu J 1991 Some estimates for a weighted L_2 projection *Mathematics of computation* **56** (194) 463-76

[7] Ciarlet P 1978 *The finite element method for elliptic problems* (Amsterdam-New York-Oxford: North-Holland Publishing Company)

[8] Carstensen C E and Merdon C 2013 Effective postprocessing for equilibration a posteriori error estimators *Numer. Math.* **123**(3) 425-59

[9] Frolov M, Neittaanmaki P and Repin S 2003 On the reliability, effetivity and robustness of a posteriori error estimation methods *Numerical Methods for Sientific Computing Variational problems and applications* Barelona: CIMNE 153-75

[10] Korneev V G 1977 *The finite element methods of high order of accuracy* (Leningrad: Leningrad State University) (in Russian)

[11] Korneev V G 2011 Simple algorithms for computation of a posteriri error bounds for numerical solutions of elliptic equations *Uchenye zapiski Kazanskogo gos. universiteta Serija Fiziko-matematicheskie nauki* **154**(4) 11-27 (In Russian)

[12] Korneev V G and Langer U 2015 *Dirichlet-Dirichlet Domain Decomposition Methods for Elliptic Problems, h and hp Finite Element Discretizations* (New Jesey-London-Singapore-Beijing: World Scientific)

[13] Li B and Zhang Z 1999 Analysis of a class of superconvergence patch recovery techniques for linear and bilinear finite elements *Numer. Methods PDE* **15** (2) 151-67

[14] Oganesian L A and Ruhovets L A 1979 *Variational-difference methods for solution of elliptic equations* (Yerevan: Publishing house of Armenian Academy of Sciences of Armenian SSR)

[15] Repin S 2000 A posteriori error estimation for variational problems with uniformly convex functionals *Math. Comp* **69** 481-500

[16] Repin S and Frolov M 2002 On a posteriori error bounds for approximate solutions of elliptic boundary value problems *Zhurnal vychislitel'noi matematiki i matematicheskoi fiziki* **42** (12) 1774-87 (in Russian)

[17] Repin S and Sauter S 2006 Functional a posteriori estimates for the reaction-diffusion problem *C. R. Math. Acad. Sci. Paris* **343** (5) 349-354

[18] Churilova M A 2014 Numerical properties of functional a posteriori bounds for stationary reaction-diffusion problem *Vestnik SPbSU Seria 1: Matematika, Mehanika, Astronomija* **1** (1) 68-78 (in Russian)

[19] Zhang Z 1996 Ultraconergence of the patch recovery technique *Mathematics of computaion* **65** (216) 1431-7

Eigenvectors method in mechanics of solids one-dimensional linear boundary problems solution

E A Korovaytseva

Institute of Mechanics, Lomonosov Moscow State University, Leninskie Gory, 1, Moscow, 119991, Russian Federation

E-mail: katrell@mail.ru

Abstract. Problems of deformed elements stress-strain state analysis for small displacements, angles of rotation and deformations are considered. Eigenvectors method is suggested for solution of corresponding linear boundary problems. The method is based on lattice functions theory and differential equations theory, allowing to obtain analytical form of integral matrices. One of the modifications of the suggested method allows obtaining solution with the precision equal to computer zero. An example illustrating advantages of eigenvectors method is represented.

1. Problem statement

For all known linear models of deformed elements the system of equations describing their deformed state includes three types of linear relations: differential equilibrium equations, differential and algebraic geometry relations and physical relations [1]. For stationary one-dimensional problems these equations can reduce to the system of linear ordinary differential equations in some vector $\mathbf{y}(x,\boldsymbol{\mu})$ of dimension m:

$$\mathbf{y}'(x,\boldsymbol{\mu}) = \mathbf{A}(x,\boldsymbol{\mu})\mathbf{y} + \mathbf{b}(x,\boldsymbol{\mu}) \tag{1}$$

Here prime mark denotes derivative with respect to x, $\mathbf{A}$ is square matrix of dimension m, vector $\mathbf{b}$ describes the loading of the deformed element, $\boldsymbol{\mu}$ is the vector of problem parameters. The form of the matrix and vectors depends on the type of the considered deformed element.

As a rule, boundary problem for the system (1) is solved on the interval $[x_1, x_2]$ with boundary conditions

$$\mathbf{C}_i\mathbf{y}_i = \mathbf{d}_i, \tag{2}$$

Here $\mathbf{y}_i = \mathbf{y}(x_i)$, $i = 1, 2$, $\mathbf{C}_i$ are matrices of dimensions $m \times m_i$ ($m_1 + m_2 = m$), $\mathbf{d}_i$ are vectors of inhomogeneous components of boundary conditions.

2. Initial parameters method of boundary problem solution

For solution of the problem (1)-(2) we shall use initial parameters method [2]. According to this method, we represent the unknown vector using its value on the left boundary of the integration interval as follows:

$$\mathbf{y}(x) = \mathbf{M}(x)\mathbf{y}_1 + \mathbf{H}(x) \tag{3}$$

In this representation matrix $\mathbf{M}(x)$ and vector $\mathbf{H}(x)$ are solutions of the following Cauchy problems:

$$\mathbf{M}'(x) = \mathbf{A}(x)\mathbf{M}(x), \quad \mathbf{M}(x_1) = \mathbf{E} \tag{4}$$

$$\mathbf{H}'(x) = \mathbf{A}(x)\mathbf{H}(x) + \mathbf{b}(x), \quad \mathbf{H}(x_1) = \mathbf{0} \tag{5}$$

Vector $\mathbf{y}(x)$ represented in such a way satisfies the system of equations (1).

Substituting vector (3) in boundary conditions (2) we obtain the system of linear algebraic equations in vector $\mathbf{y}_1$:

$$\begin{bmatrix} \mathbf{C}_1 \\ \mathbf{C}_2\mathbf{M}(x_2) \end{bmatrix} \mathbf{y}_1 = \begin{bmatrix} \mathbf{d}_1 \\ \mathbf{d}_2 - \mathbf{C}_2\mathbf{H}(x_2) \end{bmatrix} \tag{6}$$

So boundary problem (1)-(2) reduces to (m+1) Cauchy problems in m columns of matrix $\mathbf{M}(x)$ and vector $\mathbf{H}(x)$. We shall write these problems in the following common form:

$$\mathbf{y}'(x) = \mathbf{A}(x)\mathbf{y}(x) + \mathbf{b}(x), \quad \mathbf{y}(x_1) = \mathbf{y}_0 \tag{7}$$

3. Eigenvectors method of Cauchy problem solution

When solving the problem (7) we use lattice functions for all the variables. By definition [3] lattice function exists only at discrete equidistant values of independent variable x and is equal to zero between these values of the argument. Then on each period of discreteness lattice analogue of vector function representing the right hand side of the equation (7) $\mathbf{f}(x,\mathbf{y}) = \mathbf{A}(x)\mathbf{y}(x) + \mathbf{b}(x)$ can be represented by its envelope $\boldsymbol{\varphi}(\tau,\boldsymbol{\xi}) = \mathbf{const}$, where $\boldsymbol{\xi}$ is an envelope of vector function $\mathbf{y}$ on the considered period, τ is an argument of lattice function, $\tau = x_0 + (n-1)X_0$, X_0 is the period of discreteness, n is a number of the period of discreteness.

In order to construct exponential form of the problem (7) solution based on lattice functions using we transform the right hand side of differential equation (7) as follows:

$$\mathbf{y}'(x) = \mathbf{P}(x)\mathbf{y}(x) + \boldsymbol{\varphi}^*(\tau,\boldsymbol{\xi}) \tag{8}$$

Here $\mathbf{P}(x)$ is regularizing matrix of the system (7), and $\boldsymbol{\varphi}^*(\tau,\boldsymbol{\xi}) = \boldsymbol{\varphi}(\tau,\boldsymbol{\xi}) - \mathbf{P}(\tau)\boldsymbol{\xi}$.

We seek the general solution of differential equations system (8) using method of initial parameters relation (3). Now we obtain the following homogeneous and inhomogeneous equations for matrix $\mathbf{M}(x)$ and vector $\mathbf{H}(x)$:

$$\mathbf{M}'(x) = \mathbf{P}(x)\mathbf{M}(x), \quad \mathbf{M}(x_1) = \mathbf{E} \tag{9}$$

$$\mathbf{H}'(x) = \mathbf{P}(x)\mathbf{H}(x) + \boldsymbol{\varphi}^*(\tau,\boldsymbol{\xi}), \quad \mathbf{H}(x_1) = \mathbf{0} \tag{10}$$

Equation (9) has the following form of solution [4]:

$$\mathbf{M}(x) = \boldsymbol{\Omega}(x)\boldsymbol{\Omega}^T(0) \tag{11}$$

Here $\boldsymbol{\Omega}(x) = \left[\gamma_1 \exp(\lambda_1 x) \ \ \ldots \ \ \gamma_m \exp(\lambda_m x)\right]$, where γ_i are eigenvectors of the matrix $\mathbf{P}(x)$, λ_i are its eigenvalues.

Solution of the equation (10) for the vector $\mathbf{H}(x)$ has the form

$$\mathbf{H}(x) = \mathbf{M}(x)\int_{x_1}^{x} \mathbf{M}^{-1}(\zeta)\boldsymbol{\varphi}^{*}(\tau,\xi)\mathrm{d}\zeta \tag{12}$$

So the general solution of the equation (7) can be written as follows:

$$\mathbf{y}(x) = \mathbf{M}(x)\left[\mathbf{y}_1 + \int_{x_1}^{x} \mathbf{M}^{-1}(\zeta)\boldsymbol{\varphi}^{*}(\tau,\xi)\mathrm{d}\zeta\right] \tag{13}$$

Here $x \in [x_1 + iX_0,\ x_1 + (i+1)X_0],\ i \in [0\ N-1]$.

On the next lattice functions periods of discreteness we use fitting method [2]. On the ith period of problem (7) solution parameters $\mathbf{y}_1$ and $\boldsymbol{\varphi}^{*}(\tau,\xi)$ in the right side of equation (13) have the values calculated by the values $\mathbf{y}(x_1 + (i-1)X_0)$ of (13) on the previous period.

There can be two variants of the suggested method. If regularizing matrix $\mathbf{P}$ is chosen constant, we calculate matrix $\mathbf{M}$ only once on the first period of discreteness and set it constant for all the solution interval. We shall call this variant global eigenvectors method. If regularizing matrix $\mathbf{P}(x)$ depends on the argument x (as a rule, Jacobian of equations system is taken as a regularizing matrix in this case), the matrix $\mathbf{M}(x)$ is calculated on each period of discreteness. We shall call this variant local eigenvectors method.

The equation (13) is an exact analytical solution of Cauchy problem (7) on any period of discreteness. Analysis of accuracy and stability of the suggested method is a separate fundamental investigation, so in this work we shall represent testing the method on a concrete example to reveal its characteristics.

4. Example

The method was tested on a variety of mathematical physics and mechanics of deformed solid linear problems. Thorough testing didn't reveal any disadvantages of the method. In order to prove correctness of the results obtained we shall consider here the problem which has an exact analytical solution in the class of continuous functions.

We consider the following system of equations:

$$\mathbf{y}' = \mathbf{A}\mathbf{y} \tag{14}$$

Here $\mathbf{A} = \begin{pmatrix} 0 & 1 & 0 & 0 \\ 0 & 0 & 1 & 0 \\ 0 & 0 & 0 & 1 \\ -4\beta^4 & 0 & 0 & 0 \end{pmatrix}$.

Such a system with different meanings of vector $\mathbf{y}$ and parameter β is used in analysis of beams on elastic foundation, cylindrical and spherical shells. Krylov functions $K_0(x)$, $K_1(x)$, $K_2(x)$, $K_3(x)$ satisfy system (14) [1]. Analytical solution of the system (14) can be represented as follows:

$$\mathbf{y} = \mathbf{M}(x)\mathbf{C} \tag{15}$$

Here $\mathbf{C}$ is a vector of constants, $\mathbf{M}(x)$ is normalized integral matrix:

$$\mathbf{M}(x) = \begin{pmatrix} K_0(x) & \beta^{-1}K_1(x) & \beta^{-2}K_2(x) & \beta^{-3}K_3(x) \\ -4\beta K_3(x) & K_0(x) & \beta^{-1}K_1(x) & \beta^{-2}K_2(x) \\ -4\beta^2 K_2(x) & -4\beta K_3(x) & K_0(x) & \beta^{-1}K_1(x) \\ -4\beta^4 K_1(x) & -4\beta^2 K_2(x) & -4\beta K_3(x) & K_0(x) \end{pmatrix} \tag{16}$$

Method precision was tested for all the components of the matrix $\mathbf{M}(x)$. For comparison we used a number of well-known numerical methods, in particular, Runge-Kutta, Kutta-Merson with automatic step change, Bulirsch-Stoer for non-stiff systems of differential equations, and Radau, Bulirsch-Stoer and Rosenbrock for stiff systems of differential equations. When carrying out calculations with constant step we took 100 steps. For the methods with changing step we took relative tolerance equal to 10^{-3}. When estimating tolerances of normalized integral matrix calculation we used maximum matrix norms. Norms of relative tolerances of investigated methods are represented in table 1. Here matrix $\mathbf{M}(x)$ was calculated at $x=0.5$ with the value of the parameter $\beta=20$. The norms of relative tolerances for the methods used vary from $6.42\cdot10^{-3}$ for Kutta-Merson method to $4.58\cdot10^{-15}$ for the suggested in the paper local eigenvectors method.

Table 1. Relative tolerances of normalized integral matrix calculation.

Method	Norm of relative tolerance
Runge-Kutta	$3.83\cdot10^{-5}$
Kutta-Merson	$6.42\cdot10^{-3}$
Bulirsch-Stoer	$9.07\cdot10^{-7}$
Radau	$1.03\cdot10^{-5}$
Bulirsch-Stoer for stiff equations	$3.0\cdot10^{-4}$
Rosenbrock	$2.27\cdot10^{-5}$
Global eigenvectors	$1.42\cdot10^{-3}$
Local eigenvectors	$4.58\cdot10^{-15}$

It should be pointed out that methods of stiff differential equations systems solution (Radau, Bulirsch-Stoer, Rosenbrock) didn't turn out to be the best, which proves the point of view given in [5], stressing the importance of numerical experiments and their equal status with theoretical analysis of Cauchy problem solution methods characteristics.

Among the methods suggested in the paper global eigenvectors method showed accuracy slightly better than Kutta-Merson method, while local eigenvectors method allowed obtaining the solution with the accuracy equal to computer zero.

Figure 1 represents comparison of several methods (Runge-Kutta, Rosenbrock, global and local eigenvectors) relative tolerances common logarithms of calculating the function $K_0(x)=ch(x)\cdot\cos(x)$ as a component of system (14) solution. We can see that global eigenvectors method is the least precise, Rosenbrock method (as any other mentioned in the paper method of stiff differential equations solution) is insignificantly better than Runge-Kutta method, while local eigenvectors method shows essentially different order of relative tolerance, which varies from $\sim10^{-20}$ to $\sim10^{-15}$.

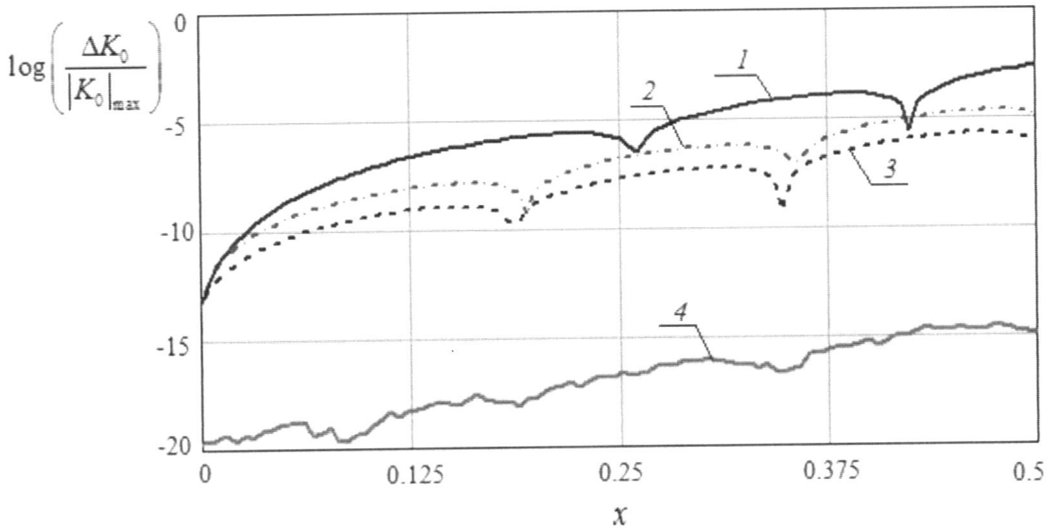

Figure 1. Comparison of relative tolerances common logarithms of function $K_0(x)$ calculation (1 – global eigenvectors method, 2 – Runge-Kutta method, 3 – Rosenbrock method, 4 – local eigenvectors method).

5. Conclusion

Eigenvectors method of deformed elements state calculation is represented. The method doesn't require any limitations for right hand side of differential equations system within the meaning [1, 2]. It allows obtaining analytical expression of the solution based on eigenvectors and eigenvalues of the equations system regularizing matrix. Local form of eigenvectors method gives the precision equal to computer zero and can be used as a basic method for solving the problems which don't have exact analytical solution.

References

[1] Abramyan B L *et al.* 1968 *Strength, Stability, Vibrations* (Moscow: Mashinostroenie) [in Russian]

[2] Biderman V L 1977 *Mechanics of Thin Structures* (Moscow: Mashinostroenie) [in Russian]

[3] Arens V D, Fedorov S M, Khitrik M S and Luchko S V 1972 *Dynamics of Rockets with Board Digital Computers Guidance Systems* (Moscow: Mashinostroenie) [in Russian]

[4] Strazheva I V and Melkumov V S 1973 *Vector-matrix Methods in Flight Mechanics* (Moscow: Mashinostroenie) [in Russian]

[5] Hall J and Watt J M 1976 *Modern Numerical Methods for Ordinary Differential Equations* (Oxford: Clarendon Press)

Study of motion of optimal bodies in the soil of grid method

V L Kotov and E Yu Linnik

Research Institute for Mechanics of the National research Lobachevsky State University of Nizhni Novgorod, Nizhny Novgorod, Russia

E-mail: ElenkaLinnik@gmail.com

Abstract. The paper presents a method of calculating the optimum forms in axisymmetric numerical method based on the Godunov and models elastoplastic soil vedium Grigoryan. Solved two problems in a certain definition of generetrix rotation of the body of a given length and radius of the base, having a minimum impedance and maximum penetration depth. Numerical calculations are carried out by a modified method of local variations, which allows to significantly reduce the number of operations at different representations of generetrix. Significantly simplify the process of searching for optimal body allows the use of a quadratic model of local interaction for preliminary assessments. It is noted the qualitative similarity of the process of convergence of numerical calculations for solving the optimization problem based on local interaction model and within the of continuum mechanics. A comparison of the optimal bodies with absolutely optimal bodies possessing the minimum resistance of penetration below which is impossible to achieve under given constraints on the geometry. It is shown that the conical striker with a variable vertex angle, which equal to the angle of the solution is absolutely optimal body of minimum resistance of penetration for each value of the velocity of implementation will have a final depth of penetration is only 12% more than the traditional body absolutely optimal maximum depth penetration.

1. Introduction

Research processes of impact and penetration of solids into the soil media using numerical methods becam widespread. The combination of known mathematical models of soils with grid methods allows us to describe a non-linear properties of the medium such as irreversible compressibility and shear strength and dimensional flow effects: the effect of free surface, cavitation, unbalanced separation and other. However, when calculating the optimal body forms, these effects are usually not counted.

Known numerical-analytical methods for determining the contact forces and the depth of penetration of projectiles into the dense, including soil media, are mainly based on local interaction hypothesis [1]. Models of local interaction (LIM) are widely used in the study of motion of bodies in soil media [2] and the search for optimal forms of penetrating bodies [3-6]. The use of different LIM modifications allowed to allocate [7] class of absolutely optimal body (AOB), which contains a general solution quite a wide range of problems of body shape optimization of penetration into the soil media. The normal to the surface of these bodies is the direction of movement constant optimum angle in the 3D formulation. Among the bodies of revolution is AOB is a circular cone with the same optimum angle, determined by velocity of the body of the solution and the strength characteristics of the soil media. Axially symmetric AOB have high elongation for soils of low strength.

To find the body forms of minimum resistance penetration for a given length of the body, at of absolutely optimal length of the cone, which is known to contain end [7], it is necessary to use numeric (grid) methods. This necessity is caused by the fact that the power to determine the characteristics of the strikers in the class of blunt bodies LIM have a maximum error.

2. Formulation of the problem and the results of the comparative analysis

We consider two classic problems of optimization – define a convex curve $r(z)$ that has a minimum full resistance, another task is to find the form of rotation of the body for maximum depth of penetration:

$$F(r)\to\min \quad \text{and} \quad S(r)=m\int_0^{V_0}\frac{VdV}{F(r)}\to\max,\ r(0)=r_0,\ r(L)=R. \tag{1}$$

Full axial resistance F of the body having a length L and radius midsection R with an end face with a radius r_0 it has the form:

$$F=\sigma_0\pi r_0^2+\pi\int_{r_0}^{R}\sigma_n dr^2+\pi\int_{r_0}^{R}\sigma_\tau\frac{\sqrt{1-\alpha^2}}{\alpha}dr^2, \tag{2}$$

$$\alpha=\frac{r'}{\sqrt{1+(r')^2}},\ r'\equiv\frac{dr(z)}{dz},$$

here the first term is due to the stresses acting on the flat front end of the body, the second and third terms - respectively normal and tangential stresses acting on the lateral surface of the body.

Stress on the body surface, penetrating at a velocity V_0 described by a quadratic model of local interaction:

$$\sigma_n=A_1V_0^2\alpha^2+C_1,\ \sigma_0=A_1V_0^2+C_1, \tag{3}$$

$$\sigma_\tau=k_f\sigma_n+C_2.$$

The constant coefficients A_1 and C_1 of binomial models (3) reflects the contribution of the inertial forces and the strength of the soil, the tangential stresses are calculated in accordance with Coulomb's law constant friction k_f and shear strength of the soil C_2.

For an approximate solution of problems forming the desired shape optimization of the optimal of the body is represented as a polyline with coordinates $r_i=r(z_i)$, $i=\overline{1,N}$, similar to [7]. The strength of the resistance to the introduction of such a body, taking into account (3) has the form:

$$F=F_0+\sum_1^N F_i,\ F_0=\pi\left(A_1V^2+C_1\right)r_0^2,\ i=\overline{1,N}, \tag{4}$$

$$F_i=\pi\left(\left(A_1V^2\alpha_i^2+C_1\right)\left(1+\tilde{k}_f\right)+\frac{\sqrt{1-\alpha_i^2}}{\alpha_i}C_2\right)\left(r_i^2-r_{i-1}^2\right),\ \tilde{k}_f=k_f\frac{\sqrt{1-\alpha_i^2}}{\alpha_i},$$

where z_i coordinates distributed across the body length in steps Δz_i, $\alpha_i=\Delta r_i\big/\sqrt{\Delta z_i^2+\Delta r_i^2}$, $\Delta r_i=r_i-r_{i-1}$, $i=\overline{1,N}$, $r_N=R$.

To solve the second problem of the expression (4) for the resistance force is converted as follows:

$$F=AV^2+C,\ A=\pi A_1r_0^2+\pi A_1\sum_{i=1}^N\alpha_i^2\left(1+\tilde{k}_f\right)\left(r_i^2-r_{i-1}^2\right),$$

$$C = \pi C_1 r_0^2 + \pi\left(1+\tilde{k}_f\right)C_1 \sum_{i=1}^{N}\left(r_i^2 - r_{i-1}^2\right) + \pi C_2 \sum_{i=1}^{N} \frac{\sqrt{1-\alpha_i^2}}{\alpha_i}\left(r_i^2 - r_{i-1}^2\right).$$

Then, the depth of penetration of the body rotation to a full stop will be equal:

$$S = \frac{m}{2A}\ln\left(\frac{A}{C}V_0^2 + 1\right). \tag{5}$$

Approximate solution of tasks (1) based on the known method of local variations involves the reduction of the original problem to the problem of unconstrained optimization. Find a set of coordinates $(r_0,...,r_{N-1})$ representing the piecewise linear forming a body of revolution, that a value $D(r_0,...,r_{N-1})$ representing the integral resistance force of penetration F (4) or the penetration depth S (5), taken with the opposite sign, reaching a minimum. The search for optimal forms reduced to solving systems of nonlinear equations of Newton's method in both formulations.

Significantly reduce the number of calculations allows the generator polynomial approximation of a Bezier based blunting, where the unknown parameters are the varying blunting radius and angles of the characteristic triangle [8].

In the paper calculations were carried out in the framework of continuum mechanics. In the cylindrical coordinate system rOz (Oz is the axis of symmetry), the Grigoryan mathematical model of the dynamic deformation of soil is formulated as a system of differential equations containing the laws of conservation of mass, momentum, and the maximum density reached in the process of active loading of soil and the plastic flow equations with the Mises plasticity condition:

$$d\rho/dt + \rho(u_{r,r} + u_{z,z}) = -(\rho u_r)/r,$$

$$\rho du_r/dt - \sigma_{rr,r} - \sigma_{rz,z} = (\sigma_{rr} - \sigma_{\theta\theta})/r,$$

$$\rho du_z/dt - \sigma_{rz,r} - \sigma_{zz,z} = (\sigma_{rz})/r, \tag{6}$$

$$d\rho_*/dt = d\rho/dt H(\rho - \rho_*)H(d\rho/dt),$$

$$D_J s_{ij} + \lambda s_{ij} = 2Ge_{ij}, \ (i, j = r, z),$$

$$s_{ij}s^{ij} \le \frac{2}{3}\sigma_T^2,$$

here t is time, ρ and $\rho*$ are the current and maximum density reached during loading, u_i, σ_{ij} , s_{ij}, and e_{ij} are the components of the velocity vector, the Cauchy stress tensor, and the stress tensor and strain rate deviators, respectively, H is a Heaviside function, D_J is the Jaumann derivative, d/dt is the total time derivative, G is the shear modulus, σ_y is the yield point; the summation is performed over repeated indices; $\lambda = 0$ in the case of elastic deformation and $\lambda > 0$ in the case of plastic deformation.

The system of differential equations (6) is closed by finite relations for the pressure p and the condition of plasticity of soil:

$$p = f_1(\rho, \rho_*)H(\rho_* - \rho)H(\rho_0 - \rho), \tag{7}$$

$$\sigma_T = f_2(p).$$

here ρ_0 is the initial density.

System (6), (7) describing the dynamic deformation of soil is supplemented with initial and boundary conditions. As shown by experiments, the penetration of blunt (in particular, spherical) bodies at an impact velocity of more than 100 m/s leads to crushing of sand particles. Then, the crushed sand particles are compacted at the head of the projectiles to form an attached cone. Traces of the interaction of soil with the cylindrical surface of the projectile and the adjoining part of the hemisphere are absent, indicating the cavitation nature of the dry sand flow over the projectiles. Therefore, the calculations are performed using an algorithm which assumes impermeability of the projectile in the normal direction and tangential dry-friction sliding of particles, in accordance with the mixed friction model:

$$u'_s = u''_s,\ q'_s = -q''_s,\ q_s = q'_s = \begin{cases} q_s, & |q_s| \le k_f |q_\xi| \\ k_f |q_\xi| \cdot sign(q_s), & |q_s| > k_f |q_\xi| \end{cases}. \quad (8)$$

where u_α and q_α are the components of the velocity and contact pressure in a local coordinate basis ($\alpha = s, \xi$); s and ξ are the tangential and normal directions; k_f is the sliding friction coefficient; primes denote the corresponding quantities on different sides of the projectile–soil interface.

Condition (8) was specified at the head of the projectile in contact with the soil, and zero normal and shear stresses were specified on the free surfaces of the soil and projectile. The outer boundaries of the computational domain of the soil were considered rigid. At the initial time, the stresses and velocity of the soil particles were set equal to zero. The projectile was assumed to be rigid and moving with constant velocity equal to the impact velocity.

Equations (6) and (7) were solved numerically using a procedure based on a modified Godunov scheme and implemented in the Dinamika-2 software. The results of previous numerical calculations of the impact and penetration of axisymmetric projectiles into to soft soils are in good agreement with experimental data, which indicates the reliability of the chosen soil model and the effectiveness of the employed mathematical apparatus and numerical method. Just necessary resistance force calculations introduction and penetration depth were carried out on the basis of local interaction model.

Is noted qualitatively similar behavior of numerical solutions in productions based on LIM and continuum mechanics, but a significant difference in the forms of optimal bodies, their resistance and depth of penetration. Accounting for internal and surface friction, as well as other features of dynamic deformation of soils, the implementation of which is not difficult in the framework of known models of soils medium to reduce the differences.

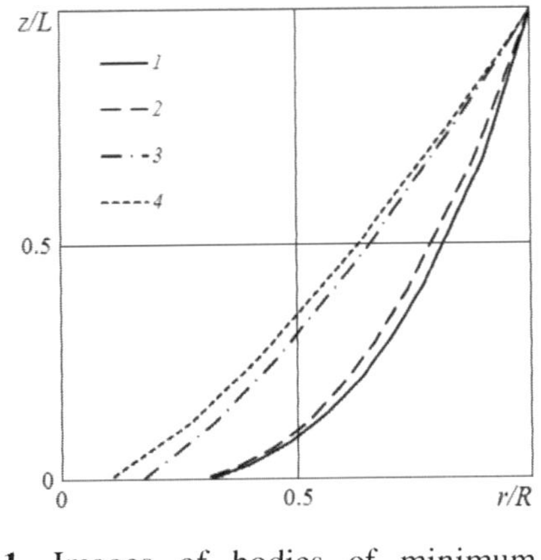

Figure 1. Images of bodies of minimum resistance of penetration of a given length L and the radius of the base R: curves 1, 2 – calculations in the formulation of continuum mechanics, curves 3, 4 – model of local interaction.

The optimal form of minimum resistance rotating bodies are shown in Figure 1, taking into account the different representations of friction. Curves 1 and 3 were obtained in the presence the internal and friction of the surface in the soil medium, curves 2 and 4 correspond to the absence of friction. It is evident that with a significant quantitative difference is maintained qualitatively correct the tendency of small changes in the forms of optimal body.

Thus, when searching for the optimal bodies advisable to use a combined approach: to carry out preliminary calculations based on local interaction model with a further refinement of the forms using known grid methods in the formulation of continuum mechanics.

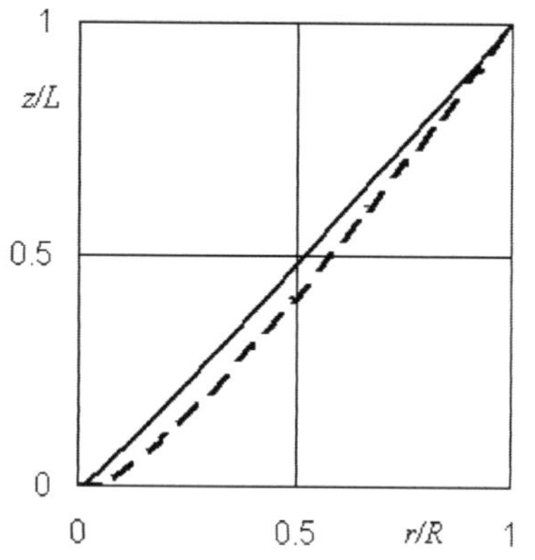

Figure 2. Images of bodies minimum resistance and maximum depth of penetration (V_0=100 m/s).

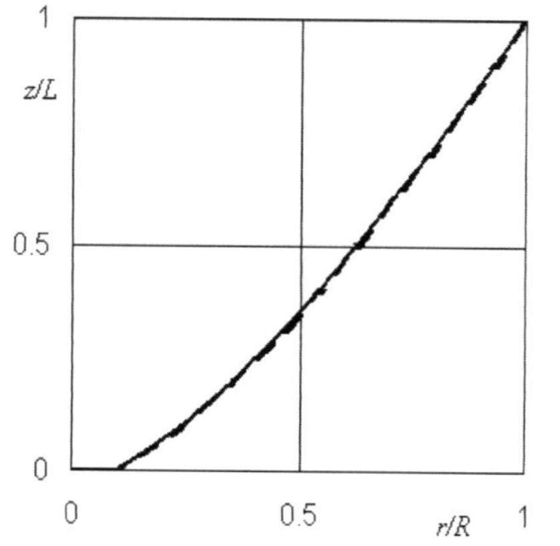

Figure 3. Images of bodies minimum resistance and maximum depth of penetration ($V_{0=}$1000 m/s).

In the comparison of optimal forms of minimum resistance rotating bodies (dashed line) and the maximum depth of penetration (solid line) for the given initial impact velocities 100 м/с (Figure 2) and 1000 м/с (Figure 3). It was established [9] that with increasing velocity appears nonzero front end, rising to a value of about R/10. There is a significant difference forms at velocities values close to critical, and the convergence of generators with increasing velocity. When the linear parameterization of forming an optimal rotation of the body with the end (truncated cone is solution of Newton was first obtained), the radius of the front end at high velocities was equal to about R/7.

Comparison of the numerical calculations the optimal forms with absolutely optimum bodies showed that the evaluation of the resistance force is absolutely the optimal bodies does not depend on the velocity. Evaluation of the resistance force of penetration of body with a maximum depth of penetration and depth of penetration of the minimum resistance of the body penetration will only decrease.

Advantages of the optimal blunt bodies of revolution occur at high velocities of penetration and are in the range of 10-20% for the given parameters of the soil medium and the body. An estimate of the ratio of the resistance of the body with the introduction of a maximum depth of penetration to the minimum value of resistance force introduction, which does not depend on velocity. It is shown that the conical striker with a variable opening angle equal to the angle of the solution absolutely optimal body of minimum resistance introduction for each value of the implementation of speed, will have a final depth of penetration is only 12% more than traditional absolutely optimal body the maximum penetration depth (at least, at high velocities).

At lower velocities it should be noted that the axially symmetric absolutely optimal body of the minimum resistance of penetration is cone only up to a certain value of an angle.

In [10] the example of solving the problem of optimizing the shape of the body, which has a maximum depth of penetration, influence of resistance and friction laws. It is shown that the best fit with the experimental data obtained by using the law of the mixed friction and a discontinuous resistance law. It was found that at high velocities the penetration of Coulomb's law of friction leads to substantially greater differences of forms and maximum depth of penetration compared with the mixed friction model.

In this paper numerically shown that at high velocities, the optimal form of the introduction of the minimum resistance of the body will contain a non-zero front end. Also, this form will be close to the optimum form of the body the maximum depth of penetration. The use of methods of investigation based on MLV to preliminary calculations, the optimal form, it is possible to achieve a substantial reduction in the amount of computing axisymmetric.

Acknowledgments
This work was supported in part by the Program for Support of Leading Scientific Schools of Russia (NSh_6944.2016.8), and also by the Russian Foundation for Basic Research (project nos. 16-38-60035-mol_a_dk, 16-08-00825-a, 16-38-00102-mol-a).

References

[1] Ben-Dor G , Dubinsky A and Elperin T 2014 Engineering models of high speed penetration into geological targets *Central European Journal Engineering* **4**(1) 1-9

[2] Kotov V L, Balandin V V, Bragov A M, Linnik E Yu and Balandin V V 2013 Using a local-interaction model to determine the resistance to penetration of projectiles into sandy soil *Journal of Applied Mechanics and Technical Physics* **54** (4) 612-621

[3] Yakunina G Ye 2005 The optimum shapes of bodies moving in a medium taking friction into account Journal of Applied Mathematics and Mechanics **69** (5) 680-692

[4] Banichuk N V and Ivanova S Yu 2007 Shape optimization of rigid body, penetrating into continuous medium *J The Problems of Strength and Plasticity* **69** 47-57

[5] Bazhenov V G, Balandin V V, Grigoryan S S and Kotov V L 2014 Analysis of models for calculating the motion of solids of revolution of minimum resistance in soil media *Journal of Applied Mathematics and Mechanics* **78** (1) 98-115

[6] Kotov V L, Linnik E Yu 2014 The analysis of models and methods of calculation forms of minimum resistance to motion in soils *X International Conference "Mesh methods for boundary value problems and applications"* 435-439.

[7] Kotov V L and Linnik E Yu 2013 Numerical computation of the form of a body of revolution with the minimal resistance to motion in a soil medium in the frame of the local interaction model *J The Problems of Strength and Plasticity* **75** (4) 296-302

[8] Linnik E Yu 2014 Numerical Calculation of the Optimum Shape of a Body of Revolution at Its Penetration into Soil Media, *Uchenye Zapiski Kazanskogo Universiteta. Fiziko-Matematicheskie Nauki* **156** (2) 95-101

[9] Kotov V L, Linnik E Yu and Tarasova A A 2015 Comparing the forms of the bodies of revolution of minimal resistance and of maximal depth of penetration into the ground *J The Problems of Strength and Plasticity* **77** (1) 85-92

[10] Linnik E Yu 2016 Some special features of determining the maximal penetration depth of optimal bodies into a sand soil *J The Problems of Strength and Plasticity* **78** 208-217

Udzawa-type iterative method with parareal preconditioner for a parabolic optimal control problem

A Lapin[1,2] **and A Romanenko**[1]

[1] Kazan (Volga region) Federal University, Kazan, Russia
[2] Tianjin University of Finance and Economics, Tianjin, China

E-mail: romart92@mail.ru

Abstract. The article deals with the optimal control problem with the parabolic equation as state problem. There are point-wise constraints on the state and control functions. The objective functional involves the observation given in the domain at each moment. The conditions for convergence Udzawa's type iterative method are given. The parareal method to inverse preconditioner is given. The results of calculations are presented.

1. Introduction

The numerical solution of parabolic control problems requires multiple solving the direct and adjoint equations (cf., e.g., [1]). The corresponding simulations can be too expensive and lead to the difficulty in obtaining "real-time solutions". To overcome this difficulty the parallel algorithms for multiprocessor workstations are usually constructed. A common way to construct a parallel algorithm is based on the decomposition of a spatial domain in elliptic problems as well as in evolutionary problems. Another approach consists in splitting the global problem into a series of independent evolutionary problems on smaller time intervals. It is so-called parareal algorithm, proposed in [2], [3] and widely used for solving different scientific and applied problems (cf., e.g., [4]-[9] and the bibliographies therein).

In this article we use the results of [10], [11] to construct a preconditioned Udzawa-type iterative method for mesh approximation of a state constrained parabolic optimal control problem. For the implementation of this method we propose a parareal algorithm of inverting the preconditioning matrix.

2. Initial problem, mesh approximation

Let $\Omega = (0,1)^n$, $n \geq 1$, with the boundary $\partial\Omega$, $Q_T = \Omega \times (0,T]$ and $\Sigma_T = \partial\Omega \times (0,T]$. A parabolic initial-boundary value problem

$$y_t - \Delta y = u \text{ in } Q_T,\ \ y = 0 \text{ on } \Sigma_T,\ \ y = 0 \text{ for } t = 0,\ x \in \Omega \qquad (1)$$

is a state problem. For $u \in L_2(Q_T)$ there exists a unique solution of problem (1) from the space

$W(Q_T) = \{y \in L_2(0,T;H_0^1(\Omega)),\ y_t \in L_2(Q_T)\}$ ([12], p.370). We define the sets of constraints for control u and state y:

$$U_{ad} = \{u \in L_2(Q_T) : |u(x,t)| \leq u_{\max}\},\quad Y_{ad} = \{y \in W(Q_T) : y_{\min} \leq y(x,t) \leq y_{\max}\} \quad \forall (x,t) \in Q_T.$$

Above $u_{max}>0$ and $-\infty \le y_{min} < 0 < y_{max} \le +\infty$. Let objective function be defined by the equality

$$J(y,u)=\frac{1}{2}\int_{Q_T}(y(x,t)-y_d(x,t))^2\,dxdt+\frac{1}{2}\int_{Q_T}u^2(x,t)dxdt \tag{2}$$

with a given observation function $y_d \in L_2(Q_T)$. Optimal control problem reads as follows:

$$\min_{(y,u)\in K} J(y,u),\qquad K=\{(y,u)\in Y_{ad}\times U_{ad} : equation(1)\ holds\}. \tag{3}$$

It is a minimization problem of quadratic functional $J(y, u)$ on a closed, bounded and convex set $K\neq\varnothing$, so, it has a unique solution (cf. [1]).

Below we suppose for the simplicity that the function y_d is continuous. We construct a finite difference approximation of problem (3) on the uniform grid $\omega_x\times\omega_\tau$ in $\overline{Q}_T$, where ω_x is a grid with a step h while $\omega_\tau=\{t_j=j\tau,\ j=0,1,\dots,M_\tau;\ M_\tau\tau=T\}$. Let V_h be the space of mesh functions defined on the grid ω_x vanishing in the boundary nodes $\partial\omega_x$ and $y_j=y(x,t_j)\in V_h$ be a mesh function on a time level $t_j=j\tau\in\omega_\tau$. Later we use the same notations both for mesh functions and the vectors of their nodal values. By N_x we denote the dimension of V_h and by $\|\cdot\|_x$ the Euclidian norm of the vectors of nodal values in this space.

Let A be well-known matrix of mesh Laplace operator on the grid ω_x with homogeneous Dirichlet boundary conditions. Approximate state problem (1) using backward Euler finite difference method:

$$\frac{y^j-y^{j-1}}{\tau}+Ay^j=u^j,\ j=1,2,\dots,M_\tau,\quad y^0=0. \tag{4}$$

The approximation of the objective function (2) on the grid $\omega_x\times\omega_\tau$ is the sum $\frac{\tau h^n}{2}\left(\sum_{j=1}^{M_\tau}\|y^j-y_d^j\|_x^2+\sum_{j=1}^{M_\tau}\|u^j\|_x^2\right)$. We scaled it to derive the following mesh objective function:

$$J(y,u)=\frac{1}{2}\sum_{j=1}^{M_\tau}\|y^j-y_d^j\|_x^2+\frac{1}{2}\sum_{j=1}^{M_\tau}\|u^j\|_x^2. \tag{5}$$

The sets of the constraints for the mesh control and state functions we define as follows

$$U_{ad}^h=\{(u^1,\dots,u^{M_\tau}): u^j\in V_h,\ |u^j|\le u_{\max}\},$$
$$Y_{ad}^h=\{(y^1,\dots,y^{M_\tau}): y^j\in V_h,\ y_{\min}\le y^j\le y_{\max}\}.$$

Now, mesh optimal control problem reads as follows:

$$\min_{(y,u)\in K_h} J(y,u),\qquad K_h=\{(y,u)\in Y_{ad}^h\times U_{ad}^h : equation(4)\ holds\}. \tag{6}$$

Problem (6) is a minimization problem of a quadratic function on a compact set, so, it has a unique solution.

Later we consider problem (6) with very big number of time levels M_τ and use a parareal algorithm for solving (4). Namely, let $\Delta t >> \tau$ be a new time step and $\omega_{\Delta t} = \{t_j = j\Delta t,\ j = 0,1,\ldots,M_{\Delta t};\ M_{\Delta t}\Delta t = T\}$. Further for the simplicity we suppose that $\Delta t = m\tau$ with $m >> 1$.

Following the ideas and terminology of the article [13], we define as a coarse propagator backward Euler finite difference scheme on the coarse grid $\omega_{\Delta t}$, while a fine propagator - backward Euler finite difference scheme on the fine grid ω_τ.

3. Iterative method for the mesh problem

Let $N = N_x M_\tau$ be the dimension of the mesh functions of the variables x and t, $E \in R^{N\times N}$ be unit matrix and $L \in R^{N\times N}$ be defined by the equality

$$(Ly)^j = \left\{\frac{y^1}{\tau}\ for\ j=1;\quad \frac{y^j - y^{j-1}}{\tau} + Ay^j\ for\ j = 2,\ldots,M_\tau\right\}.$$

Let also ψ and φ be the indicator functions of the sets Y_{ad}^h and U_{ad}^h, respectively, while $\partial\psi$ and $\partial\varphi$ be their subdifferentials. Mesh optimal control problem (5) can be rewritten as

$$\min_{Ly=u}\{J(y,u) + \psi(y) + \varphi(u)\}.$$

Lagrange function for this problem is defined by the equality

$$\Im(y,u,\lambda) = J(y,u) + \psi(y) + \varphi(u) + (\lambda, Ly - u),$$

where $(\cdot,\cdot)$ is Euclidian inner product in R^N.

Due to theory of saddle point problems ([14], p.169) a saddle point of this Lagrange function satisfies the following system:

$$\begin{pmatrix} E & 0 & L^T \\ 0 & E & -E \\ L & -E & 0 \end{pmatrix}\begin{pmatrix} y \\ u \\ \lambda \end{pmatrix} + \begin{pmatrix} \partial\psi(y) \\ \partial\varphi(u) \\ 0 \end{pmatrix} \ni \begin{pmatrix} y_d \\ 0 \\ 0 \end{pmatrix}. \qquad (7)$$

With the notations $z = (y, u)^T$, $f = (y_d, 0, 0)^T$, $\Psi(z) = (\psi(y), \varphi(u))^T$ and

$$A = diag(E\quad E),\quad B = (L\quad -E)$$

problem (7) reads as

$$\begin{pmatrix} A & B^T \\ B & 0 \end{pmatrix}\begin{pmatrix} z \\ \lambda \end{pmatrix} + \begin{pmatrix} \partial\psi(z) \\ 0 \end{pmatrix} \ni \begin{pmatrix} f \\ 0 \end{pmatrix}. \qquad (8)$$

Using the results of [15] we can prove the existence of a solution to (6) and convergence of a preconditioned Udzawa method

$$\begin{aligned} &Az^{k+1} + \partial\psi(z^{k+1}) \ni f, \\ &\frac{1}{\rho}D(\lambda^{k+1} - \lambda^k) + Bz^{k+1} = 0, \quad \rho > 0. \end{aligned} \qquad (9)$$

with preconditioning matrix $D = D^T > 0$ and iterative parameter ρ which satisfy the inequality

$$D > \frac{\rho}{2} BA^{-1}B^T = \frac{\rho}{2}(LL^T + E). \tag{10}$$

In [16] the convergence result has been proved for method (9) with the preconditioner $D = LL^T$ and iterative parameter $\rho \in (0,1)$. The detailed form of the corresponding iterative method reads as follows:

$$\begin{aligned} &u^{k+1} + \partial\varphi(u^{k+1}) \ni \lambda^k, \\ &y^{k+1} + \partial\psi(y^{k+1}) \ni y_d - L^T\lambda^k, \\ &LL^T \frac{\lambda^{k+1} - \lambda^k}{\rho} = Ly^{k+1} - u^{k+1}. \end{aligned} \tag{11}$$

Implementation of every step in (11) consists of solving the inclusions to find u^{k+1} and y^{k+1}, and solving a system of linear equations with the matrix LL^T. Since the operators $\partial\varphi$ and $\partial\psi$ have diagonal forms, vectors u^{k+1} and y^{k+1} can be found component-wise by the explicit formulae. So, the most time consuming part of the algorithm is solution of the system with the matrix LL^T. In the following section we describe a parareal algorithm of inverting matrix LL^T.

4. Parareal preconditioner

To find the solution of the third equation in system (10) we have to invert the matrix

$$L = \begin{pmatrix} \tau^{-1}E & 0 & \dots & 0 & 0 \\ -\tau^{-1}E & -\tau^{-1}E + A & \dots & 0 & 0 \\ 0 & -\tau^{-1}E & \dots & 0 & 0 \\ \vdots & \vdots & \dots & \vdots & \vdots \\ 0 & 0 & \dots & -\tau^{-1}E & -\tau^{-1}E + A \end{pmatrix}$$

and its transposed L^T. Denote by $g = Ly^{k+1} - u^{k+1}$. Then we get from (11) the equation $Lv = g$, where $v = L^T \dfrac{\lambda^{k+1} - \lambda^k}{\rho}$.

We use the following variant of parareal method for finding a solution v of the equation $Lv = g$.

Let $\tilde{v}^k$ be calculated using backward Euler finite difference scheme on a coarse grid $\omega_{\Delta t} = \{t_j = j\Delta t,\ j = 0,1,\dots,M_{\Delta t};\ M_{\Delta t}\Delta t = T\}$ with $\Delta t = m\tau,\quad m >> 1$.

Then at each time subinterval $[j\Delta t, (j+1)\Delta t],\ j = 0,1,\dots,M_{\Delta t} - 1$ we use the values of $\tilde{v}_j^k = \tilde{v}^k(t_j)$ as initial values to find a solution v^k in the points $t_0 = j\Delta t, t_1 = j\Delta t + \tau, \dots, t_m = (j+1)\Delta t,\ j = 0,1,\dots,m$ of the fine grid:

$$\frac{v^k(t_{j+1}) - v^k(t_j)}{\tau} + Av^k(t_{j+1}) = g(t_{j+1}),\quad v^k(t_0) = \tilde{v}^k(j\Delta t). \tag{12}$$

After that we define a correction vector $\delta^k = v^k - \tilde{v}^k$ in the points of coarse grid and find a new value $\tilde{v}^k$ via following sequential procedure for $t_j \in \omega_{\Delta t}$:

$$\frac{w-\tilde{v}^k(t_j)}{\Delta t}+Aw=g(t_{j+1}),\quad \tilde{v}^k(t_{j+1})=w+\tilde{\delta}^k(t_{j+1}).$$

The main feature of (12) is the possibility of its parallel implementation. Namely, we can use $M_{\Delta t}$ processors for searching v^k in $M_{\Delta t}$ subintervals in parallel manner.

Denoting $w=\dfrac{\lambda^{k+1}-\lambda^k}{\rho}$ we get $L^T z=v$. The implementation of this equation differs from the algorithm above in that all calculations start from the last time-point $t=T$ and continue backward for $j=M_{\Delta t},M_{\Delta t}-1,\ldots,1$. Finally, $\lambda^{k+1}=\lambda^k+\rho z$.

The stability of the parareal algorithm is proved in the article [17]. In addition, the results of [18] show us that suitably damping coarse schemes ensure unconditional stability of the parareal algorithm.

5. Numerical experiments

The parareal algorithm is realized with the following data. For simplicity we choose one-dimensional case, when n = 1, and there are no restrictions on functions y and u. The final time moment is T = 1, mesh size h = 0.05. Time step of the fine grid t = h, the step of the coarse grid $\tau = h^2$. The iterative parameter ρ = 0.005, the observation function $y_d(x, t) = e^t \sin(\pi x)$. The calculation stopping criterion is number of iterations in Udzawa method.

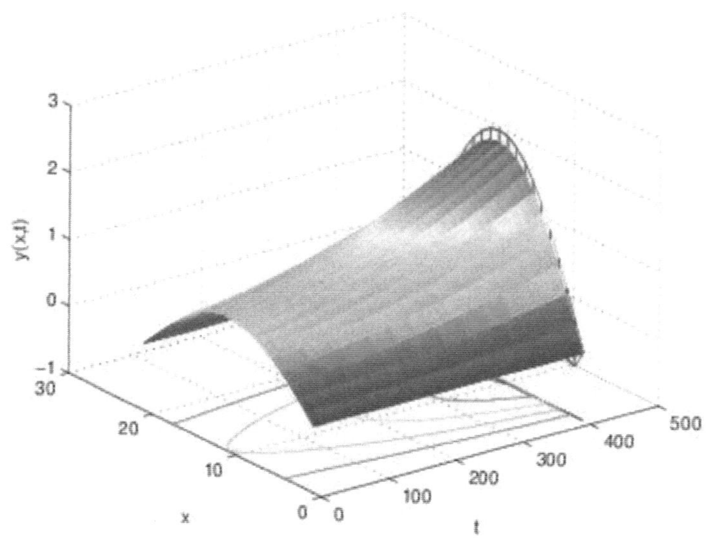

Figure 1. Shape of the state function y.

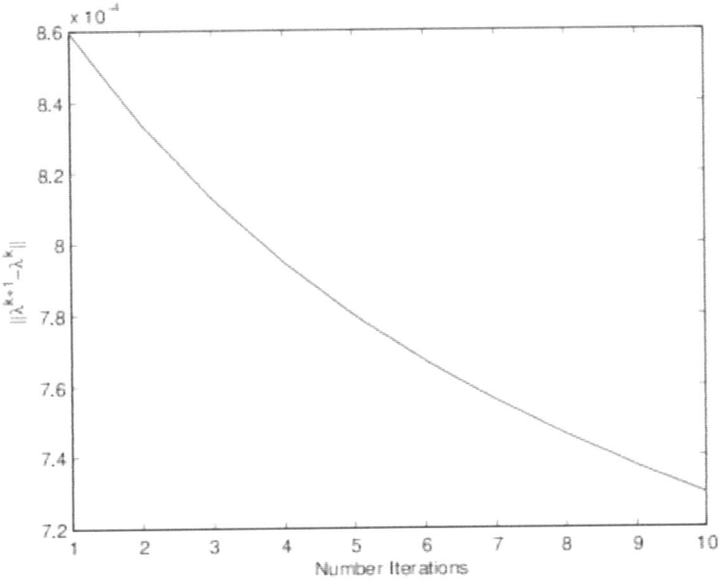

Figure 2. Residual $\left\|\lambda^{k+1}-\lambda^{k}\right\|_{L_2(Q_T)}$.

References

[1] Tiba D and Neittaanmaki P 1994 *Optimal control of nonlinear parabolic systems. Theory, algorithms and applications* (New York: Marcel Dekker)
[2] Lions J, Maday Y and Turinici G 2001 *CR Math. Acad. Sci. Paris I Math* **332** pp 661-668
[3] Bal G and Maday Y 2001 Proceedings of *the Workshop on Domain Decomposition, LNCSE Series* (Berlin: Springer-Verlag) pp 189-202
[4] Maday Y and Turinici G 2002 *CR Math. Acad. Sci. Paris I Math* **335** pp 387-392
[5] Srinivasan A and Chandra N 2005 *Parallel Computing* **31** pp 777-796
[6] Trindade J and Pereira J 2006 *Numerical Heat Transfer Part B-Fundamentals* **50** pp 25-40
[7] Samaddar D, Newman, B and Sanchez R 2010 *Journal of Computa-tional Physics* **229** pp 6558-6573
[8] Geiser J and Guttel S 2012 *Journal of Mathematical Analysis and Applications* **388** pp 873-887
[9] Samuel H 2012 *Geochemistry Geophysics Geosystems* **13** pp 1-16
[10] Lapin A, Laitinen E and Lapin S 2015 *Russian J. Numer. Analysis Math. Modeling* **30** pp 351-362
[11] Lapin A and Laitinen E 2016 *Lobachevskii J. Math.* **37** pp 561-569
[12] Quarteroni A and Valli A 1997 *Numerical approximation of partial differential equations* (Berlin: Springer)
[13] Maday Y and Turinici G 2005 *Lecture Notes in Computational Science and Engineering* **40** pp 441-448
[14] Ekeland I and Temam K 1976 *Convex analysis and variational problems* (Amsterdam: North-Holland)
[15] Lapin A 2010 *Lobachevskii J. Math.* **31** pp 309-322
[16] Lapin A and Platonov A 2016 *Proceedings of Kazan University, Physics and Mathematics Series* (in Russian) **158** pp 81-89
[17] Staff G and Ronquist E 2005 *Lecture Notes in Computational Science and Engineering* **40** pp 449-456
[18] Bal G 2005 *Lecture Notes in Computational Science and Engineering* **40** pp 425-432

Non-overlapping domain decomposition method for a variational inequality with gradient constraints

A Lapin[1,2], E Laitinen[3] and S Lapin[1,4]

[1] Kazan (Volga region) Federal University, Kazan, Russia
[2] Tianjin University of Finance and Economics, Tianjin, China
[3] University of Oulu, Oulu, Finland
[4] Washington State University, Pullman, WA, USA

E-mail: avlapine@mail.ru

Abstract. Non-overlapping domain decomposition method is applied to a variational inequality with nonlinear diffusion-convection operator and gradient constraints. The method is based on the initial approximation of the problem and its subsequent splitting into subproblems. For the resulting constrained saddle point problem block relaxation-Uzawa iterative solution method is applied.

1. Introduction

Domain decomposition methods for the variational inequalities have been investigated for a long time. The most attention has been paid to Schwarz-type iterative methods for the problems with pointwise constraints to solution [1-12]. This type of problems includes obstacle problems, some contact problems, Stefan problem on a fixed time level among others. Non-overlapping domain decomposition method has been applied to variational inequalities with constraints to a solution in the supposition that the location of free boundary is known [13-15].

Convergence of Uzawa-type iterative methods for the constrained saddle point problems has been investigated in [16-18]. First general result on the convergence of Uzawa iterative method has been proved in the article [16], where sufficient convergence condition has been formulated in terms of matrices inequality. In [17] a generalization of this result for wider class of saddle point problems and for so-called block relaxation-Uzawa iterative solution method have been investigated. These results were applied to iterative solution methods for mesh variational inequalities with gradient constraints and for mesh approximations of state and control constrained optimal control problems in the numerous subsequent articles.

In this article we apply the aforementioned results on the iterative solution methods for the constrained saddle point problems to non-overlapping domain decomposition method for variational inequalities with gradient constraints.

2. Variational inequality and its approximation

Let $\Omega \subset R^2$ with be a bounded domain with a piecewise smooth boundary $\partial\Omega$, $V = H_0^1(\Omega)$ be Sobolev space with the norm $\|u\| = \left(\int_\Omega |\nabla u|^2 dx \right)^{1/2}$. Further, c_f is a constant in Friedrichs inequality $\|u\|_{L_2(\Omega)} \le c_f \|u\| \quad \forall u \in V$.

Define the functions

$$f \in L_2(\Omega), a, b \in L_\infty(\Omega), \quad where \quad a(x) \ge a_0 > 0 \quad a.e.\, in\, \Omega, \tag{1}$$

and $g_1(\bar{t})$, $g_2(s, \bar{t})$, which for all $s \in R, \bar{t} \in R^2$ satisfy the following assumptions:

$$\begin{cases} g_1(\bar{t}) \text{ and } g_2(s,\bar{t}) \text{ are continuous}, |g_1(\bar{t})| \le c|\bar{t}|, \\ |g_2(s_1,\bar{t}) - g_2(s_2,\bar{t})| \le \beta_1 |s_1 - s_2| + \beta_2 |\bar{t}_1 - \bar{t}_2|, \\ (g_1(\bar{t}_1) - g_1(\bar{t}_2), \bar{t}_1 - \bar{t}_2) \ge \sigma_0 |\bar{t}_1 - \bar{t}_2|^2, \\ \sigma_0 > 0, a_0\sigma_0 - b\beta_1 c_f^2 - b\beta_2 c_f \equiv \sigma > 0, b = \sup_{x \in \Omega} |b(x)|. \end{cases} \tag{2}$$

Define a semilinear form $a(\cdot,\cdot): V \times V \to R$ by the equality

$$a(u,v) = \int_\Omega a(x) g_1(\nabla u) \cdot \nabla v\, dx + \int_\Omega b(x) g_2(u, \nabla u) v\, dx.$$

Due to assumptions (1), (2) the form $a(u,v)$ is continuous on $V \times V$ and uniformly monotone:

$$a(v, v-u) - a(u, v-u) \ge \sigma \|u - v\|^2.$$

Consider variational inequality

$$u \in V: \quad a(u, v-u) + \int_\Omega |\nabla v|\, dx - \int_\Omega |\nabla u|\, dx \ge \int_\Omega f(v-u)\, dx \quad \forall v \in V. \tag{3}$$

Proposition 1. *Under the assumptions (1), (2) variational inequality (3) has a unique solution.*

The proof is based on the theory of variational inequalities with monotone operators [19].

Let Ω be a polygonal domain. We approximate variational inequality (3) by using first order finite element method on a triangle grid (cf [20]-[22]). Let T_h be a conforming triangulation of $\bar{\Omega}$ into triangle finite elements e, $V_h = \{u_h \in H_0^1(\Omega) \cap C(\bar{\Omega}) : u_h \in P_1 \quad on \quad every \quad e \in T_h\}$ be the space of the continuous and piecewise linear functions, while $U_h = \{u_h \in L_2(\Omega) : u_h \in P_0 \quad on \quad every \quad e \in T_h\}$ be the space of the piecewise constant functions. Consider f_h, a_h and b_h are functions from U_h which equal to mean values of f, a and b on every $e \in T_h$. Namely, $f_h = |e|^{-1} \int_{t \in e} f(t)\, dt$ for every $e \in T_h$, $|e| = mease$, and similarly for a_h, b_h. Let us define

$$a_h(u_h, v_h) = \int_\Omega a_h g_1(\nabla u_h) \cdot \nabla v_h\, dx + \int_\Omega b_h g_2(u_h, \nabla u_h) v_h\, dx.$$

Due to assumptions (1), (2) the form $a_h(u_h, v_h)$ is continuous on $V_h \times V_h$ and uniformly monotone. Discrete variational inequality, approximating (3):

$$u_h \in V_h: \quad a_h(u_h, v_h - u_h) + \int_\Omega |\nabla v_h|\, dx - \int_\Omega |\nabla u_h|\, dx \ge \int_\Omega f_h(v_h - u_h)\, dx \quad \forall v_h \in V_h. \tag{4}$$

Proposition 2. *Under the assumptions (1), (2) variational inequality (4) has a unique solution.*

3. Domain decomposition and constructing a saddle point problem

Let us decompose the domain Ω into m subdomains $\Omega=\sum_{i=1}^{m}\Omega_i+\bigcup_{i\neq j}\Gamma_{ij}$, where every open subdomain Ω_i consists of the elements $e\in T_h$ and $\Gamma_{ij}\in\Omega$ is the general part of the boundaries of $\partial\Omega_i$ and $\partial\Omega_j$ lying inside Ω. Hereafter we use the symbol Σ for the non-intersected sets (above $\Omega_i\bigcap\Omega_j=\varnothing$ for all $i\neq j$).

By $\Gamma=\bigcup_{i\neq j}\Gamma_{ij}$ we denote the skeleton of the decomposition. The boundary of a subdomain generally consists of two parts: $\partial\Omega_i=\Gamma_{i0}+\Gamma_i$ with $\Gamma_{i0}\subset\partial\Omega$, $\Gamma_i\subset\Gamma$. The case $\Gamma_{i0}=\varnothing$ is allowed.

We use the following spaces of mesh functions: $V_{hi}=\{u_{hi}\ \textit{is the restriction of}\ u_h\in V_h\ \textit{to}\ \overline{\Omega}_i\}$, $\overline{V}_h=\{\overline{u}_h=(u_{h1},u_{h2},\dots,u_{hm}),\ u_{hi}\in V_{hi}\}$. W_h is the space of traces of functions from V_h to skeleton Γ. $K_h=\{(\overline{u}_h,s_h)\in\overline{V}_h\times W_h:\ u_{hi}=s_h\ \ on\ \ \Gamma_i,\ \ i=1,2,\dots,m\}$.

Let

$$a_{hi}(u_h,v_h)=\int_{\Omega}a_h g_1(\nabla u_h)\cdot\nabla v_h\,dx+\int_{\Omega}b_h g_2(u_h,\nabla u_h)v_h\,dx,$$

$$\overline{a}_h(\overline{u}_h,\overline{v}_h)=\sum_{i=1}^{m}a_{hi}(u_{hi},v_{hi})\quad for\quad \overline{u}_h\in\overline{V}_h.$$

Similarly define $\varphi_{hi}(u_{hi}), f_{hi}$ on the subdomains and composite functions $\overline{\varphi}_h(\overline{u}_h),\overline{f}_h$.

Consider variational inequality: find $(\overline{u}_h,s_h)\in K_h$ such that

$$\overline{a}_h(\overline{u}_h,\overline{v}_h-\overline{u}_h)+\overline{\varphi}_h(\overline{v}_h)-\overline{\varphi}_h(\overline{u}_h)\geq\overline{f}_h(\overline{v}_h-\overline{u}_h)\quad\forall(\overline{v}_h,s_h)\in K_h.\tag{5}$$

Proposition 3. *Problem (5) has a unique solution $(\overline{u}_h,s_h)\in K_h$ such that u_{hi} is the restriction to Ω_i of the solution u_h of problem (4), while s_h is the trace of u_h on the skeleton Γ.*

We put in the correspondence to a mesh function the vector of its nodal values. In particular, let $u_i\in R^{N_i}$ be the vector of nodal values of a function $u_{hi}\in V_{hi}$, and we use notation $u_i\leftrightarrow u_{hi}$ for this correspondence. Vector $u=(u_1,u_2,\dots,u_m)^T$ with $u_i\in R^{N_i}$ corresponds to $\overline{u}_h\in\overline{V}_h$ $(u\leftrightarrow\overline{u}_h)$, while $\overline{q}$ corresponds to $\overline{q}_h=(q_{1h},q_{2h})\in U_h\times U_h$.

Below by $(\cdot,\cdot)_i$ we mean the Euclidian scalar product in R^{n_i} and denote $(u,v)=\sum_{i=1}^{m}(u_i,v_i)$ for $u=(u_1,u_2,\dots,u_m)^T$ and $v=(v_1,v_2,\dots,v_m)^T$.

Let for every i=1,2,...,m the following matrices and nonlinear operators be defined by the forms on the subdomains:

$$(\tilde{L}_i u_i,q_i)_i=\int_{\Omega_i}\nabla u_{hi}(x)\cdot\overline{q}_{hi}(x)dx,\quad (M_{pi}p_i,q_i)_i=\int_{\Omega_i}\overline{p}_{hi}(x)\cdot\overline{q}_{hi}(x)dx,$$

$$(M_{ui}u_i,v_i)_i=\int_{\Omega_i}u_{hi}(x)\cdot v_{hi}(x)dx,\quad (k_{1i}(p_i),q_i)_i=\int_{\Omega_i}a_h(x)g_1(\overline{p}_h(x))\overline{q}_h(x)dx,$$

$$(k_{2i}(u_i,p_i),v_i)_i=\int_{\Omega_i}b_h(x)g_2(u_h,\overline{p}_h(x))v_h(x)dx,\quad L_i=M_{pi}^{-1}\tilde{L}_i.$$

The corresponding composite matrices and operators

$$Lu = \sum_{i=1}^{m} L_i u_i, \quad L^T k_1(Lu) = \sum_{i=1}^{m} L_i^T k_{1i}(L_i u_i), \quad k_2(u, Lu) = \sum_{i=1}^{m} k_{2i}(u_i, L_i u_i)$$

have block diagonal forms.

Using the introduced notations we get the equality

$$\bar{a}_h(\bar{u}_h, \bar{v}_h) = (k_1(Lu), Lv) + (k_2(u, Lu), v) = (N(u), v) \quad for \quad \bar{u}_h \leftrightarrow u, \bar{v}_h \leftrightarrow v$$

with block diagonal, continuous and monotone operator $N(u) = L^T k_1(Lu) + k_2(u, Lu)$. Denote by $\theta_i(L_i u_i) = \int_{\Omega_i} |\nabla u_{hi}| \, dx \quad for \quad u_i \leftrightarrow u_{hi}$, and $\theta(Lu) = \sum_{i=1}^{m} \theta_i(L_i u_i)$.

Finally, $(\bar{u}_h, s_h) \in K_h \leftrightarrow (u, s) \in K = \{R_i u_i = S_i s, \quad i = 1,2,...,m\}$, where $R_i u_i$ is the trace of u_i on Γ_i, while $S_i s$ is the restriction of s to Γ_i.

Now, variational inequality (5) can be written as the following variational inequality for the vectors of nodal values of corresponding mesh functions:

$$(u, s) \in K: \quad (N(u), v - u) + \theta(Lv) - \theta(Lu) \geq f(v - u) \quad \forall (v, t) \in K. \tag{6}$$

Obviously, this variational inequality has a unique solution because (5) has a unique solution.

Proposition 4. *A pair $(u, s) \in K$ is a solution of (6) if and only if there exists a vector μ such that the triple (u, s, μ) satisfies the following saddle point problem*:

$$\begin{pmatrix} N + L^T \circ \partial\theta \circ L & 0 & -R^T \\ 0 & 0 & S^T \\ -R & S & 0 \end{pmatrix} \begin{pmatrix} u \\ s \\ \mu \end{pmatrix} \ni \begin{pmatrix} f \\ 0 \\ 0 \end{pmatrix}, \tag{7}$$

where $R = diag(R_1, R_2, \ldots R_m)$, $S = (R_1, R_2, \ldots R_m)^T$ and $\partial\theta$ is the subdifferential of the convex function.

To apply Uzawa-type iterative method we make several equivalent transformations of saddle point problem (7). First, define the auxiliary vectors $p=Lu$ and $\lambda \in k_1(p) + \partial\theta(p)$. Then the inclusion $N(u) + L^T \partial\theta(Lu) - f \ni R^T \mu$ in the first row of (7) transforms to the system with respect to the vectors (u, p, λ): $k_2(u, p) + L^T \lambda - R^T \mu = f, \quad k_1(p) + \partial\theta(p) \ni \lambda, \quad Lu = p$.

Problem (7) turns to the following one: find $w = (u, p, s)^T$, $\eta = (\lambda, \mu)^T$ such that

$$\begin{pmatrix} A_0 & B \\ -B^T & 0 \end{pmatrix} \begin{pmatrix} w \\ \eta \end{pmatrix} \ni \begin{pmatrix} F \\ 0 \end{pmatrix}. \tag{8}$$

with right-hand side $F=(f,0,0)^T$, matrix $B=\begin{pmatrix} L & -E & 0 \\ -R & 0 & S \end{pmatrix}$, E is identity matrix, and non-linear operator $A_0(w)=(k_2(u,p),k_1(p)+\partial\theta(p),0)^T$ for $w=(u,p,s)^T$.

Since the operator A_0 is degenerate, we make further equivalent transformation of system (8) by using the equations Lu-p=0 and Ru-Ss=0. These transformations result in the following saddle point problem

$$\begin{pmatrix} A & B \\ -B^T & 0 \end{pmatrix}\begin{pmatrix} w \\ \eta \end{pmatrix} \ni \begin{pmatrix} F \\ 0 \end{pmatrix} \tag{9}$$

with $A(w)=((rL^T M_p L+r_1 R^T R)u - rL^T M_p p + k_2(u,p) - r_1 R^T Ss, k_1(p)+\partial\theta(p), -r_1 S^T Ru + r_1 S^T Ss)^T$ and positive parameters r and r_1 which are defined to ensure the uniform monotonicity of A.

Proposition 5. *If $r_1>0$ while parameter r satisfies the inequalities*

$$0<2a_0\sigma_0 - b\beta_2 c_f - 2\sqrt{a_0\sigma_0\sigma} < r < 2a_0\sigma_0 - b\beta_2 c_f + 2\sqrt{a_0\sigma_0\sigma}, \tag{10}$$

then operator A is uniformly monotone:

$$\exists \gamma_1>0:\ (A(w_1)-A(w_2),w_1-w_2)_w \geq \gamma_1\|w_1-w_2\|_w^2. \tag{11}$$

Proposition 6. *Problem (9) has a solution (u,p,s,λ,μ) with unique components (u,p,s).*

4. Iterative solution method

Now let us fix a parameter $r_1>0$ and $r_0=2\sigma_0-\beta_2 c_f$ (the middle point of the admissible interval). Define a non-linear operator $A(w,\tilde{w})$ for the vectors $w=(u,p,s)^T$ and $\tilde{w}=(\tilde{u},\tilde{p},\tilde{s})^T$ by the equality

$$A(w,\tilde{w})=((rL^T M_p L+r_1 R^T R)u - rL^T M_p p + k_2(\tilde{u},p) - r_1 R^T S\tilde{s}, k_1(p)+\partial\theta(p), -r_1 S^T Ru + r_1 S^T Ss)^T.$$

Proposition 7. *If $D_1=\begin{pmatrix} M_p^{-1} & 0 \\ 0 & E \end{pmatrix}$ and $\tau<r_0$, then there exists $\alpha>1$, $\gamma>0$ such that*

$$(A(w_1,\tilde{w}_1)-A(w_2,\tilde{w}_2),w_1-w_2)_w - \frac{\alpha\tau}{2}(D^{-1}B(w_1-w_2),B(w_1-w_2)) \geq \tag{12}$$

$$\geq \gamma\|w_1-w_2\|_w^2 + \rho(w_1-w_2) - \rho(\tilde{w}_1-\tilde{w}_2),\quad \rho(w)=0.5b\beta_1 c_f^2(M_p Lu,Lu)+0.5r_1\|Ss\|^2.$$

Block relaxation-Uzawa method with a preconditioner D for solving (9) reads as follows:

$$A(w^{k+1},w^k)\ni F,$$

$$D\frac{\eta^{k+1}-\eta^k}{\tau}+Bw^{k+1}=0. \tag{13}$$

Due to the inequalities (11) and (12) and general convergence Theorem 1 from [16] we get the following result:

Theorem 1. If $D=\begin{pmatrix} M_p^{-1} & 0 \\ 0 & E \end{pmatrix}$ and $\tau<r_0$ then iterative method (13) converges from any initial guess w^0.

The algorithm to implement method (13) reads as follows: given an initial guess (u^0,λ^0,s^0) for $k=0,1...$ calculate sequently

$$k_1(p^{k+1}) + \partial\theta(p^{k+1}) \ni \lambda^k,$$
$$(rL^T M_p L + r_1 R^T R)u^{k+1} = rL^T M_p p^{k+1} - k_2(u^k, p^{k+1}) - r_1 R^T S s^k + f - L^T \lambda^k + S^T \mu^k,$$
$$r_1 S^T S s^{k+1} = r_1 S^T R u^{k+1} - S^T \mu^k,$$
$$\lambda^{k+1} = \lambda^k + \tau M_p (p^{k+1} - L u^{k+1}),$$
$$\mu^{k+1} = \mu^k + \tau(R u^{k+1} - S s^{k+1}).$$

Thus, we have to solve on every iteration the system of inclusions to find p^{k+1} and then the systems of linear equations to find u^{k+1} and s^{k+1}. We emphasize that owing to the block diagonal form of the operators and matrices the inclusion for p^{k+1} and equation for u^{k+1} are split up into uncoupled systems, corresponding to the subdomains. Moreover, since every operator $k_{1i} + \partial\theta_i$ for $i = 1,2,\ldots,m$ has block diagonal form with 2×2 blocks, we can easily find the exact solutions of the corresponding inclusions in explicit forms.

References

[1] Kuznetsov Y and Lapin A 1988 *Soviet J. Numer. Anal. Math. Modelling* **3** pp 487-504
[2] A. Lapin 1991 *Int. Ser. Num. Math.* **99** pp 243-250
[3] Kuznetsov Y, Neittaanmaki P and Tarvainen P 1994 *Contemporary Mathematics* **157 pp** 271-277
[4] Zeng J and Zhou S 1998 *SIAM J. Numer. Anal.* **35** pp 600-616
[5] Tai X and Espedal M 1998 *Numer. Meth. for Part. Diff. Equat.* **14** pp 717-737
[6] Badea L and Wang J. 1999 *Mathematics of Computation* **69** pp 1341-1354
[7] Laitinen E, Lapin A and Pieska J 2001 *Comp. Methods in Appl. Mech. Eng.* **191** pp 295-309
[8] Tai X 2001 *East-West J. Numer. Anal.* **9** pp 233-252
[9] Laitinen E, Lapin A and Pieska J 2003 *J. Comp. Appl. Math* **154** pp 393-413
[10] Tai X 2003 *Numerische Mathematik* **93** pp 755-786
[11] Dostal Z et al. 2008 *Lecture Notes in Computational Science and Engineering. Domain Decomposition Methods in Science and Engineering* **91** pp 39-49
[12] Haslinger J et al. 2012 *Proceedings DDM-21* pp 743-750
[13] Jiang B, Bruch Jr and Sloss J 2006 *Numerical Methods for Partial Differential Equations* **22** pp 1-17
[14] Jiang B 2009 *Computer Methods in Applied Mechanics and Engineering* **198** pp 9-12
[15] Riaz S and Loghin D 2012 *Proceedings DDM-21* pp 735-743
[16] Lapin A 2010 *Lobachevskii J. Math.* **31** pp 309-322
[17] Laitinen E, Lapin A and Lapin S 2015 *Russian J. Numer. Analysis Math. Modeling* **30** pp 43-54
[18] Laitinen E, Lapin A and Lapin S 2015 *Lobachevskii J. Math.* **33** pp 364-371
[19] Ekeland I and Temam K 1976 *Convex analysis and variational problems* (Amsterdam: North-Holland)
[20] R. Glowinski R, Lions J and Tremolieres R 1981 *Numerical Analysis of Variational Inequalities* (Amsterdam: North-Holland)
[21] Ciarlet P 1978 *The finite element method for elliptic problems* (Amsterdam: North-Holland)
[22] Glowinski R 2008 *Numerical Methods for Nonlinear Variational Problems* (Berlin: Springer-Verlag)

Multilayer neural network models based on grid methods

T Lazovskaya and D Tarkhov

Peter the Great St-Petersburg Polytechnic University, 29 Politechnicheskaya Str, 195251 Saint-Petersburg, Russia

E-mail: tatianala@list.ru

Abstract. The article discusses building hybrid models relating classical numerical methods for solving ordinary and partial differential equations and the universal neural network approach being developed by D Tarkhov and A Vasilyev. The different ways of constructing multilayer neural network structures based on grid methods are considered. The technique of building a continuous approximation using one simple modification of classical schemes is presented. Introduction non-linear relationships into the classic models with and without posterior learning are investigated. The numerical experiments are conducted.

1. Introduction

Neural network approach to solving differential equations which was proposed by A.N. Vasilyev and D.A. Tarkhov was quite versatile. It could be applied without strong changes for the most extensive range of tasks [1-8]. Classical approaches to solving many of these problems require special artificial techniques. Neural network models were resistant to errors in the given data, they can be built based on the heterogeneous data which may include the results of experiments and have a number of other advantages. At the same time, the neural network approach is not free of a number of disadvantages in comparison to the classical methods of the grids, the finite elements and etc.

First of all, the neural network learning is still very resource consuming procedure. In the second place, the required size of the neural network and its learning time increase sharply with accuracy requirements. This problem is actual when the boundary value problem displays the simulated real object properties with high precision. Third, the neural network approach has not been implemented as a standard software package and it is not integrated at any type of standard package, for example, ANSYS.

In this context, the topical problem is the coupling of the neural networks and classical methods. We assume that obtained hybrid algorithms will be saved the disadvantages of each separate approach while keeping their advantages. We tried three different ways to such coupling.

The first way is that the results of classical methods are used as the additional data at training of the neural network [1,2,5].

The second way consists in the implicit grid methods transforming to the explicit methods using pre-trained neural networks. This approach has been tested for ordinary differential equations in [2].

A third way is to construct multilayer neural network structures based on the grid methods. In the above-cited works on neural network approach, we used neural networks with one hidden layer. For scaling to multilayer networks, it is necessary to determine the network structure (number of layers and the number of elements in each layer), and the initial values of the network parameters (network weights). This can be done by means evolutionary algorithms [1,6,7], and based on a grid pattern, passing from linear relations to similar nonlinear relations which are typical of neural networks. The result is an analogue of deep learning [2].

In this article, we focus on the second and third approaches due to their principal novelty.

2. Continuous approximations based on classical methods

Before turning from the classical methods to trainable neural networks, we consider one technique of constructing a continuous approximation of the solution of the problem based on the classical Euler method.

We solve the boundary problem

$$y'(t) = \begin{cases} f(t, y(t)), \ t \in [t_0, T]; \\ \quad y(t_0) = y_0. \end{cases} \tag{1}$$

We construct a solution using the explicit Euler method on the interval $[t_0, x]$ with a variable right endpoint $x \in [t_0, T]$. The step size is defined by the next expression $h(t) = \frac{x - t_0}{M}$, where M is called the layer number. Then, the approximate solution $\hat{y}(x)$ is built iteratively and coincides with $y_M(x)$ in the scheme

$$\begin{cases} t_i(x) = t_0 + h(x)i, \\ y_{i+1}(x) = y_i(x) + h(x) f(t_i(x), y_i(x)), \end{cases} \quad i = 1, \ldots, M. \tag{2}$$

Similarly, we can modify and other classical schemes – the predictor-corrector, the trapezoidal rule, Runge-Kutta, Adams methods and etc.

The result is a continuous function approximating the required solution on all of the interval $[t_0, T]$. The accuracy of the solution, obviously, depends on the layer number M of considered model. The convergence and the accuracy can be estimated by the error (residual) functional.

Such Euler and Runge-Kutta method modifications were applied to solve the problem with a parameter

$$\begin{cases} y'(0) = \alpha(\cos t - y(t)); \\ y(0) = 0, \end{cases} \tag{3}$$

where $\alpha \in [0.1, 50]$ and $t \in [0,1]$.

For large values of the parameter α, we have the classic stiff problem [3, 4] for which explicit methods are numerically unstable.

The solution quality was defined as the mean square residual Δ_{1000} at 1000 pairs of random points (x_i, α_i) uniformly distributed in the area $[0,1] \times [0.1, 50]$. When choosing the number of layers $M = 30$ a continuous solution based on Euler method was obtained with an error $\Delta_{1000} = 0.004$, the relative mean square error, in this case, was 0.5 percent.

Note that for $\alpha = 50$ while the classical scheme gives a pointwise solution oscillating with a large error in the neighborhood of the point $x = 0$, the errors of continuous approximation of the required function based on the modified methods are rather small. Figure 1 shows errors of all methods mentioned.

Decrease the number of layers to 25 significantly degrades the accuracy of solutions. This is readily illustrated by Figure 2.

Note that it is possible to obtain uniform estimates for the continuous approximate solutions on the interval based on the known estimates for the initial methods residuals at the finite point of the interval.

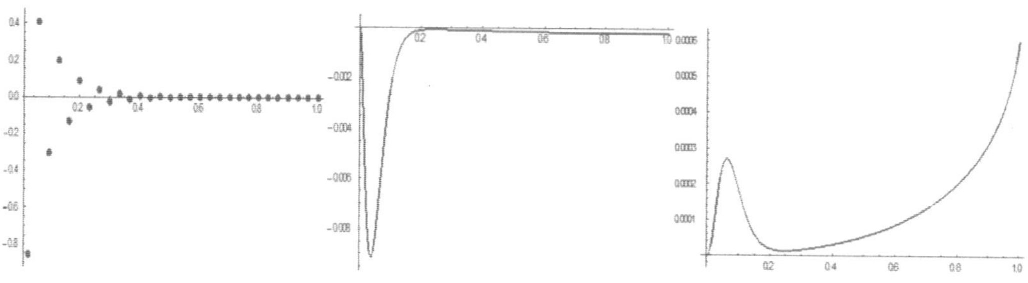

a) b) c)

Figure 1. Error graphs for the classical explicit Euler method for solving the problem (3) with $\alpha = 50$ (a) and the continuous approximations of it obtained by using the modifications of the explicit scheme of Euler method (b) and Runge-Kutta method of second-order accuracy (c) in the case of the layer number $M = 30$.

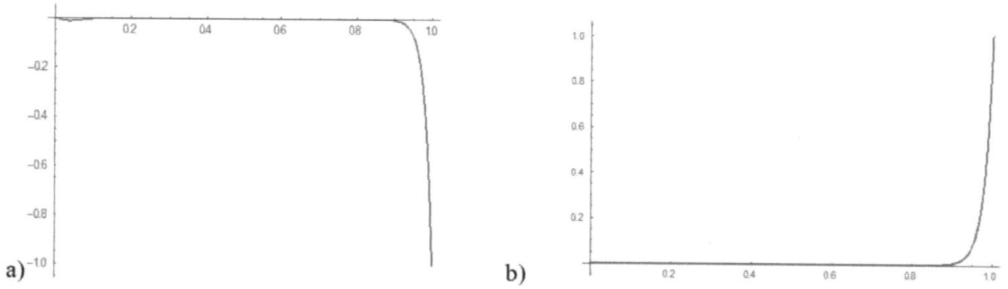

Figure 2. Error Graphs for the continuous approximation of the solution of problem (3) with $\alpha = 50$ obtained by using the modifications of the explicit scheme of Euler's method (a) and Runge-Kutta method of second-order accuracy (b) in the case of the layer number $M = 25$.

3. Introduction nonlinear relationships into the classic models

If in the previous section as a result of modification of the classical method of Euler, we obtained an approximate solution in the form of a function not associated with neural networks, then below we build in a similar manner an approximate solution of the problem (1) most closely to possible neural network.

The idea is to replace the increment linear expression with the nonlinear relationship which behaves almost linearly for small values of the argument. We can get an initial approximation in the form of the neural network choosing the functions which are typical for a neural network basis as such nonlinear dependencies.

If it is necessary to find approximate solutions more accurate (in the terms of the residual or error functional) the constructed neural network can be trained by fitting the weight during the minimization of the corresponding error functional. Below we present particular examples.

4. Nonlinear Euler method

Solving the problem (1) we transform the already modified scheme (2) of the Euler method to the next form

$$\begin{cases} t_i(x) = t_0 + h(x)i, \\ y_{i+1}(x) = y_i(x) + \mathrm{th}[h(x) f(t_i(x), y_i(x))], \end{cases} \quad i = 1, \ldots, M. \tag{4}$$

The number M still determines the layer number of the model. As will be seen, a sufficiently large value of M provides constructing the model of a good accuracy without recourse to further learning and fitting of parameters.

We continue to solve the problem (3) using the scheme (4) and build a multi-layer perceptron network. Even when the number of layers M is equal to 25 we get a quite good approximation (Figure 3).

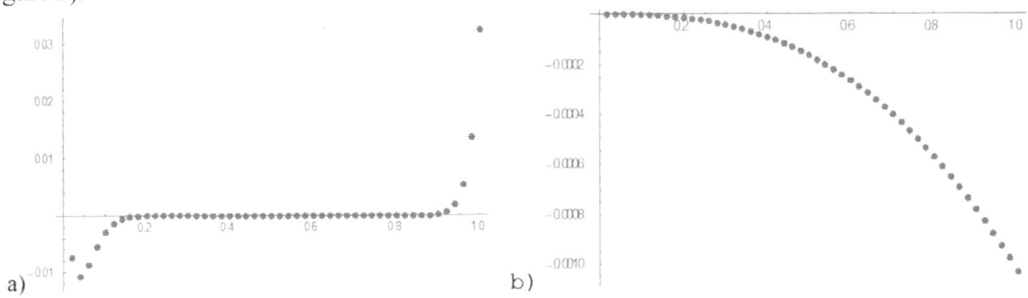

Figure 3. Error Graphs for the continuous approximation of the solution of problem (3) with $\alpha = 50$ (a) and $\alpha = 0.1$ (b) obtained by using the scheme (4) in the case of the layer number $M = 25$.

The comparison with Figure 2 shows a significantly higher accuracy of the solution obtained based on the scheme (4). With the layer number increasing the calculation time cost and the complexity of the constructed solution increase too.

5. Trainable neural networks based on the Euler method

Here, we build a multi-layer neural network by using the previous section methods and choose a small number of layers. So, we are able to train obtained network at the lowest cost. Network parameters are introduced by the addition of linear factors in the argument of the sigmoid function (th($\cdot$)) included in the solution constructed.

The three-layer parameterized neural network corresponded to the problem (3) and constructed by the scheme (4) has the following form

$$y(x,\alpha\ ;\ b,\ c,\ d) = \mathrm{th}(bx\alpha) + \mathrm{th}[cx\alpha(\cos\frac{x}{3} - \mathrm{th}(bx\alpha))] + \\ \mathrm{th}[dx\alpha(\cos\frac{2x}{3} - \mathrm{th}(bx\alpha) - \mathrm{th}[cx\alpha(\cos\frac{x}{3} - \mathrm{th}(bx\alpha))])]. \tag{5}$$

Further, the network weights *b*, *c* and *d* are fitted during the minimization of the corresponding error functional. We have used the functional in the discrete form

$$\sum_{j=1}^{m}\left(y'(x_j) + \alpha_j(y(x_j) - \cos x_j)\right)^2, \tag{6}$$

test points $(x_j, \alpha_j) \in [0;1] \times [\alpha_{\min}; \alpha_{\max}]$ are regenerated after every 3-5 training steps. We have used the normalized functional too in the form

$$\sum_{j=1}^{m}\left(y'(x_j) + \alpha_j(y(x_j) - \cos x_j)\right)^2 / \alpha_j^2. \tag{7}$$

The parameter values $b = c = d = 1/3$ from the scheme (4) were used as a quite good initial approximations. Optimization was carried out using a combination of the RProp and the Particle swarm algorithms.

Neural network learning allowed to get an acceptable accuracy in the case the significantly smaller number of layers (Figure 4). Note that the methods [3] give similar accuracy for the single-layer network with 20 neurons.

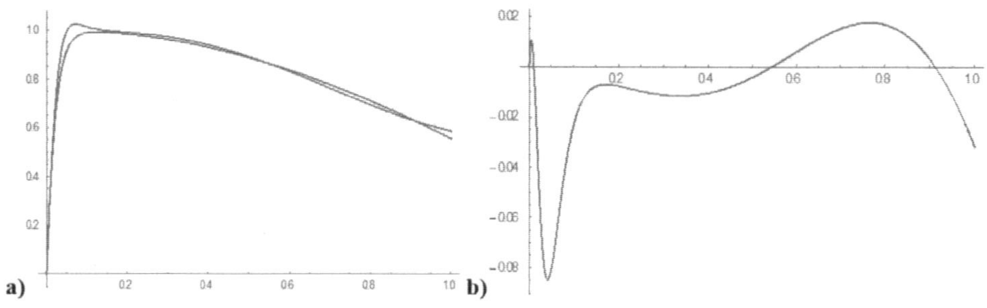

Figure 4. (a) True and approximate solution graphs precise and (b) the approximation error for stiff problem (3) in the case of $\alpha = 50$ obtained by the use of the three-layer neural network (5).

6. Nonlinear modification of the implicit Euler method vs the classical methods

We solve the problem (3) for various fixed values of the parameter α using different approaches: classic explicit and implicit (8) Euler methods, trapezoidal rule (9) and the method modified by introducing a sigmoid nonlinearity (10). We remind that depending on the alpha the problem cannot be solved by explicit classical methods. For considered problem, implicit methods equations are solved simply. Let $h = 1/M$ is a step size in terms of classical methods (M is the layer number). Corresponding schemes will be of the next forms

$$y_{i+1} = \frac{1}{1+\alpha h} y_i + \frac{\alpha h}{1+\alpha h} \cos ih; \tag{8}$$

$$y_{i+1} = \frac{1-0.5\alpha h}{1+0.5\alpha h} y_i + \frac{0.5\alpha h}{1+0.5\alpha h} \left(\cos ih + \cos(i+1)h\right); \tag{9}$$

$$y_{i+1} = \frac{1-\text{th}(\alpha h)}{1+\text{th}(\alpha h)} y_i + \frac{\text{th}(\alpha h)}{1+\text{th}(\alpha h)} \cos(i+1)h, \tag{10}$$

where $y_0 = 0,\ i = 0, \ldots, M$.

In the case of a non-stiff problem statement, all methods work equally well, the function $\text{th}(x)$ is almost linear for small values of the argument.

By choosing the number of layers $M = 30$ we consider the problem (3) in a fairly stiff statement at $\alpha = 100$. Figure 5 shows that while the explicit method diverges schemes (8)-(10) give error increase in the neighborhood of the point $x = 0$ but sigmoidal scheme accuracy is an order of magnitude more.

With further increase of the parameter α sigmoidal modification (10) shows the best result.

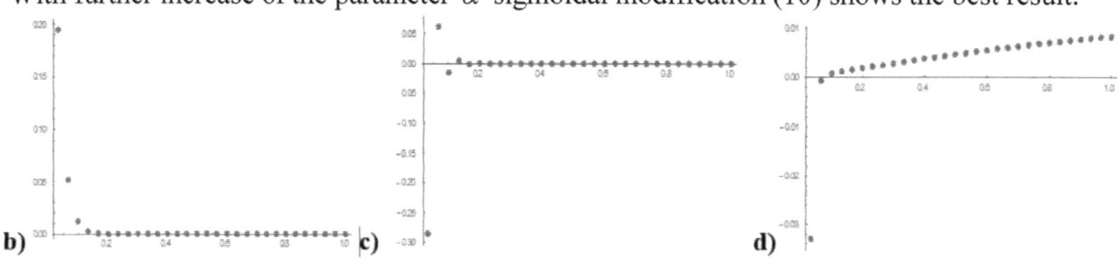

Figure 5. Error graphs of the problem (3) solution obtained based on schemes (8) - (10) in the case of $\alpha = 100$ and the layer number $M = 30$.

7. Introduction of neural network parameters to original functional object

We have already discussed above construction of continuous approximations based on classical methods. We showed how you can select the network parameters for further learning. At the same time, the interior point positions were prefixed on the assumption of the uniform decomposition of an interval. Another way to organize the algorithm it to make steps variables.

The use of explicit Euler method with one intermediate point to the problem (3) gives the following formula

$$y(x,\alpha;b) = \alpha bx\cos(bx) + \alpha(1-b)x(\cos x - \alpha bx\cos(bx)). \quad (11)$$

Figure 6 shows the acceptable result obtained by the use (11).

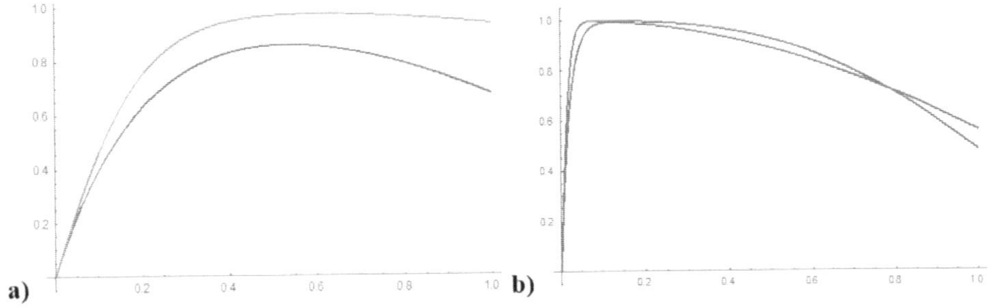

Figure 6. Graphs of the solutions of the problem (3) with the parameter values (a) $\alpha = 5$ and (b) $\alpha = 50$ obtained using the formula (11).

8. Extension of new approaches to partial differential equations

It is possible to offer different ways to extend the proposed approaches to partial differential equations. The simplest of these is the method of lines [4].

We present a simple example of this scheme for the heat equation

$$\frac{\partial y}{\partial t} = \frac{\partial^2 y}{\partial x^2}, \ (t,x) \in [0,1] \times [0,1], \quad (12)$$

with boundary conditions $y(0,x) = 1 - x^2$, $y'(t,0) = 0$ and $y(t,1) = 0$. The explicit scheme of the methods is considered in the form

$$G(i,j) = G(i,j-1) + R\Big[R\big(h\big(G(i-1,j-1) - G(i,j-1)\big)\big) + R\big(h\big(G(i+1,j-1) - G(i,j-1)\big)\big)\Big] \quad (13)$$

If $R(x) = x$ we obtain a simple explicit method. If we choose sigmoid function $R(x) = \dfrac{x}{1+|x|}$ the scheme (13) can be considered as the neural network model. Figure 7 is illustrated the results in the case when we divide the segment by 10 parts with respect to x and by 180 parts – to t.

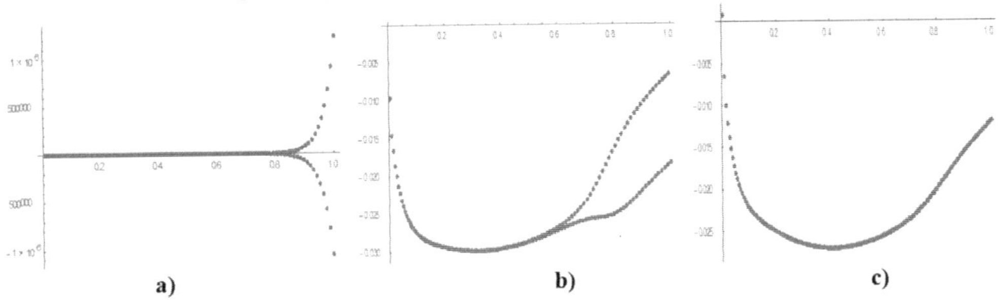

Figure 7. Error graphs at $x = 0$ of the approximate solution of the heat equation problem obtained by regular explicit scheme (a), the scheme (13) (b) and the scheme (13) with averaged solution at the current and the previous points (c).

9. Conclusions

In this paper, we made the first attempts to bridge the gap between classical numerical methods for solving differential equations and neural network approach. We have considered the possibility of constructing continuous approximations based only on classical numerical schemes, the introduction of non-linear relationships both on searching pointwise solutions and in constructing the continuous approximations. We have shown the possibility of learning of the obtained models through the introduction of the corresponding parameters. The calculations based on considered algorithms can be speed up by several orders of magnitude in the case of using specialized neuron chips.

This area is promising for the further investigation and theoretical study results.

Acknowledgments

The work is supported by the Russian Foundation for Basic Research (Grant no. 14-01-00660).

References

[1] Tarkhov D and Vasilyev A 2014 Mathematical Models of Complex Systems on the Basis of Artificial Neural Networks *Nonlinear Phenomena in Complex Systems* **17** pp 327-335

[2] LeCun Y, Bengio Y and Hinton G 2015 Deep learning *Nature* **521 553** pp 436-444

[3] Lazovskaya T and Tarkhov D 2015 Fresh approaches to the construction of parameterized neural network solutions of a stiff differential equation *St. Petersburg Polytechnical University Journal: Physics and Mathematics* http://dx.doi.org/10.1016/j.spjpm.2015.07.005

[4] Hairer E and Wanner G 2010 *Solving Ordinary Differential Equations II: Stiff and Differential-Algebraic Problems* (Springer) 614 p

[5] Budkina E, Kuznetsov E, Lazovskaya T, Leonov S, Tarkhov D and Vasilyev A 2016 Neural Network Technique in Boundary Value Problems for Ordinary Differential Equations *ISNN 2016 LNCS* **9719** pp 277-283

[6] Gorbachenko V, Lazovskaya T, Tarkhov D, Vasilyev A and Zhukov M 2016 Neural Network Technique in Some Inverse Problems of Mathematical Physics *ISNN 2016 LNCS* **9719** pp 320-316

[7] Shemyakina T, Tarkhov D and Vasilyev A 2016 Neural Network Technique for Processes Modeling in Porous Catalyst and Chemical Reactor *ISNN 2016 LNCS* **9719** pp 547-554

[8] Tarkhov D and Vasilyev A 2005 New neural network technique to the numerical solution of mathematical physics problems. I: Simple problems *Optical Memory and Neural Networks (Information Optics)* **14** pp 59-72

[9] Tarkhov D and Vasilyev A 2005 New neural network technique to the numerical solution of mathematical physics problems. II: Complicated and nonstandard problems *Optical Memory and Neural Networks (Information Optics)* **14** pp 97-122

[10] Kainov N, Shemyakina T and Tarkhov D 2014 Application of neural network modeling to identification and prediction problems in ecology data analysis for metallurgy and welding industry *Nonlinear Phenomena in Complex Systems* **17 (1)** pp 57-63

Influence of antenna dielectric cover on the parameters of the electrode microwave discharge

Yu A Lebedev, A V Tatarinov, A Yu Titov and I L Epstein

Topchiev Institute of Petrochemical Synthesis of the Russian Academy of Sciences (TIPS RAS), Moscow, Russia

E-mail: lebedev@ips.ac.ru

Abstract. One possibility of reducing the influence of contamination of the plasma volume, when initiating the microwave discharges using antennas, is to cover the antenna with a dielectric. The results of 1-D and 2-D modeling showing the impact of such dielectric cover on the parameters of a non-equilibrium plasma of an electrode microwave discharge in hydrogen at reduced pressure (the change of the electrodynamics of the discharge, charge deposition on the dielectric surface, changing the catalytic properties of the surface of the antenna) are presented in the article.

1. Introduction

One of the advantages of microwave discharges is the absence of electrodes in the discharge chamber that can lead to contamination of the plasma by the products of erosion. Because of that reason it was the electrodeless microwave systems like the waveguide, resonator and surface wave, etc systems [1] that have been widely used for plasma generation until recently. However there are conditions, when it is necessary to introduce the electrode for microwave plasma generation. This applies primarily to the initiated discharges when the microwave field from an external source is insufficient to initiate and maintain the discharge; creation of a discharge requires additional electrodes where an increase in the local field values exists [2]. In addition, there is a whole class of microwave devices where the plasma is initiated by electromagnetic vibrators of different types [3-6]. Such devices are used to obtain low pressure [7-9] and atmospheric pressure [4-6] plasma. These include widespread coaxial microwave plasma torch [10-12]. The microwave antenna systems are also applied for generating plasma inside liquids [13-16].

In these cases, there is the problem of contamination of the volume by the products of erosion of the metal electrodes. It seems that the possibility of contamination of the plasma with electrode material in microwave devices is exaggerated at least at low power. In [17] in the electrode microwave discharge in conditions typical for diamond deposition (a mixture of hydrogen with a small addition of methane, at pressure of about 15 Torr, the substrate, located at a distance of 1.5 cm from the electrode, is heated up to ~1000°C) the ESCA and Auger spectroscopy methods have not revealed even trace quantities of the electrode material in the deposited material. This is due to the fact that in the microwave plasma the electrode surface is not probed by high energy ions. In RF discharge, for example, such ions appear as a result of their acceleration in large DC fields that exist in the electric double layer at the electrode due to the rectification of the applied RF voltage (~100-300V) on the nonlinearity of double layer. In the microwave plasma the electron concentration is high enough ($\geq 10^{11}$ cm^{-3}) and thickness of a near-electrode layer is small. Under these conditions, the layer impedance is capacitive, the rectification of the microwave signal, though exists, but is not a factor leading to the occurrence of high energy ions. Therefore the microwave discharge, as a rule, is the discharge with predominant volume ionization (α-form level) and the secondary ionization (γ-processes) can be neglected. However in high energy regimes, when the temperature of the electrode becomes large, melting and even explosion of metallic electrodes are possible [8, 9], which leads to the mass transfer of the material of electrodes into a gas media. Thermal erosion of the electrode also occurs in the microwave discharges in liquids [18, 19].

One method of eliminating the influence of electrode material is coating the electrode with a heat-resistant dielectric, as it was done, e.g., in [19]. In this regard the problem of influence of such coatings on the plasma parameters arises. This influence may be related to: the deposition of charges on the dielectric surface in the plasma, changing of the electrodynamic properties of the discharge system and also change of the catalytic properties of the surface (changing of the metal to dielectric) and an associated change of the surface impact on concentration of heavy plasma components. All of these factors are analyzed in the present work, based on modeling of the electrode microwave discharge in hydrogen at pressure of 1 Torr.

2. The model

We used one- and two-dimensional models for analyzing the effect of dielectric coating at the antenna on properties of the microwave plasma.

In the present paper a two-dimensional model is applied for studying steady-state microwave field which is set inside the empty discharge chamber with the inner electrode covered with dielectric. The discharge chamber is described in detail in [7-9]. The discharge chamber is a stainless steel cylinder with a diameter of 15 cm. An electromagnetic wave with a frequency of 2.45 GHz is fed into the discharge chamber through a waveguide-coaxial adapter and is introduced into the chamber along the electrode-antenna (a copper cylindrical electrode with a diameter of 5mm). Full self-consistent two-dimensional model for the description of the discharge in hydrogen is described by the authors [20, 21] in detail. It includes the Maxwell equations, balance equations for the plasma particles and the Poisson equation, taking into account the field of charge separation.

Dielectrics with $\varepsilon = 1, 2, 3, 4$ and thicknesses d = 1, 2, 3 mm have been used for coating of the inner electrode.

A one-dimensional self-consistent spherically symmetric model includes the equation for the electromagnetic field in the quasi-static approximation. The *rms* electric component of the microwave field in the electrode system with spherical symmetry coincides with the field of a spherical capacitor. The inner electrode of such a capacitor is covered with a dielectric layer ε and the remaining discharge region is filled with plasma of dielectric permeability $\varepsilon_{pl} = \left\{(1 - n_e/n_{norm})^2 + \left(\frac{\nu}{\omega}\frac{n_e}{n_{norm}}\right)^2\right\}^{1/2}$, where $n_{norm} = m(\omega^2 + \nu^2)/4\pi e^2$, ω is a circular frequency of the microwave signal, ν is the frequency of collisions of electrons with neutral heavy particles and n_e is the electron concentration. The microwave field in the capacitor is determined from the Poisson equation. The field amplitude at the internal electrode boundary is set.

The model also includes: the Boltzmann equation for the electron energy distribution function (EEDF), recorded in the two-term approximation of the expansion of the EEDF into spherical harmonics; non-stationary balance equations for the concentrations of charged and neutral particles, and the equation for the field calculation of charge separation. The solution of the Boltzmann equation considering both the microwave and DC fields showed that the influence of the DC field on the EEDF was negligible. Therefore, when calculating the EEDF only the microwave field has been taken into account. The effect of the vibrationally excited hydrogen molecules on the EEDF and rates of different plasma chemical processes was taken into account by using the vibrational distribution function of hydrogen molecules obtained in the diffusion approximation described in [22, 23]. All processes in hydrogen plasma which are used in the model are described in [20, 21] in detail.

Boundary conditions for charged plasma particles, described by the balanced equations, are set by the expression for fluxes to the surface of the insulator covering the inner electrode and to the outer wall of the discharge camera:

$$\Gamma_{\boldsymbol{e}}\cdot\boldsymbol{n} = -\frac{1}{4}n_e \upsilon_{T_e} - (a-1)\cdot\mu_e n_e\,\mathbf{E}\cdot\mathbf{n}\cdot$$

$$-\Gamma^{\mathrm{i}}_{\mathbf{p}}\cdot\boldsymbol{n} = -\frac{1}{4}n^i_p \upsilon^i_{T_p} - a\cdot\mu^i_p n^i_p\,\mathbf{E}\cdot\mathbf{n}\cdot$$

Flows for electrons and ions of the i-th type to the wall consist of two components: diffusion and drift. Here υ_{Te} and υ^i_{Tp} are thermal velocities of electrons and ions on the wall, $\boldsymbol{n}$ is a unit vector normal to the electrode surface (normal vector). The corresponding electron temperature at the wall was computed using the Boltzmann equation, and the temperature of ions on the walls was taken equal to the gas temperature. The coefficient $a = 1$, if the DC field vector $\boldsymbol{E}$ is directed towards the wall, i.e. $\boldsymbol{E}\cdot\boldsymbol{n} > 0$, and $a = 0$, if $\boldsymbol{E}\cdot\boldsymbol{n} < 0$.

The equation for the accumulation of surface charge σ, deposited from the discharge, is solved at the dielectric-plasma boundary: $\frac{\partial\sigma}{\partial t} = -e\left(\sum_i \Gamma^{\mathrm{i}}_{\mathbf{p}} - \Gamma_{\mathbf{e}}\right)$. The boundary conditions for the potential φ, are the following: on the surface of the dielectric $\varepsilon_0\varepsilon\nabla\varphi = \sigma$; the outer wall of the working chamber is grounded.

The concentration of neutral excited particles of plasma on the wall was supposed to be zero. The flux of unexcited hydrogen atoms incident from plasma to the wall is equal to the heat flux (diffusion) multiplied by the probability of their losses due to recombination γ_{rec}: $-\Gamma_{\boldsymbol{n}}\cdot\boldsymbol{n} = -\gamma_{rec}\frac{1}{4}n_e \upsilon_{T_e}$.

The probability of the recombination loss for hydrogen atoms on the copper electrode was taken 0.05. According to [24] this value lies in the range 0.02-0.1. The probability of the recombination loss for hydrogen atoms on the dielectric surface covering the electrode (to be specific quartz was chosen as an insulator), was 0.01 [25]. The probability of the recombination loss of the hydrogen atoms on the chamber walls of stainless steel was taken equal to 0.1 [26].

The one- and two-dimensional simulations have been carried out numerically, using finite element methods implemented in the commercial package Comsol 3.5a [27] on 12-core 2.3GHz Xeon of 64 bit system with 32 GB of RAM.

The calculations were carried out at hydrogen pressure of 1 Torr and gas temperature of 600 K. In accordance with experimental results [28] the temperature was considered constant throughout the volume. In the following we present the stationary distributions of plasma parameters.

3. Results and discussions

3.1. . The change of the electrodynamic properties of the discharge system.

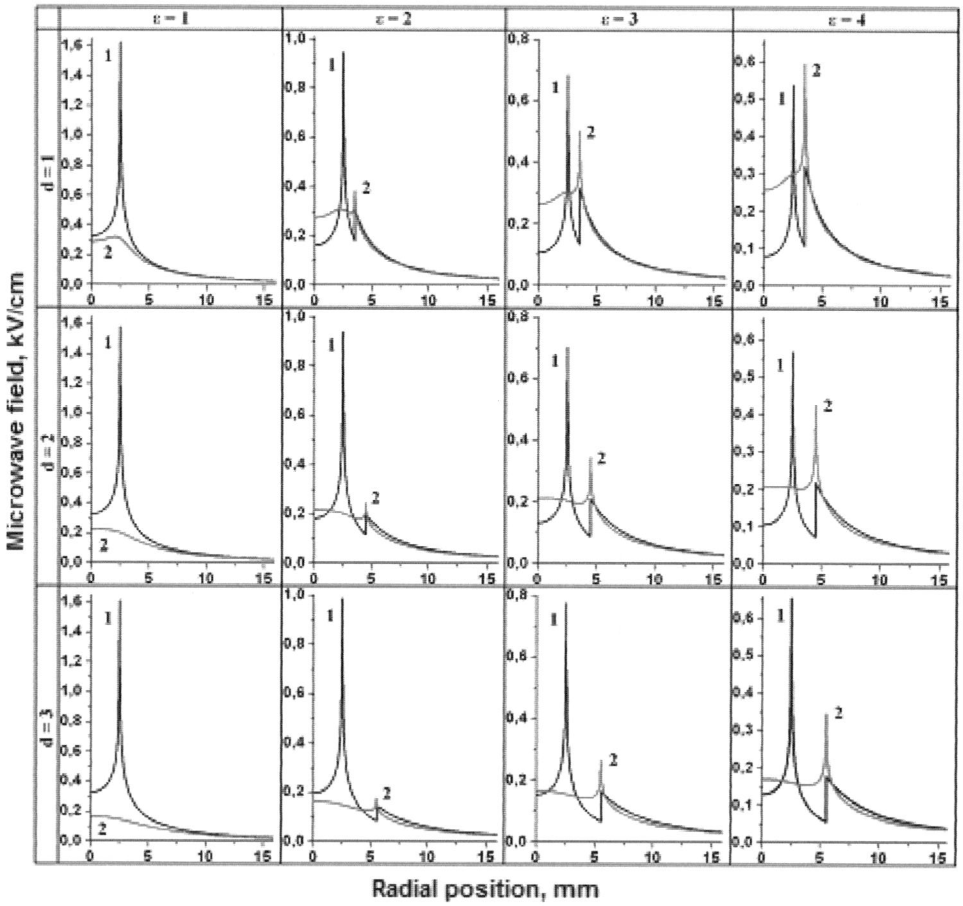

Figure 1. Radial distributions of microwave field at distance 10 microns below the metal electrode (black curves - 1) and 10 microns below the dielectric cover (red curves - 2) for different thickness of the cover and different dielectric constants (results of 2-D modeling).

The microwave energy is delivered into the discharge chamber along the antenna with the discharge ignited at its tip. A dielectric cover should lead to a change in the distribution of electromagnetic fields around the antenna. This problem has been studied in [19] by modeling of creating a microwave plasma in water. The antenna, coated with aluminum oxide, served as a central conductor of the coaxial line with a teflon insulator. The end of antenna is surrounded by water. It is shown that at the triple point, where the aluminum oxide, teflon and water are in contact, there has been a sharp increase in the intensity of the microwave field, which may lead to a discharge ignition. Note that in this case the discharge exists near the point of introduction of the antenna into the chamber. The problem on finding a place where the discharge is sustained at the body of antenna is investigated in [29] and it is shown that the in the media with high tangent loss the discharge always burns at the entrance of the antenna into the discharge chamber. In the media with low losses the discharge always burns at the end of the antenna.

In this paper we have studied the coaxial electrodynamic system (Fig. 1), where the discharge was excited by means of the antenna in hydrogen at a pressure of 1 Torr. Our previous experiments and calculations (see reviews [7-9]) have shown that the discharge at low power is sustained at the end of the antenna. Since plasma absorbs the microwave energy and represents a region with large losses, the plasma area shifts to the place of insertion of the antenna into the chamber at higher power. In this section we present a distribution of the microwave field in the discharge chamber without plasma ($\varepsilon = 1$) with a cylindrical antenna covered by a lossless dielectric of different thickness and dielectric permeability. Radial distribution of the electric field at the border of the antenna and dielectric is shown in Fig.1.

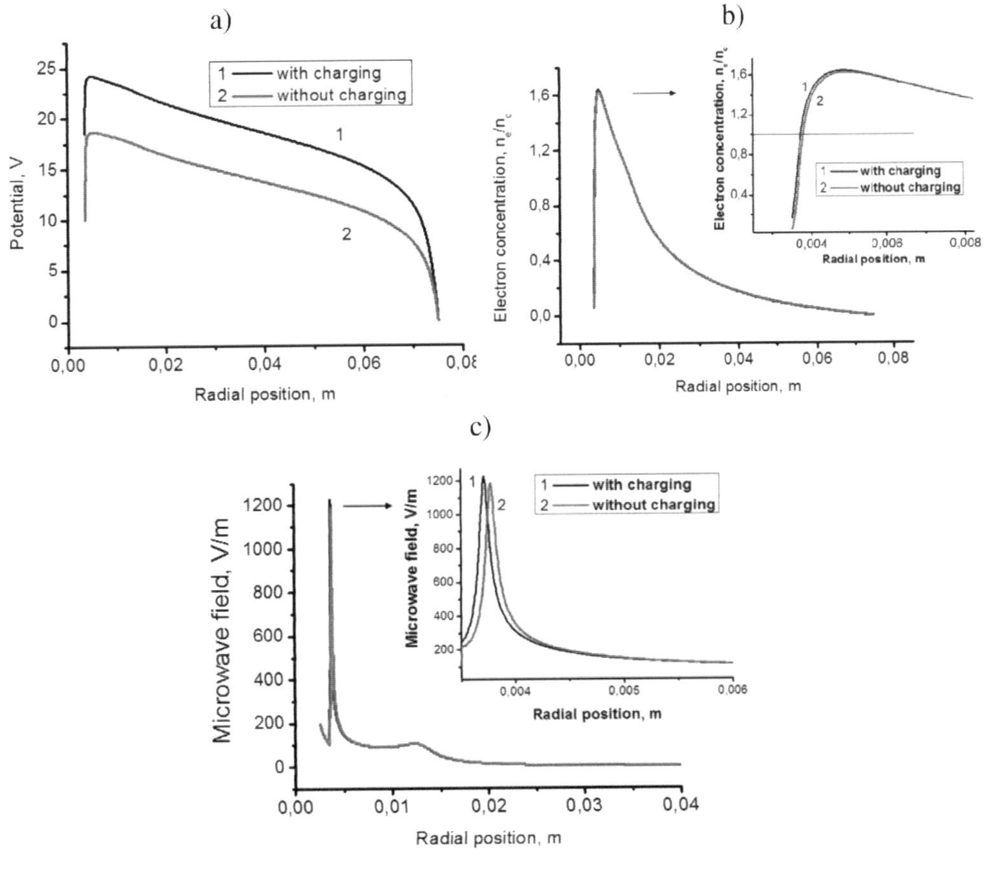

Figure 2. Radial distributions of DC potential (a), electron density (b) and microwave field (c) with 1 (black curves) and without 2 (red curves) charge deposition on quartz cover of thickness of 1 mm (results of 1-D modeling).

As it should be, on the boundary of the dielectric and gas there is a field discontinuity, the microwave field experiences a jump in accordance with the difference of dielectric permeability. It is seen also that the presence of an insulator with appropriate ε and thickness allows to obtain more uniform distribution of the field in the region under the dielectric, than without it. There is also a certain increase of the field at the edge on the end face of the dielectric.

3.2. *The deposition of charges on the dielectric surface in the plasma*

Let us consider in the framework of the one-dimensional model, the impact of deposition of charges on the dielectric-plasma interface on the plasma parameters. Since the dielectric is ideal (lossless), the influence of charging of its surface is considered only for one thickness of the dielectric (1 mm). The results of the calculations are shown in Fig.2.

Charging leads to an increase in surface potential and the potential difference between the electrode and plasma, though the profile of spatial potential distribution changes slightly (see Fig.2a).

The maximum value of the electron density is modified slightly, but the position of its maximum slightly shifts to the dielectric surface (Fig.2b). The changes are noticeable only near the surface, which occurs due to changes in the potential of the surface when it is charging. The change in the distribution of electron density near the surface leads to the fact that the region of plasma resonance approaches the surface and the magnitude of the resonance field slightly increases (Fig.2c). The position of the plasma resonance, $n_e/n_c=1$ (n_c~7×10^{11} cm^{-3}) lie on a straight line, marked in Fig.2b parallel to the x-axis.

Thus the accumulation of the charge on the surface of the electrode covered with dielectric has little effect on the plasma parameters in the volume of discharge.

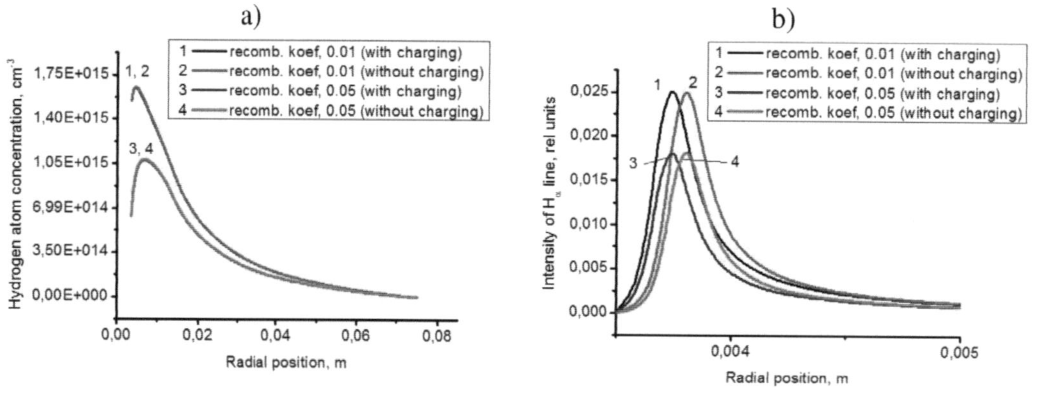

Figure 3. Radial distributions of hydrogen atom concentration (a) and H_α-line intensity (b) with (1, 3 curves) and without (2, 4 curves) charge on quartz cover of thickness 1 mm for different values of H-atom recombination coefficient (results of 1-D modeling).

3.3. *The change of catalytic properties of the surface of the antenna covered with dielectric.*

The calculations within the one-dimensional model showed that the catalytic activity affects neither the magnitude nor the distribution of the electron density and microwave field, and they coincide with those given in Fig.3b,c. Two values for coefficients of surface recombination of atoms have been used in calculations: 0.05 for copper and 0.01 for quartz. At the same time, the concentration of hydrogen atoms increases significantly at the transition from the copper to quartz. This is due to the fact that in considered conditions the main channel of loss of atoms is diffusion towards the surface with subsequent recombination. The deposition of charges on the surface of the electrode does not practically affect the concentration of the atoms (curves 1, 2 and also 3, 4). The difference is small and has a spatial scale, determined from the inset in Fig.2b. The absolute values of the emission intensity of line H_α also increase with changing recombination coefficient of hydrogen atoms from 0.05 to 0.01. Fig.3b shows the radial distribution of the emission intensity of the line H_α in the vicinity of the electrode, where the microwave field is high (see Fig.2c). Unlike for atoms the emission curves reveal the influence of the charges deposition on the electrode surface: as in the case of the microwave fields, the deposition of charges leads to a shift of the emission maximum to the surface of the electrode

It can be expected that at high pressures and high concentrations of the atoms the channel of volume losses of atoms begins to dominate and the role of the catalytic activity of the wall will decrease.

4. Conclusion

Modeling of the influence of the dielectric cover (the change in electrodynamics of the discharge, deposition of charges on the dielectric surface, changing the catalytic properties of the surface of the antenna) on plasma parameters of the microwave discharge in hydrogen at reduced pressure have shown that: (a) the use of dielectric covers of different thickness and different dielectric permeability makes it possible to control the spatial distribution of microwave field intensity near the antenna; (b) modification of electrodynamics and the deposition of charges on the dielectric surface only affect the surface layer of plasma and practically does not affect the characteristics of plasma in the plasma volume (e.g., the maximum value of the electron density); (c) catalytic properties of the dielectric surface strongly affect the maximum value of the concentration of hydrogen atoms in the case if their loss occurs due to the processes of diffusion and recombination on the surface; (d) deposition of charges on the dielectric surface leads to an increase in surface potential relative to the plasma, but the form of the spatial profile of the potential remains the same, and also leads to the shifting of the region of plasma resonance towards the surface of the electrode. It can be expected that when increasing the role of volume recombination of hydrogen atoms, the role of the recombination coefficient on the surface of the dielectric on the plasma parameters would be small.

Acknowledgments

This study was partly supported by the Russian Foundation for Basic Research, Project no. 15-08-00070.

References

[1] Lebedev Yu A 2015 Plasma Sources Sci. Technol. **24** 053001
[2] Brovkin V G, Kolesnichenko Yu F 1994 *Tech. Phys.* **64** 194-6
[3] Ganguli A, Jarwal R K, Tarey R D and Akhtar M K 1997 *IEEE Trans Plasma Sci.* **25** 1086-95
[4] Aleksandrov K V, Volkov A A, Grachev L P, Esakov I I and Severinov L G 2011 *Tech. Phys.* **56** 351-5
[5] Grachev L P, Esakov I I, Lavrov P V and Ravaev A A 2012 *Tech. Phys.* **57** 230-5
[6] Toyota H, Nomura S and Mukasa S 2013 *Int. J. Materials Sci. Appl.* **2** 83-8
[7] Lebedev Yu A, Epstein I L, Tatarinov A V and Shakhatov V A 2006 *J. Phys.: Conf. Ser.* **44** 30-9

[8] Lebedev Yu A, Mokeev M V, Tatarinov A V, Shakhatov V A and Epstein I L 2008 *J. Phys. D: Appl. Phys.* **41** 194001
[9] Lebedev Yu A, Epstein I L, Tatarinov A V and Shakhatov V A 2010 *J. Phys.: Conf. Ser.* **207** 012002
[10] Nowakowska H, Jasinski M and Mizaraczyk J 2009 *Eur. Phys. J. D* **54** 511-8
[11] Stonies R, Schermer S, Voges E and Broekaert J A C 2004 *Plasma Sources Sci. Technol.* **13** 604-11
[12] Shimizu T, Steffes B, Pompl R, et al *Plasma Process. Polym.* 2008 **5** 577-82
[13] Nomura S, Toyota H, Mukasa S, Yamashita H, Maehara T and Kawashima A 2009 J. *Appl. Phys.* **106** 073306
[14] Hattori Y, Mukasa S, Nomura S and Toyota H 2010 *J. Appl. Phys.* **107** 063305.
[15] Lebedev Yu A, Epstein I L, Shakhatov V A, Yusupova E V and Konstantinov V S 2014 *High Temp.* **52** 319-27
[16] Lebedev Yu.A., Tatarinov A.V., Epstein I.L. and Averin K.A. 2016 *Plasma Chem. Plasma Process.* **36** 535-52
[17] Bardos L, Barankova H, Lebedev Yu A, Nyberg T and Berg S 1997 *Diamond and Related Materials* **6** 224-29.
[18] Lebedev Yu A, Konstantinov V S, Yablokov M Yu, Shchegolikhin A N and Surin N M 2014 *High Energy Chem.* **48** 385-8
[19] Hattori Y, Mukasa S, Toyota H, Yamashita H and Nomura S 2012 *Surface & Coatings Technology* **206** 2140-5
[20] Lebedev Yu A, Tatarinov A V, Titov A Yu, Epstein I L, Krashevskaya G V and Yusupova E V 2014 *J. Phys. D: Appl. Phys.* **47** 335203.
[21] Lebedev Yu A, Tatarinov A V, Titov A Yu and Epstein I L 2014 Uchenye Zapiski Kazanskogo Universiteta. Seriya Fiziko-Matematicheskie Nauki **156** 120-33
[22] Capitelli M, Ferreira C, Gordiets B and Osipov A 2000 *Plasma Kinetics in Atmospheric Gases.* (Berlin: Springer)
[23] Gordiets B F, Osipov A I and Shelepin L A 1980 *Kinetic Processes in Gases and Molecular Lasers* (Moscow: Nauka)
[24] Gorse C, Capitelli M, Bacal M, Bretagne J and Lagan´a A 1987 *Chem. Phys.* **117** 177-95
[25] Lavrenko B 1973 *Recombination of Hydrogen Atoms on the Surface of Solids* (Kiev: Naukova Dumka)
[26] Kae-Nune P, Perrin J, Jolly J and Guillon 1996 *J Surface. Science.* **360** L495-8
[27] COMSOL 3.5a (http://www.comsol.com).
[28] Lebedev Yu A and Mokeev M V 2003 *Plasma Phys. Reports* **29** 226-30.
[29] Hattori Y, Mukasa S, Nomura S, Toyota H 2010 *J. Appl. Phys.* **107** 063305

Differential neural network approach in information process for prediction of roadside air pollution by peat fire

V Lozhkin[1], D Tarkhov[2], V Timofeev[1], O Lozhkina[1] and A Vasilyev[2]
[1] St. Petersburg University of State Fire Service of EMERCOM of Russia, 196105, Moskovsky, 149, St. Petersburg, Russia
[2] Peter the Great St. Petersburg Polytechnic University, 195251, Polytechnicheskaya, 29, St. Petersburg, Russia

E-mail: olojkina@yandex.ru

Abstract. The paper presents a novel differential neural network model estimating the dispersion of CO emissions from a peat fire near a highway. We have developed approaches for the optimization of the model on the base of simulated and experimental measurements of CO concentrations in the area of dispersion of the smoke cloud. The numerical solutions of the problem are presented in the form of neural network approximations by the Gaussian model and in the form of neural network approximate solutions of partial differential equations. The trained neural network model can be used for the prediction of emergency when wind speed and direction and other fire parameters are changing. The method is also recommended for the development of air quality monitoring and predicting information systems.

1. Introduction

Peat deposits are found in many places around the world, but the world's largest peatlands are the West Siberian Lowland, the Hudson Bay Lowland, and the Mackenzie River Valley [1]. The peat ecosystem is the most efficient carbon sink on the planet, for that reason, peat fires are significant sources of carbon dioxide, other greenhouse gases, carbon oxide and other pollutants, which may be a serious human health concern [1, 2]. Peat can burn deep underground for meters, even in damp conditions and in winter time under the snow layer. Peat fires are difficult to extinguish, and severe fires in peatlands can last for months. Winter peat fires are often smoldering fires that create a lot of smoke because of incomplete combustion and result in greater emissions of carbon monoxide and other harmful substances [3-6]. If a peat fire develops near a motorway, the smoke from the burning peat-bog reduces the visibility, makes the breathing difficult, affect human nervous and cardiovascular systems and may finally result in traffic accidents or in an emergency. Such an emergency, caused by a peat fire, lasted in Irkutsk Region (situated in Siberia) near the Federal Motorway P-255 "Siberia" from 26.10.15 to 15.01.16.

Atmospheric chemical transport models are widely used for developing air quality management strategies [7-13], however, the temporal and spatial estimates of emissions from wildfires, including peat fires, are problematic because of uncertainties in the size and location as well as their temporal and spatial variability [7].

Development of functional information systems allowing modeling of such situations without resolving differential equations in partial derivatives for different sets of input parameters is of great interest. The capability of neural network models to learning and optimization using the results of observations and numerical calculations as the problem parameters (e.g. wind speed and direction, the area of the fire, the intensity of emissions, etc.) is their important advantage [14-19]. Thus, the aim of the present study was to develop and investigate a neural network model for the prediction of CO concentration in the area of a peat fire considering heterogeneous information from simulated and experimental measurements for the Federal Motorway P-255 "Siberia".

2. Modeling methodology

2.1. K-theory approach

According to Berlyand [20], such parameters as instant concentrations of CO, pulsed deviations from these values and the velocity of the CO dispersion should be taken into consideration while developing an emission model of a peat fire near a motorway. The problem can be simplified by the application of the turbulent diffusion model [20-23]:

$$\frac{\partial q}{\partial t}+u\frac{\partial q}{\partial x}+v\frac{\partial q}{\partial y}+w\frac{\partial q}{\partial z}=\frac{\partial}{\partial x}k_x\frac{\partial q}{\partial x}+\frac{\partial}{\partial y}k_y\frac{\partial q}{\partial y}+\frac{\partial}{\partial z}k_z\frac{\partial q}{\partial z}-\alpha q\ , \tag{1}$$

where q is the concentration of pollutant (g/m^3); x and y are the horizontal axis (m); z is the vertical axis (m); t is the time (s); u, v, w are the components of the average CO transfer speed along the axes x, y, z, respectively, (m/s); k_x, k_y, k_z are the components of the exchange coefficient; α is the coefficient taking into account probable metabolism of CO in the atmosphere.

This dispersion approach is based on the analytical approximation of results of joint numerical integration of the equation of atmospheric diffusion and the system of equations of hydro thermodynamics for the atmospheric boundary layer and is described in the Russian normative document for the calculation of the dispersion of harmful substances in the atmospheric air – OND-86 [21] and other studies [20-23]. Using this approach, also known as K-theory, together with some approximations and assumptions [21-24], there was established that the concentration of the pollutant emitted from a non-regular superficial source, such as a peat fire, is as follow [21]:

$$C_M=\frac{AMFmn\eta}{H^2(V_1\Delta T)^{1/3}} \tag{2}$$

where C_M is the concentration of CO (g/m^3); A is the coefficient considering the temperature stratification of the atmosphere; ΔT is the temperature difference (^{0}C); M is the emission of pollutant (g/s), F is the dimensionless coefficient considering the velocity of a probable CO metabolism in the atmosphere; V_1 is the smoke flow rate from the peat fire (m^3/s); m *and* n are the experimental coefficients; η is the dimensionless coefficient considering the influence of the terrain relief; H is the height of the steady smog cloud (m).

The authors of the present paper have a long-term positive experience in the application of this model for the estimation and the forecast of transport and fire-related air pollution [24-27]. At the same time, this approach is time-consuming and does not specify inaccurate problem parameters derived from measurements. To solve these problems, we offer to apply a neural network approach.

2.2. Neural network approach

On the base of the measurements, we developed an original neural network model with parameters (weights) tuned via optimization methods [14-19]. The RProp method and the combination of cloud and RProp methods were in use. The neural network model of the complex system can gather pieces of heterogeneous information – differential equations, conservation laws, equations of state, symmetry conditions, etc. The information exchange via neural network parameters between different levels of hierarchy makes computing less resource consuming.

3. Results and discussion

3.1 Case study 1

Turbulent diffusion loses importance when modeling the transfer of the smog from the peat fire over long distances. In addition, there may happen not only a smoldering peat fire but a burning peat fire followed by the emission of hot gases [27]. We have developed an original neural network model, based on the Gaussian dispersion to estimate these physical phenomena.
Let's assume that the average cross-section of a peat fire smog, migrating towards of a motorway, is similar to the Gaussian distribution having a plume profile (Fig. 1).

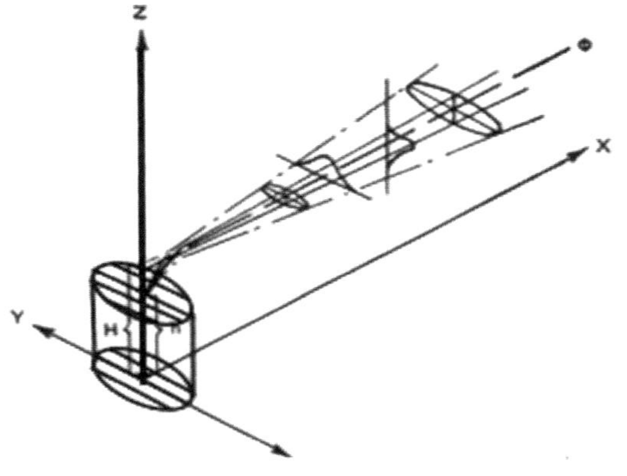

Figure 1. Schematic illustration of the transfer of CO from a peat fire: h is the height of the turbulent diffusion; H is the height of the smog.

According to the Gaussian model, the dispersion of CO concentrations in the smog obeys the normal distribution law:

$$q(t,x,y,z)=\frac{Q\exp\left(-\frac{1}{2}\left(\frac{(x-x_0-ut)^2}{(\sigma_x)^2 t}+\frac{(y-y_0-vt)^2}{(\sigma_y)^2 t}+\frac{(z-z_0-wt)^2}{(\sigma_z)^2 t}\right)\right)}{\left(\sqrt{2\pi t}\right)^3\sigma_x\sigma_y\sigma_z}, \quad (3)$$

where q is the concentration of pollutant (g/m^3); Q is the emission rate of the pollutant (g m^{-1}s^{-1}); t is the time (s), x_0, y_0, z_0 are the coordinates of the CO emission source (m); u, v, w are the components of wind speed; σ_X, σ_Y, σ_Z are the mean square deviations of the CO concentration at a time t by the axes OX, OY, OZ:

$$\sigma_x^2=\frac{2}{h}\int_0^h K_x(z)dz\ ,\ \sigma_y^2=\frac{2}{h}\int_0^h K_y(z)dz\ ,\ \sigma_z^2=\frac{2}{h}\int_0^h K_z(z)dz\ , \quad (4)$$

where h is the height of the surface layer with active turbulence, K_x, K_y, K_z are the numerical coefficients.

Using the principle of superposition, it is easy to obtain from equation (3) a formula for the concentration of CO released from a persistent peat fire:

$$q(t,x,y,z)=\int_{\Omega}\frac{Q\exp\left(-\frac{1}{2}\left(\frac{(x-x_0-ut)^2}{(\sigma_x)^2 t}+\frac{(y-y_0-vt)^2}{(\sigma_y)^2 t}+\frac{(z-z_0-wt)^2}{(\sigma_z)^2 t}\right)\right)}{\left(\sqrt{2\pi t}\right)^3\sigma_x\sigma_y\sigma_z}dx_0dy_0dz_0 \qquad (5)$$

The calculation of the integral is complex: on one hand, analytical methods will lead to huge formulas, on the other hand, numerical methods are time-consuming. To solve the problem, let's substitute the integral by a finite sum:

$$q(t,x,y,z,u,v,w)=\sum_{i=1}^{N}K_i\frac{Q\exp\left(-\frac{1}{2}\left(\frac{(x-x_0-ut)^2}{(\sigma_x)^2 t}+\frac{(y-y_0-vt)^2}{(\sigma_y)^2 t}+\frac{(z-z_0-wt)^2}{(\sigma_z)^2 t}\right)\right)}{\left(\sqrt{2\pi t}\right)^3\sigma_x\sigma_y\sigma_z}, \qquad (6)$$

where K_i are the numerical coefficients; $i = 0,1,2,3, \ldots, n$ are the integration points.

Figure 2 visualizes the joint results of experimental and simulated measurements of the peat fire-related CO concentrations (received according to formula (2)) near the Federal Motorway P-255 – "Siberia". The concentrations of CO are expressed in terms of Limit Value Units: 20 minutes CO Limit Value is 5mg/m^3. The calculations were realized using the software program Ecolog 4 (Integral Co Ltd, St. Petersburg, Russia). The results of the measured and simulated CO concentrations were later used as input data to test the proposed neural network technique. Figure 3 shows the dynamic development of the pollution in this area at the wind in the direction to the motorway (4 neurons).

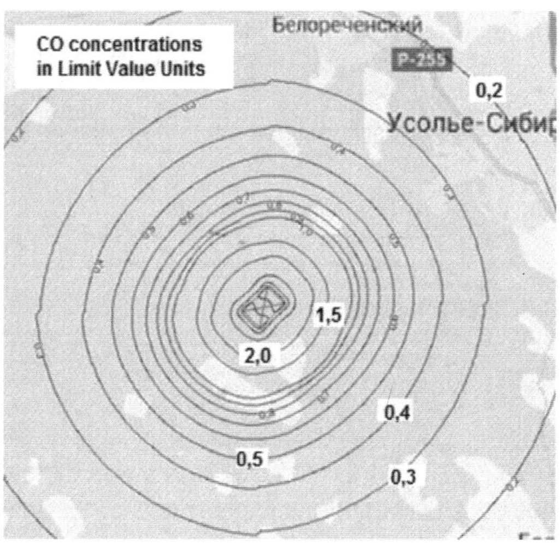

Figure 2. Visualization of the fields of CO concentrations (in terms of Limit Value Units) emitted from the peat fire near the Federal Motorway P-255 – "Siberia".

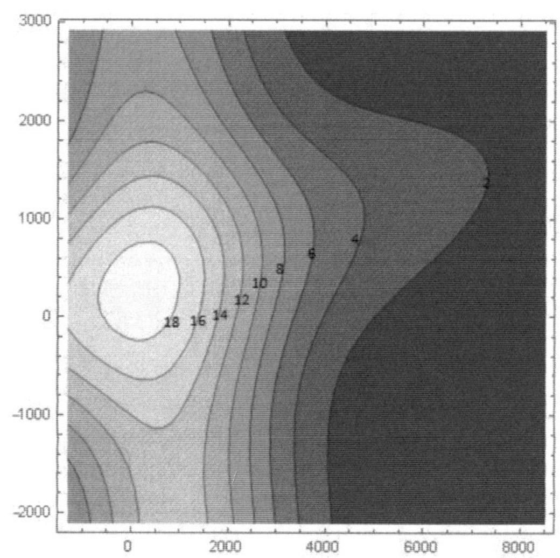

Figure 3. Isolines of the level of CO at the wind in the direction to the Motorway P-255 – "Siberia" obtained by the neural network model (4 neurons).

3.2. Case study 2

Equation (1) can be greatly simplified for the solution of the problem concerning peat fire. Changing the parameters of the problem (the wind speed, the radius of the fire, etc.), we suppose the process is quasi-stationary, in other words, the concentrations are stabilized faster than the parameters are changed. Thus, let us assume $\frac{\partial q}{\partial t}=0$.

Let us assume the z-axis pointed upward, so, w for heavy particles, having own sedimentation rate (with the sign "-"), is equal to that rate, and w for light fractions is equal to 0. Let's also assume that the average turbulent flow of pollutants at the earth's surface is small and the exchange coefficients are constant. Let us suppose further that the concentration of CO satisfies the equation:

$$u\frac{\partial q}{\partial x}+v\frac{\partial q}{\partial y}-k(\frac{\partial^2 q}{\partial x^2}+\frac{\partial^2 q}{\partial y^2})=Q\ , \quad (x,y)\in\Omega \quad , \tag{7}$$

where Q is constant in the area of peat fire ($Q=0$ out of the fire); k is the known parameter, u, v are the known wind speed components corresponding to the measurements.

We assume then the CO concentration tending to 0 zero at infinity, and we use this as boundary conditions. The solution of the problem can be simplified by the following conversions of the equation (7). Firstly, let us transfer the origin to the center of the peat fire, which we consider a superficial CO emission source of circular geometry. In the second place, let us direct the OX-axle in the direction of the wind by turning the coordinates. We eliminate the parameter k by scaling x and y, and we eliminate the parameter Q by scaling the density of pollution. The actual value of Q can be determined by measurements.

Here there are left such parameters as the radius of the area of the fire and the wind speed. In the first version of the program, they are fixed and set at the beginning together with the other input data. They can be re-changed further by training the program.

Let us solve the equation (7) in the form of a neural network model:

$$q(x,y)=\sum_{i=1}^{N}c_i\exp\left\{\left[-a_i(x-x_i)^2-b_i(y-y_i)^2\right]\right\}\ , \tag{8}$$

whose parameters a_i, b_i, c_i, x_i and y_i can be found by minimizing the error:

$$J=\sum_{j=1}^{M}\left|u\frac{\partial q(x_j^1,y_j^1)}{\partial x}-\frac{\partial^2 q(x_j^1,y_j^1)}{\partial x^2}-\frac{\partial^2 q(x_j^1,y_j^1)}{\partial y^2}-Q(x_j^1,y_j^1)\right|^2\ , \tag{9}$$

where $\{x_j^1,y_j^1\}$ are the test points in the area where the solution is built, they are updated through every 3-5 steps of the minimization J; $Q(x^1_j, y^1_j)=1$ in the area of the peat fire and $Q(x^1_j, y^1_j)=0$ at other points.

Figure 4 shows the results of calculations for a peat fire with a radius of 1; N=20, M=200; u=1.

a) b)

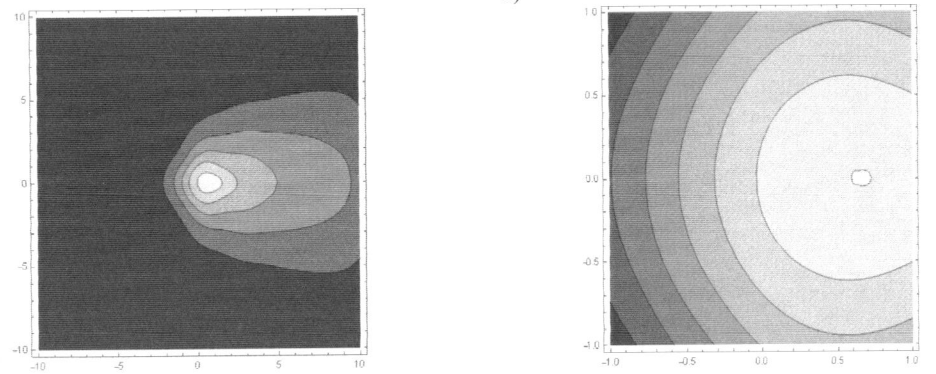

Figure 4. Lines of the gates of the neural network model $q(x, y)$ in the entire area of variables change where the solution is sought (a), and in the area of the peat fire (b).

Moreover, we have built a parametric model considering wind speed as input data. In this case, the approximate solution is sought in the form of heterogeneous neural network function:

$$q(x,y,u)=\sum_{i=1}^{N}c_i\exp\left\{\left[-a_i\left(x-x_i\right)^2-b_i\left(y-y_i\right)^2\right]\right\}th\left[d_i(u-u_i)\right], \tag{10}$$

whose parameters a_i, b_i, c_i, d_i, x_i, y_i and u_i can be found by minimizing the error:

$$J_2=\sum_{j=1}^{M}\left|u_j^1\frac{\partial q(x_j^1,y_j^1,u_j^1)}{\partial x}-\frac{\partial^2 q(x_j^1,y_j^1,u_j^1)}{\partial x^2}-\frac{\partial^2 q(x_j^1,y_j^1,u_j^1)}{\partial y^2}-Q(x_j^1,y_j^1)\right|^2, \tag{11}$$

where the points $\{x_j^1,y_j^1,u_j^1\}$ are periodically updated.

This parametric model allows predicting the level of peat fire-related air pollution at different wind directions.

Such models can be easily trained if there are measurement data in a set of points. In this case, there will be added an additional term to the error functional (9) or (11) equal to the sum of the squares of the differences of the measured values and the neural network output. Likewise, data from numerical calculations are taken into account using standard methods or packages for individual parameter sets.

Conclusions

In this paper, an original self-learning neural network modeling technique is proposed that is capable of predicting dangerous roadside air pollution by carbon monoxide released from a nearby peat fire. The model was developed and tested by investigating a real peat fire happened in the Irkutsk region near atmospheric dispersion in the form of partial differential equations, calculation data (simulated measurements) obtained on the base of the Gaussian function; experimental data (real measurements) were used to build the neural network model. The results of the study show that the approaches developed here can be recommended as a useful tool for the air quality management and forecasting as well as for the prediction and prevention of such emergencies.

Acknowledgements

The investigations were supported by the grant of the Russian Foundation for Basic Research, project № 14-01-00733 A.

References

[1] Fraser L H, Keddy P A 2005 The World's Largest Wetlands: Ecology and Conservation. Cambridge University Press, Cambridge, UK 488 p

[2] Biester H, Bindler R 2009 Modelling Past Mercury Deposition from Peat Bogs – The Influence of Peat Structure and ^{210}Pb Mobility *Working Papers of the Finnish Forest Research Institute* **128** p 483

[3] De Groot W 2012 Peatland fires and carbon emissions. Natural Resources Canada. Canadian Forest Service. Great Lakes Forestry Centre, *Frontline Express (GLFC - Sault Ste. Marie)* **50** 2p.

[4] Fokeeva E V, Safronov A N, Rakitin V S, et al 2011 Investigation of the 2010 July–August fires impact on carbon monoxide atmospheric pollution in Moscow and its outskirts, estimating of emissions *Izv. Atmos. Ocean. Phys.* **47(6)** 682-698

[5] Konecny K, Ballhorn U, et al 2016 Variable carbon losses from recurrent fires in drained tropical peatlands *Global Change Biology* **22(4)** 1469–1480

[6] Gaveau D L A, Salim M A, et al 2014 Major atmospheric emissions from peat fires in Southeast Asia during non-drought years: evidence from the 2013 Sumatran fires *Sc. Rep.* **4** 6112

[7] Pouliot G, Pierce T, et al 2005 Wildfire Emission Modeling: Integrating BlueSky and SMOKE *Proc. 14th Int. Emission Inventory Conference "Transforming Emission Inventories - Meeting Future Challenges Today" (Las Vegas 11-14 April 2005)* Ed EPA US

[8] Benson P 1992 A review of the development and application of the CALINE 3 and 4 models *Atmos. Environ.* **26 B (3)** 379-390

[9] Berkowicz R 2000 OSPM – a parameterized street pollution model *Environmental Monitoring and Assessment* **65 (2)** 323-331

[10] Hadjiiski L and Hopke P 2000 Application of artificial neural networks to modeling and prediction of ambient ozone concentrations *J. Air and Waste Manage. Assoc.* **50** 894-901

[11] Feng X et al 2015 Artificial neural networks forecasting of PM2.5 pollution using air mass trajectory based geographic model and wavelet transformation *Atmos. Environ.* **107** 118-128

[12] Kukkonen J, Partanen L et al 2003 Extensive evaluation of neural network models for the prediction of NO2 and PM10 concentrations, compared with a deterministic modeling system and measurements in central Helsinki *Atmos. Environ.* **37** 4549-4550

[13] Lu H, Hsieh J et al 2006 Prediction of daily maximum ozone concentrations from meteorological conditions using a two-stage neural network *Atmos. Res.* **81** 124-139

[14] Tarkhov D, Vasilyev A 2005 New neural network technique to the numerical solution of mathematical physics problems. II: Complicated and nonstandard problems *Optical Memory and Neural Networks (Information Optics)* **14** 97-122

[15] Vasilyev A, Tarkhov D 2009 *Neural Network Modeling. Principles. Algorithms. Applications* (SPbSPU Publishing House, Saint-Petersburg) p 528 (in Russian)

[16] Vasilyev A, Tarkhov D 2014 Mathematical Models of Complex Systems on the Basis of Artificial Neural Networks *Nonlinear Phenomena in Complex Systems* **17** 327-35

[17] Kainov N, Tarkhov D and Shemyakina T 2014 Application of Neural Network Modeling to Identification and Prediction Problems in Ecology Data Analysis for Metallurgy and Welding Industry *Nonlinear Phenomena in Complex Systems* **17** 57-63

[18] Shemyakina T, Tarkhov D, Vasilyev A 2016 *Neural Network Technique for Processes Modeling in Porous Catalyst and Chemical Reactor* ed. L. Cheng et al (Switzerland Springer International Publishing) 547–554

[19] Haykin S 2008 *Neural Networks and Learning Machines* (Prentice Hall) p 936

[20] Berlyand M E 1991 Prediction and Regulation of Air Pollution 1991 *Atmospheric Sciences Library* **14** 320 p

[21] Methodology for the calculation of the concentrations in the air of harmful substances contained in industrial emissions. OND-86 1987 Ed. Berlyand M E, Gasilina N K et al (Leningrad: Gidrometeoizdat) p 93 (In Russian)

[22] Genikhovich E L, Gracheva I G, Onikul R I, Filatova E N 2002. Air pollution modeling at urban scale - Russian experience and problems. *Water, Air & Soil Pollution: Focus* **2 (5-6)** 501-512

[23] Genikhovich E L, Sciermeier F A 1995. Comparison of the United States and Russian complex terrain diffusion models developed for regulatory applications. *Atmos. Environ.* **29 (17)** 2375-2385.

[24] Ložkin V, Ložkina O, Ušakov A 2013 Using K-theory in geographic information investigations of critical-level pollution of atmosphere in the vicinity of motor roads *World Applied Science Journal* **23 (13)** 96-100

[25] Lozhkina O, Nevmerzhitsky N, Lozhkin V 2016 Evaluation of air pollution by PM10 and PM2.5 on St. Petersburg ring road: mobile measurements and source apportionment modeling *Proc. 10th Int. Conf. on Air Quality: Science and Application (Milano 14-18 March 2016)* Ed S Finardi, A Farrow et al (Hertfordshire: University of Hertfordshire) p 176

[26] Lozhkina O V, Lozhkin V N 2015 Estimation of road transport related air pollution in Saint Petersburg using European and Russian calculation models *Transport. Res. Part D* **36** 178-189

[27] Lozhkin V, Sukhoivanov A 2000 Model for estimation of Impact of forest fires on Environment *Proc. Conf. on Improving operational performance of engines, tractors, and vehicles* Ed Nikolaenko et al (St. Petersburg: SPbGAU) 121-129.

Detached eddy simulations of side-loads in an overexpanded nozzle flow

E A Luchikhina and L E Tonkov

Institute of Mechanics Ural Branch of the Russian Academy of Sciences, ul. T. Baramzinoy, 34, Izhevsk, 426067, Russia

E-mail: tnk@udman.ru

Abstract. In this paper, we numerically examine the separated flow in an overexpanded nozzle developed during the starting phase of a rocket engine featuring the side-load effect. We use detached eddy simulation and compare our results with the existing experimental data. The calculations are based on finite volume approximation of the Navier-Stokes equations for perfect gas and pressure-based formulation with PISO coupling. The main flow properties (wall pressure levels, shock waves pattern, side load locus) are rather well reproduced. We suggest using a delayed detached eddy simulation for more accurate predictions of the flow separation.

1. Introduction

The loss of axial symmetry of a supersonic nozzle flow and, as a result, the lateral deviation of the thrust vector is inherent in any real flow. However, for a number of applications, such as first-stage rocket engines with high geometric expansion ratio of the nozzle, the loss of symmetry results in significant transient side loads. The side-load assessment is a challenging engineering problem. The side load effect has been studied for American (J-2, J-2S, J-2X), European (Vulcain), and Japanese (LE-7A) engines [1].

The start-up of a rocket engine nozzle that operates in overexpanded flow conditions is not just an important practical problem. It is interesting in terms of developing new mathematical models and numerical schemes. This class of flows has a number of specific features, such as: sub- and supersonic flow regions separated by shock waves pattern, recirculation bubble with boundary layer detachment, and reattachment on the nozzle wall. Shock-induced separation flow caused by these features plays a key role in the loss of symmetry and the side-load effect [2]. Therefore, for accurate numerical modeling of the internal gas dynamics of the nozzle, it is necessary to take into account all these processes.

On the numerical side, most research on shock patterns and side-loads in rocket nozzle with high area-ratio is based on the resolution of the Unsteady Reynolds Averaged Navier–Stokes equations (U-RANS) in conjunction with k-ω family of turbulence models [2,3,4,5]. There are very few works on using Large-Eddy Simulation (LES) to study side-loads effect in an overexpanded nozzle flow. A detailed description of the numerical experiment based on the RANS–LES method, namely Delayed Detached Eddy Simulation (DDES), can be found in [6].

It is well known that the solving RANS equations requires less computing resources, but RANS-based models are not universal, which limit their applicability in practice. Furthermore, RANS time averaging may be unsuitable for solving significantly transient problems.

Direct numerical simulation (DNS) is still difficult because of the high computational costs and its applications is limited to a small set of model examples. An alternative to DNS is eddy-resolving methods (LES, DES), which are based on the idea of spatial low-pass filtering of the original system of equations (1). Such filtering reduces the computational cost by ignoring the smallest length scales, which are the most computationally expensive to resolve. However, when considering the near-wall flows, the computational cost dramatically increases, since significant turbulence eddy structures near the wall are quite small, which in turn leads to a filter width decrease.

Detached Eddy Simulation (DES) is a promising approach to modeling the turbulent flows. In this method, "simple"" RANS-based turbulence model, for example Spalart–Allmaras (SA) model, is used for simulation of only structures with dimensions the smaller size of the filter, and larger eddies are resolved exactly.

The aim of this work is to investigate the special regime and side-loads effect using advanced DES turbulence modeling approaches.

This paper is structured as follows. Section 2 describes the computational fluid dynamics model and grid. Section 3 introduces DES-SA turbulence model in the frame of overexpanded nozzle flows. Section 4 then presents the summary of results obtained and their analysis. Section 5 concludes the paper.

2. Mathematical model and computational methodology

The computational fluid dynamics model is based on a multi-dimensional system of Navier–Stokes equations for perfect gas:

$$\begin{aligned} &\partial\rho/\partial t + \nabla\cdot\rho\mathbf{U} = 0, \\ &\partial\rho\mathbf{U}/\partial t + \nabla\cdot\rho\mathbf{U}\mathbf{U} = -\nabla p + \nabla\boldsymbol{\sigma}, \\ &\partial\rho E/\partial t + \nabla\cdot\rho\mathbf{U}E = -p(\nabla\cdot\mathbf{U}) - \nabla\cdot\mathbf{q} - \nabla\cdot(\boldsymbol{\sigma}\cdot\nabla\mathbf{U}). \end{aligned} \quad (1)$$

Here $\boldsymbol{\sigma}$ — viscosity stress tensor, $E = \frac{1}{2}\mathbf{U}^2 + c_v T$ — total energy, $\mathbf{q}$ — heat flow.

Computational domain covers the nozzle interior as well as the freestream region. Following [4], the geometric parameters of the nozzle (J-2S) were selected following: length of diverging section $l = 1.913$ m, throat radius $r_* = 0.154$ m, exit radius $r_a = 0.975$ m.

Computational block-structured grid was generated by rotating an 2-D grid (pic. 1) around the symmetry axis to become a three-dimensional grid containing 4 552 500 hexagonal cells. Let us note that the final 3-D grid is symmetric to the central axis, ensuring that the computed asymmetric flow comes from the physics and not an asymmetric grid topology.

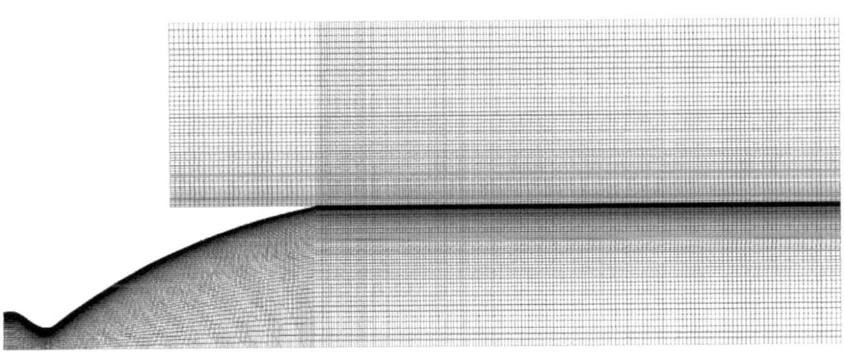

Figure 1. 2-D grid used to construct the final axisymmetric 3-D grid

No-slip and adiabatic boundary conditions are imposed on all the nozzle walls. The outlet boundary condition was set to the fixed-pressure type with ambient pressure $p_a = 10^5$ Pa. The inlet boundary condition was set to the total-pressure type with stagnation pressure $p_* = 43 \cdot 10^5$ Pa.

At walls, the modified turbulent kinematic viscosity $\tilde{\nu}$ was set to zero, at inlet boundary — $\tilde{\nu} \approx \nu$ [7]. Initial conditions was established by stagnation values into combustion chamber: $U_s = 0$, $p_s = 43 \cdot 10^5$ Pa, $T_s = 300$ K, and ambient values behind throat: $U_a = 0$, $p_a = 1 \cdot 10^5$ Pa, $T_a = 288$ K.

The calculations is based on finite volume approximation of system (1) and pressure-based formulation with PISO coupling. A second-order central-difference scheme was employed to discretize the diffusion fluxes and source terms. For the convective terms, a second-order, total-variation-diminishing, upwind-difference based scheme with Van Leer limiter function was used. The numerical scheme provides a good approximation and stability in the calculation of sub- and supersonic flows, containing shock-wave discontinuities for the wide range of Mach numbers.

3. Turbulence modelling: detached eddy simulation

The nature of the investigated flow is influenced by large eddies behind the Mach disk as well as small eddies in wall-near recirculation zones. In the frame of DES, the grid resolution requirements in the boundary layer are closer to those of RANS, thereby significantly reducing the computational cost. At the same time a DES is an LES outside the boundary layer, that allows as to resolve a turbulence structures on wide-range scales without a time averaging of (1).

From the computational point of view, the notation of RANS and LES governing equations just the same and the RANS model can be reduced to a LES subgrid-scale one by replacing the linear scale in a destruction term for the eddy viscosity. Thus, the computational model, based on the system of Navier–Stokes equations (1), includes one additional transport equation for modified eddy viscosity $\tilde{\nu}$ [8]:

$$\frac{\partial \tilde{\nu}}{\partial t} + \mathbf{U} \cdot \nabla \tilde{\nu} = \frac{1}{\sigma}\{\nabla[(\nu + \tilde{\nu})\nabla \tilde{\nu}] + c_{b2}\nabla \tilde{\nu} \cdot \nabla \tilde{\nu}\} + S. \tag{2}$$

Generation term in equation (2) defined by:

$$S = c_{b1}\tilde{S}\tilde{\nu} - c_{w1}f_w\left(\frac{\tilde{\nu}}{d}\right)^2, \tag{3}$$

where d is the linear scale of turbulence.

Relation between modified eddy viscosity and turbulence viscosity are given by: $\nu_t = f_{v1}\tilde{\nu}$ or $\mu_t = \rho f_{v1}\tilde{\nu}$.

The above formulation of the Spalart–Allmaras (SA) model (2)-(3) does not include the terms for realisation of a numerical-induced laminar-turbulent transition [7]. The values of constants and functions can be found, for example, in [8]. In order to use SA model as a LES subgrid-scale one is necessary to replace the linear scale d by

$$l_{\mathrm{DES}} = \min\{l_{\mathrm{RANS}}, C_{\mathrm{DES}}\Delta\}, \tag{4}$$

where l_{RANS} is closest wall distance, $\Delta = \sqrt[3]{\Delta_x\Delta_y\Delta_z}$ — LES filters width and $C_{\mathrm{DES}} = 0.65$.

4. Results and discussion

Following [4], we have considered the time interval $0 \leq t \leq 0.1818$ s which is a period sufficiently long to generate restricted shock separation (RSS) flow mode then transition to free shock separation (FSS) flow with near-wall recirculation zones.

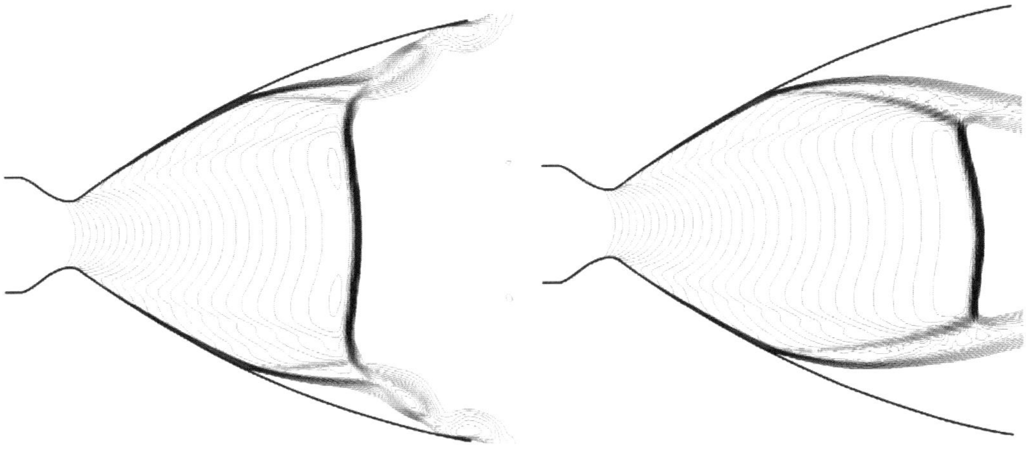

(a) (b)

Figure 2. The iso-Mach contour lines: a) $t = 0.007$s; b) $t = 0.186$s.

The iso-Mach contour lines at two time slices $t = 0.007$ s and $t = 0.186$ s are given in figure 2. One can notice loss of symmetry (figure 2, b), due to the instability of the flow in recirculation zone.

Computed by DES-SA model (1), (2)–(3) side load locus are plotted in figure 3, b versus numerical results [4] (figure 3, a). As seen in the plots, the present data are in good agreement with [4], but oscillation frequency is a slightly higher.

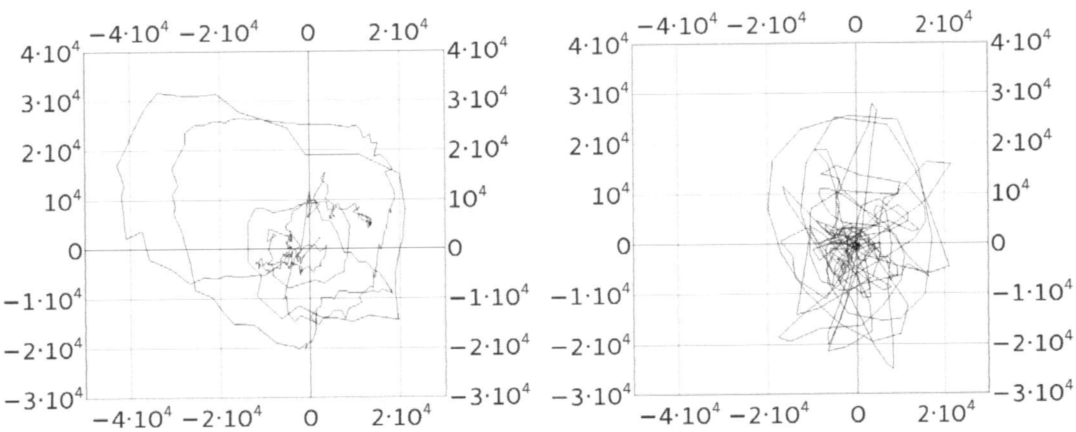

(a) (b)

Figure 3. Computed side force locus F_{xy}, N: a) numerical data [4]; b) present calculations by DES-SA model.

Figure 4 shows in comparison with [4] computed streamwise wall pressure distributions for FSS mode (a) and RSS mode (b).

The vertical lines on the plot show the set of wall pressure values in the corresponding cross-sestion and their magnitude indicating the flow is mostly asymmetric. It can also be seen that the side load effect develops at FSS mode. Examining the wall pressure distributions in figure 4, it can be seen that a calculated flow separation point located somewhat upstream than that of [4], and the maximum pressure value slightly higher. From the results it is concluded that the subgrid-scale eddy viscosity model needs to be adjusted as well as computational grid.

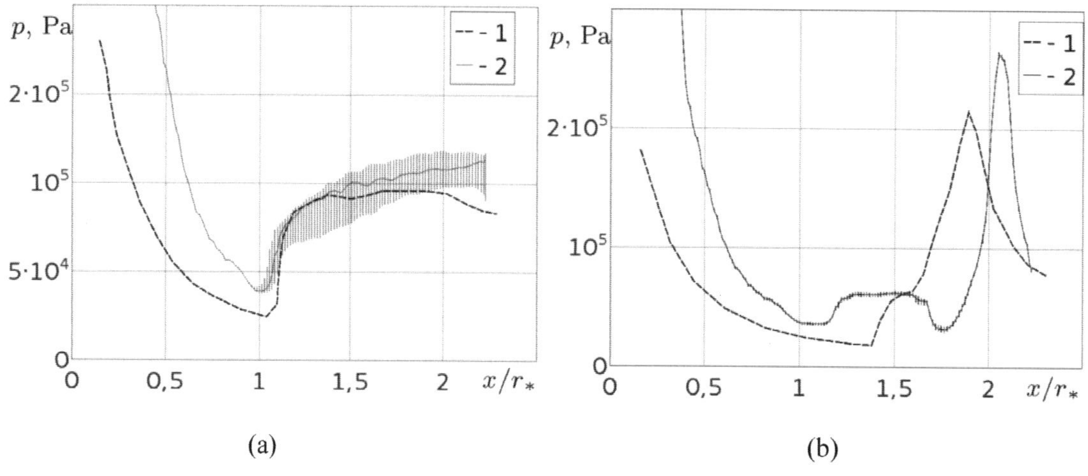

Figure 4. Computed streamwise wall pressure distributions, FSS flow mode (a); RSS flow mode (b): 1 — numerical data [4]; 2 — present calculations by DES-SA model; a) $t = 0.186$ b) $t = 0.007$.

5. Conclusions

The nozzle flow developed during the starting phase of the rocket engine was analyzed via numerical simulation using detached eddy approach (DES) in conjuncton with the Spalart–Allmaras eddy viscosity model; the results was compared with data [4]. The comparative study provide some evidence for the correctness of the mathematical model that adequately describes the main characters of this type of flows. The resulting differences indicate that additional work is needed to improve the method and computational grid resolution first of all at near-wall zone.

It is suggested that a delayed DES (DDES) approach should be used for more accurate predictions of the separation point position. Furthermore, it is necessary that a more thorough study of the eddy viscosity model parameters that are responsible for the generation-destruction balance at RANS/LES crossover region.

Acknowledgments

This work was supported in part by the RFBR under grants 14-08-00064-a and 14-01-00055-a.

References

[1] Tomita T, Sakamoto H, Onodera T, Sasaki M, Takahashi M and Tamura H 2004 *AIAA Paper* **4** 9

[2] Glushko G S, Ivanov I E and Kriukov I A 2010 *Physical and Chemical Processes In Gas Dynamics* **1** 172–179

[3] Wang T S 2009 *Shock Waves* **19** 251–264

[4] Zhao X, Bayyuk S and Zhang S 2013 Aeroelastic response of rocket nozzles to asymmetric thrust loading *Computers and Fluids* 76 pp 128–148

[5] Kuzmin I, Tonkov L and Chernova A 2015 Izv. *Inst. Mat. Inform., Udmurt. Gos. Univ.* **2(46)** 93–98

[6] Deck S 2009 *Shock Waves* **19** 239–249

[7] Garbaruk A V, Niculin D, Strelets M K, Dyadkin A, Krylov A and Stekenius K 2013 *Progress in Flight Physics* **5** 3–22

[8] Spalart P R and Allmaras S R 1994 A *La Recherche Aerospatiale* **1** 5–21

Flow past a porous cylinder in a rectangular periodic cell: Brinkman and Darcy models comparison

R F Mardanov[1], O V Soloveva[1,2] and S K Zaripov[1]

[1]Kazan Federal University, 18, Kremlyovskaya St., Kazan 420008, Russia
[2]Kazan State Power Engineering University, 51, Krasnoselskaya St., Kazan 420066, Russia

E-mail: Renat.Mardanov@kpfu.ru

Abstract. The problem of the gas suspension flow around a porous cylinder in a periodic rectangular cell within the models of the Stokes – Darcy and the Navier–Stokes – Brinkman using the boundary element and finite volumes method was solved. Streamlines of the carrier phase flow and the air and inertia-less particles capture coefficients of at varying porosities of cylinder medium and periodic cylinder packing were compared.

1. Introduction

A flow around a porous body is widely encountered in the field of gas suspension filtration. Fully or partially porous bodies can act as an aerosol filter element. To evaluate the efficiency of filters with porous elements, the flow of gas suspension in an ordered or random packing should be calculated. The flow field of the carrier phase can be found with a good accuracy in the approximation of a circular or rectangular periodic cell model [1-3]. To the best of our knowledge, there are not enough comparisons of accuracy of the predictions of various flow models in the region with mixed homogeneous and porous areas. The study of the air flow past a porous cylinder in a rectangular periodic cell on the base of two flow models is the aim of the present work.

2. The problem statement

The problem of incompressible gas flow around a porous cylinder in a periodic rectangular cell using two mathematical models is solved. The first model describes the fluid flow in a homogeneous field in the Stokes approximation and in the porous cylinder domain on the base of the Darcy law. The obtained boundary value problem is solved by the boundary element method (BEM). Within the second model a combination of the Navier–Stokes and the extended Brinkman equations for uniform and porous areas, respectively, is implemented. The solution of the model equations is obtained using the CFD code ANSYS/FLUENT by the finite volume method (FVM). Let us describe problem for both models.

2.1. The Stokes – Darcy model

Let us consider a two-dimensional flow of an incompressible viscous fluid with suspended particles of a porous cylinder with the radius R_c in the periodic rectangular cell. As the scale of a linear size and speed select the radius of the cylinder R_c and the average speed U on the left side of the cell, respectively. In dimensionless coordinates xy the rectangular periodic cell of the porosity ε has the

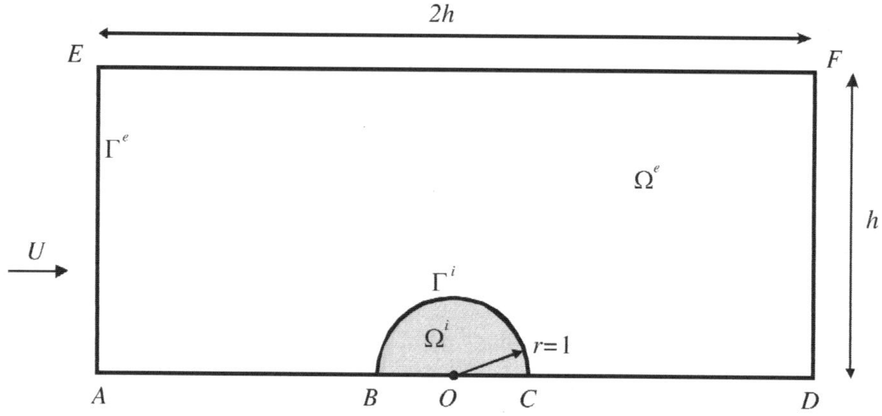

Figure 1. The geometry of the computational domain.

height $h=\sqrt{\pi/(1-\varepsilon)}/2$ and width $2h$. The computational domain Ω consists of a homogeneous area Ω^e of the external flow and a porous medium Ω^i of the internal flow inside the cylinder with the radius $r=1$: $\Omega=\Omega^e\cup\Omega^i$ (figure 1).

The external flow in the area Ω^e is described in the Stokes approximation. In this case, the stream function $\psi^e(x,y)$ of the external flow satisfies the biharmonic equation

$$\Delta^2\psi^e=0 \tag{1}$$

and the conditions on the boundaries Γ^e and Γ^i. On the lines AE and DF the periodic conditions are set

$$\psi^e(-h,y)=\psi^e(h,y),\ \ \psi^{e\prime}(-h,y)=-\psi^{e\prime}(h,y),\ \ \omega^e(-h,y)=\omega^e(h,y),\ \ \omega^{e\prime}(-h,y)=-\omega^{e\prime}(h,y), \tag{2}$$

where $\omega^e=-\Delta\psi^e$ is the vorticity (a prime denotes differentiation with respect to the outward normal to the boundary). On the top line EF

$$\psi^e=h,\ \ \omega^e=0. \tag{3}$$

On the axis-lines AB and CD – the symmetry conditions hold

$$\psi^e=0,\ \ \omega^e=0. \tag{4}$$

In the porous domain Ω^i the flow is described within the Darcy model. The stream function $\psi^i(x,y)$ of the fluid flow inside the cylinder satisfies the Laplace equation

$$\Delta\psi^i=0 \tag{5}$$

and the symmetry conditions on the line BOC

$$\psi^i=0,\ \ \omega^i=0. \tag{6}$$

On the line BC between the free space and the porous medium (the boundary of the cylinder) the conditions are

$$u_r^e=u_r^i,\ \ p^e=p^i,\ \ \frac{\partial u_\theta^e}{\partial r}=-\alpha_s S\left(u_\theta^i-u_\theta^e\right), \tag{7}$$

where u_r, u_θ are the radial and tangential components of the fluid flow velocity, indexes e and i correspond to the external and internal flow, p is the pressure, $S=R_c/\sqrt{k}$, k is the permeability of the porous medium. The dimensionless slip coefficient α_s is determined by the viscosity of the carrier medium and geometric parameters of the porous medium. The last condition in (7) is a widely known Beavers condition [4].

2.2. *The Navier–Stokes – Brinkman model*

The flow in the rectangular periodic cell in the approximation of a laminar flow of a viscous incompressible fluid is described by the extended Brinkman equations in the porous region and the Navier–Stokes equations in the homogeneous region [3, 6, 7]:

$$\nabla \cdot \bar{U} = 0, \tag{8}$$

$$\varepsilon_f^{-2} \rho \bar{U} \cdot \nabla \bar{U} = -\nabla P + \frac{\mu}{\varepsilon_f} \Delta \bar{U} - b \frac{\mu}{k} \bar{U}, \tag{9}$$

where ε_f is the porosity of the cylinder medium. The quantity b equals zero and unity outside and inside the cylinder, respectively. The conditions at external boundaries of the rectangular cell correspond to the conditions accepted in the previous model: the periodic conditions on the vertical boundaries of the cell and the symmetry conditions on the top and bottom boundaries.

3. Calculation results

The boundary value problems (1) – (7) are solved together by BEM [5]. The equations (8) – (9) are numerically integrated by FVM using ANSYS/FLUENT code. The numerical studies using the two models described were conducted for the same values of the parameters.

Figure 2 shows the flow streamlines in the periodic cell of $\varepsilon = 0.96$, calculated by two models for the two values of the parameter *S*. The solid and dashed lines correspond to the streamlines obtained by BEM and FVM respectively. In the Stokes – Darcy model α_S is assumed to be $\alpha_S = 1$ in the Beavers condition (7). To better visualize the flow in the porous cylinder more fluid flow streamlines near the symmetry axis are given. It is seen that the streamlines obtained by two models agree well away from the porous cylinder and the agreement is better for larger S. Near and on the porous cylinder there is a significant difference between two models.

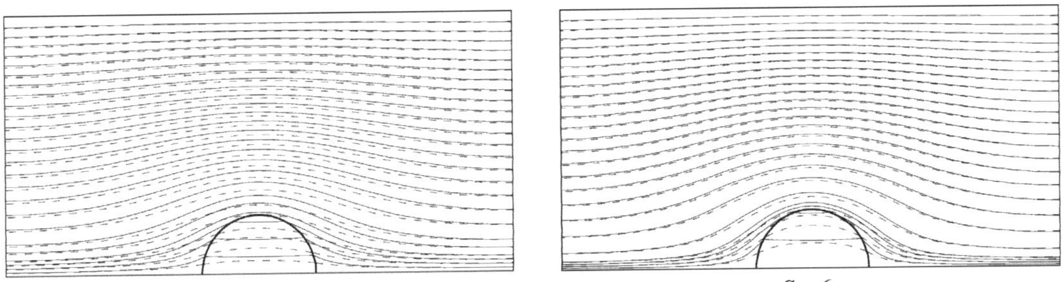

$S = 3$ $S = 6$

Figure 2. The fluid flow streamlines in the periodic cell.

To estimate the inertialess particles coefficient for the flow around the porous cylinder the carrier phase capture as a function $Q(S)$ should be calculated. The quantity Q is defined as the ratio of the initial ordinate of the limiting streamline passing through the cylinder to its radius and is found as the value of stream function of the upper point of the cylinder with the coordinates $x = 0,\ y = 1$: $Q = \psi(0,1)$. The calculated functions $Q(S)$ for the various values ε and α_S are shown in figure 3. With the growth of S (decrease in the permeability of the cylinder) the value Q decreases monotonically. The parameter α_S in the Darcy model with the Beavers boundary condition (7) significantly affects the value Q and this effect is larger for less porosity ε. An approximate model of Darcy's flow in the porous cylinder yields the low values for the considered α_S due to neglecting the air tangential velocity at the cylinder boundary.

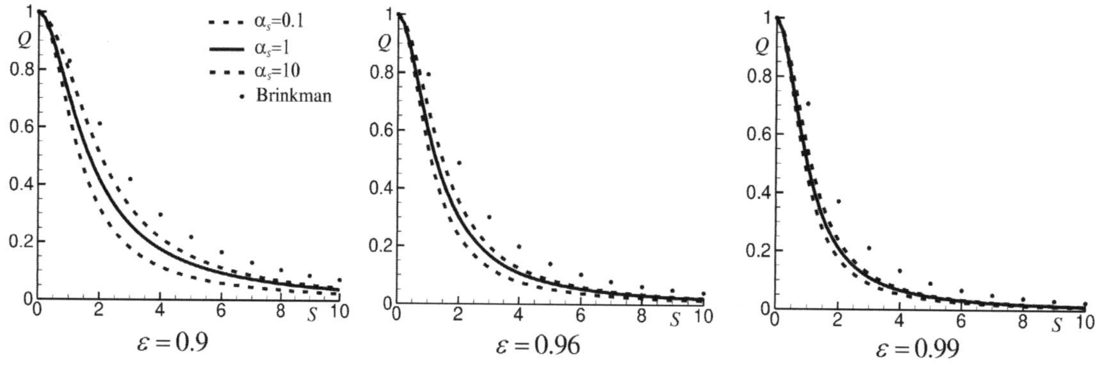

Figure 3. The coefficient of air capture $Q(S)$.

In the case of a porous fiber of an aerosol filter the air capture coefficient Q gives the efficiency of the capture of inertialess particles [3]. Using the numerically found streamfunction $\psi^e(x,y)$ we can calculate the particle deposition coefficient by the interception phenomena [8]. The contribution of the interception to the particles deposition becomes considerable for the particle sizes that are comparable with cylinder radius and depends on the ratio $\delta = r_p/R_c$, where r_p is the particle radius. The total efficiency E of capture of inertialess particles can be found as $E(\delta)=\psi(0,1+\delta)$. On the base of the analytical Kuwabara model the formula was obtained in [3]

$$E(\delta)=\psi(0,1+\delta)=(1+\delta)+C\left(a_1(1+\delta)^{-1}-a_2(1+\delta)+a_3(1+\delta)^3\right),$$
$$a_1=0.5-0.25\alpha,\ a_2=\mathrm{Ku}-0.5\alpha+0.5-\ln(1+\delta),\ a_3=-0.25\alpha, \quad (10)$$

$$C=\left(\mathrm{Ku}+2(1+\alpha)S^{-2}\right)^{-1},\ \mathrm{Ku}=-0.5\ln\alpha-0.75+\alpha-0.25\alpha^2,\ \alpha=1-\varepsilon.$$

The dependencies $E(\delta)$ for the two values of S at $\varepsilon=0.96$ and $\alpha_S=1$ are given in figure 4. In [3] it was shown that the value of the parameter $S<3$ for real porous bodies. The curve $E(\delta)$ is higher than the interception efficiency of a solid cylinder. The difference is the air capture coefficient Q. In general, the Darcy model gives low values of the particle capture rate as compared with the model of Brinkman. The curves $E(\delta)$ obtained by BEM are in a better agreement with an analytical formula (10).

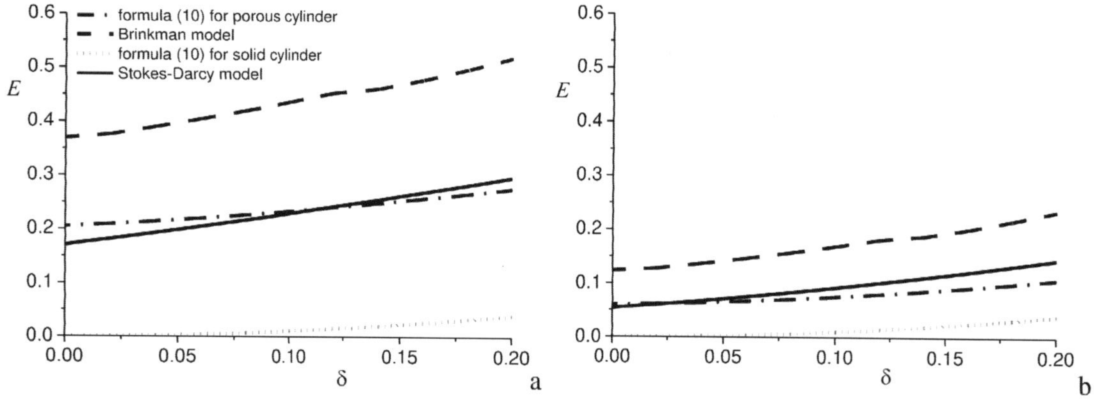

Figure. 4. The dependence $E(\delta)$ at S=3(a), 6(b).

Acknowledgements
This work is supported by the Russian Government Program of Competitive Growth of Kazan Federal University, RFBR grant № 15-01-06135 and President of Russian Federation grant for young scientists MK-6235.2015.1.

References
[1] Stechkina I B 1979 *Fluid Dyn.* **14(6)** pp 912–915
[2] Kirsch V A 2006 *Theoretical Foundations of Chemical Engineering* **40(5)** pp 465–471
[3] Zaripov S K, Solov'eva O V and Solov'ev S A 2015 *Aerosol Science and Technology* 49(6) pp 400–408
[4] Beavers G S and Joseph D D 1967 *J. of Fluid Mechanics* **30** pp 197–207
[5] Mardanov R F, Dunnett S J and Zaripov S K 2016 *Engineering Analysis with Boundary Elements* pp 54–62
[6] Bhattacharyya S, Dhinakazan S and Khalili A 2006 *Chemical Engineering Science* **61** pp 4451–4461
[7] Grigor'eva O V and Zaripov S K 2012 *Russian Aeronautics* **55(1)** pp 19–24
[8] Brown R C 1993 *Air filtration. An integrated approach to the theory and applications of fibrous filters* (Oxford; New York: Pergamon Press) p 272

Modeling of unsteady flow of viscous fluid in the channel of complex geometry

E A Marfin[1,2] and A A Abdrashitov[2]

[1] Institute of Physics, Kazan Federal University, 16a Kremlyovskaya Street, Kazan, Russia
[2] Kazan Science Centre, Russian Academy of Science, 2/31 Lobachevsky Street, Kazan, Russia

E-mail: marfin_ea@mail.ru

Abstract. The article concerns an issue of exploring the mechanism of wave influence on the process of filtration. To describe a filtration flow, the porous medium is represented as a capillary, the radius of which varies sinusoidally. In this article we are presenting the results of numerical modeling of pulsating liquid flow in a sinusoidally-shaped channel. Numerical research was conducted with the help of the program complex FlowVision. As a result of series of calculations we received fields of velocities and pressure in the axial section of flowing channel. It was found that it is the flow in narrow isthmuses between the pores that contributes to the pressure difference the most. We revealed the signs of steady-state liquid flow in the channel when imposing fluctuations of pressure in the absence of pressure gradient. The conditions of formation of such flow are revealed.

1. Introduction

Effective mastering of oil reserves is a topical and complicated task, the resolution to which requires fundamentally new approaches which would allow drastic reducing unit power expenses and material expenses. Numerous theoretical and experimental researches, which took place in laboratory or in trading, show high efficiency of wave method of increasing oil recovery [1-6]. The essence of wave method lies in the way elastic vibrations influence on productive formation and bottom-hole area of the hole. The most perspective thing to do is to combine the methods of increasing oil recovery with wave influence on productive formation [7-10]. Variety of events that take place under this influence can be summed to two factors which provide increasing the efficiency of oil production. These are acceleration of filtration process and involvement of dead oil into filtration process [11-13]. This article is an attempt to model the filtration process under wave influence on formation numerically which is aimed at studying the mechanism of acceleration of filtration flows in porous mediums when imposing elastic vibrations.

To describe a filtration flow in porous mediums it is quite common to use mathematical models in which the real porous medium is represented as a system of capillaries, cracks, packages with spheres etc. One of the most popular ones is the capillary model which consists of a beam of straight parallel capillaries, radiuses of which are allocated according to some function of allocation.

The subject of our investigation is the flow of fluid in a cylindrical sinusoidally-shaped channel when imposing a wave field. Geometrical sizes of this channel are commensurable with sizes of pores in productive formations of oil and gas fields.

2. Subject of investigation and setting the task

Geometrical model of isotropic porous medium is represented as a narrow channel, the radius of which varies sinusoidally. I.e. pores and narrow isthmuses, which connect adjacent pores, alternate inside of the channel. The sketch of the channel, in which tests were run, is depicted in Figure 1. Characteristic sizes of the channel are the following: diameter of a pore is $D=3\cdot10^{-5}$ m, diameter of a flow section (where it narrows) of the channel is $d=0{,}1\cdot D$. The surface of calculation area is a complex of flat polygons that are facets, on which boundary conditions are defined.

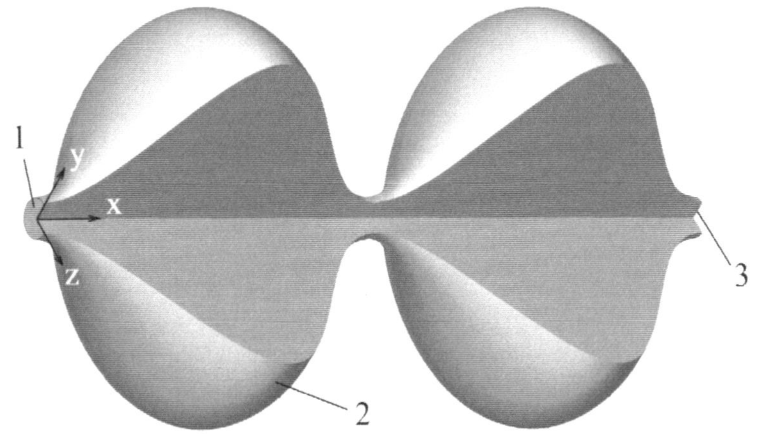

Figure 1. Calculation area. Borders: 1 – input, 2 – wall, 3 – output.

To describe hydrodynamics of the flow we used the model 'laminar liquid' as a mathematical model, which describes the flow of viscid liquid with low Mach and Reynolds numbers. The model includes Navier–Stokes equation and continuity equation, the solution to which is performed by the finite volume method with an implicit algorithm of splitting into variables with the program complex FlowVision. Computer modeling was performed for a steady-state flow as well as for the one with imposed pulsations. Pressure pulsations, which were set at the entrance to the calculation area, modeled wave influence on filtration process.

The following boundary conditions were applied to give a task:

1) On the wall as on the border of area the condition of sticking is set:

$$\mathbf{V}\big|_{w}=0 \tag{1}$$

2) At the entrance to the channel pressure-time (t) relation is set:

$$p(\mathrm{t})=p+p_0\cdot\sin(2\cdot\pi\cdot f\cdot t), \tag{2}$$

where p is for pressure difference, Pa; p_0 represents the amplitude of fluctuations, Pa; f is for frequency of fluctuations, Hz.

3) At the outflow section there is a free exit with zero pressure on the border.

The initial condition was velocity of liquid flow being equal to zero, i.e. liquid is resting and pressure in the channel is equal to the initial one.

The properties of liquid during the calculations are set by the database FlowVision, that is the clean water. Physical characteristics of liquids are the following: density is 998 kg/m^3, viscosity is 0.001 Pa·s. Initial values: temperature is 293 K, pressure is 101 kPa.

3. Results of the numerical research

To investigate influence of different factors (amplitude and frequency of fluctuations) on non-steady-state process of liquid flow in the channel, we conducted series of calculations in an identical calculation area and an identical calculation net. The amplitude of fluctuations varied from 5 to 20 kPa, frequency changed from 200 to 3000 Hz. To make the numerical experiment, a step of time was set for each series and it was equal to 1/(1000·*f*).

Figures 2 and 3 depict characteristic allocation of pressure and velocity on a plane of the channel with steady-state flow. From these figures it follows that pressure inside of the 'pore' is spread practically evenly. The pressure difference is observed in a narrow part of the channel which connects adjacent 'pores'. Analysis of velocity allocation allows concluding that maximum velocity is observed in the narrow part of the channel. Meanwhile, the flow is observed in the central part of the channel. Inside of the channel the velocity of liquid decreases to almost zero as we move away from an axis.

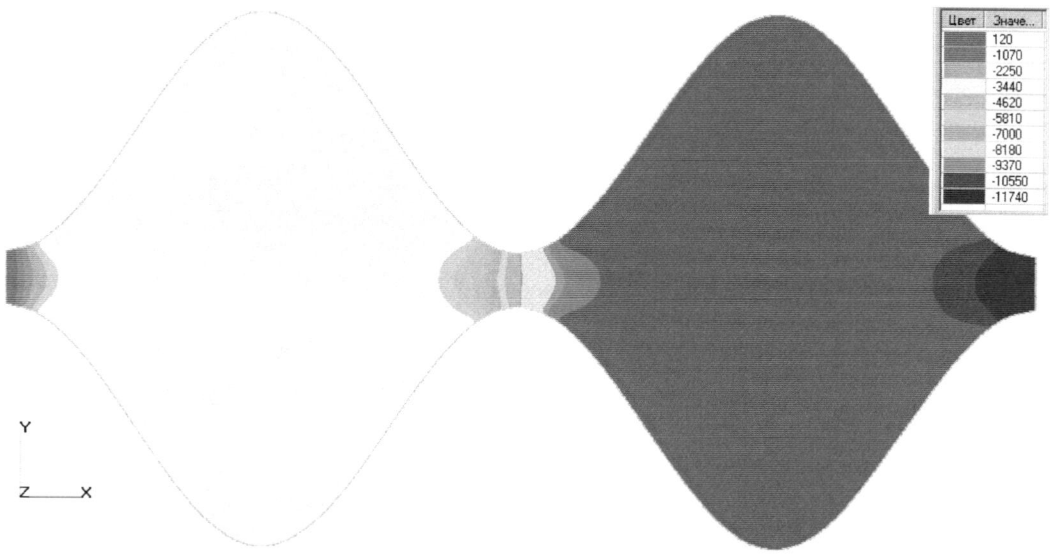

Figure 2. The field of allocation of pressure in the channel.

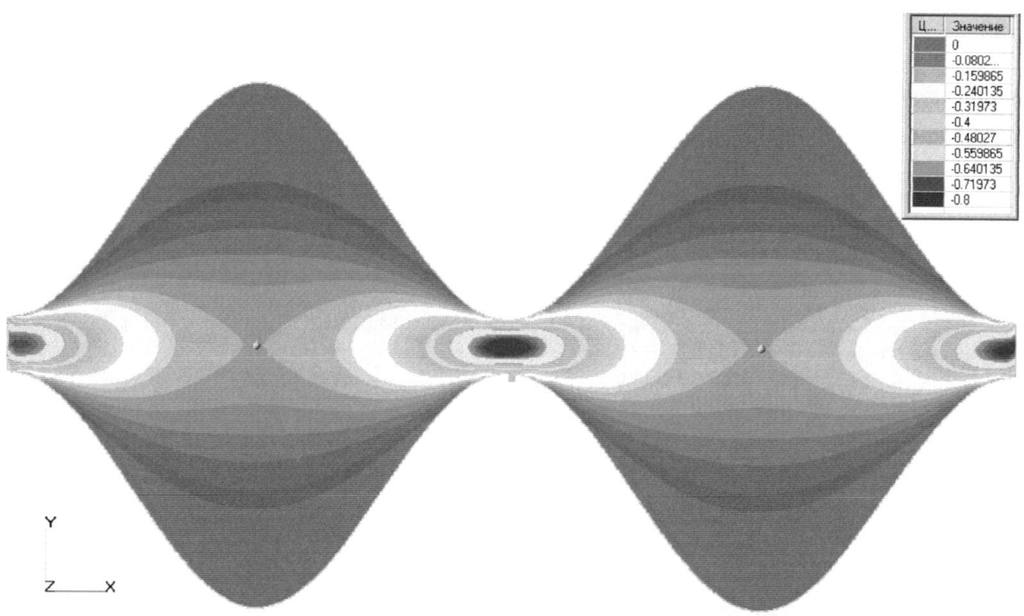

Figure 3. The field of allocation of velocity of the liquid in the channel.

For a better understanding of character of pulsating flow of liquid we took into consideration characteristics of velocity by time. The flow is the pulsating motion with a frequency of pressure pulsation at the entrance. However, a deeper analysis leads to a conclusion that pulsations of velocity are imposed on some steady-state flow (average value $\bar{u}$). The numerical value $\bar{u}$ is ten times less than the amplitude of pulsation; nevertheless it is much higher than characteristic value of velocity of filtration.

We studied the average velocity in the narrowing part on different distance from an axis of the channel. We received a non-dimensional profile of relative velocity which is typical for all the series of tests and coincides with the profile of laminar steady-state flow. The dispersion of relative velocities for different frequencies and amplitudes does not exceed 0.1%.

Processing and analyzing the results of numerical experiment allowed defining how average velocity depends on amplitude of fluctuations and frequency. Figure 4 depicts how average velocity on an axis in the narrow part of the channel depends on an amplitude and frequency of fluctuations.

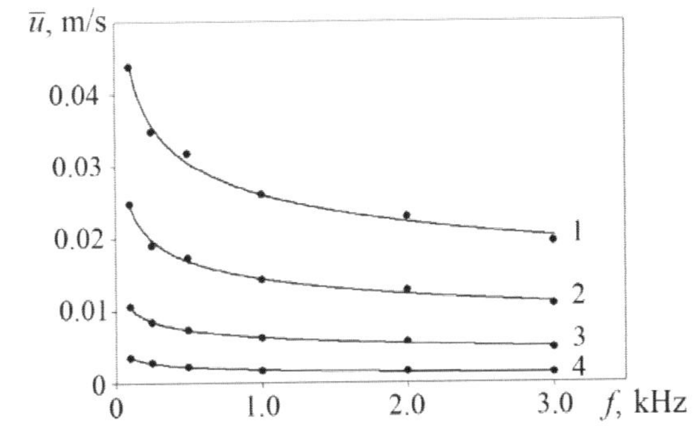

Figure 4. Dependence of average velocity of flow in the narrow channel on frequency *f* and amplitude of fluctuations p_0: 1 –20 kPa; 2 – 15 kPa; 3 – 10 kPa; 4 – 5 kPa.

To determine functional dependence of average velocity of the flow on defining factors, we inserted the following non-dimensional parameters:

$$\frac{\bar{u}}{fd} \text{ and } \frac{p_0}{\rho c f d} \tag{3}$$

where ρ represents density of the liquid, kg/m^3; c represents speed of sound in the liquid, m/s.

Figure 5 depicts interrelation of these non-dimensional parameters which is approximated by power function.

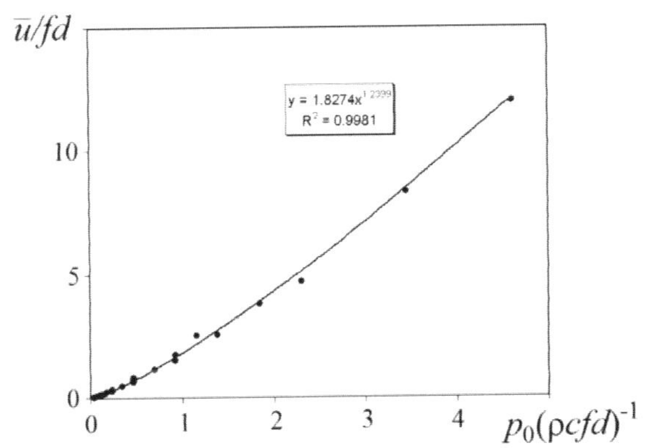

Figure 5. Interrelation of non-dimensional parameters.

From this power function it follows that average velocity is defined the following way:

$$\bar{u} = 1.83 \cdot \left(\frac{p_0}{\rho c} \right)^{1.24} \cdot (fd)^{-0.24} \tag{4}$$

The calculated data, averaged by the integer of fluctuation periods, demonstrated that the liquid flow, which is directed from the emission, is formed when imposing pressure fluctuations in the absence of steady-state flow. The velocity allocation through a section of the channel is similar to a steady-state flow during some pressure difference.

4. Conclusion

Analysis of results of the numerical research demonstrated that imposing wave influence on motionless liquid, which is located in a sinusoidally-shaped channel, leads to its pulsating motion. Average velocity is not equal to zero, i.e. steady-state flow arouses towards the allocation of fluctuations. It is found that average velocity depends on the amplitude and frequency of imposed fluctuations; we defined its functional relation. The results open a new direction for researches in the field of influence of pressure fluctuations on the flow of liquids in complex-shaped channels and its practical use for filtration in porous mediums. The results can be used for perfecting methods of wave influence on productive formations.

Acknowledgments
The reported study was funded by RFBR according to the research project №16-29-15118.

References

[1] Beresnev I A and Johnson P A 1994 *Geophysics* **59** 1000-17
[2] Marfin E A, Kravtsov Y I, Abdrashitov A A, Gataullin R N and Galimzyanova A R 2015 *Pet. Sci. Tech.* **33**(15) 1526-32
[3] Jeong C, Kallivokas L F, Kucukcoban S, Deng W and Fathi A 2015 *J. Pet. Sci. Eng.* **129** 205-220
[4] Marfin E A, Kravtsov Y I, Abdrashitov A A and Metelev I S *2015 Geomodel 2015 - 17th Sci.-Pract. Conf. on Oil and Gas Geol. Explor. and Devel.* doi: 10.3997/2214-4609.201413912
[5] Keshavarzi B, Karimi R, Najafi I, Ghazanfari M H, Amani M and Ghotbi C 2013 *Pet. Sci. Tech.* **31**:11 1159-67
[6] Marfin E A, Kravtsov Ya I, Abdrashitov A A and Gataullin R N 2014 *Georesources* **2**(57) 14-16
[7] Gataullin R N, Kravtsov Ya I and Marfin E A 2013 *Neftyanoe khozyaystvo - Oil Industry* **1** 90-93
[8] Kravtsov Ja I and Marfin E A 2011 *Georesources* **3**(39) 17-18
[9] Marfin E A and Abdrashitov A A 2014 *6th Saint Petersburg Int. Conf. and Exhib. on Geosciences 2014: Investing in the Future*. doi: 10.3997/2214-4609.20140186
[10] Marfin E A, Abdrashitov A A and Kravtsov Y I 2016 *7th EAGE Saint Petersburg Int. Conf. and Exhib.: Understanding the Harmony of the Earth's Resources Through Integration of Geosciences.* doi: 10.3997/2214-4609.201600094
[11] Manga M, Beresnev I, Emily E, Brodsky E E, Elkhoury J E, Elsworth D, Ingebritsen S E, David C, Mays D C and Wang C-Y 2012 *Rev. Geophys.* **50** RG2004
[12] Kravtsov J I, Marfin E A, Butorin E A and Gataullin R N 2009 *Georesources* **1**(29) 43-45
[13] Jankov M, Lovoll G, Knudsen H A, Maloy K J, Planet R, Toussaint R and Flekkoy E G 2010 *Transp. Porous Med.* **84**:3, 569-585

Multigrid finite element method in stress analysis of three-dimensional elastic bodies of heterogeneous structure

A D Matveev

Institute of Computational Modeling SB RAS, 660036, Akademgorodok, Krasnoyarsk, Russia

E-mail: mtv@icm.krasn.ru

Abstract. To calculate the three-dimensional elastic body of heterogeneous structure under static loading, a method of multigrid finite element is provided, when implemented on the basis of algorithms of finite element method (FEM), using homogeneous and composite three-dimensional multigrid finite elements (MFE). Peculiarities and differences of MFE from the currently available finite elements (FE) are to develop composite MFE (without increasing their dimensions), arbitrarily small basic partition of composite solids consisting of single-grid homogeneous FE of the first order can be used, i.e. in fact, to use micro approach in finite element form. These small partitions allow one to take into account in MFE, i.e. in the basic discrete models of composite solids, complex heterogeneous and microscopically inhomogeneous structure, shape, the complex nature of the loading and fixation and describe arbitrarily closely the stress and stain state by the equations of three-dimensional elastic theory without any additional simplifying hypotheses. When building the m grid FE, m of nested grids is used. The fine grid is generated by a basic partition of MFE, the other $m-1$ large grids are applied to reduce MFE dimensionality, when m is increased, MFE dimensionality becomes smaller. The procedures of developing MFE of rectangular parallelepiped, irregular shape, plate and beam types are given. MFE generate the small dimensional discrete models and numerical solutions with a high accuracy. An example of calculating the laminated plate, using three-dimensional 3-grid FE and the reference discrete model is given, with that having 2.2 milliards of FEM nodal unknowns.

1. Introduction

In current paper, to calculate the elastic bodies with heterogeneous (microscopically inhomogeneous) structure and static loading, a method of multigrid finite elements using three-dimensional composite MFE is proposed. The essence of MFE is as follows. At a basic partition (on the fine grid) of m - grid FE, $m \geq 2$, which consists of the 1st order single-grid homogeneous FE and takes into account its heterogeneous structure according to the micro approach [1], the total potential energy as a F function of many variables depending on the fine grid nodal displacements has been determined. On the other $m-1$ large grids (enclosed in the fine one), the displacement functions used to reduce the dimension of the F function are built by the FEM algorithms [2–4] that allows one to develop MFE of small dimension. To approximate the MFE displacement functions, power-law and Lagrangian polynomials of various orders and the equations of three-dimensional problem of elasticity [5], written in local Cartesian coordinate systems without any simplifying hypotheses are used [6, 7]. Lagrange polynomials are effectively applied in the construction of plate and beam MFE. The mixed multigrid discrete models used to calculate the composite solids of irregular shape have been considered.

Advantages of MFE are to take into account the heterogeneous structure of three-dimensional bodies according to the micro approach, to form the discrete models, whose dimensions are $10^4 \div 10^5$ less than the ones of the basic models, to generate computational solutions with a high rate of convergence to the exact solutions that allows one to find solutions with a small error. The calculations have shown that realization of the FEM for multigrid discrete models requires $10^4 \div 10^6$ times less of computer memory than the base models at low time cost. 3-grid FE have been developed and computationally investigated to calculate the sandwich plate.

2. The first procedure of developing composite 2-grid FE

The main principals of the given procedure (without loss of statement generality) are shown by the example of constructing a composite 2-grid FE (2gFE) V_a (of 2nd order) of a $6h \times 6h \times 6h$ cube (figure 1), $Oxyz$ - local Cartesian system coordinate. The connections between the heterogeneous structure components of 2gFE are considered to be ideal. The components are homogeneous isotropic bodies.

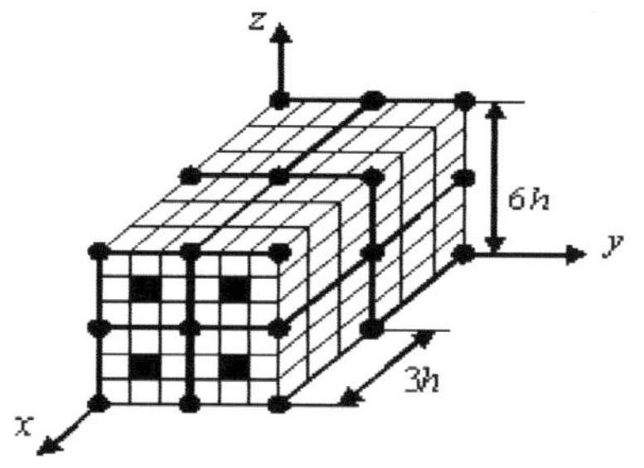

Figure 1. 2gFE V_a, V_S.

Displacement, stress and deformation functions of the components satisfy the Hooke's law, and Cauchy relations corresponding to the three-dimensional problem of elasticity theory [5], i.e. in the entire area of 2gFE, three-dimensional stress state without simplifying hypotheses is realized. 2gFE is fiber-reinforced, parallel to the axis Ox; fiber sections shown in figure 1 are shaded. To construct 2gFE, two nested grids: fine h_a and coarse H_a are used. The grid h_a is generated by a basic partition R_a of 2gFE, which consists of a homogeneous FE V_e of the 1st order in the form of cube with a side h [2–4], and takes into account the heterogeneous structure of 2gFE, its shape, loading and fixing $e = 1,...,M$; M is a total number of FE V_e. Uniform grid h_a with a step h is shown in figure 1. The uniform coarse grid H_a (has the $n_1 \times n_2 \times n_3$ dimension) with a step $3h$ whose nodes are depicted by the dots is identified on the grid h_a (in figure 1 $n_1 = n_2 = n_3 = 3$). The total potential energy Π_a of the 2gFE base partition is given as follows

$$\Pi_a = \sum_{e=1}^{M} (\frac{1}{2} \mathbf{q}_e^T [K_e] \mathbf{q}_e - \mathbf{q}_e^T \mathbf{P}_e), \tag{1}$$

where $[K_e]$ – stiffness matrix, $\mathbf{P}_e$ – nodal force vector, $\mathbf{q}_e$ – nodal displacement vector of FE V_e, T – transposition.

Using the Lagrange second order polynomials [3] on the grid H_a, displacement functions u_a, v_a, w_a of 2gFE are determined and written as follows

$$u_a = \sum_{\beta=1}^{n} N_\beta u_\beta, \quad v_a = \sum_{\beta=1}^{n} N_\beta v_\beta, \quad w_a = \sum_{\beta=1}^{n} N_\beta w_\beta, \tag{2}$$

where u_β, v_β, w_β, N_β – displacements and shape function of β - node of the grid H_a, n - a total number of grid nodes H_a, $n = n_1 n_2 n_3$ ($n = 27$ for 2gFE V_a of the second kind, figure 1).

Let $\mathbf{q}_a = \{u_1,...,u_n,\ v_1,...,v_n,\ w_1,...,w_n\}^T$ is the vector of nodal displacements of the coarse grid H_a, i.e. vector of nodal displacements of 2gFE. Using (2), $\mathbf{q}_e$ vector is expressed through $\mathbf{q}_a$ vector; as a result, the following relation is obtained

$$\mathbf{q}_e = [A_e]\mathbf{q}_a, \tag{3}$$

where $[A_e]$ – rectangular matrix, $e = 1,...,M$.

Using (3) in (1), provided that $\partial \Pi_a / \partial \mathbf{q}_a = 0$, the relation $[K_a]\mathbf{q}_a = \mathbf{F}_a$ for 2gFE V_a corresponding to its state of equilibrium is obtained, where

$$[K_a] = \sum_{e=1}^{M} [A_e^a]^T [K_e][A_e^a],\ \mathbf{F}_a = \sum_{e=1}^{M} [A_e^a]^T \mathbf{P}_e, \tag{4}$$

$[K_a]$, $\mathbf{F}_a$ – stiffness matrix and nodal force vector of 2gFE V_a.

Note that the order of 2gFE V_a equals n_a when $p = n_1 = n_2 = n_3$, where $n_a = p - 1$.

Commentary 1. The solution corresponding to H_a grid of 2gFE V_a using the formula (3) is projected onto the fine grid h_a of the 2gFE basic partition, providing the stress calculation in either FE V_e of the basic partition of 2gFE V_a, i.e. to determine stresses in any component of heterogeneous structure of three-dimensional body.

Commentary 2. When developing the approximative functions of 2gFE displacements on the H_a grid the known polynomials of the 1st, 2nd and 3d orders [2–4] can be used. It is appropriate to apply the following P_3 polynomial of the third kind [4]

$$\begin{aligned} P_3 &= a_1 + a_2 x + a_3 y + a_4 z + a_5 xy + a_6 zx + a_7 zy + a_8 xyz + a_9 x^2 + a_{10} y^2 + a_{11} z^2 + a_{12} x^2 y + \\ &+ a_{13} x^2 z + a_{14} y^2 x + a_{15} y^2 z + a_{16} z^2 x + a_{17} z^2 y + a_{18} x^3 + a_{19} y^3 + a_{20} z^3 + a_{21} x^3 y + a_{22} x^3 z + \\ &+ a_{23} y^3 x + a_{24} y^3 z + a_{25} z^3 x + a_{26} z^3 y + a_{27} x^2 yz + a_{28} y^2 xz + a_{29} z^2 xy + a_{30} x^3 yz + a_{31} y^3 xz + a_{32} z^3 xy. \end{aligned} \tag{5}$$

3. The second procedure of developing composite 2gFE

The second procedure is briefly considered (without loss of statement generality) by producing 2gFE V_S of a $6h \times 6h \times 6h$ cube form having a heterogeneous structure and located in a local Cartesian coordinate system $Oxyz$, as a 2gFE V_a (figure 1). Fine grid h_a and the basic partition R_a of 2gFE V_a are used. Superelement G is built at a basic partition R_a of 2gFE V_S using a condensation method [3]. The total potential energy Π_g of the superelement G is written as

$$\Pi_g = \frac{1}{2}\mathbf{q}_g^T [K_g]\mathbf{q}_g - \mathbf{q}_g^T \mathbf{P}_g, \tag{6}$$

where $[K_g]$ – stiffness matrix; $\mathbf{P}_g, \mathbf{q}_g$ – nodal force vector and nodal displacement vector of the superelement G.

The coarse grid H_S nested in the fine grid h_a is determined at the boundary of the superelement G. Let $\mathbf{q}_S$ is the node displacement vector of the grid H_S, i.e. the nodal displacements of 2gFE V_S.

Using the displacement function, built on a coarse grid H_S, the relation between the vectors $\mathbf{q}_S, \mathbf{q}_g$ is as follows

$$\mathbf{q}_g = [A_g^S]\mathbf{q}_S, \tag{7}$$

where $[A_g^S]$ – a rectangular matrix.

Using (7) in (6) provided that $\partial \Pi_g / \partial \mathbf{q}_S = 0$, the relation $[K_S]\mathbf{q}_S = \mathbf{F}_S$ corresponding its equilibrium state is obtained for 2gFE V_S, where $\mathbf{F}_S = [A_g^S]^T \mathbf{P}_S$, $[K_S] = [A_g^S]^T [K_g][A_g^S]$, $[K_S]$ – stiffness matrix, $\mathbf{F}_S$ – nodal force vector of 2gFE V_S.

Composite elastic 2–dimensional 2gFE of rectangular shape are projected by the procedures that are similar to those in section 2, 3 [6, 7]. When developing the composite 2gFE of triangular shape (tetragonal ones), the 1st, 2nd and 3rd order polynomials have been used [2–4].

Commentary 3. Calculations have demonstrated that 2gFE V_S (based on the second procedure) generate more exact solutions than 2gFE V_a (based on the first procedure). However, the second procedure involves the development of G superelement that is associated with the high-order matrix inversion. This increases the span time required to build 2gFE V_S.

Commentary 4. With the help of the step ratio varying of fine and coarse grids of 2gFE V_a, V_S, the solution error projected for double grid discrete models is regulated.

Commentary 5. According to (3) the vector dimension $\boldsymbol{\delta}_a$ (2gFE V_a dimension) doesn't depend on the number M i.e. on the dimension of R_a partition. Therefore, arbitrarily small basic partitions R_a comprised of FE V_e can be used to take into account the heterogeneous and microinhomogeneous structure of 2gFE V_a (see section 2). In this case, the three-dimensional stress state in FE V_e is arbitrarily closely described (without any simplifying hypotheses). With a sharp increase in the dimensions of basic 2gFE partitions, i.e. M number (dimension of superelement G, see section 3), the time spending needed to develop 2gFE V_a, V_S is dramatically increased. Therefore, it is worth using a 3-grid FE requiring less time costs to be constructed and giving rise to the discrete models of three-dimensional bodies of smaller dimension than 2gFE.

4. Developing the composite 3-grid FE

The main principals of this procedure are considered to be an example of the 3-grid FE (3gFE) V_b of the third order of $12h \times 12h \times 12h$ cube located in a local Cartesian coordinate system $Oxyz$ (figure 2). The area of 3gFE consists of 2gFE V_n^a, $n = 1,...,N$, N - the total number of 2gFE V_n^a ($6h \times 6h \times 6h$ sizes), $N = 8$ is shown in figure 2. Coarse grids of 2gFE generate 3gFE fine grid h_b. Coarse grid H_b of 3gFE is identified at h_b grid. 27 grid nodes H_b are point - marked in the figure 2.

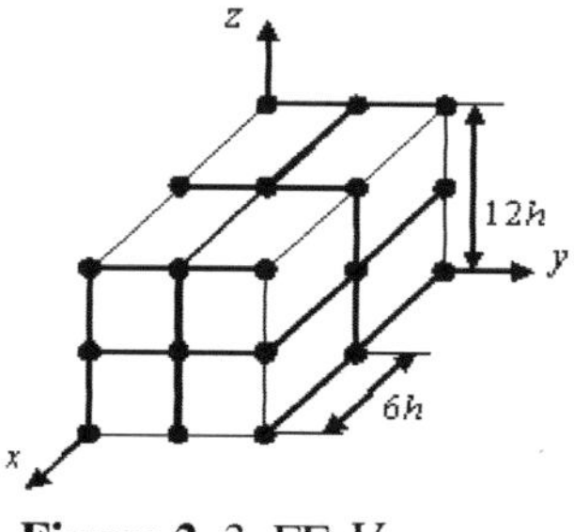

Figure 2. 3gFE V_b.

Displacement functions u_b, v_b, w_b of 3gFE built on a grid using the Lagrange polynomials [3] can be written as follows

$$u_b = \sum_{\beta=1}^{m} N_\beta q_\beta^u, \quad v_b = \sum_{\beta=1}^{m} N_\beta q_\beta^v, \quad w_b = \sum_{\beta=1}^{m} N_\beta q_\beta^w, \tag{8}$$

where q_β^u, q_β^v, q_β^w, N_β - displacements and shape function of β node of the H_b grid, $m=27$ is shown in figure 2.

To build 3gFE V_b, three typical grids are used: two grids of 2gFE V_n^a and a coarse one H_b. Total potential energy W_b of 3gFE V_b is presented by the following expression

$$W_b = \sum_{n=1}^{N} \{\frac{1}{2}(\boldsymbol{\delta}_n^a)^T [K_n^a] \boldsymbol{\delta}_n^a - (\boldsymbol{\delta}_n^a)^T \mathbf{P}_n^a\}, \tag{9}$$

where $[K_n^a]$ – stiffness matrix; $\mathbf{P}_n^a$, $\boldsymbol{\delta}_n^a$ – nodal force and displacements vectors of 2gFE V_n^a.

Let $\boldsymbol{\delta}_b$ is a nodal displacement vector of the H_b grid. Using (8), the nodal displacements of the $\boldsymbol{\delta}_n^a$ vector in terms of the nodal $\boldsymbol{\delta}_b$ vector displacement of the coarse grid H_b are expressed. As a result the following relation can be deduced

$$\boldsymbol{\delta}_n^a = [A_n^b]\boldsymbol{\delta}_b, \tag{10}$$

where $[A_n^b]$ – rectangular matrix, $n = 1,..., N$.

Using (10) in the functional (9) and minimizing it based on the displacements $\boldsymbol{\delta}_b$, $[K_b]\boldsymbol{\delta}_b = \mathbf{F}_b$ is obtained for 3gFE V_b which corresponds to its state of equilibrium, where $[K_b]$ - stiffness matrix, $\mathbf{F}_b$ - nodal force vector of the V_b 3gFE identified by the following formulas

$$[K_b] = \sum_{n=1}^{N} [A_n^b]^T [K_n^a][A_n^b], \quad \mathbf{F}_b = \sum_{n=1}^{N} [A_n^b]^T \mathbf{P}_n^a. \tag{11}$$

According to (10), the vector dimension $\boldsymbol{\delta}_b$ (i.e. the dimension of 3gFE V_b) doesn't depend on the total number N of 2gFE V_n^a, therefore, the partition of 3gFE V_b into 2gFE V_n^a can be very small (i.e. the sizes of 2gFE are arbitrarily small). In this case, 3-D stress state with no simplifying hypotheses in basic partitions of 2gFE V_n^a, i.e. in FE V_e (see section 2, 3), are accurately presented. Determination of the stresses in 3gFE V_b is as follows. Let vector $\boldsymbol{\delta}_b$ was build. According to the formula (10), vectors $\boldsymbol{\delta}_a^n$ of nodal displacements of 2gFE V_n^a were found. Stresses in FE V_e of the basic partition of 2gFE V_n^a, $n = 1,..., N$, are determined based on the algorithms in section 2.

Commentary 6. Using 3gFE, based on the procedure similar to one in section 4, 4-grid FE are developed, so m - grid FE are [8, 9], $m \geq 4$. Note that the m grid FE gives rise to the discrete model of a smaller dimension than the $m-1$ grid FE.

Commentary 7. When developing plate (beam) type MFE, Lagrange polynomials having along the Ox, Oy axes (Oy - direction) a higher order approximation than along the Oz axis, $n_1, n_2 > n_3$ (Oz, Ox - directions, $n_2 > n_1, n_3$, see section 2). To construct homogeneous MFE similar procedures, described in section 2-4, are used.

5. Mixed multigrid discrete models of composite solids

Key point of the combined application of the single - and multigrid modeling of three-dimensional inhomogeneous elastic solids of complex shape is as follows. Subregion of the body, which includes the complex boundary or a body binding, is presented by (fine) base partition, which consists of the homogeneous first order single grid FE of a cube form and takes into account the heterogeneous structure and complex body shape. The rest of the body is represented by a coarse partition containing the composite cuboid MFE. Fine and coarse partitions using the binding MFE are connected and a mixed discrete model with a less dimension than that of the basic model is obtained. Mixed multigrid discrete models are in detail reviewed in [10].

6. Developing MFE of irregular shape

6.1. The first procedure is briefly considered by the example of 2-D composite 2gFE *ABCD* of the first order given in figure 3, where the 2gFE curved boundary *AB* coincides with the boundary of body. The coarse grid nodes are point-marked. The 1-st order polynomial is applied to produce the displacement function on a coarse grid of 2gFE, written in local Cartesian coordinate systems *Oxyz* (figure 3).

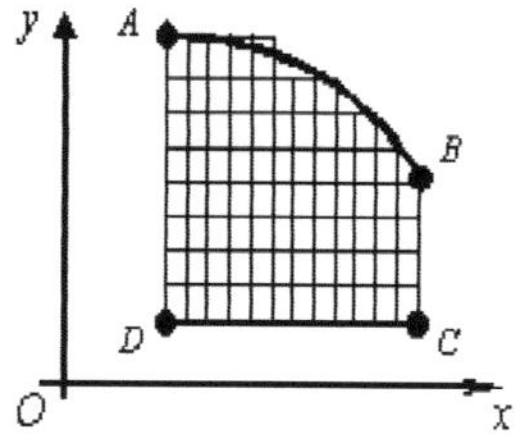

Figure 3. Two-dimensional 2gFE.

2gFE of irregular shape using the base partition is taken into consideration. The base partition grid of 2gFE is shown in figure 3. The power - law polynomials have been used for the displacement functions to be developed on 2gFE coarse grids, with some nodes of coarse grids being on the 2gFE curved boundaries. It should be noted that 2gFE (MFE) application is effective when calculated homogeneous bodies of irregular shape. Using the basic partitions of homogeneous 2gFE (MFE), their irregular shape (figure 3), the complexity of fixing and body loading have been taken into account. In addition, 2gFE (MFE) generate discrete model of homogeneous bodies whose dimensions are smaller than the ones of the basic model.

6.2. Consider the procedure of building composite MFE in the form of a right prism with the bases of irregular shape [11]. Approximating displacement functions of MFE built on coarse grids and presented as a product of power and Lagrange polynomials have been made use of. The key point of the procedure is considered by 2gFE example of triangular right prism shape with a hole shown in figure 4. The hole cross section is shaded; the coarse grid nodes are point-marked (24 nodes).

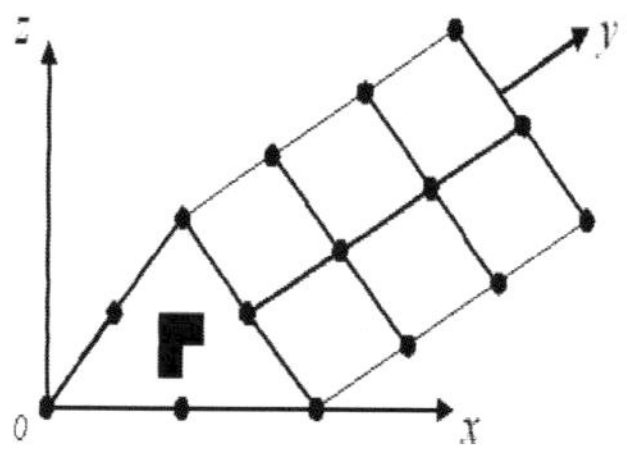

Figure 4. 2gFE of triangular right prism shape.

Basic partition of 2gFEs takes into account its heterogeneous structure and irregular shape. In the *Oxz* plane (figure 4) on the coarse grid nodes of 2gFE (6 nodes), the approximating displacement functions using the 2nd order polynomial $P_2(x,z)$ of the form

$$P_2(x,z) = a_1 + a_2 x + a_3 z + a_4 xz + a_5 x^2 + a_6 z^2 . \qquad (12)$$

Basis function φ_β for β - node of 2gFE coarse grid is represented as a product of the power and Lagrange polynomials, i.e.

$$\varphi_\beta(x,y,z) = N_\beta(x,z)L_\beta(y), \tag{13}$$

where $N_\beta(x,z)$ is a basis function of the polynomial $P_2(x,z)$ of the form (12) and $L_\beta(y)$ is Lagrange polynomial, corresponding to β- node of 2gFE coarse grid, $\beta = 1,...,N$, N is the total number of the coarse grid nodes.

Figure 4 demonstrates $N = 24$, $L_\beta(y)$ is the third order polynomial, for (13) $\beta = 1,...,24$.

7. Numerical study results

Model problem of bending the $216h_x \times 432h_y \times 6h_z$ V_0 three-dimensional rectangular six-layer plate have been considered in Cartesian reference system $Oxyz$ (figure 5) where $h_x = h_y = 3.5$, $h_z = 6$, $h = 6h_z$ is the plate thickness. For $y = 0: u = v = w = 0$. Layers are homogeneous isotropic bodies with the thickness $h_0 = h/6 = 6$. The elastic moduli (Young`s modulus) from the bottom are equal to 10, 25, 40, 55, 70, 85 respectively. The Poisson's ratio is 0.3. For $z = h$, $y \geq 324h_y$, the plate is given a uniform vertical surface load $P_z = 0.0003$.

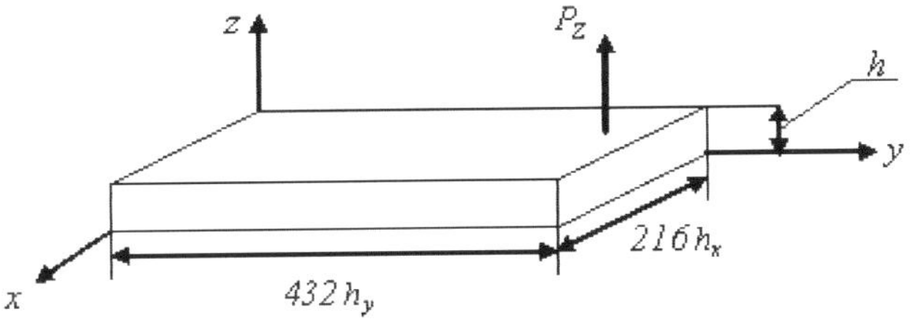

Figure 5. Sizes of V_0 plate.

The R_n^0 base discrete models of the plate consist of the first order homogeneous FE of $h_n^x \times h_n^y \times h_n^z$ size, where $h_n^x = h_x/(2n-1)$, $h_n^y = h_y/(2n-1)$, $h_n^z = h_z/(2n-1)$, $n = 1,...,6$. Grid of the R_n^0 discrete model has the $m_n^1 \times m_n^2 \times m_n^3$ dimension, where $m_n^1 = 216(2n-1)+1$, $m_n^2 = 432(2n-1)+1$, $m_n^3 = 6(2n-1)+1$, $n = 1,...,6$.

The R_n multigrid discrete models of the V_0 plate consist of $108h_n^x \times 108h_n^y \times 6h_n^z$ plate composite 3gFE (see section 3). Coarse grid of 3gFE along the Ox, Oy, Oz axes has the $36h_n^x$, $36h_n^y$, $6h_n^z$ steps, respectively. 3gFE involve the third order 2gFE of $6h_n^x \times 6h_n^y \times 6h_n^z$ sizes developed by section 2. There are 32 nodes on 2gFE coarse grid, the steps of the coarse grid in Ox, Oy, Oz- directions equal $2h_n^x$, $2h_n^y$, $2h_n^z$, respectively. To approximate the displacement functions on a 2gFE coarse grid, the 3rd order polynomial of the form (5) has been used. The calculation results are shown in Table. 1, where w_n^m, σ_n^m - the maximum displacement and equivalent stress of the R_n models, $n = 1,...,6$. Parameters δ_n^m, Δ_n^m have been found by the following formulas

$$\delta_n^m(\%) = 100\% \,|\sigma_n^m - \sigma_{n-1}^m| / \sigma_n^m, \quad \Delta_n^m(\%) = 100\% \,|w_n^m - w_{n-1}^m| / w_n^m, \; n \geq 2.$$

Stresses σ_n^m are determined according to the 4th theory of strength.

Analysis of the results has shown (table 1) the nature of the parameter changes $\Delta_n^m(\%)$, $\delta_n^m(\%)$ ($n=1,...,6$) demonstrates a high convergence rate of the maximum stresses σ_n^m and displacements w_n^m to the exact solutions.

Table 1. Maximum displacements and stresses of R_n discrete models.

R_n	R_1	R_2	R_3	R_4	R_5	R_6
w_n^m	684.854	754.407	764.893	767.851	768.131	767.838
$\Delta_n^m(\%)$	–	9.087	1.371	0.385	0.036	0.038
σ_n^m	0.686	0.948	1.071	1.125	1.153	1.166
$\delta_n^m(\%)$	–	27.637	11.485	4.800	2.428	1.115

As the $\Delta_6^m = 0.00038$, $\delta_6^m = 0.01115$ are small, the displacement $w_6^m = 767.838$ and stress $\sigma_6^m = 1.166$ can be considered to be the exact solution. There are 2270396480 nodal unknowns in the discrete base model R_6^0 of the plate (i.e. 2.2 milliards of unknown FEM), the bandwidth of equation system (ES) of FEM is equal to 955967. The dimension of R_6 multigrid model is 318384; bandwidth of ES of FEM is equal to 14699. FEM implementation for the R_6 model requires 463770 times less computer memory than for the R_6^0 base model, at small time spanding.

Acknowledgments
This work was supported by the Russian Foundation for Basic Research (project No. 14-01-00130).

References

[1] Fudzii T and Dzako M 1982 *Fracture mechanics of composite materials* (Moscow: Mir)
[2] Zenkevich O *1975 Finite element method in engineering* (Moscow: Mir)
[3] Norrie D and Vries J 1981 *Introduction to the finite element method* (Moscow: Mir)
[4] Sekulovich M 1993 Finite *Element Method* (Moscow: Stroyizdat)
[5] Samul V I 1982 *Fundamentals of the elasticity and plasticity theory* (Moscow: Higher School)
[6] Matveev A D 2000 Some approaches of the multigrid elastic finite element design *Dep. VINITI* 2990-V00
[7] Matveev A D 2004 Multigrid modeling the irregular structure composites with a low fullness coefficient *J. Appl. Mech. Tech. Phys.* **45** 440
[8] Matveev A D 2013 Construction of complex multigrid elements with microscopic structures *Proc. of XXIII All-Russian Conf. Numerical methods for solving the problems of elasticity and plasticity theory* (Novosibirsk: Parallel) pp 142-144
[9] Matveev A D 2014 Construction of complex multigrid finite elements of inhomogeneous and microinhomogeneous structure *News of Altai State University* **1/1** 80
[10] Matveev A D 2013 Mixed discrete models in the analysis of elastic three-dimensional heterogeneous solids of complex shape *PNRPU Mechanics Bulletin* **1** 182
[11] Matveev A D 2015 Calculation of three-dimensional composite beams of complex shapee using double grid finite element *The Bulletin of KrasGAU* **8** 92

Upscaling of absolute permeability for a super element model of petroleum reservoir

A B Mazo and K A Potashev

Department of Aerohydromechanics, Kazan Federal University, Kazan, Russia

E-mail: KPotashev@mail.ru

Abstract. This paper presents a new method of local upscaling of absolute permeability for super element simulation of an oil reservoir. Upscaling was performed for each block of a super element unstructured grid. For this purpose, a set of problems of a one-phase steady-state flow was solved on a fine computational grid with the initial scalar field of absolute permeability with various boundary conditions. These conditions reflect the specific variants of filtrational flow through the super element and take into account the presence or absence of boreholes in the coarse block. The resulting components of the effective permeability tensor in each super element were found from the solution of the problem of minimizing the deviations of the normal flows through the super element faces, averaged on a detailed computational grid, from those approximated on a coarse super element grid. The results of using the method are demonstrated for reservoirs with river-type absolute permeability. The method is compared with the traditional methods of local upscaling.

1. Introduction

The super element model of petroleum reservoir simulation makes it possible to essentially speed up the computation process by using a coarse computational grid with a size of blocks being the same order as the distance between the wells [1]. The absolute (AP) and relative permeability (RP) functions contained in the equations of this model are obtained as a result of upscaling of the original characteristics by the scale of the coarse computational grid. Upscaling of RP in a super element model was considered earlier [2]. The present study focuses on AP upscaling. Simplified analytical formulas of upscaling prove to be applicable to a small number of idealized types of heterogeneity. Reservoirs with a more complex structure require using other averaging methods. Mathematical averaging tools for the problems of filter theory are described in [3] and allow us to estimate the effective parameters of porous media. Numerical methods of flow-based upscaling are currently the most widespread methods in the practice of reservoir simulation [4]. Both global [5] and more popular local methods [6, 4] are used for this purpose. It is obvious that upscaling must take into account not only the grid geometry, but also the computational scheme of the problem solved on a coarse grid. This paper proposes a specialized method for local upscaling of AP for a super element model of petroleum reservoir simulation.

2. Constitutive equations

The method of AP upscaling will use the mass balance equation for Darcy's flow neglecting the compressibility, gravitational and capillary forces:

$$\operatorname{div}\mathbf{u}=0, \tag{1}$$

$$\mathbf{u}=-\sigma\nabla p. \tag{2}$$

Here $p(\mathbf{x})$ is the pressure; $\mathbf{u}$ is the flow rate; $\sigma(\mathbf{x})=k(\mathbf{x})/\mu$ is the specific conductivity of the reservoir; $k(\mathbf{x})$ is the scalar field of absolute permeability; μ is the fluid's dynamic viscosity, which will be assumed to be constant in the super element (SE) volume. The function $k(\mathbf{x})$ is defined on a geological grid, which is able to accurately reproduce the complex heterogeneous structure of the reservoir.

Equations (1)–(2) describe the flow in the domain D, bounded from above and below by the impermeable top and the bottom of the formation, and from one side – by the outer cylindrical surface, on which the hydrostatic pressure is maintained, as well as by the inner boundaries – perforated surfaces of wellbores with the radius r_w, where the values of the bottomhole pressure p^w or the summarized flow q are set.

Let us consider the super element V with the outer boundary Γ, consisting of faces Γ_j, and with the inner boundary γ. After integrating Equation (1), we obtain the relation

$$Q_\Gamma+Q_\gamma=0,$$

$$Q_\Gamma=\int_\Gamma u_n\,\mathrm{d}\Gamma=\sum_j|\Gamma_j|\,u_{nj},\quad Q_\gamma=\int_\gamma u_n\,\mathrm{d}\gamma=q,$$

where the sign for the average flow rate on the faces Γ_j is introduced:

$$\text{e, } u_n=-\sigma\,\partial p/\partial n, \tag{3}$$

$\mathbf{n}$ is the outer normal direction. Let us determine the way of expressing velocities (3) through the average pressure

$$\langle p\rangle=|V|^{-1}\int_V p\,dV, \tag{4}$$

where $|V|$ is the super element volume V.

The set of values $\langle p\rangle$ in the super elements determines the average pressure field in the reservoir. The use of a particular replenishment procedure for the grid function $\langle p\rangle$ makes it possible to determine the gradient of the average pressure field and to use the relation similar to Darcy's law for the average flow rate field (2):

$$\mathbf{U}=-\mathbf{\Sigma}\cdot\nabla\langle p\rangle.$$

Here $\mathbf{\Sigma}=\mathbf{K}/\mu$; $\mathbf{K}$ is the effective tensor of absolute permeability in SE. Taking into the last formula, we can introduce approximation for the flow rate (3):

$$u_n\approx U_n=\mathbf{U}\cdot\mathbf{n}=-(\mathbf{\Sigma}\cdot\nabla\langle p\rangle)\cdot\mathbf{n}=-(\mathbf{k}/\mu)\cdot\nabla\langle p\rangle=-(|\mathbf{k}|/\mu)(\partial\langle p\rangle/\partial\mathrm{k}), \tag{5}$$

where $\mathbf{k}=\mathbf{n}\cdot\mathbf{K}$ is defined as the vector of full permeability on the super element edge, and $\partial/\partial\mathrm{k}$ is the directional derivative, which is constructed by the colocations method [7, 2]:

$$U_n\approx-\frac{\bar{k}}{\mu}\frac{P_2-P_1}{(h_1+h_2)},\quad \bar{k}=\frac{|\mathbf{k}_1||\mathbf{k}_2|(h_1+h_2)}{|\mathbf{k}_2|h_1+|\mathbf{k}_1|h_2}$$

Here $\mathbf{k}_1=\mathbf{n}\cdot\mathbf{K}_1$ and $\mathbf{k}_2=\mathbf{n}\cdot\mathbf{K}_2$ are the full permeability vectors in the adjacent super elements with the AP effective tensors $\mathbf{K}_1$ and $\mathbf{K}_2$; P_1, P_2 are the average pressures at the distances h_1 and h_2 from the face on the directional lines k_1 and k_2, which are determined by the interpolation of values of the average pressure $\langle p\rangle$. Thus, the colocation algorithm is a technique of the average field $\langle p\rangle$ differentiation by the node values, which does not require an explicit replenishment of this function.

3. Upscaling algorithm

Coefficients of the effective AP tensors in each SE are found from the condition of the best approximation of the average normal flow rates (5). As an "exact" solution for the rates $u_{n\,j}$ on the faces Γ_j, we use the value (3), where the pressure p is obtained as the solution to the problem (1), (2) with the initial conductivity $\sigma(\mathbf{x})$ on a fine enough grid in the SE volume.

To reflect various modes of flow through the SE, the following boundary conditions are set

$$p^b\Big|_{\mathbf{x}\in\Gamma} = \sum_i \delta_i^b x^i, \quad b = 1..B, \tag{6}$$

where x^i are the coordinates at the outer boundary, δ_i^b is the Kronecker symbol, B is the number of various boundary conditions, equal to the problem dimension. For SE with the borehole γ, we set the well discharge in addition to the conditions (6):

$$\int_\gamma u_n \,\mathrm{d}\gamma = q. \tag{7}$$

The coefficients K^{ij} of the tensor $\mathbf{K}$ are defined from the following condition

$$R^2 = \frac{1}{B}\sum_{b=1}^{B} \rho^b\left(K^{ij}\right) \to \min_{K^{ij}}, \quad \rho^b\left(K^{ij}\right) = \frac{1}{M}\sum_{m=1}^{M}\left(\left|\Gamma_m\right|\left(u_{nm}^b - U_{nm}^b\left(K^{ij}\right)\right)\left[\max_{l=1..M}\left(\left|\Gamma_l\right|\left|u_{nl}^b\right|\right)\right]^{-1}\right)^2.$$

4. Testing

We will test the presented super element upscaling method (SEM) for a two-dimensional problem on the (x, y) plane. Let the reservoir with the absolute permeability $k = 0.001$ have the high-permeable $(k = 1)$ channel. We will cover the rectangular area of the reservoir by a super element grid of 50 blocks (figure 1). The ratio of the channel width to the mean radius of the super elements is 0.25. Let us arrange two production (P-1, P-2) and two injection (I-1, I-2) wells with the same rates in absolute magnitude so that the high-permeable channels connect pairwise the wells I-1 with P-1 and I-2 with P-2. We will set the constant value of the hydrostatic pressure at the external boundaries of the area.

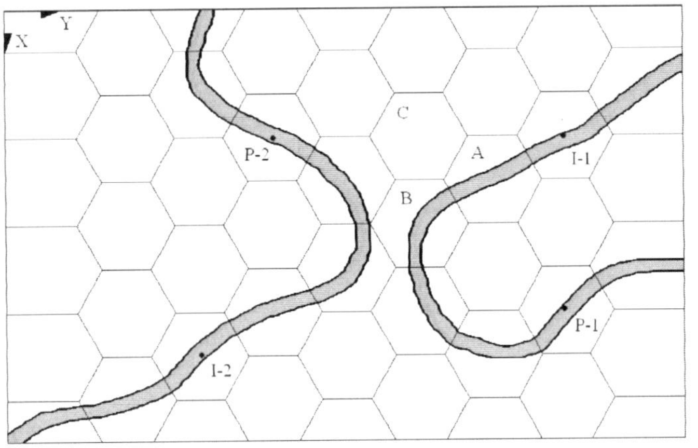

Figure 1. Scheme of the reservoir with high-permeable channels.

To evaluate the proposed upscaling method, the coefficients K^{ij} of the effective AP tensors in each SE can be found in four ways. The simplest one is averaging by the volume method (VOL). The velocity averaging (VEL) and minimal dissipation (DIS) methods [4] with periodic and linear boundary conditions are considered to be more reliable. The forth method is the presented super element upscaling method (SEM).

Figure 6 shows three typical types of SE, which differ in their location relative to the high-permeable channel: "A" demonstrates the channel which passes through the SE mainly in one direction, "B" shows the channel which changes its direction within the SE, and "C" illustrates the channel which does not pass through the SE. The AP upscaling for C-type SE gives the same results by all the listed methods: the effective tensor $\mathbf{K}$ is isotropic, and its diagonal coefficients are equal to the reservoir permeability $k = 0.001$. The AP tensor components obtained by different methods for the SE types "A" and "B" are given in table 1.

Table 1. Examples of effective tensors

Method	Super element "A"				Super element "B"			
	K^{xx}	K^{xy}	K^{yx}	K^{yy}	K^{xx}	K^{xy}	K^{yx}	K^{yy}
VOL	0.152	0.0000	0.0000	0.152	0.179	0.000	0.000	0.179
VEL	0.002	0.0015	0.0015	0.004	0.003	0.001	0.001	0.002
DIS	0.057	0.0700	0.0700	0.155	0.103	0.062	0.062	0.060
SEM	0.041	0.0820	0.0690	0.144	0.106	0.056	0.062	0.033

Let us consider VEL results. The coefficients underrated by one or two orders on average are the result of using the periodic boundary conditions, which are unacceptable for the non-periodical structure of the $k(\mathbf{x})$ field. Calculation of the coefficients K^{ij} by SEM method for super elements with boreholes was carried out following the second scenario with the boundary conditions (7). As an example, we can compare the coefficients K^{ij} for the "A"-type SE (see table 1) and the coefficients for super elements with boreholes I-1: $K^{xx} = 0.235$, $K^{xy} = 0.404$, $K^{yx} = 0.404$, $K^{yy} = 0.702$.

Let us estimate the quality of the upscaling techniques by the values of deviations:

$$r_u = \sqrt{\frac{1}{M}\sum_{i=1}^{M}\left(\left(u_{ni} - U_{ni}\right)/u_*\right)^2}, \quad u_* = \max_{k=1..M}\left|u_{nk}\right|,$$

$$r_p = \sqrt{\frac{1}{N}\sum_{i=1}^{N}\left(\left(\langle p\rangle_i - P_i\right)/\Delta P\right)^2}, \quad \Delta P = \max_{i=1..N}\langle p\rangle_i - \min_{i=1..N}\langle p\rangle_i,$$

$$R_p = \max_{i=1..N}\left|\left(\langle p\rangle_i - P_i\right)/\Delta P\right|,$$

where M, N are the total numbers of faces and blocks in the super element grid; the values u_{ni} are calculated using Equation (3) for each SE face by the numerical integration of the solution obtained on a fine grid; the values $\langle p\rangle_i$ are obtained using Equation (4) for each SE; U_{ni}, P_j are the average normal rate of flow through the super element faces and the average pressure in super elements, obtained from the solution on the course grid. The average fine grid step was about 1 % of the average coarse grid step.

The values of these deviations for the considered methods are given in table 2. The best results demonstrate solutions built using DIS and SEM methods, though the latter method leads to smaller values of residual errors.

Minimizing these area-averaged residual errors is not the only criterion of AP upscaling methods quality. Equally important is to adequately describe, using a course grid, the structure of the flow determining the interwell interference. To analyze this description, we solve the problem of imitation of tracer injection alternately in each injector. The resulting tracer distribution in the reservoir will show the trajectory of the fluid injected into the wells, and the final concentrations in the producers will correspond to the degree of their interaction with the respective injectors. The exact solution of the problem will be the solution built on a fine grid. Solution on a super element grid will be build using various upscaling methods.

Table 2. Average normal flow rate and average pressure deviations

Deviation \ Method	VOL	VEL	DIS	SEM
r_u	0.130	0.222	0.079	0.063
r_p	0.158	5.523	0.186	0.139
R_p	0.503	18.76	0.653	0.434

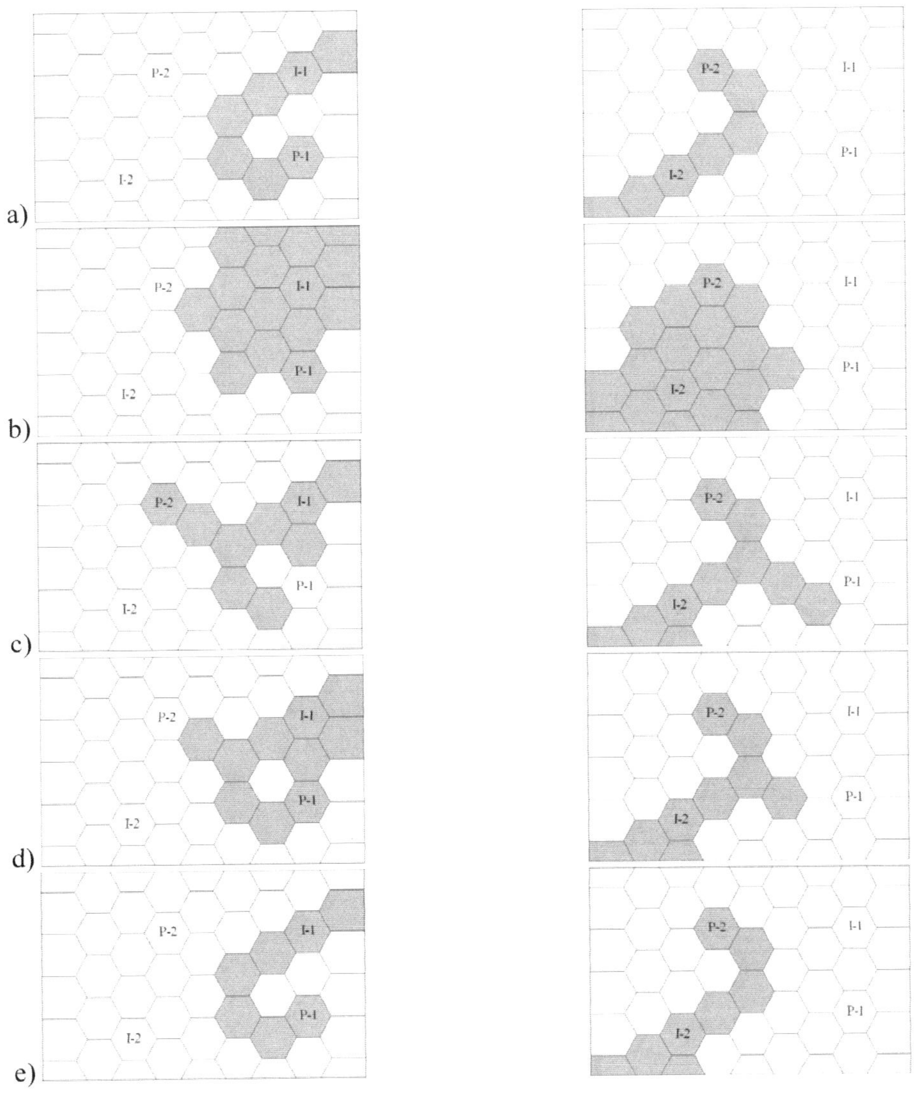

Figure 2. Tracer distribution
from the injection well I-1 (on the left) and the well I-2 (on the right):
a) exact solution, b) VEL c) VOL, d) DIS, e) SEM.

Figure 6 shows the tracer distribution in the reservoir at a fixed time moment, obtained in different ways. Superelements where the tracer concentration exceeds 20% are highlighted in color. For exact solutions, such SE were determined by numerical integration of the concentration field on a detailed grid by the areas of SE.

It can be seen that AP upscaling by VEL method with the periodic boundary conditions does not give an adequate description of the flow structure. Among other methods, the super element upscaling method ensures the closest tracer distribution in the reservoir to the actual one. Upscaling performed using VOL and DIS methods does not provide a sufficient isolation of faces separating SE with the areas of two independent high-permeable channels. Due to this fact the tracer from the injector enters the producer, but they do not interact in reality. The final concentration C of the tracer in the producer's liquid is shown in table 3.

Table 3. Final concentrations of the tracer in producers

Calculation variant	Tracer injection into the well I-1		Tracer injection into the well I-2	
	P-1	P-2	P-1	P-2
exact solution	0.375	0.000	0.000	0.498
SEM	0.291	0.025	0.039	0.440
DIS	0.233	0.144	0.084	0.372
VEL	0.164	0.275	0.148	0.294

5. Conclusion

We formulated a special method for local upscaling of absolute permeability, which makes it possible to minimize the error in flow calculation during super element simulation of oil reservoir development and accurately reflect the structure of the flow.

The comparative analysis showed a marked advantage of the presented upscaling method in comparison with the most popular methods of velocity averaging and minimum energy dissipation, used in rescaling of permeability field from a detailed geological grid to a course hydrodynamic one.

Acknowledgments

The work was financially supported by the Russian Foundation for Basic Research and the Government of the Republic of Tatarstan (nos. 15-41-02698, 15-41-02699).

References

[1] Mazo A B and Bulygin D V 2011 *Neft. Gas. Novacii* **11** 6–8 [*in russian*].
Mazo A B, Potashev K A, Kalinin E I and Bulygin D V 2013 *Mat. Model.* **25** 51 [*in russian*].
Bulygin D V, Mazo A B, Potashev K A and Kalinin E I 2013 *Geores.* **53** 31–5 [*in russian*].
Mazo A, Potashev K and Kalinin E 2015 *Proc. Earth and Planet. Sci.* **15** 482–7.

[2] Mazo A B and Potashev K A 2016 *Mat. Model.* (in print) [*in russian*].
Potashev K A 2014 *Uch. Zap. Kaz. Univ., Ser. Phys.-math. Sci.* **156** 120-34 [*in russian*].

[3] Belyaev A 2004 Usrednenie v zadachah teorii filtracii (Averaging in problems of filtration theory) (Moscow: Nauka) p 200 [*in russian*].

[4] Wu X H, Efendiev Y and Hou T Y 2002 *Discrete And Contin. Dyn. Syst.* B **2** 185–204.

[5] White C D and Horne R N 1987 *SPE J.* **16011** 209–21.
Holden L and Nielsen B F 2000 *Transp. In Porous Media* **40** 115–43.

[6] Durlofsky L J 1991 *Water Resour. Res.* **27** 699–708.

[7] Nikitin N G 2010 *Mat. Model.* **22** 131-47 [*in russian*].

Localization of microseismic events and determination of source parameters

E V Mokshin and D V Berezhnoi

Kazan Federal University, 18 Kremlyovskaya Street, Kazan, 420008, Russian Federation

E-mail: evmoks@yandex.ru

Abstract. We examine the problem of localization determining and a microseismic moment tensor of single microseismic event in the presence of strongly correlated noise. This is a typical problem occurring in monitoring of microseismic events from a daylight surface under conditions of a producing field or surface monitoring of hydraulic fracturing. We offer the solution to this problem based on the method of maximum likelihood. The article presents of decision of this problem and the results of numerical experiments. We discuss some features and problems of the proposed approach and estimate the required computing resources. We develop the problem of determination direction of fracture propagation from microseismic event.

1. Introduction

The hydraulic fracturing (HF) is one of the intensification of oil extraction technologies. HF is mechanical method of influence on the reservoir, based on increasing pressure of fluid in well higher than pore pressure, and lead to creating large fracture. During the process of fracture generation the elastic strain energy emitted, which generated seismic waves. The geophysical control based on this information and used it for monitoring of fracture geometry. This movement can be detected by seismic sensors and used to locate position of seismic sources, which occurred due to hydraulic fracture procedure and allow to evaluate space-time parameters of fracture [1].

The source mechanism of any microseismic event can be characterized by the seismic moment tensor. Aki and Richards [2] introduced the concept of the seismic moment tensor which depends on the source intensity and the medium movement orientation in the source vicinity. The full seismic moment tensor characterizes all possible stresses occurring in the vicinity of the seismic source epicenter. Tensor components M_{ij} are the values of the moments of forces applied along the i-axis when the points of force application are situated along j-axis. Thus, the seismic moment is the most general characteristic of a seismic event.

The condition of observation microseismic from the surface greatly differ than downhole observation. Even the microseismic noise level on the surface significantly higher than noise level in well, when depth of position sensor more than 500 meters. During procedure hydraulic fracturing the set of equipment is working which generate powerful noise, this noise is added to background noise. The level of signal (from source) on surface is much lower than in the case of downhole observations, due to the longer transit time of wave and intense absorption in weathering zone, especially for high frequency signal. These reasons lead to the fact that the source localization by the beamforming method (that are used in the downhole monitoring) for surface monitoring is not possible, due to too low signal-to-noise ratio and signal is not allocated visually. Thus, the surface microseismic data processing is significantly difficult problem than with a downhole data processing.

The problem of localization of microseismic sources is well-known in the radio engineering [3] as task of detection signals for multi-channel reception and solved by different methods. With regard to the problem of localization microseismic source they have different localization accuracy, robustness to noise, and computational complexity. The most commonly beamforming method and its variants is using for the recovery position of microseismic source. For example, focusing conversion method, which allows dynamically throw out the sensors which not informative. The diffraction stacking method is relatively easily to implement, however, as will be shown below, it is poorly resistant to surface noise.

There is also the method of Time Reverse Modelling [4, 5], which is resistant to surface noise and the accuracy of the localization higher than diffraction stacking method [6].

2. Maximum-likelihood estimation

From the signal theory it is known that the most effective method is a method of detecting signals based on the maximum likelihood principle. With regard to the problem of monitoring hydraulic fracturing the method of maximum likelihood is used as follows. For area of space the discrete grid defined by a increments of 5-10 meters. The nodes of this grid are the points of microseismic intensity recovery.

Then, the most probable amplitude of microseismic event is recovering for each moment of time. Solution is made by comparing the synthetic signals and field data and selecting the amplitude of microseismic events, when the function likelihood is maximized. Synthetic signals calculated in at each node on grid and model is a 3D cube of microseismic events, which include interest area of Earth (also for interval by time). This cube can be subjected to further processing - filtration, separation of individual microseismic events, their rejection by various criteria, etc. The final result of processing is a map of summarized microseismic activity corresponding to different time of phases of hydraulic fracturing.

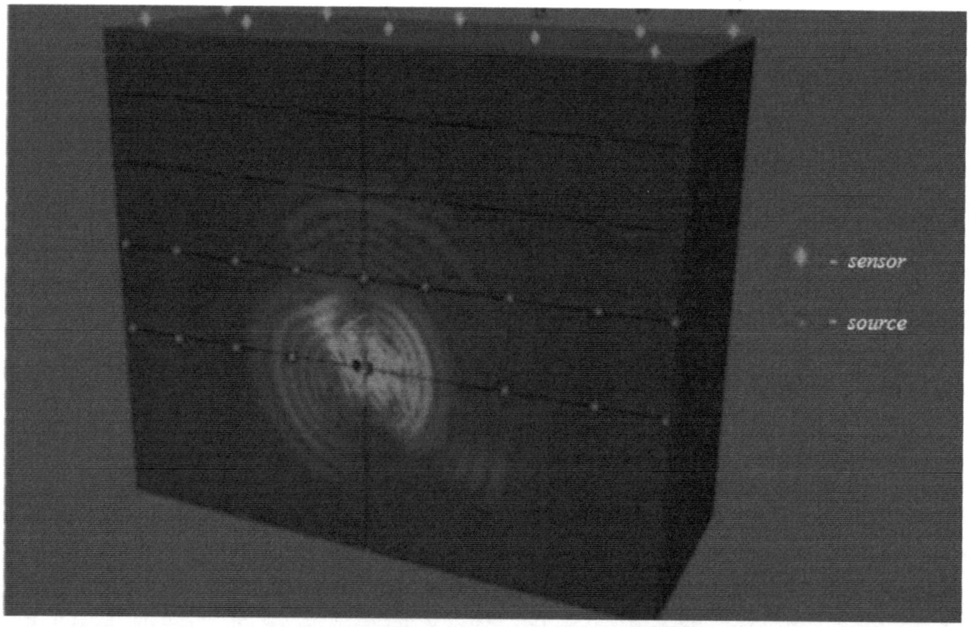

Figure 1. Example of the wave simulation.

Unlike the more simple methods such as diffraction stacking, the maximum likelihood estimation method requires knowledge of the waveform expected to sensors from microseismic source, not just the time of the first entry. In the method of diffraction stacking used ray tracing algorithm in this case is not suitable, and the only Full-wave numerical simulation is allows to solve this problem.

Full-wave numerical simulation allows the calculation of the waveform, which came to the sensor. It simulation can work with geology of any complexity, taking into account all possible waves, which generated during the seismic impulse propagating from the source to the receiver and to take into account a geometrical spreading for the actual propagation paths in inhomogeneous space (Fig. 1).

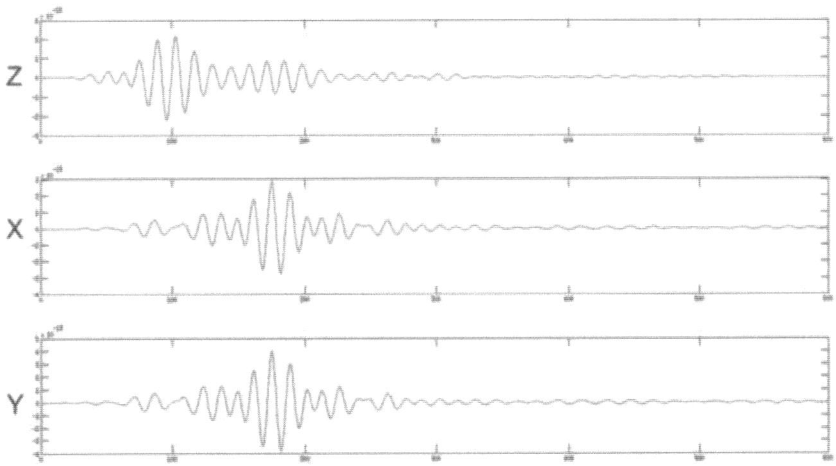

Figure 2. Calculated signals from a receiver for the three components.

Full-wave calculation is exceptionally demanding task, but it lends itself to parallelization, including advanced graphics clusters.

Using specialized software for full-wave modelling and supercomputing complexes [7-16] such a calculation is quite possible for a practical time, that making the maximum likelihood method is effective tool for the localization of weak microseism. And so he can recover the most probable values of the tensor seismic moment by time.

3. Determination of primary direction fracture propagation

Any type of underground events can be expressed in terms of seismic moment tensor. Consider seismic processes that can occur underground, which recorded at the seismic sensors. Fig. 2 shows the types of sources and the corresponding seismic moment tensors.

During hydraulic fracturing procedure occurs breakdown of formation, and fracture is growing, and this is process is source type of tensile crack. Question is of recognition this event among others.

Fig. 2 shows the ideal form of the tensor, in real cases it will also have the off-diagonal elements due to the presence of surface noise. This matrix is real and symmetric, so it can always be reduced to diagonal form by finding the eigenvalues. If we take a certain period of time, we can consider the distribution of the eigenvalues for each of the recovery tensor.

Three submitted by source type (Fig. 2) will produce different distributions of the three eigenvalues. For tensile crack will be two distributions for the first two eigenvalues, which are very close by value and one isolated distribution of the third eigenvalue. The maximum-likelihood estimation method allows to analytically evaluate the distribution and take the threshold for rejection all unlike types of events. Thus restoring only the type of source tensile crack. Monitoring these events makes it possible to identify the primary direction of fracture propagation.

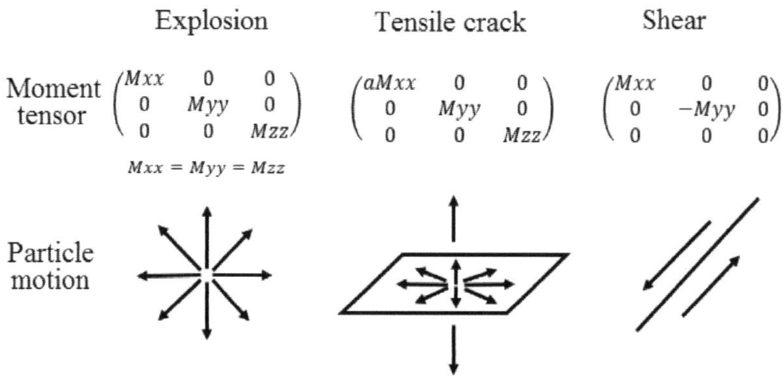

Figure 3. Type of seismic sources.

4. Conclusion

The maximum-likelihood estimation method is the most informative and robust technology of recovery signals. The article presents an approach based on the maximum likelihood method for the recovery of microseismic events, that occurs during hydraulic fracturing. The approach and shows the principle possibility of determining the priority directions of crack propagation using the maximum likelihood method.

Showed the principle possibility of determining the direction of fracture propagation by the maximum-likelihood estimation method.

Acknowledgments

The reported study was supported by Government of the Republic of Tatarstan research projects No. 15-07-05380, 15-41-02555.

References

[1] Aleksandrov S I, Gogonenkov G N and Mishin V A 2005 *Oil Industry* **5** 64–66
[2] Aki K and Richards P G 1980 *Quantitative seismology* (Freeman and Co)
[3] Howland P E 2005 *IEE Proceedings: Radar, Sonar and Navigation* **152 (3)** 105–106
[4] Levin S A 1984 *Geophysics* **49** 581–583
[5] Steiner B, Saenger E H and Schmalholz S M 2008 *Geophysical Research Letters* **35** L03307
[6] Mokshin E V, Berezhnoy D V and Birialtsev E V 2012 *Exposition Oil Gas* **2** 26-28
[7] Galimov M R and Birialtsev E V 2010 *Numerical Methods and Programming* **11** 77–93
[8] Berezhnoi D V, Sachenkov A A and Sagdatullin M K 2014 *Applied Mathematical Sciences* **8 (127)** 6341–6348
[9] Berezhnoi D V, Sachenkov A A and Sagdatullin M K 2014 *Applied Mathematical Sciences* **8 (143)** 7107–7115
[10] Badriev I B and Shagidullin R R 1995 *Journal of Mathematical Sciences* **73 (5)** 519–525
[11] Badriev I B, Makarov M V and Paimushin V N 2015 Russian Mathematics 59 (3) 66–71
[12] Sachenkov O, Kharislamova L, Shamsutdinova N, Kirillova E and Konoplev Yu 2015 *IOP Conference Series: Materials Science and Engineering* **98**
[13] Shigapova F A, Mustakimova R F, Saleeva G T and Sachenkov O A *International Journal of Applied Engineering Research* **10(24)** 44711–44714
[14] Sultanov L U 2015 *Procedia Earth and Planetary Science* **15** 119–124
[15] Davydov R L and Sultanov L U 2013 *PNRPU Mechanics Bulletin* **1** 81–93
[16] Golovanov A I and Sultanov L U 2005 *International Applied Mechanics* **41 (6)** 614–620

Numerical modeling of two-phase high speed jet with non-equilibrium solid phase crystallization

A M Molchanov, D S Yanyshev and L V Bykov

Department of aviation and space heat engineering, Moscow Aviation Institute (National Research University), Volokolamskoe shosse, 4, Moscow, A-80, GSP-3,125993, Russian Federation.

E-mail: dyanishev@gmail.com

Abstract. The main purpose of the paper is to demonstrate that the Euler approach is fully applicable to the multiphase flows with discrete phase undergoing phase transitions. It is carried out using the example of a jet flow with aluminium oxide particles non-equilibrium crystallization. The jet is strongly underexpanded. The non-equilibrium molecular effects are being taken into account. The obtained results of the simulations are in good agreement with the works of the other authors. The developed Euler approach proved itself to be the most robust in flows with complex flow geometry.

1. Introduction

Determination of parameters of high speed jets with admixture of solid phase is a very important problem in applied technical physics. Such type of flows can be seen in different systems and objects in civil area as well as special ones. And the parameters of these flows can be crucial to design.

Computational fluid dynamics today is one of the most important and powerful tools for investigation of multiphase flow. Nevertheless, it should be noted that the methods for investigating multiphase flows with phase transition by now are not quite well developed. Particle phase transition models have been usually treated using Lagrange technique. However, the handicaps of this technique are well-known and include two-way coupling difficulties, problems dealing with flows with significant particle volume fraction (>10%), necessity to analyze a great number of particle trajectories in order to maintain statistical stability of the solution, etc. On the other hand, Euler approach lacks these disadvantages, but the Euler modeling techniques for particulate flows with phase transitions for now are developed rather poorly. This provides a high motivation for the present work.

This paper proposes a method for simulation of two-phase jets with solid phase non-equilibrium crystallization. The method has the following features:

1. The mathematical model includes conservation equations for gas and solid particles, that is conservation of mass, momentum and energy. In addition, the equations for vibration gas molecular energy, conservation of species and dimensionless particle crystallization radius are solved.
2. The equations for gas and particles are solved in the Euler's form.
3. A two-way coupled solver algorithm is used.

2. Governing equations for gas in state of thermal and chemical non-equilibrium

First of all, let us write down the system of equations for gas phase.

The main system of equations includes:

1). Continuity equation

$$\frac{\partial \rho}{\partial t}+\frac{\partial}{\partial x_j}\left(\rho u_j\right)=0 \tag{1}$$

2). Momentum equation

$$\frac{\partial}{\partial t}\left(\rho u_i\right)+\frac{\partial}{\partial x_j}\left(\rho u_j u_i+\delta_{ji} p-\tau_{ij}\right)=0 \tag{2}$$

3). Total energy equation

$$\frac{\partial}{\partial t}(\rho E)+\frac{\partial}{\partial x_j}\left[\rho u_j\left(E+\frac{p}{\rho}\right)+q_{tr,j}+q_{v,j}+\sum_s V_{s,j}\rho_s h_s-u_i\tau_{ij}\right]=-Q_R \tag{3}$$

4). Vibrational energy equation for m-th vibrational mode

$$\frac{\partial}{\partial t}\left(\mathrm{E}_{v,m}\right)+\frac{\partial}{\partial x_j}\left(\mathrm{E}_{v,m}u_j+q_{v,m,j}+\mathrm{E}_{v,m}V_{m,j}\right)=S_{v,m}, \quad m=1,2,\ldots,N_M, \tag{4}$$

5). Mass conservation law for the s-th reacting mixture component

$$\frac{\partial}{\partial t}\left(\rho C_s\right)+\frac{\partial}{\partial x_j}\left(\rho C_s u_j+\rho C_s V_{s,j}\right)=\dot{w}_s, \quad s=1,2,\ldots,N_C-1 \tag{5}$$

In these equations: ρ - mixture density; u_j - j-th component of the velocity vector; p - pressure; τ_{ij} - shear stress tensor; E - specific total energy; h_s - specific enthalpy for the s-th component; ρ_s - density of the s-th component; $V_{s,j}$ - diffusion speed for the s-th; $q_{v,j}$ - vibrational energy heat flux; $q_{tr,j}$ - translational-rotational energy heat flux; Q_R - heat losses due to radiation; $\mathrm{E}_{v,m}$ - volume specific vibrational energy for m-th vibrational mode; $q_{v,m,j}$ - m-th mode vibrational energy heat flux; $S_{v,m}$ - vibrational energy source; N_M – number of vibrational energy modes; $C_s=\rho_s/\rho$ - mass fraction of the s-th component; $\dot{w}_s$ - s-th component production rate; N_C – number of gas mixture components.

In turbulent flows extra mass, energy and stress fluxes are added. For their calculation K-ε-V_n turbulence model was used [1]. For calculation of the formation rates of the chemical components system of reactions from [2] and [3] were used.

Solution methods for the gas flow system are given in many works (e.g. see [3]-[5]]) and thus are not stated here.

3. Governing equations for particles

The main case considered in this paper is crystallization process of aluminum oxide Al_2O_3 particles in underexpanded jet flow.

Polydisperse mixture of particles is represented as a set of N groups of particles, each of which is characterized by the values of the radius r_α, density $\rho_\alpha=n_\alpha\cdot m_\alpha$, velocity components $u_{\alpha,j}$ and temperature T_α.

For each of N groups (α=1,2,...), under the assumptions that the particles are spherical, chemically inert to the gaseous phase, and do not interact with each other, equations describing the motion of Al_2O_3 particles can be written as [6]

1) Particle continuity:

$$\frac{\partial \rho_\alpha}{\partial t}+\frac{\partial}{\partial x_j}\left(\rho_\alpha u_{\alpha,j}\right)=0 \tag{6}$$

2) Particle momentum:

$$\frac{\partial u_{\alpha,i}}{\partial t}+u_{\alpha,j}\frac{\partial u_{\alpha,i}}{\partial x_j}=C_f\left(u_i-u_{\alpha,i}\right) \quad (7)$$

where $C_f=\frac{3}{8}\frac{C_{D,\alpha}\rho\left|\vec{V}-\vec{V}_\alpha\right|}{r_\alpha\rho_{Al_2O_3}}$; $\rho_{Al_2O_3}$ - Al2O3 density ; $C_{D,\alpha}$ is computed using Henderson's formula [7].

3) Particle energy:

$$\frac{\partial T_\alpha}{\partial t}+u_{\alpha,j}\frac{\partial T_\alpha}{\partial x_j}=\frac{C_q}{C_S}\left(T-T_\alpha\right)+\frac{q_{phase,\alpha}}{C_S\rho_\alpha}-\frac{q_{rad,\alpha}}{C_S\rho_\alpha} \quad (8)$$

where $C_q=\frac{3Nu\cdot\lambda}{2\rho_{Al_2O_3}r_\alpha^2}$; Nu is Nusselt number [8], λ is gas mixture heat conductivity; $q_{rad,\alpha}$ is radiative heat flux.

Phase transition heat is determined the following way:

$$\frac{q_{phase,\alpha}}{\rho_\alpha}=3q_{cr}\frac{\psi^2}{r_\alpha^3}a\left(T_M-T_\alpha\right)^{1.8} \quad (9)$$

where q_{cr} - is latent heat of phase transition; T_M is onset temperature of equilibrium crystallization; ψ is dimensionless radius of crystallization ($0\le\psi\le1$, where 0 corresponds to liquid particle and 1 - to the solid one); a is constant determining the crystallization interphase velocity [9].

As it is rather hard to solve Stephan problem for each of the particles, in this paper the approximate crystallization introduced in model is used.

$$\frac{d\psi}{dt}=-\frac{a}{r_\alpha}\left(T_m-T_\alpha\right)^{1.8}F_{cr}\left(T_\alpha\right) \quad (10)$$

where $F_{cr}\left(T_\alpha\right)$ is a function determining the crystallization onset and termination (for aluminum oxide crystallization begins at temperature being equal $0.83\,T_m$ and terminates at temperature above T_m).

As the problem is considered from the Euler's point of view then the derivative in left-hand side of the equation is considered as substantial.

$$\frac{\partial\psi}{\partial t}+u_{\alpha,j}\frac{\partial\psi}{\partial x_j}=-\frac{a}{r_\alpha}\left(T_M-T_\alpha\right)^{1.8}F_{cr}\left(T_\alpha\right) \quad (11)$$

The liquid phase is assumed to transfer initially to the metastable solid γ phase. The transition process from the γ phase to the stable α phase begins as soon as any portion of the γ phase has appeared. This is described by the equation of the form $\frac{dC_\alpha}{dt}=A\exp\left(-B/T_\alpha\right)$, which yields:

$$\frac{\partial C_\alpha}{\partial t}+u_{\alpha,j}\frac{\partial C_\alpha}{\partial x_j}=A\exp\left(-B/T_\alpha\right) \quad (12)$$

where C_α is α-phase fraction; A,B are constants [10].

As the crystallization process begins and terminates at two different values of temperature, the onset function F_{cr} should depend on the history of particles along their streamlines. In Lagrange domain this is not a problem, while in the Euler domain it is rather tricky to represent F_{cr} correctly, as streamlines are not actually considered in this case.

Two ways could be considered to resolve the problem of F_{cr} non-locality. The first way to go is to actually reconstruct streamlines from the solution data and to use them to determine the value of F_{cr} at a given point. This seems to solve the problem, however, this method has all the disadvantages of the Lagrange approach.

Another way to deal with the problem is to consider F_{cr} as a transported scalar and to solve the appropriate convection equation for it.

$$\frac{\partial F_{cr}}{\partial t} + u_{\alpha,j}\frac{\partial F_{cr}}{\partial x_j} = S_{cr} \tag{13}$$

The initial value of F_{cr} is zero. The source term is zero everywhere except regions with $T_\alpha < 0.83\,T_m$. The maximum value of F_{cr} is limited by one.

In case of strong aftercombustion, the gas temperature can exceed the melting temperature of the particles and the crystallized particles can begin to melt. In this case the melted zone is determined from the following consistency relation:

$$q_{conv,\alpha} = q_{phase,\alpha} \tag{14}$$

4. Numerical scheme for the particle equations

Equations (6)-(8), (11)-(13) have a similar structure in 3d Cartesian frame (x, y, z):

$$\frac{\partial f_\alpha}{\partial t} + u_\alpha\frac{\partial f_\alpha}{\partial x} + v_\alpha\frac{\partial f_\alpha}{\partial y} + w_\alpha\frac{\partial f_\alpha}{\partial z} = S_\alpha \tag{15}$$

where $u_\alpha, v_\alpha, w_\alpha$ are the corresponding particle velocity components.

Let us consider a dimensionless coordinate frame, taking in account the shape of the computation domain:

$$\xi = \xi(x) = x / L, \quad \eta = \eta(x, y, z), \quad \zeta = \zeta(x, y, z) \tag{16}$$

where L is characteristic dimension.

In the new frame steady state form of the equation (15) will be as follows:

$$\frac{\partial f_\alpha}{\partial \xi} + \overline{v}\frac{\partial f_\alpha}{\partial \eta} + \overline{w}\frac{\partial f_\alpha}{\partial \zeta} = \overline{S}_\alpha \tag{17}$$

where $\overline{v} = L\eta_x + \dfrac{L\eta_y v_\alpha}{u_\alpha} + \dfrac{L\eta_z w_\alpha}{u_\alpha}$, $\overline{w} = L\zeta_x + \dfrac{L\zeta_y v_\alpha}{u_\alpha} + \dfrac{L\zeta_z w_\alpha}{u_\alpha}$, $\overline{S}_\alpha = \dfrac{LS_\alpha}{u_\alpha}$

The numerical solution of equation (17) is performed with the explicit-implicit predictor-corrector method (explicit for convection and implicit of the source).

Predictor step:

$$(f_\alpha)_{j,k}^{\overline{n+1}} = (f_\alpha)_{j,k}^{n} + \Delta\xi\overline{S}_\alpha^{\,n} - \frac{\Delta\xi\overline{v}}{\Delta\eta}\left[(f_\alpha)_{j+1,k}^{n} - (f_\alpha)_{j,k}^{n}\right] - \frac{\Delta\xi\overline{w}}{\Delta\zeta}\left[(f_\alpha)_{j,k+1}^{n} - (f_\alpha)_{j,k}^{n}\right] \tag{18}$$

Corrector step:

$$\begin{gathered}(f_\alpha)_{j,k}^{\overline{\overline{n+1}}} = (f_\alpha)_{j,k}^{n} + \Delta\xi\overline{S}_\alpha^{\,n+1} - \frac{\Delta\xi\overline{v}^{\overline{n+1}}}{\Delta\eta}\left[(f_\alpha)_{j,k}^{\overline{n+1}} - (f_\alpha)_{j-1,k}^{\overline{n+1}}\right] - \frac{\Delta\xi\overline{w}^{\overline{n+1}}}{\Delta\zeta}\left[(f_\alpha)_{j,k}^{\overline{n+1}} - (f_\alpha)_{j,k-1}^{\overline{n+1}}\right], \\ (f_\alpha)_{j,k}^{n+1} = 0.5\left[(f_\alpha)_{j,k}^{\overline{\overline{n+1}}} + (f_\alpha)_{j,k}^{\overline{n+1}}\right]\end{gathered} \tag{19}$$

By linearizing the source $\overline{S}_\alpha^{\,n+1} = \overline{S}_\alpha^{\,\overline{n+1}} + \left(\dfrac{\partial \overline{S}_\alpha}{\partial f_\alpha}\right)^{\overline{n+1}}\left[(f_\alpha)_{j,k}^{n+1} - (f_\alpha)_{j,k}^{\overline{n+1}}\right]$, we get the solution value at (n+1) step for ξ coordinate:

$$\left(f_\alpha\right)_{j,k}^{n+1}=\frac{0.5\left[\left(f_\alpha\right)_{j,k}^{n}+\left(f_\alpha\right)_{j,k}^{\overline{n+1}}+\Delta\left(f_\alpha\right)_{j,k}^{\overline{n+1}}-\Delta\xi\left(\frac{\partial\overline{S}_\alpha}{\partial f_\alpha}\right)^{\overline{n+1}}\left(f_\alpha\right)_{j,k}^{\overline{n+1}}\right]}{1-0.5\Delta\xi\left(\frac{\partial\overline{S}_\alpha}{\partial f_\alpha}\right)^{\overline{n+1}}} \tag{20}$$

where $\Delta\left(f_\alpha\right)_{j,k}^{\overline{n+1}}$ is the explicit function increment at the corrector step:

$$\Delta\left(f_\alpha\right)_{j,k}^{\overline{n+1}}=\Delta\xi\overline{S}_\alpha^{\overline{n+1}}-\frac{\Delta\xi\overline{v}^{\overline{n+1}}}{\Delta\eta}\left[\left(f_\alpha\right)_{j,k}^{\overline{n+1}}-\left(f_\alpha\right)_{j-1,k}^{\overline{n+1}}\right]-\frac{\Delta\xi\overline{w}^{\overline{n+1}}}{\Delta\zeta}\left[\left(f_\alpha\right)_{j,k}^{\overline{n+1}}-\left(f_\alpha\right)_{j,k-1}^{\overline{n+1}}\right] \tag{21}$$

The continuity equation has the following form in the frame (ξ,η,ζ):

$$\frac{\partial}{\partial t}\left(\frac{\rho_\alpha}{J}\right)+\frac{\partial E}{\partial\xi}+\frac{\partial F}{\partial\eta}+\frac{\partial G}{\partial\zeta}=0 \tag{22}$$

where $E=\frac{\xi_x\rho_\alpha u_\alpha}{J}$, $F=\frac{\rho_\alpha}{J}\left(\eta_x u_\alpha+\eta_y v_\alpha+\eta_z w_\alpha\right)$, $G=\frac{\rho_\alpha}{J}\left(\zeta_x u_\alpha+\zeta_y v_\alpha+\zeta_z w_\alpha\right)$ (J is the Jacobian).

In steady state:

$$\frac{\partial E}{\partial\xi}+\frac{\partial F}{\partial\eta}+\frac{\partial G}{\partial\zeta}=0 \tag{23}$$

which can be solved by explicit Korte method [5].

5. Results

The developed method was verified in three case-studies of multiphase jet flow.

5.1. Case 1

The first case was described in [8]. The jet has 0.4 atm (5.9 psia) static pressure and 2500 K static temperature at the nozzle exit (Mach number is 2.35). Three particulate groups are considered: 1.5 μm, 3 μm and 6 μm.

The results for gas and particulate temperature in comparison with the results of [8] are shown in figure 1.

Figure 2 shows axial profiles of alpha-phase fraction. These results are in good agreement with the results of [8], where crystallization of 1.5 μm particles begins approximately at a distance of 16 nozzle radii and fraction of alpha-phase reaches ~$1.6\cdot10^{-3}$ (see figures 6 and 7 of [8]).

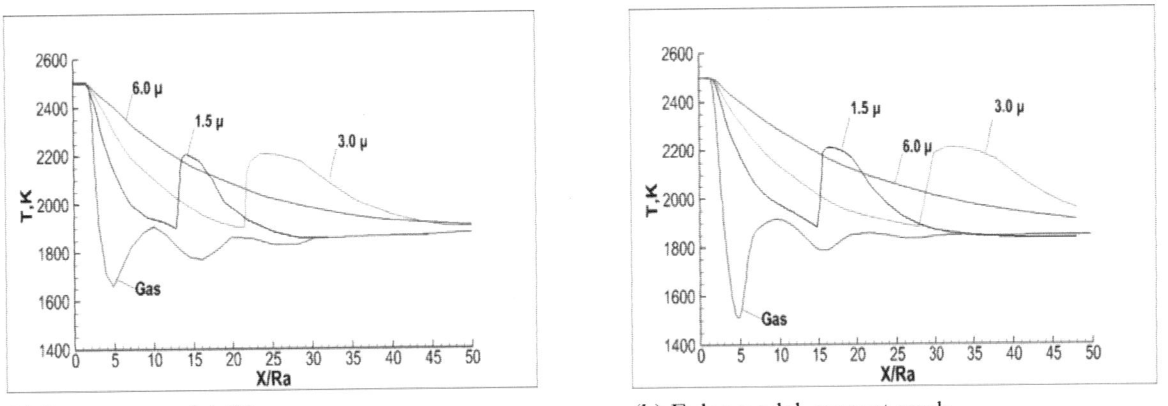

(a) Lagrange model [8] (b) Euler model, present work

Figure 1. Axial profiles of the gas and particle temperatures on the jet centerline.

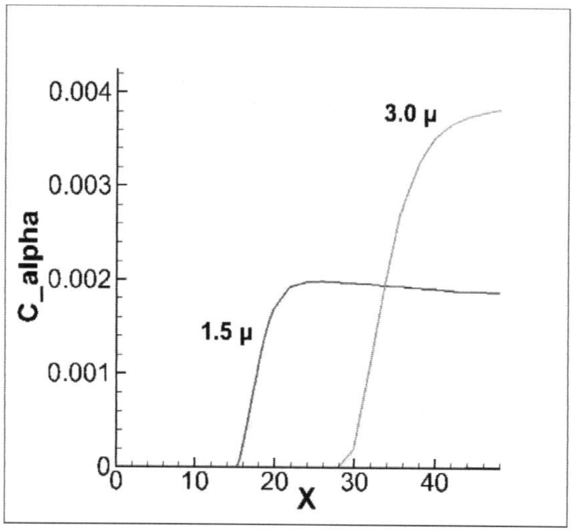

Figure 2. Axial profiles of alpha phase fraction on the jet centerline (Euler model, present work).

5.2. Case 2

The case was described in the work of F.Zavelevich et al [6]. It represents a typical solid rocket engine plume with engine thrust 12 tons at 31 km altitude.

The computation results in comparison with the results of [6] are shown in figure 3. As it can be seen, they are in good agreement.

Here one can see a very important practical result: particles "ignite" the gas mixture intensifying afterburning. In non-coupled calculations the afterburning cannot be seen.

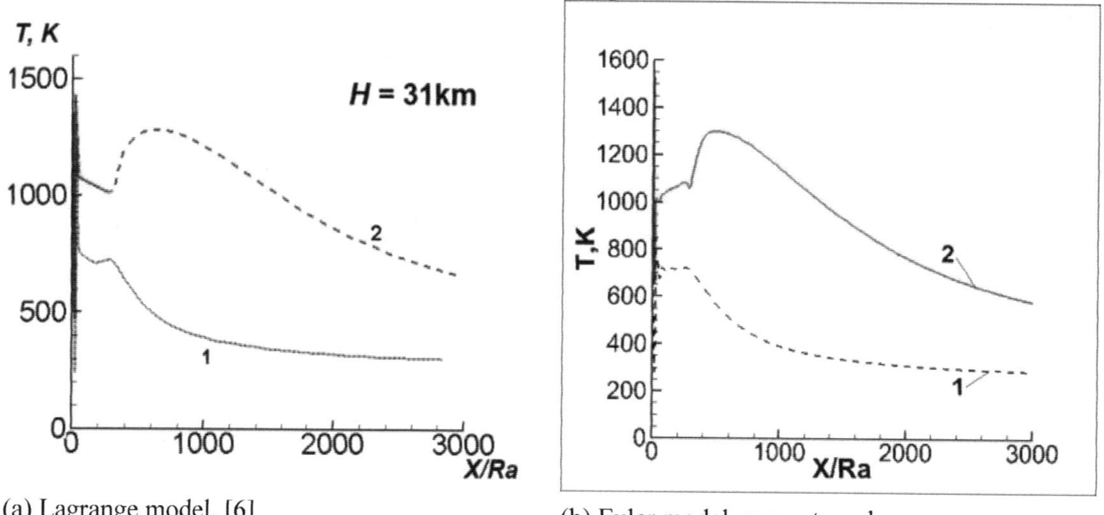

(a) Lagrange model, [6] (b) Euler model, present work

Figure 3. Axial temperature distribution in plume of typical solid rocket engine at 31 km altitude. 1- calculation results without gas-particles coupling, 2- two-way coupled calculation.

5.3. Case 3

The previous calculations were performed for 2d jets. In 3d case with cross-flow (even when cross-flow angles are insignificant) Lagrange simulations are connected with certain difficulties because particulate trajectories can become rather complex, which causes troubles with computation of particulate density. The Euler model, on the other hand, does not have this problems and thus is considered preferable for cross-flow cases.

Figure 4 shows particulate density distribution for typical solid rocket engine with 55 tons of thrust at 50 km altitude.

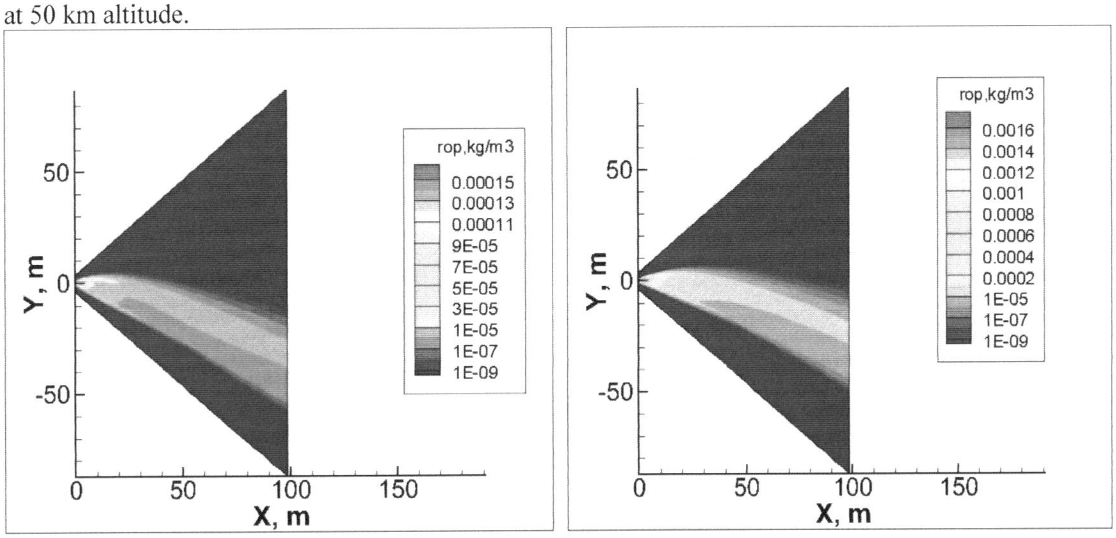

(a) 0.5 μm particles (b) 6 μm particles

Figure 4. Particulate density distribution in XY plane for jet in cross-flow (cross-flow angle is 30°).

As it can be seen, smaller particles are more carried away by the cross-flow than the larger ones.

6. Conclusions

This work has demonstrated that the Euler technique can be successfully implemented in investigation of multiphase flows with phase changes in the particulate phase.

The calculation results presented here show good agreement with the works of other authors.

The approach developed in this paper can easily be used for different types of particulate media undergoing phase changes provided that the crystallization laws are known.

References

[1] Molchanov A.M. and Bykov L.V. Three-Equation K-e-Vn Turbulence Model for High-Speed Flows. 2013, *AIAA Paper* **2013-3181**, 30p.

[2] Connaire M.O., Curran H.J., Simmie J.M., Pitz W.J., Westbrook C.K. A Comprehensive Modeling Study of Hydrogen Oxidation. *International Journal of Chemical Kinetics*, 2004, **Vol.36**, P.603-622.

[3] Molchanov A.M. Numerical Simulation of Supersonic Chemically Reacting Turbulent Jets. 2011, *AIAA Paper* **2011-3211**, 37p.

[4] MacCormack, R.W. and Pulliam, T., Assessment Of A New Numerical Procedure For Fluid Dynamics. 1998, *AIAA paper* **98-2821** Presented at the 29th AIAA Fluid Dynamics Conference, Albuquerque, New Mexico, June 15-18, 9p.

[5] Korte J.J. An explicit, upwind algorithm for solving the parabolized Navier-Stokes equations. 1991, *NASA Technical Paper* **3050**, 70p.

[6] Zavelevich F.S., Molchanov A.M., Ushakov N.N., Computation of gas and multiphase supersonic jets with nonequilibrium processes, 2015, *Journal of Thermophysics and Heat Transfer*, v.29. No.3. p.587-593.

[7] Henderson, C. B., Drag Coefficient of Spheres in Continuum and Rarefied Flows. 1976, *AIAA Journal*, Vol. 14, No. 6, June 1976, pp. 707-708.

[8] Rodionov A. V., Plastinin Yu A., Drakes J. A. , Simmons M. A., and Hicrs III R. S., Modeling of Multiphase Alumina-Loaded Jet Flow Fields.1998, *AIAA Paper* **98-3462**, 16p.

[9] Plastinin, Yu.A., Sipachev H.Ph., Karabadzhak, G.F., Khmelinin, B.A., Khlebnikov, A.G., Shishkin, Yu.N. Influence of Alumina Particles' Phase Transition on Its Radiation in the Middle Infrared and Ultraviolet Regions of Spectrum. *AIAA Paper* **2000-0735**, 21p.

[10] Rodionov A.V., A Godunov's method modification for prediction of supersonic reacting turbulent jets. 1995, *La Recherche Aerospatiale*, No 4, pp. 263-276.

Layer-by-layer ordering in parallel finite element composition on shared-memory multiprocessors

A K Novikov, N K Piminova, S P Kopysov and YA Sagdeeva

Institute of Mechanics, Ural Branch of the Russian Academy of Sciences, 34 T. Baramzinoy, Izhevsk, Russia 426067

E-mail: sc_work@mail.ru

Abstract. In this paper, we present new partitioning algorithms for unstructured meshes that prevent conflicts during parallel assembling of FEM matrices and vectors in shared memory. These algorithms use a ratio which we introduce to determine if any two mesh cells are adjacent. This adjacency ratio defines mesh layers, which are combined into domains and assigned to different parallel processes/threads. The proposed partitioning algorithms are compared with the existing algorithms on quasi-structured and unstructured meshes by the number of potential conflicts and by the load imbalance.

1. Introduction

Modern parallel computations are performed on multi-core processors. At the same time, there is an intensive use of shared memory, which imposes additional restrictions on the mesh partitioning and mesh data related to localization and access synchronization. If shared memory is used, than a bottleneck of parallel algorithms for numerical methods are assembly operations (summation).

The finite element method such operations are: assembly of global matrices and / or vectors (in the element-by-element schemes) from local matrices and vectors, respectively, the numerical integration of local matrices and vectors. During the assembly operation of the vector components or matrix entries are summed, corresponding to the vertices shared to several finite elements. In parallel assembly, simultaneous summation vector components or matrix entries for several processes / threads leads to conflicts in the shared memory, and as a result — computational errors.

To eliminate these errors, different approaches are applied: at the level of technology of parallel programming — atomic operations and critical sections, memory allocation for storing shared data [1]; algorithm level — the change element-by-element assembly scheme at nodal [2] and the introduction of additional operations — the use of different variants for ordering mesh [3]. An ordering of mesh cells are usually carried out on the basis of the graphs coloring, which is traditionally used for the ordering of unknowns in parallel solving systems of algebraic equations [4].

Coloring algorithms allow to obtain sets of unconnected mesh cells, but are NP-complexity, since they depend not only on the number of graph vertices, but also on their degrees, and the choice of the initial vertex coloring.

In order to obtain a uniform distribution of mesh cells between parallel processes/threads after cells colorization balancing procedure is necessary, and reordering the mesh for access to the memory location.

An ordering of mesh and assembly are considered as interdependent parts of the finite element composition and use variations of layer-by-layer mesh partitioning, whereby during assembly operations parallel computing processes are not treated at the same time to the finite elements containing shared vertices.

Variants of this partition based on mesh partitioning at finite elements layers with the thick in one cell and then combining these layers into domains of finite elements assigned one thread or processor core. Layers of finite elements are defined by using introduced adjacency ratio of cells by the nodes. The algorithm of layers formation is NP-complexity.

In this paper we investigate the variants of layer ordering unstructured meshes with cells of arbitrary shape. The resulting ordering is compared with known algorithms on quasi-structured and unstructured meshes by the number of potential conflicts in the assembly and the distribution uniformity of mesh cells on the mesh subdomains.

2. Layer-by-layer partitioning of finite element meshes

Finite element meshes are unstructured meshes, that is the number of a node (cell) cannot be defined by incrementing the number of his neighbor node (cell). In [5] the following adjacency ratio was introduced for element-by-element schemes of FEM

$$\omega_k \in Adj(\omega_j), \qquad if\ V(\omega_j) \cap V(\omega_k) \neq 0, \tag{1}$$

where ω_j, ω_k are the cells of the mesh Ω; $Adj(\cdot)$ is the adjacency operator; $V(\omega_j)$ and $V(\omega_k)$ are the subsets of mesh nodes that correspond to the cells ω_j and ω_k.

It is important to note that for this adjacency ratio, the cell has a large number of adjacent cells m_e, even for a quasi-structured three-dimensional mesh as shown in figure 1.

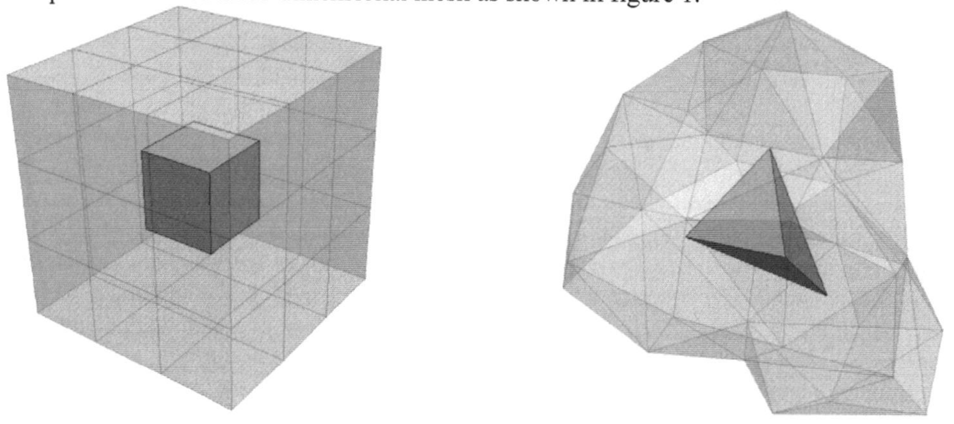

(a) (b)

Figure 1. Adjacent cells: (a) quasi-structured mesh, $m_e = 26$; (b) unstructured mesh, $m_e = 119$.

Thus, the algorithm of layer-by-layer partitioning of a finite-element mesh consists of two steps: 1) forming layers of mesh cells; 2) combining layers of mesh in subsets. Let us discuss these steps in more detail.

3. Forming layers of mesh cells

To form mesh layers by (1) the nodes are checked to belong to finite elements. Generally, unstructured meshes are stored so that the cell contains a mesh node, not the node belongs to a cell. To check which finite elements the node belongs to, two loops over mesh cells are need for $O(2 \cdot m \cdot n_e)$ operations, where m — the number of finite elements; n_e — the number of nodes in the final element.

The algorithm for generating the mesh layers [5] can be represented in the form:

(i) Set the initial layer $s_1 = \left\{\omega_j^{(i)}\right\}, j = 1,2,\dots,m_1$, where m_1 — number of cells in s_1.

(ii) Layers $s_i, i = 2,3,\dots,n_s$ determined by the expression

$$s_i = \begin{cases} Adj(s_{i-1}), & i = 2; \\ Adj(s_{i-1}) \setminus s_{i-2,} & i > 2, \end{cases}$$

where $s_i = \left\{\omega_j^{(i)}\right\}$ — a plurality of cells (finite element) in the layer i; $\Omega_j^{(i)}$ — cell with number j in the layer i; $Adj(s_i) = \bigcup_{j=1}^{m_i} Adj\left(\omega_j^{(i)}\right)$ — a set of cells adjacent to s_i; m_i — number of cells in s_i. This algorithm is also applied to the separation of disconnected subsets of cells within a layer and to the search of discontinuous layers.

If in the element-by-element schemes FEM, the existence of discontinuous layers is not essential for the method of Schur complement [6] is necessary to ensure not only the continuity of layers, but also the cells adjacency by faces. Search cells having one common vertex, and the definition of discontinuous sub-layers are carried out on the basis of the neighborhood along the faces of the cell, which is used in the determination of the previously built neighborhood on the mesh nodes (1). For each cell are considered neighbors by vertex, and excluded those with which the number of vertices in common is not equal to the number of vertices on the same face n_f (three — for tetrahedral cells and four — for hexahedral).

Determination of discontinuous layers on faces neighborhood viewpoint is as follows. Take an arbitrary cell in the current layer s_i and start searching across neighboring cells. After searching we check that all the cells were in the set of reachable vertices. If not, then all the cells trapped in a set of cells are placed in the sublayer s_j^* and the new search begins. Thus is formed a set of sublayers $\left\{s_j^*\right\}$ for a discontinuous layer s_i.

To exclude cells having one common vertex and to merge them with some layer s_i we check if this layer is discontinuous by faces. If layer has more than one sublayer, we look for the sublayer that has maximum number of cells. This sublayer is considered to be current layer s_i, and other sublayers merge with the previous layer s_{i-1}.

Examples of the layers for quasi-structured (fourth part of the membrane) and unstructured mesh (frame construction) are presented in the figures 2a and 2b, respectively.

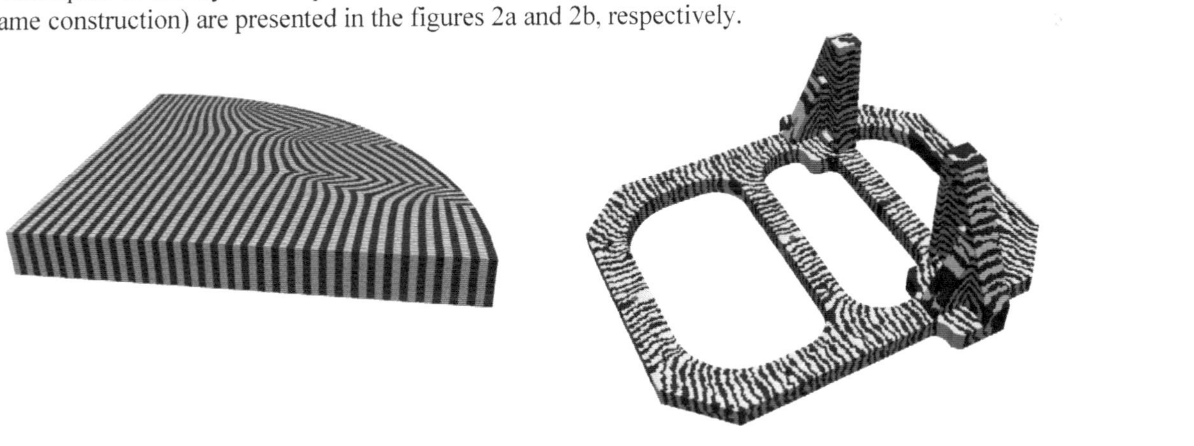

(a) (b)

Figure 2. Layers for meshes: (a) mesh of the hexagonal cells, $m = 31744$, $n_s = 64$; (b) mesh of the tetrahedral cells, $m = 485843$, $n_s = 137$.

In the case of quasi-structured mesh (variant (x, y, z_{min})) received thirty two perfectly balanced layers: $m_{max} = m_{min} = 15872$ cells. For unstructured mesh (variant (x, y_{min}, z)) the maximum number of cells in a layer — 46635 for $m = 485843$, and minimum — 17 cells, the number of layers $n_s = 47$. These variants are marked by a choice of the initial layer s_1. For example, variant (x_{min}, y, z) means that at least one vertex in a cell of the layer s_1 has the abscissa equal to x_{min}.

Further, the layers and their order are used in the formation of the mesh subdomains.

4. Combining layers of mesh in subdomains

Several variants for the formation of the subdomain $\Omega_i, i = 1,2,\dots,n_\omega$ layers from $s_j, j = 1,2,\dots,n_s$ unstructured mesh Ω was considered: block; by parity numbers of layers and their modifications.

Block variant ("block") is as follows: Combining of layers $\Omega = \bigcup_{i=1}^{n_s} s_i$; partitioning of Ω on a subdomains Ω_i, such that $|\Omega_i| \approx m / n_\omega, i = 1,2,\dots,n_\omega$, where n_ω — the number of subsets of cells or mesh subdomains (figure 3).

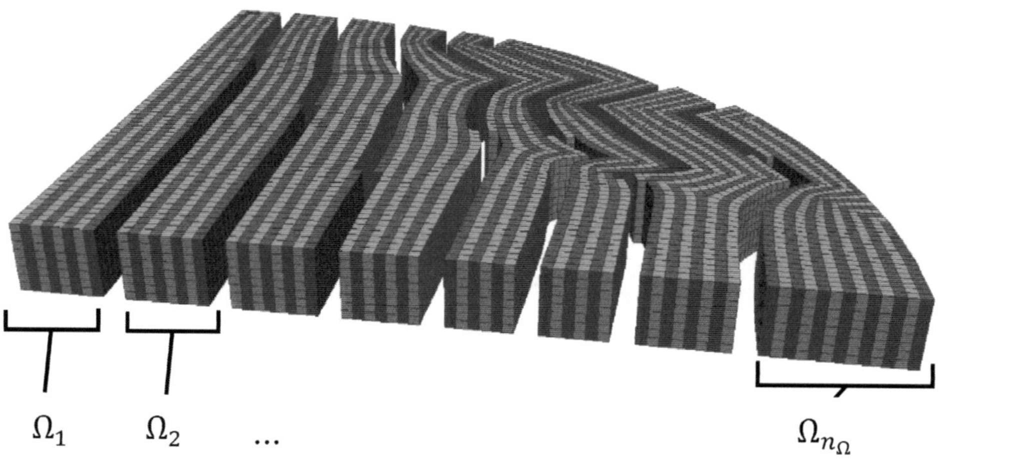

Figure 3. Subdomains are obtained in block form of the construction

As shown in figure 3, in block form, layers of unstructured mesh partitioned into parts belonging to two adjacent subdomains. For example, the subdomain Ω_1 includes seven layers $s_1, s_2, \dots, s_7$ and most of the layer s_8. Due to the division of the layers, subdomains are obtained equal by the number of cells and respectively load balancing on parallel computing processes is achieved in the future. However, the allocation of parts of the layers requires the control of the adjacency ratio. The block partition requires at least two layers in the subdomain to avoid computational errors related to the conflict with parallel access to data associated with mesh. Otherwise it is necessary to specify the order of cells in the layer, for example, in coordinate directions.

Combining of layers by the parity of numbers is as follows. First, in a subdomain $\Omega_i, i = 1,2,\dots,n_\Omega$ unite all layers with odd numbers, and then the even (figure 4a). In the subdomain ω_i is placed layers (odd-numbered or even-numbered) with the numbers $i, i + n_\Omega, i + 2n_\Omega, \dots$ for example, the subdomain Ω_1 in figure 4b.

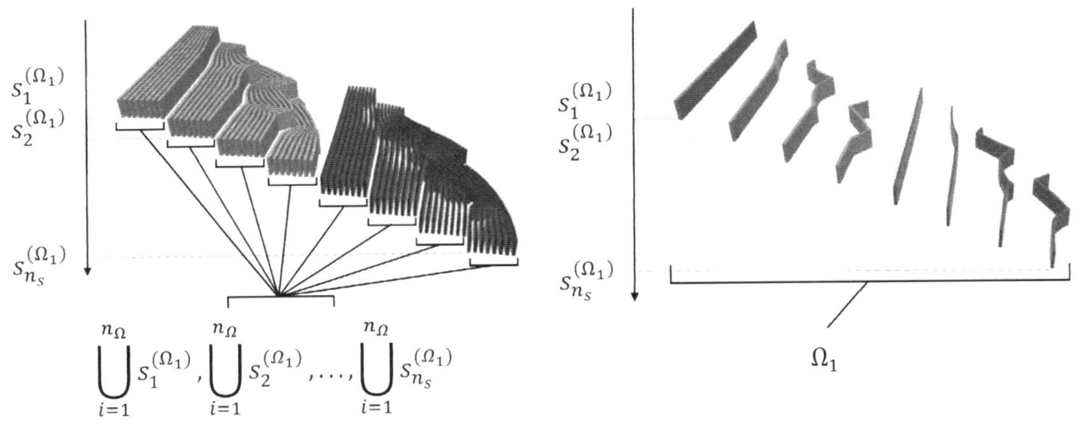

(a) (b)

Figure 4. Subsets of cells are formed by parity numbers of layers and distribution of layers on subdomains Ω_i: (a) the layers belonging to the eight subdomains ($n_\Omega = 8$), which cells will be simultaneously involved in parallel computing; (b) the sequence of layers in the subdomain Ω_1.

The results for load balance are based on the even or odd layers ("even/odd") union approach was used and providing a balanced distribution mesh cells on subdomains. In this case, the layers $s_1, s_3, s_5, \ldots$ are united in a subdomain $\Omega^{(even)}$("even") and layers $s_2, s_4, s_6, \ldots$ — into subdomain $\Omega^{(odd)}$. Further $\Omega^{(even)}$ is partitioned on n_Ω subdomains $\Omega_i^{(even)}$, such that $\left|\Omega_i^{(even)}\right| \approx \left|\Omega^{(even)}\right| / n_\Omega$ and corresponding to $\Omega^{(odd)}$ on n_Ω subdomains $\Omega_i^{(odd)}$, for which $\left|\Omega_i^{(odd)}\right| \approx \left|\Omega^{(odd)}\right| / n_\Omega$.

On figures 5 are presented obtained mesh subdomains for $n_\Omega = 8$. As can be seen from these figures, the layer-by-layer partitionig in contrast to the k-way, it allows to order domains connections (the order of colors in figure 5). In addition, in layer-by-layer variants, nodes can belong to no more than two mesh subdomains that does not guarantee, such as multi-level algorithm from the METIS library [7] for large values of n_Ω on adaptive unstructured meshes.

(a) (b) (c)

Figure 5. The partitioning of unstructured mesh into eight subdomains, $n_\Omega = 8$: (a) in block combining; (b) by combining the parity; (c) k-way algorithm from package METIS.

Discontinuous subdomains were obtained for multiply connected domain, as in the layer-by-layer so multi-level mesh partitioning. As noted earlier, in the parallel element-by-element of finite element discontinuous mesh subdomains do not affect the stability of computational algorithms, but slow memory access. In many algorithms are need to combine parts or discontinuous subdomains maintaining the total number of subdomains, or to treat these parts as independent subdomains with an increase a total number of them. The proposed ordering of layers and subdomains simplifies the procedure of combination with the neighboring subdomains, which involved a smaller number of adjacent subdomains than in [8].

Table 1 presents potential conflicts in the assembly of vector for several variants of mesh partitioning. As an estimate of possible conflicts with the parallel implementation of the operation of finite element composition is considered the number of cells to the nodes which are accessed in the adjacent subdomains of the mesh.

In variant "by generation" used the cells order obtained by mesh generation in the mesh generator. The mesh is divided into a predetermined number of subdomains n with about an equal number of cells in the subdomains $\Omega_i \approx m \ / \ n_\Omega$.

Table 1. The number of possible conflicts in the assembly of vector.

Partitioning variants						
n_Ω	"by generation"	METIS	"block"	"even/odd"		"even+odd"
				"even"	"odd"	
Quasi-structured mesh, $m = 507904$						
8	0	0	0	0	0	0
16	0	384/0	0	0	0	0
32	0	24/0	0	0	0	0
60	0	60/2	0	0	0	0
Unstructured mesh, $m = 485843$						
8	284/88	7/2	0	0	0	0
16	623/198	13/4	0	0	0	0
32	1144/352	60/25	0	0	0	0
60	2425/815	169/71	167/40	0	0	4/3

The numerator shows the number of cells, to nodes that are accessed from different mesh subdomains (threads), in the denominator — the number of cells in which the treatment of the different mesh subdomains happens to nodes with the same local (inside the cell) numbers. The estimate is presented in the denominator, more strict, based on the fact that the parallel processes (threads) begin at the same time, and access to each data cell occurs for equal period of time from any of the parallel process. If this estimate is different from zero (see table 1), then the parallel assembly of finite element vectors occurs to computational errors.

Errors of computing related to the conflicts in the shared memory can not be observed due to the ordering of cells, resulting in the mesh generation, as in the case of quasi-structured mesh (table 1, the second column). When unstructured mesh is partitioned by parity of layers ("even+odd") the occurrence of conflicts is associated with an imbalance in the number of layers of cells. For such meshes access to the mesh cells and related data is carried out in a sequence of two parallel regions OpenMP ("even/odd"), separately for the even ("even") and odd ("odd") layers. When the block partitioning conflicts are the result of discontinuity of the layers to ensure a balanced mesh partitioning and the absence of cells ordering in layers. Combining entire layers eliminates these conflicts, but leads to a much less balanced partitioning.

Acknowledgments
The reported study was partially funded by RFBR according to the research projects №14–01–00055-a, №16–01–00129-a.

References

[1] Fuhry M, Giuliani A and Krivodonova L 2014 *International Journal for Numerical Methods in Fluids* **76** 982

[2] Kopysov S P, Kuzmin I M, Nedozhogin N S, Novikov A K, Rychkov V N, Sagdeeva Y A and Tonkov L E 2014 *Computer Research and Modeling* **6** 79

[3] Komatitsch D, Micha D and Erlebacher G 2009 *J. Parallel Distr. Comp.* **69** 451

[4] Adams L and Ortega J 1982 *Int. Conf. Parallel Processing* p 53

[5] Kopysov S, Novikov A and Piminova N 2016 *Proc. of the 10th Ann. Inter. Sc. Conf. on Parallel Comp. Tech. (PCT 2016) (Chelyabinsk, Russia: South Ural State University) p 555*

[6] Kopysov S P, Kuzmin I M, Nedozhogin N S, Novikov A K and Sagdeeva Y A 2013 *Lecture Notes in Comp. Sci.* **7979** 65

[7] Karypis G and Kumar V 1998 *SIAM Journal on Scientific Computing* A **20** 359

[8] Golovchenko E and Yakobovskiy M V 2015 *Russian Supercomputing Days: Proceedings of the Int. conf. (Moscow.: Moscow State University)* p 303

The study of the wedge-shaped vibration-driven robot motion in a viscous fluid forced by different oscillation laws of the internal mass

A N Nuriev[1], **O S Zakharova**[2], **O N Zaitseva**[3] **and A I Yunusova**[4]

[1]Research Institute of Mechanics, Lobachevsky State University of Nizhnij Novgorod, Nizhnij Novgorod 603950, Russia
[2]Lobachevsky Institute of Mathematics and Mechanics, Kazan Federal University, Kazan 420008, Russia
[3]Higher Institute of Information Technologies and Information Systems, Kazan Federal University, Kazan 420008, Russia
[4]Department of Informatics and Applied Mathematics, Kazan National Technological University, Kazan 420015, Russia

nuriev_an@mail.ru

Abstract. A rectilinear motion of a two-mass system in a viscous incompressible fluid is considered. The system consists of a shell having the form of an equilateral triangular cylinder and a movable internal mass. The motion of the system as a whole is forced by longitudinal oscillations of the internal mass relative to the shell. This mechanical system simulates a vibration-driven robot, i.e. a mobile device capable to move in a resistive medium without external moving parts. Investigation of the system is carried out by a direct numerical simulation. A comparative analysis of the characteristics of the motion and flow regimes around the vibration-driven robot is carried out for different internal mass oscillation laws.

1. Introduction

Construction of vibration-driven mechanisms has long been of interest to engineers. Descriptions of numerous devices with vibrational motive power have appeared in the literature since the first half of the 20th century (see review [1]). Today the vibrational motion is a rapidly growing branch of applied mechanics and robotics.

The simplest model of a vibration-driven mechanism can be presented as a two-mass system consisting of a closed shell and movable internal mass. Devices of such architecture are often called vibration-driven robots. In a resistive medium the motion of the system as a whole is forced by longitudinal periodic oscillations of one body (internal mass) relative to another (shell). Such a principle of movement seems to be expedient for mini and micro-robots. Tightness, lack of moving external parts (screws, wheels, caterpillar tracks and etc.) are the properties, which allow using the robots for non-destructive inspections of miniature technical objects such as thin-walled pipes of small diameter, as well as in medicine, as was mentioned by several authors [2, 3].

Research of movement capabilities of vibration-driven robots were conducted earlier for medium with different resistance laws. In [4-5] the possibility of motion in an ideal fluid with related deformations of the outer shell was considered, articles [6, 7] were devoted to the study of the movement on a rough plane in the presence of a Coulomb friction, in [8] the movement on the liquid interface was studied. Movement in the Newtonian fluid was considered in [9-14].

Despite the simple architecture, the motion control of the two-mass vibration-driven system is a non-trivial problem. The movement of such device in a viscous fluid is determined by the fluid-shell interaction, which depends on the motion law of the internal mass, the geometry of the shell and the range of Reynolds number. In this paper, we consider the problems of improving the efficiency and speed of movement due to the choice of a special law of internal mass motion of the wedge-shaped vibration-driven robot at the low Reynolds number range. For these purposes, a comparative analysis of the characteristics of the motion and flow regimes around the robot are carried out for the simple harmonic law of the internal mass motion and the special two-phase law of the internal mass motion (characterized by the alternation of fast and slow motion phases), that was obtained in [13] as the solution of energy optimization problem on the basis of a simplified model of fluid-shell interaction.

Motion simulation is carried out on the basis of the OpenFOAM software package [15]. The calculations involve both original and modified modules of the package. Calculations are carried out on high performance clusters of Kazan Federal University.

2. Formulation of the problem

2.1. Two-mass system motion

Consider a system consisting of two rigid bodies. The main body (the shell) of mass M is located in a viscous incompressible fluid, while the second body of mass m (referred to below as the "internal mass") is moving within it. The longitudinal periodic motions of the internal mass relative to the body M with which the entire system moves as a whole are investigated. The velocity of the body M will be denoted by u_M, and the displacement and velocity of the internal mass relative to the body M will be denoted by s and $v = \dot{s}$ respectively. The equations of motion of the internal mass and body M in a fixed system of coordinates have the form:

$$m(\dot{u}_M + \dot{v}) = -G, \quad M\,\dot{u}_M = G + F \tag{1}$$

where F is the resistance of the medium to the motion of the body, G is the force of interaction of the internal mass and the body M. Eliminating the force G from (1) and normalizing the velocity u to the amplitude of internal mass oscillations U_0, the time t to RU_0^{-1}, where R is the characteristic size (length of the side of the shell) of the robot, we obtain the basic equation of motion of the two-mass system:

$$\dot{u}_M = -\mu_2 \dot{v} + \mu_1 \frac{R^2}{S} F \tag{2}$$

Here μ_2 is the ratio of the movable internal mass to the total mass of the robot ($\mu_2 = m/(M+m)$), μ_1 is the ratio of the mass of the viscous fluid, occupying the same volume as the robot, to the mass of the robot ($\mu_1 = M_f/(M+m)$), S is the cross-sectional area.

2.2. Fluid-structure interaction

The fluid flow around the vibration-driven robot generated by its motion is governed by the Navier-Stokes system of equations. Normalizing the spatial coordinates, time and velocity by R, RU_0^{-1}, U_0 respectively, we get the dimensionless formulation of the governing system in the following form:

$$\frac{\partial U}{\partial t} + U \cdot \nabla U = -\nabla p + \frac{1}{\mathrm{Re}} \Delta U, \quad \nabla \cdot U = 0 \tag{3}$$

Where $U=(u,v)$ is the dimensionless velocity, p is the dimensionless pressure, $\mathrm{Re}=U_0R/\nu$ is the Reynolds number.

For solving this problem numerically it is convenient to rewrite governing equations in a moving (non-inertial) coordinate system associated with the shell. To retain the governing system in the form (3), we determine a new pressure as

$$p=\tilde{p}+x\dot{w}.$$

Here the first term is the pressure in the fixed coordinate system, the second term is the inertial contribution, $\dot{w}$ is the acceleration of the moving coordinate system.

On the surface of the vibration-driven robot the no-slip conditions are fulfilled:

$$u\,|_c=v\,|_c=0. \tag{4}$$

For specifying conditions at infinity we rewrite (2) in the non-inertial coordinate system. Because the acceleration in the fixed coordinate system equals the sum of the acceleration in the moving coordinate system and acceleration of the moving coordinate system, we get

$$\dot{u}\,|_\infty=\mu_2\dot{v}-\mu_1\frac{R^2}{S}F,\ \dot{v}\,|_\infty=0 \tag{5}$$

The forces acting on the vibration-driven robot by a viscous fluid in the dimensionless formulation are calculated as

$$F_p=\int_S pnds-\int_S\bar{\sigma}\cdot nds,$$

where $\bar{\sigma}$ is the viscous stress tensor, S is the surface of the robot and n is the surface normal vector. The force vector F_p can be expanded into a vertical component F_y (lift force) and a horizontal component F_x, which consists of the drag force and inertia force. The inertial force acting on the cylinder due to the fluid acceleration can be split into two parts: the inertia force of added mass, arising due to the local acceleration near the cylinder, and the Froude-Krylov force, which is related to the pressure gradient created in the fluid by the flow oscillations. The Froude-Krylov force for the considered case can be calculated as

$$F_{fk}=\int_S x\dot{w}nds. \tag{6}$$

Taking into account (6), the infinity boundary condition (5) can be rewritten

$$\dot{u}\,|_\infty=\mu_2\dot{v}-\mu_1\frac{R^2}{S}(F_x-F_{fk}) \tag{7}$$

The system of equations (3), (4) and (7) completely describes the motion of vibration-driven robot of arbitrary shape in a viscous fluid for a given law of the internal mass motion.

3. Numerical model

3.1. Spatial and time discretization

The numerical solution is carried out using the OpenFOAM software package. The computational domain is constructed as a rectangular parallelepiped measuring 50×30×1 with an equilateral triangular cylinder in the center of it. For modelling the plane flow a three-dimensional domain is used in accordance with the software features. In chosen Cartesian coordinate system the edges of the parallelepiped are parallel to the main axes, and the plane of flow is parallel to the plane xOy.

For the discretization of the computational domain we use block-structured grids, generated by blockMesh utility of OpenFOAM. Each grid block is decomposed into disjoint hexahedral cells. The decomposition in the z-axis direction consists of one cell, which causes the flow in this direction not to be calculated due to the two-dimensional problem formulation.

The discretization of the equations of fluid flow in the OpenFOAM package is carried out by the finite volume method (FVM) in the Cartesian coordinate system. The discrete values of the velocity and pressure are localized in the cell centers of the computational grids. To calculate the volume integrals over the control volume the general Gauss procedure is used. For the approximation of the pressure gradient the linear interpolation is applied. To interpolate the variables in convective terms nonlinear NVD (normalized variable diagram) scheme "Gamma" is used. In the diffusion terms, when the discretization of the Laplace operator is carried out, the normal velocity gradients on the surface of the cells are approximated by using a symmetric second-order scheme with non-orthogonality correction (for more details see [16, 17]).

The implicit Euler scheme is used for the time discretization of the system of equations. The time step for all calculations is chosen from the condition, that the maximum Courant number does not exceed 0.1.

3.2. *Iteration scheme*

The resulting discrete problem solution is based on the PISO method implemented in the icoFoam solver of OpenFOAM. At each time iteration of the algorithm additional steps for updating the boundary conditions (9) are defined. Updating is implemented by the deferred correction scheme. The resulting iterative scheme for calculating the values of unknowns on the j-th time layer can be represented as follows:

1. The predictor for the acceleration of the moving coordinate system is calculated as

$$\dot{w}_p^j = 2\dot{w}^{j-1} - \dot{w}^{j-2}.$$

2. The boundary conditions at the input and output boundaries are updated according to (9), where the acceleration is calculated as the sum of the predictor and corrector for the old time layer, and the velocity is determined by the 2nd order upwind difference scheme:

$$\dot{u}_\infty^j = -\dot{w}_p^j + \dot{w}_c^{j-1},\; u_\infty^j = (-2\dot{u}_\infty^j dt + 4u_\infty^{j-1} - u_\infty^{j-2})/3$$

3. The discrete governing system of the fluid motion is solved by the PISO method, the force F_x is calculated.

4. The real acceleration of the system is calculated using the new value of the force:

$$\dot{w}^j = -\mu_2 \dot{v}^j + \mu_1 \frac{R^2}{S} F^j$$

5. The corrector is calculated as

$$\dot{w}_c^j = \dot{w}_p^j - \dot{w}^j.$$

We used 5 pressure correction steps in the PISO method. To solve the system of equations for the pressure, the preconditioned conjugate gradient (PCG) method with the geometric-algebraic multigrid (GAMG) preconditioner was applied. In the GAMG realization we used the Gauss-Seidel method with one and two pre- and postrelaxations for smoothing, respectively, and the faceAreaPair algorithm for agglomeration of the mesh cells. The systems of equations for the velocity components were solved by the bi-conjugate gradients (PBiCG) method with the preconditioner based on an incomplete LU factorization. All calculations were distributed according to the MPI technology and the method of decomposition of the calculation domain.

4. Results of numerical experiments

The numerical simulation was performed for the simple harmonic and the special two-phase laws of the internal mass motion (see fig. 1). Calculations were carried out in the range of Reynolds numbers $50 < \mathrm{Re} < 250$ for the following combination of control parameters: $\mu_1 = 0.06$, $\mu_2 = 0.61$, $A = H$, where $H = \sqrt{0.75}$ is the height of the triangle. Calculations were started from different initial approximations determined by different initial flow velocities. To examine the stability of motion regimes, external disturbances were included into the flow by the Martinez method [18].

The next main characteristics of the motion were considered: the average speed U_a and the motion efficiency coefficient η [19], which defines the energy consumption due to the body motion by means of an internal motive power. Using angular brackets to denote the average for a period of motion, we define them as follows:

$$U_a = \langle U \rangle,\ \eta = \frac{N_0}{N_{vbr}}$$

Here $N_0 = N(\langle U \rangle)$ is the minimum power required for the body motion with velocity U_a (the values of N_0 for different values of the parameter Re are obtained according to auxiliary calculations), $N_{vbr} = N(U)$ is the power expended on the motion of the robot with a fixed average velocity of U_a.

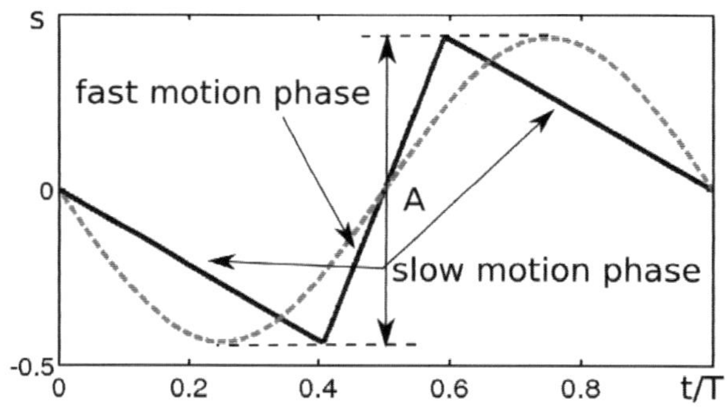

Figure 1. Harmonic (dashed line) and two-phase (solid line) laws of the internal mass motion

4.1. Harmonic law of the internal mass motion

For the harmonic law of the internal mass motion the difference between the forces on a direct and return phases of the movement, acting on the shell, causes the non zero average velocity of the whole system is achieved by asymmetrical flow around the shell with respect to the half-period.

Evolution of the motion characteristics depending on the Reynolds number for different regimes is shown in fig. 2 and fig. 3.

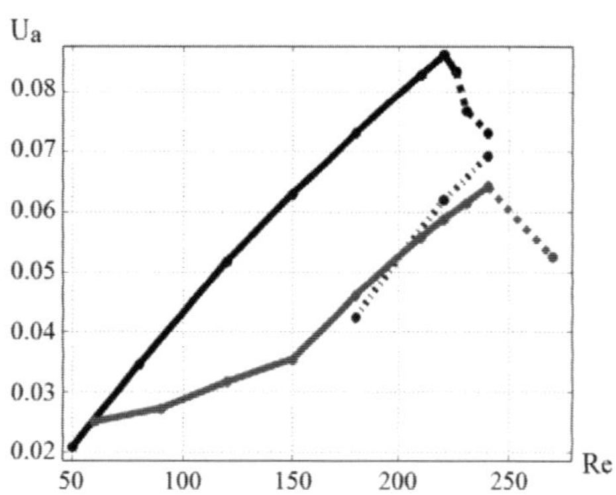

Figure 2. Average velocity of movement. Different curves represent different flow regimes: solid gray - H1, dashed gray –H2, doted – H2, solid black - TP1, dashed black - TP2.

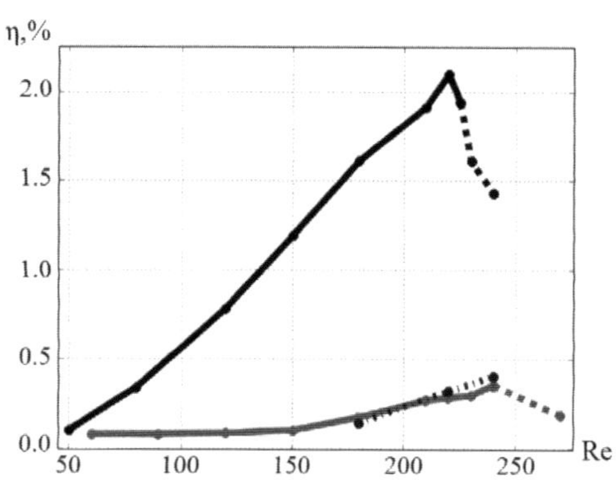

Figure 3. Efficiency of movement. Different curves represent different flow regimes: solid gray - H1, dashed gray –H2, doted – H2, solid black - TP1, dashed black - TP2.

In the area of small Reynolds numbers (Re < 170) a unique, periodic and symmetric with respect to the oscillations axis regime H1 is realized (see fig. 4). This regime corresponds to the vertex forward motion of the triangular robot. The growth of flow complexity and non-linear effects in this area lead to increase of the values of average velocity and efficiency coefficient. When Re becomes larger than

170, simultaneously with the basic regime H1 another periodic and symmetric with respect to the oscillations axis motion regime H2 appears, that implements oppositely directed base forward movement (the flow pattern is presented in fig. 4). Thus a hysteresis of motion regimes arises. Switching to one of these regimes depends on the initial parameters of the robot motion. The values of average velocity and efficiency of movement of the robot in regime H2 approximately the same as for regime S.

In the neighbourhood of Re = 240 the basic regime H1 transforms to a quasiperiodic asymmetrical regime H3 (see fig. 4). The transformation is associated with the crisis of the basic parameters of motion that is particularly noticeable on the graphs of the efficiency coefficient and average velocity (see fig. 2, 3).

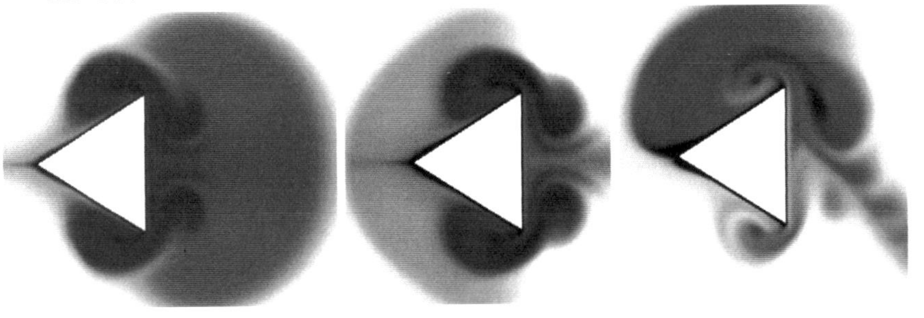

Figure 4. Flow patterns around vibration-driven robot at different motion regimes for harmonic oscillations of the internal mass. H2 (left), H1 (central), H3 (right).

4.2. Two-phase law of the internal mass motion

The structure of the considered two-phase law is shown in fig. 1. As can be seen the full period of the movement consists of two main phases: a long phase of slow motion and a short phase of fast motion, during which the internal mass moves with constant velocity. The transition between these phases is realized in a short time interval, the duration of which is less than 2% from the full period of movement. The main idea of this law is that during the slow long phase the shell of vibrating-driven robot should experience less resistance than on the fast return phase, when due to the nonlinear dependence of the resistance on the speed the forces opposing the movement of the robot increase significantly. To improve the efficiency of movement the long phase is combined with the vertex forward motion of the shell.

The two-phase motion law of the internal mass provides the only one possible direction of motion of the system (in the considered case it is vertex forward motion). In the area of small Reynolds numbers (Re < 220) a unique, periodic and symmetric with respect to the oscillations axis regime TP1 is realized (see fig. 5). In the neighbourhood of Re = 220 it transforms to a quasiperiodic asymmetrical regime TP2 (see fig. 5).

Comparison of the simulation results for different laws of the internal mass motion (see fig. 2,3) shows that the two-phase law provides a higher speed of movement (up to 50%) in the fluid, and is significantly more efficient (up to 4 times) in terms of energy consumption.

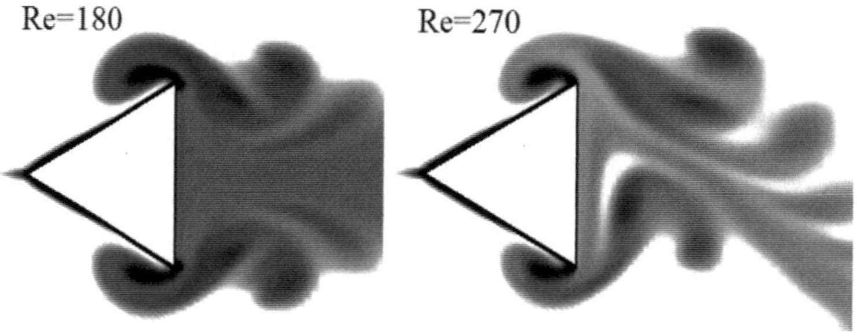

Figure 5. Flow patterns around vibration-driven robot at different motion regimes for two-phase oscillation law of the internal mass. TP1 (left), TP2 (right).

Acknowledgments
The reported study was partially supported by RSF, research project No. 15-19-10039. Numerical model was designed in the Lobachevsky State University of Nizhnij Novgorod.

References
[1] Gulia N V 1982 Inertia (Moscow: Nauka) (in Russian).
[2] Chernous'ko F L 2008 The optimal periodic motions of a two-mass system in a resistant medium *J. Appl. Math. Mech.* **72** pp. 116-125.
[3] Bolotnik N N, Figurina T Y and Chernous'ko F L 2012 Optimal control of the rectilinear motion of a two-body system in a resistive medium *J. Appl. Math. Mech.* **76** pp. 1-14.
[4] Lighthill M J 1952 On the Squirming Motion of Nearly Spherical Deformable Bodies through Liquids at Very Small Reynolds Numbers *Comm. Pure Appl. Math.* **5 (2)** pp. 109–118
[5] Saffman P G 1967 The Self-Propulsion of a Deformable Body in a Perfect Fluid *J. Fluid Mech.* **28 (2)** pp. 385–389.
[6] Ramodanov S M, Tenenev V A and Treschev D V 2012 Self-propulsion of a Body with Rigid Surface and Variable Coefficient of Lift in a Perfect Fluid *Regul. Chaotic Dyn.* **17(6)** pp. 547–558.
[7] Chernous'ko F L 2005, On the motion of a body containing a movable internal mass *Dokl. Phys.* **50** pp. 593-597.
[8] Chernous'ko F L 2006 Analysis and optimization of the motion of a body controlled by means of a movable internal mass *J. Appl. Math. Mech.* **70** pp. 819-842.
[9] Volkova L Yu and Jatsun S F 2011 Control of the Three-Mass Robot Moving in the Liquid Environment *Rus. J. Nonlin. Dyn.* **7(4)** pp. 845–857.
[10] Childress S, Spagnolie S E and Tokieda T A 2011 Bug on a Raft: Recoil Locomotion in a Viscous Fluid *J. Fluid Mech.* **669** pp. 527–556.
[11] Auziņš J, Beresņevičs V, Kaktabulis I and Kuļikovskis G. 2011 Dynamics of Water Vehicle with Internal Vibrating Gyrodrive. *Vibration Problems ICOVP 2011. Supplement: The 10th International Conference on Vibration Problems*, Czech Republic, Prague, 5-8 September.
[12] Vetchanin E 2013 The Self-propulsion of a Body with Moving Internal Masses in a Viscous Fluid *Regular and Chaotic Dynamics* **18** pp. 100–117.
[13] Egorov A G, Zakharova O S 2010 The energyoptimal motion of a vibrationdriven robot in a resistive medium *J. Appl. Math. Mech.* **74** p. 443.
[14] Egorov A G, Zakharova O S 2012 Optimal quasistationar motion of vibrationdriven robot in a viscous liquid *Izv. Vyssh. Uchebn. Zaved.* **2** pp. 57–64.
[15] Open foam (the open source cfd toolbox): User guide version 2.2.1, URL: http://www.openfoam.org/docs/user/.
[16] Nuriev A N and Zaytseva O N 2013 Solution to the problem of oscillatory motion of a cylinder in a viscous fluid in the OpenFOAM package *Heald of Kazan Technological University* **8** pp.

116-123.

[17] Egorov A G, Kamalutdinov A M, Nuriev A N and Paimushin V N 2014 Theoretical-experimental method for determining the parameters of damping based on the study of damped flexural vibrations of test specimens. 2. Aerodynamic Component of Damping *Mech. Compos. Mater.* **50** pp. 267-275.

[18] Martinez G 1979 Caractristiques dynamiques et thermiques de l'coulement autour d'un cylindre circulaire a nombres de reynolds moderes, Ph.D. thesis, I.N.P. Toulouse.

[19] Maertens A P, Triantafyllou M S and Yue D K P 2015 Efficiency of fish propulsion *Bioinspiration & Biomimetics* **10(4)**

Transient reaction of an elastic half-plane on a source of a concentrated boundary disturbance

A S Okonechnikov[1], **D V Tarlakovski**[1, 2], **A N Ul'yashina**[1] **and G V Fedotenkov**[1]

[1]Moscow Aviation Institute (National Research University), Russian Federation
[2]Lomonosov Moscow State University, Moscow, Russian Federation

E-mail: oas.mai@mail.ru

Abstract. One of the key problems in studying the non-stationary processes of solid mechanics is obtaining of influence functions. These functions serve as solutions for the problems of effect of sudden concentrated loads on a body with linear elastic properties. Knowledge of the influence functions allows us to obtain the solutions for the problems with non-mixed boundary and initial conditions in the form of quadrature formulae with the help of superposition principle, as well as get the integral governing equations for the problems with mixed boundary and initial conditions. This paper offers explicit derivations for all non-stationary surface influence functions of an elastic half-plane in a plane strain condition. It is achieved with the help of combined inverse transform of a Fourier-Laplace integral transformation. The external disturbance is both dynamic and kinematic. The derived functions in $x\tau$-domain are studied to find and describe singularities and are supplemented with graphs.

1. Problem Definition

The problem under investigation concentrates focuses on distribution of non-stationary boundary disturbance in a homogeneous isotropic elastic half-plane.

Let us introduce a coordinate system Oxz so that Oz axis is directed in the depth of the half-plane, while Ox axis corresponds with the edge of the half-plane $z=0$.

We will use dimensionless quantities system (apostrophes mark the dimensionless quantities and will be omitted in future)

$$x'=\frac{x}{L},\ z'=\frac{z}{L},\ \tau=\frac{c_1 t}{L},\ u'=\frac{u}{L},\ w'=\frac{w}{L},\ \varphi'=\frac{\varphi}{L^2},\ \psi'=\frac{\psi}{L^2},\eta=\frac{c_1}{c_2},$$
$$c_R'=\frac{c_R}{c_1},\ \sigma_{mn}'=\frac{\sigma_{mn}}{\lambda+2\mu},\ \kappa=\frac{\lambda}{\lambda+2\mu}=1-\frac{2}{\eta^2},\ c_1=\sqrt{\frac{\lambda+2\mu}{\rho}},\ c_2=\sqrt{\frac{\mu}{\rho}}, \tag{1}$$

where L – reference length, τ and t – dimensionless and dimensional time, u, w – components of displacement vectors, which match with the direction of Ox and Oz axes correspondingly; σ_{mn} $(m,n=1,3)$ – non-zero components of stress tensor, φ, ψ – scalar and vector potentials of elastic displacements; c_1, c_2 – velocities of tension and shear waves; c_R – velocity of Raleigh waves, λ, μ – Lame parameters; ρ – continuum density.

Problem definition includes [1]:

– equation of motion (dot accent represents a time derivative from now on)

$$\ddot{\varphi} = \Delta\varphi,\ \eta^2\ddot{\psi} = \Delta\psi,\ \Delta = \frac{\partial^2}{\partial x^2} + \frac{\partial^2}{\partial z^2}; \tag{2}$$

– relations between displacement vector and stress tensor components with other potentials:

$$u_1 = \frac{\partial\varphi}{\partial x} - \frac{\partial\psi}{\partial z}, u_3 = \frac{\partial\varphi}{\partial z} + \frac{\partial\psi}{\partial x}, \tag{3}$$

$$\sigma_{11} = \frac{\partial^2\varphi}{\partial x^2} + (\kappa - 1)\frac{\partial^2\psi}{\partial x\partial z} + \kappa\frac{\partial^2\varphi}{\partial z^2}, \sigma_{13} = \frac{1}{\eta^2}\left(2\frac{\partial^2\varphi}{\partial x\partial z} - \frac{\partial^2\psi}{\partial z^2} + \frac{\partial^2\psi}{\partial x^2}\right),$$
$$\sigma_{33} = \frac{\partial^2\varphi}{\partial z^2} + (1-\kappa)\frac{\partial^2\psi}{\partial x\partial z} + \kappa\frac{\partial^2\varphi}{\partial x^2}; \tag{4}$$

– initial conditions:

$$\varphi\big|_{\tau=0} = \dot{\varphi}\big|_{\tau=0} = 0,\ \psi\big|_{\tau=0} = \dot{\psi}\big|_{\tau=0} = 0; \tag{5}$$

– boundary conditions:
– no-disturbance in the infinity conditions

$$\varphi = O(1),\ \psi = O(1),\ \text{при}\ r\to\infty,\ r = \sqrt{x^2+z^2}; \tag{6}$$

– generalized conditions on the $z=0$ boundary:

$$(\alpha_1 u_1 + \beta_1\sigma_{13})\big|_{z=0} = \delta(x)\delta(\tau)\delta_{1l},$$
$$(\alpha_3 u_3 + \beta_3\sigma_{33})\big|_{z=0} = \delta(x)\delta(\tau)\delta_{3l}, \tag{7}$$

where $\delta(\bullet)$ – Dirac delta function, δ_{ml} – Kronecker symbol, $m, l = 1,3$.

Combinations of various values of $\alpha_m, \beta_m\ (m=1,3)$ parameters describe all the possible boundary conditions, while $\alpha_m^2 + \beta_m^2 \neq 0$.

At $\alpha_1 = \alpha_3 = 1,\ \beta_1 = \beta_3 = 0$ the first boundary problem can be observed.

At $\alpha_1 = \alpha_3 = 0,\ \beta_1 = \beta_3 = 1$ – we get the second boundary problem.

Displacements $u_m = G_{ml}^p(x,\tau)$ and stress $\sigma_{mk} = \Gamma_{mkl}^p(x,\tau)$, $(p=1,2;\ m,k,l=1,3)$ as solutions of problems (2)–(7) with $z=0$ will be called non-stationary surface influence functions of an elastic half-plane.

2. Construction of Surface Influence Functions

Let us apply Laplace transform over τ time to (2)–(7) and Fourier transform over x coordinate:

$$\frac{\partial^2\varphi^{FL}}{\partial z^2} - k_1^2(q^2,s^2)\varphi^{FL} = 0, \frac{\partial^2\psi^{FL}}{\partial z^2} - k_2^2(q^2,s^2)\psi^{FL} = 0, \tag{8}$$

$$k_m(q,s) = \sqrt{q+\eta_m^2 s},\ \eta_1 = 1,\ \eta_2 = \eta,$$

$$u_1^{FL} = -iq\varphi^{FL} - \frac{\partial\psi^{FL}}{\partial z}, u_3^{FL} = \frac{\partial\varphi^{FL}}{\partial z} - iq\psi^{FL},$$
$$\sigma_{11}^{FL} = -iqu_1^{FL} + \kappa\frac{\partial u_3^{FL}}{\partial z}, \sigma_{33}^{FL} = \frac{\partial u_3^{FL}}{\partial z} - iq\kappa u_1^{FL}, \tag{9}$$
$$\eta^2\sigma_{13}^{FL} = \frac{\partial u_1^{FL}}{\partial z} - iqu_3^{FL},$$

$$\varphi^{FL} = O(1), \psi^{FL} = O(1)\ \text{at}\ z\to\infty, \tag{10}$$

$$(\alpha_1 u_1^{FL} + \beta_1\sigma_{13}^{FL})\big|_{z=0} = \delta_{1l},$$
$$(\alpha_3 u_3^{FL} + \beta_3\sigma_{33}^{FL})\big|_{z=0} = \delta_{3l}. \tag{11}$$

" FL " index of function identifies its Laplace and Fourier transform, q, s – Fourier and Laplace transformation parameters correspondingly.

From (8)–(11) in the case of the first boundary problem ($p=1,\ \alpha_1=\alpha_3=1,\ \beta_1=\beta_3=0$) and the second one ($p=2,\ \alpha_1=\alpha_3=0,\ \beta_1=\beta_3=1$) we can obtain

$$\begin{gathered}G_{1l}^{pFL}(q,s)=G_{11}^{pFL}\delta_{1l}+G_{13}^{pFL}(q,s)\delta_{3l},\ G_{3l}^{pFL}(q,s)=G_{31}^{pFL}(q,s)\delta_{1l}+G_{33}^{pFL}(q,s)\delta_{3l},\\ \Gamma_{11l}^{pFL}(q,s)=\Gamma_{111}^{pFL}(q,s)\delta_{1l}+\Gamma_{113}^{pFL}(q,s)\delta_{3l},\ \Gamma_{33l}^{pFL}(q,s)=\Gamma_{331}^{pFL}(q,s)\delta_{1l}+\Gamma_{333}^{pFL}(q,s)\delta_{3l},\\ \Gamma_{13l}^{pFL}(q,s)=\Gamma_{31l}^{pFL}(q,s)=\Gamma_{131}^{pFL}(q,s)\delta_{1l}+\Gamma_{133}^{pFL}(q,s)\delta_{3l},\end{gathered} \tag{12}$$

where

$$G_{ml}^{1FL}(q,s)=\delta_{ml}\ (m,l=1,3),$$

$$\Gamma_{111}^{1FL}(q,s)=\frac{iq\left[\eta^2\kappa s^2-2R_1(q^2,s^2)\right]}{\eta^2R_1(q^2,s^2)},\ \Gamma_{333}^{1FL}(q,s)=\frac{s^2k_2(q^2,s^2)}{R_1(q^2,s^2)},$$

$$\Gamma_{133}^{1FL}(q,s)=-\frac{iq\left[2R_1(q^2,s^2)+\eta^2s^2\right]}{\eta^2R_1(q^2,s^2)},\ \Gamma_{131}^{1FL}(q,s)=\frac{s^2k_1(q^2,s^2)}{R_1(q^2,s^2)},$$

$$\Gamma_{13l}^{1FL}(q,s)=\Gamma_{31l}^{1FL}(q,s),\ \Gamma_{331}^{1FL}(q,s)=-\Gamma_{133}^{1FL}(q,s),\ \Gamma_{113}^{1FL}(q,s)=\kappa\Gamma_{333}^{1FL}(q,s),\ (l=1,3).$$

$$\Gamma_{m3l}^{2FL}=\delta_{ml}\ (m,l=1,3),$$

$$G_{11}^{2FL}(q,s)=-\frac{\eta^4k_2(q^2,s^2)s^2}{R_2(q^2,s^2)},\ G_{33}^{2FL}(q,s)=-\frac{\eta^4k_1(q^2,s^2)s^2}{R_2(q^2,s^2)},$$

$$G_{13}^{2FL}(q,s)=-G_{31}^{2FL}(q,s)=-\frac{i\eta^2q\left[\eta^2s^2+2R_1(q^2,s^2)\right]}{R_2(q^2,s^2)},$$

$$\Gamma_{111}^{2FL}(q,s)=-\frac{2iq(1+\kappa)}{\eta^2}G_{11}^{2FL}(q,s),$$

$$\Gamma_{113}^{2FL}(q,s)=-\frac{\left(\kappa\eta^2s^2-2q^2\right)\left(\eta^2s^2+2q^2\right)+4q^2k_1(q^2,s^2)k_2(q^2,s^2)}{R_2(q^2,s^2)},$$

$$R_1(q,s)=q-k_1(q,s)k_2(q,s),\ R_2(q,s)=(\eta^2s+2q)^2-4qk_1(q,s)k_2(q,s),$$

$$k_1(q,s)=\sqrt{q+s},\ k_2(q,s)=\sqrt{q+\eta^2s}.$$

3. Influence Functions in $x\tau$- Domain

As transformed functions in (12) are homogenous functions for defining the corresponding functions in $x\tau$-domain, we should use an algorithm of combined Fourier-Laplace inverse transform [5]

Among the functions in qs-domain that we study, the most interesting are $\Gamma_{mkl}^{1FL}(q,s)$, $G_{ml}^{2FL}(q,s)$, $(m,k,l=1,3)$, as well as $\Gamma_{113}^{2FL}(q,s)$. The remaining components, according to(12), are represented in $x\tau$-domain in a where they corresponds to boundary conditions in (7).

By introducing a variable substitution $\lambda=q/s$, we represent the Fourier-Laplace transformations of the functions that we are looking to obtain as:

$$\begin{gathered}\Gamma_{mkl}^{1FL}(q,s)=\gamma_1^L(s)h_{mkl}^1(\lambda),\ G_{ml}^{2FL}(q,s)=g_2^L(s)h_{ml}^2(\lambda),\\ \Gamma_{113}^{2FL}(q,s)=\gamma_2^L(s)h_{113}^2(\lambda),\ (m,k,l=1,3),\end{gathered} \tag{13}$$

where

$$g_2^L(s)=1/s,\ \gamma_1^L(s)=s,\ \gamma_2^L(s)=1,\ h_{11}^2(\lambda)=-\frac{\eta^4 k_2(\lambda^2,1)}{R_2(\lambda^2,1)},\ h_{33}^2(\lambda)=-\frac{\eta^4 k_1(\lambda^2,1)}{R_2(\lambda^2,1)},$$

$$h_{13}^2(\lambda)=-h_{31}^2(\lambda)=-\frac{i\eta^2\lambda\left[\eta^2+2R_1(\lambda^2,1)\right]}{R_2(\lambda^2,1)},\ h_{111}^1(\lambda)=\frac{i\lambda\left[\kappa-2R_1(\lambda^2,1)\right]}{\eta^2 R_1(\lambda^2,1)},$$

$$h_{333}^1(\lambda)=\frac{\kappa k_2(\lambda^2,1)}{R_1(\lambda^2,1)},\ h_{133}^1(\lambda)=-\frac{i\lambda\left[2R_1(\lambda^2,1)+\eta^2\right]}{\eta^2 R_1(\lambda^2,1)},\ h_{131}^1(\lambda)=\frac{k_1(\lambda^2,1)}{R_1(\lambda^2,1)},$$

$$h_{113}^2(\lambda)=\left[\frac{\left(\kappa\eta^2-2\lambda^2\right)\left(\eta^2+2\lambda^2\right)+4\lambda^2 k_1(\lambda^2,1)k_2(\lambda^2,1)}{R_2(\lambda^2,1)}\right],$$

$$h_{13l}^1(\lambda)=h_{31l}^1(\lambda),\ h_{331}^1(\lambda)=-h_{133}^1(\lambda),\ h_{113}^1(\lambda)=\kappa h_{333}^1(\lambda),\ (l=1,3).$$

Considering that $g_2(\tau)=H(\tau),\ \dot{g}_2(\tau)=\delta(\tau)$, where $H(\bullet)$ – Heaviside step function, as well as the properties of delta function, we obtain expressions for the analytical form [6] of $G_{ml}^2(q,s)$ function:

$$\hat{G}_{ml}^2(z,\tau)=-\frac{1}{2\pi}\dot{g}_2(\tau)*\theta_{ml}(z,\tau)=-\frac{1}{2\pi}\theta_{ml}(z,\tau),\ (m,l=1,3)$$
$$z=x+iy,\ \theta_{ij}(z,\tau)=h_{ml}\left[\lambda(z,\tau)\right]\dot{\lambda}. \tag{14}$$

Where $\lambda=\tau/iz$, $\mathrm{Re}\lambda<0$ при $(y\to+0)$; $\mathrm{Re}\lambda>0$ при $(y\to-0)$, «$*$» symbol identifies a convolution of the functions over time.

The function in the $x\tau$-domain can be depicted using the equation [7]:

$$G_{ml}^2(x,\tau)=\lim_{y\to+0}\hat{G}_{ml}^2(z,\tau)-\lim_{y\to-0}\hat{G}_{ml}^2(z,\tau)\ \ (m,l=1,3) \tag{15}$$

As $\gamma_1(\tau)=\dot{\delta}(\tau),\gamma_2(\tau)=\delta(\tau)$, the corresponding functions in the $x\tau$-domain will assume the following forms: $\Gamma_{mkl}^{1FL}(q,s)\ (m,k,l=1,3)$, $\Gamma_{113}^{2FL}(q,s)$ according to (14) should hold generalized first- and second-order derivatives correspondingly. To avoid the differentiation procedure, which can lead to complex analytical expressions for the influence functions in $x\tau$-domain, we will find the $x\tau$-domain representations of primitives of corresponding functions:

$$\Gamma_{mkl}^1(x,\tau)=\frac{d^2}{d\tau^2}\tilde{\Gamma}_{mkl}^1(x,\tau),\ \ \Gamma_{113}^2(x,\tau)=\frac{d}{d\tau}\tilde{\Gamma}_{113}^2(x,\tau), \tag{16}$$

where

$$\tilde{\Gamma}_{mkl}^1(x,\tau)=\lim_{y\to+0}\hat{\Gamma}_{mkl}^1(z,\tau)-\lim_{y\to-0}\hat{\Gamma}_{mkl}^1(z,\tau),\ \tilde{\Gamma}_{113}^2(x,\tau)=\lim_{y\to+0}\hat{\Gamma}_{113}^2(z,\tau)-\lim_{y\to-0}\hat{\Gamma}_{113}^2(z,\tau),$$

$$\hat{\Gamma}_{mkl}^1(z,\tau)=-\frac{1}{2\pi}\theta_{mkl}(z,\tau),\ \hat{\Gamma}_{113}^2(z,\tau)=-\frac{1}{2\pi}\theta_{113}(z,\tau),\ \theta_{mkl}(z,\tau)=h_{mkl}\left[\lambda(z,\tau)\right]\dot{\lambda},$$

$$\theta_{113}(z,\tau)=h_{113}\left[\lambda(z,\tau)\right]\dot{\lambda},\ z=x+iy,\ (m,k,l=1,3).$$

Here derivatives are understood in the generalized sense.

This is possible thanks to two reasons. First, the differentiation impairs the properties of the functions in the $x\tau$-domain we are looking for, as they might get singularities that are impossible to integrate. Second, the influence functions we are looking for are not the ultimate goal. They serve as the equation kernels of the corresponding governing integral operators or equations in the process of solving particular problems. Consequently, while solving we can always remove the time derivatives from these kernels, with, for example, integration by parts.

Let us find the limit values of the functions which form (15)–(16):

$$\lambda_0(x,\tau)=\lim_{y\to\pm0}\lambda(z,\tau)=-\frac{i\tau}{x} \tag{17}$$

$$\lim_{y\to\pm 0} k_m(\lambda^2,1) = \begin{cases} k_{m0}(x^2,\tau^2) & \text{at} \left|\tau/x\right| < \eta_m \\ \mp sign\, x\, k_{m0}(x^2,\tau^2) & \text{at} \left|\tau/x\right| \ge \eta_m \end{cases} \tag{18}$$
$$k_{m0}(1,\tau/x) = \sqrt{\eta_m^2 - \tau/x},$$

By substituting (17)–(18) into (14)–(15) we will obtain the function in the $x\tau$-domain with different values of $\tau/|x|$:

At $\tau/|x| < 1$

$$G^p_{ml}(x,\tau) = \Gamma^p_{mkl}(x,\tau) = 0 \;\; (m,k,l = 1,3). \tag{19}$$

At $1 < \tau/|x| < \eta$

$$G^2_{ml}(x,\tau) = G^{21}_{ml}(x,\tau),$$
$$G^{21}_{11}(x,\tau) = \frac{4\eta^2\tau^2\left(\tau^2 - \eta^2 x^2\right)}{\pi P_3(x^2,\tau^2)}\sqrt{\tau^2 - x^2},\; G^{21}_{33}(x,\tau) = \frac{\eta^2\left(\eta^2 x^2 - 2\tau^2\right)^2}{\pi P_3(x^2,\tau^2)}\sqrt{\tau^2 - x^2},$$
$$G^{21}_{13}(x,\tau) = -G^{21}_{31}(x,\tau) = -\frac{2\eta^2\tau}{\pi}\left[\frac{\left(\eta^2 x^2 - 2\tau^2\right)\sqrt{\eta^2 x^2 - \tau^2}\sqrt{\tau^2 - x^2}}{P_3(x^2,\tau^2)}\right],$$
$$\Gamma^{21}_{111}(x,\tau) = -\frac{2iq\beta_3(1+\kappa)}{\eta^2}G^{21}_{11}(x,\tau), \tag{20}$$
$$\tilde\Gamma^{21}_{113}(x,\tau) = \frac{2\tau^2\left(\eta^4 x^4 + 2\kappa\eta^4 x^4 - 4\kappa\eta^2 x^2\tau^2 - 4\tau^4\right)\sqrt{\eta^2 x^2 - \tau^2}}{\eta^2 x^4 \pi P_3(x^2,\tau^2)}\sqrt{\tau^2 - x^2},$$
$$\tilde\Gamma^{11}_{111}(x,\tau) = \frac{\tau\kappa\sqrt{\eta^2 x^2 - \tau^2}}{\pi Q_3(x^2,\tau^2)}\sqrt{\tau^2 - x^2},\;\; \tilde\Gamma^{11}_{333}(x,\tau) = \frac{\left(\eta^2 x^2 - \tau^2\right)}{\pi Q_3(x^2,\tau^2)}\sqrt{\tau^2 - x^2},$$
$$\tilde\Gamma^{11}_{133}(x,\tau) = -\frac{\tau\sqrt{\eta^2 x^2 - \tau^2}}{\pi Q_3(x^2,\tau^2)}\sqrt{\tau^2 - x^2},\;\; \tilde\Gamma^{11}_{131}(x,\tau) = -\frac{\tau^2\sqrt{\tau^2 - x^2}}{\pi Q_3(x^2,\tau^2)},$$
$$\tilde\Gamma^{11}_{13l}(x,\tau) = \tilde\Gamma^{11}_{31l}(x,\tau),\; \tilde\Gamma^{11}_{331}(x,\tau) = -\tilde\Gamma^{11}_{133}(x,\tau),\; \tilde\Gamma^{11}_{113}(x,\tau) = \kappa\tilde\Gamma^{11}_{333}(x,\tau),\;\; (l = 1,3).$$

At $\tau/|x| > \eta > 1$

$$G^2_{ml}(x,\tau) = G^{23}_{ml}(x,\tau)$$
$$G^{23}_{11}(x,\tau) = \frac{\eta^2 R_{22}(x^2,\tau^2)}{\pi P_3(x^2,\tau^2)}\sqrt{\tau^2 - \eta^2 x^2},\, G^{23}_{33}(x,\tau) = \frac{\eta^2 R_{21}(x^2,\tau^2)}{\pi P_3(x^2,\tau^2)}\sqrt{\tau^2 - x^2},$$
$$G^{23}_{13}(x,\tau) = G^{23}_{31}(x,\tau) = 0,$$
$$\Gamma^{23}_{111}(x,\tau) = -\frac{2iq\beta_3(1+\kappa)}{\eta^2}G^{23}_{11}(x,\tau),\; \tilde\Gamma^{23}_{113}(x,\tau) = 0, \tag{21}$$
$$\tilde\Gamma^{13}_{333}(x,\tau) = \frac{\sqrt{\tau^2 - \eta^2 x^2}\, R^1_{21}(x^2,\tau^2)}{\pi Q_3(x^2,\tau^2)},\; \tilde\Gamma^{13}_{131}(x,\tau) = \frac{\sqrt{\tau^2 - x^2}\, R^1_{21}(x^2,\tau^2)}{\pi Q_3(x^2,\tau^2)},$$
$$\tilde\Gamma^{13}_{111}(x,\tau) = \tilde\Gamma^{13}_{133}(x,\tau) = \tilde\Gamma^{13}_{331}(x,\tau) = 0,\; \tilde\Gamma^{13}_{113}(x,\tau) = \kappa\tilde\Gamma^{13}_{333}(x,\tau),$$
$$\tilde\Gamma^{11}_{13l}(x,\tau) = \tilde\Gamma^{11}_{31l}(x,\tau),\;\; (l = 1,3),$$

where

$$Q_3(x,\tau)=R_{21}^1(x,\tau)R_{21}^2(x,\tau),$$
$$R_{21}^1(x,\tau)=-\left(k_{10}(x,\tau)k_{20}(x,\tau)+\tau\right),\quad R_{22}^1(x,\tau)=\tau-k_{10}(x,\tau)k_{20}(x,\tau),$$
$$R_{21}(x,\tau)=\left(\eta^2x-2\tau\right)^2+4\tau\, k_{10}(x,\tau)k_{20}(x,\tau),\ R_{22}(x,\tau)=\left(\eta^2x-2\tau\right)^2-4\tau\, k_{10}(x,\tau)k_{20}(x,\tau),$$
$$R_{21}(x,\tau)R_{22}(x,\tau)=\eta^2xP_3(x,\tau).$$

Taking (19)–(21) into consideration, we rewrite the derivations for the influence functions:

$$\begin{gathered}
G_{ml}^2(x,\tau)=\sum_{r=1}^{2}G_{ml}^{2r}(x,\tau)H\left(\tau-\eta_r|x|\right),\\
G_{ml}^{22}(x,\tau)=G_{ml}^{23}(x,\tau)-G_{ml}^{21}(x,\tau),\ (m=l=1,3),\\
G_{ml}^2(x,\tau)=G_{ml}^{21}(x,\tau)H\left(\tau-|x|\right),\ \ (m,l=1,3;m\neq l)\\
\Gamma_{111}^2(x,\tau)=-\frac{2iq\beta_3(1+\kappa)}{\eta^2}G_{11}^2(x,\tau),\\
\tilde{\Gamma}_{113}^2(x,\tau)=\tilde{\Gamma}_{113}^{21}(x,\tau)H\left(\tau-|x|\right),\\
\tilde{\Gamma}_{m3l}^1(x,\tau)=\sum_{r=1}^{2}\tilde{\Gamma}_{m3l}^{1r}(x,\tau)H\left(\tau-\eta_r|x|\right),\\
\tilde{\Gamma}_{m3l}^{12}(x,\tau)=\tilde{\Gamma}_{m3l}^{13}(x,\tau)-\tilde{\Gamma}_{m3l}^{11}(x,\tau),(m=l=1,3),\\
\tilde{\Gamma}_{111}^1(x,\tau)=\tilde{\Gamma}_{111}^{11}(x,\tau)H\left(\tau-|x|\right),\ \tilde{\Gamma}_{133}^1(x,\tau)=\tilde{\Gamma}_{133}^{11}(x,\tau)H\left(\tau-|x|\right),\\
\tilde{\Gamma}_{331}^1(x,\tau)=\tilde{\Gamma}_{331}^{11}(x,\tau)H\left(\tau-|x|\right),\ \tilde{\Gamma}_{113}^1(x,\tau)=\kappa\tilde{\Gamma}_{333}^1(x,\tau),\\
\tilde{\Gamma}_{13l}^1(x,\tau)=\tilde{\Gamma}_{31l}^1(x,\tau),\ (l=1,3),
\end{gathered}\tag{22}$$

where

$$G_{11}^{22}(x,\tau)=\frac{\eta^2\left(\eta^2x^2-2\tau^2\right)^2}{\pi P_3(x^2,\tau^2)}\sqrt{\tau^2-\eta^2x^2},\ G_{33}^{22}(x,\tau)=\frac{4\eta^2\tau^2\left(\tau^2-x^2\right)}{\pi P_3(x^2,\tau^2)}\sqrt{\tau^2-\eta^2x^2},$$
$$\tilde{\Gamma}_{131}^{12}(x,\tau)=\frac{\left[\tau^2-\left(R_{21}^1(x^2,\tau^2)\right)\right]\sqrt{\tau^2-x^2}}{\pi Q_3(x^2,\tau^2)},\ \tilde{\Gamma}_{333}^{12}(x,\tau)=-\frac{\tau^2\sqrt{\tau^2-\eta^2x^2}}{\pi Q_3(x^2,\tau^2)}.$$

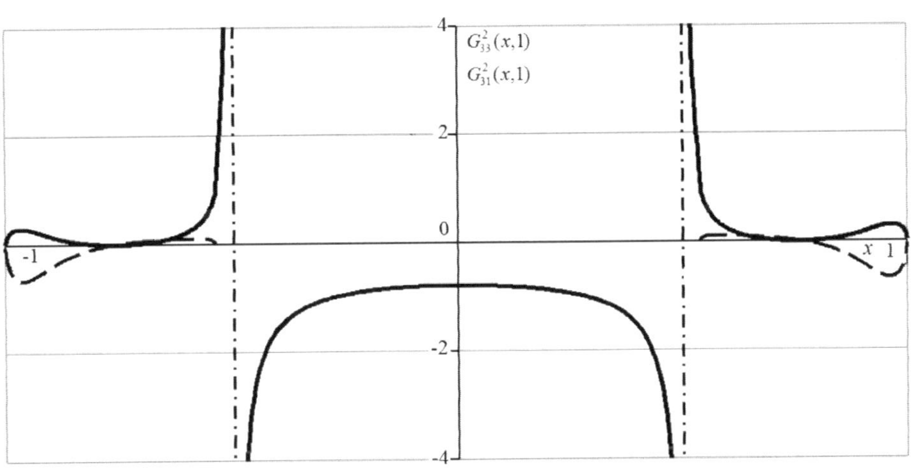

Figure 1 demonstrates influence functions, which describe the normal displacements under the effect of normal (solid line) and tangential (dashed line) load. Dot-and-dash line shows gaps corresponding to the front lines of the Raleigh waves [2].

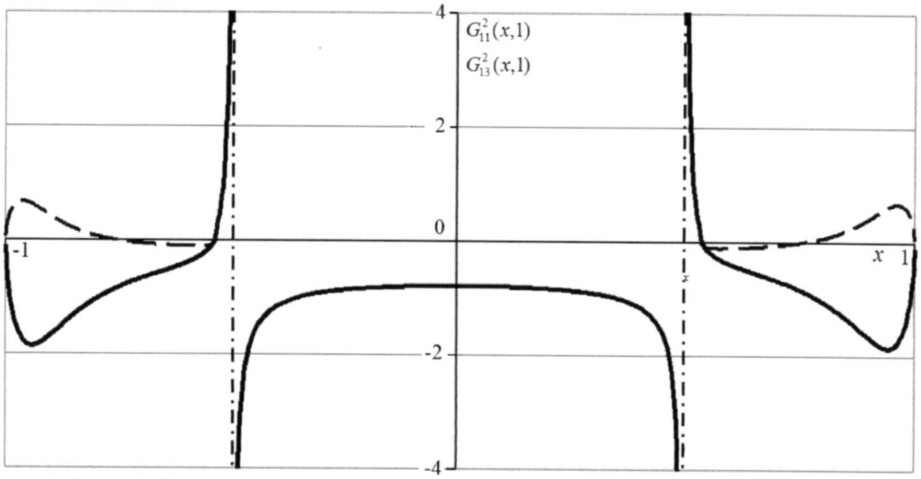

Figure 2 shows influence functions that describe the tangential displacements under the effect of tangential (solid line) and normal (dashed line) loads. Dot-and-dash line, as before, depicts the gaps in the fronts of Raleigh waves.

4. Conclusion
As a result, the influence functions for elastic half-space have been obtained. Some of these functions were applied for solving non-stationary problem of a mobile surface load [3,4], as well as in a series of works on solving the non-stationary contact problems [5-7].

Acknowledgements
The research is completed under the financial support of Russian Fund for Fundamental Research (project 16-08-00260) and grant of President of the Russian Federation for young scientists and postgraduate students (2016-2018).

References
[1] Gorshkov A G, Medvedski A L, Rabinski L N and Tarlakovski D V 2004 Waves in continua *Phizmatlit* (Moscow) 472

[2] Okonechnikov A S, Tarlakovskii D V and Fedotenkov G V 2014 Plane nonstationary problem of motion of the surface load over an elastic half-space *Journal of Mathematical Sciences* **174 (2)** 193-201. DOI 10.1007/s10958-014-2100-z.

[3] Okonechnikov A S, Tarlakovskii D V and Fedotenkov G V 2015 Nestatsionarnoe dvizhenie normal'noi sosredotochennoi nagruzki vdol' granitsy uprugoi poluploskosti *Elektronnyi zhurnal Trudy MAI* **82** https://www.mai.ru/science/trudy/published.php?ID=58453

[4] Mikhailova E Yu and Fedotenkov G V 2011 Nonstationary Axisymmetric Problem of the Impact of a Spherical Shell on an Elastic Half-Space (Initial Stage of Interaction) *Mechanics of Solids* **46** (2) 239-247. DOI: 10.3103/S0025654411020129

[5] Suvorov Ye M, Tarlakovskii D V and Fedotenkov G V 2012 The plane problem of the impact of a rigid body on a half-space modelled by a Cosserat medium *Journal of Applied Mathematics and Mechanics* **76** (5) 511–518 doi: 10.1016/j.jappmathmech.2012.11.015

[6] Tarlakovskii D V and Fedotenkov G V 2014 Two-dimensional nonstationary contact of elastic cylindrical or spherical shells *Journal of Machinery Manufacture and Reliability* **43 (2)** 145-152 doi: 10.3103/S1052618814010178

[7] Gorshkov A G and Tarlakovski D V 1995 Dynamic contact problems with motion boundaries (Moscow: Phizmatlit) 352

Discrete-layered damping model of multilayer plate with account of internal damping

V N Paimushin[1, 2] **and R K Gazizullin**[1]

[1]Kazan National Research Technical University named after A.N.Tupolev – KAI, K. Marx Str.,10, Kazan, 420111, Russian Federation
[2]Kazan Federal University, Kremlyovskaya Str.,18, Kazan, 420008, Russian Federation

E-mail: vpajmushin@mail.ru

Abstract. Construction of discrete-layered damping model of multilayer plate in small displacement and deformations with account of internal damping of layers of Thompson-Kelvin-Voight model is presented. Based on derived equations, analytical solution is given to the static damping problem of simply supported single-layer rectangular plate subjected to uniformly distributed pressure, which is applied to one of its boundary planes. Convergence to the three-dimensional case is analysed for the obtained solution with respect to the dependence on dimension of mesh in the thickness direction of plate. For thin plates, dimension reduction of the formulated problem is set on the basis of simplifying hypothesis applied for each layer.

1. Introduction

Composite materials from high strength carbon or glass fibers, which has found so far widespread utilization as construction material in manufacture of different types of products, have relatively low impact toughness, internal friction parameters and absorption of impact energy. Thereby, one of solution in improvement of specified properties of structures is application of adhesive bonding of different composite materials. Structure of those materials consists traditional composites with carbon or glass fiber reinforcements and protecting layer with various types of reinforced or unreinforced elastomers, which are inherent impact toughness and high level of energy absorption. In particular, as a protecting layer, it is expedient to use composites from high-strength and high-modulus polyethylene fibers (HSMPF) with elastomer matrixes. HSMPF differ from other high-strength fibers not only by their high level of specific mechanical properties, but also by minimum coefficient of friction, positive effect of their deformation rates on strength, sharp increase of strength in below zero temperatures and other properties. It is worth to note that due to the poor adhesion almost many plastics HSMPF is mainly used in production of cords and ropes. However, recently, there have been developed technologies that enable, for example, by irradiation with cold plasma substantially improve the adhesion properties HSMPF fibers and composites to create on their basis [1–3 et al.]. Such composites can be used merely in production of various structural parts as construction material. In particular, mentioned above specific properties make possible to recommend them in production of fuel tanks of air vehicles using as matrix of resin or polyurethane. Having relatively high rigidity in tension, they retain the flexibility to bend. To ensure the fire safety in potential penetration of tank, it is advised to use them in multilayer structure consecutively stacking unreinforced soft resin layers and rigid, in tension, elastomer matrix composite layers.

The application of elements with a multilayer structure of the described class in different structures requires the development of computational models and methods, taking into account the features of the structure and behavior of materials of the layers. Among these features, in particular, layered character of structure and relatively low stiffness in directions that do not coincide with the direction of the reinforcement can be emphasized. In addition, soft layers of such structural elements are also characterized by high values of internal friction, which induces the need accounting internal damping of material in steady dynamic damping process.

Definition of structural elements, in general, can be attributed to class of multilayered [4], consisting of rigid layers, reinforced in tangential directions and transversely soft core layers. Mechanics of deformation of multilayer structures is studied extensively in scientific literature review of which is contained in many papers [5-7 et al.]. Among them, in particular, we refer to monograph [8], in which the theoretical bases of construction of non-classical mathematical models for the dynamics of single-layer and multilayer plates and shells with variable thickness from traditional materials as well as composites with comprehensive accounting of effects of material and (or) geometrically nonlinearity during deformation, anisotropy and strength, viscoelastic properties, heterogeneity of elastic and strength parameters. For the all proposed models, numerical methods were developed on the basis of which wide range of nonlinear deformation problems as well as stability and optimum design of single-layer or composite multilayer elements and spatial structures under impulsive loadings and set the application field of classical shell theory.

Present paper is devoted to further improvements of described research directions with the main goal being construction of discrete-structural model of deformation of multilayer plate in small displacements and deformations considering the internal damping of layer materials according to Thompson-Kelvin-Voight viscoelastic model.

2. Equations of multilayer plate theory in small displacements and deformations

Let's consider plate consist of $M = N-1$ layers (fig.1), space of which $V^{[k]}$ is attributed to parametrization $\mathbf{R}^{[k]} = x_1\mathbf{e}_1 + x_2\mathbf{e}_2 + z_{[k]}\mathbf{m}, \quad -h_{[k]} \le z_{[k]} \le h_{[k]}$

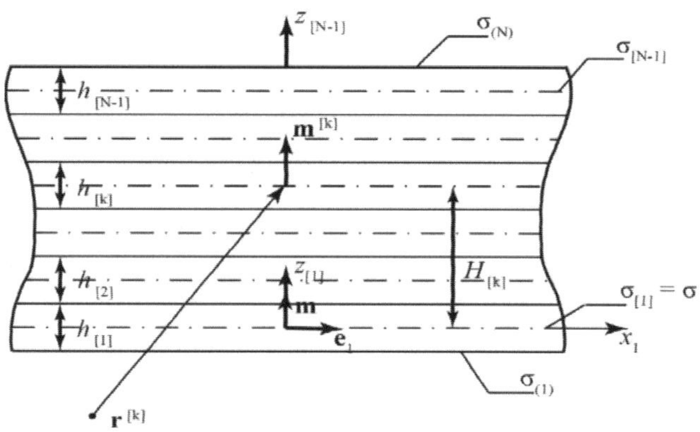

Figure 1. Multilayer plate scheme.

where layers in direction of unit vector $\mathbf{m}$ normal to the middle plane $\sigma_{[k]}$ have thickness $h_{[k]}$; x_1, x_2 – orthogonal Descartes coordinates on any plane σ, adopted as base of parametrization. Fort he description of deformation mechanics of each $[k]$-th -th layer we will use Timoshenko kinematic model [8-10 et al.], employin for the displacement vectors $\mathbf{U}^{[k]}$ the following representation

$$\mathbf{U}^{[k]} = \mathbf{u}^{[k]} + z_{[k]}\boldsymbol{\gamma}^{[k]} = u_i^{[k]}\mathbf{e}_i + w^{[k]}\mathbf{m} + z_{[k]}\left(\gamma_i^{[k]}\mathbf{e} + \gamma^{[k]}\mathbf{m}\right); k = \overline{1, N-1} \qquad (1)$$

In small deformations, translations and rotations representation (1) corresponds to tangential deformation components $\varepsilon_i^{[k]}, \gamma_{12}^{[k]}$, transverse compression component $\varepsilon_3^{[k]}$ and transverse shear components $\gamma_{13}^{[k]}$, which are calculated by formulae

$$\varepsilon_i^{[k]} = e_{ii}^{[k]} + z_{[k]}\Omega_{ii}^{[k]}, \quad \gamma_{12}^{[k]} = e_{12}^{[k]} + e_{21}^{[k]} + z_{[k]}\left(\Omega_{12}^{[k]} + \Omega_{21}^{[k]}\right); \overline{1,2}$$
$$\varepsilon_3^{[k]} = \varepsilon_{33}^{[k]} = \gamma^{[k]}, \quad \gamma_{13}^{[k]} = 2\left(e_{13}^{[k]} + z_{[k]}\Omega_{13}^{[k]}\right) \tag{2}$$

where

$$e_{ij}^{[k]} = u_{j,i}^{[k]}, \quad 2e_{i3}^{[k]} = w_{,i}^{[k]} + \gamma_i^{[k]}, \quad \Omega_{ij}^{[k]} = \gamma_{j,i}^{[k]}, \quad 2\Omega_{13}^{[k]} = \gamma_{,i}^{[k]} \tag{3}$$

Let's assume that boundary sections of plate are formed by the motion of normal $\mathbf{m}$ along the coordinate lines $x_i = x_i^-$, $x_i = x_i^+$, at the points of middle planes $\sigma_{[k]}$ of each $[k]$-th layer and their contour lines, edge forces and moments are given as

$$\mathbf{Q}_i^{[k]} = Q_{i1}^{[k]}\mathbf{e}_1 + Q_{i2}^{[k]}\mathbf{e}_2 + Q_{i3}^{[k]}\mathbf{m}, \quad \mathbf{L}_i^{[k]} = L_{i1}^{[k]}\mathbf{e}_1 + L_{i2}^{[k]}\mathbf{e}_2 + L_{i3}^{[k]}\mathbf{m}; \; k = \overline{1, N-1} \tag{4}$$

Applied to planes $\sigma_{[k]}$, and vectors of surface loads and moments

$$\mathbf{X}^{[k]} = X_i^{[k]}\mathbf{e}_i + X_3^{[k]}\mathbf{m}, \quad \mathbf{M}^{[k]} = M_i^{[k]}\mathbf{e}_i + M_3^{[k]}\mathbf{m}; \quad k = \overline{1, N-1} \tag{5}$$

are attributed to unit area of surfaces $\sigma_{[k]}$. Variation of work of these forces and moments with respect to corresponding displacements will be equal to

$$\delta A = \sum_{k=1}^{N-1}\delta A^{[k]} = \sum_{k=1}^{N-1}\left[\sum_{i=1}^{2}\int_{x_{3-i}^-}^{x_{3-i}^+}\left(Q_{ij}^{[k]}\delta u_j^{[k]} + Q_{i3}^{[k]}\delta w^{[k]} + L_{ij}^{[k]}\delta\gamma_j^{[k]} + L_{i3}^{[k]}\delta\gamma^{[k]}\right)dx_{3-i}\right|_{x_i = x_i^-}^{x_i = x_i^+} +$$
$$+\iint_\sigma\left(X_i^{[k]}\delta u_i^{[k]} + X_3^{[k]}\delta w^{(k)} + M_i^{[k]}\delta\gamma_i^{[k]} + M_3^{[k]}\delta\gamma^{[k]}\right)dx_1 dx_2 \tag{6}$$

If internal foces and moments are introduced

$$T_{\alpha\beta}^{[k]} = \int_{-h_{[k]}/2}^{h_{[k]}/2}\sigma_{\alpha\beta}^{[k]}dz_{[k]}, \quad M_{i\alpha}^{[k]} = \int_{-h_{[k]}/2}^{h_{[k]}/2}\sigma_{i\alpha}^{[k]}z_{[k]}dz_{[k]} \tag{7}$$

applied to middle plane of layers, then for the variation of strain potential energy of plate by using relationship (2), (3), we can end up with the formula below

$$\delta\Pi = \sum_{k=1}^{N-1}\delta\Pi^{[k]} = \sum_{k=1}^{N-1}\iint_\sigma\left(T_{ij}^{[k]}\delta e_{ij}^{[k]} + T_{i3}^{[k]}\delta\omega_i^{[k]} + \right.$$
$$\left. + M_{ij}^{[k]}\delta\Omega_{ij}^{[k]} + M_{i3}^{[k]}\delta\Omega_{i3}^{[k]} + T_{i3}^{[k]}\delta\gamma_i^{[k]} + T_{33}^{[k]}\delta\gamma^{[k]}\right)d\sigma \tag{8}$$

For the variation of kinetic strain energy taking into account the energy of rotations, the following expression takes place

$$\delta K = \sum_{k=1}^{N-1}\delta K^{[k]} = \sum_{k=1}^{N-1}\iint_\sigma\left(\tilde{X}_j^{[k]}\delta u_j^{[k]} + \tilde{X}_3^{[k]}\delta w^{[k]} + \tilde{Y}_j^{[k]}\delta\gamma_j^{[k]} + \tilde{Y}_3^{[k]}\delta\gamma^{[k]}\right)d\sigma \tag{9}$$

where

$$\tilde{X}_j^{[k]} = \rho_p^{[k]}h_{[k]}\ddot{u}_j^{[k]}, \quad \tilde{X}_3^{[k]} = \rho_p^{[k]}h_{[k]}\ddot{w}_j^{[k]}, \quad \tilde{Y}_j^{[k]} = \frac{\rho_p^{[k]}h_{[k]}^3}{12}\ddot{\gamma}_j^{[k]}, \quad \tilde{Y}_3^{[k]} = \frac{\rho_p^{[k]}h_{[k]}^3}{12}\ddot{\gamma}^{[k]} \tag{10}$$

Here $\rho_p^{[k]}$ – density of material of $[k]$-th layer, dots above the functions show derivations with respect to time τ.

We introduce N vectors of displacements

$$\mathbf{v}^{(k)} = u_i^{(k)}\mathbf{e}_i + w^{(k)}\mathbf{m}; \quad k = \overline{1, N} \tag{11}$$

of boundary planes points $z_{[1]}=-h_{[1]}/2$, $z_{[N-1]}=h_{[N-1]}/2$, interlayer planes and set $N-2$ conditions for interlayers of the plate according to displacement as follow

$$\mathbf{U}^{[k]}\left(z_{[k]}=-h_{[k]}/2\right)=\mathbf{v}^{(k)},\quad \mathbf{U}^{[k]}\left(z_{[k]}=h_{[k]}/2\right)=\mathbf{v}^{(k+1)};\, k=\overline{1,N-2} \tag{12}$$

Substituting representations (1) and (11) from the conditions (12), the following relations follow

$$u_i^{[k]}=\frac{u_i^{(k+1)}+u_i^{(k)}}{2},\quad w^{[k]}=\frac{w^{(k+1)}+w^{(k)}}{2},\quad \gamma_i^{[k]}=\frac{u_i^{(k+1)}-u_i^{(k)}}{h_{[k]}},\quad \gamma^{[k]}=\frac{w^{(k+1)}-w^{(k)}}{h_{[k]}} \tag{13}$$

and substitution of them into (3) we come to relations

$$\begin{gathered} e_{ij}^{[k]}=\frac{e_{ij}^{(k+1)}+e_{ij}^{(k)}}{2},\quad \omega_i^{[k]}=\frac{\omega_i^{(k+1)}+\omega_i^{(k)}}{2} \\ \Omega_{ij}^{[k]}=\frac{e_{ij}^{(k+1)}-e_{ij}^{(k)}}{h_{[k]}},\quad \Omega_{i3}^{[k]}=\frac{\omega_i^{(k+1)}-\omega_i^{(k)}}{h_{[k]}};\; k=\overline{1,N-1} \end{gathered} \tag{14}$$

in which the following kinematical relationships are introduced

$$e_{ij}^{(k)}=u_{j,i}^{(k)},\quad \omega_i^{(k)}=w_{,i}^{(k)} \tag{15}$$

Using the relationships (13), (14), we can bring expressions (6), (8) and (9) into form

$$\delta A=\sum_{k=1}^{N}\left[\sum_{i=1}^{2}\int_{x_{3-i}^{-}}^{x_{3-i}^{+}}\left(Q_{ij}^{(k)}\delta u_i^{(k)}+Q_{i3}^{(k)}\delta w^{(k)}\right)dx_{3-i}\Bigg|_{x_i=x_i^{-}}^{x_i=x_i^{+}}+\iint_{\sigma}\left(X_i^{(k)}\delta u_i^{(k)}+X_3^{(k)}\delta w^{(k)}\right)dx_1dx_2\right] \tag{16}$$

$$\delta\Pi=\sum_{k=1}^{N}\iint_{\sigma}\left(S_{ij}^{(k)}\delta e_{ij}^{(k)}+S_{i3}^{(k)}\delta\omega_i^{(k)}+N_{i3}^{(k)}\delta u_i^{(k)}+N_{33}^{(k)}\delta w^{(k)}\right)dx_1dx_2 \tag{17}$$

$$\delta K=\sum_{k=1}^{N}\iint_{\sigma}\left(\tilde{X}_i^{(k)}\delta u_i^{(k)}+\tilde{X}_3^{(k)}\delta w^{(k)}\right)dx_1dx_2 \tag{18}$$

where

$$\begin{gathered} Q_{i\alpha}^{(1)}=\frac{Q_{i\alpha}^{[1]}}{2}-\frac{L_{i\alpha}^{[1]}}{h_{[1]}},\quad Q_{i\alpha}^{(N)}=\frac{Q_{i\alpha}^{[N-1]}}{2}+\frac{L_{i\alpha}^{[N-1]}}{h_{[N-1]}} \\ Q_{i\alpha}^{(k)}=\frac{Q_{i\alpha}^{[k-1]}+Q_{i\alpha}^{[k]}}{2}+\frac{L_{i\alpha}^{[k-1]}}{h_{[k-1]}}-\frac{L_{i\alpha}^{[k]}}{h_{[k]}};\quad k=\overline{2,N-1};i=1,2;\alpha=\overline{1,3} \\ X_{\alpha}^{(1)}=\frac{X_{\alpha}^{[1]}}{2}-\frac{M_{\alpha}^{[1]}}{h_{[1]}},\quad X_{\alpha}^{(N)}=\frac{X_{\alpha}^{[N-1]}}{2}+\frac{M_{\alpha}^{[N-1]}}{h_{[N-1]}} \\ X_{\alpha}^{(k)}=\frac{X_{\alpha}^{[k-1]}+X_{\alpha}^{[k]}}{2}+\frac{M_{\alpha}^{[k-1]}}{h_{[k-1]}}-\frac{M_{\alpha}^{[k]}}{h_{[k]}};\quad \alpha=\overline{1,3};\; k=\overline{2,N-1} \\ S_{i\alpha}^{(1)}=\frac{T_{i\alpha}^{[1]}}{2}-\frac{M_{i\alpha}^{[1]}}{h_{[1]}},\quad S_{i\alpha}^{(N)}=\frac{T_{i\alpha}^{[N-1]}}{2}+\frac{M_{i\alpha}^{[N-1]}}{h_{[N-1]}} \\ S_{i\alpha}^{(k)}=\frac{T_{i\alpha}^{[k-1]}+T_{i\alpha}^{[k]}}{2}+\frac{M_{i\alpha}^{[k-1]}}{h_{[k-1]}}-\frac{M_{i\alpha}^{[k]}}{h_{[k]}};\; k=\overline{2,N-1};i=1,2;\;\alpha=\overline{1,3} \\ N_{\alpha 3}^{(1)}=-\frac{T_{\alpha 3}^{[1]}}{h_{[1]}},\quad N_{\alpha 3}^{(N)}=\frac{T_{\alpha 3}^{[N-1]}}{h_{[N-1]}},\quad N_{\alpha 3}^{(k)}=\frac{T_{\alpha 3}^{[k-1]}}{h_{[k-1]}}-\frac{T_{\alpha 3}^{[k]}}{h_{[k]}};k=\overline{2,N-1};\alpha=\overline{1,3} \end{gathered} \tag{19}$$

$$\tilde{X}_{\alpha}^{(1)} = \frac{\tilde{X}_{\alpha}^{[1]}}{2} - \frac{\tilde{Y}_{\alpha}^{[1]}}{h_{[1]}}, \ \tilde{X}_{\alpha}^{(N)} = \frac{\tilde{X}_{\alpha}^{[N-1]}}{2} + \frac{\tilde{Y}_{\alpha}^{[N-1]}}{h_{[N-1]}}$$

$$\tilde{X}_{\alpha}^{(k)} = \frac{\tilde{X}_{\alpha}^{[k-1]} + \tilde{X}_{\alpha}^{[k]}}{2} + \frac{\tilde{Y}_{\alpha}^{[k-1]}}{h_{[k-1]}} - \frac{\tilde{Y}_{\alpha}^{[k]}}{h_{[k]}}; \quad \alpha = \overline{1,3};\ k = \overline{2, N-1}$$

2.1. *Equations of equilibrium*

In static equilibrium state of the plate, variational principle of virtual displacements $\delta\Pi - \delta A = 0$ must be satisfied. The last takes the following form after substitution of expression (15) and doing standard some transformations

$$\sum_{i=1}^{2}\sum_{k=1}^{N}\left\{ \int_{x_{3-i}^{-}}^{x_{3-i}^{+}} \left[\left(S_{ij}^{(k)} - Q_{ij}^{(k)}\right)\delta u_j^{(k)} + \left(S_{i3}^{(k)} - Q_{i3}^{(k)}\right)\delta w^{(k)} \right] dx_{3-i} \Bigg|_{x_i = x_i^-}^{x_i = x_i^+} - \right. \tag{20}$$

$$\left. -\iint_{\sigma} \left(f_i^{(k)} \delta u_i^{(k)} + f_3^{(k)} \delta w^{(k)} \right) dx_1 dx_2 \right\} = 0$$

In view of the arbitrariness of variations $\delta u_i^{(k)}, \delta w^{(k)}$ from (20) follows the system of $3N$ differential equation of equilibrium

$$f_1^{(k)} = S_{11,1}^{(k)} + S_{12,2}^{(k)} - N_{13}^{(k)} + X_1^{(k)} = 0; \overrightarrow{1,2}$$
$$f_3^{(k)} = S_{13,1}^{(k)} + S_{23,2}^{(k)} - N_{33}^{(k)} + X_3^{(k)} = 0 \tag{21}$$

for which in edge sections $x_i = x_i^-, x_i = x_i^+$ boundary conditions are formulated as follow

$$S_{ij}^{(k)} = Q_{ij}^{(k)} \ when\ \delta u_j^{(k)} \neq 0,\ S_{i3}^{(k)} = Q_{i3}^{(k)} \ when\ \delta w^{(k)} \neq 0 \tag{22}$$

2.2. *Equation of motion*

From Hamilton-Ostrogradsky variational principle, system of differential equations of motion can be obtained as follow

$$f_1^{(k)} = S_{11,1}^{(k)} + S_{12,2}^{(k)} - N_{13}^{(k)} + X_1^{(k)} - \tilde{X}_1^{(k)} = 0; \overrightarrow{1,2}$$
$$f_3^{(k)} = S_{13,1}^{(k)} + S_{23,2}^{(k)} - N_{33}^{(k)} + X_3^{(k)} - \tilde{X}_3^{(k)} = 0 \tag{23}$$

In it, in contrast to (21), inertial terms $\tilde{X}_i^{(k)}$, $\tilde{X}_3^{(k)}$ appeared, which are defined by formulas (10), (19).

2.3. *Kinematic relationships*

We assume that materials of layers of shell are linear elastic and orthotropic, and orthotropy axes coincide with the selected coordinate system. For such a material, taking into account its viscoelastic properties, resulting stress and strain components in plate can be related by dependencies

$$\sigma_{\alpha 1}^{[k]} = \tilde{g}_{\alpha 1}^{[k]}\varepsilon_1^{[k]} + \tilde{g}_{\alpha 2}^{[k]}\varepsilon_2^{[k]} + \tilde{g}_{\alpha 3}^{[k]}\varepsilon_3^{[k]}, \quad \sigma_{\alpha\beta}^{[k]} = \tilde{G}_{\alpha\beta}^{[k]}\gamma_{\alpha\beta}^{[k]};$$
$$\alpha \neq \beta; \quad \alpha = \overline{1,3}; \ \beta = \overline{1,3} \tag{24}$$

Here $\tilde{g}_{\alpha\beta}^{[k]}, \tilde{G}_{\alpha\beta}^{[k]}$ – differential operators, which takes form when Thompson-Kelvin-Voight model [11-14] is used

$$\tilde{G}_{12} = \tilde{G}_{21} = G_{12}\left(1 + \frac{\delta_{12}}{\pi\omega}\frac{\partial}{\partial\tau}\right), \tilde{G}_{13} = G_{13}\left(1 + \frac{\delta_{13}}{\pi\omega}\frac{\partial}{\partial\tau}\right), \tilde{G}_{23} = G_{23}\left(1 + \frac{\delta_{23}}{\pi\omega}\frac{\partial}{\partial\tau}\right)$$

$$\tilde{g}_{11}^{[k]} = \frac{E_1^{[k]}\left(1-\nu_{23}^{[k]}\nu_{32}^{[k]}\right)}{\Delta^{[k]}}\left(1+\frac{\delta_1}{\pi\omega}\frac{\partial}{\partial\tau}\right), \tag{25}$$

$$\tilde{g}_{12}^{[k]} = \tilde{g}_{21}^{[k]} = \frac{E_1^{[k]}\left(\nu_{21}^{[k]}+\nu_{23}^{[k]}\nu_{31}^{[k]}\right)}{\Delta^{[k]}}\left(1+\frac{\delta_1}{\pi\omega}\frac{\partial}{\partial\tau}\right) = \frac{E_2^{[k]}\left(\nu_{12}^{[k]}+\nu_{13}^{[k]}\nu_{32}^{[k]}\right)}{\Delta^{[k]}}\left(1+\frac{\delta_2}{\pi\omega}\frac{\partial}{\partial\tau}\right);\quad \overrightarrow{1,2,3}$$

$$\Delta^{[k]} = 1-\nu_{12}^{[k]}\nu_{21}^{[k]}-\nu_{23}^{[k]}\nu_{32}^{[k]}-\nu_{31}^{[k]}\nu_{13}^{[k]}-2\nu_{12}^{[k]}\nu_{23}^{[k]}\nu_{31}^{[k]}$$

where $\delta_\alpha, \delta_{\alpha\beta}$ – logarithmic decrements in axial and shear deformations respectively.

Substituting expression (24) into (7) and using (2) we obtain two-dimensional elasticity relationships for forces and moments

$$T_{\alpha\alpha}^{[k]} = \tilde{C}_{\alpha 1}^{[k]}\varepsilon_{11}^{[k]} + \tilde{C}_{\alpha 2}^{[k]}\varepsilon_{22}^{[k]} + \tilde{C}_{\alpha 3}^{[k]}\varepsilon_{33}^{[k]},\quad T_{\alpha\beta}^{[k]} = 2\tilde{B}_{\alpha\beta}^{[k]}\varepsilon_{\alpha\beta}^{[k]};\ \alpha\neq\beta;\ \alpha=\overline{1,3};\ \beta=\overline{1,3}$$
$$M_{ii}^{[k]} = \tilde{D}_{i1}^{[k]}\chi_{11}^{[k]} + \tilde{D}_{i2}^{[k]}\chi_{22}^{[k]},\ M_{12}^{[k]} = 2\tilde{D}_{12}^{[k]}\chi_{12}^{[k]},\ M_{i3}^{[k]} = 2\tilde{D}_{i3}^{[k]}\Omega_{i3}^{[k]};\ i=1,2 \tag{26}$$

in which the following designations are introduced for the corresponding rigidities

$$\tilde{C}_{\alpha\beta}^{[k]} = h_{[k]}\tilde{g}_{\alpha\beta}^{[k]},\ \tilde{D}_{\alpha\beta}^{[k]} = h_{[k]}^3\tilde{g}_{\alpha\beta}^{[k]}/12,\ \tilde{B}_{\alpha\beta}^{[k]} = h_{[k]}\tilde{G}_{\alpha\beta}^{[k]},\ \tilde{D}_{12}^{[k]} = h_{[k]}^3\tilde{G}_{12}^{[k]}/12 \tag{27}$$

3. Multilayer plate under normal static pressure

Suppose that a rectangular multilayer plate with dimensions a,b is subjected to uniformly distributed monoharmonic pressures on its plane in axis $0z$, varying in time by the law (ω – circular frequency, $i=\sqrt{-1}$) $p^{(1)}=\tilde{p}^{(1)}e^{i\omega\tau}$, $p^{(N)}=\tilde{p}^{(N)}e^{i\omega\tau}$. We assume that they are applied to boundary planes $z_{[1]}=-h_{[1]}/2$, $z_{[N-1]}=h_{[N-1]}/2$, respectively. In this type of loading, the equations (22) for $[k]$-th layer, by using the relationships (26), (19), (13)-(15), which are constructed with respect to $U^{(k)}=\left\{u_i^{(k)},w^{(k)}\right\}^T$, takes form

$$\begin{aligned}
f_1^{(k)} = {} & \frac{h_{[k]}\tilde{g}_{11}^{[k]}}{6}u_{1,11}^{(k+1)} + \frac{h_{[k]}\tilde{g}_{11}^{[k]}+h_{[k-1]}\tilde{g}_{11}^{[k-1]}}{3}u_{1,11}^{(k)} + \frac{h_{[k-1]}\tilde{g}_{11}^{[k-1]}}{6}u_{1,11}^{(k-1)} + \\
& + \frac{h_{[k]}\tilde{G}_{12}^{[k]}}{6}u_{1,22}^{(k+1)} + \frac{h_{[k]}\tilde{G}_{12}^{[k]}+h_{[k-1]}\tilde{G}_{12}^{[k-1]}}{3}u_{1,22}^{(k)} + \frac{h_{[k-1]}\tilde{G}_{12}^{[k-1]}}{6}u_{1,22}^{(k-1)} + \\
& \frac{h_{[k]}\left(\tilde{g}_{12}^{[k]}+\tilde{G}_{12}^{[k]}\right)}{6}u_{2,12}^{(k+1)} + \frac{h_{[k]}\left(\tilde{g}_{12}^{[k]}+\tilde{G}_{12}^{[k]}\right)+h_{[k-1]}\left(\tilde{g}_{12}^{[k-1]}+\tilde{G}_{12}^{[k-1]}\right)}{3}u_{2,12}^{(k)} + \\
& + \frac{h_{[k-1]}\left(\tilde{g}_{12}^{[k-1]}+\tilde{G}_{12}^{[k-1]}\right)}{6}u_{2,12}^{(k-1)} + \frac{\tilde{g}_{13}^{[k]}+\tilde{G}_{13}^{[k]}}{2}w_{,1}^{(k+1)} + \\
& + \frac{\tilde{G}_{13}^{[k]}-\tilde{g}_{13}^{[k]}-\tilde{G}_{13}^{[k-1]}+\tilde{g}_{13}^{[k-1]}}{2}w_{,1}^{(k)} - \frac{\tilde{g}_{13}^{[k-1]}+\tilde{G}_{13}^{[k-1]}}{2}w_{,1}^{(k-1)} + \\
& + \frac{\tilde{G}_{13}^{[k]}}{h_{[k]}}u_1^{(k+1)} - \left(\frac{\tilde{G}_{13}^{[k]}}{h_{[k]}}+\frac{\tilde{G}_{13}^{[k-1]}}{h_{[k-1]}}\right)u_1^{(k)} + \frac{\tilde{G}_{13}^{[k-1]}}{h_{[k-1]}}u_1^{(k-1)} - \\
& -\frac{h_{[k]}\rho_p^{[k]}}{6}\ddot{u}_1^{(k+1)} - \frac{h_{[k]}\rho_p^{[k]}+h_{[k-1]}\rho_p^{[k-1]}}{3}\ddot{u}_1^{(k)} - \frac{h_{[k-1]}\rho_p^{[k-1]}}{6}\ddot{u}_1^{(k-1)} = 0;\ \overleftrightarrow{1,2}
\end{aligned} \tag{28}$$

$$
\begin{aligned}
f_3^{(k)} = & \frac{h_{[k]}\tilde{G}_{13}^{[k]}}{12} w_{,11}^{(k+1)} + \frac{5\left(h_{[k]}\tilde{G}_{13}^{[k]} + h_{[k-1]}\tilde{G}_{13}^{[k-1]}\right)}{12} w_{,11}^{(k)} + \frac{h_{[k-1]}\tilde{G}_{13}^{[k-1]}}{12} w_{,11}^{(k-1)} + \\
& + \frac{h_{[k]}\tilde{G}_{23}^{[k]}}{12} w_{,22}^{(k+1)} + \frac{5\left(h_{[k]}\tilde{G}_{23}^{[k]} + h_{[k-1]}\tilde{G}_{23}^{[k-1]}\right)}{12} w_{,22}^{(k)} + \frac{h_{[k-1]}\tilde{G}_{23}^{[k-1]}}{12} w_{,22}^{(k-1)} + \\
& + \frac{\tilde{g}_{13}^{[k]} + \tilde{G}_{13}^{[k]}}{2} u_{1,1}^{(k+1)} + \frac{\tilde{g}_{13}^{[k]} - \tilde{G}_{13}^{[k]} - \tilde{g}_{13}^{[k-1]} + \tilde{G}_{13}^{[k-1]}}{2} u_{1,1}^{(k)} - \frac{\tilde{g}_{13}^{[k-1]} + \tilde{G}_{13}^{[k-1]}}{2} u_{1,1}^{(k-1)} + \\
& + \frac{\tilde{g}_{23}^{[k]} + \tilde{G}_{23}^{[k]}}{2} u_{2,2}^{(k+1)} + \frac{\tilde{g}_{23}^{[k]} - \tilde{G}_{23}^{[k]} - \tilde{g}_{23}^{[k-1]} + \tilde{G}_{23}^{[k-1]}}{2} u_{2,2}^{(k)} - \frac{\tilde{g}_{23}^{[k-1]} + \tilde{G}_{23}^{[k-1]}}{2} u_{2,2}^{(k-1)} + \\
& + \frac{\tilde{g}_{33}^{[k]}}{h_{[k]}} w^{(k+1)} - \left(\frac{\tilde{g}_{33}^{[k]}}{h_{[k]}} + \frac{\tilde{g}_{33}^{[k-1]}}{h_{[k-1]}}\right) w^{(k)} + \frac{\tilde{g}_{33}^{[k-1]}}{h_{[k-1]}} w^{(k-1)} - \\
& - \frac{h_{[k]}\rho_p^{[k]}}{6} \ddot{w}^{(k+1)} - \frac{h_{[k]}\rho_p^{[k]} + h_{[k-1]}\rho_p^{[k-1]}}{3} \ddot{w}^{(k)} - \frac{h_{[k-1]}\rho_p^{[k-1]}}{6} \ddot{w}^{(k-1)} + p^{(k)} = 0
\end{aligned}
$$

We note that for the first and $[N-1]$-th layer corresponding equations of motion are obtained from (28) at $U^{(0)} = 0, U^{(N+1)} = 0$.

In case of simply supported edge faces of plates solutions of equation (28) should be subject to boundary conditions:

$$
\begin{aligned}
& at\ x_1 = 0, x_1 = a \quad u_{1,1}^{(k)} = 0,\ u_2^{(k)} = 0, w^{(k)} = 0 \\
& at\ x_2 = 0, x_2 = b \quad u_{2,2}^{(k)} = 0,\ u_1^{(k)} = 0, w^{(k)} = 0
\end{aligned} \tag{29}
$$

In accordance with the above assumption of monoharmonic loading, solution (28), satisfying the boundary conditions (29), will have the form

$$
\begin{aligned}
u_1^{(k)} &= e^{i\omega\tau} \sum_{m,n=1,3,\ldots} \tilde{u}_{1mn}^{(k)} \cos\lambda_m x \cdot \sin\lambda_n y, \\
u_2^{(k)} &= e^{i\omega\tau} \sum_{m,n=1,3,\ldots} \tilde{u}_{2mn}^{(k)} \sin\lambda_m x \cdot \cos\lambda_n y, \\
w^{(k)} &= e^{i\omega\tau} \sum_{m,n=1,3,\ldots} \tilde{w}_{mn}^{(k)} \sin\lambda_m x \cdot \sin\lambda_n y
\end{aligned} \tag{30}
$$

where $\lambda_m = m\pi/a$, $\lambda_n = n\pi/b$. In the virtue of (30) the pressure amplitude values represented as a Fourier series expansion

$$
\tilde{p}^{(k)} = \tilde{p}^{(k)} \sum f_{mn} \sin\lambda_m x\ \sin\lambda_n y,\ f_{mn} = \frac{16}{\pi^2 mn} \tag{31}
$$

Substituting (30) and (31) into equations of motion (28) we get system of $3N$ algebraic equations with respect to amplitude values $\tilde{u}_{i\,mn}^{(k)}$ and $\tilde{w}_{mn}^{(k)}$

$$
\begin{aligned}
f_i^{(k)} = & k_{i1}^{[k]} \tilde{u}_{1mn}^{(k+1)} + \left(\tilde{k}_{i1}^{[k]} + \tilde{k}_{i1}^{[k-1]}\right) \tilde{u}_{1mn}^{(k)} + k_{i1}^{[k-1]} \tilde{u}_{1mn}^{(k-1)} + \\
& + k_{i2}^{[k]} \tilde{u}_{2mn}^{(k+1)} + \left(\tilde{k}_{i2}^{[k]} + \tilde{k}_{i2}^{[k-1]}\right) \tilde{u}_{2mn}^{(k)} + k_{i2}^{[k-1]} \tilde{u}_{2mn}^{(k-1)} + \\
& + k_{i3}^{[k]} \tilde{w}_{mn}^{(k+1)} + \left(\tilde{k}_{i3}^{[k]} - \tilde{k}_{i3}^{[k-1]}\right) \tilde{w}_{mn}^{(k)} - k_{i3}^{[k-1]} \tilde{w}_{mn}^{(k-1)} = 0;\ i = 1, 2
\end{aligned} \tag{32}
$$

$$f_3^{(k)} = k_{31}^{[k]}\tilde{u}_{1mn}^{(k+1)} + \left(\tilde{k}_{31}^{[k]} - \tilde{k}_{31}^{[k-1]}\right)\tilde{u}_{1mn}^{(k)} - k_{31}^{[k-1]}\tilde{u}_{1mn}^{(k-1)} +$$
$$+k_{32}^{[k]}\tilde{u}_{2mn}^{(k+1)} + \left(\tilde{k}_{32}^{[k]} - \tilde{k}_{32}^{[k-1]}\right)\tilde{u}_{2mn}^{(k)} - k_{32}^{[k-1]}\tilde{u}_{2mn}^{(k-1)} +$$
$$+k_{33}^{[k]}\tilde{w}_{mn}^{(k+1)} + \left(\tilde{k}_{33}^{[k]} + \tilde{k}_{33}^{[k-1]}\right)\tilde{w}_{mn}^{(k)} + k_{33}^{[k-1]}\tilde{w}_{mn}^{(k-1)} = -\tilde{p}^{(k)} f_{mn}$$

where the following designations were introduced

$$k_{11}^{[k]} = \frac{h_{[k]}\rho_p^{[k]}\omega^2}{6} - \frac{h_{[k]}\tilde{g}_{11}^{[k]}\lambda_m^2}{6} - \frac{h_{[k]}\tilde{G}_{12}^{[k]}\lambda_n^2}{6} + \frac{\tilde{G}_{13}^{[k]}}{h_{[k]}}$$

$$\tilde{k}_{11}^{[k]} = \frac{h_{[k]}\rho_p^{[k]}\omega^2}{3} - \frac{h_{[k]}\tilde{g}_{11}^{[k]}\lambda_m^2}{3} - \frac{h_{[k]}\tilde{G}_{12}^{[k]}\lambda_n^2}{3} - \frac{\tilde{G}_{13}^{[k]}}{h_{[k]}}$$

$$\tilde{k}_{12}^{[k]} = 2k_{12}^{[k]} = \tilde{k}_{21}^{[k]} = 2k_{21}^{[k]} = -\frac{h_{[k]}\left(\tilde{g}_{12}^{[k]} + \tilde{G}_{12}^{[k]}\right)\lambda_m\lambda_n}{3}$$

$$k_{13}^{[k]} = -k_{31}^{[k]} = \frac{h_{[k]}\left(\tilde{g}_{13}^{[k]} + \tilde{G}_{13}^{[k]}\right)\lambda_m}{2}, \quad \tilde{k}_{13}^{[k]} = \tilde{k}_{31}^{[k]} = \frac{h_{[k]}\left(\tilde{G}_{13}^{[k]} - \tilde{g}_{13}^{[k]}\right)\lambda_m}{2}$$

$$k_{22}^{[k]} = \frac{h_{[k]}\rho_p^{[k]}\omega^2}{6} - \frac{h_{[k]}\tilde{G}_{12}^{[k]}\lambda_m^2}{6} - \frac{h_{[k]}\tilde{g}_{22}^{[k]}\lambda_n^2}{6} + \frac{\tilde{G}_{23}^{[k]}}{h_{[k]}}$$

$$\tilde{k}_{22}^{[k]} = \frac{h_{[k]}\rho_p^{[k]}\omega^2}{3} - \frac{h_{[k]}\tilde{G}_{12}^{[k]}\lambda_m^2}{3} - \frac{h_{[k]}\tilde{g}_{22}^{[k]}\lambda_n^2}{3} - \frac{\tilde{G}_{23}^{[k]}}{h_{[k]}} \qquad (33)$$

$$k_{23}^{[k]} = -k_{32}^{[k]} = \frac{h_{[k]}\left(\tilde{g}_{23}^{[k]} + \tilde{G}_{23}^{[k]}\right)\lambda_n}{2}, \quad \tilde{k}_{23}^{[k]} = \tilde{k}_{32}^{[k]} = \frac{h_{[k]}\left(\tilde{G}_{23}^{[k]} - \tilde{g}_{23}^{[k]}\right)\lambda_n}{2}$$

$$k_{33}^{[k]} = \frac{h_{[k]}\rho_p^{[k]}\omega^2}{6} - \frac{h_{[k]}\tilde{G}_{13}^{[k]}\lambda_m^2}{12} - \frac{h_{[k]}\tilde{G}_{23}^{[k]}\lambda_n^2}{12} + \frac{\tilde{g}_{33}^{[k]}}{h_{[k]}}$$

$$\tilde{k}_{22}^{[k]} = \frac{h_{[k]}\rho_p^{[k]}\omega^2}{3} - \frac{5h_{[k]}\tilde{G}_{13}^{[k]}\lambda_m^2}{12} - \frac{5h_{[k]}\tilde{G}_{23}^{[k]}\lambda_n^2}{12} - \frac{\tilde{g}_{33}^{[k]}}{h_{[k]}}$$

Note that under static loading of plate, the corresponding equilibrium equations for transverse bending of a multilayer plate are obtained from (32) and (33) at $\omega = 0$.

4. Results of numerical studies on the transverse bending of single-layer plate and their analysis

On the basis of the solution found, let's take into consideration the problem of determining the parameters of stress-strain state of rectangular plate subjected to uniformly distributed static pressure acting on one of its boundary planes so that $p^{(1)} = 1\ kPa$, and $p^{(N)} = 0$. We will assume that plate is made of steel and have the following values of parameters

$$E_1 = E_2 = E_3 = E = 200\,\text{GPa},\ \nu_{12} = \nu_{13} = \nu_{23} = 0.3,\ a = 480\,\text{mm},\ b = 560\,\text{mm},\ t = 3\,\text{mm}.$$

In order to approximate solutions based on the use of two-dimensional equations of the theory of plates to the solutions obtained by correspondingly use of three-dimensional equations of elasticity theory, let's consider described above single-layer plate in the form of a multilayer plate from M layers with thicknesses t/M and having equal physical and mechanical properties. Obtained solution of the problem should be regarded as the use of differential forms of the finite element method [15], in which the displacements in the layers of the plate are approximated by linear shape functions.

As an illustration, in fig.2-3, distribution of deflection w and normal stress $\sigma_x = \sigma_{11}$ through the plate thickness are introduced at the centre of the plate. Here and below dashed lines correspond to solutions, obtained at $M = 1$ (six mode variant of theory of plates and shells [16], corresponding to use in the plate Timoshenko model with transverse normal strains), results with dotted dash correspond to $M = 5$ and dotted lines to $M = 10$. These curves are compared with results, obtained on the basis of classical Kirchhoff-Love model for single-layer isotropic plate (continuous line), when the values of w and $\sigma_x = \sigma_{11}$ are calculated by formula

$$w = \sum_{m,n=1,3...} w_{mn} = \sum_{m,n=1,3...} \frac{12p^{(1)} f_{mn}\left(1-\nu^2\right)}{Et^3\left(\lambda_m^2+\lambda_n^2\right)^2}, \sigma_x = \frac{Et}{2\left(1-\nu^2\right)} \sum_{m,n=1,3...} \left(\lambda_m^2+\nu\lambda_n^2\right) w_{mn}$$

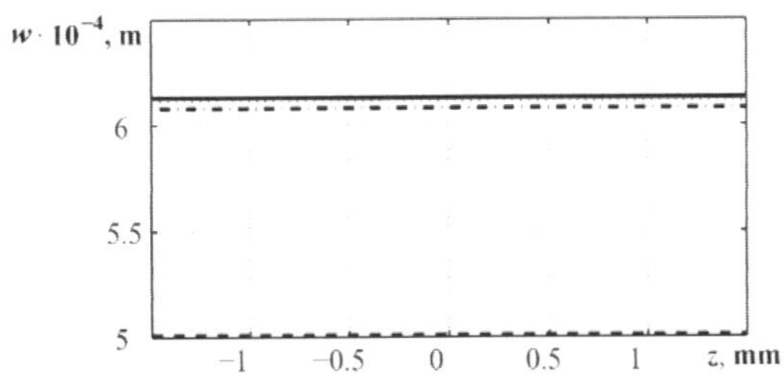

Figure 2. Relationship $w(z)$ for steel plate

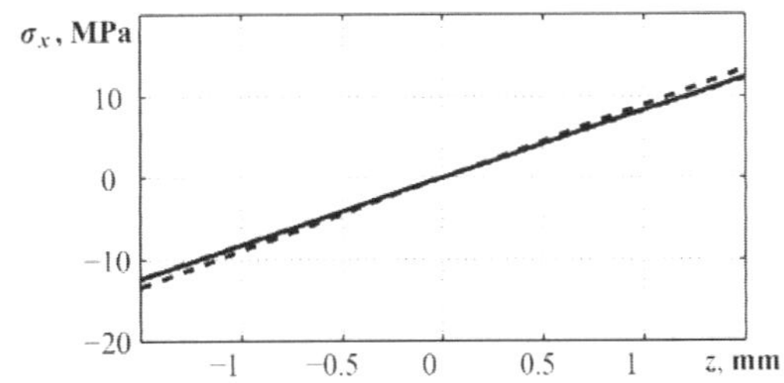

Figure 3. Relationship $\sigma_x(z)$ for steel plate

Analysing the demonstrated results, it is not difficult to see the main drawback of six-mode variant of theory, related with constant transverse normal strain of the plate, which cause [16] bending stiffness to be distorted, but, in numerical computation, cause to increase of the stiffness of discrete system: results at $M = 1$ significantly differ from the results obtained from the use of classical Kirchhoff-Love model. Results with tolerated accuracy can be obtained only increasing the number of layers in system (in this case at $M \geq 10$) and they are being closer to 3D elasticity solutions when $M > 10$.

The analysis of the equations of equilibrium and motion of the plate showed that the described disadvantage of equations with small values of M is due to the significant difference between the elasticity relations (24), (25) from the relationships, corresponding to plane stress state of the plate. Conducted computing experiments have shown that well acceptable solutions to the problem for small values of M can be obtained by taking Poisson coefficients equal to zero. For an illustration, in fig. 4-5 relationships $w = w(z)$, $\sigma_x = \sigma_x(z)$ are shown at the center of the plate, which are calculated taking $\nu_{13} = \nu_{23} = 0$. It is obvious that for the all values of parameter M results are almost identical.

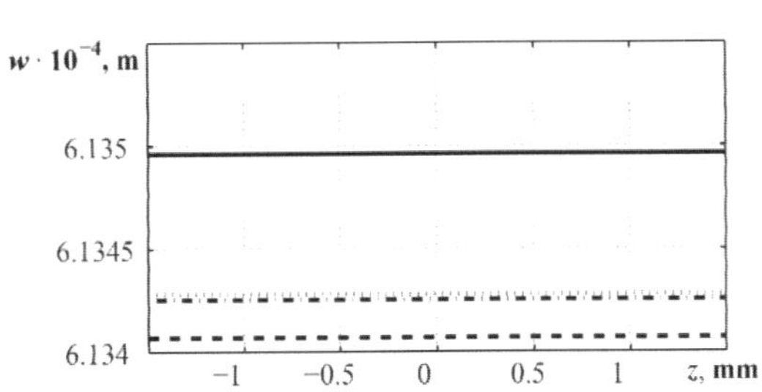

Figure 4. Relationship $w(z)$ for steel plate in case $\nu_{13} = \nu_{23} = 0$

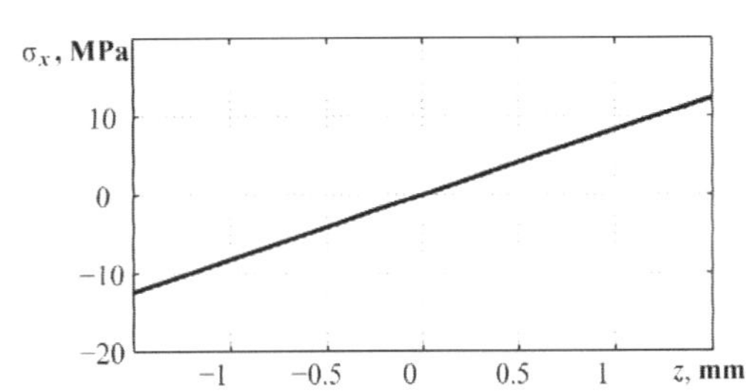

Figure 5. Relationship $\sigma_x(z)$ for steel plate in case of $\nu_{13} = \nu_{23} = 0$

Finally, through the thickness variation of transverse normal stress $\sigma_z = \sigma_{33}$ resulted at the centre of the plate is shown in fig.6., which is obtained by taking $\nu_{13} = \nu_{23} = 0$. Comparing the results, presented in fig.5 and fig.6, it can be seen that, transverse normal stress σ_{33} for the concerned thin plate is smaller than σ_{11} by three orders of its magnitude.

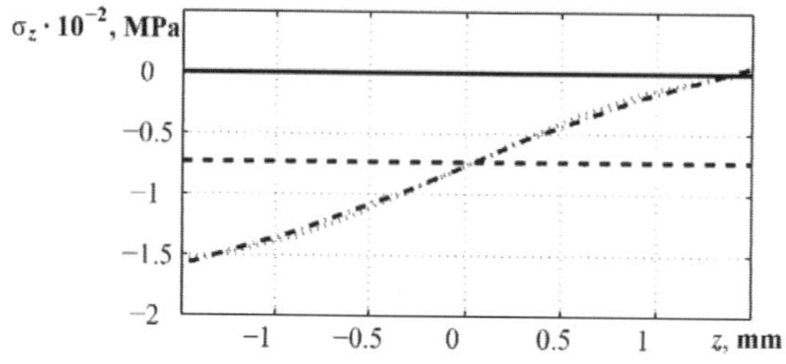

Figure 6. Relationship $\sigma_z(z)$ for steel plate in case of $\nu_{13} = \nu_{23} = 0$

Acknowledgements
This research is performed with the grant of Russian Science Foundation (project № 14-19-00667) and Russian Fond of Fundamental Research (project № 16-38-50036).

References

[1] Sergeeva E A and Abdullin I Sh 2009 Effect of high-frequency low-pressure discharge on the properties of high-strength high-modulus polyethylene fibers *Vestnik of Kazan Technological University* **2** pp 84–89
[2] Sergeeva E A, Grishanova I A, Abdulatipova L N and Ilyushina S V 2010 Optimization of low temperature plasma processing regimes of high-strength polyethylene fibers *Vestnik Kazan Technological University* **7** pp 94–98
[3] Sergeeva E A, Grishanova I A and Abdullin I Sh 2010 Influence of high capaticance plasma discharge on the physical and mechanical properties of the fibers and composite materials *Vestnik Kazan Technological University* **7** pp 109–112
[4] Bolotin V V and Novichkov Yu N 1980 *The mechanics of multilayered structures* (Moscow, Mashinostoenie Publ) p 375
[5] Dudchenko A A, Lur'e S A and Obratzov I F 1983 Anisotropic multilayered plates and shells *Mechanics of solid deformable bodies* **15** pp 3-277
[6] Grigolyuk E I and Kogan F A 1972 Current state of multilayered shell theories *Applied mechanics* **8 (6)** pp 5-17
[7] Noor A K, Burton W S and Bert Ch W 1996 Computational models for sandwich panels and shells *Applied Mechanics Reviews* **49 (3)** pp 155-199
[8] Abrosimov N A and Bazhenov V G 2002 *Nonlinear problems of the dynamics of composite structures* (Nizhni Novgorod, UNN Publ.) p 400
[9] Rickards R B and Teters G A 1974 *Stability of shells made of composite materials* (Riga, Zinatne Publ) p 310
[10] Galimov K Z 1975 *Foundations of nonlinear theory of thin shells* (Kazan, Kazan Federal University Publ) p 326
[11] Davidenkov N N 1938 On the energy absorbtion in vibration *Tech. Phys. Jour.* **8 (6)** pp 483–499
[12] Panovko Ya G 1960 *Internal friction in oscillation of elastic systems* (Moscow, FizMatGiz Publ.) p 193
[13] Sorokin E S 1960 *On the internal friction theory of the oscillations of elastic systems* (Moscow, GosStrojIzdat Publ.) p 129
[14] Pisarenko G S, Yakovlev A P and Matveev V V 1971 *Vibraiton damping properties of structural materials: Handbook* (Kiev, Naukova Dumka Publ) p 375

[15] Korneev V G and Rozin L A 1975 The differential form of the finite element method applied to the problems of elasticity theory *The proceedings in the mechanics of deformable continuum* (Moscow, Nauka Publ) pp 297-306.

[16] Bajenov V G 1981 Nonlinear dynamic problems of thin-walled structures under impulsive loads *Applied problems of strength and plasticity, Russian inter-universtiy collections, Gorkiy University* **18** pp 57–66.

IOP Conf. Series: Materials Science and Engineering **158** (2016) 012075 doi:10.1088/1757-899X/158/1/012075

Acoustic impact on the laminated plates placed between barriers

V N Paimushin[1, 2], R K Gazizullin[1] and G V Fedotenkov[3]

[1]Kazan National Research Technical University named after A.N.Tupolev – KAI, K.Marx Str.,10, Kazan, 420111, Russian Federation
[2]Kazan Federal University, Kremlyovskaya Str.,18, Kazan, 420008, Russian Federation
[3]Moscow Aviation Institute (National Research University), Volokolamskoe shosse,4, Moscow, 125993, Russian Federation

E-mail: vpajmushin@mail.ru

Abstract. On the basis of previously derived equations, analytical solutions are established on the forced vibrations of two-layer and three-layers rectangular plates hinged in an opening of absolutely rigid walls during the transmission of monoharmonic sound waves. It is assumed that the partition wall is situated between two absolutely rigid barriers, one of them by harmonic oscillation with a given displacements amplitude on the plate forms the incident sound wave, and the other is stationary and has a coating of deformable energy absorbing material with high damping properties. The behavior of acoustic environments in the spaces between the deformable plate and the barriers described by classical wave equation based on the ideal compressible fluid model. To describe the process of dynamic deformation of the energy absorbing coating of fixed barrier, two-dimensional equations of motion based on the use of models transversely soft layer are derived with a linear approximation of the displacement field in the thickness direction of the coating and taking into account the damping properties of the material and the hysteresis model for it. The influence of the physical and mechanical properties of the concerned mechanical system and the frequency of the incident sound wave on the parameters of its insulation properties of the plate, as well as on the parameters of the stress-strain state of the plate has been analyzed.

1. Introduction

In the second half of the last century, a scientific direction in mechanics was founded on the study of stationary and non-stationary interactions of acoustic waves with barriers in the form of solid deformable bodies and thin-walled structural elements. This direction continues still to attract the attention of researchers by its actuality, complexity and diversity of the phenomena inherent in the process of interaction of bodies with different physical fields. Related to this direction, issues about the aero-hydroelasticity of thin-walled shell structures have been covered in a number of monographs and reviews ([1-4], and et al.). However, they do not cover the issues about sound wave formation and the study on soundproofing problems as well as sound absorption by various deformable coatings, although in almost all publications devoted to the creation of various kinds of multi-layer constructions emphasizing that they have good soundproofing and sound absorption properties ([5, 6] et al.). Such problems of mechanics are problems acoustoelasticity, which is devoted to the rather extensive literature in the form of scientific articles ([7-10], et al.), monographs ([11-13], et al.) and review articles ([14-18] et al.).

Subject matter of this article - the problem of forced oscillations of a rectangular plate (two-layer and three-layer), hinged to an opening in absolutely rigid partitions, under the influence of monoharmonic external pressure. Without taking into account the interaction of the plate with the surrounding acoustic environments, as a rule, researches are conducted on forced vibrations of thin-walled structural elements ignoring the fact that the real structural elements are not in vacuum, but in acoustic environments. The following shows that the correct formulation of these problems requires consideration of external damping.

2. Problem statement

Let's consider multi-layer plate consist of $M = N - 1$ layers (fig.1),

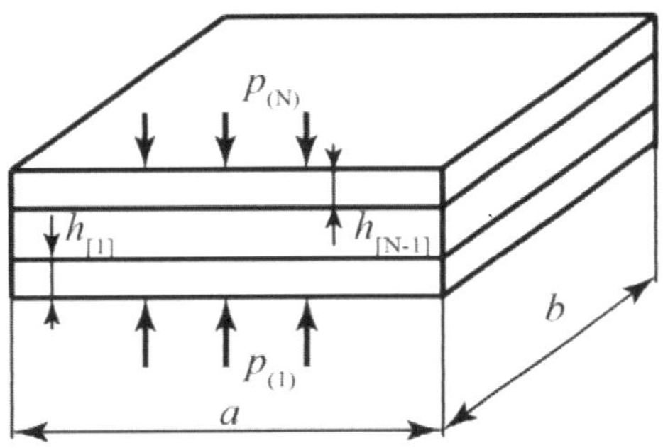

Figure 1. Multilayer plate scheme.

We assume that the plate is placed in the opening of an absolutely rigid partition separating two adjacent spaces V_1 and V_2. Coordinate plane $z = 0$ (fig.2) is compatible with the median plane of the plate. We assume that at distances $z = -l_1$ and $z = l_2$ from the plate positioned absolutely rigid barriers of infinite extent in directions x and y, coinciding with the coordinate axes of the plate x_1 and x_2, respectively. One of the barriers makes z-axis direction in the harmonic oscillation with angular frequency ω and amplitude $\tilde{U}_*$, and the second fixed coating has low-rigidity deformable layer with thickness h_c. We assume that the boundary plane of coating $z = l_2 + h_c$ is fixed and the point of the boundary plane $z = l_2$ due to the deformation of this plane undergoes the following definition of displacement

$$U = U(x, y, \tau),\ V = V(x, y, \tau),\ W = W(x, y, \tau).$$

We assume that V_s $(s = 1, 2)$ space filled with acoustic medium with density ρ_s and sound velocity c_s. Then, due to harmonic oscillations of barriers $z = -l_1$, acoustic wave pressures p_1 and p_2 will be formed in spaces V_s, which are reflected and emitted in the first medium and the radiated in the second medium.

To determine pressure values p_1^0 and p_2^0, applied to the first and N -th layers of the boundary planes of the plate, respectively, it is necessary to find solutions to the wave equation [19] (c_1, c_2 - speed of sound in the spaces V_1 and V_2)

$$\Phi_{s,xx} + \Phi_{s,yy} + \Phi_{s,zz} - \ddot{\Phi}_s / c_s^2 = 0 \tag{1}$$

written with respect to the potential velocity Φ_1, Φ_2 of sound waves, set in space V_1 and V_2 respectively. Here and throughout $s = 1, 2$, the points above the functions indicate derivatives with respect to time τ, and the subscript after the decimal point - the partial derivatives with respect to coordinates x, y and z.

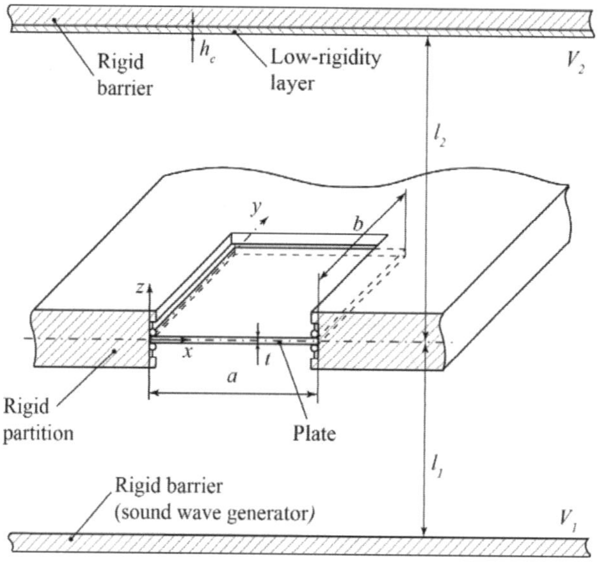

Figure 2. Plate fixing scheme.

By the functions Φ_*, Φ_s, pressure p_s and velocity components v_x^s, v_y^s, v_z^s in the spaces V_1, V_2 defined by (ρ_s - density of environments in V_1 and V_2 spaces)

$$v_x^s = \Phi_{s,x}, v_y^s = \Phi_{s,y}, v_z^s = \Phi_{s,z}, p_s = -\rho_s \dot{\Phi}_s \tag{2}$$

The process of dynamic deformation of the plate will be described by the equations of the theory of multilayer plates obtained in [20] with allowance for transverse compression and internal friction of the material of the plate on the model Thomson-Kelvin-Voigt. As the unknown components displacement components $u_1^{(k)}, u_2^{(k)}, w^{(k)}$ of the points on the facial planes of the first ($k=1$) and the last ($k=N-1$) layers, as well as points of interlayer planes are taken. In the case of simply supported plate in the opening of an absolutely rigid partitions, for these functions the following representations are valid

$$\begin{gathered} u_1^{(k)} = \sum \tilde{u}_{1mn}^{(k)} \cos \lambda_m x \cdot \sin \lambda_n y, \ u_2^{(k)} = \sum \tilde{u}_{2mn}^{(k)} \sin \lambda_m x \cdot \cos \lambda_n y, \\ w^{(k)} = \sum \tilde{w}_{mn}^{(k)} \sin \lambda_m x \cdot \sin \lambda_n y, \ \lambda_m = m\pi/a, \quad \lambda_n = n\pi/b; \end{gathered} \tag{3}$$

Hereafter, unless otherwise stated, and the summation is to be done over $m = 1,3,...$ and $n = 1,3,...$ and plate dimensions are shown in its plan view through a and b.

During the formation of the acoustic waves in space V_2 emitted by the plate under the action of pressure $p_2^+ = p_2|_{z=l_2}$, coating layers with low stiffness on fixed barrier layer will be deformed. We assume that the main stress component will be that of in the normal direction σ_z^c and the shear stresses τ_{xz}^c, τ_{yz}^c associated with the relevant components of the strain ε_z^c и $\gamma_{xz}^c, \gamma_{yz}^c$ and taking into account the viscoelastic properties of the material by Thomson-Kelvin-Voigt model with elastic relations of the following kind

$$\sigma_z^c = E_z^c D_z^c \varepsilon_z^c, \quad \tau_{xz}^c = G_c D_\gamma^c \gamma_{xz}^c, \quad \tau_{yz}^c = G_c D_\gamma^c \gamma_{yz}^c \tag{4}$$

where

$$D_z^c = 1 + \frac{\delta_z^c}{\pi\omega}\frac{\partial}{\partial\tau}, \quad D_\gamma^c = 1 + \frac{\delta_\gamma^c}{\pi\omega}\frac{\partial}{\partial\tau} \tag{5}$$

Here $E_z^c, G_c, \delta_z^c, \delta_\gamma^c$ are the corresponding elastic and damping characteristics of the coating material. Considering that $h_c << a,b$, for the displacement components in the coating we will take the approximation

$$\left(u_c, v_c, w_c\right) = \left(1 - \frac{z - l_2}{h_c}\right)\left(U, V, W\right) \tag{6}$$

allowing in necessary degree of accuracy to describe the strains $\varepsilon_z^c, \gamma_{xy}^c$ and γ_{yz}^c. Their substitution into the Cauchy relation leads to relationships

$$\gamma_{xz}^c = \left(1 - \frac{z - l_2}{h_c}\right)W_{,x} - \frac{U}{h_c}, \quad \gamma_{yz}^c = \left(1 - \frac{z - l_2}{h_c}\right)W_{,y} - \frac{V}{h_c}, \quad \varepsilon_z^c = -\frac{W}{h_c}. \tag{7}$$

When using the relations (4) - (7) for the case when the coating layer is subjected to the lateral load p_2^+, it is possible to get the system of three differential equations of motion in standard way for the coating layer

$$\begin{gathered} G_c D_\gamma^c\left(\frac{U}{h_c} - \frac{W_{,x}}{2}\right) + \frac{\rho_c h_c}{3}\ddot{U} = 0, \quad G_c D_\gamma^c\left(\frac{V}{h_c} - \frac{W_{,y}}{2}\right) + \frac{\rho_c h_c}{3}\ddot{V} = 0 \\ \frac{E_z^c}{h_c} D_z^c W + \frac{\rho_c h_c}{3}\ddot{W} - G_c D_\gamma^c\left[\frac{h_c}{3}\left(W_{,xx} + W_{,yy}\right) - \frac{1}{2}\left(U_{,x} + V_{,y}\right)\right] - p_2^+ = 0 \end{gathered} \tag{8}$$

wherein ρ_c - the density of the coating material.

By the time, solution of the equations of motion of the plate, and the equations (8) and (1) must satisfy the kinematic coupling conditions

$$\dot{w}^{(1)} = v_z^1\big|_{z=0}, \quad \dot{w}^{(N)} = v_z^2\big|_{z=0}, \dot{W} = v_z^2\big|_{z=l_2}, \quad \dot{U}_* = v_z^1\big|_{z=-l_1} \tag{9}$$

at all points of the boundary planes of the plate (the first pair of conditions (9)), and the first barrier coating.

3. Determination of aero-hydrodynamic loads acting on plate

In accordance with the assumptions made above for the functions $U_*(\tau)$ the representation of the form [21] must be taken

$$U_* = \tilde{U}_* e^{i\omega\tau} \tag{10}$$

by virtue of which, taking into account the first two terms of (9) and representation (3) solutions of equations (1) should be of the form

$$\Phi_s = e^{i\omega\tau}\sum\tilde{\Phi}_s^{mn}(z)\sin\lambda_m x\sin\lambda_n y \tag{11}$$

In turn, because of the last two terms of (9) and representation (10) - (11) for seeking solutions of equations (8) should take the form

$$\begin{gathered} U = e^{i\omega\tau}\sum\tilde{U}_{mn}\cos\lambda_m x\sin\lambda_n y, \; V = e^{i\omega\tau}\sum\tilde{V}_{mn}\sin\lambda_m x\cos\lambda_n y \\ W = e^{i\omega\tau}\sum\tilde{W}_{mn}\sin\lambda_m x\sin\lambda_n y \end{gathered} \tag{12}$$

When substituting the representations (11) into the equations (1), we arrive at the equations with respect to one-dimensional functions $\tilde{\Phi}_s^{mn}(z)$:

$$\tilde{\Phi}_{s,zz}^{mn} - \left(æ_s^{mn}\right)^2\tilde{\Phi}_s^{mn} = 0; \; \left(æ_s^{mn}\right)^2 = \lambda_m^2 + \lambda_n^2 - k_s^2, \; k_s = \omega/c_s; \; m,n = 1,3,... \tag{13}$$

Depending on the sign of $\left(æ_s^{mn}\right)^2$ the equation (13) will have a solution (A_s^{mn} and B_s^{mn} – integration constants): at $\left(æ_s^{mn}\right)^2 > 0$.

$$\tilde{\Phi}_s^{mn} = A_s^{mn} e^{æ_s^{mn} z} + B_s^{mn} e^{-æ_s^{mn} z} \tag{14}$$

and in accordance with the relations (2) and (11)

$$v_z^s = e^{i\omega\tau} \sum æ_s^{mn} \left(A_s^{mn} e^{æ_s^{mn} z} - B_s^{mn} e^{-æ_s^{mn} z} \right) \sin\lambda_m x \sin\lambda_n y \tag{15}$$

$$p_s = -i\rho_s \omega e^{i\omega\tau} \sum \left(A_s^{mn} e^{æ_s^{mn} z} + B_s^{mn} e^{-æ_s^{mn} z} \right) \sin\lambda_m x \sin\lambda_n y \tag{16}$$

And for $\left(æ_s^{mn}\right)^2 < 0$ in formulas (14) - (16), the terms $æ_s^{mn}$ should be replaced by $i\tilde{æ}_s^{mn}$, where

$$\left(\tilde{æ}_s^{mn}\right)^2 = k_s^2 - \lambda_m^2 - \lambda_n^2 \tag{17}$$

Due to the fact that modern computational tools (eg, Matlab) allow to perform operations with complex numbers, below all the conversions are only belong to the case $\left(æ_s^{mn}\right)^2 > 0$. When realizing the solutions in such computation systems, the case $\left(æ_s^{mn}\right)^2 < 0$ is taken into account automatically ($æ_s^{mn}$ can be either real or complex value).

We represent the amplitude values $\tilde{U}_*$ in the form of Fourier series

$$\tilde{U}_* = \tilde{U}_* \sum f_{mn} \sin\lambda_m x \sin\lambda_n y, \ f_{mn} = \frac{16}{\pi^2 mn} \tag{18}$$

and subject the expression (15), the representation (18) and (12) (9). As a result, after a series of transformations, we obtain dependencies

$$\begin{gathered} B_s^{mn} = b_s^{mn} \xi_s - \tilde{b}_s^{mn} \theta_s, \ A_s^{mn} = a_s^{mn} \xi_s - \tilde{a}_s^{mn} \theta_s, \\ \xi_1 = \tilde{U}_*, \ \xi_2 = \tilde{W}_{mn}, \ \theta_1 = \tilde{w}_{mn}^{(1)}, \ \theta_2 = \tilde{w}_{mn}^{(N)} \end{gathered} \tag{19}$$

where

$$\begin{gathered} a_1^{mn} = b_1^{mn} = i\frac{\omega f_{mn} e^{æ_1^{mn} l_1}}{\Delta_1^{mn}}, \ a_2^{mn} = b_2^{mn} = i\frac{\omega e^{-æ_2^{mn} l_2}}{\Delta_2^{mn}} \\ \tilde{a}_s^{mn} = i\frac{\omega e^{(-1)^{s+1} 2æ_s^{mn} l_s}}{\Delta_s^{mn}}, \ \tilde{b}_s^{mn} = i\frac{\omega}{\Delta_s^{mn}} \\ \Delta_s^{mn} = æ_s^{mn}\left(1 - e^{(-1)^{s+1} 2æ_s^{mn} l_s}\right) \end{gathered} \tag{20}$$

After introducing the expression (19) in (16), we arrive at the dependencies

$$\begin{aligned} p_s^0 = p_s\big|_{z=0} &= -i\rho_s \omega e^{i\omega\tau} \sum \left[\left(a_s^{mn} + b_s^{mn}\right)\xi_s - \left(\tilde{a}_s^{mn} + \tilde{b}_s^{mn}\right)\theta_s \right] \sin\lambda_m x \sin\lambda_n y = \\ &= e^{i\omega\tau} \sum \left(R_s^{mn} \xi_s + \tilde{R}_s^{mn} \theta_s \right) \sin\lambda_m x \sin\lambda_n y \end{aligned} \tag{21}$$

$$p_2^+ = p_2\big|_{z=l_2} = e^{i\omega\tau} \sum \left(P_2^{mn} \tilde{W}_{mn} + \tilde{P}_2^{mn} \tilde{w}_{mn}^{(N)} \right) \sin\lambda_m x \sin\lambda_n y \tag{22}$$

where for $\left(æ_s^{mn}\right)^2 > 0$

$$\begin{gathered} R_s^{mn} = \frac{2\omega^2 \rho_s f_{mn} e^{(-1)^{s+1} æ_s^{mn} l_s}}{\Delta_s^{mn}}, \ \tilde{R}_s^{mn} = -\frac{\omega^2 \rho_s \left(1 + e^{(-1)^{s+1} 2æ_s^{mn} l_s}\right)}{\Delta_s^{mn}} \\ P_2^{mn} = -\tilde{R}_2^{mn}, \ \tilde{P}_2^{mn} = -R_2^{mn} \end{gathered} \tag{23}$$

Substituting the functions (12) of the first two equations of system (8), the following algebraic dependencies are obtained

$$\tilde{U}_{mn} = A_{mn}\lambda_m \tilde{W}_{mn},\ \tilde{V}_{mn} = A_{mn}\lambda_n \tilde{W}_{mn};\ A_{mn} = \frac{G_c\left(1+i\frac{\delta_\gamma^c}{\pi}\right)}{2\left[\frac{G_c}{h_c}\left(1+i\frac{\delta_\gamma^c}{\pi}\right)-\frac{\rho_c h_c}{3}\omega^2\right]} \tag{24}$$

the use of which by virtue of (22), the third equation of the system leads to dependencies

$$\tilde{W}_{mn} = \frac{\tilde{P}_2^{mn}}{C_{mn}}\tilde{w}_{mn}^{(N)} = -\frac{R_2^{mn}}{C_{mn}}\tilde{w}_{mn}^{(N)} \tag{25}$$

where

$$C_{mn} = \frac{E_z^c}{h_c}\left(1+i\frac{\delta_z^c}{\pi}\right)-\frac{\rho_c h_c}{3}\omega^2 + G_c\left(1+i\frac{\delta_\gamma^c}{\pi}\right)\left(\lambda_m^2+\lambda_n^2\right)\left(\frac{h_c}{3}-\frac{A_{mn}}{2}\right)-P_2^{mn} \tag{26}$$

Thus, using equation (26) in accordance with (21) and (22) we obtain the dependencies

$$\tilde{p}_1^{0mn} = R_1^{mn}\tilde{U}_* + \tilde{R}_1^{mn}\tilde{w}_{mn}^{(1)} \tag{27}$$

$$\tilde{p}_2^{0mn} = R_2^{mn}\tilde{W}_{mn} + \tilde{R}_2^{mn}\tilde{w}_{mn}^{(N)} = \left(\frac{R_2^{mn}\tilde{P}_2^{mn}}{C_{mn}} + \tilde{R}_2^{mn}\right)\tilde{w}_{mn}^{(N)} \tag{28}$$

Closing the equations of motion of the plate, using (27), (28) where in accordance with adopted problem statement $\tilde{p}_{mn}^{(1)} = \tilde{p}_1^{0mn}$, $\tilde{p}_{mn}^{(N)} = \tilde{p}_2^{0mn}$, we obtain the governing system of $3N+2$ linear equations with respect to the unknowns $U^{(k)} = \left\{u_{1\,mn}^{(k)}, u_{2\,mn}^{(k)}, w_{mn}^{(k)}, \tilde{p}_{mn}^{(1)}, \tilde{p}_{mn}^{(N)}\right\}^T$. For the obtained values of $\tilde{p}_{mn}^{(1)}, \tilde{p}_{mn}^{(N)}$, it is possible to determine the amplitude values of the pressures $\tilde{p}_1^0$ and $\tilde{p}_2^0$ acting on the boundary plane of the plate

$$\tilde{p}_s^0 = \sum \tilde{p}_s^{0mn} \sin\lambda_m x \sin\lambda_n y \tag{29}$$

In this case, at the point $x = a/2$, $y = b/2$ of plate, the soundproofing parameter measured in decibels will be equal to

$$R_p^0 = -20\lg\left|\frac{\sum \tilde{p}_2^{0mn}}{\sum \tilde{p}_1^{0mn}}\right| \tag{30}$$

4. Results of computations and their analysis.

In [22], on the basis of experiments conducted in the acoustic laboratory of reverberation type, sound insulation parameters were determined for a rectangular plate made of steel and having a thickness of 20 mm, 480mm length, 560mm width. Tests were performed with both energy absorbing plate coating (rubber plate mark 2H-1-S-MBS with $t_r = 2$ mm thickness), and without it. For a qualitative comparison of the theoretical results with experimental data four computational cases are considered: the first corresponds to a single-layer steel plate with properties

$$E_p = 200\text{ GPa},\ \nu_p = 0.3,\ \rho_p = 7800\text{ kg/m}^3,\ \delta_p = 0.02;$$

the second and third - a two-layer plate having on one side a coating of rubber-like material $t_r = 2$ mm thickness and acts as a first or second layer of the plate within the established models, respectively, with the following properties of the coating

$$E_r = 5\text{ MPa},\ \nu_r = 0.36,\ \rho_r = 500\text{ kg/m}^3,\ \delta_r = 1.2;$$

fourth - three-layer plate with the outer layers are made of rubber and middle layer is from steel. It is believed that the plate is surrounded on both sides by air having the properties

$$\rho_1 = \rho_2 = \rho = 1.225\text{ kg/m}^3,\ c_1 = c_2 = c = 340\text{ м/с}$$

The calculations were performed for given size $l_1 = l_2 = 2.3\,\text{m}$, the value of the parameter $\tilde{U}_* = 0.01$ mm and the thickness of layer rigid partitions $h_c = 0.3\,\text{m}$ (fig. 2) having elastic properties, density and internal damping parameter

$$E_p = 16\,\text{GPa},\ \nu_p = 0.17,\ \rho_p = 500\ \text{kg/m}^3,\ \delta_p = 0.02$$

Each layer of the plate was divided into 20 layers. As shown by numerical experiments adopted by the mesh size in the direction of plate thickness allows to obtain numerical solutions, almost approaching to solutions on the basis of three-dimensional equations of elasticity.

In Figure 3 graphs of dependencies of the soundproofing parameter R_p^0 on the frequency $f = \omega/(2\pi)$ are illustrated for the four cases described above. Here, the calculation results for the first case shown by the solid line, the second - a bar, for the third - dashed for the fourth - the dotted dash. Almost complete coincidence of the lines for the examined four cases give evidence of that the presence energy absorbing rubber layer on the steel plate has no influence on the parameter R_p^0 of the considered mechanical system.

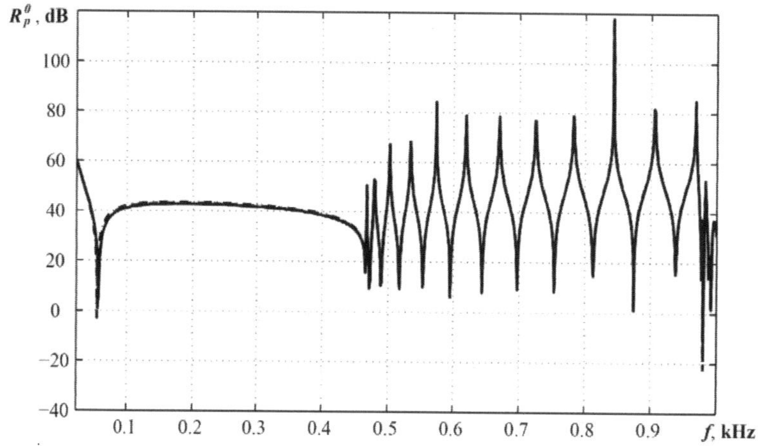

Figure 3. Relationship $R_p^0(f)$:
—— 1st case; ----- 2nd case; ······ 3rd case; -·-·- 4th case

Variation of deflection at the center of face surface of first layer of plate $\tilde{w}^{(1)}$ at two different bands of frequency f are illustrated in fig.4-5. It is obvious that there exist such frequencies of sound wave, at which significant bursts in the deflection values $\tilde{w}^{(1)}$ are observed (fig.4) due to the coincidence of the frequency of natural oscillations of the concerned mechanical system with the oscillation frequency of the incident harmonic waves. Analyzing the results, it can be seen that the addition of the energy absorbing rubber layer onto the plate leads to a substantial reduction of the deflection values at resonance frequencies f^R (fig.5). This conclusion is consistent with the results presented in [23, 24]. Along with this, there is also a slight change in the frequencies f^R themselves. We note that in fig.5 plot of dependence $\tilde{w}^{(1)} = \tilde{w}^{(1)}(f)$ for the third computation case is not listed as it fully coincided with the curve calculated for the second case.

There were also computational experiments to determine the law of variation for parameters of the stress-strain state of the plate through the thickness. For illustrative purposes, in Table 1, for the center point of the plate, the variation of the deflection values $\tilde{w}$, resulted normal stresses $\sigma_x = \sigma_{11}$, $\sigma_z = \sigma_{33}$ and shear stresses $\tau_{xz} = \sigma_{13}$ through the plate thickness are shown. The calculation results are presented for the second, third and fourth cases calculated at fixed frequency f. Results for the first computational case are not presented, as they do not differ from the results obtained on the basis of the classical Kirchhoff-Love model.

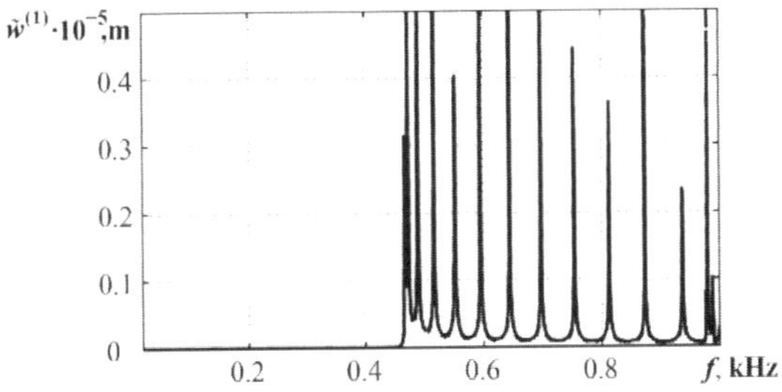

Figure 4. Relationship $\tilde{w}^{(1)}(f)$

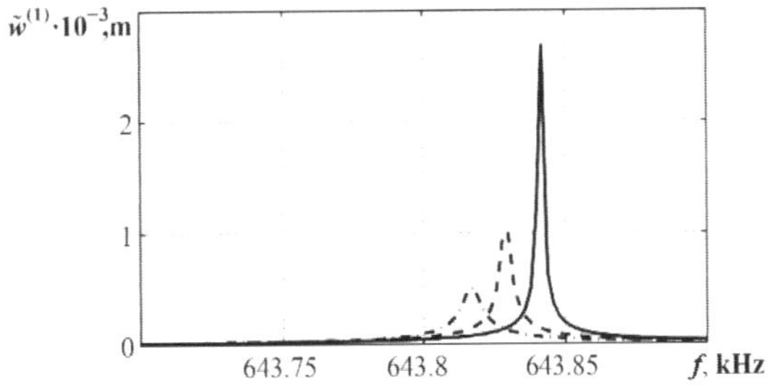

Figure 5. Relationship $\tilde{w}^{(1)}(f)$

Analyzing the graphs, it can be seen that the complicated law of variation exists for stress-strain state through the thickness of the plate, which indicates the necessity for describing the deformation mechanics of plates within the present class of problems on the basis of the refined equations [20]. It should be noted that under certain combinations of mechanical and geometrical parameters of the studied mechanical system, there are significant shifts of emerging stresses on the interfaces of the layers of materials with different mechanical properties. They, in particular, due to the complex nature of interaction of different waves propagating in the plate, including those reflected from interfaces between layers in thickness direction.

It is supposed that the results are of fundamental importance. They suggest that at high frequencies of the acoustic interaction, the damage of laminated plates of this class is likely to occur at the layer interfaces. It should be noted that at low frequencies (see e.g. $f = 50$ Hz), the main stresses in the plate are normal stresses and shear stresses. At high frequencies (see e.g. $f = 1000$ Hz) the maximum magnitude of the stress σ_z, corresponding to the deformation of the transverse compression, are comparable or even higher than the magnitudes of stress σ_x.

Table 1. The law of variation for parameters of the stress-strain state of the plate through the thickness

$f = 50$ Hz	$f = 1000$ Hz
Second computational case	
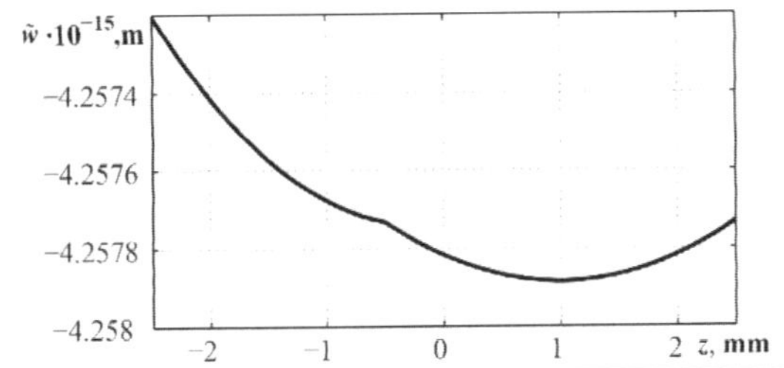	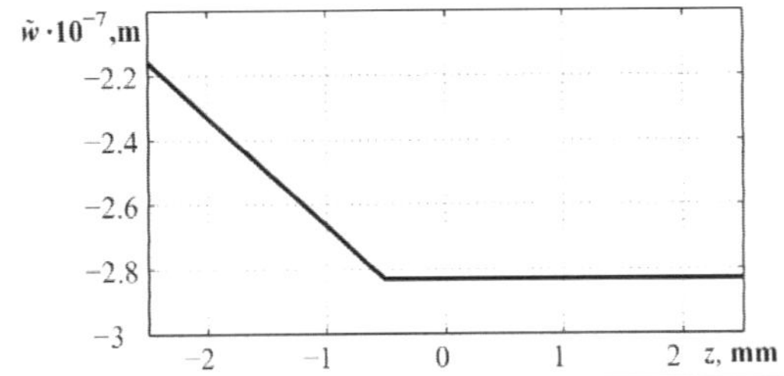

IOP Conf. Series: Materials Science and Engineering **158** (2016) 012075 doi:10.1088/1757-899X/158/1/012075

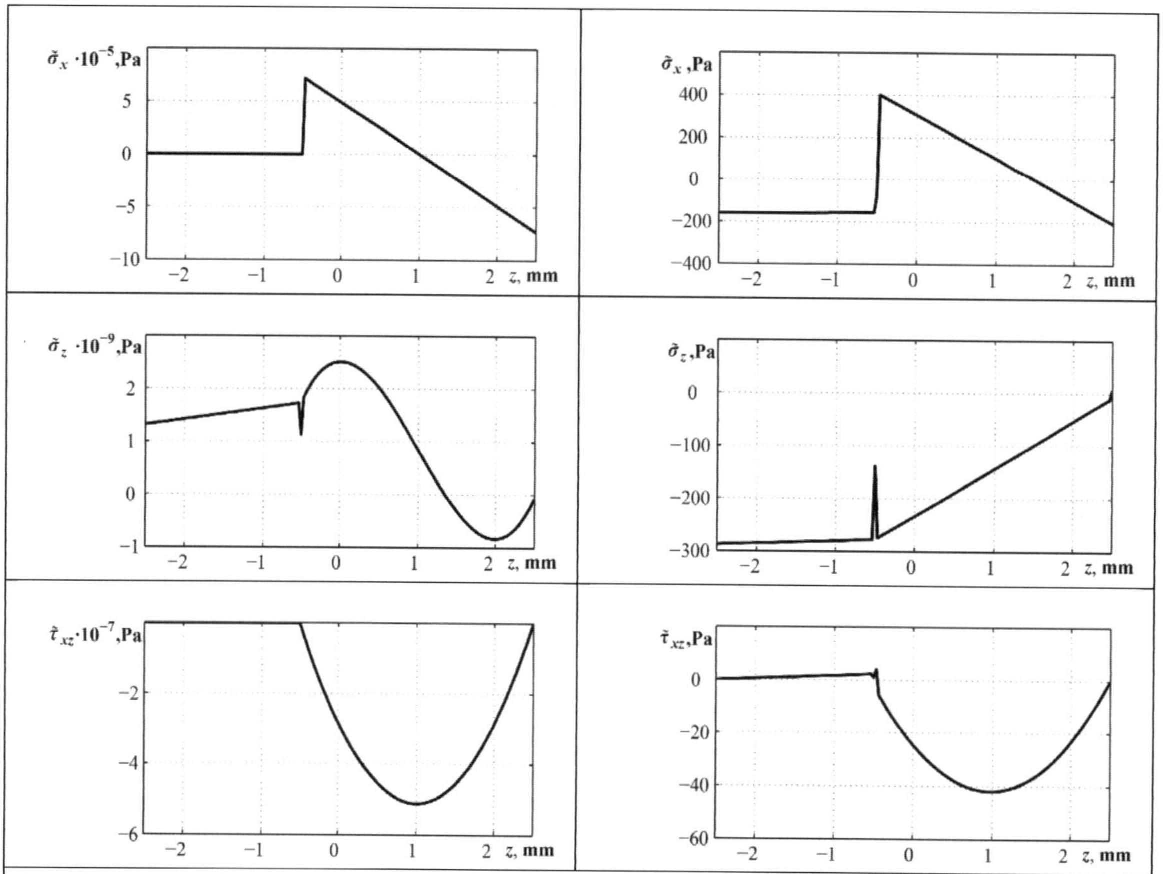

third computational case

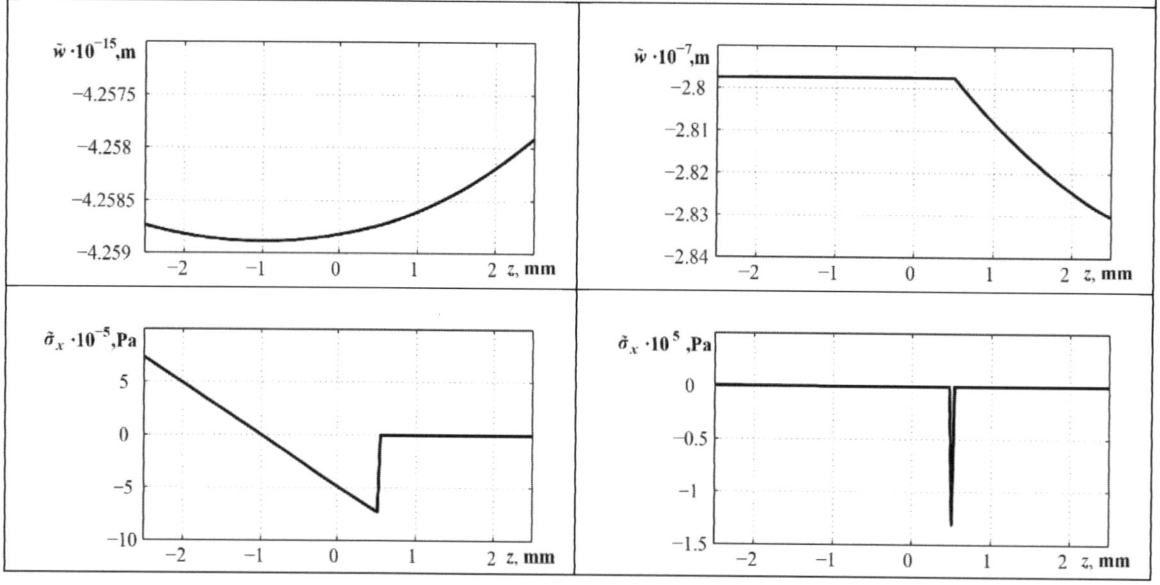

IOP Conf. Series: Materials Science and Engineering **158** (2016) 012075 doi:10.1088/1757-899X/158/1/012075

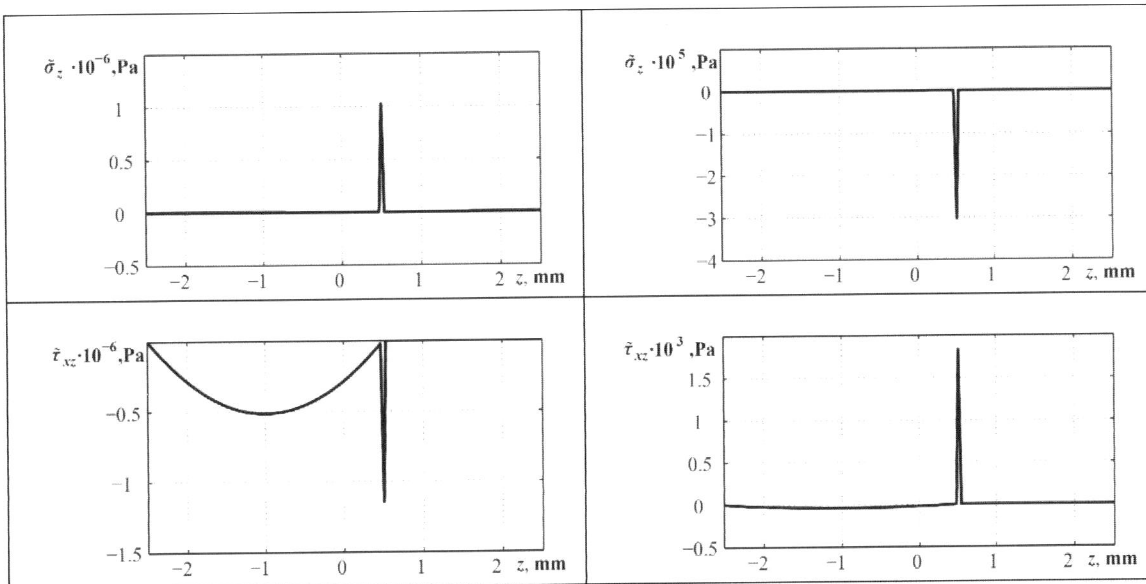

fourth computational case

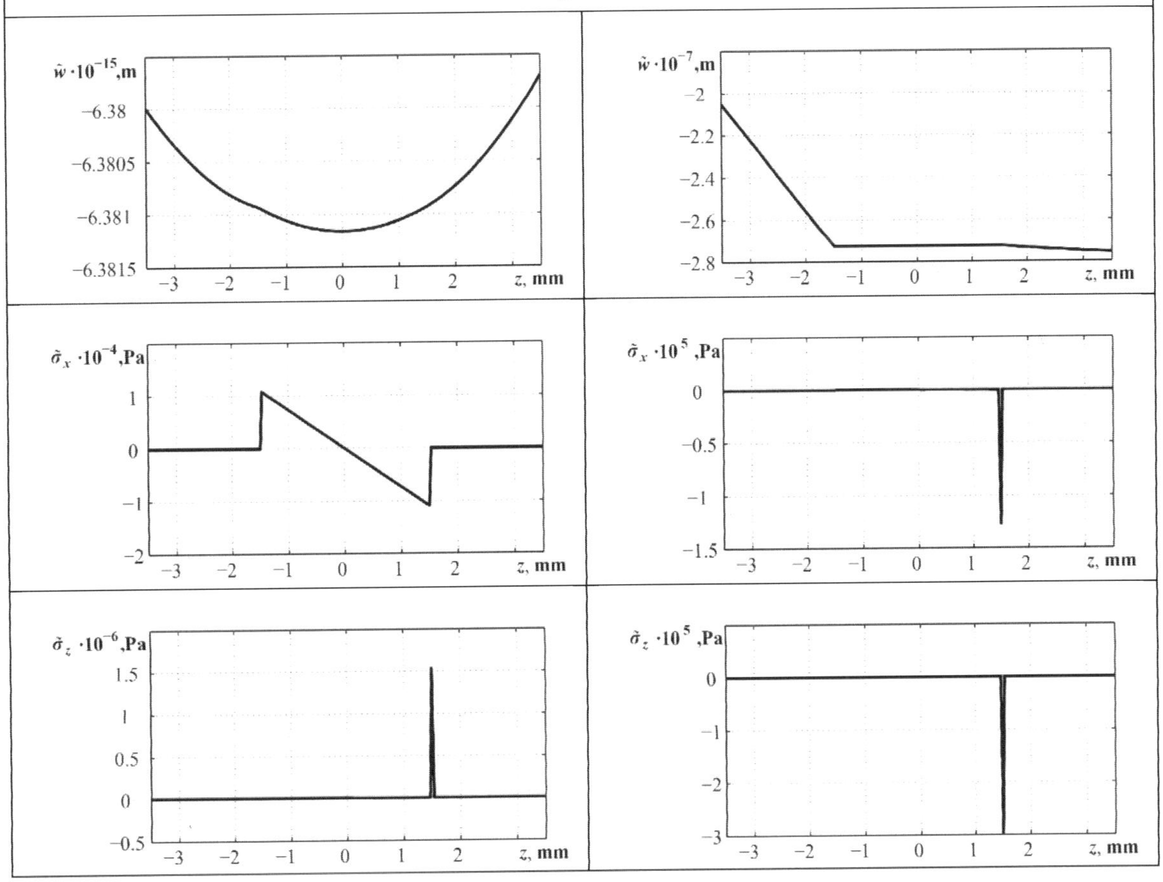

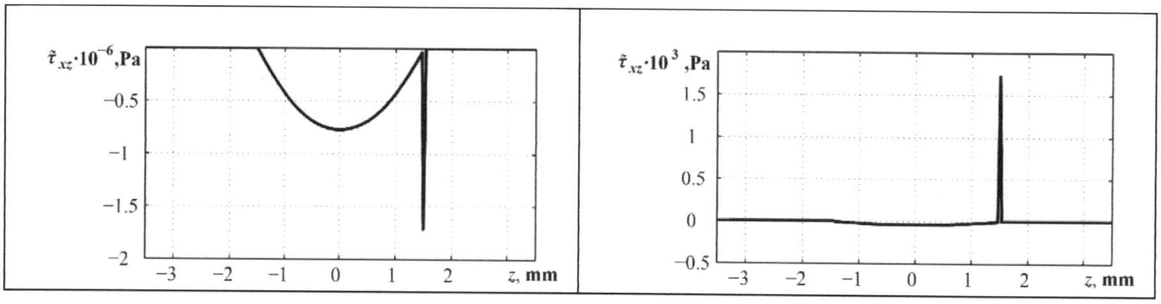

We note that stresses resulted in rubber layers σ_x, σ_y, τ_{xy}, at low frequencies are much less than the similar stresses resulted in the steel plate. In this regard, in the presented figures on the scale of the graph axes they look like zero. At high frequencies, stress shifts formed at the boundaries between the layers differing significantly by elastic characteristics are much more similar to stresses emerging in other areas.

Acknowledgements
This research is performed with the grant of Russian Science Foundation (project № 14-19-00667) and Russian Fond of Fundamental Research (project № 16-38-50036).

References

[1] Grigolyuk E I 1970 Problems of interaction with liquid membranes *Proceedings of VII All-Union conference on the theory of shells and plates, Dnepropetrovsk, 1969* (Moscow, Nauka Publ.) pp 755–778
[2] Grigolyuk E I and Gorshkov A G 1974 *Transient hydroelasticity of shells* (Leningrad, Sudostoenie Publ.) p 208
[3] Gorshkov A G 1980 Interaction of shock waves with deformable barriers *Results of science and technology. Fracture Mechanics. V.13.* (Moscow, VINITI) pp 105–186
[4] Guz A N and Kubenko V D 1982 *Computational methods of shells., V.5 The theory of unsteady aero-hydroelasticity of shells* (Kiev, Nauka. Dumka Publ.) 400 p.
[5] Stamm K and Witte H 1974 *Sandwich konstruktionen* (Wien, N. Y., Springer) p 300
[6] Endogur A I, Weinberg M V and Jerusalemsky K M 1986 *Honeycomb structures. Selection of the parameters and design* (Moscow, Mashinostroenie Publ.) p 200
[7] Pretlove A J 1965 Free vibrations of a rectangular panel backed by a closed rectangular cavity *J. Sound Vib*, V. 2, No. 3, pp 197–209
[8] Pretlove A J 1966 Forced vibrations of a rectangular panel backed by a closed rectangular cavity *J. Sound Vib*, V. 3, No. 3, pp 252–261
[9] Rdzanek Jr. W P, Rdzanek W J and Engel Z 2003 Theoretical analysis of sound radiation of an elastically supported circular plate *J. Sound Vib.* **265** pp 155–174
[10] Pico R and Gautier F 2007 The vibroacoustics of slightly distorted cylindrical shells: A model of the acoustic input impedance *J. Sound Vib.* **302** pp 18–38
[11] Hodges D H and Pierce G A 2002 Introduction to Structural Dynamics and Aeroelasticity (Cambridge: University Press) p 184
[12] Cremer L, Heckl M and Petersson B 2005 *Structure-borne Sound* (Heidelberg, Berlin, Springer) p 617
[13] Fahy F J 2007 *Sound and Structural Vibration: Radiation, Transmission and Response* (Oxford, Academic Press) p 655
[14] Campos E 2003 Review. Aeroacoustics research in Europe: The CEAS-ASC report on 2002 highlights *J. Sound Vib.* **268** pp 809–824

[15] Spyros G. Voutsinas 2007 Review. Aeroacoustics research in Europe: The CEAS-ASC report on 2005 highlights *J. Sound Vib.* **299** pp 419–459
[16] Brouwer H H and Rienstra S W 2008 Review. Aeroacoustics research in Europe: The CEAS-ASC report on 2007 highlights *J. Sound Vib.* **318** pp 625–654
[17] Juve D 2009 Aeroacoustics research in Europe: The CEAS-AS Creporton 2008 highlights *J. Sound Vib.* **328** pp 213–242
[18] Casalino D 2010 Aeroacoustics research in Europe: The CEAS-ASC report on 2009 highlights *J. Sound Vib.* **329** 4810–4828
[19] Skuchik E. 1976 *Fundamentals of acoustics. V.1.* (Moscow, Mir Publ) p 520
[20] Paimushin V N and Firsov V A 1993 *Shells made of glass. Calculation of stress-strain state* (Moscow, Mashinostroenie Publ.) p 208
[21] Paimushin V N and Gazizullin R K 2013 Research sound insulation properties of absolutely rigid plate placed on a deformable support elements between the two barriers *Uchenye Zapiski Kazanskogo Universiteta. Seriya Fiziko-Matematicheskie Nauki* **155 (3)** pp 126–141
[22] Paimushin V N, Gazizullin R K and Sharapov A A 2015 Experimental determination of the energy absorbing coating with acoustic parameters of a rectangular plate *Uchenye Zapiski Kazanskogo Universiteta. Seriya Fiziko-Matematicheskie Nauki* **157 (1)** pp 114–127
[23] Paimushin V N 2014 Dynamic Instability of a Compressible Plate Surrounded by an Acoustic Medium on Free Surfaces *Doklady Physics* **59 (3)** pp 158–162. DOI: 10.1134/S1028335814030124
[24] Paimushin V N 2015 Acoustoelastic behaviour of double-layer plates accounting internal damping of layer materials. *Procedia Engineering* **106** pp 325–335. DOI: 10.1016/j.proeng.2015.06.042

On solvability of a elliptic-parabolic problem of nonlinear filtration theory

M F Pavlova[1] **and E V Rung**[2]

[1]Department of Computational Mathematics, Institute of Computational Mathematics and Information Technologies, Kazan Federal University, 18 Kremlyovskaya Street, Kazan 420008, Russia
[2]Department of Applied Mathematics, Institute of Computational Mathematics and Information Technologies, Kazan Federal University, 18 Kremlyovskaya Street, Kazan 420008, Russia

E-mail: HelenRung@mail.ru

Abstract. Existence of a strong solution of the initial-boundary value problem modeling the process of liquid filtration in an arbitrary bounded region Ω of space R^n is proven. For determining a generalized solution, the Kirchhoff transform is used, and it is assumed that the domain of the Kirchhoff function constitutes only a part of the real axis. For proving the existence theorem, the method of semidiscretization with respect to the variable t and the Galerkin method are used.

1. Introduction

In the present work, a proof is conducted for existence of a strong solution of the initial-boundary value problem governed by the equation of nonlinear non-stationary filtration of the following type:

$$m\frac{\partial s(p)}{\partial t}+\operatorname{div} q=f(s(p)). \qquad (1)$$

Here p is pore pressure, q is filtration rate, m is porosity of the medium, $s(p)$ is saturation. We will assume that dependence of filtration rate q on p is given by the following law:

$$q=-b(s(p))(\nabla p-e), \qquad (2)$$

where the function $b(s(p))$ is relative phase permeability of the medium, $e=\rho g$, ρ is liquid density, g is gravitational force vector.

The equation under consideration is of the variable type: it is elliptic in the saturated filtration region (at $p\geq 0$), as in this case $s(p)=1$, and parabolic at $p<0$.

The problem of saturated-unsaturated filtration has been the subject of research by many authors. Most of the published works are devoted to the construction of approximate methods for solve this problem and discuss the results of numerical experiments (see, for example: [1]–[5]). Publications theoretical are few, the most important are [6]–[8], which discuss the existence of a generalized solution, and [10]–[11] devoted to the study of uniqueness of the solution.

Describe the results of [6]–[8].

First, we note that [6] is the fundamental work. Developed by the authors of this article a technique is widely used in the study of the solvability of nonlinear time-dependent problems of mathematical physics, in particular, it was used in the current work.

In [6] for a system of nonlinear equations of elliptic-parabolic type with strongly monotone space operator theorems on the existence of a generalized solution have been proved and studied its properties. This type of equation arises if in (1) by using Kirchhoff transformation we pass from an unknown function $p(x,t)$ to the function $u(x,t) = \vartheta(p(x,t))$, where

$$\vartheta(p) = \int_0^p b(s(\xi))\,d\xi. \tag{3}$$

Solvability of the equation resulted from this transformation follows from [6], provided that the domain of the Kirchhoff function ϑ was the entire axis of real numbers, and $p \leftrightarrow \vartheta$ was a one-to-one mapping.

It was later noted that the assumption that the domain of function ϑ represents the entire axis of real numbers is quite limited. We demonstrate this using conventional for applied research dependences $s(p)$ and $b(s)$:

$$s(p) = \begin{cases} (1-p)^{-\alpha}, & p \le 0 \\ 1, & p > 0, \end{cases} \qquad b(s) = s^{\beta}, \quad 0 \le s \le 1, \quad \alpha > 0, \quad \beta > 0, \quad \alpha\beta > 1. \tag{4}$$

Apparently, in this case we have $\vartheta(p) \in [-\gamma, +\infty) \quad \forall p \in R$, where $\gamma = \int_{-\infty}^{0} b(s(\xi))\,d\xi = (\alpha\beta - 1)^{-1}$.

In problems of filtration theory, the number $\int_{-\infty}^{0} b(s(\xi))\,d\xi$ is referred to as the macroscopic capillary length, which, as it was proven in [9], is always bounded.

The authors of [7], [8] consider the case when the domain of function ϑ is semi-axis of real numbers $[-\gamma, +\infty)$. In [7] is proved the existence theorem for weak solution in the one-dimensional case. In [8] the problem of saturated-unsaturated filtration in case when $b(s) = 0$, if $s \le \hat{s}$, where $\hat{s}$ is minimum irreducible wetness of soil was considered. In that work, a proof was given for existence of a solution of a regularized problem obtained from the original problem by substitution of the function $\tilde{\varphi}(u) = s(\vartheta^{(-1)}(u))$ on set $[-\gamma, -\gamma + \delta(\varepsilon)]$ (it is a zone of unbounded increase of the function $\tilde{\varphi}$) by a linear function having an angle with the axis u, the tangent of which is equal to $1/\varepsilon$, where ε is a small parameter.

The aim of the present work is to prove existence of a generalized solution of problem (1)–(2) under more general assumptions on the functions $s(p)$ and $b(s)$, which, in particular, also permit dependencies of the form (4).

2. Problem statement

We consider problem (1), (2) in bounded region Ω of space R^n with Lipschitz boundary $\Gamma = \Gamma_1 \cup \Gamma_2$ assuming that, at $t \in (0,T]$, the boundary conditions

$$q_\nu(x,t) = 0, \quad x \in \Gamma_1; \qquad p(x,t) = 0, \quad x \in \Gamma_2 \tag{5}$$

are satisfied.

Here $q_\nu(x,t) = q \cdot \nu$, where ν is outward normal to boundary Γ, $q \cdot \nu$ is scalar product of vectors q and ν in space R^n. Initial conditions are set in the form

$$(p(x,0))^- = p_0^-(x), \qquad x \in \Omega, \tag{6}$$

where $p^- = (|p| - p)/2$, p_0 is given function.

We use the Kirchhoff transform for proving the existence theorem. To eliminate the constraint $u \geq -\gamma$, for given $s(p), b(s)$, $\vartheta(p)$ we introduce the function (see (3))

$$\varphi(u) = \begin{cases} s(\vartheta^{-1}(u)), & u > -\gamma, \\ 0, & u \leq -\gamma. \end{cases} \tag{7}$$

Along with (1), (2), (5), (6) we consider the problem (which is assigned the name "extended" below) that is obtained from (1), (2), (5), (6) by replacement of equations (1), (2), boundary conditions (5) and initial conditions (6) with the relations:

$$m\frac{\partial \varphi(u)}{\partial t} + \operatorname{div} q = f(\varphi(u)), \tag{8}$$

$$q = -(\nabla u - b(\varphi(u))e), \tag{9}$$

$$q_\nu(x,t) = 0, \quad x \in \Gamma_1; \qquad u(x,t) = 0, \quad x \in \Gamma_2, \tag{10}$$

$$u_0^- = (u(x,0))^- = \vartheta(p_0^-(x)). \tag{11}$$

One can readily see that if function u is a solution of the extended problem, and $u(x,t) \geq -\gamma$ for arbitrary $(x,t) \in Q_T$, then $\vartheta^{-1}(u)$ is a solution of the problem (1), (2), (5), (6).

From this point on, we assume that the functions $s(p), b(s)$ satisfy the following conditions:

A₁. Function $s: R \to (0,1]$ is nondecreasing, continuous and piecewise-differentiable and satisfies the relation $s(p) = 1 \ \forall p \geq 0$.

A₂. Function $b: [0,1] \to [0,1]$ is nondecreasing and continuous, $b(0) = 0$.

A₃. Mapping $\vartheta: R \to [-\gamma, +\infty)$, given by formula (3), is one-to-one, where $\gamma = \int_{-\infty}^{0} b(s(\xi))\, d\xi$.

A₄. There exist constants $\kappa > 0$, $C_\varphi > 0$ such that

$$\varphi'(\eta)\varphi^{2\kappa-1}(\eta) \leq C_\varphi \qquad \forall \eta \in (-\gamma, 0). \tag{12}$$

Let us note that for dependencies (4), the condition (12) holds at $\kappa > \beta/2$ for arbitrary α and β satisfying the condition $(\alpha\beta - 1) > 0$.

3. Existence theorem

Let V be a subspace of the space $W_2^1(\Omega)$ having a zero trace on Γ_2.

Definition 1. *A generalized solution of the extended problem is a function* $u \in L_2(0,T;V)$, *for which the conditions*

$$\frac{\partial \varphi(u)}{\partial t} \in L_2(0,T;V^*), \qquad u^-(x,0) = u_0^-(x) \ \text{ almost everywhere on } \Omega$$

are satisfied, and for an arbitrary function $z \in L_2(0,T;V)$, *the equality*

$$\int_0^T < m\frac{\partial \varphi}{\partial t}, z > dt + \int_{Q_T} \nabla u \cdot \nabla z \, dxdt = \int_{Q_T} \{b(\varphi(u))(e \cdot \nabla z) + f(\varphi(u))z\} \, dxdt \tag{13}$$

holds. Here $Q_T = (0,T) \times \Omega$, V^* *is space dual to* V, $\langle v, z \rangle$ *is a value of the functional* $v \in V^*$ *on the element* $z \in V$.

Theorem 1. *Let the functions $s(p), b(s)$ and $\varphi(u)$ satisfy the conditions $\mathbf{A}_1$–$\mathbf{A}_4$. The function $f(\xi)$ is continuous on $[0,1]$ and $f(0)=0$. Then, for arbitrary $u_0 \in V$, there exists a generalized solution of the extended problem.*

For proving the theorem, we use the semidiscretization method. On the interval $[0,T]$, a uniform grid is created having the step size

$$\overline{\omega}_\tau = \{t = k\tau, 0 \le k \le M, M\tau = T\}, \qquad \omega_\tau = \overline{\omega}_\tau \setminus \{0\}.$$

Definition 2. *The function $w_\tau(t)$ is a semidiscrete solution of the extended problem, if*

$$u_\tau(t) \in V \quad \forall t \in \omega_\tau, \qquad (u_\tau(0))^- = u_0^- \text{ almost everywhere on } \Omega,$$

for any function $z \in V$ and for any $t \in \overline{\omega}_\tau \setminus \{T\}$, the equalities

$$\int_\Omega m \frac{\varphi(\hat{u}_\tau) - \varphi(u_\tau)}{\tau} z dx + \int_\Omega \nabla \hat{u}_\tau \cdot \nabla z dx = \int_{Q_T} \{b(\varphi(\hat{u}_\tau))(e \cdot \nabla z) + f(\varphi(\hat{u}_\tau))z\} dxdt \tag{14}$$

hold. Here $\hat{y} = y(t+\tau)$.

Lemma 1. *For a solution of the semidiscrete problem (14), the inequality*

$$u_\tau(x,t) \ge -\gamma \quad \text{almost everywhere on } \Omega, \quad \forall t \in \omega_\tau \tag{15}$$

holds.

Proof. By setting $z = (\gamma + \hat{w}_\tau)^-$ in equality (14), we rewrite it in the form

$$\frac{m}{\tau}\int_\Omega \varphi(\hat{u}_\tau)(\gamma + \hat{u}_\tau)^- dx - \frac{m}{\tau}\int_\Omega \varphi(u_\tau)(\gamma + \hat{u}_\tau)^- dx + \int_\Omega \nabla \hat{u}_\tau \cdot \nabla(\gamma + \hat{u}_\tau)^- dx$$
$$= \int_{Q_T} \{b(\varphi(\hat{u}_\tau))(e \cdot \nabla(\gamma + \hat{u}_\tau)^-) + f(\varphi(\hat{u}_\tau))(\gamma + \hat{u}_\tau)^-\} dxdt. \tag{16}$$

One can readily see that, by virtue of equality (7), the function $\varphi(\zeta)$ is zero for $\zeta \le -\gamma$, and the relation $(\gamma + \zeta)^- = 0$ holds for $\zeta \ge -\gamma$; therefore, $\varphi(\zeta)(\gamma + \zeta)^- = 0$ and $\zeta^+(\gamma + \zeta)^- = 0$ for arbitrary values of ζ. In addition, from conditions $b(0) = 0$ and $f(0) = 0$ it follows that $b(\varphi(\zeta)) \ne 0$ and $f(\varphi(\zeta)) \ne 0$ if and only if $\zeta > -\gamma$. Therefore, the right-hand side (16) is equal to zero.

Let us transform the third term on the left-hand side in relation (16):

$$\int_\Omega \nabla \hat{u}_\tau \cdot \nabla(\gamma + \hat{u}_\tau)^- dx = -\int_\Omega \sum_{i=1}^n \frac{\partial}{\partial x_i}(\gamma + \hat{u}_\tau)\frac{\partial}{\partial x_i}(\gamma + \hat{u}_\tau)^- dx = -\|(\gamma + \hat{u}_\tau)^-\|_1^2. \tag{17}$$

Then equality (16) takes the form

$$-\|(\gamma + \hat{u}_\tau)^-\|_1^2 - \frac{m}{\tau}\int_\Omega \varphi(u_\tau)(\gamma + \hat{u}_\tau)^- dx = 0, \tag{18}$$

here $\|v\|_1^2 = \int_\Omega |\nabla v|^2 dx$. The left-hand side in equality (18) is nonpositive; therefore, it follows from (18) that $(\gamma + \hat{u}_\tau(x,t))^- = 0$ almost everywhere in Ω $\forall t \in \omega_\tau$. The proof of the lemma is completed.

We use the Galerkin method to prove existence of a semidiscrete solution, for which a priori estimates of the kind

$$\int_\Omega \Phi(u_\tau(t'))\, dx \le C, \quad \sum_{t=0}^{t'-\tau} \tau\|\hat{u}_\tau(t)\|_1^2 \le C, \quad \sum_{t=0}^{t'-\tau} \tau\|G(\hat{u}_\tau(t))\|_1^2 \le C \qquad \forall t' \in \omega_\tau, \tag{19}$$

$$\frac{1}{k\tau}\sum_{t'=0}^{T-k\tau}\tau\int_{\Omega}\left(G(u_\tau(t'+k\tau))-G(u_\tau(t'))\right)^2 dx \le C \tag{20}$$

are satisfied, where $\Phi(u)=\int_0^u \varphi'(\xi)\xi\,d\xi$, $G(u_\tau)=\int_0^{u_\tau}\varphi'(\xi)g(\xi)\,d\xi$,

$$g(\zeta)=\begin{cases}\varphi^{2\kappa}(\zeta)\left(1+\varphi^{2\kappa}(\zeta)\right)^{-1}, & \text{if } \zeta<0,\\ 1/2, & \text{if } \zeta\ge 0.\end{cases}$$

Introducing the function $G(u)$ and obtaining the estimates represent an important stage of a theorem's proof as the estimates (19), (20) ensure existence of a converging subsequence of functions u_τ^- almost everywhere in Q_T (see [6]). For investigating non-linear problems, existence of such convergence is essential. It is allocated in the proof of the Theorem 1 a sequence of piecewise constant in t extension of solutions of a semidiscrete problem. It is proved that limit of this sequence is a generalized solution of the extended problem.

Acknowledgments
The research was supported by the Russian Foundation for research (projects 15-01-05686, 15-41-02315).

References

[1] Pop I S and Yong W-A 2002 *Numer. Math.* **92** pp 357–381
[2] Diersch H-J G and Perrochet P 1999 *Advances in Water Resources* **23** pp 271–301
[3] Radu F, Pop I S and Knabner P 2001 *Proceedings of ALGORITMY* pp 1–8
[4] Ahtareev A A and Dautov R Z 2007 Uchenye zapiski Kazanskogo universiteta. Seriya: Fiz.-mat. nauki **149** 4 pp 58–72
[5] Williams G A, Miller C T and Kelley C T 2000 *Water resources research* **36** 4 pp 923–934
[6] Alt H W and Luckhaus S 1983 *Math.Z.* **183** 8 pp 311–341
[7] Duyn van C J and Peletier L A 1982 *Arch. Ration. Mech. Anal.* **78** pp 173–198
[8] Alt H W, Luckhaus S and Visintin A 1984 *Ann. Mat. Pura ed Appl* **136** pp 303–316
[9] Morel-Seytoux H J, Meyer P D, Nachabe M, Touma J, van Genuchten M T and Lenhard R J 1996 *Water Resources Research* **32** 58 pp 1251–1258
[10] Otto F 1996 *J. Differ. Equations*, **131** 1 pp 20–38
[11] Otto F 1997 *Adv. Math. Sci. Appl.* **7** 2 pp 537–553

Application of spectral elements method to calculation of stress-strain state of anisotropic laminated shells

K A Petrovskiy[1], A V Vershinin[1] and V A Levin[2]

[1] LLC "Fidesys" Moscow State University Science Park named after M. V. University, office 402, Leninskie gory d. 1, str. 77, Moscow 119234, Russia
[2] Moscow State University, GSP-1, Leninskie gory d. 1, Moscow 119991, Russia

E-mail: constantine.89@yandex.ru

Abstract. A model of a high-order triangular spectral shell element is presented to calculate a stress-strain state of a shell construction under small deformations considering the material inhomogeneity. Consideration of shear deformations, changes in the thickness of a shell is implemented through the use of 7-parameter model of shell elements. The results of problems solving are to illustrate the possibilities of the element.

1. Introduction

Most of the materials studied in modern technology experience deformation during operation. The development of new structural materials require an availability of adequate mechanical models and analytic complexes on their basis for the assessment of the strength of structural elements made of such materials. For a numerical simulation of a wide range of physical phenomena the finite element method is one of the most powerful tools for accurate, efficient and stable approximate solutions. At present, the finite element method is widely known as the basic calculation method for the numerical simulation of problems of solid mechanics.

Since the beginning of the development of practical procedures using the finite element analysis, one of the main objective was to develop methods for the calculation of the complex shell structures. In modern practice, the shell structures are used in many branches of industry, including automotive, aviation and civil construction domains. In finite element analysis, there are several basic approaches to shell elements modeling [1].

The behavior of shells of the first type is considered as a superposition of membrane and bending behaviors. Finite elements are built by simple combining of rigidity matrices for plates bending and plane stress state. Such elements have low accuracy, since the bending and membrane behavior are combined at the nodal points only.

The second category includes elements based on the use of a particular theory of shells and discretization of a corresponding variational formulation. One of disadvantages of this approach is that if the shell theory is applicable to a specific geometry or analysis conditions only, then the composition of finite elements is subject to the same restrictions.

The third approach to shells modeling is based on the use of 3D elastic elements. Three-dimensional elastic elements are the most common elements, but their use became no longer practicable for subtle and multi-layered membranes.

At present, the most common and researched elements are the shells continuum elements and elements of the 2D theory of shells. Shell continuum elements are built by displaying two-dimensional master-element to a surface in $\mathbb{R}^3$ constituting the mid-surface of a shell. An approximated three-dimensional geometry of the shell element can usually be recovered throught the continuum approach, specifying the unit normal at each node, which is interpolated using the standard basic functions of an element. The formulation is completed through applying the relevant kinematic assumptions about the displacement field. Although no one of theories of shells is used, the resulting formulation can be defined as a shell element with properties corresponding to a shell model with shear deformations of the first-order.

Unlike the continuous shell elements, the shell elements based on 2D theory of shells, are formulated using an exact analytical description of undeformed mid-surface. In the theory of shells the basic kinematic assumptions, measures the stress and strain, constitutive equations are expressed in general curvilinear coordinates (ξ,η,ζ), which are used to describe the geometry of a three-dimensional shell. Formulations of elements of the theory of shells and elements based on continuous medium are based on the same mechanical assumptions and differ in the type of decomposition to the elements only.

In the theory of shells of Kirchhoff-Love type and in a 5-parameter shells of Reissner-Mindlin the three-dimensional constitutive equations are simplified through an assumption of stress equality to zero in the thickness direction. Return to the three-dimensional continuous medium is required if the stress and strain in the thickness direction are of great value. For example, in case of delamination of composite. To do this, an assumption of constant normal stress in the thickness direction is added to the theory of shells, and the formulation is extended to 6 parameters. But this leads to complex equations of deformation and inability to draw the explicit expression of defining relations. The difference between the distribution of stresses and strains over the thickness forced to add a linearly changing strain member, i. e. expand the formulation to 7-parameter theory. A linear deformation in the thickness direction is introduced as an additional variable and discretizated in the usual way via a finite element procedure. The main advantage of the shell 7-parameter model is the use of defining relations for the three-dimensional body with no changes [2].

For many problems of mechanics of a deformable solid the finite-element procedures of high-order offer a variety of theoretical and practical advantages as compared with finite-element methods of lower order, which for the past few decades began to dominate in researches and commercial software [3]. In particular, it is possible to avoid various forms of locking, which without adequate stabilization often spoil the finite element model of lower order in the weak formulation of the Galerkin method for elastic and inelastic solids.

Lots of disadvantages that may be encountered in the finite element model based on the principle of unconditional minimization can be avoided mainly or completely through the use of adequate polynomial of p-order when creating a finite element approximation u_{hp} inside each element. In particular, one can obtain effective finite element procedures which do not require the use of complicated special techniques, which are often required in the finite element formulation of the lower order to improve the accuracy of numerical solutions. As a result, one can use the formula of full integration and functional space of finite-element of a high order avoids any inconsistencies peculiar to a low-order approximation, which often lead to locking.

2. Spectral triangular element

The basis of spectral elements method is an integration scheme of Gauss-Lobatto [4]. It is not only a method of numerical integration. Integration points also determine a node basis, which gives the approximately diagonal matrix of the masses. This makes it possible to use the unsteady simulations which do not require the computation of inverse matrix of masses, but to obtain the optimal convergence rate. These properties allow the method of spectral elements to achieve a high order of accuracy. However, this rule of integration does not apply to triangles.

One of the alternatives to the Gauss-Lobatto-Legendre(GLL) points for triangles are the Fekete points [5]. These points are the solution of extreme problems and should be calculated numerically, but in most cases the minimization will be a simpler task than the calculation of optimal quadrature points. Fekete points on the border match with the GLL points and have natural extension to three dimensions. They were calculated for the tetrahedron, and for a wedge one can use the tensor product of Fekete points on a triangle with GLL points in the third direction. Figure 1 shows the location of the nodes in the triangular element of the 5th and 8th orders.

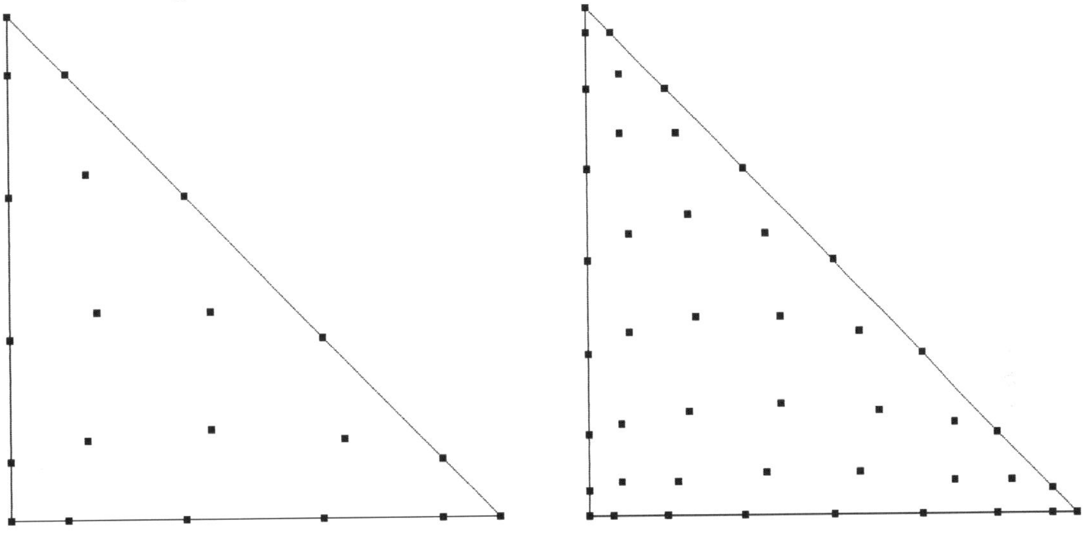

Figure 1. Location of nodal points in spectral triangle of the 5th and 8th orders

Fekete points are determined in accordance with the finite-dimensional space in which the interpolating function is built. For triangle a polynomial space of multiple two-dimensional polynomials is commonly used with a dimension $N_t = (N+1)(N+2)/2$, and in order to build interpolation polynomials exactly N_t points are required. For the selected space the Fekete points are determined by the Vandermonde matrix *V*. Let $\{D_p\}$ - is the basis for a set of polynomials of degree *N*, where $p = 1,...,N_t$, and $\{(\xi_p, \eta_p)\}$ - is a set of points in a right triangle with base and height $[-1, 1]$. Then $D_q(\xi_p, \eta_p)$ - component of (q,p) Vandermonde matrix. The Fekete points are the points maximizing the Vandermonde determinant of the matrix, where the maximum is taken over all possible sets of points in the triangle. For the selection of the orthogonal basis in the space of polynomials on the triangle the Dubiner polynomials are used.

To use the Fekete points for the spectral element method it is necessary to calculate the Lagrange interpolation functions and their derivatives [4]. Let's express these interpolation functions through Dubiner polynomials:

$$\psi_p(\xi,\eta) = \sum_{i=1}^{N_t} a_i^p D_i(\xi,\eta). \tag{1}$$

Coefficients a_i^p represent the (p,i) elements of the inverse Vandermonde matrix. Derivatives of Lagrange interpolation functions are as follows:

$$\frac{\partial \psi_p(\xi,\eta)}{\partial \xi} = \sum_{i=1}^{N_t} a_i^p \frac{\partial D_i(\xi,\eta)}{\partial \xi}, \qquad \frac{\partial \psi_p(\xi,\eta)}{\partial \eta} = \sum_{i=1}^{N_t} a_i^p \frac{\partial D_i(\xi,\eta)}{\partial \eta}. \tag{2}$$

In order to calculate integrals of some function $f(\xi,\eta,\zeta)$, that is required in the method of spectral elements, over each Ω_e element, the following formula is used:

$$\int_{\Omega_e} f(\xi,\eta,\zeta)d\Omega \simeq \int_{-1}^{1}\sum_{p=1}^{N_t} w_p J_e(\xi_p,\eta_p,\zeta) f(\xi_p,\eta_p,\zeta)d\zeta \simeq \sum_{r=0}^{N_h}\sum_{p=1}^{N_t} w_p w_r J_e(\xi_p,\eta_p,\zeta_r) f(\xi_p,\eta_p,\zeta_r),$$

where w_p - quadrature weight equal to the first Dubiner coefficient in interpolation polynomial of p order; J_e - Jacobian associated with the element mapping Ω_e to reference triangle; w_r - quadrature weight of one-dimensional Legendre polynomial of r order with (N_h+1) GLL point on the interval $[-1, 1]$.

3. Shell element model

By definition a shell is a solid with one geometric dimension considerably smaller in relation to the other two ones. Let's consider a conventional shell element Ω_e.

3.1. Shell element geometry.

The geometry of shell spectral element in the initial moment of time is described by the formula:

$$^{0}\underline{\mathbf{X}} = \sum_{k=1}^{n} \psi_k(\xi,\eta)\, {}^{0}\underline{\mathbf{X}}_{\mathrm{M}}^{k} + \frac{\zeta}{2}\sum_{k=1}^{n}\psi_k(\xi,\eta)\, {}^{0}h^{k}\, {}^{0}\underline{\mathbf{v}}_{\mathrm{n}}^{k} \quad (3)$$

where ${}^{0}\underline{\mathbf{X}}_{\mathrm{M}}$ - coordinates of a material point on the mid-surface of the shell, ${}^{0}h$ - thickness, $\underline{\mathbf{v}}_{\mathrm{n}}$ - normal to the mid-surface at that point (figure 2).

Deformed geometry at time t is described as follows:

$$^{t}\underline{\mathbf{x}} = \sum_{k=1}^{n} \psi_k(\xi,\eta)\, {}^{t}\underline{\mathbf{x}}_{\mathrm{M}}^{k} + \frac{\zeta}{2}\sum_{k=1}^{n}\psi_k(\xi,\eta)\, {}^{t}h^{k}\, {}^{t}\underline{\mathbf{v}}_{\mathrm{n}}^{k} + \zeta^2\sum_{k=1}^{n}\psi_k(\xi,\eta)\, {}^{0}h^{k}\, {}^{t}Q_{\mathrm{n}}^{k}\, {}^{t}\underline{\mathbf{v}}_{\mathrm{n}}^{k} \quad (4)$$

Here ${}^{t}\underline{\mathbf{v}}_{\mathrm{n}}$ - normal vector at the time t, ${}^{t}h$ - shell thickness at the point at the time t. ${}^{t}Q_{\mathrm{n}}$ - degree of freedom corresponding to the quadratic distribution of displacements in the direction of ${}^{t}\underline{\mathbf{v}}_{n}$ (figure 3).

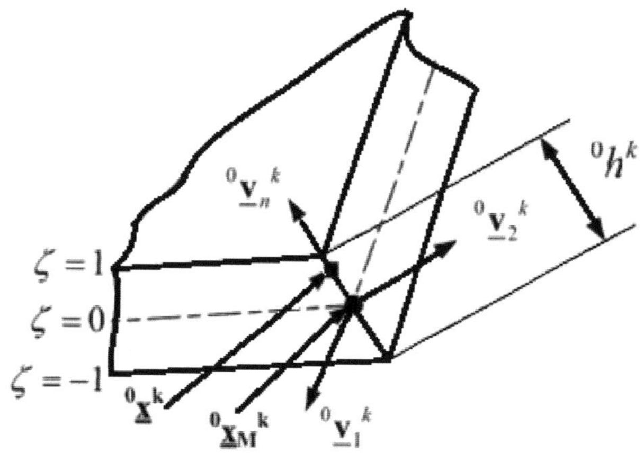

Figure2. Shell element geometry at the initial moment of time.

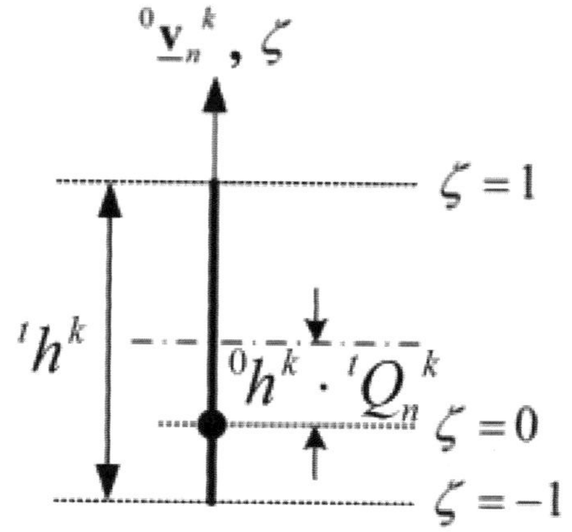

Figure3. Quadratic function of displacement in the node k.

The displacement increment from the position of the element at the time t to position at the time $t+\Delta t$ is equal to

$$\underline{\mathbf{u}} = {}^{t+\Delta t}\underline{\mathbf{x}} - {}^{t}\underline{\mathbf{x}} = \sum_{k=1}^{n}\psi_k(\xi,\eta)\left({}^{t+\Delta t}\underline{\mathbf{x}}_{\mathrm{M}}^k - {}^{t}\underline{\mathbf{x}}_{\mathrm{M}}^k\right) + \frac{\zeta}{2}\sum_{k=1}^{n}\psi_k(\xi,\eta)\left({}^{t+\Delta t}h^k\,{}^{t+\Delta t}\underline{\mathbf{v}}_{\mathrm{n}}^k - {}^{t}h^k\,{}^{t}\underline{\mathbf{v}}_{\mathrm{n}}^k\right) + \\ + \zeta^2\sum_{k=1}^{n}\psi_k(\xi,\eta)\,{}^{0}h^k\left({}^{t+\Delta t}Q_{\mathrm{n}}^k\,{}^{t+\Delta t}\underline{\mathbf{v}}_{\mathrm{n}}^k - {}^{t}Q_{\mathrm{n}}^k\,{}^{t}\underline{\mathbf{v}}_{\mathrm{n}}^k\right) \tag{5}$$

In the degrees of freedom at the time t the displacement increments can be expressed as follows:

$${}^{t+\Delta t}\underline{\mathbf{x}}_M^k - {}^{t}\underline{\mathbf{x}}_M^k = \underline{\mathbf{u}}_M^k = u_k\underline{\mathbf{e}}_1 + v_k\underline{\mathbf{e}}_2 + w_k\underline{\mathbf{e}}_3 \tag{6}$$

$${}^{t+\Delta t}h^k - {}^{t}h^k = {}^{0}h^k \cdot \Delta_a^k \tag{7}$$

$${}^{t+\Delta t}Q_{\mathrm{n}}^k - {}^{t}Q_{\mathrm{n}}^k = q_{\mathrm{n}}^k \tag{8}$$

Formula (6) describes the displacements in the global Cartesian coordinate system, (7) - change in thickness of the element, (8) - quadratic increment of displacements over the thickness.

Increment of directing vectors is expressed in the terms of rotary degree of freedom α, β, γ according to the formula

$${}^{t+\Delta t}\underline{\mathbf{v}}_i^k = \mathbf{Q}^k\,{}^{t}\underline{\mathbf{v}}_i^k, \qquad i = 1,2,n, \tag{9}$$

where $\mathbf{Q}^k$ - rotation tensor.

$$\mathbf{Q}^k = \mathbf{I} + \frac{\sin\theta}{\theta}\mathbf{S} + \frac{1}{2}\left(\frac{2\sin\left(\dfrac{\theta}{2}\right)}{\theta}\right)^2\mathbf{S}^2, \qquad \theta = \sqrt{\alpha_k^2 + \beta_k^2},$$

$$\mathbf{S} = \begin{bmatrix} 0 & 0 & \beta_k \\ 0 & 0 & -\alpha_k \\ -\beta_k & \alpha_k & 0 \end{bmatrix},$$

Here α_k, β_k - Increment of turns around the vectors ${}^{t}\underline{\mathbf{v}}_1^k$, ${}^{t}\underline{\mathbf{v}}_2^k$, respectively.

By substituting (6) - (9) to the formula (5) we obtain the increment of displacements as follows

$$\underline{\mathbf{u}}(\xi,\eta,\zeta) = \sum_{k=1}^{n}\psi_k(\xi,\eta)\underline{\mathbf{u}}_L^k \tag{10}$$

where:

$$\underline{\mathbf{u}}_L^k = \underline{\mathbf{u}}_M^k + \frac{\zeta}{2}\left[{}^{0}h^k\,{}^{t}\underline{\mathbf{v}}_n^k\Delta_a^k - {}^{t}h^k\,{}^{t}\underline{\mathbf{v}}_2^k\alpha_k + {}^{t}h^k\,{}^{t}\underline{\mathbf{v}}_1^k\beta_k\right] + \zeta^2\,{}^{0}h^k\left[{}^{t}\underline{\mathbf{v}}_n^k q_n^k - {}^{t}Q_n^k\,{}^{t}\underline{\mathbf{v}}_2^k\alpha_k + {}^{t}Q_n^k\,{}^{t}\underline{\mathbf{v}}_1^k\beta_k\right]$$

The model uses the Lagrangian formulation for small deformations. The covariant components of Green tensor of strains are defined as follows:

$${}_{0}^{t}\tilde{\varepsilon}_{ij} = \frac{1}{2}\left({}^{t}\underline{\mathbf{g}}_i \cdot {}^{t}\underline{\mathbf{g}}_j - {}^{0}\underline{\mathbf{g}}_i \cdot {}^{0}\underline{\mathbf{g}}_j\right). \tag{11}$$

In this statement of the covariant components formulation, all members higher than linear by ζ are discarded.

3.2. *Anisotropic materials*

The shells used in modern technology as structural elements are mostly natural or structurally anisotropic. Moreover, most of anisotropic shells are layered. The widespread distribution of such shells is of great interest with respect to the theory of anisotropic shells. There are many forms of relationship between strain and stress for anisotropic materials. In general, the generalized Hooke's law for homogeneous anisotropic body is as follows:

$$\begin{Bmatrix} \sigma_{xx} \\ \sigma_{yy} \\ \sigma_{zz} \\ \tau_{xy} \\ \tau_{yz} \\ \tau_{xz} \end{Bmatrix} = \begin{bmatrix} C^{1111} & C^{1122} & C^{1133} & C^{1112} & C^{1123} & C^{1113} \\ C^{1122} & C^{2222} & C^{2233} & C^{2212} & C^{2223} & C^{2213} \\ C^{1133} & C^{2233} & C^{3333} & C^{3312} & C^{3323} & C^{3313} \\ C^{1112} & C^{2212} & C^{3312} & C^{1212} & C^{1223} & C^{1213} \\ C^{1123} & C^{2223} & C^{3323} & C^{1223} & C^{2323} & C^{2313} \\ C^{1113} & C^{2213} & C^{3313} & C^{1213} & C^{2313} & C^{1313} \end{bmatrix} \begin{Bmatrix} \varepsilon_{xx} \\ \varepsilon_{yy} \\ \varepsilon_{zz} \\ \gamma_{xy} \\ \gamma_{yz} \\ \gamma_{xz} \end{Bmatrix}. \quad (12)$$

There are 21 independent elastic constants. If the body features an anisotropic elastic symmetry, the equations of generalized Hooke's law are simplified. An important class of anisotropic bodies are orthotropic bodies with three planes of elastic symmetry. In this case, relations between strain and stress may be expressed as follows:

$$\begin{Bmatrix} \varepsilon_{xx} \\ \varepsilon_{yy} \\ \varepsilon_{zz} \\ \gamma_{xy} \\ \gamma_{yz} \\ \gamma_{xz} \end{Bmatrix} = \begin{bmatrix} \frac{1}{E_x} & -\frac{\nu_{xy}}{E_y} & -\frac{\nu_{xz}}{E_z} & 0 & 0 & 0 \\ -\frac{\nu_{xy}}{E_y} & \frac{1}{E_y} & -\frac{\nu_{yz}}{E_z} & 0 & 0 & 0 \\ -\frac{\nu_{xz}}{E_z} & -\frac{\nu_{yz}}{E_z} & \frac{1}{E_z} & 0 & 0 & 0 \\ 0 & 0 & 0 & \frac{1}{G_{xy}} & 0 & 0 \\ 0 & 0 & 0 & 0 & \frac{1}{G_{yz}} & 0 \\ 0 & 0 & 0 & 0 & 0 & \frac{1}{G_{xz}} \end{bmatrix} \begin{Bmatrix} \sigma_{xx} \\ \sigma_{yy} \\ \sigma_{zz} \\ \tau_{xy} \\ \tau_{yz} \\ \tau_{xz} \end{Bmatrix}.$$

Here we are getting 9 independent parameters: E_x. E_y. E_z - elastic moduli; ν_{xy}. ν_{yz}. ν_{xz} - Poisson moduli; G_{xy}. G_{yz}. G_{xz} - shear moduli.

Layered composite materials consist of two or more materials featuring in sum the desirable properties, which do not occur when using these materials individually. Let's consider the layered shell, wherein each layer is an orthotropic material with the principal directions of elasticity $\{\hat{\mathbf{e}}_1, \hat{\mathbf{e}}_2, \hat{\mathbf{e}}_3\}$. When integrating by the thickness it is necessary to calculate the elasticity tensor C for each layer in the global coordinate system. Then, in order to write down the elasticity tensor C components for one layer in the global coordinates system $\{\mathbf{e}_1, \mathbf{e}_2, \mathbf{e}_3\}$ it is necessary to perform the conversion

$$C^{ijkl} = T_{im} T_{jn} T_{kp} T_{lq} C_{mnpq},$$

where the components T_{ij} are calculated as follows:

$$T_{ij} = \mathbf{g}^i \cdot \hat{\mathbf{e}}_j .$$

4. Numerical examples

In all the examples it is believed that each layer of the shell is made of orthotropic material, and the relationship between the components of the elastic tensor is defined by Table 1.

Table 1. Orthotropic properties adopted in the examples.

$C^{2222}/C^{1111} = 0.543103$	$C^{3333}/C^{1111} = 0.530172$

$C^{1122}/C^{1111}=0.23319$	$C^{1133}/C^{1111}=0.010776$
$C^{2233}/C^{1111}=0.098276$	$C^{1212}/C^{1111}=0.262931$
$C^{1313}/C^{1111}=0.159914$	$C^{2323}/C^{1111}=0.26681$

4.1. Shell under the uniform load, normal to the upper surface.

A simply supported shell under the uniformly distributed normal static load q_0 is considered. Numerical results presented in Table 2 are obtained for the homogeneous shell with the aspect ratio $a/b=1.0$ and a thickness $h/a=0.1$.

Table 2. Deflections and stress of homogeneous shell

	$C^{1111}w/hq_0$	σ_x/q_0	σ_y/q_0
Exact solution	-688.57	36.021	22.21
Triangular spectral shell	-703.87	36.75	22.6

4.2. Shell under the uniform load, normal to the upper surface.

A simply supported shell under the uniformly distributed normal static load q_0 is considered. The shell consists of three layers with identical top and bottom layers. The results are presented for ratio of layers $\beta=C_{\mathrm{T}}^{1111}/C_{\mathrm{M}}^{1111}$, where C_{T}^{1111} - tensor of elasticity coefficient for the top layer, and C_{M}^{1111} - the coefficient for the middle layer. Numerical results presented in Table 3 are obtained for the homogeneous shell with the aspect ratio $a/b=1.0$ and a thickness $h/a=0.1$.

Table 3. Deflections and stress of three-layer shell

$\beta=C_{\mathrm{T}}^{1111}/C_{\mathrm{M}}^{1111}$	5		10	
	Exact solution	Triangular spectral shell	Exact solution	Triangular spectral shell
$C^{1111}w/hq_0$	-258.97	-260.06	-159.38	-174.55
σ_x/q_0				
The upper surface of the top layer	60.353	60.958	65.332	71.351
The upper surface of the bottom layer	-46.426	-47.259	-48.609	-53.97
σ_y/q_0				
The upper surface of the top layer	38.491	40.124	43.566	47.252
The upper surface of the bottom layer	-30.322	-32.2	-33.756	-37.28

5. Conclusion

The 7-parameter model of triangular spectral shell element is developed. The model is intended for stress analysis of shells under small strains using the method of spectral elements. The thickness change of shell thickness is allowed, and it is assumed that strains varies linearly with the thickness of a shell. These features of the model permit one to use this model for the analysis of complex composite structures. The results of computations for some problems that can be solved analytically show a good accuracy of the proposed method.

Acknowledgements
This work was financially supported by the Ministry of Education and Science of the Russian Federation in the framework of the agreement No. 14.579.21.0112 (unique identifier of the project RFMEFI57915X0112). Investigations were carried out within the Fidesys company — a grantee of the Ministry of Education and Science of the Russian Federation.

References
[1] Zienkiewicz OC, Taylor RL and Zhu JZ 2013*The finite element method:its basis and fundamentals, Seventh edition* (Oxford: Elsevier)
[2] Kim D N and Bathe K J 2008 A 4-node 3D-shell element to model shell surface tractions and incompressible behavior*Comp. and Struct.***86**2027-41
[3] LevinV A andVershinin A V 2015 *Nonlinear computational mechanics of strength. Volume 2. Numerical methods. Parallel computing on a computer* (Moscow: Fizmatlit)
[4] Komatitsch D, Martin R, Tromp J, Taylor M A and Wingate B A 2001 Wave propagation in 2-D elastic media using a spectral element method with triangles and quadrangles *J. Comput. Acoustics* **9** 703–18
[5] Taylor M A, WingateB A and Vincent R E 2000 An algorithm for computing Fekete points in a triangle*SIAM J. Numer. Anal.* **38** pp 1707–20
[6] VallalaV P and ReddyJ N 2013 High-order spectral/hp finite element technology for shells and flows of viscous incompressible fluids *Math. and Comp. App.* **18** 152-75
[7] Srinivas S and Rao A K 1970 Bending, vibration and buckling of simply supported thick orthotropic rectangular plates and laminates *Int. J. Solids Structures* **6** 1463-81

Computational technology of multiscale modeling the gas flows in microchannels

V O Podryga

Keldysh Institute of Applied Mathematics of Russian Academy of Sciences, 125047, Russia, Moscow, Miusskaya sq., 4,

E-mail: pvictoria@list.ru

Abstract. The work is devoted to modeling the gas mixture flows in engineering microchannels under conditions of many scales of computational domain. The computational technology of using the multiscale approach combining macro - and microscopic models is presented. At macrolevel the nature of the flow and the external influence on it are considered. As a model the system of quasigasdynamic equations is selected. At microlevel the correction of gasdynamic parameters and the determination of boundary conditions are made. As a numerical model the Newton's equations and the molecular dynamics method are selected. Different algorithm types used for implementation of multiscale modeling are considered. The results of the model problems for separate stages are given.

1. Introduction

The modern computer technology makes it possible to simulate very complex systems and processes. Now the research of complex gasdynamic processes in technical micro- and nanosystems which are applied within implementation of nanotechnologies in the industry is very relevant. The paper deals with the process of gas mixture flows in the microchannels of technical systems.

The purpose of gas flows simulation tasks is determination of the optimal flow mode taking into account the geometry of installation, the properties of the gas medium and the solid wall materials. Also the fact should be taken into account that the modern technical systems can have complex multicomponent and multichannel structure. The diameters of the channels can vary from nanometers to millimeters, that leads to a problem with a wide range of Knudsen numbers (0.001 to 1 or more).

For micron and submicron sizes of the modeled system the mathematical model of gas flow can't be completely formulated within the macroscopic approach. The Navier-Stokes equations with special boundary conditions on walls or the Boltzmann equations in some approximation are usually used in such situation for description of the flow. Both methods have the advantages and the disadvantages. The solution based on the Navier-Stokes equations allows to reduce significantly computing expenses, however Knudsen number within this approach can't be more than 0.1. The solution based on the Boltzmann equation is more expensive; however the Knudsen numbers range is not limited from above. From below this range is limited to values of an order 0.01 in case of which the computational procedure on the basis of Boltzmann equation becomes unacceptable in terms of time costs.

In this paper in the range of Knudsen numbers less than 1 the system of quasigasdynamic (QGD) [1] equations is used; this system is more stable to numerical perturbations in comparison with Navier-Stokes system. The QGD equations need to be corrected and added be the boundary conditions, that is carried out using the method of molecular dynamics (MD) [2]. For Knudsen numbers of an order unity a direct molecular dynamic simulation is used. As a result, the combination of QGD equations and MD approaches covers the entire range of Knudsen numbers and allows to calculate the gasdynamic processes in systems of real microgeometry at necessary level of detailing.

The paper presents a computational technology on the use of multiscale approach combining the macroscopic description of multicomponent gas flows based the system of QGD equations and the MD models at the microlevel. Combining the models of different levels is carried out by using the method of splitting into physical processes [3]. MD acts as means of subgrid numerical correction of the flow macroparameters. This correction is applied: a) in the Knudsen layer near the channel walls, b) in the gas flow for specification of its equations of state, c) for calculating the gas kinetic coefficients and the exchange terms considering redistribution of momentum and energies between the gas mixture components.

2. Problem statement

At macroscale the system of QGD equations [1] in relaxation approximation is used. At microscale the Newton's equations [2] of classical mechanics describing the motion of separate particles and their interactions are used. It is necessary to consider the interactions: 1) the gas molecules with each other (creating the equations of state of gas mixture, implementing the mixing of components and defining the kinetic coefficients), 2) the gas molecules and the solid surface atoms (describing the boundary-layer phenomena).

The problem of binary gas mixture flow through a metal microchannel is considered as an example of applying the multiscale approach. Figure 1 shows the model geometry of the problem (a two-dimensional view of the average area). The computation of gas flow parameters in the microchannel is a research aim.

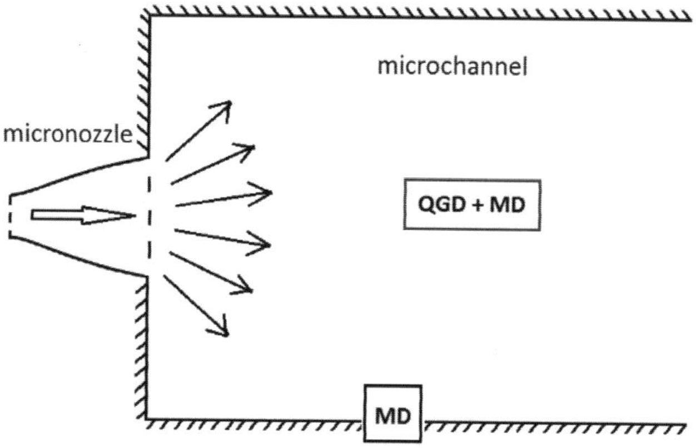

Figure1. The geometry of a model problem for the convenience provided in the form of a two-dimensional view.

Gas is supplied into the microchannel through a micronozzle at the left. At microchannel boundaries instead of usual boundary conditions the values of mass, momentum and energy fluxes calculated by the MD method are used. At the initial moment the gas medium state corresponds to an thermodynamic equilibrium state. In the central part of the microchannel (far from the walls) the simulation is performed at two scale levels (QGD and MD).

2.1. Macromodel

In the case of a gas mixture, the QGD system is written for each gas separately and has an identical form [1]; the equations include exchange terms for the momentum and the energy which are responsible for the coordination of the mixture parameters. Let us write down this system for a binary gas mixture in a form invariant under the coordinate system, together with the equations of state:

$$\frac{\partial \rho_l}{\partial t} + div\,\mathbf{W}_l^{(\rho)} = 0,\ \ l = a,b, \tag{1}$$

$$\frac{\partial}{\partial t}\rho_l u_{l,k} + div\,\mathbf{W}_l^{(\rho u_k)} = S_l^{(\rho u_k)},\ \ l = a,b,\ \ k = x,y,z, \tag{2}$$

$$\frac{\partial}{\partial t}E_l + div\,\mathbf{W}_l^{(E)} = S_l^{(E)},\ \ l = a,b, \tag{3}$$

$$E_l = \frac{1}{2}\rho_l\left|\mathbf{u}_l\right|^2 + \rho_l\varepsilon_l,\ \ P_l = Z_l\rho_l\Re_l T_l,\ \ \varepsilon_l = c_{V,l}T_l,\ \ l = a,b, \tag{4}$$

where a and b – gases of which mixture consists. Each gas has its own mass density $\rho_l = m_l n_l$, mass m_l, concentration n_l, temperature T_l and macroscopic velocity $\mathbf{u}_l$. Other parameters: P_l – partial pressures, E_l – total energy densities, ε_l – internal energies, Z_l – compressibility factors, $c_{V,l}$ – specific heat capacities at constant volume, $\Re_l = k_B / m_l$ – gas constants (k_B – Boltzmann constant). Vectors $\mathbf{W}_l^{(\rho)}$, $\mathbf{W}_l^{(\rho u_k)}$ and $\mathbf{W}_l^{(E)}$ coincide, up to the sign, with the density fluxes, fluxes of the corresponding components of the momentum density and energy density. They include QGD corrections proportional to Maxwell relaxation time, and depend on the coefficients of viscosity μ_l and thermal conductivity χ_l. The exchange terms $S_l^{(\rho u_k)}$ and $S_l^{(E)}$ take into account the momentum and energy redistribution between the gas mixture components depending on the molecular collision frequency. They contain the components of velocities and energy computed at the molecular level.

The coefficients of viscosity μ_l, thermal conductivity χ_l and compressibility Z_l, and also the specific heat capacities $c_{V,l}$ and other macroparameters for mixture components are determined either from a database of molecular calculations, or by direct MD computations.

The system of QGD equations (1)-(3) is closed by initial and boundary conditions. The initial conditions correspond to an equilibrium state of the gas medium without interactions with external factors. It is possible to consider the case of a quiescent gas medium in the entire computational domain. The densities, velocities, and total energies of the gas components are set at the channel inlet. "Soft" boundary conditions [1] are specified on free surfaces. A special microscopic system consisting of gas molecules and metal atoms is introduced near the channel walls. The boundary conditions on the wall are set as third-kind conditions describing the exchange of mass, momentum components, and energy between the gas mixture in the flow and in the near-wall layer. These conditions involve accommodation coefficients determined by tabulated physical data (which is possible for limited ranges of temperatures and pressures) or computed using the MD method. Another way of formulating the boundary conditions is the direct MD computation of density, momentum, and energy fluxes through the boundary of the near-wall layer.

2.2. Micromodel

In this problem at the microlevel away from the channel walls two types of particles (binary gas mixture) are considered, near the walls the particles constituting the wall material (e.g. metal) are added. The behavior of particles is described by Newton's equations [2]:

$$m_{l,i}\frac{d\mathbf{v}_{l,i}}{dt} = \mathbf{F}_{l,i},\ \ \mathbf{v}_{l,i} = \frac{d\mathbf{r}_{l,i}}{dt},\ \ i = 1,\ldots,N_l,\ \ l = a,b,c, \tag{5}$$

where i – particle index, l – particle kind (a and b – molecules of the first and second gases, c – metal atoms), N_l – total number of particles of kind l. The particle of kind l with index i has its mass $m_{l,i}$, position vector $\mathbf{r}_{l,i}$, velocity vector $\mathbf{v}_{l,i}$ and total force $\mathbf{F}_{l,i}$ acting on this particle.

The interaction between the particles is described by potential functions depending on the particle coordinates. The choice of interaction potential is based on comparing the mechanical properties of computer potential model and real material.

The calculation of the compressibility factor and heat capacity using the MD method is described in detail in [4], the calculation of viscosity and thermal conductivity coefficients of the mixture components using MD is described in [5].

The initial conditions at the microlevel are determined by the equilibrium or quasiequilibrium thermodynamic state of particle system at given temperature, pressure, and mean momentum. The boundary conditions at the molecular level depend on the situation to be simulated. To determine the general properties of the medium, it is sufficient to consider a distinguished three-dimensional volume of it with periodic boundary conditions on all coordinates. In the study of actual geometry microsystems, such as a microchannel, one or several directions are of finite size and shape retention of the object is achieved by the selection of potential or fixing the system. The temperature and total momentum of the system are controlled with the help of thermostats algorithms [6, 7].

3. Computational technology and examples

When solving problems the QGD system of the equations is sampled by the finite volume method [8] on the suitable structured or unstructured grids and is solved on the basis of explicit or implicit schemes on time. MD system of equations is solved by velocity Verlet scheme [9].

Simulation of problems at two scales should be done according to certain rules. It is possible to use MD calculations as directly, combining in one implementation the QGD and the MD, and indirectly, accumulating molecular database calculated earlier and using only a ready base. It is also possible the partial use of MD database and the partial direct MD modeling combined with QGD calculations. Thus, the computational technology assumes 4 types of algorithms.

Algorithms of type 1 include MD calculations of material properties and accumulation of database on these properties. Also, this type of algorithms can be used for full calculations for problems where the characteristic sizes are molecular sizes.

Algorithms of type 2 realize the calculations of gas mixture flows in technical microsystems by means of QGD equations using only the accumulated database. For receiving all necessary value range from the database the method of logarithmic interpolation is used. Conditions of pure gas or gas mixture flows can be very variable, for using this algorithm it is necessary to have access to large memory volumes for database storage and fast use these data by necessity in the calculations. Therefore calculation of MD database for the entire range of possible values can be too expensive in terms of time and information storage.

Algorithms of type 3 produce gas flows in microsystems calculations by means of QGD equations using the molecular database on boundary area and direct MD calculation in a gas flow. The boundary layer includes areas with wall particles; the wall is solid, it has structure and number of solid particles in the volume is many times more than the number of gas particles in the same volume. For this reason computations on boundaries represent one of the most expensive parts of modeling on the necessary memory capacity, operation number, and therefore on time of calculations.

Algorithms of type 4 base on direct macroscopic and microscopic simulations that are being implemented by the method of splitting into physical processes. This method allows to alternate calculations of QGD and MD equations, using MD computation as a subgrid algorithm. Direct MD calculation during computing the problem is most justified if the database on material properties is incomplete or at all is absent. In this case the algorithm of type 4 provides a complete simulation cycle by all necessary data and allows to coordinate processes on micro- and macrolevels, and also to make calculations for a certain specific set of conditions and at the same time to accumulate the database for a case of repeated computing experiments.

Let us describe more detail the algorithm of type 4.

At the microlevel all calculations are made independently in each control volume of a spatial grid, except for boundary cells. In boundary cells the nonlocal MD scheme of computations can be used.

This is especially relevant in a case of large Knudsen numbers (order of 1 and more) when QGD system loses accuracy owing to violation of the gaseous medium continuity hypothesis.

Algorithms of type 4 consist of five basic steps:

1) computing the macroparameters of gas components according to grid analogues of QGD equations excluding exchange terms in grid cells where approximation of continuous medium is valid;

2) MD calculating the exchange terms in average flow field based on local algorithms;

3) return to a macrolevel and correcting the gas momentum and total energy densities;

4) MD calculating the kinetic coefficients, the equations of state parameters and other macroparameters in average flow field, and also the parameters of boundary conditions near the computational domain boundaries and the boundaries between the different medium types;

5) return to a macrolevel and correction of gases moment and total energy densities, and also of a metal surface state.

Criterion for stopping the MD calculations is either the achievement of characteristic time of molecular system evolution (maxwellization time), or small change (for 1-2%) one or more macroparameters of molecular system (average momentum, averages kinetic and/or potential energies).

Presented computational technology allows using the appropriate type of algorithms, depending on the available computing systems and memory capacities. Described multiscale approach allows modeling multicomponent gas flows taking into account the complex geometry of the microchannels, that is provided by means of MD methods. MD calculations determine the kinetic coefficients and exchange terms necessary for computing the system of equations (1)-(3), the compressibility factors and specific heat capacities underlying the equations of state (4), and also boundary conditions.

The technique of numerical modeling is implemented in the form of parallel program tool and is verified on a set of model problems. Subsonic and supersonic flows of pure gases and a binary gas mixture were considered. The obtained calculation results agree with the known tabular data and experimental results. In work [10] the nitrogen and hydrogen mixture flow was calculated on an output from a micronozzle in the semi-open microchannel and further to the free space. The calculation results were close to data of experiments. Also in [10] the combination of the micro- and macromodels of the gas medium in a uniform research object was offered. In [11, 12] interaction of a gas flow with microchannel walls on the example of nitrogen-nickel system was considered. In [5] on an example of the same system the accommodation coefficients were obtained. These calculations represent a technique of obtaining the boundary conditions by direct MD computation. In works [12-15] methods for calculating the thermodynamic equilibrium states in metal-metal, gas-gas and gas-metal systems were developed. In particular, nickel-nickel, aluminum-aluminum, argon-argon, nitrogen-nitrogen, nitrogen-nickel systems were considered. In [4] on the example of nitrogen molecule system the technique of specifying the equations of state was considered. In [5] three methods for determining the kinetic coefficients were presented, comparing methods were produced and examples of calculating the coefficients of nitrogen molecule system were given.

4. Conclusions

For modeling the gases and their mixtures transportation through the microchannels of technical systems the multiscale approach combining the calculations on macro - and microlevels is used. At the macroscale the system of quasigasdynamic equations is used, at the microscale the method of molecular dynamics is used. For the numerical investigation the method of splitting into physical processes is applied (allowing to alternate the calculations at macro- and microlevels). The finite volume method is used for sampling the quasigasdynamic equations on the grids of different types and velocity Verlet scheme is used for implementing the Newtonian equations applied for surface atoms of the microchannel walls and for gas molecules flowing through the channel. The paper presents the computational technology of using the multiscale approach. Four types of technology algorithms are described. A set of the solved model problems is given. The obtained results were compared to experimental data and confirmed a correctness and efficiency of the provided technique.

Acknowledgements
The work is performed with financial support of the Russian Foundation for Basic Research (projects №№ 16-29-15095-ofi_m, 16-37-00417-mol-a, 16-07-00206-a).

References

[1] Elizarova T G 2009 *Quasi-Gas Dynamic Equations* (Berlin: Springer-Verlag)
[2] Frenkel D and Smit B 2002 *Understanding Molecular Simulation. From Algorithm to Applications* (New-York: Academic Press)
[3] Marchuk G I 1988 *Splitting-up Methods* (Moscow: Science)
[4] Podryga V O 2015 *Mat. Model.* **7** 80–90
[5] Podryga V O and Polyakov S V 2016 Molecular Dynamic Calculation of Gas Macroparameters in the Stream and on the Boundary *Keldysh Institute Preprints* **80**
[6] Berendsen H J C, Postma J P M, Gunsteren W F et al. 1984 *J. Chem. Phys.* **81** 3684–3690
[7] Heerman D W 1986 *Computer Simulations Methods in Theoretical Physics* (Berlin: Springer-Verlag)
[8] Eymard R, Gallouet T R and Herbin R 2000 *Handbook of Numerical Analysis* vol 7, ed P G Ciarlet and J L Lions (Amsterdam: North-Holland) p 713–1020
[9] Verlet L 1967 *Phys. Rev.* **159** 98–103
[10] Kudryashova T A, Podryga V O and Polyakov S V 2014 *Vestn. Ross. Univ. Druzhby Narodov Ser. Mat. Inf. Fiz.* **3** 154–163
[11] Podryga V O, Polyakov S V and Zhakhovskii V V 2015 *Mat. Model.* **7** 91–96
[12] Podryga V O, Polyakov S V and Puzyrkov D V 2015 *Vychisl. Metody Program.* **1** 123–138
[13] Podryga V O 2011 Math. Models Comput. Simul. **3** 381–388
[14] Podryga V O 2011 *Mat. Model.* **9** 105–119
[15] Podryga V O and Polyakov S V 2015 Math. Models Comput. Simul. **5** 456–466

Determination of muscle effort at the proximal femur rotation osteotomy

O Sachenkov[1], R Hasanov[2], P Andreev[2] and Yu Konoplev[1]

[1]Kazan Federal University, Institute of Mathematics and Mechanics, Russia, Tatarstan, Kazan, 18 Kremlevskaya St.
[2]Republican Clinical Hospital, Trauma and Orthopedic Department for Children, Russia, Tatarstan, Kazan, 138 Orenburg High Road St.

E-mail: 4works@bk.ru

Abstract. The paper formulates the problem of biomechanics of a new method for treatment of Legg–Calve–Perthes disease. Numerical calculations of the rotational flexion osteotomy have been carried out for a constructed mathematical model of the hip joint, taking into account the main set of muscles. The work presents the results of the calculations and their analysis. The results have been compared with the clinical data. The calculations of the reactive forces arising in the acetabulum and the proximal part of the femur allowed us to reveal that this reactive force changes both in value and direction. These data may be useful for assessing the stiffness of an external fixation device used in orthopedic intervention and for evaluating the compression in the joint.

1. Introduction

In surgery it is important to know biomechanical changes. Usually in orthopedic operations it is the most important to know how changes muscles. Large biomechanical changes in muscles leads to increasing the isometric force. That means that bio-system will continue changes and usually not the way surgeon wants [1]. During the rotational osteotomy, the femur is cut by a parabolic surface around the Adams arc, and the rotation of the acetabular component is performed with respect to the axis R with the help of an external fixation device [2] (see figure 1). Lengthening and shortening of the muscles lead to a redistribution of forces in the joint, and this can cause an increase in joint compression and worsen quality of health. Therefore, evaluation of the permissible angle of rotation is an important task.

2. Another section of your paper

The rotational osteotomy involves cutting of the femur by a parabolic surface around the Adams arc, and rotation of the acetabular component relative to the axis R (see figure 1) using an external fixation device. Due to the lengthening and shortening of the muscles, a redistribution of forces in the joint takes place. When studying the stress-strain state (SSS) of the hip joint (HJ) under rotational osteotomy, the list of the main muscles having a major influence was determined in the first place: mm. piriformis, rectus femoris, iliopsoas, obturator internus, gluteus minimus, medius et maximus [1, 2]. The existing parametric model of the hip joint [3–6] was used for modeling. The following anatomical characteristics of the joint and femur were taken as the main parameters in this case: the acetabular angle, the Wiberg angle, the cervical and diaphyseal angle, the length of the proximal part of the femur, the diameter of the acetabulum and the femoral head. To formulate the problem of HJ stress-strain, the following simplifications were introduced: the material of the pelvis and femur is isotropic and obeys Hooke's law. Since the rotation for tissues is instantaneous, Hooke's law is also applicable for muscles, but the muscles in this case work only in tension. The condition of frictionless contact was set on the cutting surface and the acetabulum. The rotation was simulated by kinematic boundary conditions imposed on the free surface of the proximal part cut off.

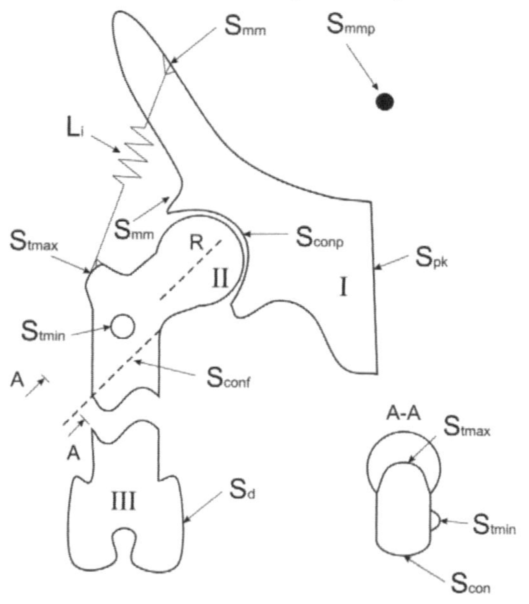

Figure 1. Calculation scheme.

The formulation considers three bodies: the pelvis – I, the proximal part of the femur – II, and the rest part of the femur – III. Let V_I, V_{II}, V_{III} denote the volumes of the respective bodies and ∂V_I, ∂V_{II}, ∂V_{II} – the boundaries of these sets (free surface), $V= V_I \cup V_{II} \cup V_{III}$. The mechanics of a system that occupies the area V in R^3 with the boundary ∂V is described within the linear elastic theory by the following system of equations:

$$\nabla \cdot \tilde{\sigma} = 0, \ \forall \mathbf{x} \in V^\circ, \tag{1}$$

$$\tilde{\varepsilon} = \frac{1}{2}(\nabla \vec{u} + (\nabla \vec{u})^T), \ \forall \mathbf{x} \in V^\circ, \tag{2}$$

$$\tilde{\sigma} = \tilde{\tilde{E}} : \tilde{\varepsilon}, \ \forall \mathbf{x} \in V^\circ, \tag{3}$$

where $V^\circ = V \cup \partial V$, $\vec{u}$ is the displacement vector, $\tilde{\sigma}$ is the stress tensor, $\tilde{\varepsilon}$ is the tensor of elastic deformations, $\tilde{\tilde{E}}$ is the tensor of elastic properties. This formulation is fulfilled for all three bodies.

Let us consider the boundary conditions and introduce a set of surfaces for this purpose. Let S_{mm} denote the surfaces of Body I, to which the muscles will be attached. Their position and size were identified according to the anatomical atlas. Those muscles which are attached to the bones absent in the model will be defined as the set of points S_{mmp}. The femoral muscles are attached to the greater trochanter – the surface S_{tmax}, and the small trochanter – the surface S_{tmin}, as well as to the distal part – the surface S_d. Contact interaction between the bodies is possible on two surfaces: the dissected parabolic surface under the small trochanter – S_{conf}, and the surface of the pelvic acetabulum and the femoral head – S_{conp}. To determine the kinematic boundary conditions, let us introduce a) the surface S_{pk} denoting the surface of the pelvis, which is attached to the sacrum, b) the surface of rotation, which is defined as $S_{rot}=\partial V_{II} \setminus S_{conf}$ (external surface of the proximal part of the femur excluding the surface being dissected), and c) ∂V_{III} denoting the external surface of the body V_{III}.

$$\vec{u}=0,\ \forall \mathbf{x}\in \partial V_{III}\cup S_{pk}, \quad (4)$$

$$\vec{u}=f(x,y,z),\ \forall \mathbf{x}\in S_{rot}, \quad (5)$$

where the function $f(x,y,z)$ defines the rotation S_{rot} relative to the axis R.

The influence of the muscles was determined by the following relation:

$$\sigma_i = E_i\cdot\varepsilon_i\cdot\frac{1+sign\varepsilon_i}{2},\ \forall \mathbf{x}\in L_i, \quad (6)$$

where L_i is the line defining the position of the i-th muscle; the attachment of all muscles was determined in the above introduced areas; all the muscles were introduced according to the anatomical atlas (figure 1 shows the illustration of only one muscle for convenience). Here σ_i and ε_i are the normal tensions and deformations in the muscle. As seen from the law, the muscles were defined by the elastic law without regard to compression; E_i is the reduced elastic modulus of the i-th muscle.

The contact interaction is defined by the expressions:

$$(\vec{r}_{II}^{\,0}-\vec{r}_{I}^{\,0})\cdot\vec{n}=(\vec{u}_{II}-\vec{u}_{I})\cdot\vec{n},\ \forall \mathbf{x}\in S_{conp} \quad (7)$$

$$(\vec{r}_{III}^{\,0}-\vec{r}_{II}^{\,0})\cdot\vec{n}=(\vec{u}_{III}-\vec{u}_{II})\cdot\vec{n},\ \forall \mathbf{x}\in S_{conf} \quad (8)$$

This technique used for calculating the contact interaction is described in detail in [3-5].

To perform a numerical investigation based on the parametric model, we built a finite-element model[6-8]. A four-node tetrahedral finite element with linear approximation was used for discretization of the volumes V_I, V_{II}, V_{III}. The characteristics of bone tissue were taken equal to E = 6 GPa, ν = 0.3. We used a one-dimensional finite element for muscle modeling [9-13]. Equation (9) was used to determine the reduced Young's modulus:

$$E=0{,}111\cdot F_{iso}\frac{kg/cm^2}{nm}, \quad (9)$$

where F_{iso} is the isometric force of the muscle.

According to the published studies on determining the maximum isometric force of a muscle, we defined the muscle stiffness and presented the results in Table 1, where $\bar{F}_{iso}$ is the mean value of the maximum isometric force of the muscle from the study [14].An additional coordinate system was introduced for rotation modeling, where the z axis coincided with the direction of the rotation axis (the line R in figure 1). The kinematic boundary conditions of rotation for the surface ∂V_{II} were set in this coordinate system.

Table 1. Values of isometric force of the muscle and the muscle stiffness.

Muscle name	$\bar{F}_{iso}$, **N**	**E, KPa**
Iliacus	615,30	56,18
Obturator externus	341,80	101,13
Obturator internus	425,80	81,18
Piriformis	175,88	196,55
Gluteus minimus	629,00	54,96
Gluteus medius	1225,15	28,22
Gluteus maximus	1747,18	19,79
Tensor fascia latae	310,75	111,24
Rectus femoris	508,75	67,95

We set the instantaneous rotation angle and then performed calculation of the model. It consisted of several stages. At the first stage, the calculations were carried out within a range of the rotation angle between –50° and 50°, where the minus sign means back rotation. As a result, we determined the elongations and efforts in the muscles.

3. Results and discussion

The paper discusses the results of the first calculation stage, which involved the determination of elongations and efforts in the muscles. For the obtained values of efforts in the muscles under rotation, we built the dependences of the force in the muscle, related to the muscle isometric force, under rotation (example of the results for two muscles is shown in figure 2).

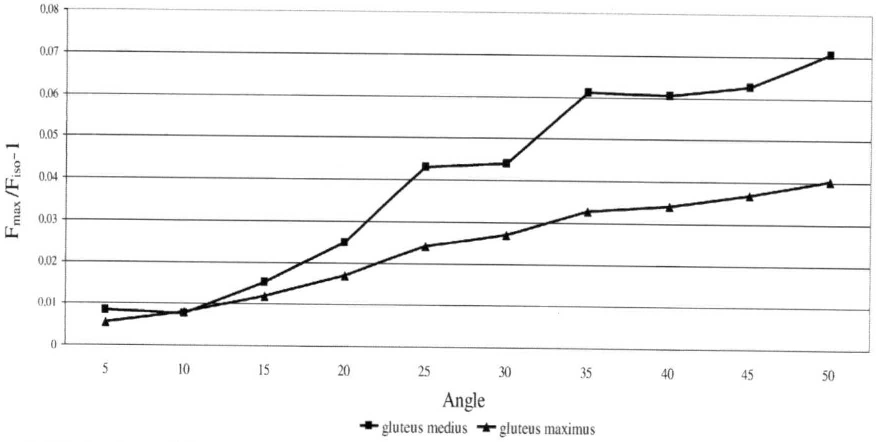

Figure 2. Variation of force in the muscle, related to the muscle isometric force, under rotation.

The analysis of the results was then carried out for all muscles that have efforts, i.e. can lengthen during rotation. The dependences of the relative (to the isometric force) increase in the force at a given angle of rotation were constructed. This dependence was nearly linear for some muscles, and therefore, it made it possible to construct a linear approximation according to the least-square method. The values calculated are shown in Table 1, which specifies the coefficients of increase in the force relative to the initial isometric force for each muscle, depending on the direction of rotation, and shows the approximation error R^2 by the least-square method.

Table 2. Coefficient of the relative grow of effort in the muscle and the approximation error.

Muscles	K	R^2
	Back rotation	
iliopsoas	0.1728	0.99
gluteus medius	0.0078	0.95
gluteus maximus	0.0041	0.98
	Forward rotation	
piriformis	1.1664	0.99
rectus femoris	Nonlinear behavior	
iliopsoas	0.0458	0.97

obturator internus	Nonlinear behavior	
gluteus minimus	0.0223	0.98
gluteus medius	0.0110	0.91
gluteus maximus	0.0041	0.95

The obtained dependences allow us to perform a qualitative assessment of the possible effort in the muscle under rotation at a given angle:

$$F(r) = F_{iso}\left(r \cdot K_m + 1\right)$$

where r is the rotation angle, F_{iso} is the muscle isometric force, K_m is the coefficient of the relative grow of effort for the muscle (see Table 2).

We obtained the picture of pressure distribution in the joint for different values of rotation. It is significant that at different angles and directions of rotation, the reactive force in the joint, which characterizes the compression, changes both in value and direction. We also determined the reactive force arising in the proximal part of the femur. Assessment of the reactive force is necessary for defining the stiffness of the external fixation device, by which rotation is carried out in practice [5].

4. Conclusions

The paper shows the formulation of the problem of modeling the biomechanics of an orthopedic method for treatment of Legg–Calve–Perthes disease. We carried out the numerical calculations of rotational flexion osteotomy for the constructed mathematical model of the hip joint, taking into account the main muscle groups. As a result of the calculations, we determined the efforts arising in the muscles at different rotations. The results were analyzed and used to obtain linear functions for estimating the efforts in the muscles depending on the angle of rotation and the isometric force. The calculations of the reactive forces arising in the acetabulum and the proximal part of the femur allowed us to reveal that this reactive force changes both in value and direction. These data may be useful for assessing the stiffness of an external fixation device used in orthopedic intervention and for evaluating the compression in the joint.

Acknowledgements

The work was partly supported by the Russian Foundation for Basic Research within the scientific project No. 15-31-20602 and No. 15-41-02555.

References

[1] Shigapova F A, Mustakimova M F, Saleeva G T and Sachenkov O A 2015 Definition of the intense and deformable jaw state under the masseters hyper *tone International Journal of Applied Engineering Research* **10** 44711-44714

[2] Galiullin R R, Sachenkov O A, Khasanov R F and Andreev P S 2015 Evaluation of external fixation device stiffness for rotary osteotomy International *Journal of Applied Engineering Research* **10** 44855-44860

[3] Sachenkov O A, Mitryaikin V I, Zaitseva T A and Konoplev Yu G 2014 Implementation of Contact Interaction in a Finite-Element Formulation *Applied Mathematical Sciences* **8** 7889 - 7897 http://dx.doi.org/10.12988/ams.2014.49769

[4] Sultanov L U 2015 Numerical Modelling of Deformation of Soil Foundations *Procedia Earth and Planetary Science* **15** 119–124 doi:10.1016/j.proeps.2015.08.029

[5] Abdrakhmanova A I, Gariffulin I R, Davydov R L, Sultanov L U and Fakhrutdinov L R 2015 Investigation of strain of solids for incompressible materials *Applied Mathematical Sciences* **9** 5907-5914 http://dx.doi.org/10.12988/ams.2015.57507

[6] Golovanov A and Sultanov L 2005 Numerical Investigation of Large Elastoplastic Strains of

Three-Dimensional Bodies International *Applied Mechanics* **41** 614-620

[7] Abdrakhmanova A I and Sultanov L U 2016 Numerical modelling of deformation of hyperelastic incompressible solids *Materials Physics and Mechanics* **26** 30-32

[8] Davydov R and Sultanov L 2013 Numerical algorithm of solving the problem of large elastic-plastic deformation by FEM *Sixth International Conference on Nonlinear Mechanics* (USA) 64-67

[9] Davydov R L and Sultanov L U 2015 Numerical algorithm for investigating large elasto-plastic deformations *Journal of Engineering Physics and Thermophysics* **88** 1280-1288 DOI 10.1007/s10891-015-1310-7

[10] Abdrakhmanova A I, Sultanov L U and Fakhrutdinov L R 2015 Calculation of Incompressible Hyperelastic Bodies *Global Journal of Pure and Applied Mathematics* **11** 5153-5162

[11] Berezhnoi D V and Paimushin V N 2011 Two formulations of elastoplastic problems and the theoretical determination of the location of neck formation in samples under tension *Journal of Applied Mathematics and Mechanics* **75** 447-462

[12] Berezhnoi D V, Paimushin V N and Shalashilin V I 2009 Studies of quality of geometrically nonlinear elasticity theory for small strains and arbitrary displacements *Mechanics of Solids* **44** 837-851

[13] Davydov R L, Sultanov L U and Kharzhavina V S 2015 Elastoplastic Model of Deformation of Three- Dimensional Bodies in Terms of Large Strains *Global Journal of Pure and Applied Mathematics* **11** 5099-5108

[14] Akulich Yu V, Podgayets R M, Scryabin V L and Sotin A V 2007 The Investigation of Stresses and Strains in the Hip Joint After Operation of Endoprosthetics *Russian journal of biomechanics* **11** 9-35

Solution for the problem of the Stokes flow past a body of arbitrary shape using conformal mapping

R R Salakhov, D F Abzalilov and S K Zaripov

Kazan Federal University, 18 Kremlevskaia Street, 420008 Kazan, Russia

E-mail: ramms_92@mail.ru

Abstract. An approach to the calculation of flow of an incompressible viscous fluid in the Stokes approximation is developed based on the finite difference method with the use of conformal mapping. The problems of flow around a circular cylinder and ellipse in a periodic circular cell have been solved. The method is particularly well suited for solving problems of flow around bodies of an arbitrary curved shape.

1. The problem statement

A flow around a body of an arbitrary shape in a circular periodic cell at low Reynolds numbers is considered (fig. 1, a). In the Stokes flow approximation, the dimensionless problem of a fluid flow in the domain *ABCD* is written in a stream function ψ – vorticity ω variables [1]:

$$\begin{aligned} \Delta\psi &= -\omega \\ \Delta\omega &= 0 \end{aligned} \tag{1}$$

On the outer boundary of the periodic cell with radius h (line *AD*), the Kuwabara conditions are adopted [2]

$$\omega = 0,\ \psi = y \tag{2}$$

On the surface of the body (line *BC*) no-slip conditions are applied

$$\psi = 0,\ \frac{\partial\psi}{\partial n} = 0 \tag{3}$$

On the lines *AB* and *CD* the symmetry conditions are taken

$$\psi = 0,\ \omega = 0 \tag{4}$$

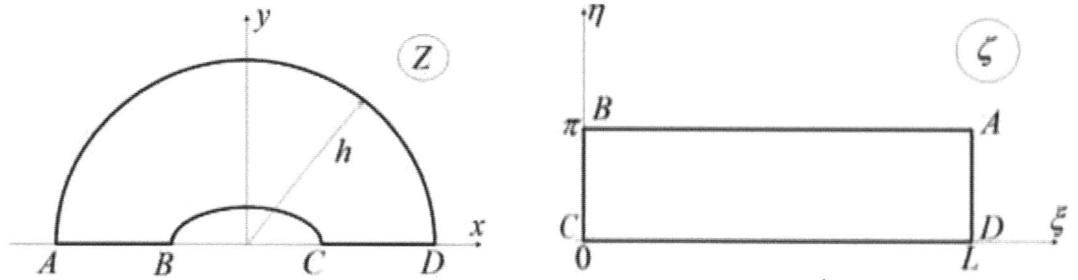

a b

Figure. 1. Computational domain in the physical (a) and canonical (b) plane.

Let us introduce the function of conformal mapping of a rectangular area in a canonical plane $\zeta = \xi + i\eta$ (Fig.1, b) to the region of flow in the physical plane $z = x + iy$ (Fig.1, a)

$$z = \sum_{k=-N}^{N} a_k e^{k\zeta} \tag{5}$$

where a_k are the real coefficients that are found from the numerical solution of the problem of conformal mapping [3]. The size of the rectangular domain is $0 \le \eta \le \pi$, $0 \le \xi \le L$ where the value L is determined in the course of the conformal mapping. Note that for the domain of an arbitrary shape the Laurent series (5) can be infinite but we use finite but large value N (the number of members of the Laurent series) to provide needed accuracy of conformal mapping.

Dividing the real and imaginary parts in (5) we obtain:

$$\begin{aligned} x &= \sum a_k e^{k\xi} \cos k\eta \\ y &= \sum a_k e^{k\xi} \sin k\eta \end{aligned} \tag{6}$$

It is known that in curvilinear coordinate system the Laplace operator has the form

$$\Delta\psi = \frac{1}{H_1 H_2}\left(\frac{\partial}{\partial\xi}\left(\frac{H_1}{H_2}\frac{\partial\psi}{\partial\xi}\right) + \frac{\partial}{\partial\eta}\left(\frac{H_2}{H_1}\frac{\partial\psi}{\partial\eta}\right)\right) \tag{7}$$

where Lame coefficients are expressed as

$$H_1 = \left(\left(\frac{\partial x}{\partial\xi}\right)^2 + \left(\frac{\partial y}{\partial\xi}\right)^2\right)^{\frac{1}{2}},\quad H_2 = \left(\left(\frac{\partial x}{\partial\eta}\right)^2 + \left(\frac{\partial y}{\partial\eta}\right)^2\right)^{\frac{1}{2}}$$

The Cauchy–Riemann conditions $\partial x / \partial\xi = \partial y / \partial\eta$, $\partial x / \partial\eta = -\partial y / \partial\xi$ are for the functions (6). In this case $H_1 = H_2$ and operator (7) has the form

$$\Delta_{x,y}\psi = H^{-1}\Delta_{\xi,\eta}\psi$$

where $H = \alpha^2 + \beta^2$, $\alpha = \partial x / \partial\xi = \sum k a_k e^{k\xi} \cos k\eta$, $\beta = \partial x / \partial\eta = -\sum k a_k e^{k\xi} \sin k\eta$

After transformation from the physical coordinates x, y to canonical ξ, η ones the equations (1) is written as:

$$\begin{aligned} \Delta\psi &= -\omega \cdot H(\xi,\eta) \\ \Delta\omega &= 0 \end{aligned} \tag{8}$$

The equations (8) are supplemented by the boundary conditions that correspond to the conditions (2–4). On the left side BC of the rectangle

$$\psi = 0,\ \partial\psi/\partial\xi = 0 \tag{9}$$

On the right side *AD*

$$\omega = 0,\ \psi = \sum a_k e^{k\xi}\sin k\eta \tag{10}$$

On the top and bottom sides *CD, AB*

$$\omega = 0,\ \psi = 0 \tag{11}$$

To solve Eq. (8) with the boundary conditions (9–11) by finite difference method we introduce the meshing of computational domain *ABCB* with nodes (ξ_j, η_i) so that $\xi_j = jh_\xi$, $j = \overline{0,M}$, $\eta_i = ih_\eta$, $i = \overline{0,N}$. We determine the discrete values of the functions $\psi_j^i = \psi(\xi_j, \eta_i)$, $\omega_j^i = \omega(\xi_j, \eta_i)$ in the nodes. Approximating the partial derivatives of order 2 by central differences, we can write Eq. (8) in a discrete form

$$\frac{\psi_{j-1}^i - 2\psi_j^i + \psi_{j+1}^i}{h_\xi^2} + \frac{\psi_j^{i-1} - 2\psi_j^i + \psi_j^{i+1}}{h_\eta^2} = -\omega_j^i \cdot H_j^i \tag{12}$$

$$\frac{\omega_{j-1}^i - 2\omega_j^i + \omega_{j+1}^i}{h_\xi^2} + \frac{\omega_j^{i-1} - 2\omega_j^i + \omega_j^{i+1}}{h_\eta^2} = 0 \tag{13}$$

where $H_j^i = H(\xi_j, \eta_i)$. The boundary conditions (9–11) on lines *BC, AD, AB*, *CD* are written as:

$$\psi_0^i = 0,\ (\psi_1^i - \psi_0^i)/h_\xi = 0 \tag{14}$$

$$\omega_M^i = 0,\ \psi_M^i = \sum a_k e^{kL}\sin k\eta_i \tag{15}$$

$$\omega_j^N = 0,\ \psi_j^N = 0 \tag{16}$$

$$\omega_j^0 = 0, \psi_j^0 = 0 \tag{17}$$

To obtain the system of algebraic equations (SAE) for ψ_j^i, ω_j^i Eq. (10–11) are written in the form

$$\frac{1}{h_\xi^2}\psi_{j-1}^i + \frac{1}{h_\xi^2}\psi_{j+1}^i - \left(\frac{2}{h_\xi^2} + \frac{2}{h_\eta^2}\right)\psi_j^i + \frac{1}{h_\eta^2}\psi_j^{i-1} + \frac{1}{h_\eta^2}\psi_j^{i+1} = -\omega_j^i \cdot H_j^i \tag{18}$$

$$\frac{1}{h_\xi^2}\omega_{j-1}^i + \frac{1}{h_\xi^2}\omega_{j+1}^i - \left(\frac{2}{h_\xi^2} + \frac{2}{h_\eta^2}\right)\omega_j^i + \frac{1}{h_\eta^2}\omega_j^{i-1} + \frac{1}{h_\eta^2}\omega_j^{i+1} = 0 \tag{19}$$

Taking into account the conditions (14–17) we will obtain SAE of order $4\times((M\text{–}2)\times(N\text{–}2))^2$. The direct solution of the SAE requires more memory but is more effective than the separate solution of (18) и (19) and use of iterations for the converged joint solution.

2. Results and discussion

As a test the flow around the cylinder in a periodic cell with the Kuwabara conditions is solved. The problem has also an analytical solution [2]

$$\psi^a(r,\theta) = \left(A\frac{1}{r} + Br + Cr\ln r + Dr^3\right)\sin\theta \tag{20}$$

IOP Conf. Series: Materials Science and Engineering **158** (2016) 012080 doi:10.1088/1757-899X/158/1/012080

$$\omega^a(r,\theta) = \frac{1}{r}\frac{\partial}{\partial r}\left(r\frac{\partial \psi^a}{\partial r}\right) + \frac{1}{r^2}\frac{\partial}{\partial r}\frac{\partial^2 \psi^a}{\partial \theta^2} = -2\left(\frac{C}{r} + 4Dr\right)\sin\theta \qquad (21)$$

where $A = 0.5\,(1-\alpha/2)/k$, $B = -0.5\,(1-\alpha)/k$, $C = 1/k, D = -0.25\alpha/k$, $\alpha = h^{-2}$, $k = \alpha - 0.25\alpha^2 - 0.5\ln\alpha - 0.75$, $r = \sqrt{x^2+y^2}$, $\theta = \mathrm{arctg}(y/x)$.

A numerical solution for M=N=100 is compared with analytical formulas (20–21). To estimate the accuracy of the developed method the absolute

$$E_\psi = \psi^a(x,y) - \psi(x,y),\; E_\omega = \omega^a(x,y) - \omega(x,y)$$

and relative errors

$$\varepsilon_\psi = \frac{\max|E_\psi(x,y)|}{\max|\psi^\circ(x,y)|},\; \varepsilon_\omega = \frac{\max|E_\omega(x,y)|}{\max|\omega^\circ(x,y)|}$$

were calculated. The distributions of the quantities $E_\psi(x,y)$ и $E_\omega(x,y)$ are shown in fig.2,a and 2,b respectively. The values of relative error are $\varepsilon_\psi = 0.003267$, $\varepsilon_\omega = 0.017285$. The conclusion can be made that the developed method gives high accuracy.

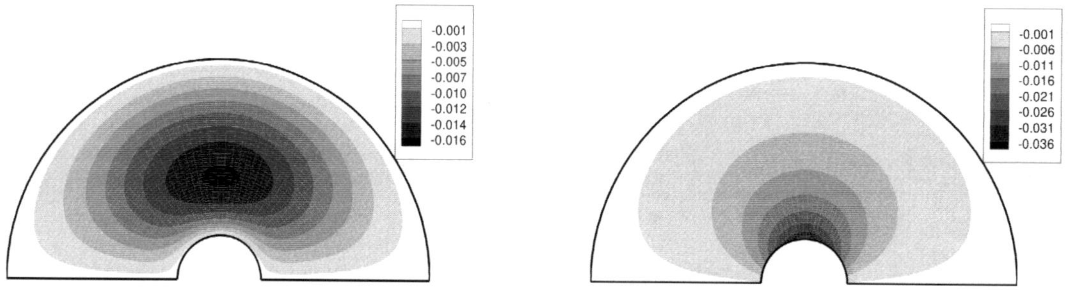

a b

Figure 2. The distributions of $E_\psi(x,y)$, $E_\omega(x,y)$

The problem of fluid flow around an elliptical cylinder with half-axes $a_1 = 0.5$, $a_2 = 1$ and radius of outer boundary h=5 has been solved. Examples of calculated streamlines and vorticity isolines in the physical and canonical planes are given in fig.3.

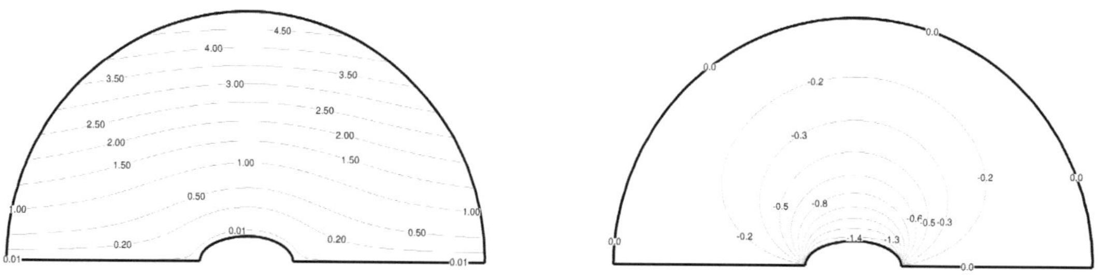

a b

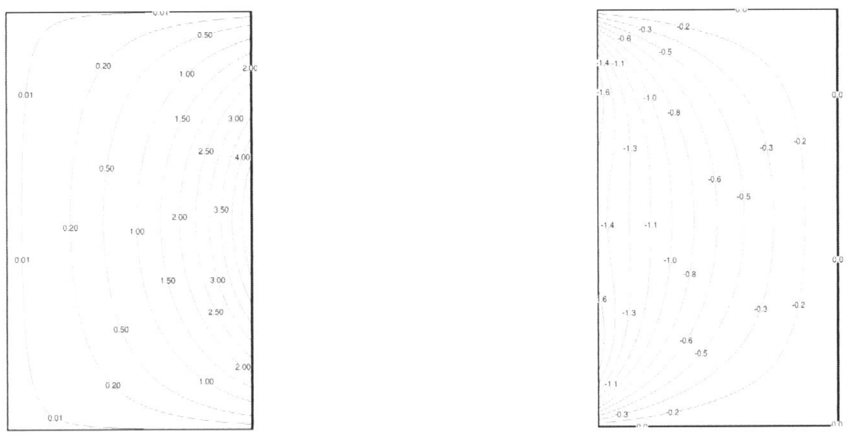

c d

Figure 3. The streamlines (a, c) and vorticity (b, d) isolines in the physical (a, b) and canonical (c, d) planes.

Acknowledgments

This investigation was conducted in the frameworks of the Russian Government Program of Competitive Growth at Kazan Federal University and supported by the Russian Foundation for Basic Research (grants No. 15-01-0613, No. 16-51-10024).

References

[1] Mardanov R F, Dunnett S J and Zaripov S K 2016 *Eng Anal Boundary Elem.* **68**, 54–62.

[2] Kuwabara S 1959 *J. of Phys. Soc. of Japan.* **14**, 527–532.

[3] Abzalilov D F and Shirokova E A 2016 *Complex Variables and Elliptic Equations. DOI: 10.1080/17476933.2016.1227978.*

Numerical scheme for non-linear model of supercritical fluid extraction from polydisperse ground plant material: single transport system

A Salamatin

Kazan Federal University, Department of Aerohydromechanics, Kremlyovskaya str., 18, Kazan 420008, Russia

E-mail: arthur.salamatin2@gmail.com

Abstract. Numerical algorithm is developed for modelling non-linear mass transfer process in supercritical fluid extraction (SFE). The ground raw material is considered as polydisperse, characterized by discrete number of effective particle fractions. Two continuous interacting counterparts separated by permeable membrane are distinguished in plant material build-up. The apoplast plays role of transport channels during extraction, and symplast contains extractable oil. The complete SFE model is non-linear as a result of non-linearity of oil dissolution kinetics. The computational scheme is based on the finite-volume approximation method and Thomas elimination procedure. The resulting system of algebraic equations is solved iteratively. Special attention is paid to polydisperse substrates, when particle scale characteristics of all fractions interact with each other through pore phase concentration on the vessel scale. Stability of the developed algorithm is demonstrated in numerical tests. Special iterative procedure guarantees a monotonic decrease of oil content in individual particles of substrate. It is also shown that in the limit of the so-called shrinking core approach the number of mesh nodes on a particle scale should be increased.

1. Problem formulation

Supercritical Fluid Extraction (SFE) – is a technological approach to extract natural compounds from particles of ground plant material. Conventionally, the complete model of the process consists of two submodels [1], external and internal. The first accounts for convective solvent flow through a porous media, composed of polydisperse ground plant material particles. After [1, 2], external mass balance is described in one-dimensional quasi-stationary approximation, and master-equation takes the following form

$$\frac{\partial c}{\partial \zeta}=\sum_{i=1}^{n}\frac{\mathrm{Sh}_i}{a_i^2}(\theta_{b,i}-c)f_i,\quad c\big|_{\zeta=0}=0. \tag{1}$$

Here, $0 \le c \le 1$ is the solute concentration in the pore phase, ζ – axial coordinate along the vessel, varying from zero at the inlet to normalized vessel height H at the outlet, f_i and Sh_i are volume fraction and Sherwood number, corresponding to i-th fraction of plant material, a_i – dimensionless radius of spherical particle, $1 \le i \le n$. Solute concentration θ_b at the particle surface is determined from the solute flux continuity condition (4). All Sherwood numbers Sh_i are assumed to be tending to infinity.

The internal submodel is of particular importance for correct SFE modeling and understanding. The one considered here contains two similarity criteria, Θ and M. The relative solvent dissolution capacity

Θ characterizes the raw material. The second criterion, $M \sim a^2$, distinguishes particle fraction (of size a) and varies from zero to infinity, what corresponds to various internal extraction regimes. The case of vegetable and fatty oil extraction is considered in this work. So, the solvent dissolution capacity $\Theta \rightarrow 0$.

The detailed description of plant material structure leads to two mass balance equations

$$\frac{\partial x_s}{\partial \tau} + (1+M)(\theta_s - \theta_a) = 0, \tag{2}$$

$$6M(\theta_s - \theta_a) + \Delta\theta_a = 0, \tag{3}$$

with boundary conditions

$$\frac{\partial \theta_a}{\partial r}\Big|_{r=0} = 0, \quad \theta_a\Big|_{r=1} = \theta_b, \quad -\frac{\partial \theta_a}{\partial r}\Big|_{r=1} = \mathrm{Sh}(\theta_b - c) \tag{4}$$

for normalized solute concentration $0 \le \theta_a \le 1$ within transport channels, and initial condition

$$x_s\Big|_{t=0} = 1 \tag{5}$$

for normalized oil density $0 \le x_s \le 1$ which accounts for extractable oil inclusions, as well as dissolved solute inside cells. Here, $0 \le r \le 1$ is the normalized radial coordinate inside spherical particles, and τ – dimensionless extraction time related to a particular particle fraction

$$\tau_i = \frac{1+M_i}{1+M_1}\tau_1 \equiv A_i\tau_1 .$$

Common SFE process time t is associated with the first fraction, $t = \tau_1$.

The normalized intra-cell solute concentration $0 \le \theta_s \le 1$ depends on x_s, and is determined by the process of oil inclusions dissolution in the solvent. The dissolution model considered here leads to the following non-linear expression

$$\theta_s = \min\{1, x_s / \Theta\}. \tag{6}$$

The internal submodel, Eqs. (2)-(6), should be solved for every i-th particle fraction. All characteristics are attributed to a particular particle fraction i, the index is omitted where there is no ambiguity.

The moving boundary $r = R$ appears inside each individual particle and is controlled by the assumed dissolution model (6) at relatively low solvent dissolution capacity, $\Theta < 1$. The boundary separates internal core with drop-phase oil, $x_s > \Theta$, from outer transport zone where all oil is completely dissolved in the solvent, $x_s < \Theta$. Position of this boundary is unknown *a priopi* and must be determined in the course of modeling. Obviously, at the beginning of the process $R(a, \zeta, 0) \equiv 1$, since no solute is extracted from substrate.

Eqs. (1)-(6), formulated above, form a complete simplified model of the SFE process. They consider intra-particle solute diffusion in quasi-stationary approximation and account only for one transport system. Nevertheless, these equations reveal the principal peculiarities and problems encountered in numerical SFE modeling.

2. Numerical algorithm

Let us introduce uniform spatial meshes $\zeta_j = h_\zeta j$, $h_\zeta = H / m_\zeta$, $j = 0, 1... m_\zeta$ and $r_l = h_r l$, $h_r = 1 / m_r$, $l = 0, 1... m_r$ in the extractor column and particles respectively. Unknown discrete analogues c_{jk}, $x_{s,ijlk}$, $\theta_{a,ijlk}$, $\theta_{s,ijlk}$, and $\theta_{b,ijlk}$ of process characteristics, c, x_s, θ_a, θ_s, and θ_b, are calculated successively at time moments $t_k = h_t k$, $k = 1, 2...$ according to the following algorithm. The subscripts j, l, k as well as i are omitted further if there is no ambiguity.

The following finite-difference scheme is deduced on the basis of well-developed integral interpolation (or finite volume) method [3, 4]

$$\frac{c_j - c_{j-1}}{h_\zeta} = \sum_{i=1}^{n} \frac{\mathrm{Sh}_i}{a_i^2}(\theta_{b,ij} - c_j) f_i, \quad c_0 = 0, \tag{7}$$

$$\frac{x_s - \breve{x}_s}{Ah_t} + (1+\mathrm{M})\left(\theta_s - \theta_a\right) = 0, \tag{8}$$

$$2\mathrm{M}\left(r_{l+1/2}^3 - r_{l-1/2}^3\right)\left(\theta_{s,l} - \theta_{a,l}\right) + r_{l+1/2}^2 \frac{\theta_{a,l+1} - \theta_{a,l}}{h_r} - r_{l-1/2}^2 \frac{\theta_{a,l} - \theta_{a,l-1}}{h_r} = 0, \tag{9}$$

$$\mathrm{M}h_r\left(\theta_{s,0} - \theta_{a,0}\right) + \frac{\theta_{a,1} - \theta_{a,0}}{h_r} = 0, \tag{10}$$

$$2\mathrm{M}\left(1 - r_{m_r-1/2}^3\right)\left(\theta_{s,m_r} - \theta_{a,m_r}\right) - \mathrm{Sh}(\theta_{a,m_r} - c) - r_{m_r-1/2}^2 \frac{\theta_{a,m_r} - \theta_{a,m_r-1}}{h_r} = 0. \tag{11}$$

Here, vector $\breve{x}_s = x_{s,k-1}$ is known from previous (k – 1)-th time layer, and algebraic expression (6) remains unchanged.

The computational scheme is based on Thomas elimination procedure [5]. First, the sweep run through Eqs. (8)-(11), and (6) delivers coefficients α and β for the linear relation between θ_b and c

$$\theta_b = \alpha c + \beta .$$

Next, c is found after substitution of the latter expression into Eq. (7). It allows to perform the second step of Thomas procedure – backward substitution, and complete the calculation of the solute characteristics from the internal model. According to Eq. (7), these parameters are sequentially determined for every cross-section ζ_j along the fluid flow. Internal submodel could be solved separately for every particle size only at ζ_0, where $c_0 = 0$ is fixed.

The coefficients α and β are found after linearization of Eq. (6), and Thomas algorithm is then applied to solve the internal-model equations. As it is already mentioned, there is a moving boundary R in every particle, that separates the internal oil saturated core and outer transport zone. The minimum of the right-hand side of Eq. (6) is well defined in every zone, and complete internal submodel is linear for $r < R$ and $r > R$. Current discrete boundary position R_k is ascribed to mesh points r_l, and it is determined iteratively for every particle fraction. The initial guess is $R^{(1)} = R_{k-1}$, since R monotonically decreases with time, and

$$\theta_s = \begin{cases} 1, & \text{if } r \le R^{(1)}; \\ x_s / \Theta, & \text{if } r > R^{(1)}. \end{cases} \tag{12}$$

The corresponding vectors $x_s^{(1)}$, and $\theta_s^{(1)}$, as well as $R^{(1)}$ have to be checked for consistency with relation (12) in the neighbourhood of $r = R^{(1)}$. For any set I of particle fractions which are inconsistent with the above conditions the guess for R should be decreased, $R_I^{(2)} = R_I^{(1)} - h_r$. Boundary positions in other fractions, $i \notin I$, are not modified. With the new guess $R^{(2)}$ another iteration is executed. The iterative procedure is continued till Eq. (6) is satisfied for every particle fraction i, i.e. the set I turns out empty. Then, unknown values could be determined at the next vessel cross-section, ζ_{j+1}, according to the same procedure.

Eventually, the solution of simultaneous algebraic Eqs. (6)-(11) could be found after a finite number of iterations for any polydisperse packed bed, since the core radius $0 \le R \le 1$, and it monotonically decreases in the course of extraction.

3. Single particle washed by pure solvent

Let us consider a single particle washed by pure solvent, at $c = 0$. Under this assumption internal submodel (8)-(11), and (6) can be solved independently to study the convergence of the computational algorithm.

Various extraction regimes take place on the particle scale during extraction, as predicted by the model, Fig. 1. At small Θ, they are primarily determined by M. As one can see, a boundary layer develops in the neighborhood of $r = R$, and its thickness decreases with M $\to \infty$. Thus, it could not be adequately resolved by the mesh at constant m_r. A noticeable concentration c deviation from the exact value can occur if this fact is disregarded, as demonstrated in the next section. Still, the scheme remains stable, and integral extraction process characteristic, Overall Extraction Curve (OEC) [1, 2, 6-9], is insensitive to this deviation.

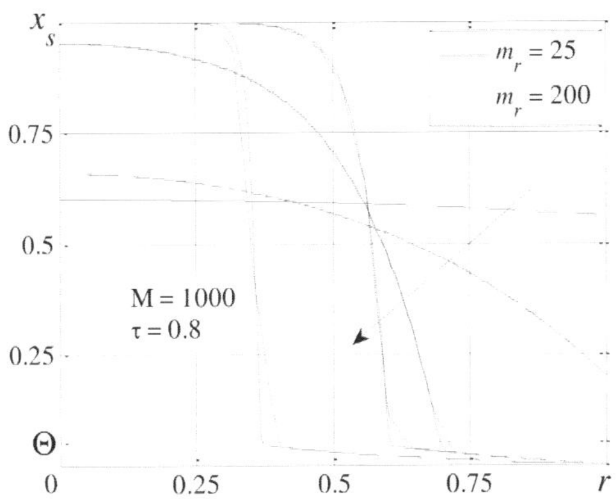

Figure 1. $x_s(r)$ at different M; Θ = 0.05. Arrow indicates lg M increase from −1 to 2 at step 1, and lines are obtained at τ = 0.5. Label corresponds to the fifth pair of curves.

The limit of M → ∞ is termed as the Shrinking Core (SC) pattern [2, 6, 7]. The model (6)-(11) at high values of M is of particular interest from the computational point of view, since x_s curve is not smooth and the jump develops in the neighborhood of $r = R$ at these conditions. The case of M = 100 and Θ = 0.05 is analyzed in Figs. 2, and 3. The curves are plotted for m_r = 25 (dashed lines), and 200 (solid lines), and demonstrate convergence of the computational algorithm on the particle scale. Arrows indicate direction of time t increase from 0.2 to 0.8 at step 0.2, and solution does not depend on h_t when it is below 0.01. The scheme convergence is apparent, and a relatively small number m_r ~ 100 of mesh points is sufficient to obtain a reliable solution in this case.

4. Monodisperse packed beds

The scheme (6)-(11) convergence is studied in this section for monodisperse packed bed, at n = 1. This simple approximation is commonly used in the kinetics data interpretation [7-9]. It is important from the computational point of view that, unlike in polydisperse packed beds, the moving internal oil-core interfaces in different particles are locally identical, and internal moving boundaries R do not mutually affect each other. So, a unique problem (6)-(11) solution exists, and it could be found after a finite number of iterations.

It is found that vessel scale mesh at m_ζ ~ 100 is well sufficient for a wide spectrum of process parameters, and it is independent of H-value, while time step h_t must be normally chosen on the order of 0.001. Special attention should be paid to the m_r impact onto external submodel solution. Inadequate spatial resolution of non-smooth distributions of internal characteristics, x_s, and θ_s on the particle scale at high M can noticeably affect the pore phase concentration c, as demonstrated in Fig. 4. Here, curves correspond to H = 20, Θ = 0.05, and extreme value of M = 1000, and are calculated at a relatively small m_r = 25. However, the computational errors become negligible with increase in m_r, at m_r = 200, and vanish with time, and they practically do not affect OEC. Hence, the major computational inaccuracy develops when only a particle superficial layer is under extraction. The convergence of the solvent flow characteristics is demonstrated in Fig. 5 for the same values of process and scheme parameters. Thus, the required accuracy of the numerical solution is mainly controlled by the quality of computations on the particle scale with the c-errors inherited via Eq. (11).

5. Bidisperse packed beds

Here, the convergence of computational algorithm is studied in a particular case of polydisperse packs, bidisperse ground plant material, for n = 2. This model of packed beds is used for detailed interpretation of SFE kinetics data [2].

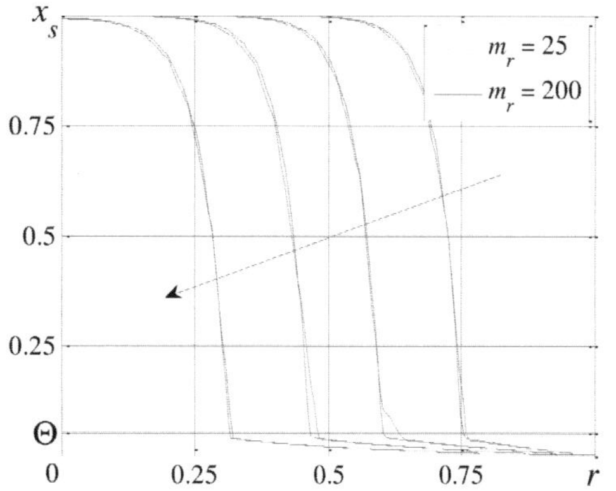

Figure 2. $x_s(r)$ dependence at different time moments t.

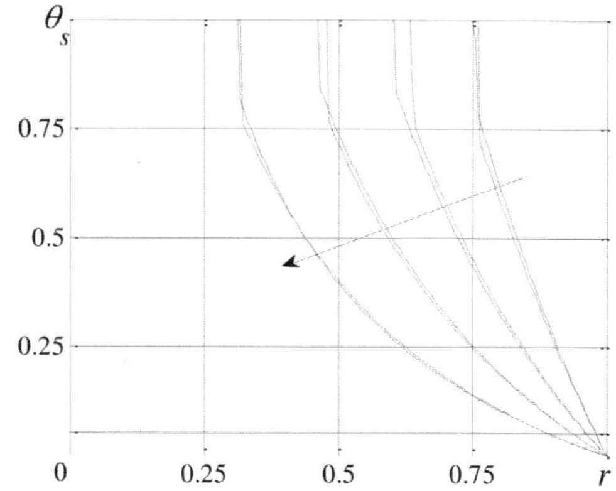

Figure 3. $\theta_s(r)$ dependence at different time moments t.

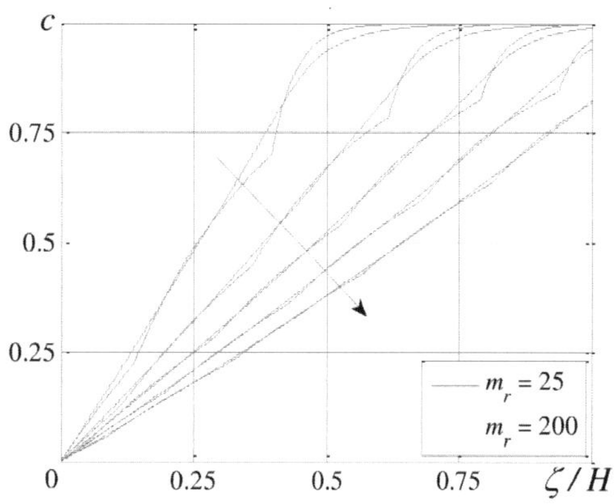

Figure 4. $c(\zeta)$ at different moments. Arrow indicate t increase from 0.02 to 0.1 at step 0.02.

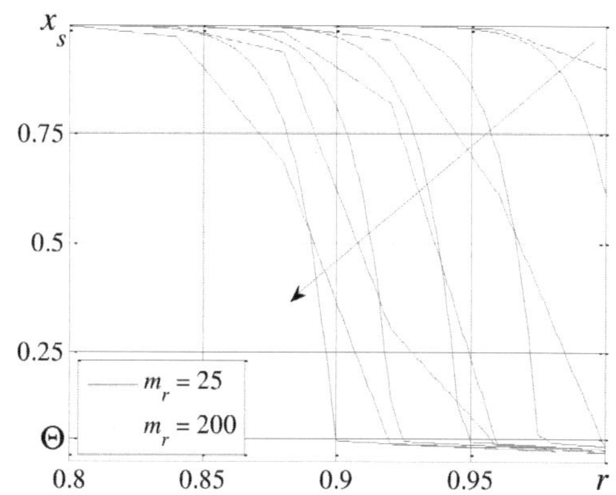

Figure 5. $x_s(r, \zeta = H / 2)$ at different moments. Arrow indicate t increase from 0.02 to 0.1 at step 0.02.

In bidisperse substrates (as well as polydisperse, at $n > 2$), moving oil-core boundaries R_i locally interact with each other through the pore phase concentration c, and should be calculated simultaneously for each fraction in every cross-section ζ. Physically consistent computational algorithm to solve the problem is described above in Sec. 2, and explicitly accounts that all R_{ij} do not increase in the course of extraction. It allows to determine a unique solution for a given set of SFE parameters and mesh characteristics.

Conventionally, bidisperse packed bed contains particles of two distinct size modes, $a_1 << a_2$, and it is set hereinafter that $M_1 = M_2 / 100$. This situation is typical for laboratory-scale experiments [2, 8, 9]. Accordingly, the substrate is characterized by two sets of spatial and time scales which manifest themselves in respective profiles of pore solute concentration versus ζ and t as illustrated by Figs. 6, and 7. Here, $\Theta = 0.05$, $M_2 = 100$, $f_1 = 0.2$, $H = 2000$, $m_\zeta = 200$, and 25 (for solid and dashed lines), $h_t = 0.03$. Arrow indicates t increase from 8 to 32 at step 8. The convergence on the particle scale is provided by $m_r = 200$.

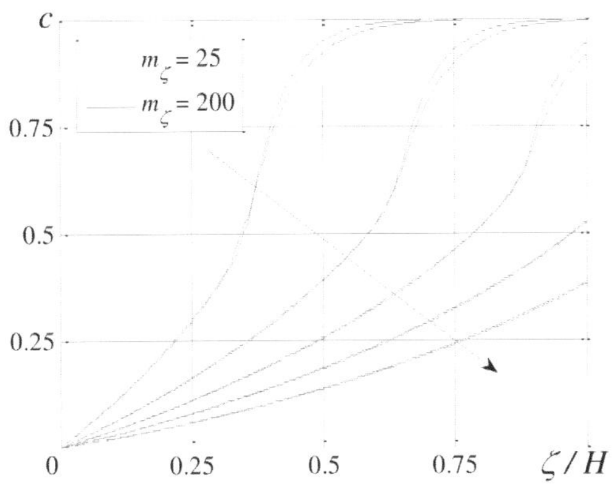

Figure 6. Pore phase concentration $c(\zeta)$ at different time moments. Bidisperse packed bed.

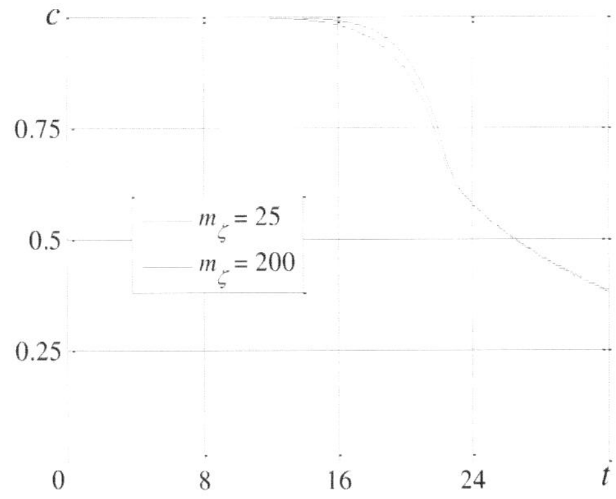

Figure 7. Outlet pore phase concentration $c(t, \zeta = H)$. Bidisperse packed bed.

6. Conclusion

The finite difference algorithm is developed to solve numerically the non-linear SFE model with the detailed description of ground plant material structure on the particle scale, and packed bed polydispercity on the vessel scale.

The scheme convergence is analyzed and demonstrated on both scales. Mesh characteristics m_ζ, and h_t are specified to provide high accuracy of simulations in a wide range of process parameters. Yet, the universal estimates for the mesh detalization, m_r, on the particle scale do not exist, since the solution of the internal submodel is not smooth at $M \to \infty$. Nevertheless, the overall integral characteristic, OEC, remains rather stable. It is found that m_r could be varied up to 200 for M up to 100.

The developed iterative computational scheme is shown to be stable and provides realistic monotonous decrease of all R_{ij} during extraction, and, therefore, could be considered as physically consistent. Apparently, a scheme with better convergence characteristics could be developed if current R-value is explicitly accounted in the quadrature formula used in finite volume method. However, it leads to a substantial complication of numerical scheme.

Acknowledgments

This study was supported by the Russian Foundation for Basic Research and by the Academy of Science of Republic of Tatarstan through the grant No. 15-41-02542 r_povolzhe_a, and grant No. 16-31-00007 mol_a.

References

[1] Egorov A G, Salamatin A A, and Maksudov R N 2014 *Theor. Found. Chem. Eng.* **48** 39
[2] Egorov A G and Salamatin A A 2015 *Chem. Eng. Technol.* **38** 1203
[3] Patankar S V 1980 *Numerical Heat Transfer and Fluid Flow* (New York: Hemisphere)
[4] Samarskii A A 2001 *The Theory of Difference Schemes* (New York: Hemisphere)
[5] Conte S D and deBoor C 1972 *Elementary Numerical Analysis* (New York: McGraw-Hill)
[6] Salamatin A A and Egorov A G 2015 *J. Supercrit. Fluids* **105** 35
[7] Goto M, Roy B C, and Hirose T 1996 *J. Supercrit. Fluids* **9** 128
[8] Salgin U and Korkmaz H 2011 *J. Supercrit. Fluids* **58** 239
[9] Ozkal S G, Yener M E and Bayindirli L. 2005 *J. Supercrit. Fluids* **35** 119

IOP Conf. Series: Materials Science and Engineering **158** (2016) 012082 doi:10.1088/1757-899X/158/1/012082

Construction and use of numerical-analytical approximating functions

M N Serazutdinov

Kazan national research technological university, 68 Karl Marx street, Kazan, 420015, Russian Federation

E-mail: serazmn@mail.ru

Abstract. The article goes over the methodology of constructing numerical-analytical approximating functions, satisfying the given boundary conditions for the function of its derivatives in the circuit areas of various shapes. The methodology is based on presenting the unknown function as a series in a complete set of functions that do not satisfy the given boundary conditions on the contour of the area, but additionally numerically defined near the contour to satisfy the boundary conditions. The additional definition of the functions near the area contour is performed numerically based on finite-difference relations. The main advantage of the stated method is the ability to build a relatively simple approximating functions satisfying the given boundary conditions on the boundary of complex shaped areas. The examples of applying the described method for solving the boundary value problem of a plate of different shapes. The possibility of using numerical-analytical functions for solving boundary value problems that contain higher derivatives up to fourth order is shown.

1. Common relations

Let's determine that the boundary problem is formulated in the variational form we need to find the function $w(x, y)$ from the following condition:

$$\iint_{\Omega} \Phi(\mathrm{w})dA = 0. \qquad (1)$$

Let's present $w(x, y)$ as series:

$$\mathrm{w} = \sum_{i}^{I} \mathrm{c_i}\, \phi_i(x, y), \qquad (2)$$

where c_i – known coefficients, $\phi_i(x, y)$ – functions that satisfy the boundary conditions.

We will present the methodology of constructing $\phi_i(x, y)$, satisfying conditions set on the contour of the areas of varios shapes. These functions can be used for getting solutions of theory of elasticity and the theory of thin shells problems in the domains of different shapes [1].

For simplicity, we will consider the cases when the contour Γ of the area Ω is made up of line segments (Figure 1,a) and the boundary conditions are set

$$w = 0,\ a_1 w + a_2 \frac{\partial w}{\partial n} + a_3 \frac{\partial^2 w}{\partial n^2} = 0,\ (x,\ y) \in \Gamma .$$

where $a_j(x, y)$ – defined functions $(j = \overline{1, 3})$, n – direction of normal to the contour.

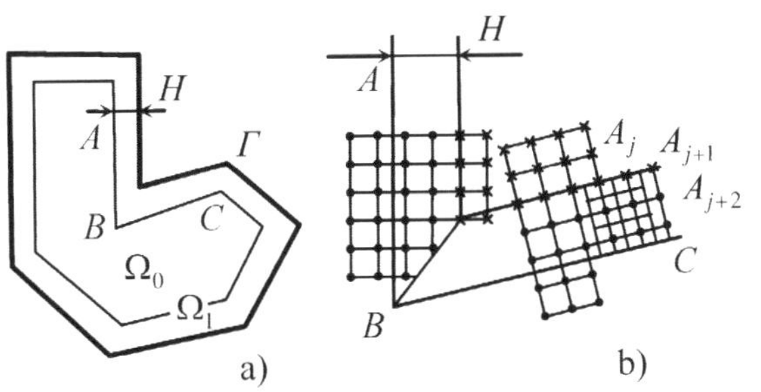

Figure 1. The area Ω and one of the possible variants of the grid.

Dividing area Ω int sub-areas Ω_0 и and Ω_1. Area Ω_1 is bordered with the contour Γ (Figure 1,a) and has a relatively small width.

Function $\phi_i(x,y)$ of series (2) we will define in the following way:

$$\phi_i(x,y) = \begin{cases} \psi_i(x,y), & x,y \in \Omega_0; \\ \psi_i^*(x,y), & x,y \in \Omega_1. \end{cases}$$

Here $\{\psi_i(x,y)\}$ – some complete system of functions, $\psi_i^*(x,y)$ – functions selected in a way for $\phi_i(x,y)$ to satisfy the boundary conditions.

We determine that $\psi_i^*(x,y)$ are discrete functions. The values $\psi_i^*(x,y)$ will be set in the grid nodes in a way to satisfy the required conditions of smoothness $\phi_i(x,y)$ and boundary conditions. The grid is chosen in the area Ω_1 and its surroundings in a way that makes it possible to satisfy the conditions at the border Γ with the use of finite-difference ratios.

One of the possible variants of the grid is shown on figure 1,b. The grid is chosen in the area Ω_1 and outside the contour points are introduced. Number of series of outside the contour points depends on the order of the derivative function w included in (1), and from the finite-difference formulas that are being used. In the nods (x_i, y_i) marked with the • sign (Figure 1,b), $\psi_i^*(x_j,y_j) = \psi_i(x_j,y_j)$ is determined. In the nodes marked with ×, values of the function ψ_i^* are found from the boundary conditions, written in finite-difference form.

The following algorithm of finding the function (2) is used. Condition (1) is presented in the way:

$$\iint_{\Omega_1} \Phi(w) dA + \iint_{\Omega_0} \Phi(w) dA = 0. \tag{3}$$

In order to calculate $\iint_{\Omega_1} \Phi(w) dA$ area $\iint_{\Omega_1} \Phi(w) dA$ divided into J elements. Grid nodes are located in the center or on the border of the elements (Figure 1,b). Using the formula of numerical integration, condition (3) can be written down as follows:

$$\sum_{j=1}^{J} \Phi(w(x_j, y_j)) A_j + \iint_{\Omega_0} \Phi(w) dA = 0,$$

where A_j – weighting coefficients of the numerical integration formulas.

Depending on the type of the condition (1) and the difference formulas used, two or more elements (two or more grid points) are located on the area width. When using the central differences, that approximate derivative functions with an error $O(h^2)$, two or three rows of the grid nodes should be placed on the width of area Ω_1. If (1) includes a higher derivative of the fourth order, three rows of the grid nodes are introduced. If (1) includes a higher derivative of the second order, two rows of the grid nodes are introduced.

Selecting the grid on the part of the area Ω_1, in a case when the condition (1) includes the higher derivatives of the fourth order and the central difference formulas with the error $O(h^2)$ are used, is shown on figure 1,b.

Circles mark the points where $\psi_i^*(x_j, y_j) = \phi_i^*(x_j, y_j)$ is determined, crosses – the points in which the values $\phi_i(x, y)$ are determined numerically from the boundary conditions. With this choice of number of elements and values $\phi_i(x, y)$ the continuity of numerical-analytical function $\phi_i(x, y)$ and its derivatives in grid nodes is ensured.

In the area Ω_0 function $w(x, y)$ is an analytical function, therefore differentiation and integration can be carried out numerically or analytically for determination of the second term of condition (3).

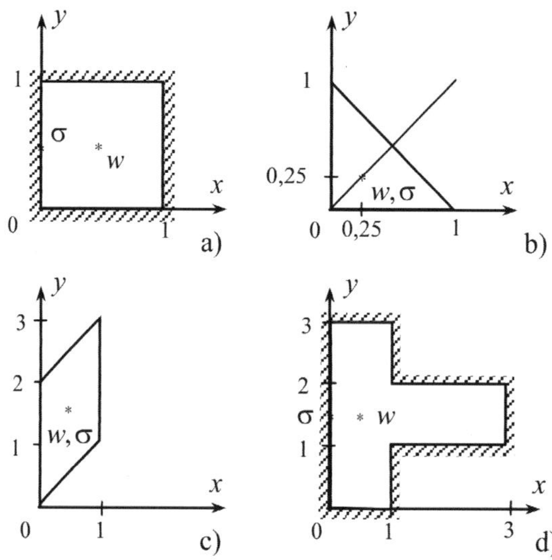

Figure 2. Plates of different shapes.

2. Results and calculations

Presenting the results of numerical-analytical functions use to solve the problem of bending of thin plates. Deflection function $w(x, y)$ of the plate is determined from the equation

$$D\nabla^2\nabla^2 w = q, \qquad x, y \in \Omega.$$

The boundary conditions of jamming are defined on the plate contour Γ

$$w = 0, \ \frac{\partial w}{\partial n} = 0, \ x, y \in \Gamma, \tag{4}$$

or of simple support condition

$$w = 0, \ \frac{\partial^2 w}{\partial n^2} = 0, \ x, y \in \Gamma. \tag{5}$$

We represent $w(x, y)$ as series (2) in which $\phi_i(x, y)$ will be defined by the methodology decribed here. To illustrate the possibilities of this methodology, the accuracy of the results, we will give an approximate solution of the problem using the Bubnov-Galerkin and Ritz methods. To determine the coefficients C_i of the series (2) with Bubnov-Galerkin method the following condition is used:

$$\iint_{\Omega} \left(\nabla^2\nabla^2 w - q\right) \phi_k ds = 0, \ k = \overline{1, I}.$$

While using Ritz method, – the condition is:

$$\iint_{\Omega} \delta\{\ D(1-\nu)[\ (\frac{\partial^2 w}{\partial x^2})^2 - \frac{\partial^2 w}{\partial x^2}\cdot\frac{\partial^2 w}{\partial y^2}\ ]+[\ 0{,}5D(\nabla^2 w)^2 - q w\ ]\ \}\, dA = 0\ .$$

Power functions $x^m y^n$ that make up a complete polynomial of degree S have been used as functions $\psi_i(x,y)$. Derivatives in the area Ω_1 were calculated using central finite-difference formulas with an error $O(h^2)$. The integration was carried out numerically according to the formula of rectangles. For the solution based on Bubnov-Galerkin method 3 rectangular elements were located on the width of area Ω_1. For the Ritz method – 2 rectangular elements. The lengths of the sides of the elements during the integration were equal to H_0 for the internal components and $H_0/2$ for elements bordering with the contour Γ ($H_0 = H/2{,}5$ in the Bubnov-Galerkin method; $H_0 = H/1{,}5$ in the Ritz method). In the vicinity of the indirect corners of the plate non-rectangular elements were used in the integration.

The functions ψ_i^* on the contour Γ and on the points outside countour to area Ω were determined from the boundary conditions.

Solution of the problem of bending plates of different shapes (Figure 2) was obtained. On the contour of the plates shown on figure 2,a,d conditions (4) were set and for the plates on figure 2,b,c – conditions (5).

Table 1. Solution based on Bubnov-Galerkin method.

Fig.	H_0	S	$\frac{w}{\Delta w\%}$	$\frac{\sigma}{\Delta\sigma\%}$	Fig.	H_0	S	$\frac{w}{\Delta w\%}$	$\frac{\sigma}{\Delta\sigma\%}$
2,a	0,04	6	$\frac{0,458}{0,9}$	$\frac{10,85}{6,8}$	2,c	0,08	8	$\frac{3,35}{0,5}$	$\frac{17,44}{3,6}$
	0,04	8	$\frac{0,488}{1,9}$	$\frac{10,05}{1}$		0,08	10	$\frac{3,38}{0}$	$\frac{17,4}{3,6}$
2,b	0,04	6	$\frac{0,218}{0,1}$	$\frac{3,54}{1,1}$	2,d	0,12	10	$\frac{1,99}{2,4}$	$\frac{23,6}{3,3}$
	0,04	8	$\frac{0,219}{0,3}$	$\frac{3,58}{0,1}$		0,12	14	$\frac{2,1}{3,1}$	$\frac{22,04}{6,1}$

Table 2. Solution based on Ritz method.

Fig.	H_0	S	$\frac{w}{\Delta w\%}$	$\frac{\sigma}{\Delta\sigma\%}$	Fig.	H_0	S	$\frac{w}{\Delta w\%}$	$\frac{\sigma}{\Delta\sigma\%}$
2,a	0,04	6	$\frac{0,46}{1,4}$	$\frac{10,4}{2,5}$	2,c	0,08	8	$\frac{3,26}{3,5}$	$\frac{16,7}{0,8}$
	0,04	8	$\frac{0,458}{0,9}$	$\frac{10,08}{0,7}$		0,08	10	$\frac{3,27}{3,1}$	$\frac{16,95}{0,7}$
2,b	0,04	6	$\frac{0,219}{0,3}$	$\frac{3,55}{0,8}$	2,d	0,06	12	$\frac{1,97}{3,4}$	$\frac{25,4}{8}$
	0,04	8	$\frac{0,22}{0,8}$	$\frac{3,59}{0,2}$		0,06	14	$\frac{2,1}{3}$	$\frac{23,3}{0,7}$

Numeric data obtained for $q = 360D$, $\nu = 0,3$ (for the plate on figure 2,c, $\nu = 0,2$). Tables 1, 2 show the values w and $\sigma = \left| w^{*}_{,xx} + \nu\, w^{*}_{,yy} \right| / (2\,(1-\nu^2))$ at the points indicated by asterisks on figure 2. Also presented, $\Delta w\%$, $\Delta\sigma\%$ – the percentage difference between the obtained and known results [2–4]. Letters a, b, c, d marke data related to the plate s shown on Figure 2, a, b, c, d.

References

[1] Serazutdinov M N and Garifullin M F 1991 Shells of complex shape *Soviet Applied Mechanics* **27** pp 1077–1082

[2] Petukhov N P 1974 Calculation of bending plates with complex contour outline *Works of a seminar according to the theory of shells* (Kazan) **5** pp 30–34

[3] Smirnov V A 1983 Remark on the accuracy of calculation of a triangular plate *Structural Mechanics and Analysis of Constructions* **5** pp 15–17

[4] Timoshenko S P and Woinowsky-Krieger S 2009 *Plates and Shells* (Moscow. URSS) p 635

IOP Conf. Series: Materials Science and Engineering **158** (2016) 012083 doi:10.1088/1757-899X/158/1/012083

Investigation of deformation at a centrifugal compressor rotor in process of interference on shaft

M R Shamim and D V Berezhnoi

Kazan Federal University, 18 Kremlyovskaya Street, Kazan, 420008, Russian Federation

E-mail: Shamimta@aol.com

Abstract. In this paper, according to the finite element method, we had implemented "master-slave" method of contact interaction in elastic deformable bodies, with consider of the friction in the contact zone. We had compiled the orientation of solving extremum problems with inequality restrictions, projection algorithm, which called "the closest point projection algorithm". Finally, an example, had brought to show the calculation of the rotor nozzle centrifugal compressor on the shaft with interference.

1. Introduction

To calculate the stress-strain state interacting elements of three-dimensional structures, which are under complex loading commonly used techniques based on various numerical methods [1-9].The most common calculations are based on the finite element discretization of the object [10-19]. Contact problem had investigated with the finite element implementation method in feature. To execution the contact conditions in finite element implementation, mainly had used method of Lagrange multipliers, penalty method, the augmented Lagrange method, direct solution of the variation problem inequality restrictions methods of quadratic programming. The common point of all these three algorithms is one algorithm which is to find the contact zone. Thus, became justified the orientation of solving extremum problems with inequality restrictions, projection algorithm, which called "the closest point projection algorithm". Current algorithm allows to build the contact elements, based on access that calls master-slave. Formulation of variational form using considered approaches problem appears as nonlinear and solution reaches by iterative methods of solution. Mentioned methods of solution to most cases require the first derivative from total received system of nonlinear equations, that named coherent system of tangent matrix. This is required primarily to obtain quadratic rate of convergence and in some arguments even if exist linear velocity of convergence of iterative process. In solving the contact problem by finite element method the most adversity is in no-penetration mode, and also additional kinematic conditions in the case that problems with friction on general unknown border.

2. Appointment of contact interaction of two bodies problem

By considering the contact of two bodies A and B which have a surface with smoothness C^1, i.e., like bodies, normal to the surface of which is continuous. We assume that if exist a common point C for two surfaces and the total normal both of bodies A and B are place in contact. Many of these points labeled as zones of contact S_C. The contact area, as well as applied stress is unknown and our mission is to determine it from the solution. Base of the variational formulation of the contact problem is on satisfaction from non-penetration of bodies. To get that we consider the penetration functions as contacting bodies A and B. On the body B we choose a point C and project it on the surface of the body. Distance which is the shortest distance between the bodies, would determines the function of penetration. We define on the body A (Fig. 1) normal $\boldsymbol{n}$.

Let:

$$p = (\boldsymbol{r}_s - \boldsymbol{r}) \cdot \boldsymbol{n}$$

given demonstration allows us to dispose the function of the contact body, if $p > 0$ there is not penetration (there is no contact between bodies), if $p \leq 0$ one body A will penetrate into next body, adequacy condition $p = 0$ become non-penetrate one body to another.

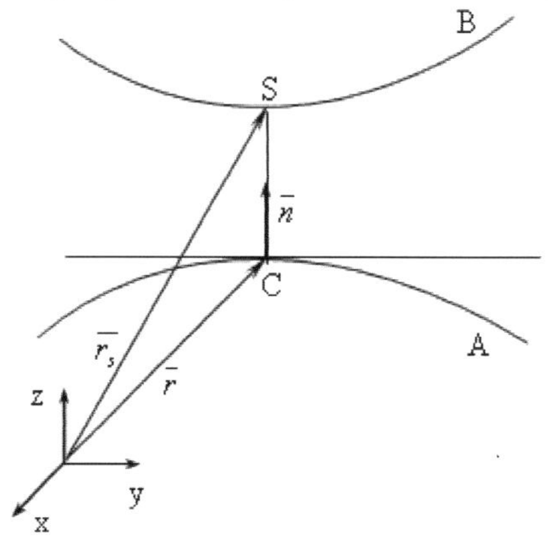

Figure 1. Contact scheme.

Figure 2. Finite element model.

The body B, in terms of which we will check the penetration modes, calls "slave" body and point – "slave" point S. The body A, on the surface of which we project "slave" point, calls the "master" body and its surface – "master" surface. Methods of solution in contact problems, established on the task "slave" points and determination of the value of penetration p in the "master" surface, which called "master-slave" algorithm. Usually nodes in finite element act as "slave" points and "master" point C determines by the closest point projection algorithm.

3. The scheme for solving the problem

The global scheme for solving the contact problem based on geometrical and physical nonlinear will present as follows.

We show general variational equation as:

$$\delta W + \delta W_c + \delta W_g + \delta W_p = 0,$$

where δW - refers to the entire structure, δW_c describes contact interaction, δW_g describes geometrically nonlinear deformation, δW_p - physically nonlinear deformation.

Next is to perform the congregate of global stiffness matrix after congregation variational the equation takes the form:

$$\{\delta q\}^T \cdot (\{R\} + \{R_c\} + \{R_g\} + \{R_p\}) = 0.$$

As result, we use Newton's method of the vector function $\boldsymbol{F}$ to solve nonlinear equation.

$$\boldsymbol{F} = \{R\} + \{R_c\} + \{R_g\} + \{R_p\},$$

where $\boldsymbol{F} = \boldsymbol{F}(\{q\})$, $\{q\}$ is the global vector of nodal displacements.

Accordingly, to specifying the initial approximation $\{q^{(0)}\}$, we construct an iterative procedure:

$$\{\Delta q\} = -\boldsymbol{F}'(\{q\})^{-1} \cdot \boldsymbol{F}(\{q\}),\ \{q\}^{(n+1)} = \{q\}^{(n)} + \{\Delta q\}^{(n)}.$$

The derivative functions of residuals give the tangent stiffness matrix:

$$\boldsymbol{F}'(u) = \frac{dF}{d\{u\}} = [K] = [K_{ep}] + [K_c] + [K_g],$$

where $[K]$ is the global stiffness matrix of structure, $[K_c] = [K_c^T] + [K_c^N]$ - is the global stiffness matrix contacts, gathered from local matrix on the contact elements, comprising part of the normal $[K_c^N]$ and tangential contact $[K_c^T]$, $[K_{ep}]$ - "elastic-plastic" stiffness matrix, $[K_g]$ - matrix of geometric stiffness. The transition to variational problem with algebraic discretization is performed by finite element method.

4. Numerical example

To describe rotor of a centrifugal compressor here used to apply different types of connection the wheels to shaft. The key seat in a shaft, splined, connection through interference and combinations are the most common element in this field. In this paper we consider the connection with interference. Goal of this work is to build the estimated compound volume modelled on interference, with consider to possible geometric variations on the contact surfaces, and evaluation while the rotor is at deformed state. To ensure the immobility connection between the impeller shaft in the design lays the necessary diametric interference. The average (nominal) contacts pressure, given by the attenuation interference in centrifugal forces, should be as when the frictional force exceeding external shear forces. Such torque and axial force appears in determination of external shear for centrifugal wheel. In stress-strain state in the hub of wheel it's necessary to calculate the optimal value of the diametrical interference, in which had provided by a fixed connection to the shaft of the wheel, and appearance of contact stress does not exceed the permissible limits. In the hub part we apply minimum yield strength of the shaft and permissible limit stress for the wheel part. As rule, the wheel has made from material with higher strength and mechanical properties. So if we use steel for the manufacture of wheels we provide amount of yield strength 835 MPa, the yield strength amount for steel to be used for the manufacture of shafts, does not exceed 450 MPa. Because of economic reasons if aim of manufacture of steel shaft is not to be much rigid so we can use from of less strength properties one. The mechanical properties of the shaft material are selected to become strength and durability condition at the area of key seat compound from the applied torque. Compression stresses in contact zone of the shaft with the wheel is not criteria. Under the influence of centrifugal forces occur weakening interference in the connection. Due to complex geometry of the wheel and its variable rigidity in the radial direction, as rule a significant decrease in interference occurs on the belt on the opposite side to the neck. To ensure the immobility of the connection at wheel with the shaft at the operating amount of rotation, it is necessary to increase the diametrical interference. In this case, calculations show that a fixed connection contact stresses can reach until the yield strength of the wheel material, i.e. so at contact zone on the shaft, occurs plastic deformation. It is considered, that, it's confirmed in practice, that the increased level of the stress state at the shaft in the contact zone does not affect the performance of the rotor entirely. This is due to the massiveness interfaced details and redistribution of stresses. Concerns to deformed state of the rotor, the assumption of uniform distribution of contact pressure in the circumferential direction in axisymmetric calculations show that will occur bending at shaft. Arise cases when after nozzle of wheels on the shaft change happen actually at geometry of rotor (increased values of the shaft radial beats at the control points). One possible reason for this situation may be compression stress tension at the zone of landing surface on the shaft. As part of this computational model is considered when the contact stress has a different amount in size in the circumferential direction, i.e., was considered irregular compression of the shaft from interference by making the contact surfaces with the maximum possible deviations indicated in construction documentation.

In practice, the conjunction surfaces are never one hundred percent. The best result is achieved will a conical under landing pressure after numerous compression and overflow on checked interfaced surfaces after each removing and with further elaboration of them. In cylindrical parts, check and modify of usage in the manufacture of rotors of centrifugal compressors is so difficult, so we don't conduct these operations in practice. For a particular test task to simulate wheel, in which observe irregular compression shaft in the circumferential direction was modeled as follows: interference was set for two sectors in each zone, located at an angle of 180° relative to each other.

Thus, it was modeled theoretical situation when interface surfaces of disc and the shaft on each circlet is 50%, and where there is no interference coupling is completely absent. In this arrangement the surfaces which interference in that had chosen from maximum possible impact condition on shaft bending. The calculation was based on the method which has presented in [20-23]. The results of calculation have shown in Figure 1. As seen in Figure 1, the assumption is that bending may occur due to irregular compression of the shaft. Result in sample the maximum shaft deflection was 0.01 mm.

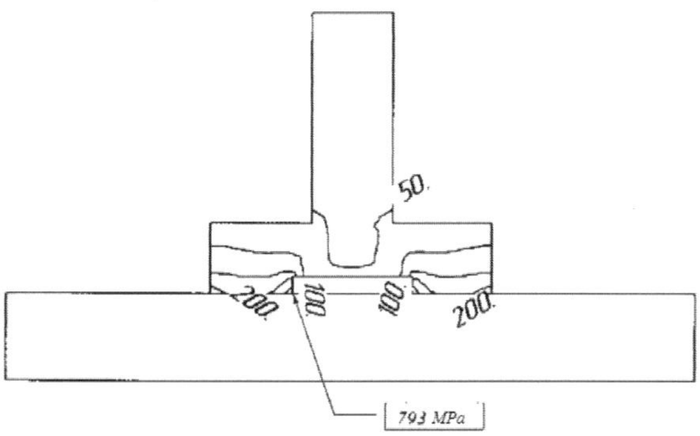

Figure 3. Contour stress intensity in connection with a disc shaft at the diameter 120 mm interference

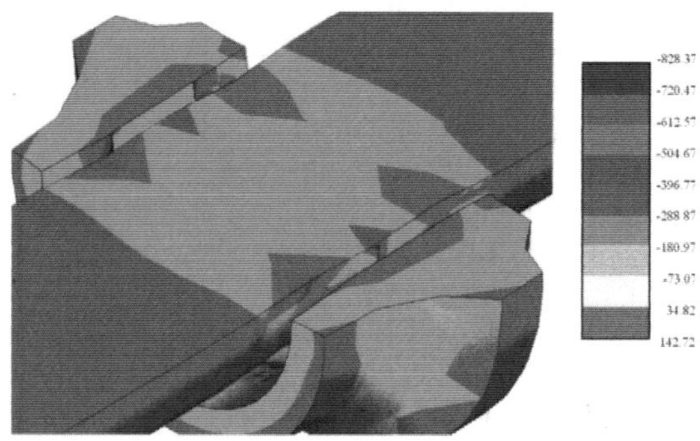

Figure 4. Radial stress (MPa) at a variable interference (250-300 micrometers) rotor

In practical terms, insignificant bending of the shaft after appearance of nozzle on the wheel, should not cause serious concerns from the normal operation of the rotor entirely. In diametric dimensions in the denomination situation by leaving a small stock on the shaft, and then having deduced nozzle on wheel, this problem solves. However, this solution is valid only for a fixed rotor. As mentioned above, during the operation there is a weakening interference until the opening of one of the girdles, with the result that will happen redistribution of contact stresses, which can again cause the rotor bending and lead to increase in vibration. As a specific example to illustrate the effect of variable interference, considered for blower rotor, during which production was set by this fact to change the geometry of the shaft after nozzle occurs on the wheel. Material properties: yield strength - 835 MPa, modulus of elasticity - 200000 MPa, density - 7800 kg/m^3. Shaft Material: material yield strength - 735 MPa, modulus of elasticity - 200000 MPa, density - 7800 kg/m^3. The range of diametrical interference within tolerance - (250-300) microns. It should be noted that tolerance at cylindrical shape at shaft 0.005 mm set just only in the areas which compression located. On the landing surfaces of shafts and wheels these parameters are not requirements. In this example, we assume that the connection has made with dimensions that can have a variable interference in the circumferential direction in the range of 250 to 300 microns. Finite element mesh model is shown in figure 3. Maximum interference of 300 microns and minimum interference 250 microns were set to sectors that provide the highest possible bending moment on the shaft. Figure 2 shows the radial displacement of the shaft with a variable interference (250-300) microns. As had seen from the calculation of the maximum shaft deflection is 0.0013 mm at the permissible value for the drawing - 0.01 mm.

5. Conclusion

Notice that in calculations used to be conducted various degree of sampling design consideration, committed similar results, indicate that the accuracy and stability of the solutions. From obtained results, we can conclude that finite element method is effective solutions for contact problems. It allows a single software implementation considers a fairly extensive category of problems with different conditions of contact interactions. The method is indifferent with respect to the geometry of the contacting parts, complicated boundary conditions and volume load. Material properties would be chosen nonhomogeneous and anisotropic, and the relationship between stress and deformation is nonlinear. All these factors insignificantly complicate the algorithms and problem solving. Comparison of the results at several problems with accurate and approximate methods, shows the reliability and accuracy of the results obtained at a relatively low complexity, and is far to become complete in possibilities of program. Valuable is the fact that as result of the calculation not only to obtain the contact pressure, but also considered stress-strained state parts in entire volume.

Acknowledgments

The reported study was supported by Government of the Republic of Tatarstan research projects No. 15-41-02555 and RFBR research projects No. 15-31-20602.

References

[1] Badriev I B, Banderov V V, Garipova G Z, Makarov M V and Shagidullin R R 2015 *Applied Mathematical Sciences* **9 (82)** 4095–4102
[2] Badriev I B and Shagidullin R R 1995 *Journal of Mathematical Sciences* **73 (5)** 519–525
[3] Badriev I B, Makarov M V and Paimushin V N 2015 Russian Mathematics 59 (3) 66–71
[4] Badriev I B, Banderov V V, Makarov M V and Paimushin V N 2015 Applied Mathematical Sciences 9 (77-80) 3887–3895
[5] Badriev I B, Makarov M V and Paimushin V N 2016 Procedia Engineering 150 1050–1055
[6] Badriev I B, Makarov M V and Paimushin V N 2016 Procedia Engineering 150 1056–1062
[7] Sachenkov O, Kharislamova L, Shamsutdinova N, Kirillova E and Konoplev Yu 2015 *IOP Conference Series: Materials Science and Engineering* **98**
[8] Shigapova F A, Mustakimova R F, Saleeva G T and Sachenkov O A *International Journal of Applied Engineering Research* **10 (24)** 44711–44714
[9] Sachenkov O A, Mitryaikin V I, Zaitseva T A and Konoplev Yu G 2014 *Applied Mathematical Sciences* **8 (159)** 7889–7897
[10] Sultanov L U 2015 *Procedia Earth and Planetary Science* **15** 119–124
[11] Davydov R L and Sultanov L U 2013 *PNRPU Mechanics Bulletin* **1** 81–93
[12] Golovanov A I and Sultanov L U 2005 *International Applied Mechanics* **41 (6)** 614–620
[13] Sultanov L U 2014 *Applied Mathematical Sciences* **8 (143)** 7117–7124
[14] Sultanov L U and Davydov R L and 2014 *Applied Mathematical Sciences* **8 (60)** 2991–2996
[15] Abdrakhmanova A I and Sultanov L U 2016 *Materials Physics and Mechanics* **26 (1)** 30–32
[16] Davydov R L and Sultanov L U 2015 *Journal of Engineering Physics and Thermophysics* **88 (5)** 1280–1288
[17] Sultanov L U and Fakhrutdinov L R 2013 *Magazine of Civil Engineering* **44(9)** 69–74
[18] Sultanov L U and Davydov R L 2013 *Magazine of Civil Engineering* **44(9)** 64–68
[19] Sultanov L U 2016 *Lobachevskii Journal of Mathematics* **37 (6)** 787–793
[20] Sagdatullin M and Berezhnoi D 2014 *Applied Mathematical Sciences* **8 (35)** 1731–1738
[21] Sagdatullin M and Berezhnoi D 2014 *Applied Mathematical Sciences* **8 (60)** 2965–2972
[22] Berezhnoi D V, Sachenkov A A and Sagdatullin M K 2014 *Applied Mathematical Sciences* **8 (127)** 6341–6348
[23] Berezhnoi D V, Sachenkov A A and Sagdatullin M K 2014 *Applied Mathematical Sciences* **8 (143)** 7107–7115

Method of the speed up calculations by the local Courant number for the filtration problem

A I Shangaraeva

Kazan Federal University, 18 Kremlevskaya Street, Kazan, 420008, Russia

AIShangaraeva@kpfu.ru

Abstract. We study the process of multiphase fluid-water flow in a heterogeneous formation in the presence of both producing and injection wells. The quasi-three-dimensional reservoir model is considered, comprising several seams, which characteristics are considered as constant thickness. Perforated wells are also can be presented. The main feature of the problem is that there are appear the filtration areas with high gradients in surroundings of wells. An approximate method is based on a preliminary finite-difference approximation of the problem. For its numerical realization there is proposed the algorithm, which allows to speed up the calculation time for the transfer equation using the explicit scheme, which is optimized taking into account the spatial inhomogeneity of the calculation time step.

1. Introduction

The filtration process is considered in the oil reservoir, which has a significant predominance of the size in the horizontal direction over the vertical. Along the strike the productive strata is limited by a cylindrical surface with a vertical generatrix. Along the vertical the reservoir is divided into the system of the interlayer, which has differ properties - not intersecting layers of variable thickness bordering on each other or separated by a cross connection with a low permeability. There are necessary to set certain transmission conditions or impermeability at these boundaries. The characteristics of the thickness are constant, so we can identify each interlayer with its middle surface. There are the system of the well, from a mathematical point of view, their presence is means the existence of boundaries relatively small size, which have the appropriate conditions. Wells are modeled as the sources/sinks. Note that the mathematical simulation of a multiphase filtration of high-viscous fluid in [1-10] is carried out.

2. Statement a problem

The problem of two-phase filtration is solved considering heterogeneity of associated oil and water according to field. This problem is described using the given equation to determine the reservoir pressure $P(x,y,z,t)$:

$$\operatorname{div}(\sigma \operatorname{grad} P) - \beta^* \frac{\partial P}{\partial t} = 0, \tag{1}$$

where $\sigma(x,y,z,t)$ is the reservoir transmissibility, $\beta^* = m(s\beta_1 + (1-s)\beta_2) + \beta_c$ is the reservoir and liquids elastocapacity, m – porosity. The equation (1) will be named as the piezoconductivity equation.

The water saturation equation $s = s_1(x, y, z, t)$ is given as the second equation of the system:

$$\operatorname{div}(f\sigma \operatorname{grad} P) - \beta_1^* \frac{\partial P}{\partial t} - m\frac{\partial s}{\partial t} = 0 \tag{2}$$

where $f(s) = k_1^*(s)/(\mu_1 k^*(s))$ is the J-function (Buckley–Leverett function), $\beta_1^* = s(m\beta_1 + \beta_c)$ is the reservoir and water elastocapacity.

The equations are valid in a multiply connected domain Ω (in the reservoir) with the boundary $\Gamma = \Gamma_H + \sum_{k=1}^{N} \Gamma_k$, $(k = 1...N)$, where Γ is the external boundary (the cylindric surface) of the filtration area, Γ_H is the top and bottom of the reservoir, Γ_k is the boundary (the cylindric surface) of the well number *k*, and N is the total amount of production and injection wells.

Initial and boundary conditions. Initially, the reservoir is in equilibrium. The fluid movement starts with the opening of its wells. At the initial time the distribution of pressure and saturation are known.

$$P = \varphi_0(x, y, z), (x, y, z) \in \Omega \tag{3}$$

$$s = s_0(x, y, z), (x, y, z) \in \Omega. \tag{4}$$

Either the total flow rate is set to wells:

$$\int_{z_H}^{z_B} \oint_{\Gamma_k} \sigma \frac{\partial P}{\partial n} d\gamma dz = q_k, (k = 1, ..., N_q) \tag{5}$$

or bottomhole pressure on the perforated intervals:

$$P = c_k(t), (k = N_q + 1, ..., N) \tag{6}$$

where N_q is the amount of the wells with the determined rate.

The pressure distribution is given on the external reservoir boundary:

$$P = \varphi(x, y, z), (x, y, z) \in \Gamma \tag{7}$$

or the condition of impermeability can be given:

$$\frac{\partial P}{\partial n} = 0, (x, y, z) \in \Gamma \tag{8}$$

And the top and bottom are impermeable:

$$\frac{\partial P}{\partial n} = 0, (x, y, z) \in \Gamma \tag{9}$$

Furthermore, the saturation is known in injection wells (on the sources):

$$s = s^*, (x, y, z) \in \Gamma_k, (k = 1, ..., N_1) \tag{10}$$

where s^* is the cutoff water saturation, N_1 is the amount injection wells.

The saturation is given on the part of the external contour $\gamma(t)$:

$$s = s_{ext}(x, y, z), (x, y, z) \in \gamma(t) \in \Gamma \tag{11}$$

where $\vec{n}$ is the external normal to the Γ, s_{ext} is the water saturation distribution along the outer contour.

Thus, it is required to find the functions $P(x, y, z, t)$ and $s(x, y, z, t), (x, y, z) \in \Omega, t \in [0, T]$, that satisfy to the equations (1), (2) and additional conditions (3) – (11).

3. Approximation of the problem

The regular grid is constructed in the filtration area for the numerical solution. The layers of this grid situated on the middle of the reservoir surface.

$$\omega_h = \left\{(x_i; y_j; z_k) : x_i = x_0 + ih;\ y_j = y_0 + jh;\ z_k = z_k\left(x_i; y_j\right)\right\},$$
$$i = \overline{0, N_x};\ j = \overline{0, N_y};\ k = \overline{0, N_z},$$

where $z_k\left(x_i; y_j\right)$ match with the vertical coordinates of the interlayer middle over the $\left(x_i; y_j\right)$ point.

The difference schemes for the equations (1) and (2), integrated over the volume of the elementary grid cells are of the form:

$$\begin{aligned}
&\beta_{i,j,k} V_{i,j,k} \frac{p_{i,j,k}^{N+1} - p_{i,j,k}^{N}}{t} - \\
&-h(k\lambda H)_{i+1/2,j,k} \frac{\left(p_{i+1,j,k} - p_{i,j,k}\right)}{h} + h(k\lambda H)_{i-1/2,j,k} \frac{\left(p_{i,j,k} - p_{i-1,j,k}\right)}{h} - \\
&-h(k\lambda H)_{i,j+1/2,k} \frac{\left(p_{i,j+1,k} - p_{i,j,k}\right)}{h} + h(k\lambda H)_{i,j-1/2,k} \frac{\left(p_{i,j,k} - p_{i,j-1,k}\right)}{h} - \\
&-h^2(k\lambda)_{i,j,k+1/2} \frac{\left(p_{i,j,k+1} - p_{i,j,k}\right)}{\tilde{H}_{i,j,k+1}} + h^2(k\lambda)_{i,j,k-1/2} \frac{\left(p_{i,j,k} - p_{i,j,k-1}\right)}{\tilde{H}_{i,j,k-1}} = -q_{i,j,k}
\end{aligned} \tag{12}$$

$$\begin{aligned}
&m_{i,j,k} V_{i,j,k} \frac{s_{i,j,k}^{n+1} - s_{i,j,k}^{n}}{\tau} = -q_{wi,j,k} - \tilde{\beta}_{i,j,k} V_{i,j,k} \frac{p_{i,j,k}^{N+1} - p_{i,j,k}^{N}}{t} - \\
&-f_{i+1/2,j,k} u_{i+1/2,j,k} + f_{i-1/2,j,k} u_{i-1/2,j,k} - \\
&-f_{i,j+1/2,k} u_{i,j+1/2,k} + f_{i,j-1/2,k} u_{i,j-1/2,k} - \\
&-f_{i,j,k+1/2} u_{i,j,k+1/2} + f_{i,j,k-1/2} u_{i,j,k-1/2}
\end{aligned} \tag{13}$$

where t – is the time step for the pressure calculation (the high index N), τ – is the time step for saturation calculation – the part of the time step t (the high index n), $\tilde{\beta} = ms\left(\beta_w + \beta_m\right)$ – is the elastocapacity, $H_{i,j,k}$ – is the local thickness of the k-layer of the reservoir, $\tilde{H}_{i,j,k+1/2} = z_{k+1}\left(x_i, y_j\right) - z_k\left(x_i, y_j\right)$ – is the distance between grid layers on vertical side, $V_{i,j,k} = h^2 H_{i,j,k}$ – is the volume of the grid; $q_{wi,j,k}$ – the amount of liquid from the cell through the corresponding well part; $q_{wi,j,k}$ – is amount of water from the cell. The amount filtration flows from are approximated in the following way:

$$\begin{aligned}
u_{i+1/2,j,k} &= -h(k\lambda H)_{i+1/2,j,k} \frac{\left(p_{i+1,j,k} - p_{i,j,k}\right)}{h}, \\
u_{i,j+1/2,k} &= -h(k\lambda H)_{i,j+1/2,k} \frac{\left(p_{i,j+1,k} - p_{i,j,k}\right)}{h}, \\
u_{i,j,k+1/2} &= -h^2(k\lambda)_{i,j,k+1/2} \frac{\left(p_{i,j,k+1} - p_{i,j,k}\right)}{\tilde{H}_{i,j,k+1}}.
\end{aligned} \tag{14}$$

The calculations are as follows: at first is one step at a time by the scheme (12). Then, according to the scheme (13) is made of r time steps, wherein $r\tau = t$ is the multiplicator, which is described below for different cells.

The approximation coefficients and taking into account features in surroundings of the wells is carried out in a standard way for the corresponding members of these types of equations. We note only that the J-function in half the nodes defined upstream. For example, according to the direction of the axis:

$$f_{i+1/2,j,k} = \begin{cases} f_{i,j,k}, & u_{i+1/2,j,k} \geq 0, \\ f_{i+1,j,k}, & u_{i+1/2,j,k} < 0. \end{cases} \tag{15}$$

For cells that contain wells difference approximation is different. You can use Peaceman approach [11]. In the right-hand sides of equations (12) and (13) are added, respectively, the terms:
– $q_l, l = l(i,j,k) \in L$ for perforated sections for a given production rate (for perforated sections assume $q_l = 0, l = l(i,j,k) \notin L$) or $P_l = P_{l\_well}, l = l(i,j,k) \in L$ for non-perforated sections with a given bottom hole pressure;
– $q_l f_l, l = l(i,j,k) \in L$ for perforated sections for a given production rate (for non-perforated $q_l = 0, l = l(i,j,k) \notin L$);
or more complicated variants of the approximation, taking into account the increasing number of cells (see., eg, [11]).

To take into account the features condition of formation in the near-well zone may use the concept of skin-factor, or a variant of effective corrections field reservoir parameters [11].

4. Algorithm of solving the problem

The IMPES method is using for solving the resulting system of difference equations, when the implicit scheme is using for pressure equation and the explicit scheme is using for saturation equation.

The reservoir pressure can be found from the (12) using a fully implicit scheme ($P_{i,j,k} = P^{N+1}_{i,j,k}$) with the use of the multigrid algorithms. As a result, in each pressure calculation step it is possible to determine approximate values velocities of flow through the respective faces (see (14)). Saturation will be determined by (13) and the explicit scheme with a known field of filtration flows $\mathbf{u}$. The Courant number for this equation is equal to:

$$C = \tau \cdot F \cdot \max\left(\frac{\mathbf{u}}{mV}\right) \tag{16}$$

where the maximum should be taken as all relations of flows through the cell borders $\mathbf{u}$ to the pore volume mV and the value F is given by the formula:

$$F = \max_s f'(s). \tag{17}$$

The flows have the greatest impact on the value of time steps in small areas near the high performance wells. In most parts of the reservoir to satisfy the condition of stability is allowed to make calculations with a much larger time steps. The term of local Courant number is introduced:

$$C_{i,j,k} = F_{i,j,k} \frac{u_{i,j,k}}{m_{i,j,k}} \frac{\tau_{i,j,k}}{V_{i,j,k}}, \tag{18}$$

where $u_{i,j,k} = \max\left(u_{i\pm1/2,j,k}, u_{i,j\pm1/2,k}, u_{i,j,k\pm1/2}\right)$.

The stability condition will be satisfied if the time step will be chosen so, that the local Courant number (18) was less than one for each cell:

$$\tau_{i,j,k} < \frac{m_{i,j,k} V_{i,j,k}}{u_{i,j,k} F_{i,j,k}}.$$

However, the direct setting of time steps in each cell can not be realized, because it is necessary the agreed calculation of the flows component field to fulfill the conditions of conservativeness.

Here the algorithm of multiple matching the number of recalculations is proposed.

1. The time step is selected, equal to the minimum of steps for each cell from the condition (18)
2. The smallest time step among all of the cells is selected.

$$\tau_{\min} = \min_{i,j,k} \tau_{i,j,k}.$$

3. The coefficient of step increase is calculated.

$$K_{i,j,k} = \frac{\tau_{i,j,k}}{\tau_{\min}}.$$

4. The coefficients for the cells are rounded towards zero to the number of which constitutes a whole integer degree of the "d". The number of small steps is determined, it will be used for recalculate the cell:

$$NS_{i,j,k} = d^{\left[\log_d K_{i,j,k}\right]} \tag{19}$$

5. The time step is calculated for each cell:

$$\tilde{\tau}_{i,j,k} = NS_{i,j,k} \cdot \tau_{\min} \tag{20}$$

6. The total saturation calculation is made with the number of steps, corresponding to the minimum step. But particular cell is " recalculated " for every step, and with its own "period" – one time for the $NS_{i,j,k}$ steps.

7. The period of recalculation of J-function on the each boundary of the cell is equal to the maximal value from periods of recalculation of the neighbor cells. It is obvious that in this case the balance of the transferred value in the cells is always complied, it means that the scheme is conservative.

5. Numerical experiment

The comparison of the results and calculation time using suggested algorithm was made in one-dimensional, two-dimensional and three-dimensional cases. During the test number "*d*" was equal two.

One dimensional case demonstrated the good agreement between the calculation results with the similar solution for the plane-parallel filtration. For the radially symmetric case it was compared the behavior of saturation front, which obtained using both algorithms (with acceleration and without).

6. Conclusion

It was found that using the suggested algorithm allows to reduce the calculation time of oil saturation field of about three times. Note that with increasing the size of the calculation grid the efficiency of the algorithm a bit improved. As a result the calculations by several million cells give a significant gain in real-time. The comparison with potentially possible "ideal" step allows us to hope that the improvement of the algorithm implementation will reduce the computation time is still about 3 times.

Acknowledgments

This work was partially supported by Russian Science Foundation, project no. 15-41-02315.

References

[1] Badriev I B, Zadvornov O A and Lyashko A D 2004 A study of variable step iterative methods for variational inequalities of the second kind *Differential Equations* **40** (7) 971-83.

[2] Badriev I B 2013 On the Solving of Variational Inequalities of Stationary Problems of Two-Phase Flow in Porous Media *Applied Mechanics and Materials* **392** 183-7 DOI: 10.4028/www.scientific.net/AMM.392.183

[3] Badriev I B, Zadvornov O A and Saddek A M 2001 Convergence Analysis of Iterative Methods for Some Variational Inequalities with Pseudomonotone Operators *Differential Equations* **37** (7) 934-42. DOI: 10.1023/A:1011901503460

[4] Badriev I B and Karchevskii M M 1989 Convergence of the iterative Uzawa method for the solution of the stationary problem of seepage theory with a limit gradient *Journal of Soviet Mathematics* **45** (4) 1302-9 DOI: 10.1007/BF01097083

[5] Badriev I B 1983 Difference schemes for nonlinear problems of filtration theory with a discontinuous law *Soviet Mathematics* **27** (5) 1–11

[6] Badriev I B and Nechaeva L A 2013 Mathematical simulation of steady filtration with multivalued law *PNRPU Mechanics Bulletin* (3) 37–65

[7] Badriev I and Banderov V 2014 Numerical method for solving variation problems in mathematical physics *Applied Mechanics and Materials* **668-669** 1094-97. DOI: 10.4028/www.scientific.net/AMM.668-669.1094
[8] Badriev I B 1989 Application of duality methods to the analysis of stationary seepage problems with a discontinuous seepage law *Journal of Soviet Mathematics* **45** (4) 1310-14. DOI: 10.1007/BF01097084
[9] Badriev I B 1989 Regularization of the nonlinear problem of seepage theory with a discontinuous law *Journal of Soviet Mathematics* **44** (5) 681-91. DOI: 10.1007/BF01095183
[10] Badriev I B and Zadvornov O A 2003 A decomposition method for variational inequalities of the second kind with strongly inverse-monotone operators *Differential Equations* **39** (7) 936-44. DOI: 10.1023/B:DIEQ.0000009189.91279.93
[11] Shangaraeva A I and Shevchenko D V 2015 Speed up of the oil saturation numerical algorithm for the plane-parallel filtration *Applied Mathematical Sciences* **9** (150) 7467–74

IOP Conf. Series: Materials Science and Engineering **158** (2016) 012085 doi:10.1088/1757-899X/158/1/012085

Methods of construction and study of Frankl system self-similar solutions in the hyperbolic case

T Shemyakina[1] and S Alekseenkко[2]

[1]Peter the Great Saint-Petersburg Polytechnical University,
29 Politechnicheskaya Street 195251 Saint-Petersburg, Russia,
[2]Nizhny Novgorod State Technical University n a R E Alekseev,
24 Minin Street 603950 Nizhny Novgorod, Russia

E-mail: sh_tat@mail.ru

Abstract. Self-similar solution of the Frankl system in the hyperbolic case was found. The Frankl system is a system of mixed type equations. Under certain conditions, it describes a model of the membrane theory of shells. The Frankl system describes a stationary irrotational motion of an ideal gas in the transition vicinity from subsonic to supersonic speeds. We find a sufficient condition on the initial data that guarantees existence of a global classical solution continued from a local solution. The proof of the nonlocal solvability of the problem in the original variables is based on the additional argument method. It allowed justify and construct a numerical solution. Numerical experiments were carried out for model examples of the Frankl system.

1. Introduction

The main object of research is the Frankl system:

$$\begin{cases} \dfrac{\partial u(x,y)}{\partial y} - P(x,y,u(x,y),v(x,y))\dfrac{\partial v(x,y)}{\partial x} = 0, \\ \dfrac{\partial u(x,y)}{\partial x} \pm Q(x,y,u(x,y),v(x,y))\dfrac{\partial v(x,y)}{\partial y} = 0, \end{cases}$$

where $u(x,y), v(x,y)$ are unknown functions;

$$P(x,y,u(x,y),v(x,y)) \geq p_0 > 0, \quad p_0, q_0 \text{ are constant},$$
$$Q(x,y,u(x,y),v(x,y)) \geq q_0 > 0.$$

The Frankl system is a system of two quasi-linear differential equations of the first order. It is a system of mixed type. Equations with opposite signs represent a system of elliptic type. Equations with the same signs represent a system of hyperbolic type.

For the first time the famous physicist F. I. Frankl [1] has combined the study of the system of equations of mixed type with the study of stationary problems of transonic gas dynamics. Some particular cases of the Frankl system can be found in neutron transport theory, theory of thermoelasticity, in describing of the distribution of electrons in the electric field of the sprite, and so on. Bibliography for mixed type equations can be found in many monographs, in particular J. M. Krikunov [2]. S. A. Chaplygin, A. V. Bitsadze, F. Tricomi, S. V. Falkovich, B. L. Rozhdestvensky and

N. N. Yanenko are the famous scientists. They studied problems for systems of mixed type equations in particular cases, when the equation system coefficients P, Q possess values of the constants or independent variables. Most often, the study came to the study of differential equations of the second order. Mathematical theory to study a model system of mixed type equations is systematically presented in the monograph by T. V. Chekmarev [3]. He considered the application of this theory to the solution of problems for mixed type systems.

A variety of approaches to the study of mixed systems of differential equations have their advantages and disadvantages. As a rule, the method of research is to convert the nonlinear equations to linear equations in the hodograph plane. Then a linear system of equations is solved. But in the general case, a return to the original variables is a more difficult task than the original task. Application of method of characteristics for quasi-linear equation systems introduces a superposition of unknown functions in the relevant integral equations. In the method of characteristics inverse transformation existence is a condition for original problem solvability. In many cases the problem of inverse function determination is so difficult that it is not solved. The existence of inverse variable transformations is taken as a condition.

To overcome these difficulties we investigate the problem using the method of additional argument [4 – 6]. We consider the Frankl system in quite a general form. The method we propose does not replace other known methods. It amplifies them and allows to determine more precisely the conditions for the solvability in the original variables. In this work [6] we found sufficient conditions for a nonlocal solvability of the Cauchy problem for a system of two quasi-linear differential equations of first order. At that, the system of equations is represented in characteristic form. However, for the Frankl system with arbitrarily given initial data, conditions for the existence of the global solution established in work [6], are not fulfilled.

In this paper, we prove the global existence of self-similar solutions of the Frankl system in hyperbolic case, subject to the coordination of the initial data. We have done numerical calculations for model problems of the Frankl system. We found numerical Frankl system solution by grid method. We found self-similar solution of the Cauchy problem by the method of additional argument. We have built a numerical solution by the method of successive approximations.

2. The statement of the problem

We consider the Frankl system in the hyperbolic case:

$$\begin{cases} \dfrac{\partial u(x,y)}{\partial y} - P(u(x,y),v(x,y))\dfrac{\partial v(x,y)}{\partial x} = 0, \\ \dfrac{\partial u(x,y)}{\partial x} - Q(u(x,y),v(x,y))\dfrac{\partial v(x,y)}{\partial y} = 0. \end{cases} \tag{1}$$

The Cauchy problem for the system of equations (1) is:

$$u(0,x) = \varphi(x),\ v(0,x) = \psi(x),\ x \in (-\infty,\infty),\ \varphi(x),\psi(x) \in \bar{C}^2(R^1) \tag{2}$$

Formulation of the problem is as follows. Find an unknown functions $u(x,y), v(x,y)$, satisfying the system of equations (1) and the initial conditions (2) in the domain:

$$\Omega_Y = \{(x,y) : x \in (-\infty,\infty),\ y \in [0,Y], Y > 0\}.$$

3. Method of problem solving

The problem (1), (2) is solved by the method of additional argument. The starting system is transformed to the system of symmetric quasi-linear equations with help of Riemann invariants:

$$z_1(x,y) = \frac{\partial u(x,y)}{\partial y} + b\big(u(x,y),v(x,y)\big)\frac{\partial v(x,y)}{\partial y},$$

$$z_2(x,y) = \frac{\partial u(x,y)}{\partial y} - b\big(u(x,y),v(x,y)\big)\frac{\partial v(x,y)}{\partial y}.$$

The result is a characteristic shape in the form of a system of differential equations for functions $z_1(x,y)$, $z_2(x,y)$, $u(x,y)$, $v(x,y)$:

$$\begin{cases} \dfrac{\partial z_1(x,y)}{\partial y} - e(u(x,y),v(x,y))\dfrac{\partial z_1(x,y)}{\partial x} = z_1(x,y)\cdot F_1(u(x,y),v(x,y),z_1(x,y),z_2(x,y)), \\ \dfrac{\partial z_2(x,y)}{\partial y} + e(u(x,y),v(x,y))\dfrac{\partial z_2(x,y)}{\partial x} = z_2(x,y)\cdot F_2(u(x,y),v(x,y),z_1(x,y),z_2(x,y)), \\ \dfrac{\partial u(x,y)}{\partial y} - e(u(x,y),v(x,y))\dfrac{\partial u(x,y)}{\partial x} = G_1(z_2(x,y)), \\ \dfrac{\partial v(x,y)}{\partial y} - e(u(x,y),v(x,y))\dfrac{\partial v(x,y)}{\partial x} = G_2(u(x,y),v(x,y),z_2(x,y)), \end{cases}$$

with the initial conditions:

$$\begin{cases} z_1(x,0) = \Phi_1(x) = P(\varphi(x),\psi(x))\psi'(x) + e(\varphi(x),\psi(x))\varphi'(x), \\ z_2(x,0) = \Phi_2(x) = P(\varphi(x),\psi(x))\psi'(x) - e(\varphi(x),\psi(x))\varphi'(x), \\ u(x,0) = \varphi(x), \\ v(x,0) = \psi(x), \end{cases}$$

where $e(u(x,y),v(x,y)) = \sqrt{\dfrac{P(u(x,y),v(x,y))}{Q(u(x,y)v(x,y))}}$,

$b(u(x,y),v(x,y)) = \sqrt{P(u(x,y),v(x,y))\cdot Q(u(x,y)v(x,y))}$.

We define the coordination of initial conditions on demand:

$$z_2(x,0) = 0, \quad P(\varphi(x),\psi(x))\psi'(x) - e(\varphi(x),\psi(x))\varphi'(x) = 0,$$

then the Cauchy problem can be represented as:

$$\begin{cases} \dfrac{\partial z_2(x,y)}{\partial y} + e(u(x,y),v(x,y))\dfrac{\partial z_2(x,y)}{\partial x} = z_2(x,y)\cdot F_2(u(x,y),v(x,y),z_1(x,y),z_2(x,y)). \\ z_2(x,0) = 0, \end{cases}$$

There exists a trivial solution $z_2(x,y) \equiv 0$.

Thus, we get the coordination of the initial data for the Frankl system:

$$\varphi'(x) = b(\varphi(x),\psi(x))\psi'(x) \tag{3}$$

The characteristic form is converted to the following form:

$$\begin{cases} \dfrac{\partial u(x,y)}{\partial y} - e(u(x,y),v(x,y))\dfrac{\partial u(x,y)}{\partial x} = 0, \ u(x,0) = \varphi(x), \\ \dfrac{\partial v(x,y)}{\partial y} - e(u(x,y),v(x,y))\dfrac{\partial v(x,y)}{\partial x} = 0, \ v(x,0) = \psi(x), \\ \dfrac{\partial z_1(x,y)}{\partial y} - e(u(x,y),v(x,y))\dfrac{\partial z_1(x,y)}{\partial x} = z_1^2(x,y)\cdot E(u(x,y),v(x,y)), \\ z_1(x,0) = \Phi_1(x) = 2P(\varphi(x),\psi(x))\psi'(x) = 2e(\varphi(x),\psi(x))\varphi'(x), \end{cases}$$

where $E(u(x,y),v(x,y)) = \dfrac{1}{2e(u,v)}\cdot\dfrac{\partial e(u,v)}{\partial u} + \dfrac{1}{2P(u,v)}\cdot\dfrac{\partial e(u,v)}{\partial v} > 0$.

In the domain of $\Omega_{Ys} = \{(x,y,s):\ x\in(-\infty,\infty),\ \ 0\le s\le y\le Y,\ Y>0\}$ we consider the extended characteristic system, in accordance with the method of additional argument:

$$\begin{cases} \dfrac{d\eta(x,y,s)}{ds} = -e_1(u_1(x,y,s),v_1(x,y,s)), & \eta(x,y,y) = x, \\ \dfrac{du_1(x,y,s)}{ds} = 0, & u_1(x,y,0) = \varphi(\eta(x,y,0)), \\ \dfrac{dv_1(x,y,s)}{ds} = 0, & v_1(x,y,0) = \psi(\eta(x,y,0)), \\ \dfrac{dw(x,y,s)}{ds} = w^2(x,y,s)E(u_1(x,y,s),v_1(x,y,s)), & w(x,y,0) = \Phi(\eta(x,y,0)). \end{cases} \quad (4)$$

Unknown functions $\eta(x,y,s)$, $u_1(x,y,s)$, $v_1(x,y,s)$, $w(x,y,s)$ depend not only on the original independent variables x,y, but they depend on an additional argument s.

In the framework of an additional argument in [4–5] we determined the conditions of the local existence of a bounded solution for the Frankl system in the hyperbolic case.

Using $P(u(x,y),v(x,y)) \ge p_0 > 0$, $Q(u(x,y),v(x,y)) \ge q_0 > 0$, $\varphi(x),\psi(x) \in \bar{C}^2(R^1)$, $P(u,v) \in \bar{C}^{2,2}(V_e)$, $Q(u,v) \in \bar{C}^{2,2}(V_e)$, $V_e = \{(u,v):\ u \in [-K,K],\ \ v \in [-K,K],\ K > 0\}$, we get: the Frankl system (1) with initial conditions (2) has a unique bounded on the set $x \in R^1$ solution $u(x,y) \in \bar{C}^{2,1}(\Omega_{Y0})$, $v(x,y) \in \bar{C}^{2,1}(\Omega_{Y0})$.

The functions $u(x,y)$, $v(x,y)$ are determined from the constructed system of equations (4) with $s = y$. The initial conditions (2) are used for $0 \le y \le Y_0 \le Y$ exactly where the constant Y_0 is determined from the initial data.

We have $u(x,y) = u_1(x,y,y)$, $v(x,y) = v_1(x,y,y)$, $z_1(x,y) = w(x,y,y)$.

Integrating (4) with respect to s, we get the solution of system of ordinary differential equations of the first order:

$$\begin{cases} \eta(x,y,s) = x + (y-s)e_1(u_1(x,y,y),v_1(x,y,y)), \\ u_1(x,y,s) = \varphi(x + y\cdot e_1(u_1(x,y,y),v_1(x,y,y)), \\ v_1(x,y,s) = \psi(x + y\cdot e_1(u_1(x,y,y),v_1(x,y,y)), \\ w(x,y,s) = \dfrac{\Phi(x + y\cdot e_1(u_1,v_1))}{1 - s\cdot E(u_1,v_1)\cdot\Phi(x + y\cdot e_1(u_1,v_1))}. \end{cases}$$

We found the solution of the characteristic form as follows:

$$\begin{cases} u(x,y) = \varphi(x + y\cdot e(u(x,y),v(x,y)), \\ v(x,y) = \psi(x + y\cdot e(u(x,y),v(x,y)), \\ z_1(x,y) = \dfrac{\Phi_1(x + y\cdot e(u,v))}{1 - y\cdot E(u,v)\cdot\Phi_1(x + y\cdot e(u,v))}. \end{cases}$$

Thus, the problem is reduced to the study of global solvability of a system of two algebraic equations:

$$\begin{cases} u(x,y) = \varphi(x + ye(u(x,y),v(x,y))), \\ v(x,y) = \psi(x + ye(u(x,y),v(x,y)). \end{cases} \quad (5)$$

The standard method of successive approximations is used to prove the existence of continuous constrained solutions together with derivatives of the first order.

We introduce the notation: $C_\Phi = \max\{\sup\limits_{(-\infty,\infty)}|\Phi(x)|, \sup\limits_{(-\infty,\infty)}|\Phi'(x)|\}$,

$$C_\varphi = \max\{\sup_{(-\infty,\infty)}|\varphi(x)|, \sup_{(-\infty,\infty)}|\varphi'(x)|\},\ C_\psi = \max\{\sup_{(-\infty,\infty)}|\psi(x)|, \sup_{(-\infty,\infty)}|\psi'(x)|\},$$

$$N_u = \max\{\sup_{\Omega_Y}\left|\frac{\partial u}{\partial x}\right|, \sup_{\Omega_Y}\left|\frac{\partial u}{\partial y}\right|\},\quad N_v = \max\{\sup_{\Omega_Y}\left|\frac{\partial v}{\partial x}\right|, \sup_{\Omega_Y}\left|\frac{\partial v}{\partial y}\right|\}.$$

Solutions $u(x,y), v(x,y)$ have an estimations:

$$|u(x,y)| \le C_\varphi,\quad |v(x,y)| \le C_\psi. \tag{6}$$

We demand the implementation of inequalities:

$\Phi_1(x+y\cdot e(u,v)) = 2P(\varphi,\psi)\psi'(x+y\cdot e(u,v)) \le 0$,

$\Phi_1(x+y\cdot e(u,v)) = 2e(\varphi,\psi)\varphi'(x+y\cdot e(u,v)) \le 0$,

not to a function $z_1(x,y)$ become infinite.

Then for the derivatives of initial functions, we obtain the following conditions:

$$\varphi'(x) \le 0,\quad \psi'(x) \le 0. \tag{7}$$

Derivatives of functions $u(x,y), v(x,y)$ have the following form:

$$\frac{\partial u}{\partial x} = \varphi'\cdot\left(1-y\left(\varphi'\frac{\partial e}{\partial u}+\psi'\frac{\partial e}{\partial v}\right)\right)^{-1},\quad \frac{\partial u}{\partial y} = \varphi'\cdot e\cdot\left(1-y\left(\varphi'\frac{\partial e}{\partial u}+\psi'\frac{\partial e}{\partial v}\right)\right)^{-1},$$

$$\frac{\partial v}{\partial x} = \psi'\cdot\left(1-y\left(\varphi'\frac{\partial e}{\partial u}+\psi'\frac{\partial e}{\partial v}\right)\right)^{-1}\quad \frac{\partial v}{\partial y} = \psi'\cdot e\cdot\left(1-y\left(\varphi'\frac{\partial e}{\partial u}+\psi'\frac{\partial e}{\partial v}\right)\right)^{-1}.$$

We obtain the estimates of derivatives of functions $u(x,y), v(x,y)$:

$$\left|\frac{\partial u(x,y)}{\partial x}\right| \le N_u,\quad \left|\frac{\partial u(x,y)}{\partial y}\right| \le N_u,\quad \left|\frac{\partial v(x,y)}{\partial x}\right| \le N_v,\quad \left|\frac{\partial v(x,y)}{\partial y}\right| \le N_v, \tag{8}$$

then we obtain the main condition:

$$\varphi'(x)\frac{\partial e}{\partial u} + \psi'(x)\frac{\partial e}{\partial v} \le 0. \tag{9}$$

Global assessments received (6), (8) for the functions $u(x,y), v(x,y)$ provide an opportunity to extend the solution on any given interval $[0,Y]$. We take as initial values $u(x,Y_0), v(x,Y_0)$ and we will extend the solution on some interval $[Y_0,Y_1]$. Then, we take as initial values $u(x,Y_1), v(x,Y_1)$ and we will extend the solution to the interval $[Y_1,Y_2]$. The range of solvability interval will not decrease, as it is defined by global assessments (6), (8). As a result, the solution $u(x,y), v(x,y)$ for finite number of steps can be extended to the entire specified period $[0,Y_k], Y_k \le Y$.

The uniqueness of the solution $u(x,y), v(x,y)$ is proved by using similar global assessments, which allowed establishing the convergence of successive approximations in the works of authors [4 - 6].

Eventually we get the following results.

Theorem. *Suppose the functions* $P(u(x,y),v(x,y)) \ge p_0 > 0$, $Q(u(x,y),v(x,y)) \ge q_0 > 0$, p_0, q_0 are constant, $P(u,v) \in \bar{C}^{2,2}(V_e)$, $Q(u,v) \in \bar{C}^{2,2}(V_e)$.

Let the functions $\varphi(x), \psi(x) \in \bar{C}^2(R^1)$ *satisfy the conditions:*

1. $\varphi'(x) = b(\varphi(x),\psi(x))\psi'(x)$ $\left(b(u,v) = \sqrt{P(u,v)\cdot Q(u,v)}\right)$,
2. $\varphi'(x) \le 0$, $\psi'(x) \le 0$,
3. $\varphi'(x)\frac{\partial e}{\partial u} + \psi'(x)\frac{\partial e}{\partial v} \le 0$.

Then for any Y the Cauchy problem (1), (2) has a unique bounded on the set $x \in R^1$ solution $u(x,y) \in \bar{C}^{2,1}(\Omega_Y)$, $v(x,y) \in \bar{C}^{2,1}(\Omega_Y)$.

Points to note. Functions $u(x,y)$, $v(x,y)$ become infinite, if both of the inequalities:

$$\varphi'(x)>0,\quad \psi'(x)>0,\ \varphi'(x)\frac{\partial e}{\partial u}+\psi'(x)\frac{\partial e}{\partial v}>0.$$

4. Calculations

Numerical calculations were carried out for many modeling tasks of the Frankl system , in particular [7 – 9]. In works [8 – 9] we studied the problem of modeling processes in the environment.

In this article, numerical calculations are carried out for model problem Frankl system (1), (2) with the following data:

$$P(u,v)=v^2,\ Q(u,v)=1.0,\ e(u,v)=v,\quad b(u,v)=v,$$

$$\varphi(x)=1.0-0.5x+0.125x^2,\quad \psi(x)=1.0-0.5x,\ x\in[0.0,1.0],\quad y\in[0.0,1.0],$$

The number of dividing points portioning the interval of variables set x: $m=6$, y: $n=11$.

Analytical solution of the problem has the following form:

$$u(x,y)=\frac{(x-2)^2+(y+2)^2}{2(y+2)^2},\ v(x,y)=\frac{(2-x)}{(y+2)}$$

The first experiment. The problem was solved by grid method (Godunov scheme).
Boundary conditions have the form:

$$u(x_0,y)=\varphi_1(y)=0.5+\frac{2}{(y+2)^2},\ u(x_m,y)=\varphi_2(y)=0.5+\frac{2}{(y+2)^2},$$

$$v(x_0,y)=\psi_1(y)=\frac{2}{(y+2)},\ v(x_m,y)=\psi_2(y)=\frac{1}{(y+2)}.$$

We have error of approximation of the numerical solution to the exact solution

$$\delta u=0.3750,\quad \delta v=0.2955.$$

The second experiment. The problem was solved by the method of successive approximations according to the formulas:

$$\begin{cases} u(x,y)=\varphi(x+ye(u(x,y),v(x,y))), \\ v(x,y)=\psi(x+ye(u(x,y),v(x,y))). \end{cases}$$

We have error of approximation of the numerical solution to the exact solution:

$$itr=1,\quad \delta u=0.2778,\quad \delta v=0.3333;$$

$$itr=2,\quad \delta u=0.0972,\quad \delta v=0.1667;$$

$$itr=3,\quad \delta u=0.0590,\quad \delta v=0.0833.$$

The third experiment. The problem is solved by the hodograph method.

$$\begin{cases} x=\frac{1}{2}c_1u^2+\frac{1}{4}c_1v^4+\frac{1}{3}c_2v^3+c_3u+c_5, \\ y=c_1uv+c_2u+c_3v+c_4. \end{cases}$$

where the coefficients c_i are const, $i=1,2,3,4,5$.

In this example we found constants, but the formulas of the functions $u(x,y)$, $v(x,y)$ are very tedious. We do not show them.

5. Conclusion

In the present work we proposed one more method to study the Frankl system of hyperbolic case. This is the method of additional argument. This method made it possible to prove the global existence of self-similar solution, subject to the approval of the initial data. Requirements for the initial data are quite simple. In addition, we found the algorithm for finding the numerical solution of the problem. Numerical calculations were performed on model problems of the Frankl system. Application of additional argument method has simplified numerical calculations and increased accuracy of estimation evaluation. In further studies we suggest the use of hybrid methods, based on methods of grids and additional argument.

Acknowledgments
This work was supported by the Russian Foundation for Basic Research, project № 14-01-00660-A

References

[1] Frankl F 1973 *Selected works on gas dynamics* (Moscow: Nauka) p 712
[2] Krikunov Yu 1986 *Boundary value problem for model equations of mixed type* (Kazan: Kazan State University) p 148
[3] Chekmarev T 1995 *The system of equations of mixed type* (Nizhny Novgorod: Nizhny Novgorod State Technical University) p 200
[4] Alekseenko S, Shemyakina T and Kruts K 2006 Local existence of a bounded solution of the Frankl system in the hyperbolic case *Study on integro-differential equations* (Bishkek: Ilim). **35** pp 142–147
[5] Shemyakina T 2012 The theorem on existence of a bounded solution of the Cauchy problem for the Frankl system of hyperbolic type *St. Petersburg State Polytechnical University Journal. Physics and Mathematics* **2 (146)** pp 130–131
[6] Alekseenko S, Shemyakina T and Dontsova M 2013 Nonlocal resolvability conditions for systems of the first order partial differential equations *St. Petersburg State Polytechnical University Journal. Physics and Mathematics* **3 (177)** pp 190–201
[7] Shemyakina T 2012 Examples of solutions of the Cauchy problem for some variants of the Frankl system of hyperbolic type *Proceedings of the IX International Conference on Nonequilibrium Processes in Nozzles and Jets (NPNJ' 2012)* (Moscow: MAI) pp 525–528
[8] Kainov N, Tarkhov D and Shemyakina T 2014 Application of Neural Network Modeling to Identification and Prediction Problems in Ecology Data Analysis for Metallurgy and Welding Industry *Nonlinear Phenomena in Complex Systems* **17** pp 57-63
[9] Vasilyev A, Tarkhov D and Shemyakina T *A neural network approach to problems of mathematical physics* (SPb: Nestor-History) p 260

Investigation of the influence of heated catalyst feeding system on the intensity of temperature-dependent chemical reaction in the fluidized bed apparatus

O V Soloveva[1,3] and S A Solovyev[2,3]

[1]Institute of Chemistry, Kazan Federal University, Kremlevskaja st. 29, Kazan 420111, Russia
[2]Institute of Mathematics and Mechanics, Kazan Federal University, Kremlevskaja st. 35, Kazan 420008, Russia
[3]Institute of Heat Power Engineering, Kazan State Power Engineering University, Krasnoselskaja st. 51, Kazan 420066, Russia

E-mail: serguei_s349@mail.ru

Abstract. A mathematical model was developed and a numerical study of operation parameters of the fluidized bed apparatus for temperature-dependent processes was performed. Fields of catalyst concentration and temperature fields were obtained. The circulation flow analysis was carried out. The effect of the influence of heated catalyst feeder on the efficiency of apparatus heating was analyzed. The change of the circulating gas flows and catalyst structures due to changes in the heated catalyst feeder was shown. The influence of the catalyst fractional composition on the efficiency of apparatus heating was studied.

1. Introduction

Fluidized bed apparatus are widely used in petrochemical industry. Advantages of industrial using of the fluidized bed apparatus are constructive simplicity at elevated heat and mass transfer and mixing of chemical elements in different aggregation states.

There are a lot of works devoted to the design of reactors (e.g. [1]), and also the generalization of models for fluidized bed mathematical simulation (e.g. [2]). Despite the fact that these apparatus are used in the industry for over fifty years, interest in the study of the properties and possibilities of the fluidized bed using is not exhausted. Each structural element contributes to the efficiency of apparatus. The works, considering individual apparatus singularities, regularly appear in recent years. Changes in the character of the movement and concentration of particles along the wall with deflectors are shown in [3]. In [4, 5] the numerical simulation analysis of the cylindrical apparatus projections on the walls influence on the hydrodynamics of the fluidized bed was carried out. Experimental studies with internal elements in a liquid-solid fluidized bed are presented in [6]. Investigations of fluidized bed processes in the column with an inner cylinder were experimentally made in [7]. Modelling and comparative analysis of two large fluidized bed blocks with various gas raw feeders were performed in [8].

In this paper, a mathematical model was developed and a numerical study of the fluidized bed operation parameters of apparatus design for the dehydrogenation of isobutane to isobutylene, passing with the heat absorption, was performed. Catalyst concentration fields and temperature fields were obtained. The circulation flow analysis was carried out. The influence of structural modifications (changes in the heated catalyst feeder) on the heating efficiency was analyzed. The change of the circulating gas flow and catalyst due to changes of apparatus design was shown. The influence of the catalyst structural composition on the efficiency of apparatus heating, and consequently, on the reaction rate was studied.

2. Problem formulation and solution method

In the present work we study the fluidized bed cylindrical apparatus, designed for reactions with the heat absorption in the presence of a catalyst. In the lower part there is a gas feeder at the temperature of 550 ^{O}C. There are microspherical catalyst particles with the diameter of 20-200 μm in the apparatus. Along the axis is located a vertical pipe for the feeding of the heated catalyst regenerated at 650 ^{O}C. In the middle of the reactor, in the working area, are located angular falling through type grids with the free-sectional area of approximately 30% (figure 1, a).

A vertical pipe for the heated catalyst feeding is located in the center of the feeder, as will be shown below, can have a significant impact on the catalyst circulation. Another version of the feeder layout is therefore considered. There is no a vertical pipe in the center. A catalyst feeder simulates a pipe, embedded on the side of the apparatus, and also provides the catalyst flow in the center of the apparatus (figure 1, b).

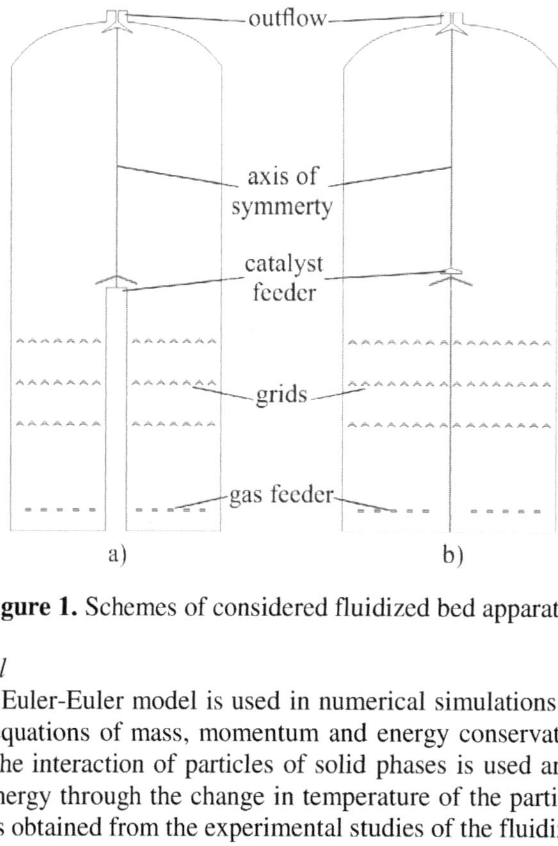

Figure 1. Schemes of considered fluidized bed apparatus.

2.1. Fluidized bed model

Continuous multi-phase Euler-Euler model is used in numerical simulations of the fluidized bed. For each of the phases the equations of mass, momentum and energy conservation are implemented. To take into consideration the interaction of particles of solid phases is used an equation that takes into account the change of energy through the change in temperature of the particles. To close the system are used the relationships obtained from the experimental studies of the fluidized bed.

Conservation of mass

$$\frac{\partial \alpha_i \rho_i}{\partial t} + \nabla \cdot (\alpha_i \rho_i \vec{v}_i) = 0, \tag{1}$$

were α_i is the volume fraction of the i-th phase, ρ_i is the real density, $\vec{v}$ is the velocity.

Conservation of momentum

$$\frac{\partial \alpha_i \rho_i \vec{v}_i}{\partial t} + \nabla \cdot \left(\alpha_i \rho_i \vec{v}_i \vec{v}_i\right) = -\alpha_i \nabla p + \nabla \cdot \bar{\bar{\tau}}_i + \alpha_i \rho_i \vec{g} + \sum_j \vec{R}_{ij}, \quad (2)$$

where p is the pressure, $\bar{\bar{\tau}}_i = \alpha_i \mu_i \left(\nabla \vec{v}_i + \nabla \vec{v}_i^T\right) + \alpha_i \left(\lambda_i - (2/3)\mu_i\right) \nabla \cdot \vec{v}_i \bar{\bar{I}}$ is the stress tensor, μ_i, λ_i are shear and bulk viscosity, $\bar{\bar{I}}$ is a unity tensor, $\vec{R}_{ij} = K_{ij}\left(\vec{v}_i - \vec{v}_j\right)$ is the force of interphase interaction.

Conservation of energy

$$\frac{\partial \alpha_i \rho_i h_i}{\partial t} + \nabla \cdot \left(\alpha_i \rho_i \vec{v}_i h_i\right) = \alpha_i \frac{\partial p_i}{\partial t} + \bar{\bar{\tau}}_i : \vec{v}_i + \sum_j Q_{ij}. \quad (3)$$

where h_i is the enthalpy, Q_{ij} is the heat exchange intensity, p_i is the solid phase granule pressure.

To account for the interaction of solid phases particles is used an equation that considers the change of energy through a change in temperature of the particles [22]:

$$\frac{3}{2}\left[\frac{\partial}{\partial t}\left(\alpha_i \rho_i \Theta_i\right) + \nabla \cdot \left(\alpha_i \rho_i \vec{v}_i \Theta_i\right)\right] = \left(-p_i \bar{\bar{I}} + \bar{\bar{\tau}}_i\right) : \nabla \vec{v}_i + \nabla \cdot \left(k_{\Theta_i} \nabla \Theta_i\right) - \gamma_{\Theta_i} + \phi_{ij}. \quad (4)$$

where k_{Θ_i} is the diffusion coefficient of the granules, Θ_i is the temperature of the solid phase granules, γ_{Θ_i} is the energy dissipation due to the particles collision, ϕ_{ij} is the energy exchange between *i*-th solid phase and *j*-th liquid (gaseous) phase.

The recorded equations system is unclosed. To close the system use the relationships obtained from the experimental studies of the fluidized bed. The interphases interaction coefficient is determined experimentally and depends on the type of the interacting phases. For two types of phase: liquid (gaseous) phase (i-th) – solid granular phase (j-th) model used in [9], in the case of two solid granular phases is the model of [10]. The energy dissipation γ_{Θ_i} due to the particles collision is written by the model obtained in [11]. Coefficient of the heat exchange intensity is determined for the interaction of two phases by the relation, given in [12]. The motion considered in the reactor at the accepted loads of the raw gas feeding and parameters of the used catalyst is turbulent. In the model accepted for the calculations a dispersion $\kappa - \varepsilon$ model of turbulence was used in which the motion of "secondary" solid granular phases is generated on the background of the turbulent motion of "primary" gas phase.

2.2. *CFD solver*

In this paper differential equations that describe the hydrodynamic and thermal processes in the computational model were solved in CFD ANSYS Fluent for unsteady flow regime. The whole computational domain was divided into elements of a triangular or rectangular shape in different subregions, the dimensions of which are sufficient to determine the characteristic factors of the investigated phenomenon. For a given initial distribution of the catalyst in the reactor, after a while the solution reached quasi-stationary regime, at which characteristic hydrodynamic and thermal pictures were calculated.

Boundary conditions were set on all elements of each constructed model of apparatus in accordance with the mechanism of work and the solver used. Since the reactor model assumes rotationally symmetric, then on the axis of the constructed model conditions of axial symmetry «axis» were set. In all non-permeable surfaces «wall» conditions were accepted. In models of raw gas feeder nozzle and in the outlet of the centrally located pipe of the catalyst feed were set «velocity-inlet» (the value of the flow rate) conditions. In the area at the top of the apparatus model there were «outflow» (free flow yield) conditions.

3. Results and discussions

From the analysis of the fractional composition of the catalyst used in the reactor two following models are assumed: monodisperse model with a mean particle size of 80 μm and two-fractional model with a large particle size of 80 μm and with small particles (15% of the total volume of particles) size of 40 μm. The gas feeding rate in the calculations corresponds to the average velocity over the apparatus section of 0.3m/s.

3.1. A case with central pipe

Figure 2 shows the characteristic pattern of catalyst particles concentration fields: monodisperse and polydisperse (two fractional). Concentration scale varies from zero to a dense bed catalyst concentration (60% of the volume fraction of the catalyst material considering its voidage). This picture corresponds to the instantaneous state. The monodisperse representation of the fluidized bed particle size is chosen based on that the average velocity of the gas is not enough for the catalyst granules ablation. This ensures that there is the upper boundary of the layer. In turn, the fine particles rise up with the gas flows and circulate throughout the apparatus working area.

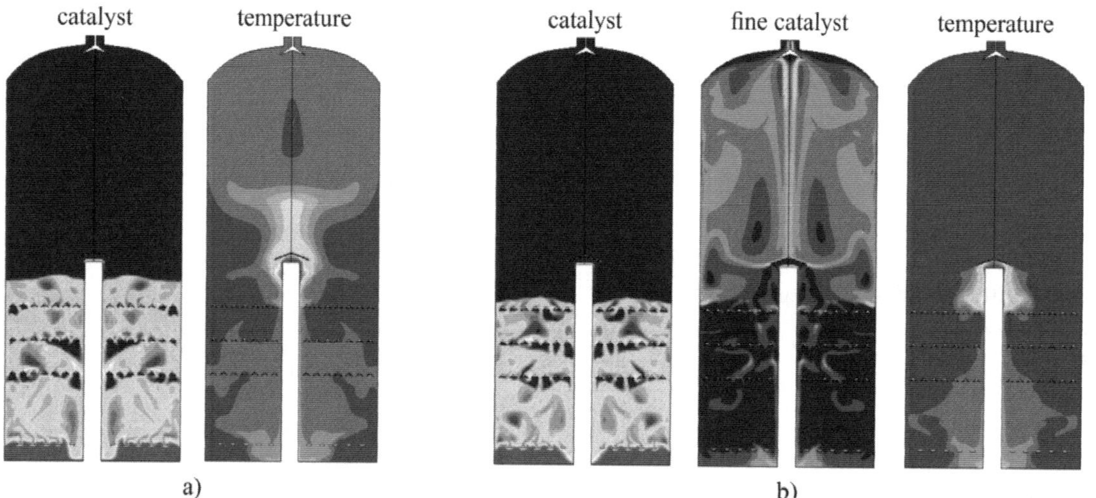

Figure 2. The catalyst concentration field and the temperature filed for the case with a centerline pipe: monodisperse (a), two-fractional (b).

Also, the figure shows the temperature fields. The flow rate of the studied chemical reaction depends on the temperature and occurs with heat absorption. In order to maintain a temperature sufficient for the reaction hot new catalyst is continuously fed into reactor. It is seen, that in the case of monodispersed catalyst the area in the upper part of the apparatus becomes more heated. This is due to the fact that the feed of heated catalyst carried under the base layer and the rising up gas flows take most of the heat from the received catalyst particles. In turn, the heated gas is in a catalyst-free zone, where the flow of chemical reaction is unlikely.

In the presence of small particles the pattern of the temperature field is different. Bottom and middle part of the apparatus occur more heated. This is due to the wall effect along a centerline pipe. Circulating fine particles press the incoming hot catalyst to the pipe, which thin layer flows into the bottom zone of the apparatus, where is a large amount of coarse catalyst. Thus, the presence of fine particles has a positive effect on the heating of apparatus zones and on the chemical reaction rate.

3.2. A case without central pipe

Let us consider the case, when the pipe for the heated catalyst feeding is not vertical, but, for example, is situated from the side. This eliminates the wall effect of the catalyst rolling down. Figure 3 shows the field of catalyst concentration and temperature fields. Almost there is no difference between two concentration patterns from the version with centerline pipe.

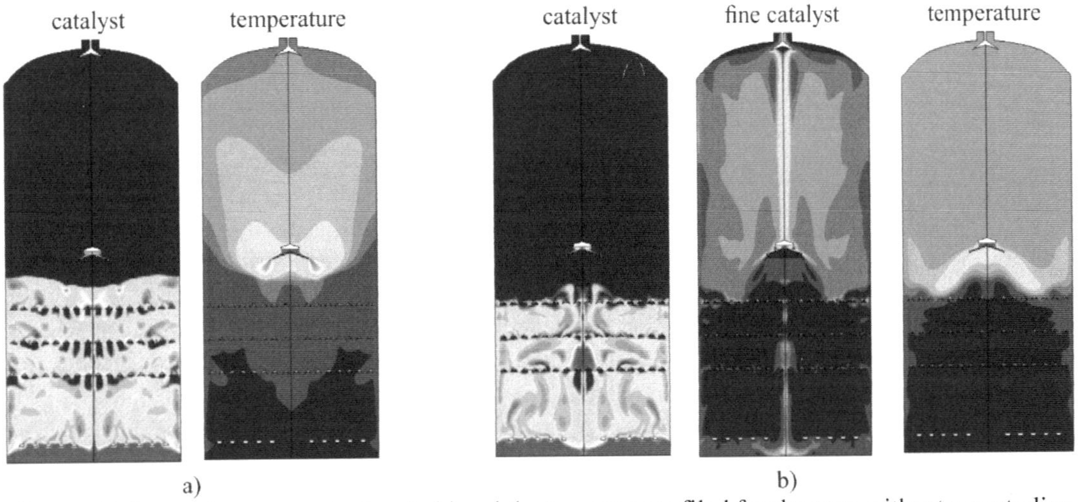

Figure 3. The catalyst concentration field and the temperature filed for the case without a centerline pipe: monodisperse (a), two-fractional (b).

However, when considering the temperature fields there are significant differences. The received catalyst gives off heat to the rising gas, as in the case of monodisperse catalyst, or falls into a zone with a sparse fine fraction in the case of two-fractional catalyst. As a result, larger catalyst particles located at the bottom of the apparatus are heated very slowly, which adversely affects the chemical reaction.

4. Conclusion

Thus, indirect structural feature as the location of the pipe to the point of heated catalyst feeding can have a significant impact on the heating of the whole apparatus. Slight movement of the catalyst along pipe (in comparison with the global circulation streams in the apparatus) can have an impact on the processes of heat transfer. In the temperature-dependent processes (chemical reactors) it can be crucial for the apparatus efficiency.

This work is supported by the Russian Government Program of Competitive Growth of Kazan Federal University.

References

[1] Kunii D, Levenspiel O 1991 *Fluidization Engineering* Butterworth-Heinemann p 498
[2] Gidaspow D 1994 *Multiphase Flow and Fluidization* Boston: Academic Press p 467
[3] Gorliz M R, Grace J R 2002 *International Journal of Heat and Mass Transfer* **45** pp 1149-1154
[4] Samruamphianskun T, Piumsomboon P, Chalermsinsuwan B 2012 *Chemical Engeneering Journal* **210** pp 237–251
[5] Chalermsinsuwana B, Samruamphianskun T, Piumsomboon P 2014 *Chemical Engineering Research and Design* **92** p 2479
[6] Piovano S, Salierno G L, Montmany E, D'Agostino M, Maestri M., Cassanello M 2015 *Chemical Engeneering and Technology* **38(3)** p 423
[7] Gujjula R, Mangadoddy N 2015 *Canadian Journal of Chemical Engineering* **93** p 1380
[8] Solovyev S A, Egorov A G, Lamberov A A, Egorova S R, Kataev A N 2016 *Catalysis in Industry* **8 (1)** pp 48-55
[9] Syamlal M and O'Brien T J 1989 *AIChE Symp. Series* **85** pp 22–31

[10] Syamlal M 1987 *The Particle-Particle Drag Term in a Multiparticle Model of Fluidization* National Technical Information Service, *VA, DOE/MC/21353-2373, NTIS/DE87006500*
[11] Lun C K K, Savage S B, Jeffrey D J 1984 *J. Fluid Mech.* **140** pp 223–256
[12] Ranz W E, Marshall W R 1952 *Chem. Eng. Prog.* **48 (4)** pp 173–180

Approximation of operator eigenvalue problems in a Hilbert space

S I Solovyev

Department of Computational Mathematics, Institute of Computational Mathematics and Information Technologies, Kazan (Volga Region) Federal University, 35 Kremlevskaya Street, Kazan, 420008, Russia

E-mail: sergei.solovyev@kpfu.ru

Abstract. The eigenvalue problem for a compact symmetric positive definite operator in an infinite-dimensional Hilbert space is approximated by an operator eigenvalue problem in finite-dimensional subspace. Error estimates for the approximate eigenvalues and eigenelements are established. These results can be applied for investigating the finite element method with numerical integration for differential eigenvalue problems.

Introduction

The eigenvalue problem for a compact symmetric positive definite operator in an infinite-dimensional Hilbert space is considered. This problem has a non-increasing sequence of positive eigenvalues of finite multiplicity with a limit point at zero, which corresponds to complete orthonormal system of eigenelements. The Hilbert space and the compact operator are the data of the problem. The original problem is approximated by an operator eigenvalue problem in finite-dimensional subspace. New error estimates for the approximate eigenvalues and eigenelements via the errors contributed in the data of the problem are established. These results can be applied to derive optimal order estimates for the accuracy of finite element method with numerical integration for differential eigenvalue problems.

Eigenvalue problems arise in various applications [1–16]. Mesh methods for solving differential eigenvalue problems were studied in [17–29]. The theoretical results obtained in the present paper develop the results of the paper [30] for operator eigenvalue problems.

1. Statement of the problem

Let V be a real infinite-dimensional Hilbert space with the scalar product $(.,.)$ and the norm $\|\cdot\|$, and let $\mathbb{R}$ be the real line. Define a linear compact symmetric positive definite operator T mapping V into V. By $\|\cdot\|$ we also denote the norm of operators mapping V into V.

Let us state the eigenvalue problem: find $\lambda \in \mathbb{R}$ and $u \in V \setminus \{0\}$ such that

$$Tu = \lambda u. \tag{1}$$

The number λ that satisfies (1) is called an eigenvalue, and the element u is called an eigenelement of problem (1) corresponding to λ. The set $U(\lambda)$ that consists of the eigenelements corresponding to the eigenvalue λ and the zero element is a closed subspace in V, which is called the eigensubspace corresponding to the eigenvalue λ. The dimension of this subspace is called the multiplicity of the eigenvalue λ.

Theorem 1. Problem (1) has a non-increasing sequence of positive eigenvalues λ_i, $i=1,2,...,$ with finite multiplicities and with limit point at zero, which are numbered with regard to their multiplicities; i.e. $\lambda_1 \geq \lambda_2 \geq ... \geq \lambda_i \geq ...,$ $\lambda_i \to 0$ as $i \to \infty$. To these eigenvalues, there corresponds an orthonormal system of eigenelements u_i, $i=1,2,...,$ such that $(u_i,u_j)=\delta_{ij}$, $i,j=1,2,...$ The elements u_i, $i=1,2,...,$ form a complete system in the space V. For any element $v \in V$, the following representations are valid

$$Tv = \sum_{i=1}^{\infty} \lambda_i (v,u_i) u_i, \quad v = \sum_{i=1}^{\infty} (v,u_i) u_i.$$

Proof. These results are well known and can be found, e.g., in [31].

2. Approximation of the problem

Let V_h be real finite-dimensional subspaces of the Hilbert space V, $N_h = \dim V_h < \infty$. By P_h we denote the operator of orthogonal projection onto the subspace V_h. Suppose that the limit density condition of the family of subspaces V_h in the space V is satisfied; i.e. for any element $v \in V$ the relation $\|v - P_h v\| \to 0$ as $h \to 0$ holds. The limit density condition implies $N_h \to \infty$ as $h \to 0$. Define the family of linear compact symmetric positive definite operator T_h mapping V_h into V_h such that $\|(T_h - P_h T)P_h\| \to 0$ as $h \to 0$.

Problem (1) is approximated by the finite-dimensional problem: find $\lambda^h \in \mathbb{R}$ and $u^h \in V_h \setminus \{0\}$ such that

$$T_h u^h = \lambda^h u^h. \tag{2}$$

Theorem 2. Problem (2) has a non-increasing sequence of positive eigenvalues λ_i^h, $i=1,2,...,N_h$, with finite multiplicities, which are numbered with regard to their multiplicities; i.e. $\lambda_1^h \geq \lambda_2^h \geq ... \geq \lambda_{N_h}^h$. To these eigenvalues, there corresponds an orthonormal system of eigenelements u_i^h, $i=1,2,...,N_h$, such that $(u_i^h,u_j^h)=\delta_{ij}$, $i,j=1,2,...,N_h$. The elements u_i^h, $i=1,2,...,N_h$, form a complete system in the space V_h. For any element $v^h \in V_h$, the following representations are valid

$$T_h v^h = \sum_{i=1}^{N_h} \lambda_i^h (v^h,u_i^h) u_i^h, \quad v^h = \sum_{i=1}^{N_h} (v^h,u_i^h) u_i^h.$$

Proof. These results can be proved by analogy with Theorem 1.

Denote by I the identical operator in the space V, and set $P^h = I - P_h$. Since T is compact, $\|P^h T\| \to 0$ as $h \to 0$.

Theorem 3. The following estimate is valid

$$|\lambda_k - \lambda_k^h| \leq 2\|P^h T\| + \|(T_h - P_h T)P_h\|.$$

Proof. By Weyl's theorem [31], we get

$$|\lambda_k - \lambda_k^h| \leq \|T - T_h P_h\| \leq \|T - P_h T P_h\| + \|(T_h - P_h T)P_h\| =$$

$$= \|P^h T - P_h T P^h\| + \|(T_h - P_h T)P_h\| \leq 2\|P^h T\| + \|(T_h - P_h T)P_h\|.$$

This proves Theorem 3.

Theorem 4. The convergence $\lambda_k^h \to \lambda_k$ as $h \to 0$ holds. Each sequence of normalized eigenelement u_k^h, $\|u_k^h\| = 1$, contains a subsequence converging to the eigenelement u_k corresponding to the eigenvalue λ_k.

Proof. These results follow from Theorem 3.

3. Error of approximate eigenelements

Let λ_k be an eigenvalue of problem (1) with multiplicity r such that $\lambda_{k-1} > \lambda_k = \lambda_{k+1} = \ldots = \lambda_{k+r-1} > \lambda_{k+r}$, U_k is the eigensubspace corresponding to the eigenvalue λ_k, $U_k^h = \operatorname{span}\{u_k^h, u_{k+1}^h, \ldots, u_{k+r-1}^h\}$. Denote by Q_k and Q_{kh} the operators of orthogonal projection onto the subspaces U_k and U_k^h, respectively, $Q^k = I - Q_k$, $Q^{kh} = I - Q_{kh}$. Denote by $\mathbb{N} = \{1, 2, \ldots\}$ the set of positive integers. Put $J_k = \{k, k+1, \ldots, k+r-1\}$, $J^k = \mathbb{N} \setminus J_k$,

$$\varepsilon_h(u) = \frac{\|P^h T u\|}{\|u\|}, \quad \delta_h(u) = \frac{\|(T_h - P_h T) P_h u\|}{\|u\|},$$

$$\rho_k = \max\left\{\frac{1}{|\lambda_k|}, \max_{j \in J^k} \frac{1}{|\lambda_k - \lambda_j|}\right\}, \quad \rho_i^h = \max\left\{\frac{1}{|\lambda_i^h|}, \max_{j \in J^k} \frac{1}{|\lambda_i^h - \lambda_j^h|}\right\},$$

$$a_i^h = \frac{1 + \rho_k \|P^h T\|}{|\lambda_k|\left(1 - \rho_k |\lambda_i^h - \lambda_k|\right)}, \quad b_i^h = \frac{\rho_k}{1 - \rho_k |\lambda_i^h - \lambda_k|},$$

$$c_i^h = \frac{1 + \rho_i^h \|P^h T\|}{|\lambda_k|\left(1 - \rho_i^h |\lambda_i^h - \lambda_k|\right)}, \quad d_i^h = \frac{\rho_i^h}{1 - \rho_i^h |\lambda_i^h - \lambda_k|},$$

where $a_i^h \to |\lambda_k|^{-1}$, $b_i^h \to \rho_k$, $c_i^h \to |\lambda_k|^{-1}$, $d_i^h \to \rho_k$, as $h \to 0$, $i \in J_k$, $u \in V$.

Theorem 5. Let u_i^h be an eigenelement of problem (2) corresponding to λ_i^h, $\|u_i^h\| = 1$, $1 - \rho_k |\lambda_i^h - \lambda_k| > 0$, $i \in J_k$. Then

$$\|u_i^h - Q_k u_i^h\| \le a_i^h \varepsilon_h(u_i^h) + b_i^h \delta_h(u_i^h).$$

Proof. According to the equality

$$(T - \lambda_k I) u_i^h = \left(\lambda_i^h - \lambda_k\right) u_i^h - (T_h - P_h T) u_i^h + P^h T u_i^h, \tag{3}$$

we obtain

$$\left(Q^k T - \lambda_k I\right) Q^k u_i^h = \left(\lambda_i^h - \lambda_k\right) Q^k u_i^h - Q^k (T_h - P_h T) u_i^h + Q^k P^h T u_i^h.$$

Hence we derive the inequality

$$\|Q^k u_i^h\| \le \rho_k |\lambda_i^h - \lambda_k| \|Q^k u_i^h\| + \rho_k \|(T_h - P_h T) u_i^h\| + \left\|\left(Q^k T - \lambda_k I\right)^{-1} Q^k P^h T u_i^h\right\|,$$

which implies the desired estimate. Thus, the theorem is proved.

Theorem 6. Let $u \in U_k$ be an eigenelement of problem (1) corresponding to λ_k, $\|u\|=1$, $1-\rho_i^h\left|\lambda_i^h-\lambda_k\right|>0$, $i \in J_k$. Then

$$\|u-Q_{kh}u\| \le c_i^h \varepsilon_h(u) + d_i^h \delta_h(u).$$

Proof. This result can be proved by analogy with Theorem 5.

4. Error of approximate eigenvalues

Set

$$\Delta_h(u) = \frac{\left|\left((T_h - P_h T) P_h u, P_h u\right)\right|}{\|u\|^2},$$

$$\xi_i^h = a_i^h \varepsilon_h(u_i^h) + b_i^h \delta_h(u_i^h), \quad \zeta_i^h = c_i^h \varepsilon_h(u) + d_i^h \delta_h(u),$$

$$p_i^h = \frac{1-\rho_k\left|\lambda_i^h-\lambda_k\right|}{1+\rho_k\left\|P^h T\right\|}, \quad q_i^h = \frac{1-\rho_i^h\left|\lambda_i^h-\lambda_k\right|}{1+\rho_i^h\left\|P^h T\right\|},$$

where $\xi_i^h \to 0$, $\zeta_i^h \to 0$, $p_i^h \to 1$, $q_i^h \to 1$, as $h \to 0$, $i \in J_k$, $u \in V$.

Theorem 7. Let u_i^h be an eigenelement of problem (2) corresponding to λ_i^h, $\|u_i^h\|=1$, $1-\rho_k\left|\lambda_i^h-\lambda_k\right|>0$, $\xi_i^h<1$, $i \in J_k$. Then

$$\left|\lambda_i^h-\lambda_k\right| \le \Delta_h(Q_k u_i^h) + \frac{\xi_i^h}{\sqrt{1-\left(\xi_i^h\right)^2}}\left(p_i^h \varepsilon_h(Q_k u_i^h) + \delta_h(Q_k u_i^h)\right).$$

Proof. Using equality (3) and the relations

$$\left((T-\lambda_k I)u_i^h, Q_k u_i^h\right) = \left(u_i^h, (T-\lambda_k I) Q_k u_i^h\right) = 0, \quad \left(u_i^h, Q_k u_i^h\right) = \left\|Q_k u_i^h\right\|^2,$$

$$\left(P^h T u_i^h, Q_k u_i^h\right) = \left(P^h T u_i^h, \lambda_k^{-1} T Q_k u_i^h\right) = \lambda_k^{-1}\left(P^h T u_i^h, T Q_k u_i^h\right),$$

$$\left\|Q_k u_i^h\right\| = \sqrt{\left\|u_i^h\right\|^2 - \left\|Q^k u_i^h\right\|^2} \ge \sqrt{1-\left(\xi_i^h\right)^2}\left\|u_i^h\right\|,$$

$$\left((T_h - P_h T) u_i^h, Q_k u_i^h\right) = \left(u_i^h, (T_h - P_h T) P_h Q_k u_i^h\right) =$$

$$= \left(P_h Q_k u_i^h, (T_h - P_h T) P_h Q_k u_i^h\right) + \left(u_i^h - Q_k u_i^h, (T_h - P_h T) P_h Q_k u_i^h\right) =$$

$$= \left((T_h - P_h T) P_h Q_k u_i^h, P_h Q_k u_i^h\right) + \left((T_h - P_h T) P_h Q_k u_i^h, u_i^h - Q_k u_i^h\right),$$

we get the relations

$$\left(\lambda_i^h - \lambda_k\right)\left\|Q_k u_i^h\right\|^2 = \left((T_h - P_h T) u_i^h, Q_k u_i^h\right) - \left(P^h T u_i^h, Q_k u_i^h\right) =$$

$$= \lambda_k^{-1}\left(P^h T u_i^h, P^h T Q_k u_i^h\right) + \left((T_h - P_h T) P_h Q_k u_i^h, P_h Q_k u_i^h\right) + \left((T_h - P_h T) P_h Q_k u_i^h, u_i^h - Q_k u_i^h\right),$$

which imply the desired estimate:

$$\left|\lambda_i^h - \lambda_k\right| \le \frac{1}{|\lambda_k|} \frac{\|P^h T u_i^h\|}{\sqrt{1-\left(\xi_i^h\right)^2}\|u_i^h\|} \frac{\|P^h T Q_k u_i^h\|}{\|Q_k u_i^h\|} + \frac{\left|\left((T_h - P_h T)P_h Q_k u_i^h, P_h Q_k u_i^h\right)\right|}{\|Q_k u_i^h\|^2} +$$

$$+\frac{\|(T_h - P_h T)P_h Q_k u_i^h\|}{\|Q_k u_i^h\|} + \frac{\|u_i^h - Q_k u_i^h\|}{\sqrt{1-\left(\xi_i^h\right)^2}\|u_i^h\|} \le \Delta_h(Q_k u_i^h) + \frac{\xi_i^h}{\sqrt{1-\left(\xi_i^h\right)^2}}\left(p_i^h \varepsilon_h(Q_k u_i^h) + \delta_h(Q_k u_i^h)\right).$$

Thus, the theorem is proved.

Theorem 8. Let $u_i^h = Q_{kh} u$ be an eigenelement of problem (2) corresponding to λ_i^h, $u \in U_k$ $\|u\| = 1$, $1 - \rho_i^h\left|\lambda_i^h - \lambda_k\right| > 0$, $\zeta_i^h < 1$, $i \in J_k$. Then

$$\left|\lambda_i^h - \lambda_k\right| \le \Delta_h(u_i^h) + \frac{\zeta_i^h}{\sqrt{1-\left(\zeta_i^h\right)^2}}\left(q_i^h \varepsilon_h(u_i^h) + \delta_h(u_i^h)\right).$$

Proof. Theorem 8 can be proved by analogy with Theorem 7.

Acknowledgments
This work was supported by Russian Science Foundation, project no. 16-11-10299.

References

[1] Zhigalko Yu P and Solov'ev S I 2001 Natural oscillations of a beam with a harmonic oscillator *Russ. Math.* **45** (10) 33–5

[2] Solov'ev S I 2016 Eigenvibrations of a beam with elastically attached load *Lobachevskii J. Math.* **37** (5) 597–609

[3] Solov'ev S I 2003 Eigenvibrations of a plate with elastically attached load *Preprint* SFB393/03-06 Technische Universität Chemnitz

[4] Solov'ev S I 2003 Vibrations of plates with masses *Preprint* SFB393/03-18 Technische Universität Chemnitz

[5] Solov'ev S I 2011 *Nonlinear Eigenvalue Problems. Approximate Methods* (Saarbrücken: Lambert Acad. Publ.) [in Russian]

[6] Goolin A V and Kartyshov S V 1993 Numerical study of stability and nonlinear eigenvalue problems *Surv. Math. Ind.* **3** 29–48

[7] Betcke T, Higham N J, Mehrmann V, Schröder C and Tisseur F 2013 NLEVP: A collection of nonlinear eigenvalue problems *ACM Trans. Math. Software* **39** (2) Article number 7

[8] Kozlov V A, Maz'ya V G and Rossmann J 2001 *Spectral Problems Associated with Corner Singularities of Solutions to Elliptic Equations* (Providence: American Mathematical Society)

[9] Apel Th, Sändig A-M and Solov'ev S I 2002 Computation of 3D vertex singularities for linear elasticity: Error estimates for a finite element method on graded meshes *Math. Model. Numer. Anal.* **36** (6) 1043–70

[10] Lyashko A D and Solov'ev S I 1991 Fourier method of solution of FE systems with Hermite elements for Poisson equation *Sov. J. Numer. Anal. Math. Modelling* **6** (2) 121–9

[11] Solov'ev S I 1994 Fast direct methods of solving finite-element grid schemes with bicubic elements for the Poisson equation *J. Math. Sciences* **71** (6) 2799–804

[12] Solov'ev S I 1995 A fast direct method of solving Hermitian fourth-order finite-element schemes for the Poisson equation *J. Math. Sciences* **74** (6) 1371–6

[13] Karchevskii E M and Solov'ev S I 2000 Investigation of a spectral problem for the Helmholtz operator on the plane *Differ. Equations* **36** (4) 631–4

[14] Zheltukhin V S, Solovyev P S and Chebakova V Yu 2015 Boundary conditions for electron

balance equation in the stationary high-frequency induction discharges *Res. J. Appl. Sci.* **10** (10) 658–62

[15] Dautov R Z, Lyashko A D and Solov'ev S I 1994 The bisection method for symmetric eigenvalue problems with a parameter entering nonlinearly *Russ. J. Numer. Anal. Math. Modelling* **9** (5), 417–427

[16] Solov'ev S I 2006 Preconditioned iterative methods for a class of nonlinear eigenvalue problems *Linear Algebra Appl.* **415** (1) 210–29

[17] Solov'ev S I 1992 The error of the Bubnov–Galerkin method with perturbations for symmetric spectral problems with a non-linearly occurring parameter *Comput. Math. Math. Phys.* **32** (5) 579–93

[18] Solov'ev S I 1997 The finite element method for symmetric nonlinear eigenvalue problems *Comput. Math. Math. Phys.* **37** (11) 1269–76

[19] Dautov R Z, Lyashko A D and Solov'ev S I 1991 Convergence of the Bubnov–Galerkin method with perturbations for symmetric spectral problems with parameter entering nonlinearly *Differ. Equations* **27** (7) 799–806

[20] Solovyev S I 1994 Superconvergence of finite element approximations of eigenfunctions *Differ. Equations* **30** (7) 1138–46

[21] Solov'ev S I 2002 Superconvergence of finite element approximations to eigenspaces *Differ. Equations* **38** (5) 752–3

[22] Solovyev S I 2010 Approximation of variational eigenvalue problems *Differ. Equations* **46** (7) 1030–41

[23] Solovyev S I 2011 Approximation of positive semidefinite spectral problems *Differ. Equations* **47** (8) 1188–96

[24] Solovyev S I 2012 Approximation of sign-indefinite spectral problems *Differ. Equations* **48** (7) 1028–41

[25] Solovyev S I 2013 Approximation of differential eigenvalue problems *Differ. Equations* **49** (7) 908–16

[26] Zheltukhin V S, Solovyev S I, Solovyev P S and Chebakova V Yu 2014 Computation of the minimum eigenvalue for a nonlinear Sturm–Liouville problem *Lobachevskii J. Math.* **35** (4) 416–26

[27] Solovyev S I 2014 Approximation of differential eigenvalue problems with a nonlinear dependence on the parameter *Differ. Equations* **50** (7) 947–54

[28] Solovyev S I 2015 Approximation of nonlinear spectral problems in a Hilbert space *Differ. Equations* **51** (7) 934–47

[29] Solovyev S I 2015 Finite element approximation with numerical integration for differential eigenvalue problems *Appl. Numer. Math.* **93** 206–14

[30] Vainikko G M 1965 Evaluation of the error of the Bubnov–Galerkin method in an eigenvalue problem *USSR Comput. Math. Math. Phys.* **5** (4) 1–31

[31] Riesz F and Sz-Nagy B 1972 *Leçons d'Analyse Fonctionelle* (Budapest: Académiai Kiadó)

Modelling of large deformations of elastoplastic solids using FEM

L U Sultanov

Kazan Federal University, 18 Kremlyovskaya Street, Kazan, 420008, Russian Federation

E-mail: Lenar.Sultanov@kpfu.ru

Abstract. The paper is devoted to development of technique for numerical studies of finite elastoplastic deformations. The kinematics of the medium is described by the deformation gradient tensor. Incremental method based on the equation of principle of virtual work is used. The von Mises criterion of plasticity is applied. The constitutive relationships are obtained for the Cauchy stress rate tensor. Computer implementation is based on finite elements method.

1. Introduction

The numerical solution of the problem of large elastic-plastic deformations is considered in this paper. There are many publications dedicated to solution of nonlinear problems, for example [1–8], which deals with the geometrically and physically nonlinear problem. They are distinguished by the difference in the formulations and solution of the resulting algebraic problems.

2. Solution algorithm

Kinematics of finite deformations is described using a gradient of deformations [9–10], which can be represented in the next form:

$$(F)=(F_e)\cdot(F_p),$$

where (F_e) – the tensor of the gradient of elastic deformations, (F_p) – the tensor of the gradient of plastic deformations.

The stress state is described by the Cauchy stress tensor [9, 11]. Physical ratio of the elastic deformation will get from the equation of the second law of thermodynamics for isothermal process:

$$(\Sigma)\cdot\cdot(d)-\dot{W}\geq 0,$$

where $W=W(B_e)$ is the function of potential energy of elastic deformation, $(B_e)=(F_e)\cdot(F)_e^T$.

$$(\Sigma)=\frac{2}{J}(B_e)\cdot\left(\frac{\partial W}{\partial B_e}\right),$$

The Cauchy stress rate tensor is defined as

$$(\dot{\Sigma})=(\Lambda_\Sigma)\cdot\cdot(d)+(h)\cdot(\Sigma)+(\Sigma)\cdot(h)^T-(\Sigma)I_{1d},$$

where $(\Lambda_\Sigma)=\frac{4}{J}(B_e)\cdot\left(\frac{\partial^2 W}{\partial B_e \partial B_e}\right)\cdot(B_e)$, $(h)=(\dot{F})\cdot(F^{-1})$, $(d)=\frac{1}{2}\left[(h)+(h)^T\right]$, $I_{1d}=tr(d)=\frac{\dot{J}}{J}$.

The von Mises yield criterion is used:

$$\Phi_p=\sigma_i-\sigma_T\le 0,$$

where $\sigma_i=\left(\frac{3}{2}(\Sigma')\cdot\cdot(\Sigma')\right)^{\frac{1}{2}}$, σ_T – yield stress. For the plastic deformation rate must hold flow rule:

$$(F_e)\cdot(d_p)\cdot(F_e^{-1})=\dot{\lambda}\left(\frac{\partial\Phi_p}{\partial\Sigma}\right). \quad (1)$$

Here $\dot{\lambda}$ is the consistency parameter, $(d_p)=\frac{1}{2}\left[(h_p)+(h_p)^T\right]$, $(h_p)=(\dot{F}_p)\cdot(F_p^{-1})$.

For the velocity gradients

$$(d)=\frac{1}{2}\left[(h)+(h)^T\right]=Sym\left[(h_e)+(F_e)\cdot(h_p)\cdot(F_e^{-1})\right]. \quad (2)$$

Assuming that plastic deformation does not cause the volume change, i.e. $(h_p)=(d_p)$, the expression (1) can be written in the following form:

$$(d)=Sym\left((h_e)+(F_e)\cdot(h_p)\cdot(F_e^{-1})\right)=Sym(h_e)+Sym\left((F_e)\cdot(h_p)\cdot(F_e^{-1})\right)=$$
$$=Sym(h_e)+Sym\left((F_e)\cdot(d_p)\cdot(F_e^{-1})\right)=(d_e)+(d_p^*),$$

where $(d_e)=\frac{1}{2}\left[(h_e)+(h_e)^T\right]$, $(d_p^*)=Sym\left((F_e)\cdot(d_p)\cdot(F_e^{-1})\right)=(F_e)\cdot(d_p)\cdot(F_e^{-1})$.

Then, using (1), the plastic velocity gradients can be written as

$$(d_p^*)=\dot{\lambda}\left(\frac{\partial\Phi_p}{\partial\Sigma}\right)=\dot{\lambda}\frac{\partial\sigma_i}{\partial\Sigma'}=\frac{3}{2}\dot{\lambda}\frac{\Sigma'}{\sigma_i}.$$

The research algorithm is based on an Update Lagrange formulation. The principle of virtual work in terms of the virtual velocity is used [12–15]:

$$\int_V(\Sigma)\cdot\cdot(\delta d)dV=\int_V\vec{f}\cdot\delta\vec{\upsilon}dV+\int_{S_\sigma}\vec{t}_n\cdot\delta\vec{\upsilon}dS.$$

Return mapping algorithm is applied to solve plastic problem [12, 16–19].The trial stress on kth step is defined as

$$({}^{k+1}\tilde{\Sigma})=({}^{k}\Sigma)+({}^{k+1}\Delta\Sigma)$$

The plastic flow $\dot{\lambda}$ can be found from equation

$$({}^{k+1}\Sigma)+\frac{3}{2}\frac{\dot{\lambda}}{\sigma_i}(\Lambda_\Sigma)\cdot\cdot({}^{k+1}\Sigma')=({}^{k+1}\tilde{\Sigma}). \quad (3)$$

The equation (3) is solved using the Newton method [20, 21]. Then, we use iterative refinement. The numerical implementation is based on the finite element method. An 8-node brick element is used.

3. Numerical example

As an example the potential of elastic deformation is considered:

$$W=\frac{\lambda+2\mu}{8}(I_{1B}-3)^2+\mu(I_{1B}-3)-\frac{\mu}{2}(I_{2B}-3),$$

where λ, μ are Lame parameters.

The problem of deforming of the pipe under internal pressure is considered (Figure 1). The material data for isotropic elasticity and the von Mises yield condition are given as follows: the inner radius a=1 cm, the outer radius b=2 cm, the radius of the plastic zone c=1.5 cm (Figure 2), $E = 2$ MPa, $\mu = 0.3$, p/σ_T=0.7208.

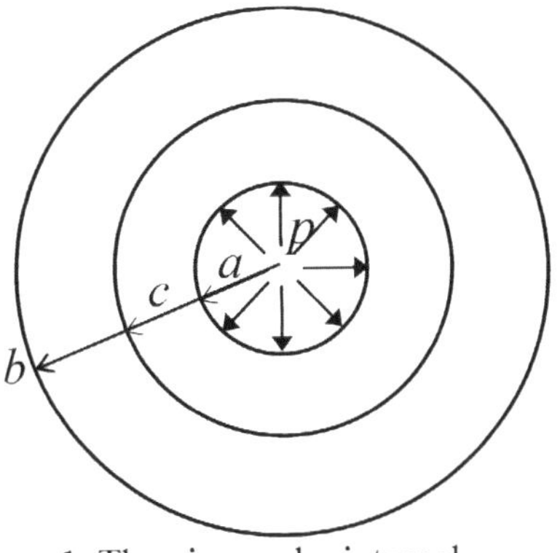

Figure 1. The pipe under internal pressure.

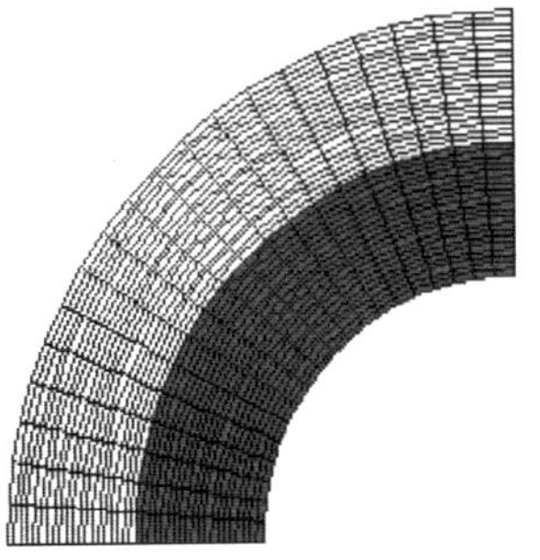

Figure 2. Deformed state of pipe.

4. Conclusion

The algorithm for solving elastoplastic problems is built. The problem of deformation of the pipe under internal pressure is solved. The obtained results have quite good agreement with the results of other authors.

Acknowledgements

The reported study was supported by Government of the Republic of Tatarstan research projects No. 15-41-02557.

References

[1] Badriev I B, Banderov V V, Garipova G Z, Makarov M V and Shagidullin R R 2015 *Applied Mathematical Sciences* **9 (81-84)** 4095–4102
[2] Badriev I B, Makarov M V and Paimushin V N 2016 *Procedia Engineering* **150** 1050–1055
[3] Badriev I B, Garipova G Z, Makarov M V, Paimushin V N and Khabibullin R F 2015 *Lobachevskii Journal of Mathematics* **36 (4)** 474–481. doi: 10.1134/S1995080215040216
[4] Berezhnoi D V, Paimushin V N and Shalashilin V I 2009 *Mechanics of Solids* **44 (6)** 837–851
[5] Berezhnoi D V and Paimushin V N 2011 *Journal of Applied Mathematics and Mechanics* **75 (4)** 447–462
[6] Sagdatullin M K and Berezhnoi D V 2014 *Applied Mathematical Sciences* **8 (35)** 1731–1738
[7] Sachenkov O A, Mitryaikin V I, Zaitseva T A and Konoplev Yu G 2014 *Applied Mathematical Sciences* **8 (159)** 7889–7897
[8] Sultanov L U 2015 *Procedia Earth and Planetary Science* **15** 119–124
[9] Sultanov L U 2016 *Lobachevskii Journal of Mathematics* **37 (6)** 784–790
[10] Sultanov L U and Fakhrutdinov L R 2013 *Magazine of Civil Engineering* **44(9)** 69–74
[11] Davydov R L and Sultanov L U 2013 *PNRPU Mechanics Bulletin* **1** 81–93
[12] Golovanov A I and Sultanov L U 2005 *International Applied Mechanics* **41 (6)** 614–620
[13] Sultanov L U 2014 *Applied Mathematical Sciences* **8 (143)** 7117–7124
[14] Sultanov L U and Davydov R L and 2014 *Applied Mathematical Sciences* **8 (60)** 2991–2996
[15] Abdrakhmanova A I and Sultanov L U 2016 *Materials Physics and Mechanics* **26 (1)** 30–32
[16] Davydov R and Sultanov L 2013 *Sixth International Conference on Nonlinear Mechanics (ICNM-VI)* 64–67

[17] Eidel B, Gruttmann F 2003 *Computational Materials Science* **28** 732–742
[18] Taylor L M and Becher E B 1983 *Computer Methods in Applied Mechanics and Engineering* **41 (3)** 251–277
[19] *Yamada Y* 1982 *Computer Methods in Applied Mechanics and Engineering* **33 (1-3)** 417–437
[20] Davydov R L and Sultanov L U 2015 *Journal of Engineering Physics and Thermophysics* **88 (5)** 1280–1288
[21] Sultanov L U and Davydov R L 2013 *Magazine of Civil Engineering* **44(9)** 64–68

Investigation of oscillations of the elastic bodies with joined a mass–spring–damper systems

L U Sultanov

Kazan Federal University, 18 Kremlyovskaya Street, Kazan, 420008, Russian Federation

E-mail: Lenar.Sultanov@kpfu.ru

Abstract. The problem of free oscillations of elastic systems joined with a simple mechanical system consisting of mass, springs and dampers is considered. The differential equation is obtained. The solution of the received equation is based on series. The oscillation of rectangular plate is considered.

1. Introduction

The work is devoted to the analysis of vibrations of elastic systems with joined mass, springs and dampers at the point. These problems are relevant at modern technology, since real shell structures are characterized by inclusions of masses, springs or dampers. The effect of joined masses, springs or dampers is expressed in the appearance of qualitatively new types of dynamic deformations that is a result of the excitation and interaction of the various bending shapes of the elastic systems.

2. Mathematical model

Consider the problem of free oscillations of elastic systems connected at a finite number of points with a simple mechanical system consisting of mass (M), springs (S) and dampers (D) (Figure 1).

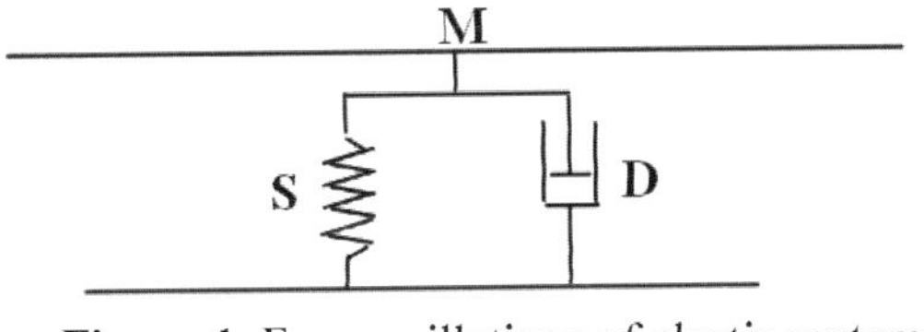

Figure 1. Free oscillations of elastic systems.

Give the differential equation which describes the free vibrations of elastic systems (beams, plates, shells) connected with MSD-systems:

$$L(w)+aL_1(\ddot{w})+\sum_j L_2\left(b\ddot{w}+c\dot{w}+ew\right)\delta\left(P-P_j'\right)=0. \tag{1}$$

Here $w(P,t)$ – the displacements, L, L_1, L_2 – static operators, a, b, c, e – constants, δ – Dirac delta function, P is an arbitrary point, P_j' is the point of joined of MSD systems.

Model (1) is constructed on the basis of the classical Kirchhoff theory. Takes into account only the transverse component of the inertia forces are not taken into account the internal friction in the material. The characteristics of all MSD systems are the similar. To complete this mathematical model to the equation (1) it is necessary to add appropriate boundary conditions.

We rewrite (1) in the form

$$L(w)+aL_1(\ddot{w})=-\sum_j\left[L_2\left(b\ddot{w}+c\dot{w}+ew\right)\right]_{P=P_j'}\delta\left(P-P_j'\right). \tag{2}$$

And we will seek its solution in the class of functions describing the harmonic damped oscillations. To do this, put

$$w(P,t)=W(P)\exp\left(i\omega^0 t\right) \tag{3}$$

where i is the imaginary unit, $\omega^0=\omega+i\mu$, ω – frequency oscillations, μ – damping factor. Substituting (3) into (2), we get

$$L(w)-a\left(\omega^0\right)^2 L_1(W)=-\sum_j\left[b\ddot{w}+c\ddot{w}+ew\right]_{P=P_j'}\left[L_2(W)\right]\delta\left(P-P_j'\right). \tag{4}$$

Decompose the function $W(P)$ and Dirac delta function $\delta\left(P-P_j'\right)$ in the system of orthogonal functions $F_{mn}(P)$ describing the forms of natural oscillations of elastic systems:

$$\begin{aligned} W(P)&=\sum_{m,n}^{\infty}W_{mn}F_{mn}(P),\\ \delta(P-P_j')&=\sum_{m,n}^{\infty}\Delta_{mn}(P_j')F_{mn}(P). \end{aligned} \tag{5}$$

After substituting (5) into (4) we have

$$A_{mn}a\left[\omega_0^2-(\omega^0)^2\right]W_{mn}=\sum_j\left[b(\omega^0)^2-ic\omega^0-e\right]\Delta_{mn}\left(P_j'\right)\left[L_2(W)\right]_{P=P_j'}$$

from which

$$W_{mn}\left(P_j'\right)=\frac{\sum_j\left[b(\omega^0)^2-ic\omega^0-e\right]\Delta_{mn}\left(P_j'\right)\left[L_2(W)\right]_{P=P_j'}}{a\left[\omega_{0mn}^2-(\omega^0)\right]A_{mn}}. \tag{6}$$

Here ω_{0mn} is the natural frequency of the elastic system, A_{mn} – eigenvalues determined by the equation $L_1(F)-AF=0$, where $F=F_{mn}$.

Now the expression of the unknown function $W(P)$ can be written in the form

$$W(P)=\sum_{m,n}W_{mn}\left(P_j'\right)F_{mn}(P). \tag{7}$$

Where the coefficients $W_{mn}\left(P_j'\right)$ is given by (6). Will act on both parts of the equality (7) the operator L_2. Then

$$L_2(W)=\sum_{m,n}W_{mn}\left(P_j'\right)L_2\left(F_{mn}\right). \tag{8}$$

Equation (8) should be performed at any point in the region definition of the function $W(P)$. Demand its implementation in the points of attachment of MSD systems, the resulting homogeneous system of linear equations relative to the unknown $\left[L_2(W)\right]_{P=P_j'}$:

$$\left[L_2(W)\right]_{P=P_j'}-\sum_{j,m,n}\frac{\left[b(\omega^0)^2-ic\omega^0-e\right]\Delta_{mn}(P_j')\left[L_2(F_{mn})\right]_{P=P_j'}\left[L_2(W)\right]_{P=P_j'}}{a\left[\omega_{0mn}^2-(\omega^0)^2\right]A_{mn}}=0.$$

The condition of existence of nontrivial solutions of this system are the equality to zero of its determinant

$$\det\left(\omega^0\right)=0. \tag{9}$$

Solving equation (9) we will find frequency ω and damping factor μ.

3. Example

Let's consider a rectangular plate with sides a along the x-axis and b on the y pivotally supported on all edges. Combining the direction of the x and y axes with the principal axes, by the usual arguments come to the equation of oscillations of an orthotropic plate with an attached MSD system at point (x', y')

$$D_1\frac{\partial^4 w}{\partial x^4}+2D_3\frac{\partial^4 w}{\partial x^2\partial y^2}+D_2\frac{\partial^4 w}{\partial y^4}+\rho h\frac{\partial^2 w}{\partial t^2}+\left[M\frac{\partial^2 w}{\partial t^2}+\alpha\frac{\partial w}{\partial t}+\beta w\right]\delta(x-x',y-y')=0. \tag{10}$$

That is, in this problem, operators take the form

$$L(w)=D_1\frac{\partial^4 w}{\partial x^4}+2D_3\frac{\partial^4 w}{\partial x^2\partial y^2}+D_2\frac{\partial^4 w}{\partial y^4},\ L_1(w)=1,\ L_2(w)=1.$$

And constants $a=\rho h$, $b=M$, $c=\alpha$, $e=\beta$, where M is the magnitude of the concentrated mass, α – viscosity damper coefficient, β – spring stiffness, ρ is the density of plate, h – plate thickness.

We write in accordance with boundary conditions of the orthogonal function

$$F_{mn}(P)=Sin\frac{m\pi x}{a}Sin\frac{n\pi y}{b} \tag{11}$$

and

$$\Delta_{mn}(P')=\frac{4}{ab}Sin\frac{m\pi x'}{a}Sin\frac{n\pi y'}{b} \tag{12}$$

Then the frequency equation in this case will take the form of

$$\sum_m^\infty\sum_n^\infty 4\frac{\left[M\left(\omega^0\right)^2-i\alpha\omega^0-\beta\right]F_{mn}^2(x',y')}{M_p\left[p_{mn}^2-\left(\omega^0\right)^2\right]}=1. \tag{13}$$

where $M_p=abh\rho$ – mass of the plate, and p_{mn} – frequency oscillation, defined as

$$p_{mn}^2=\frac{\pi^4 a}{b^3 M_p}D_2\left[\frac{D_1}{D_2}\left(\frac{mb}{a}\right)^4+2\frac{D_3}{D_2}n^2\left(\frac{mb}{a}\right)^2+n^4\right] \tag{14}$$

The equation (13) is an algebraic equation with respect to unknown ω^0, generally speaking, infinitely high degree. In specific calculations, there is a natural truncation of the series, standing in the left side of the equation, and how it will end.

Let us turn to dimensionless quantities

$$p_{mn}^* = \frac{p_{mn}}{p_{11}},\ \omega^* = \frac{\omega}{p_{11}},\ M^* = \frac{M}{M_p},\ \beta^* = \frac{\beta}{M_p p_{11}^2}.$$

Then the equation (2.4) takes the form

$$\sum_m^\infty \sum_n^\infty 4 \frac{\left[M^*(\omega^*)^2 - iq\omega^* - \beta^*\right] F_{mn}^2(x',y')}{\left[(p_{mn}^*)^2 - (\omega^*)^2\right]} = 1. \tag{15}$$

The solution of the reduced equation (15), we obtain the frequency spectrum of the vibrations of the plate.

Fixing such parameters

$$\frac{D_1}{D_2} = 20,\ \frac{D_3}{D_2} = 2{,}196,\ \frac{a}{b} = 1,\ x' = 0{,}2a,\ y' = 0{,}2b$$

and having considered the various cases of MSD systems, we present graphs of the $\omega^* = \omega^*(q)$, characterizing, respectively, the oscillation frequency and the viscosity of the damper.

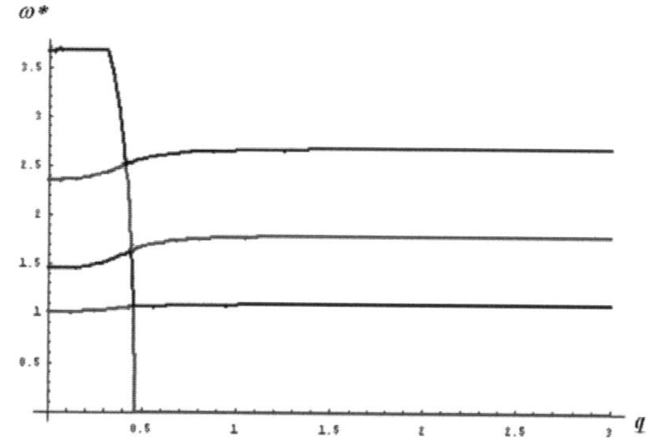

Figure 2. Graphics frequency–viscosity, $M^* = 0, \beta^* = 0$.

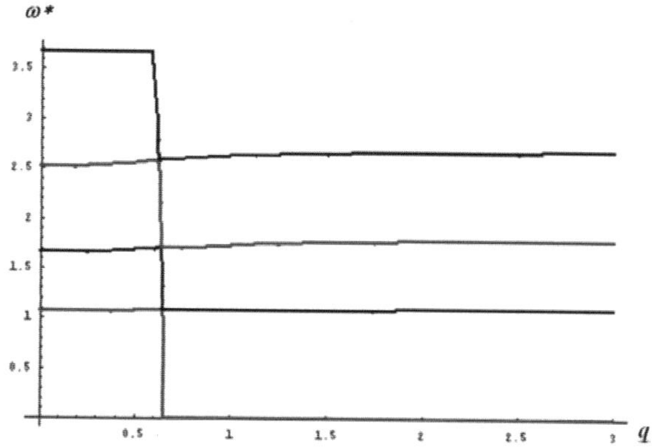

Figure 3. Graphics frequency–viscosity, $M^* = 0, \beta^* = 1$.

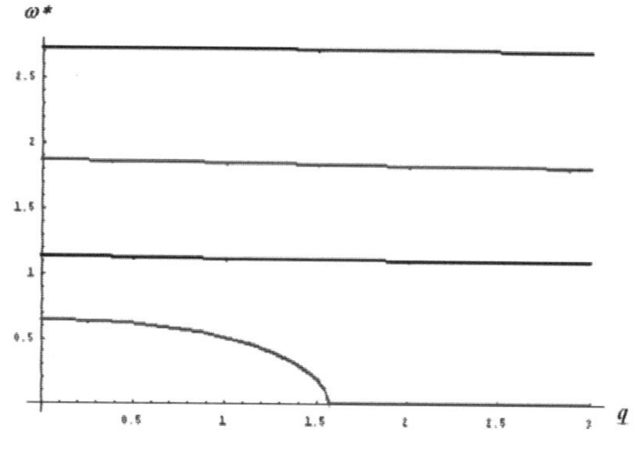

Figure 4. Graphics frequency–viscosity, $M^* = 1, \beta^* = 0$.

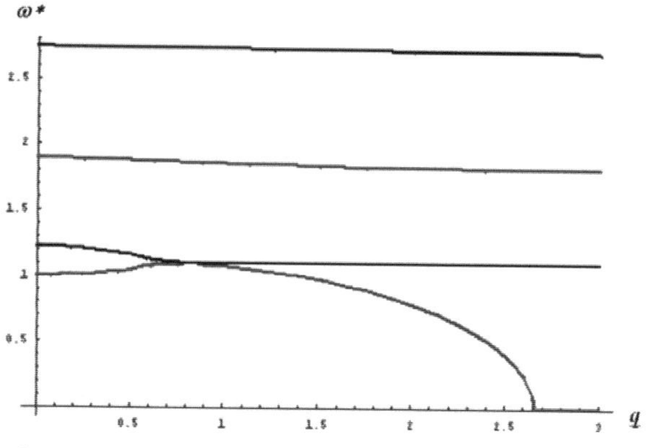

Figure 5. Graphics frequency–viscosity, $M^* = 1, \beta^* = 1$.

The graphs show that for all cases of MSD systems there are some values of the parameter $q = q^*(M^*, \beta^*)$ when $q \geq q^*$ the plate oscillates as if the attachment point of the MSD systems are fixed. The mass M^* and stiffness β^* affect the character of the frequency ω^* when $q \leq q^*$.

4. Conclusion
The paper is dedicated to a solution of the problem of oscillations of elastic system with joined mass, spring and damper. The mathematical model is obtained. The problem of vibration of a plate with attached MSD system is considered. The frequency range for different values of masses and spring stiffness is received.

Acknowledgements
This work was supported by the Russian Science Foundation (project 16-11-10299).

References
[1] Vol'mir A S 1972 *Nelineynaya dinamika plastinok i obolochek* (Moscow: Nauka) 432
[2] Grigorenko A Ya *Prikladnaya Mekhanika* **9 (10)** 87–90
[3] Kubenko V D, Koval'chuk P S and Krasnopol'skaya T S 1984 *Nelineynoye vzaimodeystviye form izgibnykh kolebaniy tsilindricheskikh obolochek* (Kiev, Nauk. Dumka) 220
[4] Kubenko V D, Koval'chuk P S and Podchasov N P 1989 *Nelineynyye kolebaniya tsilindricheskikh obolochek* (Kiev: Vyshcha shk) 208
[5] Kozlov S V 1981 *Prikladnaya Mekhanika* **17 (2)** 46–52

IOP Conf. Series: Materials Science and Engineering **158** (2016) 012090 doi:10.1088/1757-899X/158/1/012090

Calculation of elastic-plastic deformations by FEM

L U Sultanov

Kazan Federal University, 18 Kremlyovskaya Street, Kazan, 420008, Russian Federation

E-mail: Lenar.Sultanov@kpfu.ru

Abstract. The article is devoted to elasto-plastic analysis for finite deformations, large displacements and rotations. An incremental method is used. The stressed state is represented by Cauchy stress and objective Jaumann rate of Cauchy stress. The von Mises yield criterion and radial return method are applied.

1. Introduction

There are many publications, where solutions of nonlinear problems of a solid mechanic are discussed, for example [1–17]. In this paper the algorithm of numerical solution of the problem of large elastic-plastic deformations is considered [1–8, 13–15]. Kinematics of a medium is described by the deformation rate tensor, the stress state is determined by the Cauchy stress tensor. The theory of flow is used for describing plastic deformation. The total deformation rate is represented as a sum of elastic and plastic parts. The solution algorithm is based on an Update Lagrange formulation. The principle of virtual work in terms of the virtual velocity is used. The numerical implementation is based on the method of finite elements (FEM) is used.

2. Kinematics

We assume that at the current time process of deformation represents the continua with elastoplastic properties. This corresponds to the Euler approach, widely used in continuum mechanics. In line with this, the kinematics of deformation of the body will be determined using the following tensors [22–26]: velocity gradient tensor $\mathbf{h}=h_{ij}\mathbf{e}_i\mathbf{e}_j$, deformation rate tensor $\mathbf{d}=\frac{1}{2}\left[\mathbf{h}+\mathbf{h}^T\right]=d_{ij}\mathbf{e}_i\mathbf{e}_j$, rate of rotation tensor $\boldsymbol{\omega}=\frac{1}{2}\left[\mathbf{h}-\mathbf{h}^T\right]=\omega_{ij}\mathbf{e}_i\mathbf{e}_j$.

3. Stress–strain relationship

The state of a stress in a body is defined by Cauchy stress tensor $\boldsymbol{\Sigma}=\sigma_{ij}\mathbf{e}_i\mathbf{e}_j$. Also we use objective Jaumann rate of Cauchy stress $\boldsymbol{\Sigma}^J$ [22–24]

$$\boldsymbol{\Sigma}^J=\dot{\boldsymbol{\Sigma}}-\boldsymbol{\omega}\cdot\boldsymbol{\Sigma}+\boldsymbol{\Sigma}\cdot\boldsymbol{\omega}, \tag{1}$$

where $\dot{\boldsymbol{\Sigma}}$ is the rate of Cauchy stress.

Consider isotropic material and denote $d_0 = \frac{1}{3}d_{ii}$, $\sigma_0^J = \frac{1}{3}\sigma_{ii}^J$, $d'_{ij} = d_{ij} - \delta_{ij}d_0$, $\sigma_{ij}'^J = \sigma_{ij}^J - \delta_{ij}\sigma_0^J$, where δ_{ij} is Kronecker delta.

We adopt an additive split of the deformation of rate into elastic deformation of rate and plastic deformation of rate [23, 24]. Stress-strain relationship for elastic deformation is written as

$$\mathbf{\Sigma}'^J = 2G\mathbf{d}',\ \mathbf{\Sigma}_0^J = 3K\mathbf{d}_0, \tag{2}$$

where $G = \dfrac{E}{2(1+\mu)}$, $K = \dfrac{E}{3(1-2\mu)}$, E – Young's modulus, μ – Poisson's ratio. In this case the elastic stress-strain relationships will satisfy to indifference principle.

4. Algorithm of computation

Process of deformation is represented as a sequence of equilibrium states [23–27]. Conversion from previous state to next one proceeds by increment load. On each step of a loading the geometry and stress state are updated. Techniques of computation is consist in a development of algorithm of calculation of the state (l+1) when the state l is known. Consider the equation of the virtual powers in an actual configuration [31–33]

$$\iiint_\Omega \mathbf{\Sigma} : \delta\mathbf{d}d\Omega = \iiint_\Omega \mathbf{Q}\cdot\delta\mathbf{v}d\Omega + \iint_{S^\sigma} \mathbf{P}\cdot\delta\mathbf{v}dS, \tag{3}$$

where Ω is the current volume, S^σ is the surface on which the force $\mathbf{P}$ is applied, $\mathbf{Q}$ is the body force vector. The basic equation is obtained from the equation (3) using follow relations

$$\frac{d}{dt}\iiint_\Omega \mathbf{\Sigma} : \delta\mathbf{d}d\Omega = \iiint_\Omega \left[\dot{\mathbf{\Sigma}} : \delta\mathbf{d} + \mathbf{\Sigma} : \delta\dot{\mathbf{d}} + \frac{\dot{J}}{J}\mathbf{\Sigma} : \delta\mathbf{d}\right] d\Omega,$$

$$\frac{d}{dt}\iiint_\Omega \mathbf{Q}\cdot\delta\mathbf{v}d\Omega = \iiint_\Omega \left(\dot{\mathbf{Q}} + \mathbf{Q}\frac{\dot{J}}{J}\right)\cdot\delta\mathbf{v}d\Omega,$$

$$\frac{d}{dt}\iint_{S^\sigma} \mathbf{P}\cdot\delta\mathbf{v}dS = \iint_{S^\sigma}\left[\dot{\mathbf{P}} + \mathbf{P}\frac{\dot{J}}{J}\right]\cdot\delta\mathbf{v}dS,$$

where $\delta\dot{\mathbf{d}} = -\frac{1}{2}\left[\mathbf{h}^T\cdot\delta\mathbf{h}^T + \mathbf{h}\cdot\delta\mathbf{h}\right]$, $\frac{\dot{J}}{J} = 3d_0 = div\mathbf{v}$, where $\mathbf{v}$ – the velocity of a particle.

The von Mises yield criterion and radial return method are applied to find the current state [23, 25, 27]. Since the derived stress-strained state does not satisfy basic equation (3), then we carry out iteration procedure for increasing the accuracy current state. This procedure is based on the introduction of the "additional stresses" on virtual deformation of the rate. In this case the final equation will be

$$\begin{aligned}
&\iiint_{\Omega_l}\Big\{2G\,{}^l\mathbf{d}'^m : \delta^l\mathbf{d}' + 9K\,{}^l\mathbf{d}_0^m : \delta^l\mathbf{d}_0 + {}^l\mathbf{\Sigma} : \left[\delta^l\mathbf{d}\cdot{}^l\boldsymbol{\omega}^m - {}^l\boldsymbol{\omega}^m\cdot\delta^l\mathbf{d} + \delta^l\dot{\mathbf{d}}^m\right] + \\
&+\frac{{}^l\dot{j}^m}{{}^lJ^m}{}^l\mathbf{\Sigma} : \delta^l\mathbf{d}\Big\}d\Omega - \iiint_{\Omega_l}\frac{{}^l\dot{j}^m}{{}^lJ^m}{}^l\mathbf{Q}\cdot\delta^l\mathbf{v}d\Omega - \iint_{S_l^\sigma}{}^l\mathbf{P}\frac{{}^l\dot{j}^m}{{}^lJ^m}\cdot\delta^l\mathbf{v}dS = \\
&= \iiint_{\Omega_l}{}^l\dot{\mathbf{Q}}\cdot\delta^l\mathbf{v}d\Omega + \iint_{S_l^\sigma}{}^l\dot{\mathbf{P}}\cdot\delta^l\mathbf{v}dS - \\
&-\left\{\iiint_{\Omega_l}[{}^l\mathbf{\Sigma} : \delta^l\mathbf{d} - {}^l\mathbf{Q}\cdot\delta^l\mathbf{v}]d\Omega + \iint_{S_l^\sigma}{}^l\mathbf{P}\cdot\delta^l\mathbf{v}dS\right\}\Delta t^{-1} + \iiint_{\Omega_l}{}^l\mathbf{\Sigma}_\partial^m\cdot\delta^l\mathbf{d}d\Omega,
\end{aligned} \tag{4}$$

where m is the number of iteration, l is the number of step.

For quasi-static problems will take

$$^{l}\mathbf{v}^{m} = \frac{^{l+1}\mathbf{R}^{m} - {}^{l}\mathbf{R}}{\Delta t} = \frac{\Delta^{l}\mathbf{u}^{m}}{\Delta t}.$$

And let Δt=1. In the result we get from (4), the equations for the increments of displacements $\Delta^{l}\mathbf{u}^{m}$. After solving the general system of equations (4), we define tensors $^{l}\mathbf{d}^{m}$, $^{l}\boldsymbol{\omega}^{m}$, using the relations (1, 2) find the rate of Cauchy stress $^{l}\dot{\boldsymbol{\Sigma}}^{m}$ and trial stress tensor

$$^{l+1}\boldsymbol{\Sigma}_0 = {}^{l}\boldsymbol{\Sigma}_0 + {}^{l}\dot{\boldsymbol{\Sigma}}_0\Delta t,$$

$$\tilde{\boldsymbol{\Sigma}}'^{m} = {}^{l}\boldsymbol{\Sigma}' + {}^{l}\dot{\boldsymbol{\Sigma}}'^{m}\Delta t,$$

check the yield condition

$$\tilde{\sigma}_i^{m} = \sqrt{\frac{3}{2}\tilde{\sigma}_{ij}'^{m}\tilde{\sigma}_{ij}'^{m}} \le \sigma_{\mathrm{T}},$$

and if $\tilde{\sigma}_i^{m} > \sigma_{\mathrm{T}}$ the radial return method with an iterative refinement of the current mode of deformation is applied:

$$\boldsymbol{\Sigma}'^{m+1} = \tilde{\boldsymbol{\Sigma}}'^{m}\frac{\sigma_{\mathrm{T}}}{\tilde{\sigma}_i^{m}}, \quad \boldsymbol{\Sigma}_{\partial}^{m+1} = \boldsymbol{\Sigma}'^{m+1} - \tilde{\boldsymbol{\Sigma}}'^{m}. \tag{5}$$

Using relations (5) the general equation (4) is solved. If $\boldsymbol{\Sigma}_{\partial}^{m+1} \le \mathrm{tol}$ then the next configuration $^{l+1}y^{i} = {}^{l}y^{i} + \Delta^{l}u^{i}$ and stress state $^{l+1}\boldsymbol{\Sigma} = \boldsymbol{\Sigma}^{m}$ are defined.

5. Numerical example

The numerical computational is based on eight node isoparametric hexahedron element [21].

The problem of bending of a rectangular cross-section beam is considered. The beam is rigidly fixed on the one side, and on the other loaded with a bending moment. The length of the beam l=50 cm, height h=1 cm, width b=0.125 cm, E=1000 kg/cm^2, Poisson's ratio is=0. The value of the moment at which the beam is bent into a ring, was taken from the analytical solution of the $M=2\cdot\pi$kg cm (figure 1).

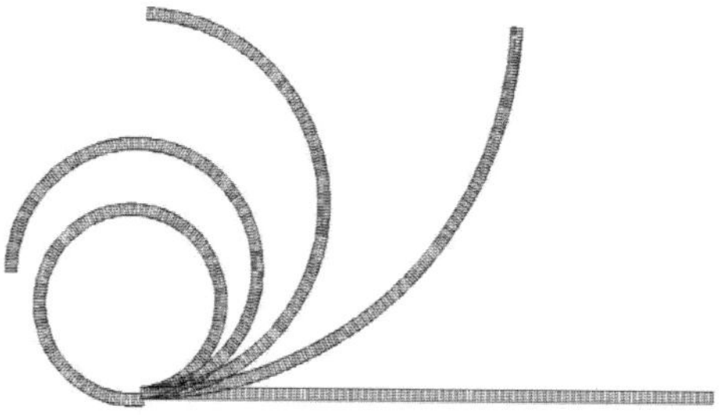

Figure 1. Different stages of the bending of a beam.

The problem of elastoplastic deformation rigidly clamped at both ends of the beam under the action of distributed load q=27 kg/cm^2. The length of the beam l=25 cm, height h=1 cm, width b=0.125 cm, E=20000 kg/ cm^2, $\mu=0$, $\sigma_{\mathrm{T}} = 750\,\mathrm{kg/cm^2}$ (figure 2, 3).

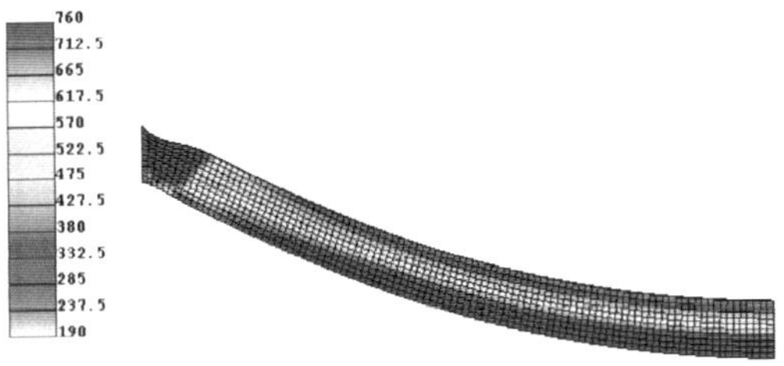

Figure 2. Equivalent stress on deformed structures.

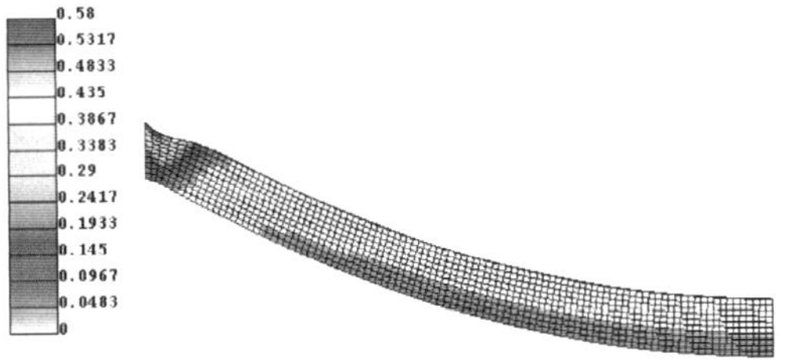

Figure 3. Equivalent plastic strain on deformed structures.

6. Conclusion

A method of numerical investigation of elastic-plastic solids with finite deformations is considered. For plasticity the von Mises criterion is applied. Incremental loading procedure is used, where allowing the variation equation is derived from the principle of virtual work in the current configuration. For the simulation of plastic deformation the radial return method with iterative refinement of the current stress-strain state is applied. The numerical discretization is based on the finite element method.

Acknowledgements

This work was supported by the Russian Science Foundation (project 16-11-10299).

References

[1] Eidel B and Gruttmann F 2003 *Computational Materials Science* **28** 732–742
[2] Taylor L M and Becher E B 1983 *Computer Methods in Applied Mechanics and Engineering* **41 (3)** 251–277
[3] *Yamada Y* 1982 *Computer Methods in Applied Mechanics and Engineering* **33 (1-3)** 417–437
[4] Eterovic A L and Bathe K –J 1990 *International Journal for Numerical Methods in Engineering* **30 (6)** 1099–1114
[5] Fish J and Sheek K 2000 *Computer Methods in Applied Mechanics and Engineering* **190** 75–93
[6] Simo J S Taylor R L and Pister K S 1985 *Computer Methods in Applied Mechanics and Engineering* **51** 177–208
[7] Kanok-Nukulchai W and Wong W K 1988 *Computers and Structures* **30 (4)** 967–974
[8] Gouveia B P P A, Rodrigues J M C and Martins P A F 1998 *Journal of Materials Processing Technology* **80-81** 647-652
[9] Badriev I B, Banderov V V, Garipova G Z, Makarov M V and Shagidullin R R 2015 *Applied Mathematical Sciences* **9 (82)** 4095–4102
[10] Badriev I B, Banderov V V, Makarov M V, Paimushin V N 2015 *Applied Mathematical Sciences* **9(82)** 3887-3895
[11] Badriev I B, Makarov M V and Paimushin V N 2016 *Procedia Engineering* **150(3)** 1050–1055
[12] Badriev I B, Garipova G Z, Makarov M V and Paymushin V N 2015 *Research Journal of Applied Sciences* **10(8)** 428–435
[13] Golovanov A I, Konoplev Yu G and Sultanov L U 2008 *Uchenye Zapiski Kazanskogo Universiteta. Seriya Fiziko-Matematicheskie Nauki* **150 (3)** 122–132
[14] Golovanov A I, Konoplev Yu G and Sultanov L U 2010 *Uchenye Zapiski Kazanskogo Universiteta. Seriya Fiziko-Matematicheskie Nauki* **151 (3)** 108–120
[15] Paimushin V N 2008 *Journal of Applied Mathematics and Mechanics* **72 (5)** 597–610
[16] Sachenkov O A, Mitryaikin V I, Zaitseva T A and Konoplev Yu G 2014 *Applied Mathematical Sciences* **8 (159)** 7889–7897
[17] Sachenkov O, Kharislamova L, Shamsutdinova N, Kirillova E and Konoplev Yu 2015 *IOP Conference Series: Materials Science and Engineering* **98**

[18] Sultanov L U 2015 *Procedia Earth and Planetary Science* **15** 119–124
[19] Shigapova F A, Mustakimova R F, Saleeva G T and Sachenkov O A *International Journal of Applied Engineering Research* **10 (24)** 44711–44714
[20] Sagdatullin M and Berezhnoi D 2014 *Applied Mathematical Sciences* **8 (35)** 1731–1738
[21] Sagdatullin M and Berezhnoi D 2014 *Applied Mathematical Sciences* **8 (60)** 2965–2972
[22] Bonet J and Wood R D 1997 *Nonlinear continuum mechanics for finite element analysis* (Cambridge University Press) 283
[23] Golovanov A I and Sultanov L U 2005 *International Applied Mechanics* **41 (6)** 614–620
[24] Golovanov A I and Sultanov L U 2008 *Russian Aeronautics* **51 (4)** 2008 362–368
[25] Davydov R L and Sultanov L U 2013 *PNRPU Mechanics Bulletin* **1** 81–93
[26] Sultanov L U 2014 *Applied Mathematical Sciences* **8 (143)** 7117–7124
[27] Sultanov L U and Davydov R L and 2014 *Applied Mathematical Sciences* **8 (60)** 2991–2996
[28] Abdrakhmanova A I and Sultanov L U 2016 *Materials Physics and Mechanics* **26 (1)** 30–32
[29] Davydov R L and Sultanov L U 2013 *Sixth International Conference on Nonlinear Mechanics (ICNM-VI)* 64–67
[30] Davydov R L and Sultanov L U 2015 *Journal of Engineering Physics and Thermophysics* **88 (5)** 1280–1288
[31] Sultanov L U and Fakhrutdinov L R 2013 *Magazine of Civil Engineering* **44 (9)** 69–74
[32] Sultanov L U and Davydov R L 2013 *Magazine of Civil Engineering* **44 (9)** 64–68
[33] Sultanov L U 2016 *Lobachevskii Journal of Mathematics* **37 (6)** 787–793

Iterative method for the numerical solution of a system of integral equations for the heat conduction initial boundary value problem

N N Svetushkov

Applied Mathematics and Physics Department, Moscow Aviation Institute (National Research University), Volokolamskoe shosse 4, Moscow, Russia.

E-mail: svt.n.n@mail.ru

Abstract. The paper deals with a numerical algorithm to reduce the overall system of integral equations describing the heat transfer process at any geometrically complex area (both two-dimensional and three-dimensional), to the iterative solution of a system of independent one-dimensional integral equations. This approach has been called "string method" and has been used to solve a number of applications, including the problem of the detonation wave front for the calculation of heat loads in pulse detonation engines. In this approach "the strings" are a set of limited segments parallel to the coordinate axes, into which the whole solving area is divided (similar to the way the strings are arranged in a tennis racket). Unlike other grid methods where often for finding solutions, the values of the desired function in the region located around a specific central point here in each iteration step is determined by the solution throughout the length of the one-dimensional "string", which connects the two end points and set them values and determine the temperature distribution along all the strings in the first step of an iterative procedure.

1. Introduction

Integral equations, on contrary to the differential, have a number of advantages, which primarily include the ability to solve them with the help of iterative algorithms and evaluate the accuracy of the approximate solution with the vector of errors or discrepancy. With regard to integral equations of heat conduction additional advantage is no need to coordinate continuity of temperature and heat flux at the interface between different media. Thus, the integral equation simplifies considerably the numerical problem definition temperature fields in geometrically complicated articles having non-uniform switching, which, among others, can be attributed to composite materials (prepregs). Modes of heat treatment of products from such materials greatly affect their distortion and ultimately structural strength, particularly for products spatially extended. In this regard, more relevant than the direct problem of determining the temperature fields in the blanks for a given external thermal effects, and the inverse problem - finding such thermal effects, which would ensure a more uniform curing and thus a more regular product structure thickness.

2. A general view of integral equations for the of heat conduction initial boundary value problem

IOP Conf. Series: Materials Science and Engineering **158** (2016) 012091 doi:10.1088/1757-899X/158/1/012091

The author has developed an approach that allowed to bring integral equations describing the process of unsteady heat conduction [1-5], as in the two-dimensional and three-dimensional cases for bodies of arbitrary geometry. Consider the conventional two-dimensional formulation of the non-stationary heat conduction problems with heat sources:

$$c\rho\partial u/\partial t=\partial/\partial x(k_x\partial u/\partial x)+ \partial/\partial y(k_y\partial u/\partial y)+q, \quad (1)$$

with initial and boundary conditions in the case when set boundary heat flows:

$$u(x,y,t)|_{t=0}= u_0,$$
$$W|_\Gamma = w_\Gamma, \quad (2)$$

where: $u=u(x,y,t)$ – the temperature function, $u_0=u_0(x,y)$ – the initial temperature function, $k_x= k_x(x,y,u)$ – the thermal conductivity along the axis x, $k_y=k_y(x,y,u)$ – the thermal conductivity along the axis y, $c=c(x,y,u)$ – the specific heat, $\rho=\rho(x,y,u)$ – the density, depending on the spatial variable and temperature and $q=q(x,y,t)$ – the distribution function of the heat sources (power density), w_Γ - external heat flux density, defined in every point of the boundary of solving area.

The figure below shows the solving area, the initial and boundary conditions:

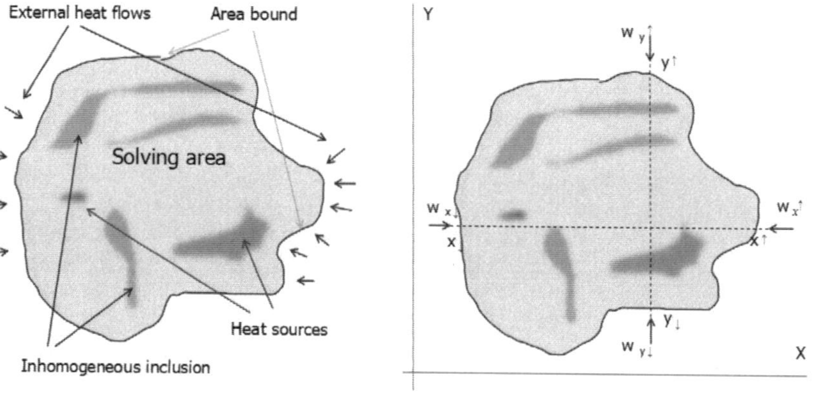

Figure 1. An a illustration of two-dimensional heat conduction initial-boundary value problem

Integral equations are based on the fact that instead of the unknown function of temperature are introduced new functions (two in case of two-dimensional and three when three-dimensional), which are physically meaningful spatial derivative of the heat flux for each of the coordinate axes:

$$S_x=[c\rho]^{-1}\ \partial/\partial x(k_x\partial u/\partial x) - f_y,$$
$$S_y=[c\rho]^{-1}\ \partial/\partial y(k_y\partial u/\partial y) - f_x. \quad (3)$$

For the two-dimensional case we have the following integral equation containing the unknown function as mentioned above:

$$u_0+\int_0 S_x dt+\int_0 S_y dt + \int_0\{ [\int_\downarrow{}^\uparrow c\rho dy]^{-1} \int_\downarrow{}^\uparrow c\rho S_y dy + [\int_\downarrow{}^\uparrow c\rho dy]^{-1} (w^\uparrow{}_y - w_{\downarrow y}) \}dt+\int_0 [c\rho]^{-1} q\, dt =$$
$$=0.5\{\int_\downarrow [k_x]^{-1} r^\uparrow{}_x - \int^\uparrow [k_x]^{-1} r^\uparrow{}_x\}\int_\downarrow c\rho S_x\, dxdx - 0.5\{\int_\downarrow [k_x]^{-1} r_{\downarrow x} - \int^\uparrow [k_x]^{-1} r_{\downarrow x}\} \int^\uparrow c\rho S_x\, dxdx +$$
$$+0.5 w_{\downarrow x}\{ \int_\downarrow [k_x]^{-1} r^\uparrow{}_x\, dx - \int^\uparrow [k_x]^{-1} r^\uparrow{}_x\, dx\} + 0.5 w^\uparrow{}_x \{ \int_\downarrow [k_x]^{-1} r_{\downarrow x}\, dx - \int^\uparrow [k_x]^{-1} r_{\downarrow x}\, dx\}; \quad (4)$$
$$u_0+\int_0 S_y dt+\int_0 S_x dt + \int_0\{ [\int_\downarrow{}^\uparrow c\rho dx]^{-1} \int_\downarrow{}^\uparrow c\rho S_x dx + [\int_\downarrow{}^\uparrow c\rho dx]^{-1} (w^\uparrow{}_x - w_{\downarrow x})\}dt + \int_0 [c\rho]^{-1} q\, dt =$$

$$=0.5\{\int_{\downarrow}[k_y]^{-1}r^{\uparrow}_{y}-\int^{\uparrow}[k_y]^{-1}r^{\uparrow}_{y}\}\int_{\downarrow}c\rho S_y\,dydy-0.5\{\int_{\downarrow}[k_y]^{-1}r_{\downarrow y}-\int^{\uparrow}[k_y]^{-1}r_{\downarrow y}\}\int^{\uparrow}c\rho S_y\,dydy+$$
$$+0.5w_{\downarrow y}\{\int_{\downarrow}[k_y]^{-1}r^{\uparrow}_{y}dy-\int^{\uparrow}[k_y]^{-1}r^{\uparrow}_{y}dy\}+0.5w^{\uparrow}_{y}\{\int_{\downarrow}[k_y]^{-1}r_{\downarrow y}\,dy-\int^{\uparrow}[k_y]^{-1}r_{\downarrow y}\,dy\};$$

where:

$$\begin{aligned} r^{\uparrow}_{x} &= [\int_{\downarrow}^{\uparrow}c\rho dx]^{-1}\int^{\uparrow}c\rho dx,\\ r_{\downarrow x} &= [\int_{\downarrow}^{\uparrow}c\rho dx]^{-1}\int_{\downarrow}c\rho dx,\\ r^{\uparrow}_{y} &= [\int_{\downarrow}^{\uparrow}c\rho dy]^{-1}\int^{\uparrow}c\rho dy,\\ r_{\downarrow y} &= [\int_{\downarrow}^{\uparrow}c\rho dy]^{-1}\int_{\downarrow}c\rho dy. \end{aligned} \tag{5}$$

There designations were introduced for a more transparent recording of integral equations:

$$\begin{aligned} &\int_0^t g(t)dt \equiv \int_0 g\,dt,\\ &\int_{\underline{x}}^{x} f(x)dx \equiv \int_{\downarrow} f\,dx,\\ &\int_{x}^{\bar{x}} f(x)dx \equiv \int^{\uparrow} f\,dx,\\ &\int_{\underline{x}}^{\bar{x}} f(x)dx \equiv \int^{\uparrow}_{\downarrow} f\,dx. \end{aligned} \tag{6}$$

After solving the resulting system of integral equations and determining the unknown functions S_x and S_y, the temperature distribution function can be calculated by simple integration:

$$u=u_0+\int_0 S_x dt+\int_0 S_y dt+\int_0 f_x dt+\int_0 f_y dt+\int_0 q/c\rho d, \tag{7}$$

where:

$$\begin{aligned} f_y &=[\int_{\downarrow}^{\uparrow}c\rho dx]^{-1}[\int_{\downarrow}^{\uparrow}c\rho S_x dx+(w^{\uparrow}_{x}-w_{\downarrow x})],\\ f_x &=[\int_{\downarrow}^{\uparrow}c\rho dy]^{-1}[\int_{\downarrow}^{\uparrow}c\rho S_y dy+(w^{\uparrow}_{y}-w_{\downarrow y})]. \end{aligned} \tag{8}$$

3. Principles of the string method in two dimension area

The string method was set up as an iterative procedure in order to find the solution of integral equations (4) through its opening. This approach allows solving (on each iteration step) integral set of heat conduction equations for one-dimensional problems.

Common views of the resulting set of «strings», describing the original problem in a two-dimensional formulation, are shown below in figure 2:

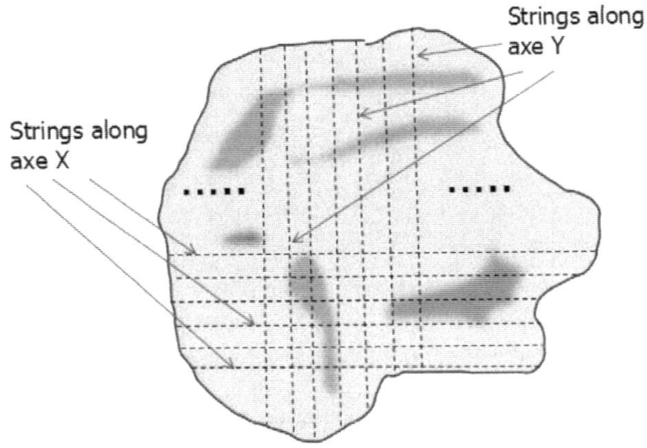

Figure 2. Strings along axes

In order to have clear principles of the method of strings, we rewrite the integral equation (4) in the form in which terms relating to "crosstalk" between these equations, moved to the right side:

$$
\begin{aligned}
&\int_0 S_x dt - 0.5\{\int_\downarrow [k_x]^{-1} r^\uparrow_x - \int^\uparrow [k_x]^{-1} r^\uparrow_x\}\int_\downarrow c\rho S_x\, dxdx + 0.5\{\int_\downarrow [k_x]^{-1} r_{\downarrow x} - \int^\uparrow [k_x]^{-1} r_{\downarrow x}\}\int^\uparrow c\rho S_x\, dxdx = \\
&= \boldsymbol{-\int_0 S_y dt - \int_0 [\int_\downarrow^\uparrow c\rho dy]^{-1} \int_\downarrow^\uparrow c\rho S_y dy dt - \int_0 [\int_\downarrow^\uparrow c\rho dy]^{-1} (w^\uparrow_y - w_{\downarrow y}) dt} + \\
&+ 0.5 w_{\downarrow x}\{\int_\downarrow [k_x]^{-1} r^\uparrow_x\, dx - \int^\uparrow [k_x]^{-1} r^\uparrow_x\, dx\} + 0.5 w^\uparrow_x\{\int_\downarrow [k_x]^{-1} r_{\downarrow x}\, dx - \int^\uparrow [k_x]^{-1} r_{\downarrow x}\, dx\} - \\
&- u_0 - \int_0 [c\rho]^{-1} q\, dt; \\
&\int_0 S_y dt - 0.5\{\int_\downarrow [k_y]^{-1} r^\uparrow_y - \int^\uparrow [k_y]^{-1} r^\uparrow_y\}\int_\downarrow c\rho S_y\, dydy + 0.5\{\int_\downarrow [k_y]^{-1} r_{\downarrow y} - \int^\uparrow [k_y]^{-1} r_{\downarrow y}\}\int^\uparrow c\rho S_y\, dydy = \\
&= \boldsymbol{-\int_0 S_x dt - \int_0 [\int_\downarrow^\uparrow c\rho dx]^{-1} \int_\downarrow^\uparrow c\rho S_x dx dt - \int_0 [\int_\downarrow^\uparrow c\rho dx]^{-1} (w^\uparrow_x - w_{\downarrow x}) dt} + \\
&+ 0.5 w_{\downarrow y}\{\int_\downarrow [k_y]^{-1} r^\uparrow_y dy - \int^\uparrow [k_y]^{-1} r^\uparrow_y dy\} + 0.5 w^\uparrow_y\{\int_\downarrow [k_y]^{-1} r_{\downarrow y}\, dy - \int^\uparrow [k_y]^{-1} r_{\downarrow y}\, dy\} - \\
&- u_0 - \int_0 [c\rho]^{-1} q\, dt;
\end{aligned}
\tag{9}
$$

If bold members in (9) are set up to zero, we obtain exactly the one-dimensional integral equation of heat conduction. Therefore, we can consider the iterative procedure, on each step assumed to be known (taken from the previous step) of S_y function (for the first equation of the system) and S_x (for the second equation), then the solution of (4) is reduced to the decision of an independent set of integral equations for dimensional problems with a new right-hand side.

4. Iterative method of solving one-dimensional heat conduction integral equations

Solution of one-dimensional integral equations also preferably performed using iterative procedures for the same reasons - the limited properties of the integral operator, which allows to evaluate the accuracy of the approximate solution. The best known of the methods used are the method of successive approximations and the method of steepest descent. The main drawback of these methods is that often the first few iterations of these methods give a good approximation, and their further use or lead to divergence or convergence to substantially zero iterative procedure. This is due to the fact that these methods work well only in the case of systems with diagonally dominant. The proposed method for solving the one-dimensional integral equation is based on the summation of a geometric progression, and can be used for solving systems for which there is no diagonal predominance.

Let us consider a one-dimensional integral equation of heat conduction in the form of an operator equation in a Hilbert space:

$$
\begin{gathered}
H = H([x_\downarrow, x^\uparrow] \mathrm{x} [0, T]) \\
(f, g) = \int_H fg dw \\
\|f\|_H^2 = (f, f);
\end{gathered}
\tag{10}
$$

and:

$$Ax = b \tag{11}$$

where A is a bounded operator on the entire Hilbert space H:

$$A: H \rightarrow H,\ \|A\|_H = C < \infty. \tag{12}$$

Definition. The linear system (11) with the vector of the right side of b is called the fixed-sign, if all the elements of vectors $b, Ab, \ldots A^n b, \ldots$ have the same sign:

$$sign\ A^n b = sign\ b,\ \ n = 0, 1, 2, \ldots \tag{13}$$

Theorem. If the linear system is a fixed sign, its solution can be found as the sum of a convergent series of geometric progression:

$$x = \lambda b + (E - \lambda A)\lambda b + (E - \lambda A)^2 \lambda b + .. + (E - \lambda A)^n \lambda b + ... \quad (14)$$

Note that if b is an eigenvector of A with eigenvalue m, then we get a common geometric progression, which can ensure the convergence selection parameter smaller than $\lambda < 1/m$. Therefore, the proof is based on the decomposition of the vector on the right part of the eigenvectors A. In order to ensure the convergence of exponentially as convergence parameter selected value of $\lambda = 1/m_{max}$, where m_{max} maximum eigenvalue of integral operator. Due to the limited conditions of the operator and fixed sign, the maximum eigenvalue of a positive and of course: $0 < m_{max} < ||A||_H$

5. Conclusions

Description of initial boundary value problem of heat conduction in the form of integral equations to solve them allows you to use iterative algorithms, which on one hand can effectively parallelize the calculation process, but on the other - to control the accuracy of the calculations for the error function (residual). The proposed approach can be used to solve a variety of problems in which the accuracy of the calculation is a critical condition for the calculations - such as the modeling of heat treatment processes, products from metals and alloys, forecast heat load on the combustion chamber wall of aircraft and rocket engines, and others.

References

[1] Svetushkov N N 2014 Metod strun v zadachah mnogomernoj nestacionarnoj teploprovodnosti *Informacionnye tehnologii* (Moskva: Novye tehnologii №12) pp 14-19

[2] Svetushkov N N 2015 Integral approach for the numerical modeling quenching process of forming rolls *CHT-15: 6th International Symposium on Advances in Computational Heat Transfer* (USA: Rutgers University) ID 259

[3] Svetushkov N N 2015 Uravnenija parabolicheskogo tipa dlja modelirovanija techenij i teplovyh nagruzok v uslovijah detonacionnogo gorenija *Materialy XIX mezhdunarodnoj konferencii po vychislitel'noj mehanike i sovremennym prikladnym programmnym sistemam (VMSPPS'2015)* (Krym: Alushta) pp 106-107

[4] Svetushkov N.N. 2016 Uproshhennaja matematicheskaja model' rascheta temperaturnyh polej pri detonacionnom gorenii *Informacionnye tehnologii* (Moskva: Novye tehnologii t **21** №9)

[5] Svetushkov N.N. 2016 Primenenie integral'nogo metoda strun dlja rascheta teplovyh nagruzok na ohlazhdaemye stenki soplovoj chasti raketnogo dvigatelja *Informacionnye tehnologii* (Moskva: Novye tehnologii t 22 №1) pp 32-36

IOP Conf. Series: Materials Science and Engineering **158** (2016) 012092 doi:10.1088/1757-899X/158/1/012092

On the existence of solutions of one nonlinear boundary-value problem for shallow shells of Timoshenko type with simply supported edges

S N Timergaliev[1] **and L S Kharasova**[2]

[1]Department of higher mathematics KGASU, Republic of Tatarstan, Kazan, Zelenaya street, 1, Russia, 420043
[2]Department of mathematics Naber,ezhnye Chelny Institute, Kazan Federal University, Republic of Tatarstan, Naberezhnye Chelny, 68/19, Russia, 423812

E-mail: kharasova.liya@mail.ru

Abstract. Solvability of one system of nonlinear second order partial differential equations with given initial conditions is considered in an arbitrary field. Reduction of the initial system of equations to one nonlinear operator equation is used to study the problem. The solvability is established with the use of the principle of contracting mappings. The method used in these studies is based on the integral representations for the displacements. These representations are constructed with the use of general solutions to the inhomogeneous Cauchy-Riemann equation.

1. Introduction

Let us introduce in the plane bounded domain Ω and consider a system of nonlinear differential equations in the form

$$\begin{gathered}
w_{1\alpha^1\alpha^1} + \mu_1 w_{1\alpha^2\alpha^2} + \mu_2 w_{2\alpha^1\alpha^2} = f_1, \\
\mu_1 w_{2\alpha^1\alpha^1} + w_{2\alpha^2\alpha^2} + \mu_2 w_{1\alpha^1\alpha^2} = f_2, \\
k^2\mu_1\left(w_{3\alpha^1\alpha^1} + w_{3\alpha^2\alpha^2} + \psi_{1\alpha^1} + \psi_{2\alpha^2}\right) + k_3 w_{1\alpha^1} + k_4 w_{2\alpha^2} - k_5 w_3 + \\
+ k_3 w_{3\alpha^1}^2/2 + k_4 w_{3\alpha^2}^2/2 + \beta_2\left[\left(T^{\lambda\mu} w_{3\alpha^\lambda}\right)_{\alpha^\mu} + R^3\right] = 0 \\
\psi_{1\alpha^1\alpha^1} + \mu_1\psi_{1\alpha^2\alpha^2} + \mu_2\psi_{2\alpha^1\alpha^2} = g_1 + k_0\psi_1, \\
\mu_1\psi_{2\alpha^1\alpha^1} + \psi_{2\alpha^2\alpha^2} + \mu_2\psi_{1\alpha^1\alpha^2} = g_2 + k_0\psi_2,
\end{gathered} \tag{1}$$

under the following conditions at the boundary Γ:

$$w_1 = w_3 = \psi_1 = 0, \tag{2}$$

$$\mu_1\left(w_{1\alpha^2} + w_{2\alpha^1}\right)(t)\,d\alpha^2/ds - \left(\mu w_{1\alpha^1} + w_{2\alpha^2}\right)(t)\,d\alpha^1/ds = \varphi_1(w_3)(t), \tag{3}$$

$$\mu_1\left(\psi_{1\alpha^2} + \psi_{2\alpha^1}\right)(t)\,d\alpha^2/ds - \left(\mu\psi_{1\alpha^1} + \psi_{2\alpha^2}\right)(t)\,d\alpha^1/ds = \varphi_2(t). \tag{4}$$

In (1)–(4) the following notations are used:

$$f_j \equiv f_j(w_3) = k_{j+2} w_{3\alpha^j} - w_{3\alpha^j} w_{3\alpha^j\alpha^j} - \mu_2 w_{3\alpha^{3-j}} w_{3\alpha^1\alpha^2} - \mu_1 w_{3\alpha^j} w_{3\alpha^{3-j}\alpha^{3-j}} - \beta_2 R^j,$$

$$g_j \equiv g_j(w_3) = k_0 w_{3\alpha^j} - \beta_1 L^j, j = 1,2, \mu_1 = (1-\mu)/2, \mu_2 = (1+\mu)/2,$$
$$\varphi_1(w_3)(t) = \beta_2 P^1(s) + \left[\mu w^2_{3\alpha^1}(t)/2 + w^2_{3\alpha^2}(t)/2\right] d\alpha^2/ds - \mu_1 w_{3\alpha^1}(t) w_{3\alpha^2}(t) d\alpha^1/ds, \quad (5)$$
$$\varphi_2(t) = \beta_1 N^2(s), t = t(s) = \alpha^1(s) + i\alpha^1(s) \in \Gamma, k_3 = k_1 + \mu k_2, k_4 = k_2 + \mu k_1,$$
$$k_5 = k_1^2 + k_2^2 + 2\mu k_1 k_2, k_0 = 6k^2(1-\mu)/h^2, \beta_1 = 12(1-\mu^2)/(h^3 E), \beta_2 = (1-\mu^2)/(Eh).$$

The system (1) together with the boundary conditions (2)–(4) describes the state of equilibrium isotropic elastic homogeneous shell with simply supported edges within the framework of Timoshenko shear model [1, pp. 168-170, 269]. Here $T^{\lambda\mu}$ are stresses $(\lambda, \mu = \overline{1,3})$; $w_j (j = 1,2)$ and w_3 are tangential and normal displacements of the points of S_0; $\psi_i (i = 1,2)$ are rotation angles of normal cross-section of S_0; $R^j(j = \overline{1,3}), L^k(k = 1,2), N^2, P^2$ are components of the external forces acting on the shell; $\mu = const$ is the Poisson coefficient, $E = const$ is Young's modulus, $k_1, k_2 = const$ are principal curvatures; $k^2 = const$ is the shear coefficient; $h = const$ is the shell width; α^1, α^2 are the Cartesian coordinates of the points in the domain Ω.

Problem (1)-(4). Find a solution to system (1) under boundary conditions (2)-(4).

There are a number of works devoted to the solvability of nonlinear problems in the framework of the Timoshenko displacement model [2-7]. For this purpose the theory of problems Rimann-Hilbert for holomorphic functions in the unit circle is used. Therefore, the field assumed from the beginning the unit circle [2-6], or conformal mappings on the unit circle [7]. At the present time, on the unit circle existence theorems of solutions of nonlinear problems for Timoshenko-type shell with rigidly clamped edges [2], with free edges [3] and with simply supported edges [4-6] are obtained. In [7] the system (1) is studied for shells of Timoshenko type with free edges in an arbitrary field Ω. The method of works [3], [4], [7] is developing on the case of arbitrary elastic shell with simply supported edges in this paper.

Consider boundary-value problem (1)-(4) in a generalized formulation. Let the following conditions hold true: (a) Ω is a simply connected domain with the boundary $\Gamma \in C^1_\beta$; (б) external forces $R^i(i = \overline{1,3}), L^k(k = 1,2) \in L_p(\Omega)$, $N^2, P^2 \in C_\beta(\Gamma)$; in what follows $p > 2, 0 < \beta < 1$.

Definition. The vector of generalized displacements $a = (w_1, w_2, w_3, \psi_1, \psi_2) \in W^{(2)}_p(\Omega), p > 2,$ is a generalized solution to the problem (1)-(4) if the vector satisfies almost everywhere the equations of system (1) and it satisfies boundary conditions (2)–(4) in pointwise fashion.

Here $W^{(2)}_p(\Omega)$ is a Sobolev space. Let us note that due to embedding theorems for Sobolev spaces $W^{(2)}_p(\Omega)$ with $p > 2$, the generalized solution α belongs to $C^1_\beta(\overline{\Omega})$. In what follos $\alpha = (p-2)/p$.

2. Solution to problem (1)-(4) with respect to tangential displacements and angles of rotation

Let us consider the first two equations in (1) and initially assume that w_3 is fixed. In terms of the complex function $\omega = w_{1\alpha^1} + w_{2\alpha^2} + i\mu_1(w_{2\alpha^1} - w_{1\alpha^2})$ these equations can be represented in the form

$$\omega_{\bar z} = f, \quad (6)$$

where $f = (f_1 + f_2)/2, \omega_{\bar z} = (\omega_{\alpha^1} + i\omega_{\alpha^2})/2, z = \alpha^1 + i\alpha^2.$

Equation (6) is an inhomogeneous Cauchy–Riemann equation. It has general solution [8]:

$$\omega(z) = \Phi_1(z) + Tf(z), \quad Tf = -\frac{1}{\pi}\iint_\Omega \frac{f(\zeta)}{\zeta - z} d\xi d\eta, \zeta = \xi + i\eta, \quad (7)$$

where $\Phi_1(z)$ is an arbitrary holomorphic function that belongs to the space $C_\alpha(\overline{\Omega})$.

It is well-known [8,pp.41,53], that Tf is a completely continuous operator which acts in $L_p(\Omega), p>2,\ C_\alpha^k(\overline{\Omega})$. It maps these spaces into $C_\alpha(\overline{\Omega})$ and $C_\alpha^{k+1}(\overline{\Omega})$, respectively. Besides, there exist the generalized derivatives [8,pp.33-34,53-67]

$$\frac{\partial Tf}{\partial \bar{z}} = f, \frac{\partial Tf}{\partial z} \equiv Sf = -\frac{1}{\pi}\iint_\Omega \frac{f(\zeta)}{(\zeta - z)^2} d\xi d\eta, \tag{8}$$

where the integral exists in the principal value sense of Cauchy (almost everywhere when $f \in L_p(\Omega), p>1$) and Sf is a linear bounded operator in $L_p(\Omega),\ C_\alpha^k(\overline{\Omega})$.

With the function $\omega_0(z) = w_2 + iw_1$ relation (7) can be also rewritten in the form of an inhomogeneous Cauchy–Riemann equation

$$\omega_{0\bar{z}} = i(d_1\omega + d_2\overline{\omega}) \equiv id[\omega], d_j = (\mu_1 + (-1)^j)/(4\mu_1), j = 1,2, \tag{9}$$

The general solution of this equation is

$$\omega_0(z) = \Phi_2(z) + iTd[\Phi_1 + Tf](z) \tag{10}$$

where $\Phi_2(z)$ is an arbitrary holomorphic function of the class $C_\alpha^1(\overline{\Omega})$.

Thus, for fixed w_3 the general solution of the two first equations (1) is of the form (10) and contains two arbitrary holomorphic functions $\Phi_j(z), j = 1,2$. We define these functions so that tangential displacements w_1, w_2 will satisfy boundary conditions (2), (3). First, we find $\Phi_2(z)$ from the condition $w_1 = 0$ on Γ. We have a Rimann-Hilbert problem for the holomorphic function $\Phi_2(z)$ with the boundary condition

$$\mathrm{Re}[i\Phi_2(t)] = \mathrm{Re}\, Td[\omega](t), t \in \Gamma. \tag{11}$$

Let $z = \varphi(\zeta)$ is conformal mapping of the unit disk $\overline{K}: |\zeta| \le 1$ to the area $\overline{\Omega}$. It is known that if condition (a) a function $\varphi(\zeta)$ belongs to the space $C_\beta^1(\overline{K})$, [8, p.25]. Under the conditions (11) will hold a replacement $t \to \varphi(t)$, $\Phi_2(\varphi(t)) \to \Phi_2(t)$, leaving for new variables to the previous notation. As a result, we have a Rimann-Hilbert problem for the holomorphic function $\Phi_2(z)$ of the unit disk K with the boundary condition

$$\mathrm{Re}[i\Phi_2(t)] = \mathrm{Re}\, Td[\omega](\varphi(t)), t \in \partial K: |t| = 1. \tag{12}$$

Then the solution of the Riemann-Hilbert problem (12) has the form [9, p.253]

$$\Phi_2(z) = -\frac{1}{2\pi}\int_{\partial K} \mathrm{Re}\, Td[\Phi_1 + Tf](\varphi(t))\frac{t+z}{t-z}\frac{dt}{t} + c_0, z \in \overline{K}, \tag{13}$$

where c_0 is a arbitrary real constant.

We differentiate relation (13) with respect to z, we find

$$\Phi_2'(z) = -\frac{1}{\pi\varphi'(z)}\int_{\partial K} \mathrm{Re}\, Td[\Phi_1 + Tf](\varphi(t))\frac{dt}{(t-z)^2}. \tag{14}$$

We substitute relations for the tangential displacements w_1, w_2 from (10) into (3). Hence, boundary conditions (3) take the form

$$\mathrm{Re}\{t'\Phi(t)\} = h(t), t' = dt/ds, t \in \Gamma, \tag{15}$$

where

$$h(t) = l(w_3)(t) + \mathrm{Re}\{t' Sd[\Phi_1]^+(t)\} - \mathrm{Re}\{\mu_3 \bar{t}'\Phi_1(t)\}/2, \mu_3 = (1+\mu)/(2(1-\mu)), \tag{16}$$

$$l(w_3)(t) = \frac{\varphi_1(w_3)(t)}{\mu - 1} + \mathrm{Re}\{t' Sd[Tf]^+(t) - \mu_3\, d\alpha^1/ds\, \mathrm{Re}\, Tf(t)\} \equiv l[f(w_3);\varphi_1(w_3)].$$

Via the $Sd[\Phi_1]^+(t)$ means the limit of the function $Sd[\Phi_1](z)$ as $z \to t \in \Gamma$ from the interior of the domain Ω. $\Phi(t)$ the boundary value of holomorphic functions in Ω

$$\Phi(z) = i\Phi_2'(z) + \mu_3\,\Phi_1(z)/2 \tag{17}$$

Thus, for the $\Phi(z)$ we have a Rimann-Hilbert problem with the boundary condition (15) in an arbitrary field Ω. Using conformal mapping, we have a Rimann-Hilbert problem for the holomorphic function $\Phi(z)\varphi'(z)$ in the unit disk K with the boundary condition

$$\mathrm{Re}[t'\varphi'(t)\Phi(t)] = h(\varphi(t))|\varphi'(t)|, t \in \partial K : |t| = 1, \tag{18}$$

where $t' = dt/d\sigma, d\sigma$ is part of an arc of a circle ∂K.

The index of problem (18) equals -1. Therefore, the solution of this problem is [9, p.253]

$$\Phi(z) = -\frac{1}{\pi\varphi'(z)}\int_{\partial K}\frac{h(\varphi(t))|\varphi'(t)|}{t-z}\frac{dt}{t}, z \in \overline{K}, \tag{19}$$

and the solvability condition

$$\int_{\partial K}\frac{h(\varphi(t))|\varphi'(t)|}{t}dt = 0 \tag{20}$$

of problem should be fulfilled. Which can be converted to the form

$$\int_{\Gamma} P^2(s)ds + \iint_{\Omega} R^2 d\alpha^1 d\alpha^2 = 0. \tag{21}$$

We find $\Phi_1(z)$ from the relation (17). For this purpose $\Phi(z)$ and $\Phi_2'(z)$ replace it with expressions from (19) and (14). Taking into consideration $h(\varphi(t))$ in (16) and equality

$$Sd[\Phi_1]^+(\varphi(t)) = \frac{1}{2}\bar{t}^2\frac{\overline{\varphi'(t)}}{\varphi'(t)}d[\Phi_1(t)] +$$

$$+\frac{d_1}{2\pi i}\bar{t}\int_{\partial K}\frac{\overline{\varphi_0(\tau,t)}}{[\varphi_0(\tau,t)]^2}\frac{\varphi'(\tau)\bar{\tau}}{\tau - t}\Phi_1(\tau)d\tau + \frac{d_2}{2\pi i}\int_{\partial K}\frac{\overline{\varphi'(\tau)}}{\varphi_0(\tau,t)}\frac{\bar{\tau}^2}{\tau - t}\overline{\Phi_1(\tau)}d\tau, \tag{22}$$

where $\varphi_0(\tau,t) = [\varphi(\tau) - \varphi(t)]/(\tau - t)$ to $\tau \neq t$ and $\varphi_0(t,t) = \varphi'(t)$ to $\tau = t$, $d[\Phi_1(t)], d_j$ are defined in (9), using the Cauchy formula for $\Phi_1(z)$ and (4.7), (8.8a) from [8, p.28, 55], we have

$$\Phi_1(z) = \Phi_1[l(w_3)](z), z \in K, \tag{23}$$

where

$$\Phi_1[l(w_3)](z) = 2(\mu - 1)S_{\partial K}(\mathrm{Re}\,Td[Tf](\varphi(\tau)))(z) + \frac{(\mu - 1)}{\pi}\int_{\partial K}\frac{l(w_3)(\varphi(t))|\varphi'(t)|}{t(t-z)}dt, \tag{24}$$

$$S_{\partial K}f(z) = \frac{1}{2\pi i}\int_{\partial K}\frac{f(t)}{(t-z)^2}dt,$$

$l(w_3)$ is defined in (16).

Let $\zeta = \psi(z)$ reverse function to $z = \varphi(\zeta)$. It is known that $\psi(z) \in C^1_\beta(\overline{\Omega})$ [8, p.25]. Therefore, $\Phi_1(\psi(z)) \in W^{(1)}_p(\Omega)$, $2 < p < 2/(1-\beta)$.

We substitute expression (23) into (13) to obtain

$$\Phi_2(\psi(z)) = \Phi_2[l(w_3)](\psi(z)) + c_0, z \in \Omega, \tag{25}$$

$$\Phi_2[l(w_3)](\psi(z)) = -\frac{1}{2\pi}\int_{\partial K}(\mathrm{Re}\,Td[\Phi_1[l(w_3)]](t) + \mathrm{Re}\,Td[Tf](t))\frac{t+\psi(z)}{t-\psi(z)}\frac{dt}{t}.$$

Consider tangential displacements w_1 and w_2 that satisfy the first two equations (1) and conditions (2), (3). Upon substituting (23), (25) into (10) and assuming that condition (21) is true, we obtain

$$\omega_0(z) = H_0 w_3(z) + c_0, z \in \Omega, \tag{26}$$

$$H_0 w_3(z) \equiv H_0[f(w_3); l(w_3)](z) = \Phi_2[l(w_3)](\psi(z)) + iTd[\Phi_1[l(w_3)](\psi(\zeta)) + Tf(w_3)(\zeta)](z).$$

We now turn to functions ψ_1, ψ_2 in the last two equations (1). These functions should satisfy boundary conditions (2), (4).

Let us note that the structure of left-hand sides in the last two equations (1) coincides with the structure of left-hand sides in boundary conditions (2) and (4). Relations for tangential displacements differ only in the right-hand sides. Therefore at fixed right-hand sides for rotation angles we obtain

$$\psi = \psi_2 + i\psi_1 = H_0[g + \tilde{\psi}; l[g + \tilde{\psi}; \varphi_2]] + c_1, \tag{27}$$

$$g \equiv g(w_3) = (g_1 + ig_2)/2, \tilde{\psi} = k_0(\psi_1 + i\psi_2)/2,$$

where $g_j (j = 1,2)$ are defined in (5), $H_0[f; g]$ is defined in (26), $l[g + \tilde{\psi}; \varphi_2]$ is defined in (16); c_1 an arbitrary real constant.

As this takes place, the condition of solvability, similar to (20), can be reduced to the form

$$\beta_1\left(\int_\Gamma N^2(s)ds + \iint_\Omega L^2 d\alpha^1 d\alpha^2\right) - k_0 \iint_\Omega \psi_2 d\alpha^1 d\alpha^2 = 0, \tag{28}$$

where N^2, L^2 are components of external load.

Since $H_0[g + \tilde{\psi}; l[g + \tilde{\psi}; \varphi_2]] = H_0[g; \varphi_2] + H_0[\tilde{\psi}; 0]$, then (27) can be written in the form

$$\psi - K_0\psi = H_0[g; \varphi_2] + c_1, K_0\psi = H_0[\tilde{\psi}; 0]. \tag{29}$$

It is obvious that $K_0\psi$ are linear completely continuous operators in $W_p^{(2)}(\Omega)$, $2 < p < 2/(1-\beta)$. Let us show that the homogeneous equation $\psi - K_0\psi = 0$ has only the trivial solution in $W_p^{(2)}(\Omega)$, $2 < p < 2/(1-\beta)$. Suppose the contrary: $\psi = \psi_2 + i\psi_1 \in W_p^{(2)}(\Omega)$, $2 < p < 2/(1-\beta)$, is a nonzero solution satisfying condition (28) with $N^2 = L^2 = 0$. Obviously, the function $\psi = \psi_2 + i\psi_1$ satisfies the last two equations in system (1) with $g_1 = g_2 = 0$, the homogeneous boundary conditions $\psi_1 = 0$, and conditions (4) with $\varphi_2(t) = 0$. We multiply the last two relations in (1) by ψ_1 and ψ_2, respectively, integrate the resulting relations over the domain Ω, and add them. In view of the boundary conditions, we obtain the relation

$$\iint_\Omega \{\mu_1(\psi_{1\alpha^2} + \psi_{2\alpha^1})^2 + \mu(\psi_{1\alpha^1} + \psi_{2\alpha^2})^2 + (1-\mu)(\psi^2{}_{1\alpha^1} + \psi^2{}_{2\alpha^2}) + k_0(\psi_1^2 + \psi_2^2)\} d\alpha^1 d\alpha^2 = 0,$$

which implies that $\psi_1 = \psi_2 = 0$ in $\overline{\Omega}$.. Then there exists an inverse operator $(I-K)^{-1}$, which is bounded in $W_p^{(2)}(\Omega)$, $2 < p < 2/(1-\beta)$, and whose application to Eq. (29) gives the relation

$$\psi = (I-K)^{-1}(H_0[g; \varphi_2] + c_1). \tag{30}$$

Note that the function $\psi_* = (I-K)^{-1}c_1$ satisfies the last two equations in system (1) with $g_1 = g_2 = 0$ and the homogeneous boundary conditions, of the above. Therefore, $\psi_* = 0$ in $\overline{\Omega}$. Then from Eq. (30), we derive the unique representation via the deflection for the rotation angles,

$$\psi \equiv \psi(w_3) = (I-K)^{-1} H_0[g(w_3); \varphi_2]. \tag{31}$$

Condition (28) for the rotation angles (31) is identically satisfied. To justify this fact, it suffices to substitute the expression (31) into the last equation in system (1) and integrate the resulting relation over Ω with regard to the boundary condition (4).

3. Reduction of system (1) to a single equation for the deflection and its study

The functions w_1, w_2, ψ_1, ψ_2 included in the third equation of the system (1), we replace them with expressions from (26), (31). As a result, we obtain a nonlinear second-order partial differential equation for the deflection,

$$w_{3\alpha^1\alpha^1} + w_{3\alpha^2\alpha^2} + K_1 w_3 + G_1 w_3 = 0,$$

which is equivalent to the equation

$$w_3 + Kw_3 + Gw_3 = 0, \tag{32}$$

where

$$Kw_3 = \iint_\Omega H(\varsigma, z) K_1 w_3(\varsigma) d\xi d\eta, Gw_3 = \iint_\Omega H(\varsigma, z) G_1 w_3(\varsigma) d\xi d\eta; \tag{33}$$

$H(\varsigma, z)$ is the harmonic Green function of the Dirichlet problem for the domain Ω, $K_1 w_3$ is a linear compact operator and $G_1 w_3$ is a nonlinear bounded operator from $W_p^{(2)}(\Omega)$ into $L_p(\Omega)$, $2 < p < 2/(1-\beta)$.

Then from (33) we find that Kw_3 is a linear compact operator and Gw_3 is a nonlinear bounded operator in $W_p^{(2)}(\Omega)$, $2 < p < 2/(1-\beta)$; further , the estimate

$$\left\| G_* w_3^1 - G_* w_3^2 \right\|_{W_p^{(2)}(\Omega)} \le c[q_0 + (1+r)r] \left\| w_3^1 - w_3^2 \right\|_{W_p^{(2)}(\Omega)}, \tag{34}$$

$$q_0 = \sum_{\lambda,\mu=1}^{2} \left\| T^{\lambda\mu}(0) \right\|_{C(\overline{\Omega})} + \sum_{\lambda=1}^{2} \left\| R^\lambda \right\|_{L_p(\Omega)},$$

$$T^{\lambda\mu}(0) \equiv T^{\lambda\mu}(a(0)), \quad a(0) = (w_1(0), w_2(0), 0, \psi_1(0), \psi_2(0))$$

holds for arbitrary $w_3^j (j = 1,2) \in W_p^{(2)}(\Omega)$ that belong to the ball $\|w_3\|_{W_p^{(2)}} < r$.

Let us show that the equation

$$w_3 + Kw_3 = 0 \tag{35}$$

has only the trivial solution in $W_p^{(2)}(\Omega)$, $2 < p < 2/(1-\beta)$. Let $w_3 \in W_p^{(2)}(\Omega)$, $2 < p < 2/(1-\beta)$, is a nonzero solution of Eq. (35). By relations (26), (31), to this solution, there correspond tangential displacements $w_j(w_3)$ and rotation angles $\psi_j(w_3), j = 1,2$. They satisfying the system (1) with $R_1 = R_2 = R_3 = L_1 = L_2 = 0$, in which the nonlinear terms do not exist, and the homogeneous boundary conditions (2), (3) with $\varphi_1(t) = 0$ and (4) with $\varphi_2(t) = 0$. The each of identities in (1) we multiply by $w_1, w_2, w_3, \psi_1, \psi_2$, respectively, integrate over the domain Ω, and sum the resulting relations. Then, by integrating by parts in the resulting relations, together with boundary conditions, imply that $w_3 = 0$ in $\overline{\Omega}$. Consequently, Eq. (35) has only the zero solution in $W_p^{(2)}(\Omega)$, $2 < p < 2/(1-\beta)$. Then there exists an inverse operator $(I+K)^{-1}$ bounded in $W_p^{(2)}(\Omega)$, $2 < p < 2/(1-\beta)$, which permits one to reduce Eq. (32) to the equivalent equation

$$w_3 + G_* w_3 = 0, \quad G_* w_3 = (I+K)^{-1} Gw_3, \tag{36}$$

where $G_* w_3$ is a nonlinear bounded operator in $W_p^{(2)}(\Omega)$, $2 < p < 2/(1-\beta)$; moreover, the estimate $\left\| G_* w_3^1 - G_* w_3^2 \right\|_{W_p^{(2)}(\Omega)} \le q_* \left\| w_3^1 - w_3^2 \right\|_{W_p^{(2)}(\Omega)}$, $q_* = c\left\| (I+K)^{-1} \right\|_{W_p^{(2)}(\Omega)} [q_0 + (1+r)r]$ holds for arbitrary $w_3^j (j = 1,2) \in W_p^{(2)}(\Omega)$ that lie in the ball $\|w_3\|_{W_p^{(2)}} < r$.

Suppose that the radius r of the ball and the external forces acting on the shell satisfy the conditions

$$q_* < 1, \quad \left\| G_*(0) \right\|_{W_p^{(2)}(\Omega)} < (1 - q_*) r. \tag{37}$$

Under these conditions, for Eq. (36), one can use the contraction mapping principle [10, p. 146]. The fair following the main

Theorem. Let conditions (a), (б) in Section 1 be fulfilled and inequality (37) holds. Then condition (21) is necessary and sufficient for the solvability of the geometrically nonlinear equilibrium problem for shallow elastic isotropic homogeneous shells of the Timoshenko type under the boundary conditions (2)–(4). Then the problem has generalized solution $a = (w_1, w_2, w_3, \psi_1, \psi_2) \in W_p^{(2)}(\Omega)$, $2 < p < 2/(1 - \beta)$. Components w_1, w_3, ψ_1, ψ_2 are uniquely defined and component w_2 depends on constant.

References

[1] Galimov K Z 1975 *Principles of the Nonlinear Theory of Thin Shells* (Kazan: Kazan Univ. Press)
[2] Timergaliev S N 2012 *Differential Equations* (Moscow: Russian Academy of Sciences) **48 (3)** pp 458-463
[3] Timergaliev S N 2014 *Russian Mathematics* (Kazan: Kazan Federal University) **58 (3)** pp 31–46
[4] Timergaliev S N, Uglov A N and Kharasova L S 2015 *Russian Mathematics* (Kazan: Kazan Federal University) **59 (5)** pp 41–51
[5] Ahmadiev M G, Timergaliev S N and Kharasova L S 2016 *Journal of Siberian Federal University Mathematics & Physics* (Krasnoyarsk: Siberian Federal University) 9 (2) pp 131-143
[6] Timergaliev S N and Kharasova L S 2016 *Differential Equations* (Moscow: Russian Academy of Sciences) **52 (5)** pp 651-664
[7] Timergaliev S N 2015 *Differential Equations* (Moscow: Russian Academy of Sciences) **51 (3)** pp 373–386
[8] Vekua I N 1988 *Generalized Analytic Function* (Moscow: Nauka)
[9] Gakhov F D 1963 *Boundary-Value Problems* (Moscow: Fizmatgiz)
[10] Krasnosel'skii M A 1956 *Topological Methods in the Theory of Nonlinear Integral Equations* (Moscow: Gostekhizdat)

Computational experiment on the numerical solution of some inverse problems of mathematical physics

V I Vasil'ev, A M Kardashevsky and PV Sivtsev

North-Eastern Federal University, 58 Belinskogo, 677000 Yakutsk, Russia

Email: vasvasil@mail.ru

Abstract. In this article the computational experiment on the numerical solution of the most popular linear inverse problems for equations of mathematical physics are presented. The discretization of retrospective inverse problem for parabolic equation is performed using difference scheme with non-positive weight multiplier. Similar difference scheme is also used for the numerical solution of Cauchy problem for two-dimensional Laplace equation. The results of computational experiment, performed on model problems with exact solution, including ones with randomly perturbed input data are presented and discussed.

1. Introduction

We have sufficiently wide class of inverse problems for equations with partial derivative, which are recently used for mathematical description of various processes. The important feature of inverse problems arising from the processing of the experimental results is that the initial information is defined approximately, because used measurement devices have a certain level of error. Therefore, methods of solution of these inverse problems must be resistant to small errors of input data.

Inverse problems of mathematical physics, as a rule, belong to the class of ill-posed problems in the classical sense. The ill-posedness is primarily caused by lack of continuous dependence of solutions on the initial data. In this case it is necessary to narrow the class of feasible solutions and to use a special regularization procedure to find sustainable solutions. In addition to the existence and stability of solutions a question of uniqueness is also relevant, which study answers the question of whether available information is sufficient for unique identification of the desired characteristics of the studied object or process.

Construction of mathematical models of various processes leading to inverse problems of mathematical physics, their theoretical study and design of numerical methods are summarized in the monographs [1-3]. Retrospective inverse heat transfer problem is one of the most popular inverse problems of mathematical physics, for which there is no continuous dependence of solutions on the initial data. However, while narrowing the class of solutions we achieve stability - this problem is conditionally correct [1]. In paper [4] a method for solution of retrospective inverse problem for a parabolic equation is proposed. The Cauchy problem for an elliptic equation, as shown by Hadamard, is the first example of incorrect inverse problem, which is unstable in the classical sense. In work [5] as it is proved by M.M. Lavrentiev [6], this problem is conventionally well-posed in the class of sufficiently smooth functions.

2. Retrospective heat transfer problem

In rectangular $(a,b)\times[0,T)$ we search the function $u(x,t)$, which is a solution of one-dimensional parabolic equation of second order

$$\frac{\partial u}{\partial t}-\frac{\partial}{\partial x}\left(k(x)\frac{\partial u}{\partial x}\right)=0,\ \ x\in(a,b)=\Omega,\ \ 0\le t<T. \tag{1}$$

Suppose that homogeneous Dirichlet boundary conditions are given

$$u(0,t)=0,\ u(l,t)=0,0<t<T. \tag{2}$$

We formulate a retrospective inverse problem, where the required function at the final moment of time is set instead of the initial condition

$$u(x,T)=\varphi(x),x\in[a,b]=\bar{\Omega}. \tag{3}$$

Let the coefficient $k(x)$ from differential equation be sufficiently smooth, positive and bounded function, i.e. it satisfies the condition $0<k_1\le k(x)\le k_2<\infty$. Then the assigned problem (1)–(3) in the class of bounded functions is conditionally well-posed [1]. It should be notedthat for $k(x)\equiv 1, a=-\infty, b=\infty$ the equation (1) has automodel solution

$$u(x,t)=\frac{1}{\sqrt{\beta+4t}}e^{-\frac{(x-x_0)^2}{\beta+4t}},\beta>0, \tag{4}$$

where $x_0\in\Omega$. This function is sufficiently smooth and limited and can be subsequently used to establish the numerical accuracy of the proposed method.

In domain $\Omega_T=(a,b)\times[0,T)$ let us introduce a uniform rectangular mesh

$$\omega_{h\tau}=\omega_h\times\omega_\tau,$$
$$\omega_h=\{x_i=a+ih, i=1,\dots,N-1; h=(b-a)/N\},$$
$$\omega_\tau=\{t_j=j\tau, j=1,\dots,M-1; \tau=T/M\}.$$

On the set of grid functions $y\in H$, which are $y(x)=0,\ x\notin\omega_h$, we define a grid operator as

$$\mathrm{A}y=-(a(x)y_{\bar{x}})_x, x\in\omega_h$$

as for example, setting $a(x)=k(x-0.5h)$.

In finite-dimensional Hilbert grid space H we introduce inner product and norm as

$$(y,v)=\sum_{x\in\omega}yvh,\qquad \|y\|=\sqrt{(y,y)}.$$

It is known that [7] in Hilbert space H the operator A is self-adjoint, positive definite ($A=A^*>0$) and there is two-sided inequality $8k_1/l^2\le\|A\|<4k_2/h^2$, which means it is positive definite and limited. In the rectangular space-time mesh $\omega_{h\tau}$ we approximate the problem using implicit difference scheme with weight factor σ:

$$\frac{y_{m+1}-y_m}{\tau}-A((1-\sigma)y_{m+1}+\sigma y_m)=0, m=M-1,\dots,1,0, \tag{5}$$

$$y_M(x)=\varphi(x),\quad x\in\bar{\omega}_h. \tag{6}$$

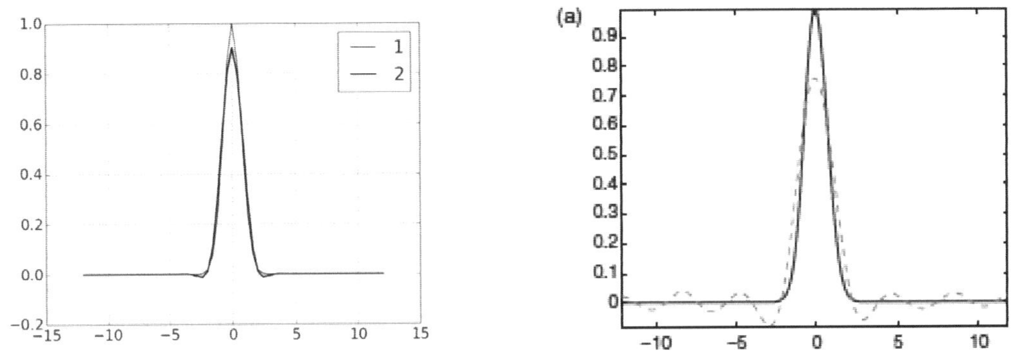

Figure 1. Results of the proposed difference method (on the left side) and results from paper [8] (on the right side)

Difference scheme (5)–(6) on solution of non-classical problem (1)–(3) has the first order of approximation by τ and the second order by h. For its solution at σ=0 we have following apriori estimate

$$y_0 \le e^{2CM}\varphi, \tag{7}$$

where $C=2k_2\tau / h^2$ is Courant number of explicit scheme for heat transfer equation, M is amount of time layers. The estimate (7) implies that the satisfactory results on this scheme can be achieved for not too large time intervals (T) and not too detailed spatial grids.

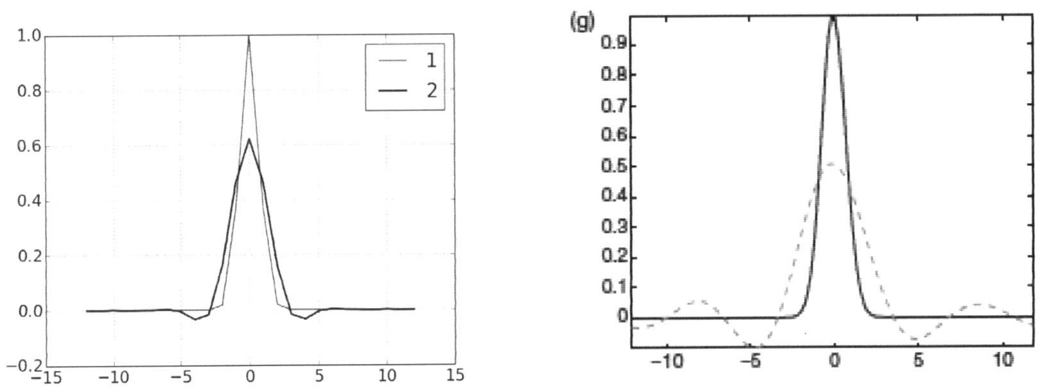

Figure 2. Results of the proposed difference method (on the left side) and results from paper [8] (on the right side)

The results of numerical calculations are carried out by using implicit difference scheme (5)–(6) for σ=0 on problem with the exact automodel solution (4) with following initial data: $k(x)\equiv 1, a=-12, b=12$. It should be noted that since we perform calculations downward by time,computations are carried out by explicit formulas [8]. Figure 1 and Figure 2 demonstrate the results of calculations obtained respectively for T=1, N=60, M=500 and T=3, N=24, M=300. On illustration 1 there is a graph of the exact solution and on 2 there is a graph of numerical solution on the selected spatial grid (on the left side) and similar graphs from paper [8] (on the right side). The graphs show that the accuracy of results obtained by implicit scheme on a fairly rough time-space grids do not concede to the accuracy of results presented in paper [8] and received by modified methods of regularization of A.N. Tikhonov.

The typical results of solution of retrospective inverse problem under conditions, where additional condition $\varphi(x)$ is set with an error γ, are presented. In the experiments the additional condition was perturbed as follows:

$$\varphi_\gamma(x) = \varphi(x) + \gamma \cdot \text{rand}(-1,1), x \in [a,b],$$

where rand$(-1,1)$ is random variable uniformly distributed on the segment$[a,b]$. To achieve smoother solution as smoothing operator we take K-time used three-point formula (filter):

$$\varphi_i^{k+1} = \left(\varphi_{i+1}^k + 4\varphi_i^k + \varphi_{i-1}^k\right)/6,\ i=1,\dots,N-1,\ k=0,\dots,K-1;$$

$$\varphi_{\gamma i} = \varphi_i^K, i=1,\dots,N-1.$$

On the Figure 3 the results obtained for following input data: N=60, m=5, T=1, $a=-12$, b=12 are presented. The left graph is obtained for γ=0.01, K=12 and the right for γ=0.01, K=24. The graphs show that growth of K leads to stronger smoothing of the solution.

Let us now present results of computations by difference scheme (5)–(6) with negative weight factor. On the left side on Figure 4 the calculation results for $\sigma=-2, T=1, a=-12, b=12, N=120, M=85$ are presented. On the right side there are results with presence of perturbation δ=0.01, K=24, N=120, M=8. The remaining data is the same. On the left figure: 1 is a graph of the desired exact solution, 2 is a graph of additional condition (function $\varphi(x)$), 3 is a graph of restored initial condition. On the right figure: 1 is desired exact solution smoothed by a filter, 2 is perturbed additional condition, 3 is found initial condition. The graphs illustrate that the choice of negative weight factor allows to calculate on more detailed spatial grid. In case when additional condition is defined with errors the algorithm gives more accurate results.

Note that in [1] has shown that the use of negative multiplier is equivalent to using pseudoparabolic regularization with parameter $\delta = -\sigma\tau$ so in our case δ=1/12.

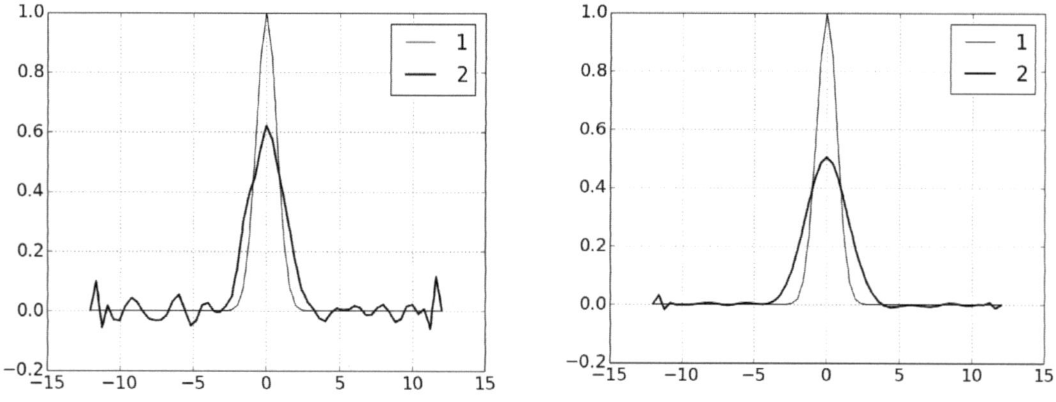

Figure 3. On the left side there are results for K=12, and on the right side there are results for K=24

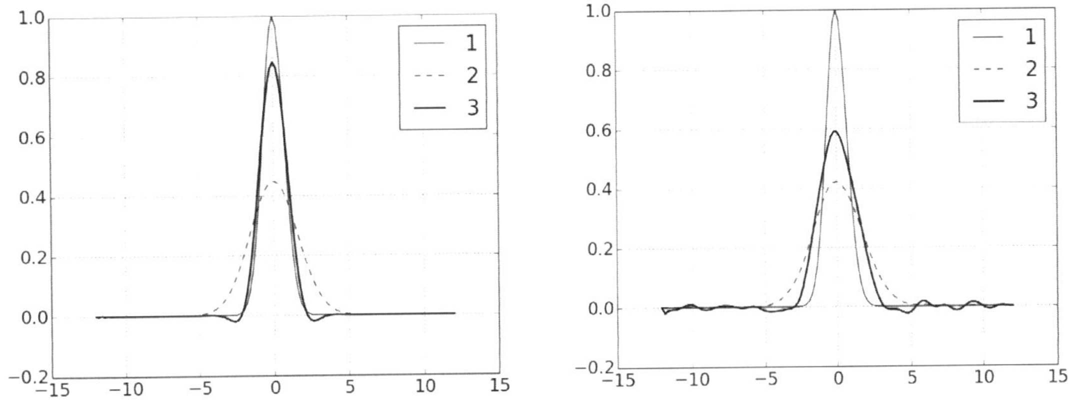

Figure 4. On the left side there are results for K=12, and on the right side there are results for K=24

3. Cauchy problem for Laplace equation

Suppose that it is required to find a function $u(x, y)$ that satisfies two-dimensional Laplace equation on rectangular $(0,L_1)\times[0,L_2)$

$$\frac{\partial^2 u}{\partial x^2}+\frac{\partial^2 u}{\partial y^2}=0, \quad (x, y)\in\Omega \tag{8}$$

and following conditions

$$u(0,y)=0, u(L_1, y)=0,\ 0<y\le L_2, \tag{9}$$

$$u(x,0)=\varphi(x), \frac{\partial u}{\partial y}(x,0)=\phi(x), 0\le x\le L_1. \tag{10}$$

This problem is a particular case of Cauchy problem for elliptic equation. On this problem the function $u(x, y)$ should be defined in domain Ω including upper boundary. But at lower boundary two conditions are set. It is conventionally well-posed on the class of smooth solutions [2].

It should be noted that problem (8)–(10) resembles the initial-boundary value problem for the wave equation. There are two initial conditions for variable y and a negative definite operator for variable x. Thus, we are dealing with the problem similar to a retrospective inverse problem, but for hyperbolic equation. This fact suggests applying a similar difference scheme.

On rectangular mesh

$$\omega=\omega_h\times\omega_\tau,$$

$$\omega_h=\{x_i=ih, i=1,\ldots,N-1; h=L_1/N\},$$

$$\omega_\tau=\{t_j=j\tau, j=1,\ldots,M-1; \tau=L_2/M\}.$$

We assign to a problem (8)–(10) the difference scheme with non-positive weight factor

$$\frac{y_{m+1}-2y_m+y_{m-1}}{\tau^2}-A(\sigma y_{m+1}+(1-2\sigma)y_m+\sigma y_{m-1})=0, \qquad m=2,3,\ldots,M, \tag{11}$$

$$y_0(x)=\phi(x),\ y_1(x)=y_0(x)+\tau\varphi(x)-\frac{\tau^2}{2}Ay_0, \qquad x\in\bar{\omega}_h. \tag{12}$$

Here the grid operator A is defined as follows

$$Ay=-y_{\bar{x}x}, \quad x\in\omega_h.$$

The relative error of the numerical method is calculated according to formula

$$R=\sqrt{\frac{\sum_{i=0}^{n}\left(y_{iM}-u(x_i,L_2)\right)^2}{\sum_{i=0}^{n}\left(u(x_i,L_2)\right)^2}}. \tag{13}$$

On Figure 5 we present the results of numerical calculations by the proposed difference scheme (11)–(12) for $\sigma=0, L_1=1, L_2=1$ on the problem with exact solution

$$u(x,y)=\pi^2\sin(\pi x)e^{\pi(y-1)}. \tag{14}$$

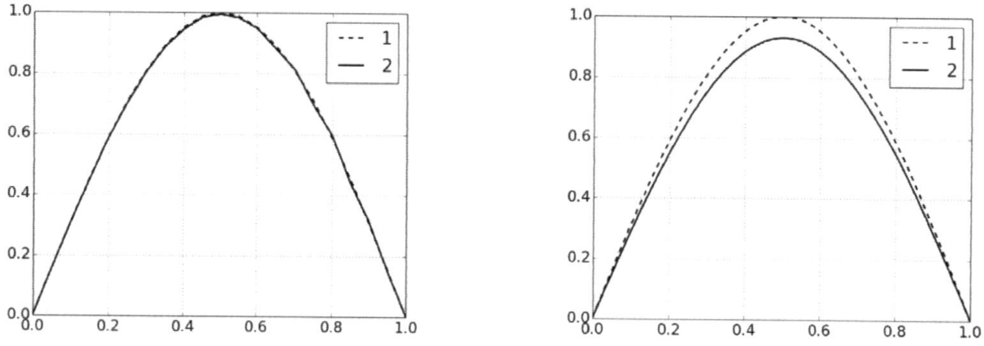

Figure 5. The results for $N = 20$, $M = 28$, $R = 0.01255$ (on the left side), $N = 100$, $M = 5$, $R = 0.07022$(on the right side)

The figures show results obtained for maximum possible values allowing stability of computational algorithm. The left figure shows the computational results obtained for $N=20; M=28$ and the right for $N=100; M=5$. On presented figures: 1 is graph of exact solution, 2 is graph of numerical solution on the selected grid. The graphs demonstrate that the results obtained by the proposed difference scheme provide a good accuracy on the gridcoarse by variable x (the left graph). Increasing the number of partitions by x leads to increasing of mesh step by y, so the accuracy fails (the right graph). It should be noted that for large N calculations will be stable at low values of L_2.

Now we present the results of calculations carried out by the difference scheme (11)–(12) with a negative weight factor on the example of the exact solution of (12). Figure 6 shows that the use of a negative weighting factor increases the number of partitions of the grid by the variable y. This leads to accuracy increasing.

It should be noted that like in the previous problem, as shown in paper [1] the use of the negative factor is equivalent to using of regularization difference scheme with regularization parameter $\delta=-\sigma\tau^2$ or associated with quasi-inversion method for approximate solution of ill-posed problem.

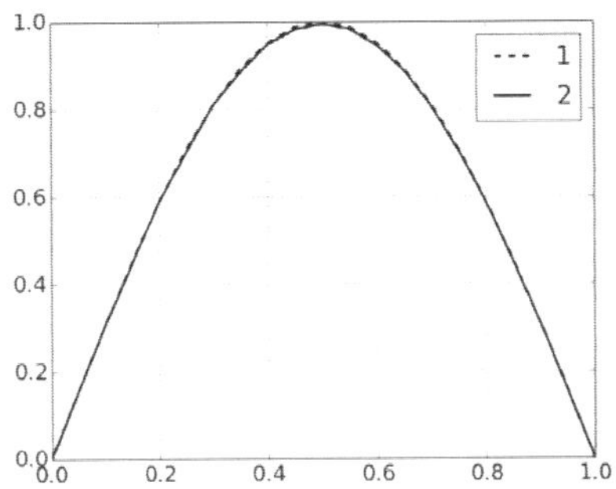

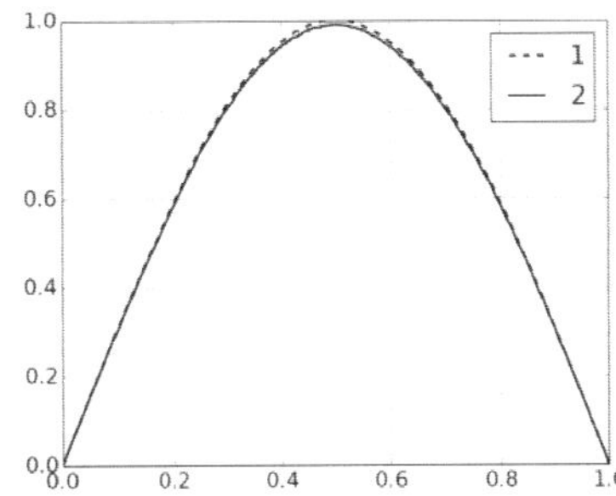

Figure 6. The results for $N = 20$, $M = 100$, $R = 0.00608$ (on the left side), $N = 100$, M = 50, $R = 0.01115$ (on the right side)

4. Conclusion

Direct solution of a retrospective heat transfer problem by using difference scheme with a negative weight factor gives good results even in the presence of random errors in additional conditions at the final moment of time. Numerical solutions of the Cauchy problem by difference scheme with a negative weight factor gives satisfactory results with proper selection of mesh size.

Acknowledgments

Authors express gratitude to professor P.N. Vabishchevich for constructive remarks and fruitful discussions. The work is performed with partial financial support of RFBR (project 15-31-20856).

References

[1] Samarskii A A and Vabishchevich P N 2007 *Numerical methods for solving inverse problems of mathematical physics* (De Gruyter)

[2] Kabanikhin S I 2011 *Inverse and Ill-Posed Problems. Theory and Applications* (Germany: De Gruyter)

[3] Lattès R and Lions J L 1969 The Method of Quasi-reversibility *Applications to Partial Differential Equations* (New York: American Elsevier)

[4] Samarskii A A, Vabishchevich P N and Vasil'ev VI 1997 Iterative Solution of a Retrospective Inverse Problem of Heat Conduction *Matematicheskoe Modelirovanie* **9 (5)** 119–127

[5] Hadamard J 1932 *Le problem de Cauchy et les equations aux derives partielles lineares hyperboliques* (Paris: Herman)

[6] Lavrent'ev M M 1955 O zadache Koshi dlja uravnenija Laplasa *Dokl. AN SSSR* **102 (2)** 205–206

[7] Samarskii A A 2001 *The theory of difference schemes* (Marcel Dekker)

[8] Zhenyu Zhao, Zehong Meng 2011 A modified Tikhonov regularization method for a backward heat equation *Inverse Problems in Science and Engineering* **19 (8)** 1175–1182

Calculation of the effective thermal properties of the composites based on the finite element solutions of the boundary value problems

I I Vdovichenko[1,2], M Ya Yakovlev[2], A V Vershinin[1,2] and V A Levin[1]

[1]Lomonosov Moscow State University, Faculty of Mechanics and Mathematics, Russia, 119991, Moscow, GSP-1, 1 Leninskiye Gory, Main Building

[2]Fidesys LLC, Office 402, 1 bld. 77, MSU Science Park, Leninskie Gory, Moscow 119234

E-mail: vdovichenko@cae-fidesys.com

Abstract. One of the key problems of mechanics of composite materials is an estimation of effective properties of composite materials. This article describes the algorithms for numerical evaluation of the effective thermal conductivity and thermal expansion of composites. An algorithm of effective thermal conductivity evaluation is based on sequential solution of boundary problems of thermal conductivity with different boundary conditions (in the form of the temperature on the boundary) on representative volume element (RVE) of composite with subsequent averaging of the resulting vector field of heat flux. An algorithm of effective thermal expansion evaluation is based on the solution of the boundary problem of elasticity (considering the thermal expansion) on a RVE of composite material with subsequent averaging of a resulting strain tensor field. Numerical calculations were performed with the help of Fidesys Composite software module of CAE Fidesys using the finite element method. The article presents the results of numerical calculations of the effective coefficients of thermal conductivity and thermoelasticity for two types of composites (single-layer fiber and particulate materials) in comparison with the analytical estimates. The comparison leads to the conclusion about the correctness of algorithms and program developed.

1. Introduction

At present, in many areas of technology various composite materials (composites) are widely used - artificially created heterogeneous solid materials consisting of two or more components with a clear boundary between them. The composites feature high strength and stiffness combined with lightness and resistance to mechanical, thermal and chemical influence. When combining several components of the composite a material is formed with a set of characteristics that reflect not only the properties of the components, but sometimes some new properties which the components do not possess individually.

When working with composite materials, there is a problem of averaging of their mechanical or physical properties. With the known geometry of the composite and the known properties of the components it is necessary to calculate the effective (i.e. averaged) properties of the composite material [1, 2, 3]. There are different solutions of this problem - both analytic, using formulas, and numerical, using modern computational tools.

This article discusses the method of numerical evaluation of effective thermal properties of composites based on the numerical solution of boundary problems on a representative volume element (RVE) of composite material. Calculations on a RVE were performed using the finite element analysis method [4, 5] with the help of CAE Fidesys [6].

2. Method of the effective properties calculating

Since we are talking about calculations on a RVE, one should give a definition of this concept first. The representative volume element (RVE) of composite material is the minimal volume of the material, which can be used for some experiments and measurements, on the basis of which it is possible to draw conclusions about the behavior of the material as a whole. If the composite material features an irregular structure, its RVE should contain a sufficient volume of each component to in order to havbe a possibility to average the properties of the entire material. if a composite features a periodic structure - then a cell of periodicity is an RVE.

2.1. Estimation of effective coefficients of thermal conductivity

We begin with the definition of the effective thermal conductivity. Let's call the homogeneous material**as the effective (averaged) material** (from the point of view of thermal conductivity), if it meets the following condition: if we consider a RVE of the initial composite material and fill the exact same volume of with this homogeneous material, then the average (averaged by volume) vectors of the heat flow in these two volumes will be equal at the same temperatures at the boundaries of the volumes. Let's call the thermal conductivity of this material as the **effective thermal conductivity**.

Let's describe a methodology for estimation of the effective thermal conductivity of the composite material using this definition. For a RVE V in the form of a rectangular parallelepiped sized $2A$x$2B$x$2C$ we will solve a certain number of boundary problems of thermal conductivity [7] (for steady thermal state with no internal heat sources):

$$\nabla \cdot q = 0 \tag{1}$$

with *nonperiodic* boundary conditions in a form of predetermined temperature on the boundary

$$T\big|_{\Gamma} = \nabla T^{e} \cdot r \tag{2}$$

or with *periodic* boundary conditions in the form of bonds to temperatures of corresponding to each other points on the opposite faces of a RVE

$$\begin{aligned} T_i - T_{-i} &= 2A \cdot (\nabla T^{e})_1 \\ T_j - T_{-j} &= 2B \cdot (\nabla T^{e})_2 \\ T_k - T_{-k} &= 2C \cdot (\nabla T^{e})_3 \end{aligned} \tag{3}$$

where ∇ is gradient operator, q is heat flux vector, T is temperature, r is radius vector of a point on the boundary. For *periodic* boundary conditions (i, $-i$), (j, $-j$), (k, $-k$) are pairs of points corresponding to each other on opposite sides of the RVE.

Here ∇T^{e} is a kind of effective temperature gradient on a RVE, which we are setting up. Each boundary thermal conductivity problem corresponds to separate effective temperature gradient. For *periodic* boundary conditions $(\nabla T^{e})_1$, $(\nabla T^{e})_2$, $(\nabla T^{e})_3$ are components of the gradient.

As a result of each task solution, we will get a vector distribution of the heat flux q over the RVE. Let's average it by the volume according to the formula

$$q^{e} = \frac{1}{V}\int_{V} q dV \tag{4}$$

and we will obtain an efficient vector of heat flux q_e. The effective thermal conductivity of the composite will be evaluated in the form of relation of q_e, obtained by us as a result of the calculation, to the effective temperature gradient ∇T^e, which we have defined as an input - in a form of Fourier law for thermal conductivity:

$$q_i^e = -\lambda_{ij}\left(\nabla T^e\right)_j \tag{5}$$

In order to calculate the effective thermal conductivity coefficients, which we are interested in, we will solve three boundary thermal conductivity problems corresponding to the following temperature gradients:

1) $\nabla T^e = (\delta;0;0) \Rightarrow \lambda_{i1} = -\dfrac{q_i^e}{\delta}, i = 1,2,3$

2) $\nabla T^e = (0;\delta;0) \Rightarrow \lambda_{i2} = -\dfrac{q_i^e}{\delta}, i = 1,2,3$

3) $\nabla T^e = (0;0;\delta) \Rightarrow \lambda_{i3} = -\dfrac{q_i^e}{\delta}, i = 1,2,3$

Thus, by solving these three problems, we calculate all the effective thermal conductivity coefficients of the composite material.

2.2. Estimation of the effective coefficients of thermal expansion

First, let's define the effective thermal expansion. Let's call the effective (as averaging) material (in terms of thermal expansion) such a homogeneous material which meets the following condition: if we consider a RVE of the initial composite material and fill the exact same volume with this homogeneous material, then the average (averaged by volume) thermal strain of these volumes will be equal at the same temperature distribution over the volumes. The coefficients of thermal expansion of the material will be called as effective coefficients of thermal expansion.

Let's describe a methodology for estimation of the effective thermal conductivity of the composite material using this definition. For a RVE V_0 in the form of a rectangular parallelepiped sized $2A$x$2B$x$2C$ wewill solve a certain number of elastic boundary value problems [8] considering the thermal expansion:

$$\begin{gathered} \nabla \cdot \sigma^{el} = 0 \\ \sigma^{el} = \sigma - \sigma^{th} \end{gathered} \tag{6}$$

In (6) σ is total stress tensor, σ^{el} is mechanic (elastic) stress tensor, σ^{th} is thermal stress tensor.

At the same time *nonperiodic* or *periodic* boundary conditions are applied to the RVE, and the entire volume is heated to ΔT. Under the influence of temperature the volume deforms (freely expands). By averaging the strain tensor over the volume (in the final state after deformation) we will obtain an effective strain tensor:

$$\varepsilon^e = \frac{1}{V}\int_V \varepsilon dV \tag{7}$$

Let's find the effective thermal expansion

$$\varepsilon_{ij}^{th} = \alpha_{ij}\Delta T \tag{8}$$

where ε^{th} is thermal strains tensor. A value of ΔT, to which a RVE was heated, was set up by us; and the effective strain tensor resulting from this heating was calculated. Effective thermal expansion coefficients are calculated according to the formula:

$$\alpha_{ij} = \frac{\varepsilon_{ij}^{e}}{\Delta T} \tag{9}$$

Boundary conditions applied to the RVE, should be discussed separately. *Nonperiodic* boundary conditions in this problem are zero pressure on the border of the volume, which allows it to expand freely as the temperature increases. *Periodic* boundary conditions also allow the volume to expand freely, with only one restriction: the RVE which was the periodicitycell prior to the deformation of the composite, should remain to be the periodicity cell after deformation.

For this purpose a pair of points (1, 2) is fixed, the first point lies on the edge $x = A$, the second point lies on the edge $x = -A$, and their projections match each other (these could be the corner points). And for the displacement of each pair of points $(i, -i)$, the first of which lies on the edge $x = A$, the second one lies on the edge $x = -A$ a relationship is put on:

$$\bar{u}_1 - \bar{u}_2 = \bar{u}_i - \bar{u}_{-i} \tag{10}$$

where u is the displacement vector of a point. Once again, we note that pairs of points $(i, -i)$ in (10) are varied, passing across the surface of edges $x = A$ and $x = -A$; and a pair of points (1, 2) - is a fixed one and the same for all relationships (10).

Relationships similar to (10) are put on points of edges $y = B$ and $y = -B$ and edges $z = C$ and $z = -C$. These boundary conditions allow the model to change its volume freely while remaining the periodicity cell of the composite material by the form.

3. Examples of calculations

The described methods of numerical evaluation of the effective thermal conductivity and thermal expansion coefficients are implemented in a Fidesys Composite software module of CAE Fidesys. Boundary problems of elasticity and thermal conductivity are solved with the help of computational kernels CAE Fidesys using the finite element analysis method. For example, let's give the numerical solution of the problem of effective evaluation of thermal conductivity and thermal elasticity coefficients for the two types of composites as compared with analytical solutions.

3.1. Single-layer fiber composite

A single-layer fiber composite material is a two component composite material, wherein each layer of matrix is reinforced by unidirectional continuous fibers of the same thickness and the same spacing (Fig. 1). Effective *thermal conductivity* coefficients of such composite material can be estimated using the analytical formulas [9] (in which it is assumed that the fibers are directed along the axis X):

$$\lambda_x = \lambda_f \gamma_f + \lambda_m \gamma_m$$

$$\lambda_y = \lambda_z \approx \lambda_m \frac{1 + \gamma_f + \gamma_m \dfrac{\lambda_m}{\lambda_f}}{\gamma_m + (1 + \gamma_f)\dfrac{\lambda_m}{\lambda_f}} \tag{11}$$

where λ_f and λ_m are thermal conductivity coefficients of the matrix and fibers, respectively, γ_f and γ_m - volumetric concentration of the fibers and the matrix, respectively.

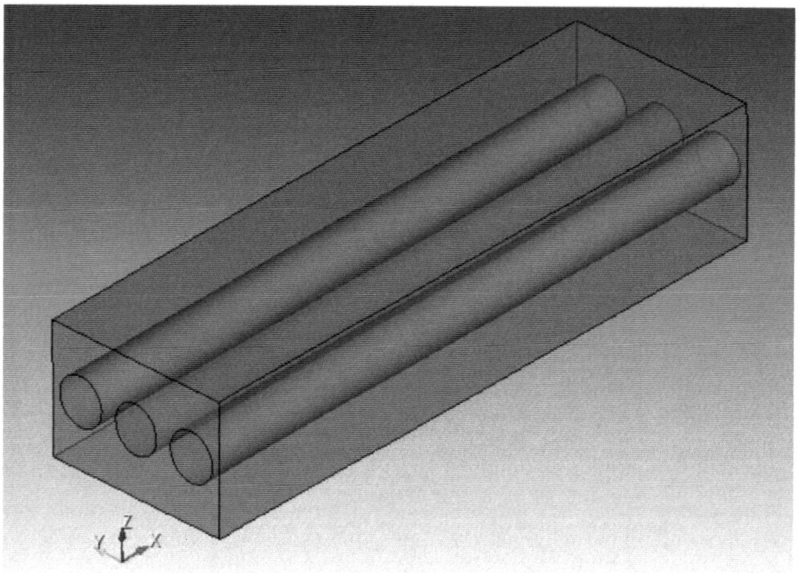

Fiqure 1. Single-layer fiber composite material

Effective *coefficients of thermal expansion* for such composite material can be estimated using formulas [9]:

$$\alpha_x = \frac{\alpha_m E_m \gamma_m + \alpha_f E_f \gamma_f}{E_m \gamma_m + E_f \gamma_f}$$

$$\alpha_y = \alpha_z = \frac{\alpha_f + 2\alpha_m - 2\nu_f \dfrac{\alpha_m - \alpha_f}{1 + \dfrac{E_f \gamma_f}{E_m \gamma_m}} + 2\nu_m s \dfrac{\alpha_m - \alpha_f}{1 + \dfrac{E_m \gamma_m}{E_f \gamma_f}}}{1 + 2s} \quad (12)$$

where $s = 2\left(t_f / d - 1\right)$.

In the formula (12) α_f and α_m are coefficients of thermal expansion of the fibers and the matrix, respectively, γ_f and γ_m are volumetric concentration of the fibers and the matrix, respectively, E_f and E_m are Young's modulus fibers and the matrix, respectively; ν_f and ν_m are Poisson's ratios of the fibers and the matrix, respectively; d is fiber diameter, t_f is fibers spacing (distance between adjacent fibers).

A single-layer fiber composite material with fibers concentration of 10% was considered. Fiber properties: Young's modulus of 2,000 MPa, Poisson's ratio of 0.2, the thermal expansion coefficient of 0.1 K^{-1}, the thermal conductivity of 10 W/(m·K). Matrix properties: Young's modulus of 2 MPa, the Poisson's ratio of 0.3, the thermal expansion coefficient of 0.001 K^{-1}, thermal conductivity coefficient of 2 W/(m·K). Analytical and numerical values of thermal conductivity are listed in Table 1.

Table 1. Coefficients of thermal conductivity of single-layer fiber composite material.

	Analytical, W/(m·K)	Numerical, W/(m·K)
λ_x	2.8	2.78952
λ_y	2.28571	2.27496
λ_z	2.28571	2.27484

Analytical and numerical values of the thermal expansion coefficients are listed in Table 2.

Table 2. Coefficients of thermal expansion of single-layer fiber composite.

	Analytical, K^{-1}	Numerical, K^{-1}
α_x	0.099117	0.099097
α_y	0.005296	0.005087
α_z	0.005296	0.005086

As one can see in the Table 1, the thermal conductivity coefficients for the fiber composite obtained numerically are equal to analytical values with very good accuracy (the error is less than 1%). Table 2 shows that the numerical values of the coefficients of thermal expansion match the analytical with error of not more than 4%, which is also a good accuracy.

3.2. The particulate composite reinforced by spherical inclusions

The particulate composite is a matrix reinforced by individual filler particles. For simplicity, let's consider the numerical calculation model of such a composite in which the filler particles have the same spherical shape, the same size and are arranged in a periodic manner in the martix, and the bulk concentration of filler in the composite is small. Effective *coefficient of thermal conductivity* of such composite can be estimated by the analytical formula [9]:

$$\lambda = \lambda_m \frac{1-2k\gamma_i}{1+k\gamma_i}, \text{ where } k = \frac{\lambda_m - \lambda_i}{2\lambda_m + \lambda_i} \tag{13}$$

where λ_i and λ_m are thermal conductivity coefficient of filler inclusions and the the matrix, respectively, γ_i and γ_m - volumetric concentration of filler inclusions and the matrix, respectively.

Effective *coefficient of thermal expansion* of such composite can be estimated by the formula [9]:

$$\alpha = \bar{\alpha} - \frac{\gamma_m \gamma_i (\alpha_m - \alpha_i)(K_i - K_m)}{\bar{K} + K_i \dfrac{1+\nu_m}{2(1-2\nu_m)}}, \text{ where } \begin{array}{l} \bar{\alpha} = \alpha_i \gamma_i + \alpha_m \gamma_m \\ \bar{K} = K_i \gamma_i + K_m \gamma_m \end{array} \tag{14}$$

In the formula (14) α_i and α_m are coefficients of thermal expansion of inclusions and the matrix, respectively; γ_i and γ_m are volumetric concentration of inclusions and the matrix, respectively; K_i and K_m are volumetric compression modules of inclusions and the matrix, respectively; ν_i and ν_m are Poisson's ratios of inclusions and the matrix, respectively.

A particulate composite with concentration of spherical inclusions of 5% was considered. Fiber properties: Young's modulus of 10 MPa, Poisson's ratio of 0.25, the thermal expansion coefficient of 0.1 K^{-1}, thermal conductivity coefficient of 10 W/(m·K). Matrix properties: Young's modulus of 1 MPa, the Poisson's ratio of 0.4, the thermal expansion coefficient of 0.001 K^{-1}, thermal conductivity coefficient of 2 W/(m·K). Analytical and numerical values of the thermal conductivity coefficient are shown in Table 3.

Table 3. Coefficient of thermal conductivity of particulate composite material.

	Analytical, W/(m·K)	Numerical, W/(m·K)
λ	2.17642	2.17519

Analytical and numerical values of the thermal expansion coefficient are listed in Table 4.

Table 4. Coefficient of thermal expansion of particulate composite material.

	Analytical, K^{-1}	Numerical, K^{-1}
α	0.006879	0.007077

As one can see in the Table 3, the thermal conductivity coefficient of the particulate composite material obtained numerically match the analytical value with very good accuracy (error is less than 1%). The numerical values of thermal expansion coefficients match the analytical values with the accuracy of no more than 3%, which also shows good accuracy of numerical solution.

4. Conclusion

The article presents the developed algorithm of calculation of effective thermal properties of composite materials on the basis of solving of boundary value problems of thermal conductivity and elastic boundary value problem for a RVE with the subsequent averaging of the results. The algorithm is implemented in a Fidesys Composite software module of CAE Fidesys. The comparison with the known analytical solutions was conducted, which allowed to draw a conclusion about the correctness of algorithms and the program developed. In the future we plan to generalize the approach to the case of nonlinear thermal conductivity and thermo-elasticity models.

Acknowledgments

The research for this article was performed in FIDESYS LLC and was financially supported by Russian Ministry of Education and Science (project №14.579.21.0112, project ID RFMEFI57915X0112).

5. References

[1] Levin V A, Zingerman K M, Vershinin A V, Yakovlev M Ya 2015 Numerical analysis of effective mechanical properties of rubber-cord composites under finite strains *Compos. Struct.* **131** pp 25–36.

[2] Vershinin A V, Levin V A, Zingerman K M, Sboychakov A M, Yakovlev M Ya 2015 Software for estimation of second order effective material properties of porous samples with geometrical and physical nonlinearity accounted for *Adv. Eng. Softw.* **86** pp 80–84.

[3] Levin V A, Vdovichenko I I, Vershinin A V, Yakovlev M Ya, Zingerman K M 2016 Numerical estimation of effective mechanical properties for reinforced Plexiglas in the two-dimensional case *Model. Simulat. Eng.* Article ID 9010576 10 pp. Available online: http://www.hindawi.com/journals/mse/2016/9010576/

[4] Zienkiewicz O C, Taylor R L 2000 *The finite element method. Vol. 1. The basis.* (Butterworth-Heinemann: Oxford, United Kingdom) 707 pp.

[5] Zienkiewicz O C, Taylor R L 2000 *The finite element method. Vol. 2. Solid mechanics.* (Butterworth-Heinemann: Oxford, United Kingdom) 479 pp.

[6] Fidesys LLC official website. Available online: http://cae-fidesys.com

[7] Lykov A V 1967 *Theory of Thermal Conductivity* [in Russian] (Moscow: Vysshaya Shkola).

[8] Lurie A I 1990 *Non-Linear Theory of Elasticity* (North-Holland: Amsterdam, Netherlands) 617 pp.

[9] Vishnyakov L R et al. 1985 *Composites* [in Russian] (Kiev: Naukova Dumka) 294 pp.

Numerical solution of moving boundary problem for deposition process in solid fuel gas generator

V M Volokhov, S O Dorofeenko, M S Sharov and P D Toktaliev
IPCP RAS, Chernogolovka, Russia
E-mail: dorofeenko@icp.ac.ru

Abstract. Moving boundary problem in application to process of depositions formation in gas generator are considered. Gas generator, as a part of fuel preparation system of high-speed vehicle, convert solid fuel into multicomponent multiphase mixture, which further burned down in combustion chamber. Mathematical model of two-phase "gas-solid particles" flow, including Navier-Stokes equations for turbulent flow in gas generator and mass, impulse conservations laws for elementary depositions layer are proposed. Verification of proposed mathematical model for depositions mass in gas generator conditions is done. Further possible improvements of proposed model, based on more detail accounting of particle-wall interaction and wall's surface adhesion properties are analyzed.

1. Introduction

In moving boundary problem or similar Stefan problem, position of interfacial surface are considered and often this position can't be determined a priori due the dependency of some other variable. Existing of locally changing in time and space interfacial surface is one of the main difficulties of such type problems and as a rule these problems have multidisciplinary and non-one dimensional nature. In particular, the great interest at nowadays is attached to systems, in which transition from solid to gas phase exist, example of such systems is burning surface of solid rocket fuel [1, 2, 3], and systems with gas-solid transition, for example, solid deposition formation in fuel lines of propulsion [4] or heat exchanger [5, 6] or in problems with icing [7]. Mathematical modelling of mentioned systems requires a description of a number of phenomena such as turbulent flow in complex geometries, heat transfer, chemical kinetics including combustion, discrete particles propagation, interaction of particles with solid surfaces and others. Modern CFD tools, providing advanced platform for solution of Navier-Stokes equations, can potentially take into account many of physical effects of the continuum media, though obtaining of physically correct solution often is a challenge. There are roughly two branches of methods within CFD, which can capture multiphase flow physics with moving interfacial surface. First branch is methods with Eulerian computational meshes, for example, immersed boundaries method, and second branch is dynamic meshes methods, which moves and recalculate mesh nodes positions in a way, that one or more of the mesh family lines always lie on the interfacial surface through the flow iteration. Last branch used in current work, two of the advantages of this branch of methods are in the absence of the special treatment of boundary conditions and relatively easy control of the near-wall mesh layer resolution.

In current work moving boundary problem in application to process of depositions formation

(a) (b)

Figure 1. Depositions formation in gas generator: (a) device state before test; (b) after test.

in gas generator are considered. Gas generator convert solid fuel into multicomponent multiphase mixture, which further burned down in combustion chamber. Two stages fuel combustion scheme improve operational characteristics of engine, but wherein burning of gas generator solid fuels can be accompanied by significant amount of condensed products, whose total mass fraction may reach about 5% from overall fuel mass in typical gas generator conditions, and about 50% for equilibrium conditions. Amount and properties of discrete phase essentially depends on fuel composition and conditions of gasification. Interaction of chemically active discrete phase particles with walls of gas generator under medium temperatures $T \approx 800$ K leads to deposition of solid fraction and appearance of evolving with time interfacial surface [4]. Further depositions growth results in changes in heat transfer at walls and flow blockage in critical sections of gas generator, as shown at figure 1. Besides that, the depositions formation reduces the energy efficiency of the engine and could lead to a performance degradation.

The aim of this paper is to develop mathematical model of solid depositions formation process in solid fuel gas generator and to estimate by the proposed model typical full blockage time of the flow.

2. Mathematical model

Solid fuel gas generator mathematical model includes incompressible viscous fluid flow description in the channel with deformable boundary. Flow was considered in quasi-stationary formulation with time marching procedure: at first, velocity field in the channel with a certain geometry was obtaining, then deformation of boundary connected with deposition formation was computing, and after that, calculation was repeating for the modified channel geometry. Thus, gas flow governing equation system is of the form

$$\nabla \cdot \mathbf{U} = 0, \tag{1}$$

$$\nabla \cdot (\mathbf{U}\mathbf{U}) = -\nabla p + \nabla \cdot (\nu \nabla \mathbf{U}), \tag{2}$$

where $\mathbf{U}$ is flow velocity, p is kinematic pressure, ν is gas kinematic viscosity. Typical Reynolds number evaluation shows that there is developed turbulent flow in gas generator, hence equations (1)–(2) was calculated in Reynolds-averaged approach (RANS) with k-ω-SST turbulence model [8] for the computational cost reduction purposes. k-ω-SST turbulence model combines k-ω model features in near-wall regions and k-ϵ model features in free flow, hence allows to obtain more precise solution in entire channel in comparison with each model individually.

Depositions formation occurs mostly due to convectional settling of particles from the flow. Therefore, depositions layer growth rate depends on mean velocity of the flow, which brings

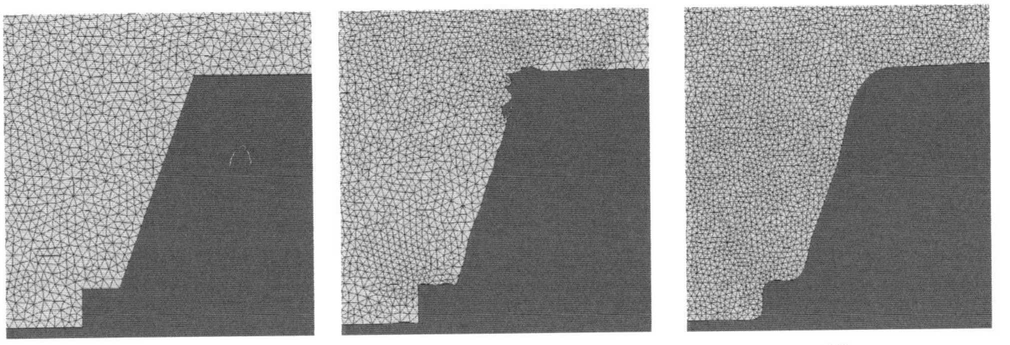

(a) (b) (c)

Figure 2. Depositions layer growth and roughness smoothing with Savitsky-Goley filter: (a) initial surface structure; (b) geometry deformation due to depositions formation; (c) deformed boundary after Savitsky-Goley filter smoothing with window $n = 7$.

settling particles and on turbulent pulsations which may increase probability of collision of near-wall particle and surface in comparison with the laminar flow. Probability of particle being captured by surface during the contact defined by many factors such as collision velocity, incidence angle, properties and features of surface, e.g. liquid film presence, etc. Problem of different factors influence estimation goes beyond the scope of the paper. Inertialess particles ($\mathrm{Sk} \ll 1$) was considered as a first approximation of discrete phase model under the assumption that dispersed phase volume flow rate is small in comparison with carrying flow rate, $G_p \sim 0.01 G_g$, and the influence of dispersed phase on the flow is negligible. Then depositions growth rate v is proportional to normal flow velocity in the near-wall region:

$$v = k_d \mathbf{U}_w \cdot \mathbf{n}, \tag{3}$$

where $\mathbf{U}_w$ is gas flow velocity in the near-wall layer, $\mathbf{n}$ is surface normal, k_d is coefficient of probability of particle being captured by surface during collision.

Expression (3) defines velocity of computational mesh boundary nodes displacement during transition from one time step to the next one. This displacement corresponds to deposition growth (fig. 2). Time step $\Delta t = 10^{-4}$ was defined as lesser than typical convectional time and typical turbulent mixing time under the assumption that depositions growth is determined mostly by mass transport processes in gas. Discretization of numerical domain for modelling leads to forming of small-scaled highly indented surface (fig. 2(b)), that occurs even when nodes displacement velocity is smooth function as in (3). Local boundary irregularity increasing can lead to degeneracy of computational mesh because of negative volume cells formation. Smoothing procedures was applied to the interfacial surface for convergence improvement and avoiding of negative volume cells formation. Boundary-forming polygon was processed with Savitsky-Goley filter with window $n <= 9$ after boundary nodes displacement. Obtained boundary was used as initial geometry for the next step mesh generation (fig. 2(c)).

Solution of discrete equations was carried out by SIMPLE method on unstructured mesh. Gradient reconstruction was performed with the 2nd order approximation scheme. Resulting set of linear algebraic equations was solved with Gauss-Zeidel method and hybrid multigrid technology GAMG.

3. Simulation results and discussion

Model gas generator consists of two sequential cylindrical sections with diameter $D = 0.045$ m and $d = 0.019$ m, and conic junction between them as shown at fig. 3(a). Computational domain

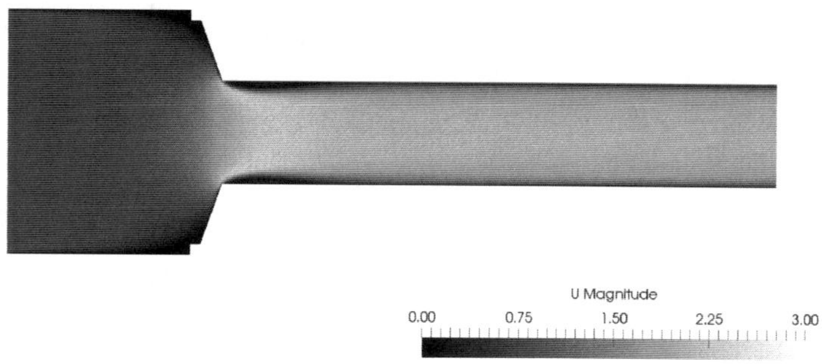

(a)

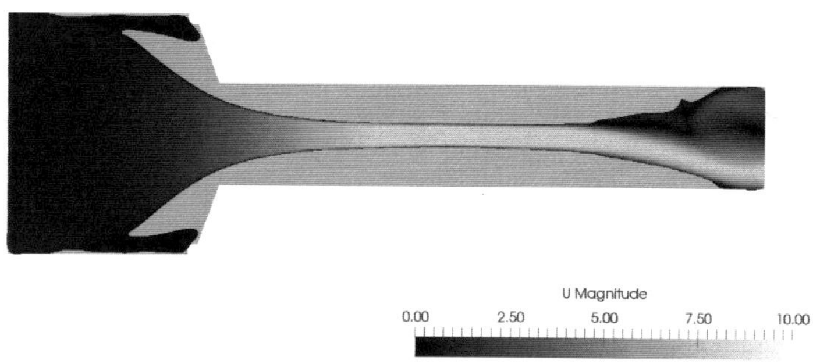

(b)

Figure 3. Flow field in the gas generator with initial geometry (a) and deformed by the depositions geometry (b).

geometry is a copy of experimental model gas generator simplified by removing small constructive elements (device fragments pictures presented at fig. 1). The problem was considered in a 2D formulation for computational cost decrease purposes. Computational domain inlet boundary (left boundary at fig. 3(a)) coincide with solid fuel charge location, which is the source of gasification products. The location of this boundary was fixed, plain gas flow velocity profile $U_x = 1$ m/s was set on this boundary. The outlet boundary with soft boundary conditions was located on the distance of $5.5d$ downstream from channel step. Other boundaries was considered as solid surfaces with no-slip wall conditions.

Velocity field in non-deformed by depositions gas generator channel is presented on fig. 3(a). Uniform inlet flow forms two low-intensive zones with significant gas residence time in throat area of gas generator. Local flow acceleration with flow compression and two elongated recirculation zones occurs in a channel throat after the corner. Maximum reversed flow velocity is about 0.5 m/s, length of recirculation zone is about d. With depositions layers growth these recirculation zones tend to decrease in size and completely disappear eventually. After that flow without separation is developed, as shown at fig. 3(b). For initial stages of flow development outlet velocity profile has typical turbulent form with a large constant velocity flow core.

Calculation results in deformed by depositions gas generator channel are presented on fig. 3(b). Solid depositions are formed mostly in cylindrical part of channel with diameter

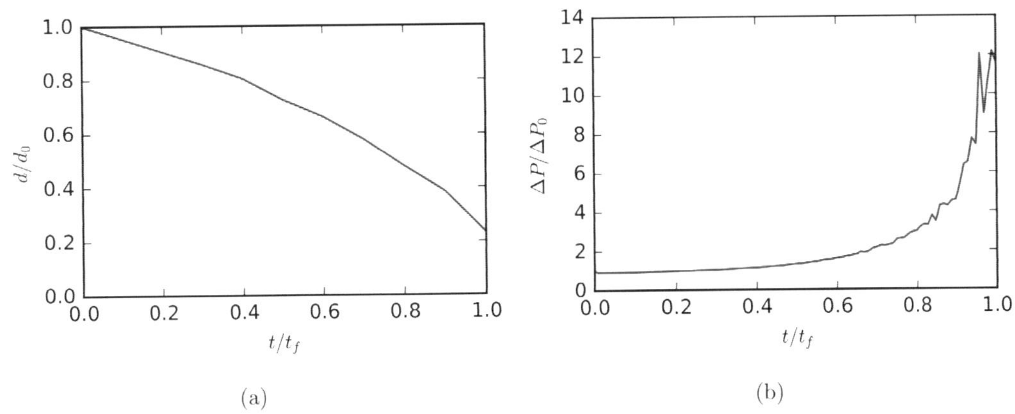

Figure 4. Influence of the depositions growth on gas generator parameters: diameter of the flow section (a), total pressure loss (b).

d, solid depositions layers have conic form in inlet and outlet parts of the channel. However, flow in the inlet part differs from the outlet part considerably: there are typical stable confuser flow and unstable flow with separation correspondingly. Further at time depositions cone angle at the outlet part of the channel exceeds critical value (8°) with the following stability loss, after that maximum velocity zone of the flow starts to oscillate. Flow instability leads to formation of inhomogeneities at the depositions layer surface. "Pockets" are forming in the cylindrical depositions layer, this process could lead to computational domain connectivity increasing, for example, if parts of depositions layer leaved the solid surface or if cavities formed in the depositions layer volume. With growth of depositions layer velocity magnitude in the paraxial zone of the channel is increased and for fig. 3(b) it equals to 11.2 m/s. For near blockage time the mass of the depositions in the inlet part of the gas generator with the diameter D is insignificant and depositions layer has wave-like form.

There are many important characteristics of gas generator, among them are the flow section of long cylindrical part and overall device hydraulic resistance. Time dependences for minimal channel flow section and total pressure losses in comparison with their initial values are presented on fig. 4. Calculation results show that minimum diameter of flow section decreases monotonously in time and at for shown at 4 time moment equals to 4.4 mm (approximately five times smaller than initial value). Further depositions growth leads to full blockage of the gas generator channel, and the corresponding moment in time could be estimated at fig. 4(a) as abscissa value at intersection point between the curve and abscissa axis. Minimal flow section locates at the distance of about $0.7d$ from the beginning of the smaller cylindrical part of the gas generator and moves downstream with the growing of depositions layer until the distance of $2.5d$. Monotony of $d(t)$ dependence can be explained by absence of global flow separation in the flow domain. Time dependence behavior of total pressure losses differs from minimal flow section one. Curve at fig. 4(b) could be divided into three parts: initial part until $t/t_f \approx 0.5$ with approximately constant pressure loss value, rapid growth part until $t/t_f \approx 0.9$, and last part at $t/t_f > 0.9$ with significant pressure oscillations related to flow instability. Net increase of relative total pressure losses equal to 12 for near blockage time.

Experimental data for solid fuel gas generator with a similar geometry at operational time 9.51s show about 2.6% reduction of relative minimal flow section, while numerical simulation results show value in 2.1% for the same parameter. Experimental growth rate of the depositions layer corresponds to coefficient value of $k_d = 521$. The main reasons of differences between experimental and numerical results are in discrepancy of quantitative parameters of depositions growth mechanism and neglecting of heat effects, which could alter adhesive properties of particles and surface. Nevertheless, proposed mathematical model can qualitatively describe solid depositions formation at gas generator walls and can estimate depositions influence on flowfield. Approximation of frozen particles can be in natural way replaced by lagrangian massive particles model and that should allow to simulate influence of particle inertia and weight distribution on depositions formation. Capturing gas-solid surface with dynamic mesh approach allow in a direct way to extend model for accounting thermo-physical properties of solid surfaces, that can be do by wall boundary conditions modification in computational cells with depositions. Future improvements of proposed mathematical model can include mesh deformation small-scale smoothing procedure, which should give correct correlation between physical inhomogeneities scales of depositions surface and computational mesh ones (filter window width and depositions roughness scale correspondence); exact formulation for mass conservation law for solid depositions (this property could be expressed using geometrical conversation laws [9]).

4. Conclusion

In this paper description of proposed mathematical model of two-phase flow "gas-solid particles", including Navier-Stokes equations for turbulent flow and mass, impulse conservations laws for elementary depositions layer are given. Despite the simplicity of the proposed model, it allow with sufficient for engineering applications accuracy describe evolution of interfacial surface. Comparison between numerical and experimental data show, that discrepancy don't exceed 40% for total depositions mass. Further possible improvements of proposed model, based on more detail accounting of particle-wall interaction and wall's surface adhesion properties are analyzed.

Acknowledgments

The activity is a part of the work "Supercomputer simulation of physical and chemical processes in the high-speed direct-flow propulsion jet engine of the hypersonic aircraft on solid fuels" supported by Russian Science Foundation (project no. 15-11-30012).

References

[1] Szmelter J and Ortiz P 2007 *Proceedings of the Institution of Mechanical Engineers, Part G: Journal of Aerospace Engineering* **221** 429–439
[2] Jiao X 2007 *Journal of Computational Physics* **220** 612–625
[3] Gueyffier D, Roux F, Fabignon Y, Chaineray G, Lupoglazoff N, Vuillot F and Alauzet F 2014 *50th AIAA Joint Propulsion Conf.*
[4] Myakochin A and Yanovskiy L 2001 *Formation of the deposits in fuel propulsion systems and methods of their suppresion* (Moscow Aviation Instititute Press) in Russian
[5] Pérez M, Vakkilainen E and Hyppänen T 2015 *Fuel* **158** 139–151
[6] Tomeczek J and Wacławiak K 2009 *Fuel* **88** 1466–1471
[7] Kinzel M, Sarofeen C, Noack R and Kreeger R 2010 *28th AIAA Applied Aerodynamics Conf.*
[8] Menter F 1994 *AIAA Journal* **32** 1598–1605
[9] Thomas P and Lombard C 1979 *AIAA Journal* **17** 1030–1037

Computer modeling of capillary flow with superimposed pulsations

A E Yaganova[1] and E A Marfin[1,2]

[1] Institute of Physics, Kazan Federal University, 16a Kremlyovskaya Street, Kazan, Russia
[2] Kazan Science Centre, Russian Academy of Science, 2/31 Lobachevsky Street, Kazan, Russia

E-mail: marfin_ea@mail.ru

Abstract. Increasing efficiency of methods of oil production can be achieved by the influence of elastic vibrations. It is a well-known fact that shift viscosity of oil changes under the effect of elastic vibrations. This change depends on properties of the oil and exposure mode. Existing approaches to the research of the way wave exposure impacts on viscosity are based on measuring it after the processing. This article concerns development of methods to measure viscosity of liquid right during its exposure to elastic vibrations. The suggested approach is based on combining numerical and natural experiments. We investigated the pulsating flow of viscid liquid in a capillary numerically in this article. We received allocations of fields of average velocity and pressure in a capillary. It is demonstrated that imposed pulsations in a capillary do not impact on hydrodynamics of the flow. We offered the scheme of an experimental installation for a research of the impact that wave exposure has on the viscosity of liquids. The installation is based on a capillary viscometer.

1. Introduction

Current period of development of oilfields in the main oil-producing regions is characterized by the increase in proportion of reserves, which are hard to reach, and ineffective deposits. Therefore specialists of oil branch do not lose interest in practical use of new effective technologies, which would provide stable maintenance and growth of oil production in complicated geological and trading conditions. Applying wave method has a prospect in technologies of increasing efficiency of oil production [1-10]. Its essence lies in intensification of processes inside of a stratum by its exposure to elastic vibrations. However, there are questions to be answered, which concern applying it. One of the biggest ones is the investigation of an impact, which wave action has on the medium, and its mechanisms.

There are different ideas of mechanisms of the exposure [11-14]. One of them is changing permeability of a porous collector; another one is the release of the dissolved gas; also there is one which lies in decreasing viscosity of formation fluids, to name but a few. However, despite significant results in understanding basic mechanisms of the exposure, there is still a vacancy in our knowledge when it comes to the way elastic vibrations impact on viscosity of liquids. Changes of rheological behavior of liquids, which are characterized by existence of viscoelastic and viscoplastic properties of non-Newtonian liquids, were found in an experiment by authors of the article [14]. In particular, change of oil shift viscosity under the action of elastic vibrations was observed (figure 1). It was established that shift viscosity decreases by 20-30% right after the exposure and after some time (5-6 hours and more) it restores fully or partially.

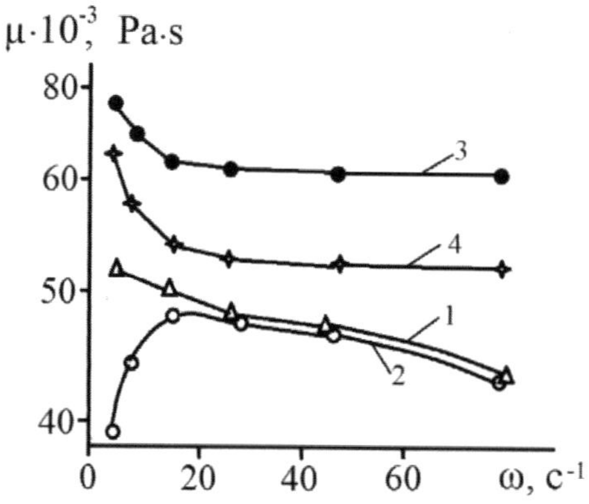

Figure 1. Dependence of dynamic viscosity of oil μ on frequency of rotor rotation ω with different time *t* after the acoustic treatment [14]: 1 – before the treatment; 2 – *t*=0; 3 – *t*=22 h; 4 – *t*=120 h.

The presented data do not give an answer to the question of the value of oil viscosity right in the middle of the exposure. It is important to know because in some cases wave impact on formation is performed permanently in the process of oil production. The main gap in our knowledge is absence of tools that would allow observing change of viscosity right in the process of the exposure. Thus, the aim of the article is development of method of measuring viscosity of liquid in the process of acoustic influence. To achieve this goal it is necessary to investigate hydrodynamics of the liquid in existing devices for measuring viscosity with imposed fluctuations.

2. Setting the task

Essence of the offered approach is the following. Wave exposure of liquid is performed right in the process of measuring it with viscometer. We can observe change of readings on the device which will testify some changes in rheological properties of the liquid. However, to isolate the change of viscosity from the other ones, it is necessary to know how elastic vibrations influence hydrodynamics of the liquid in a flow channel of the device. This task can be resolved by numerical modeling of the flow of liquid.

The flow of liquid in a capillary is used as a viscometer in this article. The modeling is based on methods of computing hydrodynamics, including the numerical solution to Navier–Stokes equations and continuity equations. Calculation is performed in steady state and with imposed fluctuations and is followed by the analysis of the results. In numerical modeling properties of the medium stay permanent, i.e. the impact of wave exposure does not affect rheological properties of the medium, thereby we investigate only exposure of hydrodynamics of the flow to elastic vibrations.

The three-dimensional geometric model of a flow passage of a capillary is devised to prepare for modeling in the system SolidWorks, which is intended for solid-state parametrical associative modeling of complex pieces. Modeling was carried out using the program complex for research of flows of liquids and gases FlowVision. Calculation area was internal flow area of a capillary with the one that adjoins at the entrance and the exit of a capillary (figure 2). In our case the needle of a medical syringe was used as a capillary; geometric sizes of the needle were the following: L=42 mm, R=0.51 mm. The liquid that we investigated was glycerol at 20 °C.

Results of calculations were compared to known analytical relation that is Poiseuille equation, which describes steady-state motion of the liquid:

$$Q = \pi\rho \frac{P_1 - P_2}{8\mu L} R^4 \qquad (1)$$

where Q represents volume flow of liquid, m^3/s; μ represents viscosity coefficient, Pa·s; ρ represents density of the liquid, kg/m^3; P_1 and P_2 are pressures at the entrance and the exit respectively, Pa; L is length of a capillary, m; R is radius of a capillary, m.

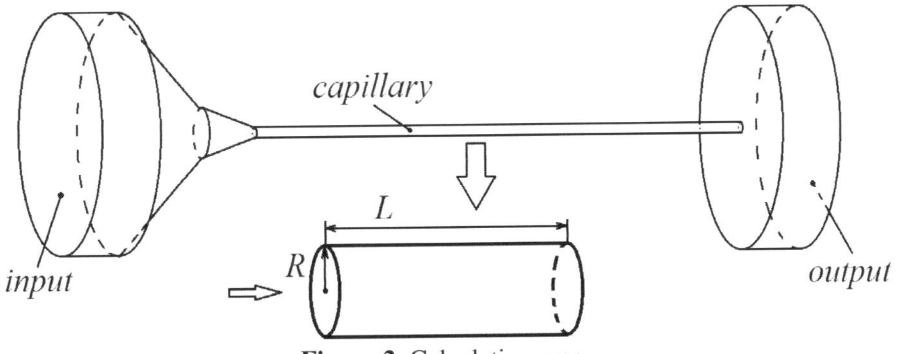

Figure 2. Calculation area.

3. Results of the modeling

Comparison of the results of the numerical experiment with the laminar steady-state flow to the one with imposed pulsations of the flow was carried out according to the change of average pressure difference on a region of the capillary as well as the change of velocity profile in a capillary section.

As a result of numerical modeling of steady-state flow we received the allocation of velocity through a capillary section (figure 3). Direction of the flow is from the left to the right. Value of velocity is established on 800 iteration of calculation.

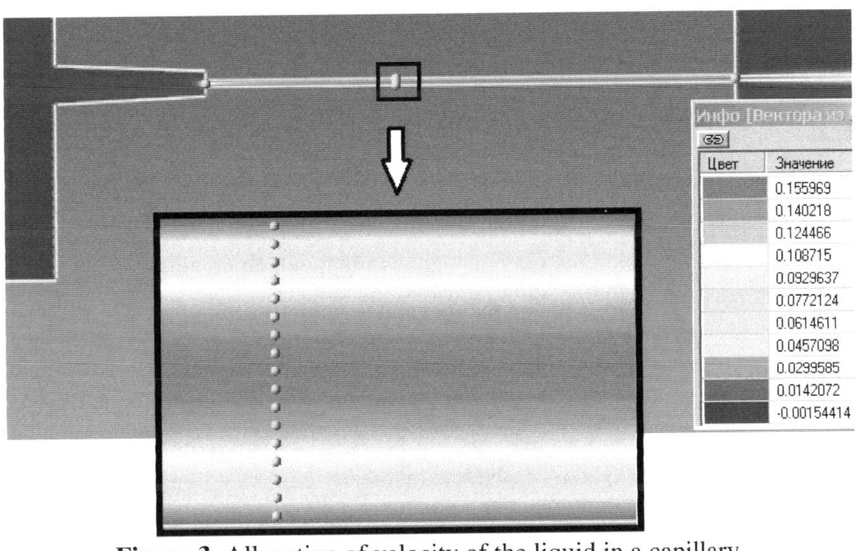

Figure 3. Allocation of velocity of the liquid in a capillary.

At the next stage of numerical research the velocity with imposed pulsations of different frequencies in the range from 1 kHz to 10 kHz was set at the entrance to a capillary. Step of time was $1.25 \cdot 10^{-5}$ s. Measurements of velocity were carried out at different points along the radius of the capillary (figure 4). Value of velocity was noted at every point. Typical change of velocity by time is depicted in figure 5. It should be mentioned that pulsations of velocity were imposed on the average velocity of the steady-state flow. Dependence of average velocity on distance to the axis of a capillary is depicted in figure 5. Such allocation of profile fits the theory that says the farther away from the wall a point is located, the higher velocity of the flow is.

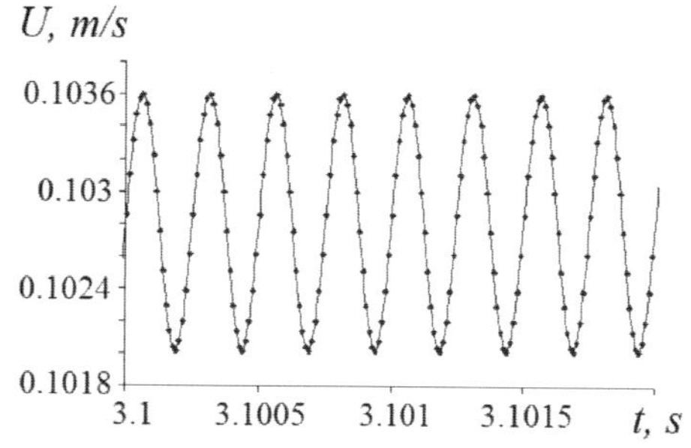

Figure 4. Change of velocity of the flow in a capillary.

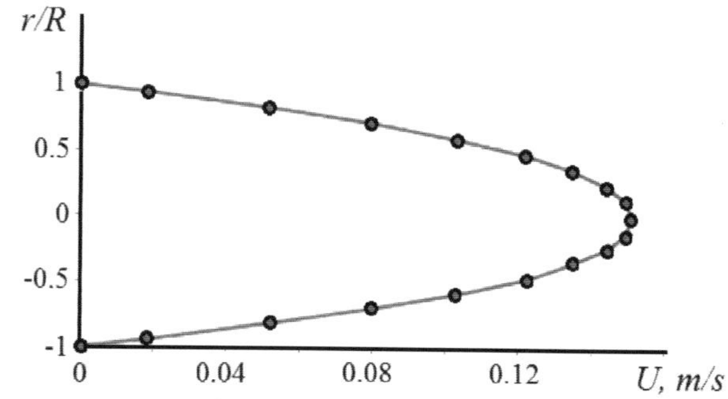

Figure 5. Profiles of velocity of the flow in a capillary (the line is for stationary conditions, the points are for when applying fluctuations 2 kHz).

Difference between average velocities of the steady-state flow, the flow with imposed pulsations and theoretical numbers calculated according to equation (1) is less than 0.4%, which is less than the error of the numerical experiment. Thus, we can conclude that presence of pressure pulsations in a capillary does not affect hydrodynamics of the flow. It means that for research of the way wave exposure impacts on viscosity of liquid we can use the capillary method, installation of which includes a device for generation of elastic fluctuations. We developed a scheme for an experimental installation for a research of the impact that wave exposure has on viscosity of liquid, which is based on a capillary viscometer. The scheme is depicted in figure 6. The installation includes the following components: a capillary that the tested liquid comes through; a fluctuations oscillator, which carries out wave exposure; a system of measuring the flow rate through a capillary and pressure difference in it.

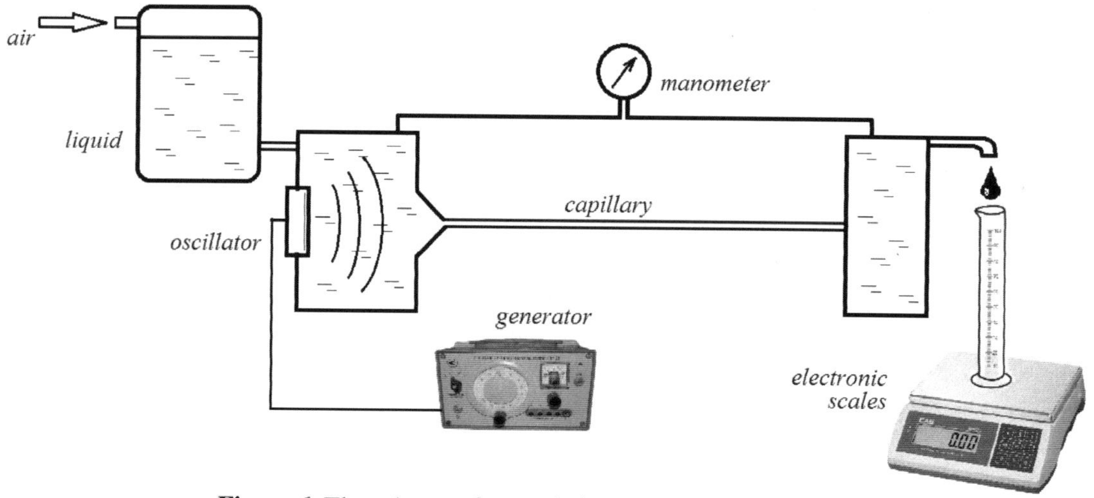

Figure 6. The scheme of upgraded experimental installation.

4. Conclusion

According to the results of the research it was established that imposed pulsations do not affect hydrodynamics of the flow of viscid liquid in a capillary. Thus, the method based on applying capillary viscometer can serve as a foundation for a way to research the impact of wave exposure on viscosity of the liquid. We developed a scheme for an experimental installation for measuring viscosity of the liquid right in the process of exposing it to elastic waves.

References

[1] Beresnev I A and Johnson P A 1994 *Geophysics* **59** 1000-17

[2] Marfin E A, Kravtsov Y I, Abdrashitov A A, Gataullin R N and Galimzyanova A R 2015 *Pet. Sci. Tech.* **33**(15) 1526-32

[3] Jeong C, Kallivokas L F, Kucukcoban S, Deng W and Fathi A 2015 *J. Pet. Sci. Eng.* **129** 205-220

[4] Marfin E A and Kravtsov Ya I 2005 *Izvestiya Akademii Nauk. Energetika* **6** 108-113

[5] Jankov M, Lovoll G, Knudsen H A, Maloy K J, Planet R, Toussaint R and Flekkoy E G 2010 *Transp. Porous Med.* **84**:3, 569-58

[6] Marfin E A, Kravtsov Ya I, Abdrashitov A A and Gataullin R N 2014 *Georesources* **2**(57) 14-16

[7] Marfin E A, Kravtsov Y I, Abdrashitov A A and Metelev I S *2015 Geomodel 2015 - 17th Sci.-Pract. Conf. on Oil and Gas Geol. Explor. and Devel.* doi: 10.3997/2214-4609.201413912

[8] Gataullin R N, Kravtsov Ya I and Marfin E A 2013 *Neftyanoe khozyaystvo - Oil Industry* **1** 90-93

[9] Marfin E A and Abdrashitov A A 2014 *6th Saint Petersburg Int. Conf. and Exhib. on Geosciences 2014: Investing in the Future*. doi: 10.3997/2214-4609.20140186

[10] Marfin E A, Abdrashitov A A and Kravtsov Y I 2016. *7th EAGE Saint Petersburg Int. Conf. and Exhib.: Understanding the Harmony of the Earth's Resources Through Integration of Geosciences.* doi: 10.3997/2214-4609.201600094

[11] Kravtsov J I, Marfin E A, Butorin E A and Gataullin R N 2009 *Georesources* **1**(29) 43-45

[12] Manga M, Beresnev I, Emily E, Brodsky E E, Elkhoury J E, Elsworth D, Ingebritsen S E, David C, Mays D C and Wang C-Y 2012 *Rev. Geophys.* **50** RG2004

[13] Hamidi H, Mohammadian E, Junin R, Rafati R, Manan M, Azdarpour A and Junid M 2014 *Ultrasonics* **54**:2 655-662

[14] Sokolov A V and Simkin E M 1981 Study of influence of acoustic treatment on rheological properties of some oils. *Topics in nonlinear geophysics* (*Voprosi nelineinoy geofiziki*) (Moscow: All-Union Research Institute of Nuclear Geophysics and Geochemistry) pp 137–142

One cutting plane algorithm using auxiliary functions

I Ya Zabotin and K E Kazaeva

Kazan (Volga Region) Federal University, 18, Kremlyovskaya st., Kazan, 420008, tel.: (843)233-71-56

E-mail: iyazabotin@mail.ru

Abstract. We propose an algorithm for solving a convex programming problem from the class of cutting methods. The algorithm is characterized by the construction of approximations using some auxiliary functions, instead of the objective function. Each auxiliary function bases on the exterior penalty function. In proposed algorithm the admissible set and the epigraph of each auxiliary function are embedded into polyhedral sets. In connection with the above, the iteration points are found by solving linear programming problems. We discuss the implementation of the algorithm and prove its convergence.

1. Introduction

In the projecting of complex technical systems solving one-criterion and multi-criteria optimization problems is often a necessity. Some examples of the optimization problem statements related to the synthesis of electronic systems and their individual subsystems can be found in [1, 2]. Often such problem statements have the form of mathematical programming problems.

Cutting plane methods form a well-known class of mathematical programming problem solving methods (e.g., [3 - 6]). Part of this class of methods uses the polyhedral approximation of the epigraph of the objective function (e.g., [3, 6 - 9]). Proposed cutting algorithm precisely refers to this group. This algorithm differs from the known ones in using approximation of the auxiliary function's epigraphs. These auxiliary functions are constructed in the form of the sum of the objective function and external penalties for the constraint region of the original problem.

2. Problem setting.

We solve the problem

$$\min \{ f(x) : x \in D \}, \quad (1)$$

D – a convex bounded closed set in the n-dimensional Euclidean space R_n, $f(x)$ – a convex function in R_n.

We set $f^* = \min \{ f(x): x \in D \}$, $X^* = \{ x \in D: f(x) = f^* \}$, $\text{epi}(g, G) = \{ (x, \gamma) \in R_{n+1} : x \in G, \gamma \geq g(x) \}$, where $G \subset R_n$, $g(x)$ – the function defined in R_n. Denote by $W(z, Q) = \{a \in R_n : \langle a, u - z \rangle \leq 0 \ \forall u \in Q\}$ the bunch of normalized generally support vectors for the set $Q \subset R_{n+1}$ at the point $z \in R_{n+1}$, int$Q$ – interior of the set Q and $K = \{0, 1, \ldots\}$.

3. The cutting algorithm and discussion

The proposed solution algorithm for the problem (1) generates a sequence of approximations $\{x_k\}$, $k \in K$, by the following rule.

Choose point

$$v \in \text{int epi}(f, D).$$

Set a convex penalty function $P_0(x)$ with the conditions $P_0(x) = 0$ for all $x \in D$ and $P_0(x) > 0$ for all $x \notin D$. Put

$$F_0(x) = f(x) + P_0(x).$$

Construct a convex bounded closed set $D_0 \subset R_n$ and convex closed set $M_0 \subset R_{n+1}$ such that

$$D \subset D_0, \text{ epi}(F_0, R_n) \subset M_0.$$

Fix a number

$$\gamma \leq \min \{ f(x): x \in D_0 \} = f_0^*$$

and a positive numerical sequence $\{\Delta_k\}$, $k \in K$. Set $i = 0$, $k = 0$.

1. Find a solution $u_i = (y_i, \gamma_i)$, where $y_i \in R_n$, $\gamma_i \in R_1$, of the following problem

$$\min \{ \gamma: (x, \gamma) \in M_i, x \in D_0, \gamma \geq \overline{\gamma} \}. \quad (2)$$

If $u_i \in \text{epi}(f, D)$, then $y_i \in X^*$ - a solution of problem (1), and the process is over.

2. If

$$F_i(y_i) - \gamma_i > \Delta_k, \quad (3)$$

then put

$$P_{i+1}(x) = P_i(x),\ F_{i+1}(x) = F_i(x),$$

and go to step 4. Otherwise set a convex penalty function $P_{i+1}(x)$ such that $P_{i+1}(x) = 0$ for all $x \in D$ and $P_{i+1}(x) > P_i(x)$ for all $x \notin$ D. Put

$$F_{i+1}(x) = f(x) + P_{i+1}(x),$$

and go to step 3.

3. Let $i_k = i$,

$$x_k = y_{i_k},\ \sigma_k = \gamma_{i_k}, \quad (4)$$

increase the value of k by one.

4. Find a point $v_i \in R_{n+1}$ as the intersection point of the segment $[v, u_i]$ and the boundary of the set $\text{epi}(F_i, R_n)$, choose a vector $a_i \in W(v_i, \text{epi}(F_i, R_n))$.
5. Let

$$M_{i+1} = M_i \cap \{u \in R_{n+1}: \langle a_i, u - v_i \rangle \leq 0\}.$$

Increase the value of i by one and go to step 1.

Let us make some remarks concerning the algorithm.

If the sets D_0 and M_0 are polyhedral, then the problems of constructing approximations u_i (2) are linear programming problems, for all $i \in K$. If put $M_0 = R_{n+1}$ then the pair $(y_0, \overline{\gamma})$, where $y_0 \in D_0$, can be a solution of problem (2) for $i = 0$.

Let $x^* \in X^*$. As $x^* \in D_0$ and the choice of number $\overline{\gamma}$ can be made by the inequalities $f^* \geq f_0^* \geq \overline{\gamma}$, it is easy to prove that $(x^*, f^*) \in M_i$, $i \in K$, i.e. the admissible set of the problem (2) is not empty. The last inclusion implies the inequality

$$\gamma_i \leq f^*,\ i \in K. \quad (5)$$

The proof of the following criterion of optimality (from step 1) is based on this inequality.

Theorem 1. If the inclusion $u_i \in \text{epi}(f, D)$ occurs for some $i \in K$, then $y_i \in X^*$.

Notice that some methods of specifying penalty functions can be found, for example, in [10, 11].

We move on to the investigation of the convergence of the algorithm. Firstly note the boundness of the sequences $\{u_i\}$ and $\{v_i\}$ constructed by the algorithm, because the inclusions $y_i \in D_0$ and the inequalities (5) are implemented for the points (y_i, γ_i).

Show that with a sequence $\{u_i\}$, $i \in K$, the algorithm constructs sequences $\{x_k\}$, $\{\sigma_k\}$, $k \in K$, too.

Lemma 1. If the sequence $\{u_i\}$, $i \in K$ is constructed by the described algorithm, there is an index $i_k \in K$ which satisfies (4) for every $k \in K$.

Proof. Fix an index $k \in K$ and show the existence of such index $i_k \in K$ for which the inequality $F_{i_k}(y_{i_k}) - \gamma_{i_k} \leq \Delta_k$ holds. Thereby the equalities (4) will be proved for the selected $k \in K$.

Assume the contrary, i.e. (3) is executed for the fixed Δ_k and for all $i \in K$. Then, according to the step 2 of the algorithm, we have the equalities $P_i(x) = P_0(x)$, $F_i(x) = F_0(x)$ for all $i > 0$. It means that

$$F_0(y_i) - \gamma_i > \Delta_k \; \forall \; i \in K. \tag{6}$$

From the sequence $\{u_i\}$, $i \in K$, distinguish a convergent subsequence $\{u_i\}$, $i \in K' \subset K$, and let $u' = (y', \gamma')$ be its limit point. Then referring to inequalities (6)

$$F_0(y') - \gamma' \geq \Delta_k. \tag{7}$$

Prove that the following equality holds for this subsequence:

$$\lim_{i \in K'} || v_i - u_i || = 0 \tag{8}$$

By the choice of the points v_i, there is such $\mathcal{T}_i \in [0, 1)$ for each $i \in K$ that

$$v_i = u_i + \mathcal{T}_i(v - u_i). \tag{9}$$

Fix $i', i'' \in K'$ so that $i'' > i'$. By the construction $M_{i''} \subset M_{i'}$ and consequently $a_{i'} \in W(v_{i'}, M_{i''})$. In view of inclusion $u_{i''} \in M_{i''}$ we have the inequality $<a_{i'}, u_{i''} - v_{i'}> \leq 0$. Hence from (9) there is the inequality

$$<a_{i'}, u_{i'} - u_{i''}> \geq \mathcal{T}_{i'} <a_{i'}, u_{i'} - v>. \tag{10}$$

As $v \in \text{int epi}(f, D)$ and $v_i \notin \text{int epi}(f, D)$, $i \in K$, by lemma 1 from [12] there is a number $\delta > 0$ such that $<a_i, v - v_i> \leq -\delta$ for all $i \in K$. By the equality (9) and the inequality $0 \leq \mathcal{T}_i < 1$, $i \in K$, we have the inequality $<a_i, v - u_i> \leq -\delta$, $i \in K$. Then, from (10) we obtain $<a_{i'}, u_{i'} - u_{i''}> \geq \mathcal{T}_{i'}\,\delta$ or $\| u_{i'} - u_{i''} \| \geq \mathcal{T}_{i'}\,\delta$. Due to convergence of the sequence $\{u_i\}$, $i \in K'$, the limit relation $\mathcal{T}_i \to 0$, $i \in K'$, follows from the last inequality. By the boundness of $\{ \| v - u_i \| \}$, $i \in K'$, using (9), we obtain the required equality (8).

Then, by equality (8) and by the inclusion $v_i \in \text{epi}(F_0, R_n)$, $i \in K'$, there is the inclusion $u' \in \text{epi}(F_0, R_n)$, i.e.

$$F_0(y') \leq \gamma'. \tag{11}$$

Contrariwise, $\text{epi}(F_0, R_n) \subset M_i$, $i \in K$. It means that the point $(y_i, F_0(y_i))$, $i \in K$, is a permissible solution of the problem (2) for each $i \in K$. Then the inequality $\gamma_i \leq F_0(y_i)$ holds for the solution (y_i, γ_i) of the problem (2) for each $i \in K$. Passing in this inequality to the limit $i \in K'$ we get $\gamma' \leq F_0(y')$. Adding (11) we obtain equality $F_0(y') = \gamma'$ which contradicts (7). QED

Theorem 2. Let the functions $P_i(x)$, $i \in K$, are chosen in the algorithm on condition

$$\lim_{i \in K} P_i(x) = +\infty \quad \forall \, x \notin D, \tag{12}$$

$\{ (x_k, \sigma_k) \}$, $k \in K' \subset K$, is a convergent subsequence of the sequence $\{ (x_k, \sigma_k) \}$, $k \in K$, and $\overline{u} = (\overline{x}, \overline{\sigma})$ is its limit point. Then

$\overline{x} \in X^*, \overline{\sigma} \in f^*.$

Proof. Remind that according to (4) $(x_k, \sigma_k) = (y_{i_k}, \gamma_{i_k}) = u_{i_k}$, $k \in K$. By the method of justification of equality (8) it is proved that

$$\lim_{k \in K'} || v_{i_k} - u_{i_k} || = 0 \qquad (13)$$

Distinguish a convergent subsequence $\{v_{i_k}\}$, $k \in K'' \subset K'$, of the sequence $\{v_{i_k}\}$, $k \in K'$. Let $\overline{v} = (\overline{w}, \overline{\alpha})$, where $\overline{w} \in R_n$ and $\overline{\alpha} \in R_1$ is its limit point. It is not difficult to check the validity of inequality $f(\overline{w}) \le \overline{\alpha}$ for this point. Moreover, considering the terms of (12) the inclusion $w \in D$ is proved. Hence, $v \in \mathrm{epi}(f, D)$. Then from (13) we have the inclusion $u \in \mathrm{epi}(f, D)$, i.e. $\overline{x} \in D$ и $f(\overline{x}) \le \overline{\sigma}$. But according to (5) $\sigma \le f^*$. Consequently, we get the inequalities $f(\overline{x}) \le \overline{\sigma} \le f^* \le f(\overline{x})$ from which the assertion of theorem follows.

Remark that the convergence theorem of the algorithm is proved without any additional requirements for the choice of the sequence $\{\Delta_k\}$, $k \in K$. In particular it is permitted to put $\Delta_k = \Delta > 0$ for all $k \in K$. If Δ is arbitrarily large, the inequality $F_i(y_i) - \gamma_i < \Delta$ holds for all $i \in K$. In this case the functions $P_{i+1}(x)$ differ from $P_i(x)$ and $x_k = y_i$, $k \in K$.

If we set the sequence $\{\Delta_k\}$ with the condition $\Delta_k \to 0$, $k \to \infty$, algorithm can be considered as the implementation of the penalty function method, where iteration points are found on the set of auxiliary functions D_0 from the condition of the approximate minimum ([10]. c. 380).

References

[1] Zabotin I Ya and Kobchikov A V 1990 On the Choice of the Optimal Ratio between the Masses of the Electronic System Blocks (in Russian) *Kazan university* vol **18** pp 64 – 70

[2] Zabotin I Ya and Kobchikov A V 1991 On the Realization of Some Methods Applied to Design Optimization Problems *Avtomat. i Telemekh.* no **1** pp 169 – 172

[3] Bulatov V P 1977 *Embedding methods in optimization problems* (in Russian) (Novosibirsk Nauka) 161 p.

[4] Zabotin I Ya and Yarullun R S 2013 One approach to constructing cutting algorithms with dropping of cutting planes Russian Math *Iz. VUZ* Allerton Press Inc. vol **57** no 3 pp 60–64

[5] Zabotin I Ya and Yarullun R S 2013 A Cutting Method for Finding Discrete Minimax with Dropping of Cutting Planes *Lobachevskii Journal of Mathematics* vol **35** no 2 pp 157 - 163

[6] Nesterov Yu E 2010 *Introduction to convex optimization* (in Russian). (Moscow: MCCME) p 274

[7] Zabotin I Ya, Shulgina O N and Yarullun R S 2014 A Cutting Method and Construction of Mixed Minimization Algorithms on Its Basis (in Russian) *Uchenye Zapiski Kazanskogo Universiteta* Ser. Fiz.-Matem. Nauki vol **156** no 4 pp 14 – 24.

[8] Zabotin I Ya and Yarullun R S 2015 A Cutting-Plane Method Based on Epigraph Approximation with Discarding the Cutting Planes *Automation and Remote Control* vol **76** no 11 pp 1966–1975

[9] Polyak B T 1983 *Vvedenie v optimizatsiyu (Introduction to Optimization)* (Moscow: Nauka) p 384

[10] Vasil'ev F P 2011 *Optimization methods* (in Russian) (Moscow: MCCME) p 620

[11] Konnov I V 2013 *Nonlinear optimization and variational inequalities* (in Russian) (Kazan: Kazan university) p 508

Zabotin I Ya 2011 Some embedding-cutting algorithms for mathematical programming problems (in Russian) *Izv. Irkutsk. Gos. Univ*. Ser. Matem. vol **4** no 2 pp 91–101

IOP Conf. Series: Materials Science and Engineering **158** (2016) 012098 doi:10.1088/1757-899X/158/1/012098

A minimization method on the basis of embedding the feasible set and the epigraph

I Ya Zabotin, O N Shulgina and R S Yarullin

Data mining and operation research department, Institute of Computational Mathematics and Information Technologies, Kazan (Volga Region) Federal University, Kazan, Kremlyovskaya str., 18, Russia

E-mail: IYaZabotin@mail.ru

Abstract. We propose a conditional minimization method of the convex nonsmooth function which belongs to the class of cutting-plane methods. During constructing iteration points a feasible set and an epigraph of the objective function are approximated by the polyhedral sets. In this connection, auxiliary problems of constructing iteration points are linear programming problems. In optimization process there is some opportunity of updating sets which approximate the epigraph. These updates are performed by periodically dropping of cutting planes which form embedding sets. Convergence of the proposed method is proved, some realizations of the method are discussed.

1. Introduction

Some algorithms from a class of cutting methods (e. g. [1 — 9]) are used to solve many applied optimization problems. Examples of these problems which are successfully solved by the mentioned methods can be found in [10 — 12]. Cutting methods are chosen for solving optimization problems, because there are some possibilities of estimating the proximity of the value of the objective function at the current iteration point to the optimal value.

Usually during practical implementations of the cutting methods there is some problem of accumulating cutting planes which form approximating sets. In this work the proposed method belongs to the mentioned class of methods and is characterized by using operation of simultaneous approximating the epigraph and the feasible set. The main feature of the method is possibility of updating embedding sets due to discarding any cutting planes.

2. Problem Settings

A problem

$$min\{ f(x) : x \in D\}, \quad (1)$$

is solved by the proposed method, where $f(x)$ is a convex function defined in n-dimensional Euclidian space R_n, and a set $D \subset R_n$ is closed and convex.

Suppose $f^* = \min\{f(x): x \in D\} > -\infty$, $X^* = \{x \in D : f(x) = f^*\}$, $x^* \in X^*$,

$$epi\,(f, R_n) = \{(x, \gamma) \in R_{n+1} : x \in R_n, \gamma \geq f(x)\},$$

and let $int\ Q$ be an interior of the set Q, $W(\bar{u}, Q)$ be a cone of the normalized generalized-support vectors at the point $\bar{u}$ for the set Q, assign $K = \{0, 1, \dots\}$.

3. Minimization Method

The proposed method constructs an auxiliary sequence $\{\boldsymbol{u_i}\}$, $\boldsymbol{i \in K}$, and a base sequence of approximations $\{\boldsymbol{x_k}\}$, $\boldsymbol{k \in K}$, for solving problem (1) by the following rule. Select points

$$\boldsymbol{v' \in int\,D,\quad v'' \in int\,epi\,(f, R_n)}.$$

Choose a convex closed bounded set $\boldsymbol{M_0 \subset R_n}$ and a convex closed set $\boldsymbol{G_0 \in R_{n+1}}$ such that

$$\boldsymbol{x^* \in M_0,\quad epi(f, R_n) \subset G_0}.$$

Define numbers $\boldsymbol{\bar{\gamma}}$, $\boldsymbol{\varepsilon_k}$, $\boldsymbol{k \in K}$, according to conditions

$$\boldsymbol{\bar{\gamma} \le min\{\, f(x) : x \in M_0\}},$$

$$\boldsymbol{\varepsilon_k > 0,\ k \in K,\ \varepsilon_k \to 0,\ k \to \infty}. \tag{2}$$

Assign $\boldsymbol{i = 0}$, $\boldsymbol{k = 0}$.

1. Find a solution $\boldsymbol{u_i = (y_i, \gamma_i)}$ of the following problem

$$\boldsymbol{\min\{\, \gamma : (x, \gamma) \in G_i, x \in M_k, \gamma \ge \bar{\gamma}\}}, \tag{3}$$

where $\boldsymbol{y_i \in R_n}$, $\boldsymbol{\gamma_i \in R_1}$.

2. Find $\boldsymbol{\bar{u}_i = (\bar{y}_i, \bar{\gamma}_i) \in R_{n+1}}$ as an intersection point of the segment $[\boldsymbol{v''}, \boldsymbol{u_i}]$ with the border of the set $\boldsymbol{epi\,(f, R_n)}$. If

$$\boldsymbol{\bar{u}_i = u_i,\quad y_i \in D}, \tag{4}$$

then $\boldsymbol{y_i \in X^*}$, and the process of solving problem (1) is finished.

3. If the inequality

$$\boldsymbol{||\bar{u}_i - u_i|| > \varepsilon_k} \tag{5}$$

is defined, then assign

$$\boldsymbol{G_{i+1} = S_i \cap \{u \in R_{n+1} : \langle b_i, u - \bar{u}_i \rangle \le 0\}}, \tag{6}$$

where

$$\boldsymbol{b_i \in W(\bar{u}_i, epi(f, R_n))}, \tag{7}$$

$$\boldsymbol{S_i = G_i}, \tag{8}$$

and increment $\boldsymbol{i}$ by one, go to Step 1. Otherwise go to Step 4.

4. Assign $\boldsymbol{i_k = i}$,

$$\boldsymbol{x_k = y_{i_k},\quad \sigma_k = \gamma_{i_k}}, \tag{9}$$

and construct a set $\boldsymbol{G_{i+1}}$ in accordance with (6), (7), where $\boldsymbol{S_i}$ is a convex closed set such that

$$epi(f, R_n) \subset S_i. \tag{10}$$

5. If $x_k \in D$, then assign $M_{k+1} = M_k$. Otherwise find $\overline{x}_k$ as a point situated in the intersection

of the segment $[v', x_k]$ with the border of the set D, and assign

$$M_{k+1} = M_k \cap \{x \in R_n : \langle a_k, x - \overline{x}_k \rangle \leq 0\},$$

where $a_k \in W(\overline{x}_k, D)$.

6. Increment i and k by one, and go to Step 1.

Lets make some remarks about the proposed method. Firstly, prove stopping criteria of the method which is represented at Step 2.

Theorem 1. Suppose that expressions (4) is defined for some number $i \in K$. Then the point y_i is a solution of problem (1).

Proof. In accordance with ways of choosing sets M_0, G_0, condition (10) and approach of constructing cutting planes it is easy to prove by induction that the feasible set of problem (3) contains the point (x^*, f^*) for all $i \in K$, $k \in K$. Consequently, the solution (y_i, γ_i) of this problem satisfies the inequality

$$\gamma_i \leq f^* \tag{11}$$

for any $i \in K$, $k \in K$.

Assume that the number $i \in K$ is selected according to (4). Then $\overline{y}_i = y_i$, $\overline{\gamma}_i = \gamma_i = f(\overline{y}_i)$, and, consequently, $f(y_i) = \gamma_i$ is determined. But from (4) it follows that $f(y_i) \geq f^*$. On the other hand, in view of (11) we have $f(y_i) \leq f^*$. Thus, the equation $f(y_i) = f^*$ is defined, and the theorem is proved.

Further, pay attention that there are some possibilities to update approximation sets G_{i+1} at iterations with numbers $i = i_k$, and this is convenient from the practical viewpoint. These updates are performed on the basis of constructing sets S_{i_k}. Namely, it will be shown below that for each $k \in K$ during constructing $\{u_i\}$, $i \in K$, the number $i = i_k$ will be fixed such that

$$||\overline{u}_i - u_i|| \leq \varepsilon_k. \tag{12}$$

Then in view of (10) we can assign, for example, $S_{i_k} = G_{r_i}$, where $0 \leq r_i \leq i_k - 1$. In this case we can discard some cutting planes which are constructed to the iteration i_k. On the other hand, the set S_{i_k} can be constructed on the basis of the last cutting planes, in particular, by getting active planes at the point u_{i_k}.

Note that in view of (10) we can assign $S_{i_k} = G_{i_k}$ for each $k \in K$. Then the equality $S_i = G_i$ is determined in (6) independently of conditions (5), (12), and updates of approximating sets do not occur.

Lemma 1. Suppose that the sequence $\{u_i\}$, $i \in K$, is constructed while S_i is defined in accordance with (8) for all $i \in K$. Then

$$\lim_{i \in K} ||\overline{u}_i - u_i|| = 0.$$

Proof. Assume the converse. Then there is a subsequence $\{u_i\}$, $i \in K' \subset K$, such that

$$||\bar{u}_i - u_i|| \geq \Delta > 0 \quad \forall i \in K'. \tag{13}$$

Select a convergence subsequence $\{u_i\}$, $i \in K'' \subset K'$, from the sequence $\{u_i\}$, $i \in K'$. Suppose i, $p_i \in K''$ such that $p_i > i$. From (6), (8) it follows that $G_{p_i} \subset G_i$. But $u_{p_i} \in G_{p_i}$, and, moreover, $b_i \in W(\bar{u}_i, G_{p_i})$. Consequently, we have $\langle b_i, u_{p_i} - \bar{u}_i \rangle \leq 0$. Since

$$\bar{u}_l = u_l + \alpha_l (v'' - u_l) \tag{14}$$

is defined for all $l \in K$, where $\alpha_l \in (0, 1)$, then from (14) and the last inequality it follows that

$$\langle b_i, u_i - u_{p_i} \rangle \geq \alpha_i \langle b_i, u_i - v'' \rangle.$$

In view of $v'' \in int\ epi\ (f, R_n)$ it is not difficult to prove existence of the number $\sigma > 0$ such that

$$\langle b_l, v'' - u_l \rangle \leq -\sigma$$

for all $l \in K$. Therefore, we get $\langle b_i, u_i - u_{p_i} \rangle \geq \alpha_i \sigma$ or $||u_i - u_{p_i}|| \geq \alpha_i \sigma$. Further, since the sequence $\{u_i\}$, $i \in K''$, is convergence, then $\alpha_i \to 0$, $i \in K''$, is determined. Thus, from equality (14) for $l = i$ it follows that

$$||\bar{u}_i - u_i|| \to 0, \quad i \in K''.$$

This limit expression contradicts to (13). The lemma is proved.

Now on the basis of Lemma 1 lets prove that according to (9) the sequence $\{(x_k, \sigma_k)\}$ will be constructed with the sequence $\{u_i\}$.

Lemma 2. Suppose that the sequence $\{u_i\}$, $i \in K$, is constructed by the proposed method. Then there exist a number $i = i_k$ for each $k \in K$ such that (12) is defined.

Proof. Assume $k \in K$. Lets obtain inequality (12) for some $i = i_k$. Suppose the converse, i.e. inequality (5) is determined for all $i \in K$. Then sets S_i are given by (8) for all $i \in K$, and the statement of Lemma 1 contradicts to inequality (5).

Lemma 3. Let $\{x_k\}$, $k \in K' \subset K$, be a convergence subsequence of the sequence $\{x_k\}$, $k \in K$, and $\bar{x}$ be its limit point. Then the inclusion

$$\bar{x} \in D \tag{15}$$

is defined.

Proof. If the inclusion $x_k \in D$ is defined for infinitely many numbers $k \in K'$, then in view of closedness of the set D statement (15) is obviously obtained. That's why suppose that $x_k \notin D$ is given for all numbers $k \in K'$ such that $k \geq N \in K'$.

Note that

$$\bar{x}_k = x_k + \theta_k (v' - x_k), \quad k \in K', \quad k \geq N, \tag{16}$$

where $\theta_k \in (0, 1)$. Lets choose numbers $l, p_l \in K'$ such that $p_l > l \geq N$. Since $M_{p_l} \subset M_l$, then $a_l \in W(\bar{x}_l, M_{p_l})$. But from (9), (13) it follows that $x_{p_l} \in M_{p_l}$, consequently, we have

$$\langle a_l, x_{p_l} - \bar{x}_l \rangle \leq 0,$$

and in view of (16) the inequality

$$\langle a_l, x_l - x_{p_l}\rangle \geq \theta_l \langle a_l, x_l - v'\rangle \tag{17}$$

is defined. Since $x_k \notin D$, $k \geq N$, $k \in K'$ and $v' \in int\, D$, then there exist a number $\delta > 0$ such that $\langle a_k, v' - x_k\rangle \leq -\delta$ for all $k \geq N$, $k \in K'$. Further, taking into account (17) and $||a_k|| = 1$ the inequality

$$||x_l - x_{p_l}|| \geq \theta_k \delta$$

is defined for any l, $p_l \in K'$ such that $p_l > l \geq N$. From this inequality and convergence of the sequence $\{x_k\}$, $k \in K'$, it follows that $\theta_k \to 0$, $k \in K'$. Then in accordance with (16) and boundness of the sequence $\{||v' - x_k||\}$, $k \in K'$, we get

$$||\overline{x}_k - x_k|| \to 0, \quad k \in K'. \tag{18}$$

Now lets select a convergence subsequence $\{\overline{x}_k\}$, $k \in K'' \subset K'$, from the sequence $\{\overline{x}_k\}$, $k \in K'$, and let $\widetilde{x}$ be its limit point. Then in view of (18) we have $\widetilde{x} = \overline{x}$, and from closedness of the set D it follows that inclusion (15) is defined. The lemma is proved.

At last, lets obtain the theorem of convergence of the proposed method.

Theorem 2. The following expressions

$$\overline{x} \in X^*, \quad \overline{\sigma} = f^*$$

are defined for any limit point $(\overline{x}, \overline{\sigma})$ of the sequence $\{(x_k, \sigma_k)\}$, $k \in K$, which is constructed by the proposed method.

Proof. Let $(\overline{x}, \overline{\sigma})$ be a limit point of the convergence subsequence $\{(x_k, \sigma_k)\}$, $k \in K' \subset K$, selected from the sequence $\{(x_k, \sigma_k)\}$, $k \in K$. According to Lemma 3 the inclusion

$$\overline{x} \in D$$

is defined, i.e.

$$f(\overline{x}) \geq f^*. \tag{19}$$

Choose a convergence subsequence $\{\overline{u}_{i_k}\}$, $k \in K'' \subset K$ from the sequence $\{\overline{u}_{i_k}\}$, $k \in K'$, and let $\overline{u}$ be its limit point. Since the set $epi\,(f, R_n)$ is closed, then

$$\overline{u} \in epi\,(f, R_n). \tag{20}$$

In accordance with condition (2) of constructing points ε_k and equations $u_{i_k} = (x_k, \sigma_k)$ which is obtained from inequalities $||\overline{u}_{i_k} - u_{i_k}|| \leq \varepsilon_k$, $k \in K''$, we have

$$\overline{u} = (\overline{x}, \overline{\sigma}).$$

Then in view of (20) the inclusion $(\overline{x}, \overline{\sigma}) \in epi(f, R_n)$ is determined, i.e.

$$f(\overline{x}) \leq \overline{\sigma}. \tag{21}$$

But according to (11) $\sigma_k \leq f^*$, $k \in K''$, and $\overline{\sigma} \leq f^*$. Then from inequalities (19), (21) it follows that $f(\overline{x}) = \overline{\sigma} = f^*$. The theorem is proved.

References

[1] Bulatov V P, Khamisov O V 2007 Cutting methods in E^{n+1} for global optimization of a class of functions *Zhurn. Vychisl. Matem. i Matem. Fiz.* **47** pp 1830—1842.

[2] Haarala N, Miettinen K, Makela M M 2007 Globally convergent limited memory bundle method for large-scale nonsmooth optimization *Mathematical Programming* **109** pp 181—205

[3] Kelley J E 1960 The cutting-plane method for solving convex programs *SIAMJ.* **8** pp. 703—712

[4] Kiwiel K C 2006 A proximal bundle method with approximate subgradient linearizations *SIAM J.* **16** pp 1007—1023

[5] Lemarechal C, Nemirovskii A, Nesterov Yu 1995 New variants of bundle methods *Math. Program.* **69** pp 111—148

[6] Sagastizabal C 2013 Composite proximal bundle method *Mathematical Programming* **140** pp 189—233

[7] Zabotin I Ya, Yarullin R S 2013 A cutting method for finding discrete minimax with dropping of cutting planes *Lobachevskii Journal of Mathematics* **35** pp 157—163

[8] Zabotin I Ya, Yarullun R S 2013 One approach to constructing cutting algorithms with dropping of cutting planes *Russian Math. (Iz.VUZ)* **57** pp 60—64

[9] Zabotin I Ya, Yarullin R S 2015 Cutting plane method based on epigraph approximation with discarding the cutting planes *Automation and Remote Control* **76** pp 1578—1587

[10] Bulatov V P 1977 Embedding methods in optimization problems *(Novosibirsk, Nauka)* p 161

[11] Belih T I, Bulatov V P 2008 Methods of supported cons and simplexes in the convex programming and their applications in some phisical-chemistry systems *Zhurn. vichisl. mat. i mat. kibern* **48** pp 1952—1967

[12] Gibbs J W 1961 On the equilibrium of heterogeneous substance *(Dover Publications, Inc., New York)*

IOP Conf. Series: Materials Science and Engineering **158** (2016) 012099 doi:10.1088/1757-899X/158/1/012099

Domain splitting algorithms for the Li-ion battery simulation

O Iliev[1] and P E Zakharov[2]

[1] Senior Scientist, Fraunhofer ITWM, Kaiserslautern, Germany
[2] Associate Professor-Researcher, NEFU, Yakutsk, Russia

Email: zapetch@gmail.com

Abstract. Numerical simulation of electrochemical processes in rechargeable batteries has important applications in an energy technology. In this paper we have developed and compared three domain splitting algorithms for the Li-ion battery simulation. Li-ion battery simulation is based on microscopic model, which contains nonlinear equations for Li-ion concentration and potential. On the interface of electrodes and electrolyte the Lithium ions intercalation are described by nonlinear equation. This nonlinear interface condition affects the Newton's method iterations and computation time. To simplify numerical simulations we use domain splitting algorithms, which split the original problem into three independent subproblems in two electrodes and electrolyte. We investigate the numerical convergence and efficiency of the algorithms on a 2D model problem.

1. Introduction

Microstructures and electrochemical processes are important in microscopic battery simulations [1]. We consider the problem in three subdomians: anode, electrolyte and cathode $\Omega = \Omega_a \cup \Omega_e \cup \Omega_c$. We denote the external boundary of anode, electrolyte, cathode by $\Gamma_a, \Gamma_e, \Gamma_c$, respectively and the interface boundary of anode and cathode by Γ_a^I and Γ_c^I.

Mathematical model of electrochemical processes in Li-ion battery are described by equations for Li-ion concentration and potential [2,3]. For each subdomain the equations of concentration c_a, c_e, c_c are defined. The indices $i = a, e, c$ define belonging function of subdomain: anode, electrolyte, cathode respectively. For simplicity we use dimensionless concentrations

$$c_a = \frac{\tilde{c}_a}{C_a}, \qquad c_e = \frac{\tilde{c}_e}{C_e}, \qquad c_c = \frac{\tilde{c}_c}{C_c},$$

where $\tilde{c}_i, i = a, e, c$ is the absolute value of concentrations, C_a, C_c are the maximum value of concentrations in electrodes, $C_e$is the mean value of concentration in electrolyte. For the dimensionless concentrations we chose different dividers because it is more convenient for comparison and solving of independent problems.

We define equations for electrical potential φ_a, φ_c in the electrodes and electrochemical potential φ_e in the electrolyte. The current densities in electrodes are

$$\begin{aligned} \boldsymbol{j}_a &= -\kappa_a \nabla \varphi_a, \qquad \boldsymbol{x} \in \Omega_a, \\ \boldsymbol{j}_c &= -\kappa_c \nabla \varphi_c, \qquad \boldsymbol{x} \in \Omega_c, \end{aligned}$$

and the current density in electrolyte is

$$\boldsymbol{j}_e = -\kappa_e \nabla \varphi_e + \kappa_e (1 - t_+) \frac{RT}{F} \nabla \ln c_e , \qquad \boldsymbol{x} \in \Omega_e,$$

where $\kappa_i, i = a, e, c$ is the conductivity, t_+ is the transport number of Lithium ions, R is the universal gas constant, T is the temperature, F is the Faraday constant.

The current densities in all subdomains are solenoidal vector fields

$$\nabla \boldsymbol{j}_i = 0, \qquad \boldsymbol{x} \in \Omega_i, \qquad i = a, e, c. \tag{1}$$

The concentration fluxes in the electrodes are

$$\boldsymbol{N}_a = -C_a D_a \nabla c_a, \qquad \boldsymbol{x} \in \Omega_a,$$
$$\boldsymbol{N}_c = -C_c D_c \nabla c_c, \qquad \boldsymbol{x} \in \Omega_c,$$

and in the electrolyte is

$$\boldsymbol{N}_e = -C_e D_e \nabla c_e + \frac{t_+}{F} \boldsymbol{j}_e, \qquad \boldsymbol{x} \in \Omega_e,$$

where $D_i, i = a, e, c$ is the diffusion coefficients. The concentration fluxes satisfy to the continuity equation without sources

$$C_i \frac{\partial c_i}{\partial t} + \nabla \boldsymbol{N}_i = 0, \ \boldsymbol{x} \in \Omega_i, \ i = a, e, c. \tag{2}$$

Intercalation processes occur on the contact of electrode and electrolyte. The current density of the intercalation on interface is described by the Butler-Volmer equation:

$$i_i = 2 i_{i,0} \sinh\left(\frac{F}{2RT} \eta_i\right), \ i = a, c, \tag{3}$$

where $i_{i,0}$ is the exchange current density

$$i_{i,0} = \sqrt{C_e C_i} k_i \sqrt{c_e c_i (1 - c_i)}, \qquad i = a, c,$$

and k_i is the reaction coefficient, η_i is the overpotential

$$\eta_i = \varphi_i - \varphi_e - U_{i,0}(c_i), \qquad i = a, c,$$

where $U_{i,0}(c_i)$ is the open circuit potential. The current density and the concentration flux on interface satisfy following conditions

$$\begin{aligned} \boldsymbol{j}_a \cdot \boldsymbol{n}_a &= -\boldsymbol{j}_e \cdot \boldsymbol{n}_e = i_a(c_a, \varphi_a, c_e, \varphi_e), & \boldsymbol{x} \in \Gamma_a^I, \\ \boldsymbol{N}_a \cdot \boldsymbol{n}_a &= -\boldsymbol{N}_e \cdot \boldsymbol{n}_e = \frac{1}{F} i_a(c_a, \varphi_a, c_e, \varphi_e), & \boldsymbol{x} \in \Gamma_a^I, \\ \boldsymbol{j}_c \cdot \boldsymbol{n}_c &= -\boldsymbol{j}_e \cdot \boldsymbol{n}_e = i_c(c_e, \varphi_e, c_c, \varphi_c), & \boldsymbol{x} \in \Gamma_c^I, \\ \boldsymbol{N}_c \cdot \boldsymbol{n}_c &= -\boldsymbol{N}_e \cdot \boldsymbol{n}_e = \frac{1}{F} i_c(c_e, \varphi_e, c_c, \varphi_c), & \boldsymbol{x} \in \Gamma_c^I, \end{aligned} \tag{4}$$

where $\boldsymbol{n}_i$ is the outside normal of subdomains.

For the concentration we set condition of non-permeability through outer boundary of all subdomains

$$\boldsymbol{N}_i \cdot \boldsymbol{n}_i = 0, \ \boldsymbol{x} \in \Gamma_i, \ i = a, e, c. \tag{5}$$

We fix the potential on the anode outer boundary

$$\varphi_a = \varphi_D(t), \ \boldsymbol{x} \in \Gamma_a, \tag{6}$$

and set the current density on the cathode outer boundary

$$\boldsymbol{j}_c \cdot \boldsymbol{n}_c = g(t), \ \boldsymbol{x} \in \Gamma_c. \tag{7}$$

There is no current through the electrolyte outer boundary

$$\boldsymbol{j}_e \cdot \boldsymbol{n}_e = 0, \ \boldsymbol{x} \in \Gamma_e. \tag{8}$$

The initial concentrations and potentials are given as constants

$$c_i = c_i^0, \quad \varphi_i = \varphi_i^0, \quad t = 0, \quad \boldsymbol{x} \in \Omega_i, \quad i = a, e, c. \tag{9}$$

The initial potentials are calculated from the interface conditions

$$\begin{aligned} \varphi_a^0 &= U_{a,0}(c_a^0), \\ \varphi_e^0 &= \frac{2RT}{F} \operatorname{arsinh}\left(\frac{[\Gamma_c] g(0)}{2[\Gamma_a^I] i_{a,0}(c_a^0, c_e^0)}\right), \\ \varphi_c^0 &= \varphi_e^0 + U_{c,0}(c_c^0) - \frac{2RT}{F} \operatorname{arsinh}\left(\frac{[\Gamma_c] g(0)}{2[\Gamma_c^I] i_{c,0}(c_c^0, c_e^0)}\right), \end{aligned} \tag{10}$$

where $[\Gamma_c]$ is the area of outer boundary Γ_c, $[\Gamma_a^I]$, $[\Gamma_c^I]$ is the areas of interface boundaries of anode and electrolyte. We chose these initial conditions for potentials because the initial values must be significantly close to solution for Newton's method.

2. Discretization

To discretize by time we use the implicit scheme with uniform step τ and denote the time $t^n = n\tau, n = 0,\dots,N, N = T/\tau$. For brevity, the function on current time layer we denote $c_i = c_i^{n+1}$, and on previous time layer $\breve{c}_i = c_i^n$.

For spatial approximation we use a finite element methods. Due to concentration and potential discontinues on the interface we consider problem on electrodes and electrolyte domains and connect by interface condition. We formulate the variational problem by presenting the equation in the weak formulation. Taking into account the interface (4) and the boundary conditions (5), (7), (8), the equations (1), (2) are written in the weak formulation

$$F_{a,c} = \int_{\Omega_a} \left(C_a \frac{c_a - \breve{c}_a}{\tau} v_a - \boldsymbol{N}_a \cdot \nabla v_a\right) \mathrm{d}\boldsymbol{x} + \int_{\Gamma_a^I} \frac{i_a}{F} v_a \mathrm{d}s,$$

$$F_{a,\varphi} = -\int_{\Omega_a} \boldsymbol{j}_a \cdot \nabla q_a \mathrm{d}\boldsymbol{x} + \int_{\Gamma_a^I} i_a q_a \mathrm{d}s,$$

$$F_{e,c} = \int_{\Omega_e} \left(C_e \frac{c_e - \breve{c}_e}{\tau} v_e - \boldsymbol{N}_e \cdot \nabla v_e\right) \mathrm{d}\boldsymbol{x} - \int_{\Gamma_a^I} \frac{i_a}{F} v_e \mathrm{d}s - \int_{\Gamma_c^I} \frac{i_c}{F} v_e \mathrm{d}s,$$

$$F_{e,\varphi} = -\int_{\Omega_e} \boldsymbol{j}_e \cdot \nabla q_e \mathrm{d}\boldsymbol{x} - \int_{\Gamma_a^I} i_a q_e \mathrm{d}s - \int_{\Gamma_c^I} i_c q_e \mathrm{d}s,$$

$$F_{c,c} = \int_{\Omega_c} \left(C_c \frac{c_c - \breve{c}_c}{\tau} v_c - \boldsymbol{N}_c \cdot \nabla v_c\right) \mathrm{d}\boldsymbol{x} + \int_{\Gamma_c^I} \frac{i_c}{F} v_c \mathrm{d}s,$$

$$F_{c,\varphi} = -\int_{\Omega_c} \boldsymbol{j}_c \cdot \nabla q_c \mathrm{d}\boldsymbol{x} + \int_{\Gamma_c^I} i_c q_c \mathrm{d}s + \int_{\Gamma_c} g q_c \mathrm{d}s,$$

where $v_i, q_i, i = a, e, c$ are the test functions.

Nonlinear variational problem is as follows:
Find functions $(c_a, \varphi_a, c_e, \varphi_e, c_c, \varphi_c) \in C_a \times \Phi_a \times C_e \times \Phi_e \times C_c \times \Phi_c$ such that

$$F_{a,c} + F_{a,\varphi} + F_{e,c} + F_{e,\varphi} + F_{c,c} + F_{c,\varphi} = 0,$$

$$\forall (v_a, q_a, v_e, q_e, v_c, q_c) \in V_a \times Q_a \times V_e \times Q_e \times V_c \times Q_c,$$

where

$$C_a = H^1(\Omega_a), \Phi_a = \{\varphi_a \in H^1(\Omega_a): \varphi_a = \varphi_D, x \in \Gamma_a\},$$

$$C_e = \Phi_e = H^1(\Omega_e),$$

$$C_c = \Phi_c = H^1(\Omega_c),$$

$$V_a = H^1(\Omega_a), Q_a = \{q_a \in H^1(\Omega_a): q_a = 0, x \in \Gamma_a\},$$

$$V_e = Q_e = H^1(\Omega_e),$$

$$V_c = Q_c = H^1(\Omega_c),$$

where $H^1(\Omega_i)$ are Sobolev spaces for subdomains Ω_i. Since the problem is considered in several subdomains and has two unknown functions, we define the general norm as follows:

$$\| (c_a, \varphi_a, c_e, \varphi_e, c_c, \varphi_c) \| = (\| c_a \|^2 + \| \varphi_a \|^2 + \| c_e \|^2 + \| \varphi_e \|^2 + \| c_c \|^2 + \| \varphi_c \|^2)^{\frac{1}{2}},$$

where function norm in subdomains

$$\| u_i \|^2 = \int_{\Omega_i} u_i^2 \mathrm{d}\boldsymbol{x}, \qquad i = a, e, c.$$

To get a linear variational problem from the nonlinear problem we use Newton's method [4]. We introduce the following notation for the unknown of iteration $k+1$

$$c_i^{k+1} = c_i^k + \delta c_i^k, \qquad \varphi_i^{k+1} = \varphi_i^k + \delta\varphi_i^k, \qquad i = a, e, c,$$

where c_i^k, φ_i^k are the known concentrations and potentials of the previous iteration, $\delta c_i^k, \delta\varphi_i^k$ are the correction functions, which we want to find. Also we introduce the following auxiliary notations

$$
\begin{aligned}
\delta \boldsymbol{N}_i^k &= \frac{\partial \boldsymbol{N}_i}{\partial c_i}^k \delta c_i^k + \frac{\partial \boldsymbol{N}_i}{\partial \varphi_i}^k \delta \varphi_i^k, \; i = a, e, c, \\
\delta \boldsymbol{j}_i^k &= \frac{\partial \boldsymbol{j}_i}{\partial c_i}^k \delta c_i^k + \frac{\partial \boldsymbol{j}_i}{\partial \varphi_i}^k \delta \varphi_i^k, \; i = a, e, c,
\end{aligned}
\tag{11}
$$

$$
\begin{aligned}
\delta i_a^k &= \frac{\partial i_a}{\partial c_a}^k \delta c_a^k + \frac{\partial i_a}{\partial \varphi_a}^k \delta \varphi_a^k + \frac{\partial i_a}{\partial c_e}^k \delta c_e^k + \frac{\partial i_a}{\partial \varphi_e}^k \delta \varphi_e^k, \\
\delta i_c^k &= \frac{\partial i_c}{\partial c_e}^k \delta c_e^k + \frac{\partial i_c}{\partial \varphi_e}^k \delta \varphi_e^k + \frac{\partial i_c}{\partial c_c}^k \delta c_c^k + \frac{\partial i_c}{\partial \varphi_c}^k \delta \varphi_c^k.
\end{aligned}
\tag{12}
$$

Using the introduced notations we write equations for the corrections

$$
\begin{aligned}
F_{a,c}^{k+1} &= \int_{\Omega_a} (C_a\, \delta c_a^k/\tau\, v_a - \delta \boldsymbol{N}_a^k \cdot \nabla v_a)\, \mathrm{d}\boldsymbol{x} + \int_{\Gamma_a^I} \delta i_a^k/F\, v_a \mathrm{d}s + F_{a,c}^k, \\
F_{a,\varphi}^{k+1} &= -\int_{\Omega_a} \delta \boldsymbol{j}_a^k \cdot \nabla q_a \mathrm{d}\boldsymbol{x} + \int_{\Gamma_a^I} \delta i_a^k q_a \mathrm{d}s + F_{a,\varphi}^k, \\
F_{e,c}^{k+1} &= \int_{\Omega_e} (C_e\, \delta c_e^k/\tau\, v_e - \delta N_e^k \cdot \nabla v_e)\, \mathrm{d}\boldsymbol{x} - \int_{\Gamma_a^I} \delta i_a^k/F\, v_e \mathrm{d}s - \int_{\Gamma_c^I} \delta i_c^k/F\, v_e \mathrm{d}s + F_{e,c}^k, \\
F_{e,\varphi}^{k+1} &= -\int_{\Omega_e} \delta j_e^k \cdot \nabla q_e \mathrm{d}\boldsymbol{x} - \int_{\Gamma_a^I} \delta i_a^k q_e \mathrm{d}s - \int_{\Gamma_c^I} \delta i_c^k q_e \mathrm{d}s + F_{e,\varphi}^k, \\
F_{c,c}^{k+1} &= \int_{\Omega_c} (C_c\, \delta c_c^k/\tau\, v_c - \delta N_c^k \cdot \nabla v_c)\, \mathrm{d}\boldsymbol{x} + \int_{\Gamma_c^I} \delta i_c^k/F\, v_c \mathrm{d}s + F_{c,c}^k, \\
F_{c,\varphi}^{k+1} &= -\int_{\Omega_c} \delta j_c^k \cdot \nabla q_c \mathrm{d}\boldsymbol{x} + \int_{\Gamma_c^I} \delta i_c^k q_c \mathrm{d}s + F_{c,\varphi}^k,
\end{aligned}
$$

where

$$
F_{i,c}^k = F_{i,c}, \qquad F_{i,\varphi}^k = F_{i,\varphi}, \qquad i = a, e, c.
$$

Then the linear variational problem for the corrections will be as follows: Find functions correction $(\delta c_a, \delta \varphi_a, \delta c_e, \delta \varphi_e, \delta c_c, \delta \varphi_c) \in V_a \times Q_a \times V_e \times Q_e \times V_c \times Q_c$ such that

$$
\begin{gathered}
F_{a,c}^{k+1} + F_{a,\varphi}^{k+1} + F_{e,c}^{k+1} + F_{e,\varphi}^{k+1} + F_{c,c}^{k+1} + F_{c,\varphi}^{k+1} = 0 \\
\forall (v_a, q_a, v_e, q_e, v_c, q_c) \in V_a \times Q_a \times V_e \times Q_e \times V_c \times Q_c,
\end{gathered}
$$

where

$$
\begin{gathered}
V_a = H^1(\Omega_a), \;\; Q_a = \{q_a \in H^1(\Omega_a), q_a = 0, x \in \Gamma_a\}, \\
V_e = Q_e = H^1(\Omega_e), \\
V_c = Q_c = H^1(\Omega_c).
\end{gathered}
$$

As initial approximations for Newton's method we use the values from the previous time layer. The Newton iterations are performed while the relative norm of corrections $\| \delta(c,\varphi) \| / \| (c,\varphi) \|$ is less than relative tolerance $t_r = 10^{-6}$, or absolute norm of corrections $\| \delta(c,\varphi) \|$ is less than absolute tolerance $t_a = 10^{-20}$. In the transition from continuous subdomains $\Omega_i, i = a, e, c$ to discrete subdomains $\Omega_i^h, i = a, e, c$ we use Lagrange elements of the 1st order.

3. Algorithms

To reduce the calculation time and to simplify the problem we use domain splitting algorithms. Due to the relatively small time derivative of functions in the electrolyte, we assume that: we can do not change the functions in electrolyte over few time layers. Using this assumption we consider several algorithms of splitting into independent subproblems.

We assume that the original equation in the electrolyte can be replaced by the following equations

$$
\frac{c_e - \breve{c}_e}{\tau} = 0, \; \frac{\varphi_e - \breve{\varphi}_e}{\tau} = 0,
\tag{13}
$$

then the interface conditions (4) for electrodes can be replaced with the following conditions

$$
\begin{aligned}
\boldsymbol{j}_a \cdot \boldsymbol{n}_a &= i_a(c_a, \varphi_a, \breve{c}_e, \breve{\varphi}_e), && \boldsymbol{x} \in \Gamma_a^I, \\
\boldsymbol{N}_a \cdot \boldsymbol{n}_a &= \tfrac{1}{F} i_a(c_a, \varphi_a, \breve{c}_e, \breve{\varphi}_e), && \boldsymbol{x} \in \Gamma_a^I, \\
\boldsymbol{j}_c \cdot \boldsymbol{n}_c &= i_c(\breve{c}_e, \breve{\varphi}_e, c_c, \varphi_c), && \boldsymbol{x} \in \Gamma_c^I, \\
\boldsymbol{N}_c \cdot \boldsymbol{n}_c &= \tfrac{1}{F} i_c(\breve{c}_e, \breve{\varphi}_e, c_c, \varphi_c), && \boldsymbol{x} \in \Gamma_c^I.
\end{aligned}
\tag{14}
$$

If we use interface conditions (14) then the original nonlinear variational problem is split into two independent subproblems in electrodes: Find the functions $(c_i, \varphi_i) \in C_i \times \Phi_i$ such that

$$F_{i,c} + F_{i,\varphi} = 0, \qquad \forall (v_i, q_i) \in V_i \times Q_i,$$

for each electrodes subdomain $i = a, c$. In Newton's method for each subproblem expressions (12) are replaced by the following expressions

$$\delta i_a = \frac{\partial i_a}{\partial c_a}^k \delta c_a + \frac{\partial i_a}{\partial \varphi_a}^k \delta \varphi_a,$$

$$\delta i_c = \frac{\partial i_c}{\partial c_c}^k \delta c_c + \frac{\partial i_c}{\partial \varphi_c}^k \delta \varphi_c.$$

Then the linear variation subproblems for corrections in electrodes are as follows: Find correction functions $(\delta c_i, \delta \varphi_i) \in V_i \times Q_i$ such that

$$F_{i,c}^{k+1} + F_{i,\varphi}^{k+1} = 0, \qquad \forall (v_i, q_i) \in V_i \times Q_i,$$

for each electrodes subdomain $i = a, c$.
Below we describe the first domain splitting algorithm of the problem simulation

Algorithm 1. The functions are fixed in electrolyte.

while $t < t_f$ **do**
  repeat
    $t = t + \tau,\ Solve(\Omega^h)$
  until $\| \frac{c_e - \check{c}_e}{\tau} \| < \delta$
  for $i = 1$ **to** N **do**
    $t = t + \tau,\ Solve(\Omega_a^h),\ Solve(\Omega_c^h),\ c_e = \check{c}_e,\ \varphi_e = \check{\varphi}_e$
  end for
end while

In the algorithm 1 firstly for each time layer we solve the complete system while the concentration change is sufficiently small in electrolyte. When the time derivative of the concentration in electrolyte is less than δ we solve the separated subproblems. In next time layers we solve the subproblems in electrodes and fix solution in electrolyte. We repeat these steps until the simulation ends.
We propose the second domain splitting algorithm, which based on reformulating the original boundary condition on anode. We use Neumann BC instead of Dirichlet BC on anode boundary. Also we don't fix functions in electrolyte and solve independent problem in electrolyte.
The absolute value of potential do not affect the mathematical model, since the only potentials difference and gradient are used. Thus if not considering the Dirichlet boundary condition, then for all potentials we can add any constant that does not affect the equations. We introduce the new unknown depending on the potential of the original problem

$$\phi_i = \varphi_i + \phi_D, \qquad i = a, e, c, \qquad \boldsymbol{x} \in \Omega_i, \qquad \phi_D \in R.$$

Instead of the Dirichlet boundary conditions (6), we introduce the Neumann boundary condition

$$\boldsymbol{j}_a \cdot n_a = g(t), \qquad \boldsymbol{x} \in \Gamma_a,$$

and choose the constant

$$\phi_D = \phi_a - \varphi_D, \qquad \boldsymbol{x} \in \Gamma.$$

Then we can remove the explicit definition of the Dirichlet condition, but we select the appropriate constant C which satisfies this condition. Note we assume that potential is the same on whole anode outer boundary due to high conductivity of contacting battery shell.
Under the new boundary condition in variational problems we change only the function spaces

$$\Phi_a = \{\phi_a \in H^1(\Omega_a)\}, Q_a = \{q_a \in H^1(\Omega_a)\}.$$

The solution in electrolyte is computed using the concentrations and potentials of electrodes from the current time layer, which are known after solution in electrodes.
Nonlinear variational problem for subproblem in electrolyte is as follows: Find functions $(c_e, \varphi_e) \in C_e \times \Phi_e$ such that

$$F_{e,c} + F_{e,\varphi} = 0, \qquad \forall (v_e, q_e) \in V_e \times Q_e.$$

In Newton's method for subproblem in electrolyte we need to replace the expressions (12) by the following expressions

$$\delta i_a = \frac{\partial i_a}{\partial c_e}^k \delta c_e + \frac{\partial i_a}{\partial \varphi_e}^k \delta \varphi_e,$$

$$\delta i_c = \frac{\partial i_c}{\partial c_e}^k \delta c_c + \frac{\partial i_c}{\partial \varphi_e}^k \delta \varphi_e.$$

Then linear variation subproblem for the corrections in electrolyte is as follows: Find correction functions $(\delta c_e, \delta \varphi_e) \in V_e \times Q_e$ such that

$$F_{e,c}^{k+1} + F_{e,\varphi}^{k+1} = 0, \qquad \forall (v_e, q_e) \in V_e \times Q_e.$$

The second algorithm is the same as the algorithm 1, the only difference is that we solve subproblem in electrolyte.

Algorithm 2. The solution is separated into subproblems.

while $t < t_f$ **do**
  repeat
    $t = t + \tau,\ Solve(\Omega^h)$
  until $\| \frac{c_e - \tilde{c}_e}{\tau} \| < \delta$
  for $i = 1$ **to** N **do**
    $t = t + \tau,\ Solve(\Omega_a^h),\ Solve(\Omega_c^h),\ Solve(\Omega_e^h)$
  end for
end while

With the third algorithm we do not need to solve the complete system, i.e. we solve only split subproblems. Each time layer we solve subproblems in electrodes, and solve the subproblem in electrolyte using results of electrodes subproblem.

Algorithm 3. Domain splitting scheme.

while $t < t_f$ **do**
  $t = t + \tau,\ Solve(\Omega_a^h),\ Solve(\Omega_c^h),\ Solve(\Omega_e^h)$
end while

Nonlinear and linear variational problems are the same as in the algorithm 2. We modify and substantially simplify the solution algorithm. In algorithm 3 we do not have parameters δ and N. This algorithm represent the domain splitting scheme.

4. Numerical experiments

We consider 2D model Li-ion battery problem with simple domain.

Table 1. Problem parameters.

Symbol	Parameter	Value	Unit
R	Universal gas constant	8.3144621	$J/(K\text{mol})$
T	Temperature	300	K
F	Faraday constant	96485.33	As/mol
C_e	Average concentration in eletrolite	0.001	mol/cm^3
t_+	Transport number of lithium ions,	0.2	–
κ_e	Electrolyte conductivity	0.001	S/cm
D_e	Electrolyte diffusion coefficient	10^{-7}	cm^2/s
C_a	Maximum concentration in anode	0.02	mol/cm^3
k_a	Anode reaction rate	0.002	$Acm^{2.5}/mol^{1.5}$

$U_{a,0}(c_a)$	Anode open circuit voltage	$e^{-10\backslash c_a}$	V
κ_a	Anode conductivity	10	S/cm
D_a	Anode diffusion coefficient	$5 \cdot 10^{-10}$	cm^2/s
C_c	Maximum concentration in cathode	0.02	mol/cm^3
k_c	Cathode reaction rate	0.2	$Acm^{2.5}/mol^{1.5}$
$U_{c,0}(c_c)$	Cathode open circuit voltage	$4 - 100(c_c - 0.5)^5$	V
κ_c	Cathode conductivity	1	S/cm
D_c	Cathode diffusion coefficient	10^{-9}	cm^2/s

Physical constants and properties of the battery components are given in Table 1. The boundary value for the anode potential is assigned as constant $\varphi_D = U_{i,0}(c_i^0)$. We set the current density at the cathode outer boundary as fixed number $g = \alpha C_{rate}$, where $\alpha \in \mathbf{R}$, C_{rate} is the current density at which the battery is theoretically fully charged or discharged in 1 hour, depending on the sign

$$C_{rate} = \frac{C_c F[\Omega_c]}{3600[\Gamma_c]},$$

where $[\Omega_c]$ is the size of domain Ω_c. In all calculations we used $\alpha = -1$, which means we simulate the battery charging. We take the initial conditions for the concentrations as follows

$$c_a^0 = 0.1, \qquad c_e^0 = 1, \qquad c_c^0 = 0.9,$$

and the potentials we calculated using the formula (10). The simulation end time is $T = 2900\ s$.

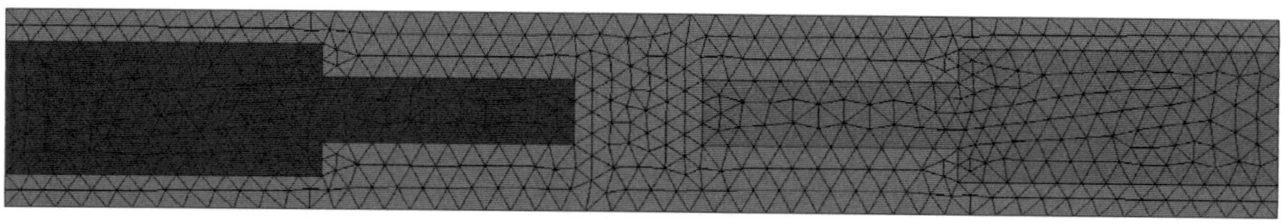

Figure 1. Computational mesh with 610 nodes. Cell size is $1\mu m$. Domain size is $50\mu m \times 7.5\mu m$.

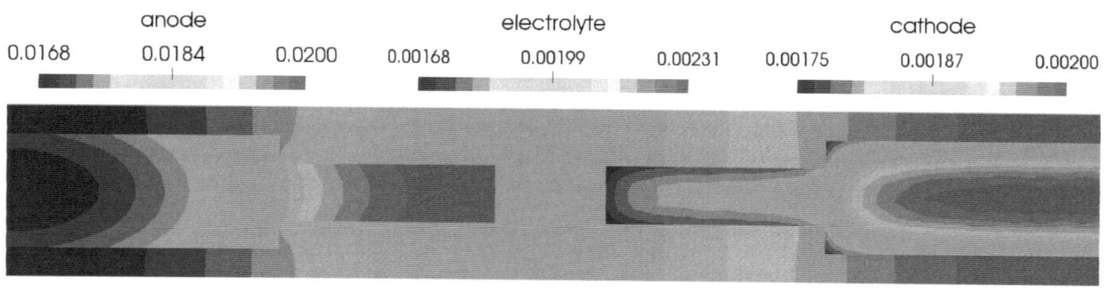

Figure 2. Concentrations in $t = 2900\ s$ with separate legends, mol/cm^3.

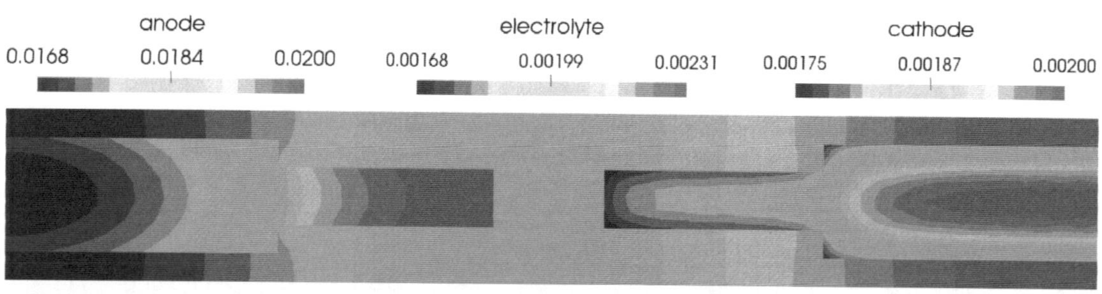

Figure 3. Potentials in $t = 2900\ s$ with separate legends, mol/cm^3.

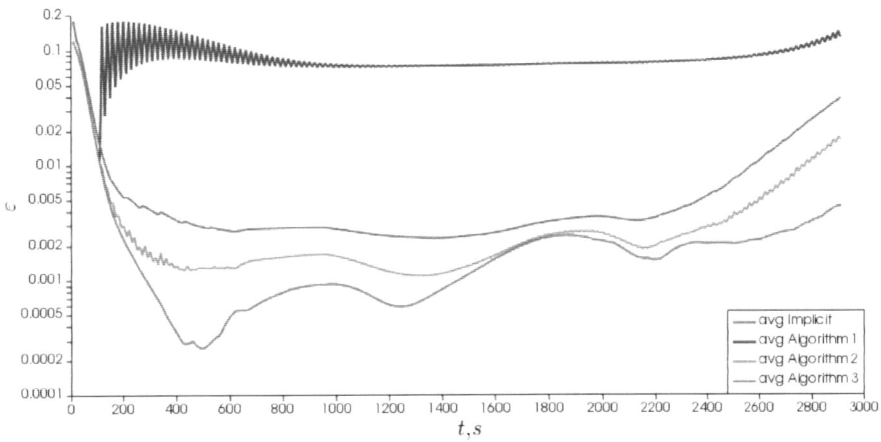

Figure 4. Errors norm comparison of the algorithms and implicit scheme.

In Figure 1 we show the computational mesh which has double unknowns on interface boundary. Numerical problem is solved in μm scale and domains size is $[\Omega] = 375\mu m^2$. Example of solution is showed in Figures 2 and 3. Numerical experiments with varying mesh size h, time step τ, parameters N and δ showed that the algorithms are numerically converge.

In the algorithms comparison we use error norm of solution in following form

$$\varepsilon = \| (c_a - c_a^*, \varphi_a - \varphi_a^*, c_e - c_e^*, \varphi_e - \varphi_e^*, c_c - c_c^*, \varphi_c - \varphi_c^*) \|,$$

where $\| (c_a, \varphi_a, c_e, \varphi_e, c_c, \varphi_c) \|$ is numerical solution with $\tau = 10\, s$ and $\| (c_a^*, \varphi_a^*, c_e^*, \varphi_e^*, c_c^*, \varphi_c^*) \|$ is the reference solution with $\tau = 2.5\, s$. Note that the general norm of constant function with value one

$$\| I \| = \| (I_a, I_a, I_e, I_e, I_c, I_c) \| = (2[\Omega_a]^2 + 2[\Omega_e]^2 + 2[\Omega_c]^2)^{\frac{1}{2}},$$

where $I_i = 1, x \in \Omega_i, i = a, e, c$. Then for our mesh $\| I \| \approx 332.6\mu m^2$

5. Conclusion

The main aim of this article is to develop and compare domain splitting algorithms for the problem describing electro-chemical processes in Li-ion batteries. We have developed and compared three domain splitting algorithms based on splitting by interface boundary. Our results showed that the algorithm 3 (splitting scheme) error is comparable with implicit scheme error (see Figure 4). The algorithms 1 and 2 computationally more complex than the algorithm 3 due to the solution of full system. The algorithm 3 is fully decouple the general problem into 3 independent subproblems and preserve precise accuracy, which can be comparable with the implicit scheme accuracy.

References

[1] Broussely M, Biensan Ph, Bonhomme F, Blanchard Ph, Herreyre S, Nechev K and Staniewicz RJ 2005 Main aging mechanisms in li ion batteries *Journal of Power Sources*, **146** 90–6

[2] Thomas KE, Newman J and Darling RM 2002 Mathematical modeling of lithium batteries *Kluwer Acad. Publ.* 345–92

[3] Less GB, Seo JH, Han S, Sastry AM, Zausch J, Latz A, Schmidt S, Wieser C, Kehrwald D and Fell S 2012 Micro-scale modeling of li-ion batteries: Parameterization and validation *Journal of The Electrochemical Society* **159(6)** A697–704

[4] Kelley CT 2003 *Solving Nonlinear Equations with Newton's Method* SIAM

Numerical studies on the performance of an aerosol respirator with faceseal leakage

S K Zaripov[1], I T Mukhametzanov[1] and S A Grinshpun[2,1]

[1]Institute of Environmental Sciences, Kazan Federal University, Kremlevskay str.,18, 420008, Russia
[2]Center for Health-Related Aerosol Studies, University of Cincinnati, USA

E-mail: shamil.zaripov@kpfu.ru

Abstract. We studied the efficiency of a facepiece filtering respirator (FFR) in presence of a measurable faceseal leakage using the previously developed model of a spherical sampler with porous layer. In our earlier study, the model was validated for a specific filter permeability value. In this follow-up study, we investigated the effect of permeability on the overall respirator performance accounting for the faceseal leakage. The Total Inward Leakage (*TIL*) was calculated as a function of the leakage-to-filter surface ratio and the particle diameter. A good correlation was found between the theoretical and experimental *TIL* values. The *TIL* value was shown to increase and the effect of particle size on *TIL* to decrease as the leakage-to-filter surface ratio grows. The model confirmed that within the most penetrating particle size range (~50 nm) and at relatively low leakage-to-filter surface ratios, an FFR performs better (*TIL* is lower) when the filter has a lower permeability which should be anticipated as long as the flow through the filter represents the dominant particle penetration pathway. An increase in leak size causes the *TIL* to rise; furthermore, under certain leakage-to-filter surface ratios, *TIL* for ultrafine particles becomes essentially independent on the filter properties due to a greater contribution of the aerosol flow through the faceseal leakage. In contrast to the ultrafine fraction, the larger particles (e.g., 800 nm) entering a typical high- or medium-quality respirator filter are almost fully collected by the filter medium regardless of its permeability; at the same time, the fraction penetrated through the leakage appears to be permeability-dependent: higher permeability generally results in a lower pressure drop through the filter which increases the air flow through the filter at the expense of the leakage flow. The latter reduces the leakage effect thus improving the overall respiratory protection level. The findings of this study provide valuable information for developing new respirators with a predictable actual workplace protection factor.

1. Introduction

Facepiece filtering respirators (FFRs) are used to protect workers from hazardous aerosol exposures. The efficiency of an FFR is generally dependent on the properties of its filter. It has been reported that the protection level offered by a respirator may substantially decrease in the presence of a faceseal leakage due to significant penetration of particle that bypass the filter [1], [2]. The particle flux through a faceseal leakage depends on the particle diameter, leak size, breathing flow rate and permeability of the filter medium. In our recent study [3], we developed a model for predicting the Total Inward Leakage (*TIL*) of an FFR. This model utilizes the previously validated fluid dynamic approach to an aerosol flow created by a spherical sampler with suction through a porous layer. In the quoted investigation, the particle penetration into an FFR was calculated for different ratios of the leak area to filter surface area at constant filter permeability which corresponded to a protection level of 95%. The present effort extends our earlier investigation [3] by considering the filter permeability as a variable that makes the model capable of predicting the efficiency of respirators equipped with various filter materials.

2. The problem statement

An idealized model of human head used in this investigation involves a spherical geometry with a circular opening for air suction (Fig.1). The sphere and opening radii are denoted by R_h and R_i respectively. The respirator filter is represented by a porous layer located upstream of the orifice. It is a spherical segment of height H_f and radius R_f. An annular peripheral opening between the sphere and the segment simulates the faceseal leakage of width W_l. The air movement in the breathing zone is described by a 2D axisymmetric steady flow passing through the porous layer and the peripheral opening. The air velocity U_0 far enough from the sphere is assumed constant.

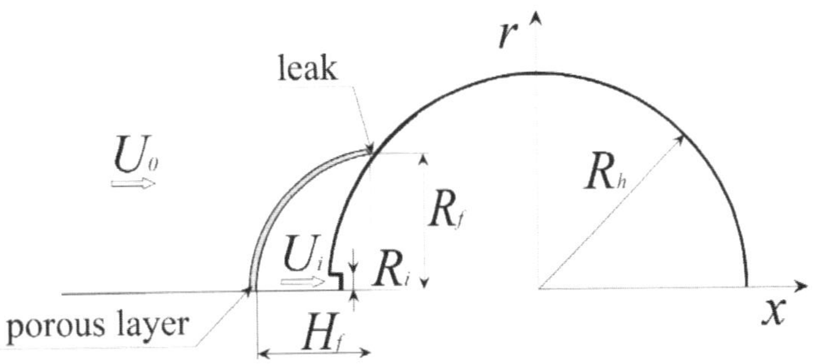

Figure 1. Schematics and major parameters included in the model.

The influence of aerosol particles on air flow is neglected due to their small concentration. The air motion in the free space and inside the porous medium (filter) approximated by a laminar viscous flow of an incompressible gas is described as follows [4]:

$$\nabla \cdot \overline{U} = 0 \tag{1}$$

$$\rho \overline{U} \cdot \nabla \overline{U} = -\varepsilon_f^2 \nabla P + \varepsilon_f \mu \Delta \overline{U} - a \varepsilon_f^2 \frac{\mu}{k} \overline{U} \tag{2}$$

In Eqs. (1)–(2), $\overline{U}$ is the air velocity; μ and ρ are the coefficients of dynamic viscosity and air density; P is the pressure; and ε_f is the medium porosity. The coefficient a is defined as being equal to zero outside the porous medium and equal to one inside the layer. Last term in Eq. (2) takes into account the additional air drag in the porous medium. In the latter case, it yields to the extended Darcy-Brinkman equations. To formulate a boundary value problem, the following boundary conditions were adopted for Eqs. (1)–(2): in the left part of the outer circular boundary of the computational domain, the air velocity is U_0; in the right part, the pressure is assumed to be equal to atmospheric pressure; the no-slip conditions are assumed on the sphere. The air velocity at the suction opening cross-section is defined as inhalation velocity U_i. The axial symmetry condition is applied along the x-axis.

The *TIL* is calculated as follows [3]:

$$TIL = \frac{Q_f \eta_f + Q_l \eta_l}{Q_i} \tag{3}$$

where Q_f and Q_l are the flow rates of air passing through the filter medium and the peripheral opening, respectively; $Q_i = U_i \pi R_i^2 = Q_f + Q_l$ is the air breathing flow rate; η_f and η_l are the particle penetrations through the filter layer and the peripheral leak, respectively.

As particles move within filter medium, the large ones are deposited on the filter fibers due to mechanisms such as inertia, interception and gravity, and the smaller ones are deposited due to diffusion and electrostatics. The particle penetration η_f through the filter medium with a thickness L is calculated as follows [5]:

$$\eta_f = \exp\left(-\frac{4\alpha E_{fiber} L}{\pi d_{fiber}(1-\alpha)}\right) \tag{4}$$

where α is the fiber packing density, d_{fiber} is the diameter of a filter fiber, and E_{fiber} is the collection efficiency of a fiber. If the aerosol particles are small enough to make the inertial and gravitational deposition negligibly small, E_{fiber} can be calculated as

$$E_{fiber} = 1-(1-E_r)(1-E_d)(1-E_q) \tag{5}$$

where E_r, E_d, and E_q are the collection efficiencies of a fiber by the action of interception, diffusion and electrostatic deposition, respectively. Given a very small depth of the opening (a typical faceseal leakage), no internal particle losses are assumed inside which suggests $\eta_l = 1$.

3. Results and discussion

The following geometrical parameters were selected for the numerical study: R_h=0.09 m, R_i = 0.007 m, R_f =0.057 m, and H_f=0.05 m. The filter surface area and the porous layer depth were S_f=0.018 m^2 and L=0.003 m, respectively. These values represent a typical European FFP2 or American N95 FFR. An ambient air velocity of U_0=0.1 m/s was chosen. The peripheral leakage area S_l (determined by W_l) was a study variable. The dimensions of the calculation domain were established so that the influence of the sphere on the ambient flow can be neglected: R_b= 20 m.

For the filter medium, the fiber diameter value was chosen to be 10.07 μm, in accordance to [3]. Three filter penetration values (referring to particles in the most penetrating size range) were considered: η_f = 0.1, 0.05, and 0.01. These represent filter media with collection efficiencies of 90%, 95% and 99%, respectively. Subsequently, the fiber packing densities were calculated to be α = 0.0479, 0.069, and 0.0888, respectively, producing permeability values of k=2.12×10^{-10} m^2, 9.55×10^{-11} m^2, and 5.41×10^{-11} m^2, respectively. In this calculation, we conservatively assumed the inhalation rate corresponding to a high workload.

To numerically solve the fluid flow equations [Eqs.(1)–(2)], ANSYS/FLUENT code was deployed. The second order upwind scheme was applied for the discretization of the air motion equations. The unstructured grid with an increased density in the vicinity of the boundary between the free space and the porous medium was utilized to obtain a numerical solution convergence with a residual of ~10^{-10}. The flow rate Q_l through leakage was determined from this numerical solution. The value of the flow rate Q_f through the filter was calculated as $Q_f = Q_i - Q_l$.

The data produced by the above model using a correction factor of 2 [3] were compared to the experimental data reported earlier in [1], [6], and [7] on *TIL* as a function of the leakage-to-filter surface ratio, S_l/S_f, obtained at Q_i=30 L min^{-1} and the porous layer permeability of k=$9.55\cdot10^{-11}$ m^2. The results of this comparison are shown in Fig. 2 for two particle diameters. It is seen that the model is capable of predicting the experimentally obtained *TIL* values. For both particle sizes, the value of *TIL* increases as S_l/S_f increases. Solid horizontal lines represent the value of *TIL* = 0.05 (which corresponds to η_f= 0.05 representing a commonly used N95 FFR). The *TIL* value exceeds 0.05 at S_l/S_f in excess of approximately 5×10^{-4}. A good agreement is seen between the theoretical findings [curves

$TIL(S_l/S_f)$] and the published experimental data. For $d_p = 50$ nm, as S_l/S_f decreases down to zero (sealed respirator, i.e. no leakage) *TIL* becomes constant (dotted line) and fully determined by the filter medium properties and the breathing condition. It is noted that the experiments and the model produced plateaus at different $TIL(S_l/S_f)$ values at S_l/S_f approaching zero. It can be explained given that the model assumes a steady inhalation flow while the experimental design adopted in [6] and [7] established a cyclic breathing.

Figure 3 presents $TIL(d_p)$ for k=2.12×10^{-10}, 9.55×10^{-11}, and 5.41×10^{-11} at different S_l/S_f and Q_i=30 L min^{-1}. At S_l/S_f = 0 and 8×10^{-5}, the peak *TIL* was observed for d_p=50 nm that falls in the range of the most penetrating particle sizes for fibrous respirator filters [8]. As S_l/S_f increases, *TIL* increases as well, and the effect of particle size on *TIL* becomes rather weak due to increase of the flow rate through the leak. It is also seen that at S_l/S_f=0 the $TIL(d_p)$ curve lies higher for a higher permeability, i.e. the model produces a well anticipated result for perfectly fit respirator facepiece (no faceseal leakage): a filter with lower permeability makes the respirator perform better. At the same time, as S_l/S_f increases (see S_l/S_f=8×10^{-4} and greater, Fig. 3), the model predicts that a better performance, i.e. a lower *TIL*, corresponds to a higher permeability suggesting an increasing flow rate through the leakage. For S_l/S_f as high as 4×10^{-3}, *TIL* exceeds 0.1 for all permeabilities considered in this study. This allows us to conclude that filter permeability does not produce a major effect on the respirator performance once the non-dimensional ratio of S_l/S_f exceeds a certain value.

Figure 4 presents the data predicted by our model in a form of *TIL* as a function of S_l/S_f for the same *k*-values, Q_i=30 *L* min^{-1}, and two very distinct particle sizes, 50 and 800 nm. The trends are similar, and a reasonable quantitative agreement is achieved for different permeabilities, except for the case of smaller particles (d_p = 50 nm) penetrating at relatively low S_l/S_f (as explained above). For the particles as large as d_p = 800 nm, a lower $TIL(S_l/S_f)$ curve corresponds to a larger *k* in the entire tested range of S_l/S_f which may seem counter-intuitive. The following explanation is suggested. In contrast to ultrafine particles, the larger ones entering a typical high- or medium-quality respirator filter are almost fully collected by the filter medium regardless of its permeability while the leakage penetration is permeability-dependent: higher permeability is generally associated with a lower pressure drop through the filter thus increasing Q_f at the expense of Q_l. Lowering Q_l reduces the leakage effect thus improving the overall respiratory protection level.

The results of this study help assess relative contributions of the two particle penetration pathways – the filter and the faceseal leakage. The findings provide valuable information for the development of new respirators with a predictable actual workplace protection factor.

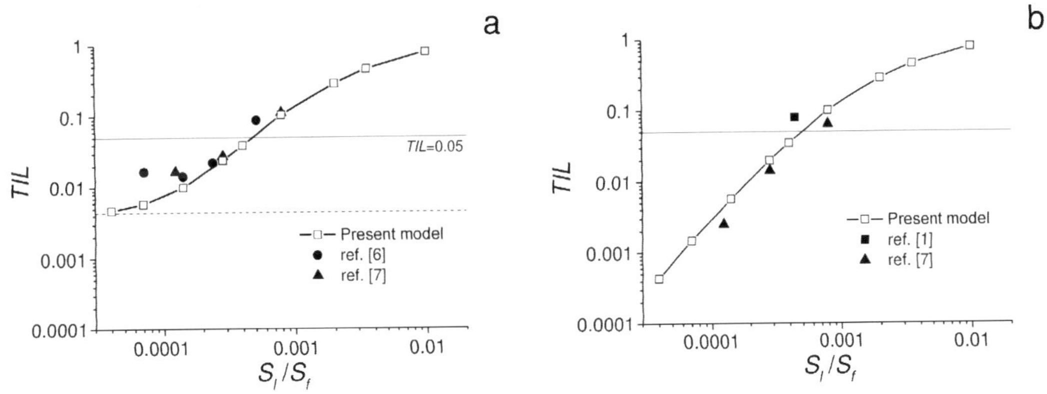

Figure 2. *TIL* as a function of S_l/S_f for d_p = 50 nm (a) and 800 nm (b). The solid horizontal lines represent the value *TIL* = 0.051 the dotted horizontal lines represent a "perfect fit" respirator (with no faceseal leakage).

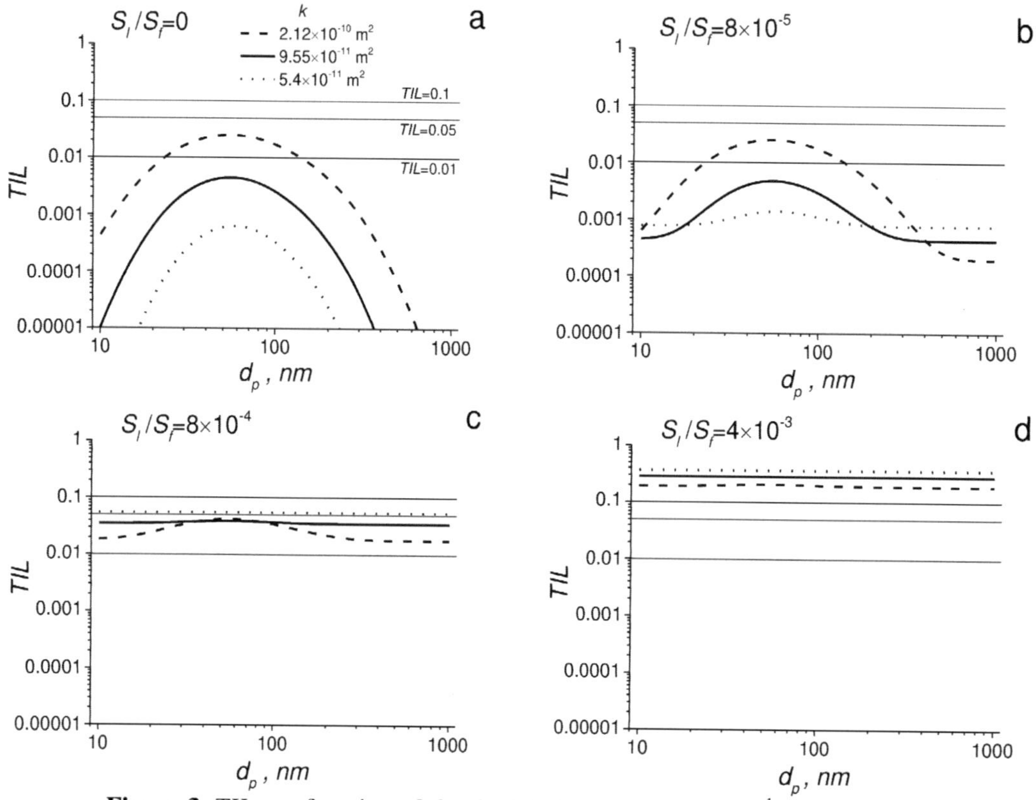

Figure 3. *TIL* as a function of d_p calculated for Q_i = 30 L min^{-1} and various S_l/S_f.

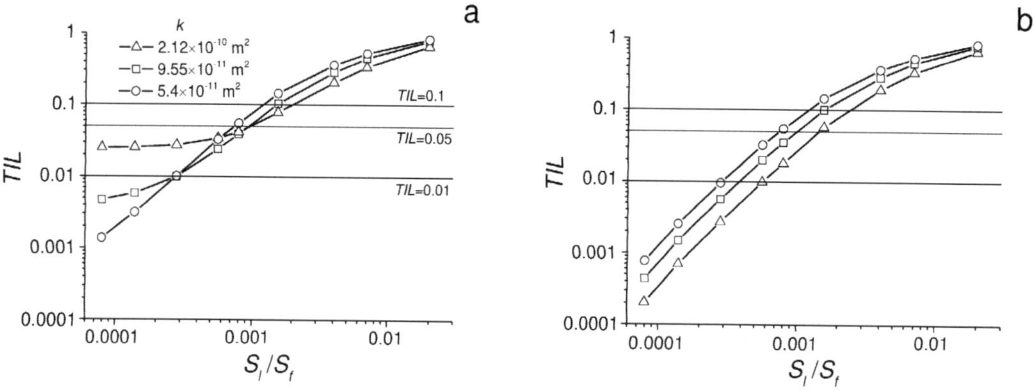

Figure 4. *TIL* as a function of S_l/S_f calculated for Q_i = 30 L min^{-1} for d_p = 50 nm (a) and 800 nm (b).

Acknowledgement

This investigation was conducted in the frameworks of the Russian Government Program of Competitive Growth at Kazan Federal University and supported by the Russian Foundation for Basic Research (grant No. 15-01-0613) and through a grant from the President of the Russian Federation (MK-6235.2015.1).

References

[1] Chen CC and Willeke K 1992 *Am Ind Hyg Assoc J.* **53** (9) 533–9

[2] Grinshpun SA, Haruta H, Eninger RM, Reponen T, McKay RT and Lee S-A. J 2009 *Occup Environ Hyg.* **6** (10) 593–603

[3] Mukhametzanov IT, Grinshpun SA, Zaripov SK and Gilfanov AK 2016 *Aerosol Air Qual Res*, in press.

[4] Bhattacharyya S, Dhinakaran S and Khalili A 2006 *Chem Eng Sci.* **61** (13) 4451–61

[5] Pich J 1966 *Aerosol Sci.* 223–83

[6] Rengasamy S and Eimer BC 2011 *Ann Occup Hyg* **55** (3) 253–63

[7] Rengasamy S and Eimer BC 2012 *Ann Occup Hyg.* **56** (5) 568–80

[8] Bałazy A, Toivola M, Adhikari A, Sivasubramani SK, Reponen T and Grinshpun SA 2006 *Am J Infect Control.* **34** (2) 51–7

IOP Conf. Series: Materials Science and Engineering **158** (2016) 012101 doi:10.1088/1757-899X/158/1/012101

Polar-symmetric problem of elastic diffusion for isotropic multi-component plane

A V Zemskov[1] and D V Tarlakovskii[2]

[1]Moscow Aviation Institute, Moscow, Russia
[2]Institute of Mechanics Lomonosov, Moscow State University, Moscow, Russia

E-mail: azemskov1975@mail.ru

Abstract. The paper considers a polar-symmetric problem of finding a stress strain condition of a plane influenced by non-stationary volume elastic diffusion disturbances. The mathematical model is based on a connected system of equations of elastic diffusion in a polar coordinate system.

The solution of the problem is sought in an integral for and presented in the form of convolutions of Green's function with the right side of equation of motion and mass transfer. Laplace time and Hankel's radial coordinate transformations are used to find the Green's functions. The inverse Laplace transform is done analytically by residue. The inverse Hankel's transform is done numerically by quadrature formulas.

1. Introduction

Depending on the geometry of the area, sometimes, it is practical to consider mechanics problem in curvilinear coordinates. For instance, for plane polar-symmetric problems it is convenient to construct a mathematical model in polar coordinates, if the problem solution area is a circle, ring or spherical cavity. Besides, the same case is with a polar-symmetric problem for a plane as an extreme case for a circle problem when its radius tends to infinity. All these problems in a polar coordinate system will be one-dimensional which may simplify their solution.

On other hand, for solution of such problems one can use an approach close to an approach to the solutions of similar problems in a Cartesian coordinate system. Specifically, it is convenient to search for a solution of an initially boundary value problem by making use of the Laplace transformation and expansions by eigenfunctions. This will allow for reducing the initial system of equations of motion and mass transfer to a system of linear algebraic equations. Correspondingly, the solution in the Laplace transforms will be a rational function which allows us to come to originals by residues.

Here the main problem is finding a system of proper functions which are solutions of the corresponding Sturm-Liouville problem. The simplest solution of this problem is found within a Cartesian coordinate system. With certain boundary conditions such functions can be represented by sinus and cosine [1]. For curvilinear coordinates there are also specific solutions [2-4] the form of which will depend on both selected boundary conditions and the geometry of the area.

Further on, a polar-symmetrical problem for an isotropic plane will be considered where a Bessel function of the first kind can be used as a proper function. The medium is an N-component solid solution where non-stationary volumetric disturbance cause coupled elastic and diffusion fields.

2. Statement of the problem

The equations describing coupled elastic and diffusion processes in an isotropic N-component solid solution in an arbitrary curvilinear coordinate system will be written as follows [5]:

$$\rho\frac{\partial^2\mathbf{u}}{\partial t^2}=(\lambda+\mu)\operatorname{graddiv}\mathbf{u}+\mu\Delta\mathbf{u}-\sum_{j=1}^{n}\alpha^{(j)}\operatorname{grad}\eta^{(j)}+\mathbf{F},$$

$$\frac{\partial\eta^{(q)}}{\partial t}=D^{(q)}\Delta\eta^{(q)}-\Lambda^{(q)}\Delta(\operatorname{div}\mathbf{u})+F^{(q)},\quad q=\overline{1,N}.$$

where t is time; $\mathbf{u}=\{u_1,u_2,u_3\}$ is displacement vector; $\eta^{(j)}=n^{(j)}-n_0^{(j)}$ is molar concentration increment of a j diffusant $n^{(j)}$ with reference to its initial concentration $n_0^{(j)}$; λ, and μ is elastic Lame constant; ρ is density of the environment; $\alpha^{(j)}$ is a coefficient of volume expansion due to mass transfer; $D^{(j)}$ is a coefficient of self-diffusion; $\Lambda^{(j)}=n_0D^{(j)}\alpha^{(j)}/(RT_0)$; R is absolute gas constant; T_0 - absolute ambient temperature, $\mathbf{F}$ is mass forces vector, $F^{(q)}$ is density of internal sources of mass transfer.

In case of polar symmetry the physical fields and mass forces vector will be written as follows:

$$\mathbf{u}=\{u_r(r),0,0\},\quad \eta^{(j)}=\eta^{(j)}(r),\quad \mathbf{F}=\{F_r(r),0,0\},$$

where r is a radial coordinate.

Correspondingly, the equations of motion and mass transfer for an isotropic plane will take the following forms:

$$\rho\frac{\partial^2u_r}{\partial t^2}=(\lambda+2\mu)\left(\frac{\partial^2u_r}{\partial r^2}+\frac{1}{r}\frac{\partial u_r}{\partial r}-\frac{u_r}{r^2}\right)-\sum_{j=1}^{N}\alpha^{(j)}\frac{\partial\eta^{(j)}}{\partial r}+F_r,$$

$$\frac{\partial\eta^{(q)}}{\partial t}=-\Lambda^{(q)}\left(\frac{\partial^3u_r}{\partial r}+\frac{2}{r}\frac{\partial^2u_r}{\partial r^2}-\frac{1}{r^2}\frac{\partial u_r}{\partial r}+\frac{u_r}{r^3}\right)+D^{(q)}\left(\frac{\partial^2\eta^{(q)}}{\partial r^2}+\frac{1}{r}\frac{\partial\eta^{(q)}}{\partial r}\right)+F^{(q)}.$$

Further on it is convenient to switch to non-dimensional numbers (with the same symbols they are marked with "*" which is omitted later on) as follows:

$$r^*=\frac{r}{L},\quad u=\frac{u_r}{L},\quad \tau=\frac{Ct}{L},\quad \eta_q=\frac{\eta^{(q)}}{n_0^{(q)}},\quad C^2=\frac{\lambda+2\mu}{\rho},$$

$$\alpha_q=\frac{n_0^{(q)}\alpha^{(q)}}{\lambda+2\mu},\quad D_q=\frac{D^{(q)}}{CL},\quad \Lambda_q=\frac{\Lambda^{(q)}}{n_0^{(q)}CL},\quad F_1=\frac{F_rL}{\lambda+2\mu},\quad F_{q+1}=\frac{F^{(q)}L}{Cn_0^{(q)}},$$

As a result, we will obtain (prime mark means a derivative with respect to r and period means a derivative with respect to non-dimensional time τ)

$$\ddot{u}=u''+\frac{u'}{r}-\frac{u}{r^2}-\sum_{q=1}^{N}\alpha_q\eta_q'+F_1,$$
$$\dot{\eta}_q=-\Lambda_q\left(u'''+\frac{2u''}{r}-\frac{u'}{r^2}+\frac{u}{r^3}\right)+D_q\left(\eta_q''+\frac{\eta_q'}{r}\right)+F_{q+1}. \quad (1)$$

The formulation of the problem is concluded by the initial conditions which will be assumed to be null.

3. Solution method

The solution of the problem is sought in an integrated form. Assume that $G_{nm}(r,\tau),\ n,m=\overline{1,N+1}$ - are Green's function of the subject problem, i.e. the solution of the following Cauchy problems:

$$\left(G''_{1m} + \frac{G'_{1m}}{r} - \frac{G_{1m}}{r^2}\right) - \sum_{j=1}^{N} \alpha_j G'_{j+1,m} + \delta_{1m}\delta(r-\xi)\delta(\tau) = \ddot{G}_{1m},$$
$$-\Lambda_q\left(G'''_{1m} + \frac{2G''_{1m}}{r} - \frac{G'_{1m}}{r^2} + \frac{G_{1m}}{r^3}\right) + D_q\left(G''_{q+1,m} + \frac{G'_{q+1,m}}{r}\right) + \delta_{q+1,m}\delta(r-\xi)\delta(\tau) = \dot{G}_{q+1,m}, \tag{2}$$
$$G_{1m}\big|_{\tau=0} = \dot{G}_{1m}\big|_{\tau=0} = G_{q+1,m}\big|_{\tau=0} = 0.$$

Thus, the solution of the problem (1) will take the form (asterisks will mean convolution by time τ and radius r):

$$u = \sum_{m=1}^{N+1} G_{1m}(r,\tau) ** f_m(r,\tau), \quad \eta_q = \sum_{m=1}^{N+1} G_{q+1,m}(r,\tau) ** f_m(r,\tau), \tag{3}$$

To find Greens functions let us employ the Laplace and Hankel transformation. The literature [6, 7] offer the following formulas for Hankel transformation from derivatives

$$H_k[f'] = -pH_{k-1}[f] - (1-k)H_k\left[\frac{f}{r}\right], \quad H_k\left[f' + (1-k)\frac{f}{r}\right] = -pH_{k-1}[f],$$

$$H_k\left[f'' + \frac{f'}{r} - \frac{k^2}{r^2}f\right] = -p^2 H_k[f], \quad H_k\left[f'' + \frac{f'}{r}\right] = -p^2 H_k[f] + k^2 H_k\left[\frac{f}{r^2}\right]$$

Here operator H_k (and the corresponding index further on) means Hankel of k order; p means a parameter of Hankel transformation; $f = f(r)$.

Let us present that

$$H_0\left[f''' + \frac{2f''}{r} - \frac{f'}{r^2} + \frac{f}{r^3}\right] = -p^3 H_1[f].$$

As a matter of fact

$$H_0\left[f''' + \frac{2f''}{r} - \frac{f'}{r^2} + \frac{f}{r^3}\right] = H_0\left[\left(f'' + \frac{f'}{r} - \frac{f}{r^2}\right)'\right] + H_0\left[\frac{1}{r}\left(f'' + \frac{f'}{r} - \frac{f}{r^2}\right)\right].$$

Consider the augend:

$$H_0\left[\left(f'' + \frac{f'}{r} - \frac{1}{r^2}f\right)'\right] = \int_0^\infty \left(f'' + \frac{f'}{r} - \frac{f}{r^2}\right)' J_0(pr)r\,dr = \int_0^\infty J_0(pr)r\,d\left(f'' + \frac{f'}{r} - \frac{f}{r^2}\right) =$$
$$= J_0(pr)r\left(f'' + \frac{f'}{r} - \frac{f}{r^2}\right)\Bigg|_0^\infty - \int_0^\infty \left(f'' + \frac{f'}{r} - \frac{f}{r^2}\right)\frac{d}{dr}[J_0(pr)r]dr =$$
$$= -\int_0^\infty \left(f'' + \frac{f'}{r} - \frac{f}{r^2}\right)[J_0(pr) + prJ_0'(pr)]dr = -H_0\left[\frac{1}{r}\left(f'' + \frac{f'}{r} - \frac{f}{r^2}\right)\right] +$$
$$+ p\int_0^\infty \left(f'' + \frac{f'}{r} - \frac{f}{r^2}\right) rJ_1(pr)dr = -H_0\left[\frac{1}{r}\left(f'' + \frac{f'}{r} - \frac{f}{r^2}\right)\right] - p^3 H_1[f].$$

In this case $J_0(x)$, $J_1(x)$ are Bessel functions of the first kind, zero and first order correspondingly. Thus,

$$H_0\left[f''' + \frac{2f''}{r} - \frac{f'}{r^2} + \frac{f}{r^3}\right] = -H_0\left[\frac{1}{r}\left(f'' + \frac{f'}{r} - \frac{f}{r^2}\right)\right] - p^3 H_1[f] +$$
$$+ H_0\left[\frac{1}{r}\left(f'' + \frac{f'}{r} - \frac{f}{r^2}\right)\right] = -p^3 H_1[f].$$

Which was to be proved. Now we apply the Laplace transformation to the equations (2), then, a first order Hankel transformation to the first equation (2), and a zero order Hankel transformation, to the second equation, which results into (s is a parameter of Laplace transformation, index L is a Laplace transform)

$$\begin{gathered}\left(p^2+s^2\right)G_{1m}^{H_1L}-\sum_{j=1}^{N}\alpha_j pG_{j+1,m}^{H_0L}=f_{1m}^{H_1L},\\ -\Lambda_q p^3 G_{1m}^{H_1L}+\left(D_q p^2+s\right)G_{q+1,m}^{H_0L}=f_{q+1,m}^{H_0L}.\end{gathered}\tag{4}$$

Where

$$G_{1m}^{H_1L}=\int_0^\infty rJ_1(pr)dr\int_0^\infty G_{1m}(r,\tau)e^{-s\tau}d\tau,$$

$$G_{q+1,m}^{H_0L}=\int_0^\infty rJ_0(pr)dr\int_0^\infty G_{q+1,m}(r,\tau)e^{-s\tau}d\tau,$$

$$f_{1m}^{H_1L}=\delta_{1m}\int_0^\infty rJ_1(pr)dr\int_0^\infty \delta(r-\xi)\delta(\tau)e^{-s\tau}d\tau=\delta_{1m}\xi J_1(p\xi),$$

$$f_{q+1}^{H_0L}=\delta_{q+1,m}\int_0^\infty rJ_0(pr)dr\int_0^\infty \delta(r-\xi)\delta(\tau)e^{-s\tau}d\tau=\delta_{q+1,m}\xi J_0(p\xi).$$

The solution of the system (4) takes the following form:

$$\begin{gathered}G_{11}^{H_1L}(p,\xi,s)=\frac{P_{11}(p,s)}{P(p,s)}\xi J_1(p\xi),\quad G_{1,q+1}^{H_1L}(p,\xi,s)=\frac{P_{1,q+1}(p,s)}{P(p,s)}\xi J_0(p\xi)\\ G_{q+1,1}^{H_0L}(p,\xi,s)=\frac{P_{q+1,1}(p,s)}{P(p,s)}\xi J_1(p\xi),\\ G_{q+1,q+1}^{H_0L}(p,\xi,s)=\left[\frac{1}{s+D_q p^2}+\frac{P_{q+1,q+1}(p,s)}{Q_q(p,s)}\right]\xi J_0(p\xi),\\ G_{q+1,p+1}^{H_0L}(p,\xi,s)=\frac{P_{q+1,p+1}(p,s)}{P(p,s)}\xi J_0(p\xi),\quad p=\overline{1,N},\quad p\neq q.\end{gathered}\tag{5}$$

where

$$P(p,s)=\left(s^2+p^2\right)\prod_{j=1}^{N}\left(s+D_j p^2\right)-\sum_{j=1}^{N}\alpha_j\Lambda_j p^4\prod_{r=1,r\neq j}^{N}\left(s+D_r p^2\right),\tag{6}$$

$$Q_q(p,s)=P(p,s)\left(s+D_q p^2\right),$$

$$P_{11}(p,s)=\prod_{j=1}^{N}\left(s+D_j p^2\right),\quad P_{1,q+1}(p,s)=\alpha_q p\prod_{j\neq q}^{N}\left(s+D_j p^2\right),\tag{7}$$

$$\begin{gathered}P_{q+1,1}(p,s)=\Lambda_q p^3\prod_{j=1,j\neq q}^{N}\left(s+D_j p^2\right),\\ P_{q+1,p+1}(p,s)=\alpha_p\Lambda_q p^4\prod_{j=1,j\neq\{p,q\}}^{N}\left(s+D_j p^2\right),\quad p=\overline{1,N},\ p\neq q,\\ P_{q+1,q+1}(p,s)=\alpha_q\Lambda_q p^4\prod_{j=1,j\neq q}^{N}\left(s+D_j p^2\right)\end{gathered}\tag{8}$$

Note. The formulas (6) - (8) must assume:

- if $N=1$

$$\prod_{j=1, j\neq q}^{1}\left(s + D_j\lambda_n^2\right) = 1,$$

- if $N = 2$

$$\prod_{j=1, j\neq p,q}^{2}\left(s + D_j\lambda_n^2\right) = 1.$$

The polynomials $P(p,s)$ and $Q_q(p,s)$ have the same structure as those of the similar polynomials in [1]. Therefore, the originals of Laplace transformed Green functions (5) will be written as follows:

$$G_{11}^{H_1}(p,\xi,\tau) = \tilde{G}_{11}(p,\tau)\xi J_1(p\xi), \quad G_{1,q+1}^{H_1}(p,\xi,\tau) = \tilde{G}_{1,q+1}(p,\tau)\xi J_0(p\xi),$$
$$G_{q+1,1}^{H_0}(p,\xi,\tau) = \tilde{G}_{q+1,1}(p,\tau)\xi J_1(p\xi),$$
$$G_{q+1,p+1}^{H_0}(p,\xi,\tau) = \tilde{G}_{q+1,p+1}(p,\tau)\xi J_0(p\xi),$$

$$\tilde{G}_{1m}(p,\tau) = e^{\gamma\tau}\left[A_{1m}^{(1)}\cos\beta\tau - A_{1m}^{(2)}\sin\beta\tau\right] + \sum_{j=1}^{N} A_{1m}^{(j+2)} e^{s_{j+2}\tau}, \quad m = \overline{1, N+1},$$

$$\tilde{G}_{q+1,p}(p,\tau) = e^{\gamma\tau}\left[A_{q+1,p}^{(1)}\cos\beta\tau - A_{q+1,p}^{(2)}\sin\beta\tau\right] + \sum_{j=1}^{N} A_{q+1,p}^{(j+2)} e^{s_{j+2}\tau}, \; p = \overline{1, N},$$

$$\tilde{G}_{q+1,q+1}(p,\tau) = e^{\gamma\tau}\left[A_{q+1,q+1}^{(1)}\cos\beta\tau - A_{q+1,q+1}^{(2)}\sin\beta\tau\right] + \sum_{j=1}^{N} A_{q+1,q+1}^{(j+2)} e^{s_{j+2}\tau} + A_{q+1,q+1}^{(N+3)} e^{-D_q p^2\tau},$$

where s_1 and $s_2 = \bar{s}_1$ are complex conjugate values, and s_{j+2}, $j = \overline{1,N}$ are real zeros of a polynomial $P(p,s)$. Meanwhile, $\beta = \operatorname{Re} s_1 < 0$, $\gamma = \operatorname{Im} s_1$, $s_{j+2} < 0$, $j = \overline{1,N}$. The coefficients $A_{nm}^{(l)}, l = \overline{1, N+2}$ are found as follows:

$$A_{1m}^{(1)} = 2\operatorname{Re}\frac{P_{1m}(\lambda, s_1)}{P'(\lambda, s_1)}, \; A_{1m}^{(2)} = 2\operatorname{Im}\frac{P_{1m}(\lambda, s_1)}{P'(\lambda, s_1)}, \; A_{1m}^{(j+2)} = \frac{P_{1m}(\lambda, s_{j+2})}{P'(\lambda, s_{j+2})},$$

$$A_{q+1,p}^{(1)} = 2\operatorname{Re}\frac{P_{q+1,p}(\lambda, s_1)}{P'(\lambda, s_1)}, \; A_{q+1,p}^{(2)} = 2\operatorname{Im}\frac{P_{q+1,p}(\lambda, s_1)}{P'(\lambda, s_1)}, \; A_{q+1,p}^{(j+2)} = \frac{P_{q+1,p}(\lambda, s_{j+2})}{P'(\lambda, s_{j+2})},$$

$$A_{q+1,q+1}^{(1)} = 2\operatorname{Re}\frac{P_{q+1,q+1}(\lambda, s_1)}{Q_q'(\lambda, s_1)}, \; A_{q+1,q+1}^{(2)} = 2\operatorname{Im}\frac{P_{q+1,q+1}(\lambda, s_1)}{Q_q'(\lambda, s_1)},$$

$$A_{q+1,q+1}^{(j+2)} = \frac{P_{q+1,q+1}(\lambda, s_{j+2})}{Q_q'(\lambda, s_{j+2})}, \; A_{q+1,q+1}^{(N+3)} = \frac{P_{q+1,q+1}(\lambda, -D_q\lambda^2)}{Q_q'(\lambda, -D_q\lambda^2)}.$$

The inverse Hankel transformation is found as follows:

$$G_{11}(r,\xi,\tau) = \int_0^\infty G_{11}^{H_1}(p,\xi,\tau) p J_1(pr) dp = \int_0^\infty \tilde{G}_{11}(p,\tau)\xi J_1(p\xi) p J_1(pr) dp,$$

$$G_{1,q+1}(r,\xi,\tau) = \int_0^\infty G_{1,q+1}^{H_1}(p,\xi,\tau) p J_1(pr) dp = \int_0^\infty \tilde{G}_{1,q+1}(p,\tau)\xi J_0(p\xi) p J_1(pr) dp,$$

$$G_{q+1,1}(r,\xi,\tau) = \int_0^\infty G_{q+1,1}^{H_0}(p,\xi,\tau) p J_0(pr) dp = \int_0^\infty \tilde{G}_{q+1,1}(p,\tau)\xi J_1(p\xi) p J_0(pr) dp,$$

$$G_{q+1,k+1}(r,\xi,\tau) = \int_0^\infty G_{q+1,k+1}^{H_0}(p,\xi,\tau) p J_0(pr) dp = \int_0^\infty \tilde{G}_{q+1,k+1}(p,\tau)\xi J_0(p\xi) p J_0(pr) dp.$$

To find solutions u and η_q, first, let us find convolution by time in (3). In this case

$$G_{11} ** f_1 = \int_0^\infty d\xi \int_0^\tau dt \int_0^\infty \tilde{G}_{11}(p,t)\xi J_1(p\xi) p J_1(pr) f_1(\xi, \tau - t) dp = \int_0^\infty S_{11}(p,\tau) p J_1(pr) dp, \tag{9}$$

and in the same way

$$G_{1,q+1} ** f_{q+1} = \int_0^\infty S_{1,q+1}(p,\tau) p J_1(pr) dp,$$

$$G_{q+1,m} ** f_1 = \int_0^\infty S_{q+1,m}(p,\tau) p J_0(pr) dp, \quad m = \overline{1, N+1},$$

where

$$\begin{aligned}
S_{11}(p,\tau) &= \int_0^\infty \xi J_1(p\xi) d\xi \int_0^\tau \tilde{G}_{11}(p,t) f_1(\xi, \tau - t) dt, \\
S_{1,q+1}(p,\tau) &= \int_0^\infty \xi J_0(p\xi) d\xi \int_0^\tau \tilde{G}_{1,q+1}(p,t) f_{q+1}(\xi, \tau - t) dt, \\
S_{q+1,1}(p,\tau) &= \int_0^\infty \xi J_1(p\xi) d\xi \int_0^\tau \tilde{G}_{q+1,1}(p,t) f_1(\xi, \tau - t) dt, \\
S_{q+1,k+1}(p,\tau) &= \int_0^\infty \xi J_0(p\xi) d\xi \int_0^\tau \tilde{G}_{q+1,k+1}(p,t) f_{k+1}(\xi, \tau - t) dt,
\end{aligned} \tag{10}$$

Here the functions $S_{mn}(p,\tau)$ can be found analytically with integration table [8]. The remaining integrals in (9) are found numerically.

4. Example

As an example let us consider a single component plane made of aluminum with the following mechanical properties [9]

$$\lambda + 2\mu = 1.26 \cdot 10^{11} \frac{H}{м^2}, \quad T_0 = 800\,K, \quad \rho = 2700 \frac{кг}{м^3}, \quad D = 7.73 \cdot 10^{-14} \frac{м^2}{c}, \quad L = 1\,м.$$

Set

$$f_1(r,\tau) = e^{-\varepsilon r} H(\tau), \quad f_2(r,\tau) = 0, \quad \mathrm{Re}\,\varepsilon > 0.$$

Then time convolutions (3) will give the following expressions for functions $S_{nm}(p,\tau)$ in (10):

$$S_{11}(p,\tau) = \frac{p}{(\varepsilon^2 + p^2)^{3/2}} \left[A_{11}^{(1)} I_1(\gamma,\beta,\tau) - A_{11}^{(2)} I_2(\gamma,\beta,\tau) + A_{11}^{(3)} I_3(s_3,\tau) \right]$$

$$S_{21}(p,\tau) = \frac{\varepsilon}{(\varepsilon^2 + p^2)^{3/2}} \left[A_{21}^{(1)} I_1(\gamma,\beta,\tau) - A_{21}^{(2)} I_2(\gamma,\beta,\tau) + A_{21}^{(3)} I_3(s_3,\tau) \right]$$

where I_m, $m = \overline{1,3}$ are found by as follows [1,8]:

$$I_1(\gamma,\beta,\tau) = \frac{1}{\gamma^2 + \beta^2} \left[e^{\gamma\tau} (\beta \sin \beta\tau + \gamma \cos \beta\tau) - \gamma \right]$$

$$I_2(\gamma,\beta,\tau) = \frac{1}{\gamma^2 + \beta^2} \left[e^{\gamma\tau} (\gamma \sin \beta\tau - \beta \cos \beta\tau) + \beta \right] \quad I_3(s,\tau) = \frac{e^{s\tau} - 1}{s}.$$

The inversed Hankel transforms are calculated numerically. The results of calculations are presented in Figures below.

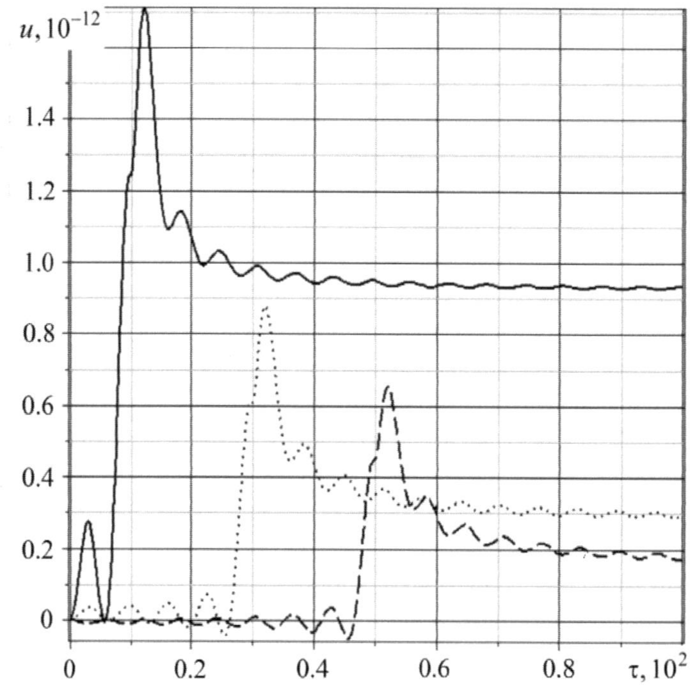

Figure 1. Time-displacement relation $u(x,\tau)$ where $r = 10$ (full line), $r = 30$ (dotted line), $r = 50$ (dash line)

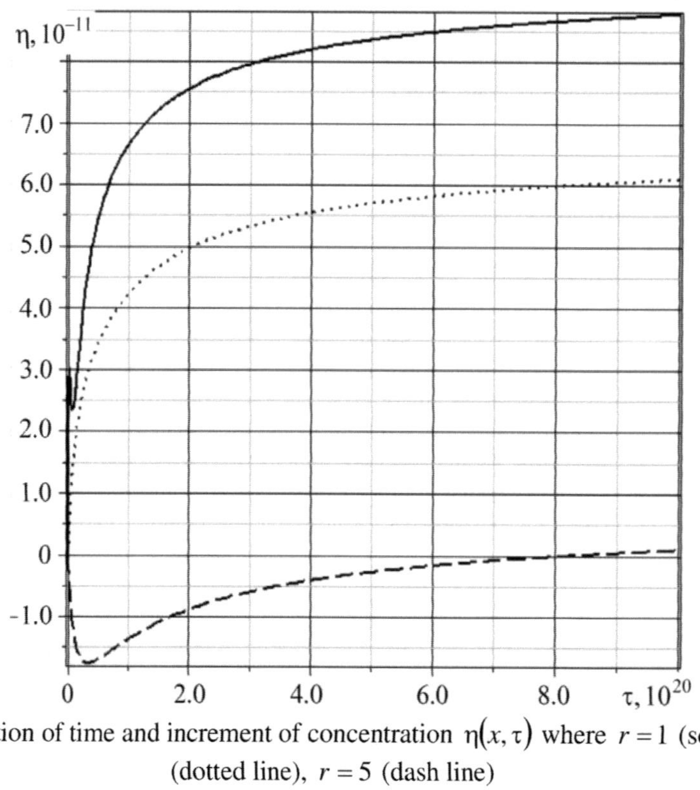

Figure 2. Relation of time and increment of concentration $\eta(x,\tau)$ where $r = 1$ (solid line), $r = 3$ (dotted line), $r = 5$ (dash line)

The calculations demonstrate that volumetric disturbances set in the right side of the equation of motion induce both mechanical displacements (Fig. 1) and mass transfer (Fig.2).

Conclusion

This paper introduces an algorithm of solution of a polar-symmetric non-stationary problem of elastic diffusion for an isotropic plane. There are dominant functions found allowing for identifying displacement fields and increment of ambient component concentration by preset volumetric disturbances. The work of the algorithm is exemplified by an effect of link of mechanic and diffusion field. The results of calculations are portrayed in graphs of relations of the desired fields and time in various points of a plane.

References

[1] Davydov S A, Zemskov A V, Tarlakovskii D V 2015 *Lobachevskii Journal of Mathematics* **36**,4 503.

[2] Sneddon I 1951 *Fourier transforms* (New York –Toronto – London).

[3] *Slepyan L.I., Yakovlev Y.S. Integralniye preobrazovaniya v nestatsionarnykh zadachakh mekhaniki.- L.: Sudostroyeniye, 1980, -344 s.* (Slepyan L.I., Yakovlev Y.S., Integral transformations in non-stationary problems of mechanics, in Russian)

[4] Tranter C J 1966 *Integral transforms in mathematical physics* (London: Methuen & Co. Ltd. New York: John Wiley & Sons, Inc)

[5] Zemskov A V, Tarlakovskiy D V *2015 Journal of Applied Mechanics and Technical Physics,* **56**,6 1023

[6] *Gorshkov A.G., Medvedevsky A.L., Rabinsky L.N., Tarlakovsky D.V. Volny v sploshnykh sredakh. Uchebnoye posobiye*: dlya vuzov. – M.: Phizmatlit, 2004. – 472 s. (Gorshkov A.G., Medvedevsky A.L., Rabinsky L.N., Tarlakovsky D.V., Waves in continuous mediums, in Russian)

[7] Piessens R. 2000 *The Hankel Transform*: (Boca Raton: CRC Press LLC)

[8] *Prudnikov A.P., Brychkov Y.A., Marichev O.I., Integraly I ryady. Tom 1. Elementarniye funktsii. – M.: Nauka, 1981, 797 s.* (Prudnikov A.P., Brychkov Y.A., Marichev O.I., Integrals and serials. Volume 1. Elementary functions. In Russian).

[9] *Fizicheskiye velichiny: Sprovochnik / Babichev A.P., Babushkina N.A., Bratkovsky A.M., I dr. Pod obshchey redaktsiyey Grigorieva I.S., Meylikhova I.Z. – M.: Energoatomizdat, 1991.* (Physical quantities: reference manual, edited by Grigorieva I.S., Meylikhova I.Z. In Russian)

Third type boundary conditions for steady state ambipolar diffusion equation

V S Zheltukhin[1], **S I Solov'ev**[2], **P S Solov'ev**[2], **V Yu Chebakova**[2] **and A M Sidorov**[2]

[1]Kazan State Technological University, 68 Karl Marx Street, Kazan, 420015, Russia
[2]Kazan Federal University, 18 Kremlevskaya Street, Kazan, 420008, Russia

E-mail: vchebakova@mail.ru

Abstract. A third type boundary condition for boundary value problem of electron balance in inductively coupled radio-frequency (ICRF) discharge is described. It is assumed that directions of diffusion and drift flows on boundary through the positive charge layer are coincided. Herein we study the effect of different boundary conditions on the parameters of ICRF discharge by computational solution of eigenvalue problems. It is showed that electronic temperature at boundary conditions derived in this study differ from the electronic temperature at boundary condition which mean a complete recombination of charged particles on the wall of gas discharge chamber (DC). The derived boundary conditions can be used for modelling ICRF discharges to determine optimal parameters of plasmatrons.

1. Introduction

ICRF discharge is widely used in different technological plasma processes. Modes of material processing are extremely sensitive to the main parameters of ICRF discharge. Mathematical modelling is necessary for more efficient and better choice of design solutions RF plasma equipment as some technological characteristics of plasma either evade direct measurements, or such measurements are labor-consuming [1-5].

2. Boundary condition derivation

Magnetic field in inductively plasmatron is directed lengthwise the coil axis, and force electric lines are the circle and plasma density is high; in other words, we can assume that current is vortex and the volume charge does not accumulate in quasineutral plasma, and relatively diffusion is ambipolar. ICRF discharges is modeled at local or non-local of electron distribution function which depends on ratio of electron energy relaxation length with skin-layer thickness. Herein we provide a mathematical model of ICRF discharge in case of non-local function of distribution in total discharge volume.

Evaluation of elementary processes taking place in low pressure ICRF plasma is showed that time of steady state stabilization exceeds a lot an electromagnetic field oscillation period. Correspondingly, fluctuations of plasma properties near average values, taken per the variation period, are negligible. With a glance to this hypothesis, balance equation of electrons in plasma of low pressure ICRF discharge takes the following form $-\frac{1}{r}\frac{\mathrm{d}}{\mathrm{d}r}\left(rD_a\frac{\mathrm{d}n}{\mathrm{d}r}+rV_a n\right)=\nu_i n-\beta n^2$, when $0<r<R$. Here R is the

radius, is a coefficient of ambipolar diffusion, β is a recombination coefficient, ν_i is an ionization frequency, V_a is gas flow velocity. It is supposed that D_a and ν_i are functions of electron temperature. Symmetry axis of the chamber stipulates condition that means continuity of flow of the charged particles. The border of the discharge including surface of the discharge chamber (DC) usually stipulates condition that means complete recombination of the charged particles on DC walls. But this condition results in infinite value of electron velocity near DC wall that is in conflict with discharge physics. It is known that double electric layer forms near the surface of a solid in plasma s and such solid has negatively potential to plasma. Therefore a condition of equality of diffusion flow from plasma to drift flow through a double layer is more correctly than the condition of complete recombination on DC walls. Evaluations of elementary processes in plasma of low pressure ICRF discharge are showed that a the last particle collisions take place approximately at the average free path from the wall surface. At law pressure the double layer thickness is more less than the mean free path of electrons. So we suppose that double layer is collisionless. Let us assume that deviation of the electron energy distribution function from Maxwellian is low at the first approximation. In this case, only those electrons can achieve the surface, which have velocity directed to the surface of electrode and energy of which is sufficient to overcome potential barrier. So, we can point microscopic density of electron and ion flux on the surface as flux density of corresponding particles, energy of which allows overcoming difference of potentials between the wall surface and potential of plasma at the border of plasma and double layer [6]. From the condition of continuity of particle flows, it follows that microscopic values of density of charged particle flux shall be equal to the corresponding values of flux density, when $x = X_b$:

$$U_e = -\frac{\mu_e}{\mu_e+\mu_i}\frac{J_b}{en_b} - \frac{D_a}{n_b}\left(\frac{\mathrm{d}n}{\mathrm{d}x}\right)_b, \quad U_i = -\frac{\mu_i}{\mu_e+\mu_i}\frac{J_b}{en_b} - \frac{D_a}{n_b}\left(\frac{\mathrm{d}n}{\mathrm{d}x}\right)_b$$

where n_b is density of particles at the border of plasma and double layer, μ_e, μ_i is electron and ion mobility ratios, J_b - density of charged particle flux.

As a result, we get the system

$$\begin{cases} -\dfrac{n_b\overline{C_e}}{4}h_e = -\dfrac{\mu_e}{\mu_e+\mu_i}\dfrac{J_b}{en_b} - \dfrac{D_a}{n_b}\left(\dfrac{\mathrm{d}n}{\mathrm{d}x}\right)_b, \\ -\dfrac{n_b\overline{C_i}}{4}h_i = -\dfrac{\mu_i}{\mu_e+\mu_i}\dfrac{J_b}{en_b} - \dfrac{D_a}{n_b}\left(\dfrac{\mathrm{d}n}{\mathrm{d}x}\right)_b. \end{cases}$$

After having divided the first equation of the system by μ_e, the second – by μ_i and after having summed them, as a result we get the following boundary condition:

$$D_a\left(\frac{\mathrm{d}n}{\mathrm{d}x}\right)_b = n_b\left(\frac{\mu_i\overline{C_e}h_e + \mu_e\overline{C_i}h_i}{4(\mu_e+\mu_i)}\right) \tag{1}$$

Here $\overline{C_{e,i}} = (8kT_{e,i}/\pi m_{e,i})^{1/2}$ are the average velocity of electrons and ions, correspondingly is, $T_{e,i}$ - temperature of electrons and ions, $m_{e,i}$ - weight of electrons and ions correspondingly, functions h_e, h_i are determined by the following way:

$$h_e = \begin{cases} \exp\left[\dfrac{e(\phi_b-\phi_p)}{kT_e}\right] & when\ \phi_b \ge \phi_p \\ 1 & when\ \phi_b \le \phi_p \end{cases}, \quad h_+ = \begin{cases} \exp\left[\dfrac{e(\phi_b-\phi_p)}{kT_e}\right] & when\ \phi_b \le \phi_p \\ 1 & when\ \phi_b \ge \phi_p \end{cases}$$

Here ϕ_b is potential at the surface, is potential of plasma at the border of plasma and double layer.

Now let us determine difference of potentials between border of plasma and surface, which is necessary to calculate $h_{e,i}$.

Let us consider the surface as an electrode with instantaneous value of the potential to plasma (at lower boundary of ambipolar zone), which is equal to V_p. In this case the constant potential of plasma to the electrode $\overline{V_p}$ can be calculated from dependences of average instantaneous potential fall through the layer for period of E-vibrations on a shift of fluctuating motion of electrons which equal to thickness of positive charge layer and on the difference $\phi_p - \phi_b = \dfrac{9kTe}{4e}$.

3. Numerical modelling

The paper work [7-8] describes balance problem of the charged particles of ICRF discharge, which is non-linear problem of eigenvalues, although a spectral parameter is absent in explicit form; as well this paper work gives explanation for origin and nature of the spectral parameter. It is shown that, as eigenvalues of differential problem can be in full accounted due to equation ratios and boundary conditions, so the sole free parameter is the value of electronic temperature in the center of the discharge, which meets some balance between ambipolar diffusion ratio, ionization frequency, recombination ratio and radius of plasmatron. This balance sets condition, necessary to support stationary low-pressure ICRF discharge. It is obvious that, by analogy with linear case, the eigenvalues of non-linear problem, forms function from electronic temperature too, but condition to calculate such case cannot be taken out in similar form. The paper work [9] describes the condition, which assumes existence of minimal eigenvalue that satisfies positive eigenfunction of non-linear problem toward eigenvalues with the ratios, which depend on spectral parameter aimed to determine concentration of the charged particles of the stable ICRF discharge. To resolve stationary and non-stationary non-linear problems, we can use methods proposed in [10-13].

Mathematical model, which takes into account the processes described in this study was implemented was realized numerically. We compared results of numerical calculations used to resolve model boundary value problems as including condition of equality to zero concentration on the wall of the gas discharge chamber and including the condition (1), by means of numerical modelling. Both problems are non-linear eigenvalue problems, spectral parameter of which is maximum value of the electronic temperature owing to dependencies D_a and v_i. It has been established that distribution of the charged particles concentration, which is rated to 1, under use of genus first boundary condition and genus third boundary condition, differs obviously near boundary of gas discharge chamber. But maximum value of electronic temperature, obtained due to resolving problem with third genus boundary condition, is far less then maximum value of electronic temperature obtained due to resolving the problem with the condition of complete recombination of the charged particles. As a result, the boundary condition, set on DC wall, provides more significant impact on maximum value of electronic temperature in the interelectrode interval, than on distribution of the charged particles concentration. This is caused owing to non-linearity of mathematical model of ICRF discharge and dependency of spectral parameter from electronic temperature, which in its turn can be presented as function set obviously from the least eigenvalue.

Acknowledgments

The work is performed according to the Russian Government Program of Competitive Growth of Kazan Federal University and this work was supported by Russian Science Foundation, project no. 16-11-10299.

References

[1] Abdullin I Sh, Zheltukhin V S and Kashapov N F 2000 *Radio-Frequency Plasma-Jet Processing of Materials at Reduced Pressures: Theory and Practice of Applications* (Kazan: Izd. Kazan. Univ.)[in Russian]

[2] Chebert P and Braithwaite N 2011 *Physics of radio-frequency plasmas* (Cambridge: Cambridge Univ. Press)

[3] Fridman A 2008 *Plasma chemistry* (Cambridge: Cambridge Univ. Press)
[4] Lebedev Y A, Tatarinov A V, Epstein I L and Averin K A 2015 Mathematical modeling of the gas bubbles in the microwave discharge of boiling n-heptane *Contemporary Engineering Sciences* **8** (21) 1057-65.
[5] Chebakova V Ju, Zheltukhin V S and Dubrovin V T 2016 Problem of radio-frequency discharge at atmospheric pressure in local statement *Applied Mathematical Sciences* **10** (21) 1013-22.
[6] Micher M and Kruger CH 1976 *Partially ionized gases* (Moscow: Mir)
[7] Zheltukhin V S 1999 On the solvability of a nonlinear spectral problem in the theory of high-frequency discharges of reduced pressure *Russ. Math.* **43** (5) 24–9
[8] Zheltukhin V S 2005 On conditions for the solvability of a system of boundary value problem in the theory of a high-frequency plasma of reduced pressure *Russ. Math.* **49** (1) 47–52
[9] Zheltukhin V S, Solov'ev S I, Solov'ev P S and Chebakova V Yu 2014 Computation of the minimum eigenvalue for a nonlinear Sturm–Liouville problem *Lobachevskii J. Math.***35** (4) 416–26
[10] Badriev I B and Banderov V V 2014 Numerical method for solving variation problems in mathematical physics *Mechanical components and control engineering III* **668-669** 1094-97
[11] Badriev I B 1983 Difference-schemes for linear-problems of the filtration theory with discontinuous law*Izvestiya Vysshikh Uchebnykh Zavedenii Matematika* **5** 3-12
[12] Solov'ev S I 1997 The finite element method for symmetric nonlinear eigenvalue problems *Comput. Math. Math. Phys* **37** (11) 1269–76
[13] Solov'ev S I 2014 Approximation of differential eigenvalue problems with a nonlinear dependence on the parameter *Differ. Equations* **50** (7) 947–54

Existence of solutions for electron balance problem in the stationary radio-frequency induction discharges

V S Zheltukhin[1], **S I Solovyev**[2], **P S Solovyev**[2] **and V Yu Chebakova**[2]

[1]Kazan State Technological University, 68 Karl Marx Street, Kazan, 420015, Russia
[2]Kazan Federal University, 18 Kremlevskaya Street, Kazan, 420008, Russia

E-mail: sergei.solovyev@kpfu.ru

Abstract. A sufficient condition for the existence of a minimal eigenvalue corresponding to a positive eigenfunction of an eigenvalue problem with nonlinear dependence on the parameter for a second order ordinary differential equation is established. The initial problem is approximated by the finite element method. Error estimates for the approximate minimal eigenvalue and corresponding positive eigenfunction are derived. Problems of this form arise in modelling the plasma of a radio-frequency discharge at reduced pressure.

1. Statement of the problem

In the present paper we investigate the problem of finding the minimal eigenvalue $\lambda \in \Lambda$, $\Lambda = [0,\infty)$, corresponding to a positive eigenfunction $u(x)$, $x \in \Omega$, $\Omega = (0,\pi)$, $\bar{\Omega} = [0,\pi]$, of the following eigenvalue problem

$$\begin{aligned} -(p(\lambda s(x))u')' &= r(\lambda s(x))u, \quad x \in \Omega, \\ u(0) &= u(\pi) = 0. \end{aligned} \tag{1}$$

Assume that $p(\mu)$, $r(\mu)$, $\mu \in \Lambda$, and $s(x)$, $x \in \bar{\Omega}$, are continuous positive functions, $p'(\mu)$, $r'(\mu)$, $\mu \in \Lambda$, are continuous functions, $p(\mu)$, $\mu \in \Lambda$, is nondecreasing bounded, $r(\mu)$, $\mu \in \Lambda$, is nondecreasing unbounded.

For fixed $\mu \in \Lambda$, by $\gamma(\mu)$ we denote the minimal simple eigenvalue corresponding to a positive eigenfunction $u(x) = u_\mu(x)$, $x \in \Omega$, of the parametric linear eigenvalue problem

$$\begin{aligned} -(p(\mu s(x))u')' &= \gamma(\mu) r(\mu s(x))u, \quad x \in \Omega, \\ u(0) &= u(\pi) = 0. \end{aligned} \tag{2}$$

Then the minimal eigenvalue λ of problem (1) is the minimal root of the following equation

$$\gamma(\mu) = 1, \quad \mu \in \Lambda. \tag{3}$$

In Section 2, assuming the condition $p(\xi s_1) > r(\xi s_2)$ for some $\xi \in \Lambda$, where s_1 and s_2 are the minimum and maximum of the function $s(x)$, $x \in \bar{\Omega}$, we prove the existence of a minimal simple eigenvalue of problem (1) being a minimal root of equation (3). In Section 3, we define a mesh scheme of the finite element method for solving problem (1). We prove the existence of a minimal simple approximate eigenvalue corresponding to a positive eigenfunction under the condition $p(\xi s_1) > r(\xi s_2)$ for some $\xi \in \Lambda$ and establish error estimates for the approximate minimal eigenvalue and corresponding positive eigenfunction. These results develop the results of the paper [1].

Problems of the form (1) arise in modelling the plasma of a radio-frequency discharge at reduced pressure. The sufficient condition obtained in the paper defines a condition necessary for maintaining a stationary inductive coupled radio-frequency discharge at reduced pressure [1–5].

Eigenvalue problems with nonlinear dependence on the parameter arise in various fields of science and technology [6–18]. Numerical algorithms for solving matrix nonlinear eigenvalue problems were constructed and investigated in [10,19–23]. Mesh methods for solving differential eigenvalue problems with nonlinear dependence on the spectral parameter were studied in [24–28]. The theoretical basis for the study of eigenvalue problems with nonlinear dependence on the parameter is results obtained for linear eigenvalue problems [29–35].

2. Existence of solutions

Let $H = L_2(\Omega)$ denotes the real Lebesgue space with norm $|.|_0$. By $V = \{v:\ v,\ v' \in H,\ v(0) = v(\pi) = 0\}$ we denote the real Sobolev space with norm $|.|_1$. Here we use the notation from [1]. For fixed $\mu \in \Lambda$, we introduce the bilinear forms

$$a(\mu,u,v) = \int_0^\pi p(\mu s(x))u'v'\mathrm{d}x, \quad b(\mu,u,v) = \int_0^\pi r(\mu s(x))uv\mathrm{d}x,$$

for $u,v \in V$ and the Rayleigh functional $R(\mu,v) = a(\mu,v,v)/b(\mu,v,v)$ for any $v \in V\setminus\{0\}$. Put $K = \{v:\ v \in V,\ v(x) > 0,\ x \in \Omega\}$.

The differential eigenvalue problem (1) is equivalent to the variational nonlinear eigenvalue problem: find minimal $\lambda \in \Lambda$ and $u \in K,\ b(\lambda,u,u) = 1,$ such that

$$a(\lambda,u,v) = b(\lambda,u,v) \quad \forall v \in V. \tag{4}$$

For fixed $\mu \in \Lambda$, the differential eigenvalue problem (2) is equivalent to the variational linear eigenvalue problem: find minimal $\gamma(\mu)$ and $u = u_\mu \in K,\ b(\mu,u,u) = 1,$ such that

$$a(\mu,u,v) = \gamma(\mu)b(\mu,u,v) \quad \forall v \in V. \tag{5}$$

Theorem 1. Suppose that $p(\xi s_1) > r(\xi s_2)$ for some $\xi \in \Lambda$, where s_1 and s_2 are the minimum and maximum of the function $s(x),\ x \in \bar{\Omega}$. Then there exists a minimal simple eigenvalue of problem (4) corresponding to a positive eigenfunction.

Proof. According to the variational characterization for the minimal simple eigenvalue of problem (5), we derive

$$\gamma(\xi) = \min_{v\in V\setminus\{0\}} R(\xi,v) = \min_{v\in V\setminus\{0\}} \frac{\int_0^\pi p(\xi s(x))(v')^2\mathrm{d}x}{\int_0^\pi r(\xi s(x))v^2\mathrm{d}x} \ge \frac{p(\xi s_1)}{r(\xi s_2)} \min_{v\in V\setminus\{0\}} \frac{\int_0^\pi (v')^2\mathrm{d}x}{\int_0^\pi v^2\mathrm{d}x} = \frac{p(\xi s_1)}{r(\xi s_2)} > 1.$$

By [1], $\gamma(\mu)$, $\mu\in\Lambda$, is the continuous function, and $\gamma(\mu)\to 0$ as $\mu\to\infty$. This implies the existence of a minimal root of equation (3), which defines the minimal eigenvalue λ of problem (4) corresponding to a positive eigenfunction. Thus, the theorem is proved.

3. Approximation of solutions

Define the partition of the interval $[0,\pi]$ by equidistant points $x_i = ih$, $i=0,1,\ldots,N$, into the elements $e_i=(x_{i-1},x_i)$, $i=1,2,\ldots,N$, $h=\pi/N$. By V_h denote the subspace of the space V consisting of continuous functions v^h linear on each element e_i, $i=1,2,\ldots,N$. Set $K_h=\{v^h:\ v^h\in V_h,\ v^h(x)>0,\ x\in\Omega\}$.

The variational nonlinear eigenvalue problem (4) is approximated by the following finite-dimensional nonlinear eigenvalue problem: find minimal $\lambda^h\in\Lambda$ and $u^h\in K_h$, $b(\lambda^h,u^h,u^h)=1$, such that

$$a(\lambda^h,u^h,v^h)=b(\lambda^h,u^h,v^h)\quad \forall v^h\in V_h. \tag{6}$$

For fixed $\mu\in\Lambda$, the variational linear eigenvalue problem (5) is approximated by the following finite-dimensional linear eigenvalue problem: find minimal $\gamma^h(\mu)$ and $u^h=u^h_\mu\in K_h$, $b(\mu,u^h,u^h)=1$, such that

$$a(\mu,u^h,v^h)=\gamma^h(\mu)b(\mu,u^h,v^h)\quad \forall v^h\in V_h. \tag{7}$$

Then the minimal eigenvalue λ^h of problem (6) is the minimal root of the following equation

$$\gamma^h(\mu)=1,\quad \mu\in\Lambda. \tag{8}$$

Theorem 2. Suppose that $p(\xi s_1)>r(\xi s_2)$ for some $\xi\in\Lambda$, where s_1 and s_2 are the minimum and maximum of the function $s(x)$, $x\in\bar{\Omega}$. Then there exists a minimal simple eigenvalue of problem (6) corresponding to a positive eigenfunction.

Proof. Using variational characterization for the minimal simple eigenvalue of problem (7), we get

$$\gamma^h(\xi)=\min_{v^h\in V_h\setminus\{0\}} R(\xi,v^h)=\min_{v^h\in V_h\setminus\{0\}}\frac{\int_0^\pi p(\xi s(x))((v^h)')^2dx}{\int_0^\pi r(\xi s(x))(v^h)^2dx}\geq\frac{p(\xi s_1)}{r(\xi s_2)}\min_{v\in V\setminus\{0\}}\frac{\int_0^\pi (v')^2dx}{\int_0^\pi v^2dx}=\frac{p(\xi s_1)}{r(\xi s_2)}>1.$$

According to [1], $\gamma^h(\mu)$, $\mu\in\Lambda$, is the continuous function, and $\gamma^h(\mu)\to 0$ as $\mu\to\infty$. This implies the existence of a minimal root of equation (8), which defines the minimal eigenvalue λ^h of problem (6) corresponding to a positive eigenfunction. Thus, the theorem is proved.

By c we denote various positive constants independent of h. For fixed $\mu\in\Lambda$, we introduce an operator $P_h(\mu):\ V\to V_h$ by the rule $a(\mu,u-P_h(\mu)u,v^h)=0$ for any $v^h\in V_h$, where $u\in V$, $|u-P_h(\mu)u|_0\leq ch^2$. Put $P_h=P_h(\lambda)$, $\|v\|^2_{b(\mu)}=b(\mu,v,v)$, $v\in V$, $\mu\in\Lambda$.

By $\gamma_i^h(\mu)$, $u_i^h(\mu)=u_i^h(\mu,x)$, $x\in\bar{\Omega}$, $\mu\in\Lambda$, $i=1,2,\ldots,N$, we denote eigenvalues and eigenfunctions satisfying (7) and such that

$$\gamma_1^h(\mu)<\gamma_2^h(\mu)<\ldots<\gamma_N^h(\mu),$$

$$a(\mu,u_i^h(\mu),u_j^h(\mu))=\gamma_i^h(\mu)\delta_{ij},\quad b(\mu,u_i^h(\mu),u_j^h(\mu))=\delta_{ij},\quad i,j=1,2,...,N,$$

$u_1^h(\mu)=u_\mu^h\in K_h$, $\gamma_1^h(\mu)=\gamma^h(\mu)$, the functions $u_i^h(\mu)$, $\mu\in\Lambda$, $i=1,2,...,N$, form a complete system in the space V_h. For fixed $\mu\in\Lambda$ and sufficiently small h, the estimates hold:

$$0\le\gamma_i^h(\mu)-\gamma_i(\mu)\le ch^2,\quad i=1,2,\quad |u_\mu^h-u_\mu|_0\le ch^2.$$

Theorem 3. Let λ be the minimal simple eigenvalue of problem (4) corresponding to the positive eigenfunction u, and let λ^h be the minimal simple eigenvalue of problem (6) corresponding to the positive eigenfunction u^h. Assume that $\gamma'(\lambda)\ne 0$. Then the following error estimates hold $0\le\lambda^h-\lambda\le ch^2$, $|u^h-u|_0\le ch^2$, for sufficiently small h.

Proof. First estimate follows from relations

$$c_1(\lambda^h-\lambda)\le-(\gamma^h(\xi^h))'(\lambda^h-\lambda)=\gamma^h(\lambda)-\gamma^h(\lambda^h)=\gamma^h(\lambda)-\gamma(\lambda)\le c_2h^2$$

for some ξ^h and sufficiently small h.

Let us prove second estimate. Set $\beta_i^h=b(\lambda^h,P_hu,y_i^h)$, $i=1,2,...,N$, where $y_i^h=u_i^h(\lambda^h)$, $i=1,2,...,N$. Since elements y_i^h, $i=1,2,...,N$, form an orthonormal basis in the space V_h, it follows that the element $P_hu\in V_h$ can be represented in the form $P_hu=\beta_1^hy_1^h+w_1^h$, where $w_1^h=\beta_2^hy_2^h+...+\beta_N^hy_N^h$. The inequality $\gamma_2(\lambda)-\gamma_1(\lambda)>0$ implies that $\gamma_2^h(\lambda^h)-1\ge c$ for sufficiently small h. Denote

$$\zeta_h(u)=\sup_{v^h\in V_h\setminus\{0\}}\frac{|a(\lambda^h,P_hu,v^h)-\lambda b(\lambda^h,P_hu,v^h)|}{|v^h|_1}.$$

Then $\zeta_h(u)\le ch^2$.

To show the estimate $|w_1^h|_1\le ch^2$ for sufficiently small h, we note that

$$a(\lambda^h,P_hu,w_1^h)=a(\lambda^h,w_1^h,w_1^h),$$

$$b(\lambda^h,P_hu,w_1^h)=b(\lambda^h,w_1^h,w_1^h),$$

$$a(\lambda^h,w_1^h,w_1^h)\ge\gamma_2^h(\lambda^h)b(\lambda^h,w_1^h,w_1^h).$$

Hence we get the relations

$$|w_1^h|_1\,\zeta_h(u)\ge a(\lambda^h,P_hu,w_1^h)-b(\lambda^h,P_hu,w_1^h)=a(\lambda^h,w_1^h,w_1^h)-b(\lambda^h,w_1^h,w_1^h)\ge$$

$$\ge\frac{\gamma_2^h(\lambda^h)-1}{\gamma_2^h(\lambda^h)}a(\lambda^h,w_1^h,w_1^h)\ge c^{-1}|w_1^h|_1^2,$$

which imply the desired estimate: $|w_1^h|_1\le c\zeta_h(u)\le ch^2$. Therefore $|P_hu-\beta_1^hy_1^h|_0\le ch^2$.

Moreover, we have

$$\beta_1^h=\|\beta_1^hy_1^h\|_{b(\lambda^h)}\le\|u\|_{b(\lambda)}+\Big|\|u\|_{b(\lambda)}-\|u\|_{b(\lambda^h)}\Big|+\|u-\beta_1^hy_1^h\|_{b(\lambda^h)}\le 1+ch^2,$$

$$\beta_1^h = \| \beta_1^h y_1^h \|_{b(\lambda^h)} \geq \| u \|_{b(\lambda)} - \left| \| u \|_{b(\lambda)} - \| u \|_{b(\lambda^h)} \right| - \| u - \beta_1^h y_1^h \|_{b(\lambda^h)} \geq 1 - ch^2,$$

since

$$\left| \| u \|_{b(\lambda)} - \| u \|_{b(\lambda^h)} \right| \leq \frac{\| u \|_{b(\lambda)}^2 - \| u \|_{b(\lambda^h)}^2}{\| u \|_{b(\lambda)} + \| u \|_{b(\lambda^h)}} \leq c(\lambda^h - \lambda) \leq ch^2,$$

$$\| u - \beta_1^h y_1^h \|_{b(\lambda^h)} \leq \sqrt{\beta_2}\, | u - \beta_1^h y_1^h |_0 \leq \sqrt{\beta_2} (| u - P_h u |_0 + | P_h u - \beta_1^h y_1^h |_0) \leq ch^2,$$

for $\beta_2 = r(\lambda s_1 + 1)$ and sufficiently small h. Consequently, we derive $| 1 - \beta_1^h | \leq ch^2$.

As a result, we conclude

$$\sqrt{\beta_1}\, | u^h - u |_0 \leq \| u - y_1^h \|_{b(\lambda^h)} \leq \| u - \beta_1^h y_1^h \|_{b(\lambda^h)} + \| y_1^h - \beta_1^h y_1^h \|_{b(\lambda^h)} =$$

$$= \| u - \beta_1^h y_1^h \|_{b(\lambda^h)} + | 1 - \beta_1^h | \leq ch^2,$$

where $\beta_1 = r(0)$. This completes the proof of the theorem.

Acknowledgments
This work was supported by Russian Science Foundation, project no. 16-11-10299.

References

[1] Zheltukhin V S, Solovyev S I, Solovyev P S and Chebakova V Yu 2014 Computation of the minimum eigenvalue for a nonlinear Sturm–Liouville problem *Lobachevskii J. Math.* **35** (4) 416–26

[2] Zheltukhin V S, Solovyev P S and Chebakova V Yu 2015 Boundary conditions for electron balance equation in the stationary high-frequency induction discharges *Res. J. Appl. Sci.* **10** (10) 658–62

[3] Abdullin I Sh, Zheltukhin V S and Kashapov N F 2000 *Radio-Frequency Plasma-Jet Processing of Materials at Reduced Pressures: Theory and Practice of Applications* (Kazan: Izd. Kazan. Univ.) [in Russian]

[4] Zheltukhin V S 1999 On the solvability of a nonlinear spectral problem in the theory of high-frequency discharges of reduced pressure *Russ. Math.* **43** (5) 24–9

[5] Zheltukhin V S 2005 On conditions for the solvability of a system of boundary value problem in the theory of a high-frequency plasma of reduced pressure *Russ. Math.* **49** (1) 47–52

[6] Zhigalko Yu P and Solovyev S I 2001 Natural oscillations of a beam with a harmonic oscillator *Russ. Math.* **45** (10) 33–5

[7] Solovyev S I 2016 Eigenvibrations of a beam with elastically attached load *Lobachevskii J. Math.* **37** (5) 597–609

[8] Solovyev S I 2003 Eigenvibrations of a plate with elastically attached load *Preprint* SFB393/03-06 Technische Universität Chemnitz

[9] Solovyev S I 2003 Vibrations of plates with masses *Preprint* SFB393/03-18 Technische Universität Chemnitz

[10] Solovyev S I 2011 *Nonlinear Eigenvalue Problems. Approximate Methods* (Saarbrücken: Lambert Acad. Publ.) [in Russian]

[11] Goolin A V and Kartyshov S V 1993 Numerical study of stability and nonlinear eigenvalue problems *Surv. Math. Ind.* **3** 29–48

[12] Betcke T, Higham N J, Mehrmann V, Schröder C and Tisseur F 2013 NLEVP: A collection of nonlinear eigenvalue problems *ACM Trans. Math. Software* **39** (2) Article number 7

[13] Kozlov V A, Mazyya V G and Rossmann J 2001 *Spectral Problems Associated with Corner Singularities of Solutions to Elliptic Equations* (Providence: American Mathematical Society)
[14] Apel Th, Sändig A-M and Solovyev S I 2002 Computation of 3D vertex singularities for linear elasticity: Error estimates for a finite element method on graded meshes *Math. Model. Numer. Anal.* **36** (6) 1043–70
[15] Lyashko A D and Solovyev S I 1991 Fourier method of solution of FE systems with Hermite elements for Poisson equation *Sov. J. Numer. Anal. Math. Modelling* **6** (2) 121–9
[16] Solovyev S I 1994 Fast direct methods of solving finite-element grid schemes with bicubic elements for the Poisson equation *J. Math. Sciences* **71** (6) 2799–804
[17] Solovyev S I 1995 A fast direct method of solving Hermitian fourth-order finite-element schemes for the Poisson equation *J. Math. Sciences* **74** (6) 1371–6
[18] Karchevskii E M and Solovyev S I 2000 Investigation of a spectral problem for the Helmholtz operator on the plane *Differ. Equations* **36** (4) 631–4
[19] Dautov R Z, Lyashko A D and Solovyev S I 1994 The bisection method for symmetric eigenvalue problems with a parameter entering nonlinearly *Russ. J. Numer. Anal. Math. Modelling* **9** (5), 417–27
[20] Solovyev S I 2006 Preconditioned iterative methods for a class of nonlinear eigenvalue problems *Linear Algebra Appl.* **415** (1) 210–29
[21] Ruhe A 1973 Algorithms for the nonlinear eigenvalue problem *SIAM J. Numer. Anal.* **10** 674–89
[22] Mehrmann V and Voss H 2004 Nonlinear eigenvalue problems: a challenge for modern eigenvalue methods *GAMM–Mit.* **27** 1029–51
[23] Beyn W-J 2012 An integral method for solving nonlinear eigenvalue problems *Linear Algebra Appl.* **436** (10) 3839–63
[24] Solovyev S I 1992 The error of the Bubnov–Galerkin method with perturbations for symmetric spectral problems with a non-linearly occurring parameter *Comput. Math. Math. Phys.* **32** (5) 579–93
[25] Solovyev S I 1997 The finite element method for symmetric nonlinear eigenvalue problems *Comput. Math. Math. Phys.* **37** (11) 1269–76
[26] Dautov R Z, Lyashko A D and Solovyev S I 1991 Convergence of the Bubnov–Galerkin method with perturbations for symmetric spectral problems with parameter entering nonlinearly *Differ. Equations* **27** (7) 799–806
[27] Solovyev S I 2014 Approximation of differential eigenvalue problems with a nonlinear dependence on the parameter *Differ. Equations* **50** (7) 947–54
[28] Solovyev S I 2015 Approximation of nonlinear spectral problems in a Hilbert space *Differ. Equations* **51** (7) 934–47
[29] Solovyev S I 1994 Superconvergence of finite element approximations of eigenfunctions *Differ. Equations* **30** (7) 1138–46
[30] Solovyev S I 2002 Superconvergence of finite element approximations to eigenspaces *Differ. Equations* **38** (5) 752–3
[31] Solovyev S I 2010 Approximation of variational eigenvalue problems *Differ. Equations* **46** (7) 1030–41
[32] Solovyev S I 2011 Approximation of positive semidefinite spectral problems *Differ. Equations* **47** (8) 1188–96
[33] Solovyev S I 2012 Approximation of sign-indefinite spectral problems *Differ. Equations* **48** (7) 1028–41
[34] Solovyev S I 2013 Approximation of differential eigenvalue problems *Differ. Equations* **49** (7) 908–16
[35] Solovyev S I 2015 Finite element approximation with numerical integration for differential eigenvalue problems *Appl. Numer. Math.* **93** 206–14

An approach for verification of finite-element analysis in nonlinear elasticity under large strains

K M Zingerman [1] **A V Vershinin** [2] **and V A Levin** [3]

[1] Tver State University, 33 Zhelyabov Street, Tver 170100, Russian Federation

[2] Fidesys LLC, Office 402, 1 bld. 77, MSU Science Park, Leninskie Gory, Moscow 119234, Russian Federation

[3] Lomonosov Moscow State University, Faculty of Mechanics and Mathematics, Russia, 119991, Moscow, GSP-1, 1 Leninskiye Gory, Main Building, Russian Federation

E-mail: zingerman@rambler.ru

Abstract. An approach to verification of finite-element calculations of stress-strain state of nonlinear elastic bodies under large deformations is suggested. The problems that may be reduced to one-dimensional ones using a semi-inverse method are taken as test problems. An example of such a test problem is the Lame problem for a cylinder. Generally, this problem for compressible hyperelastic materials has no exact analytical solution, but it can be reduced to a boundary value problem for an ordinary second-order nonlinear differential equation, and in some cases - to the Cauchy problem. A numerical solution of this problem can be used as a test one for finite element calculations carried out in three-dimensional statement. Some results of such verification (finite element calculations were performed using the Fidesys CAE-system) are presented.

1. Introduction

In finite element simulation of the phenomena taking place in deformable solids, it is necessary to verify the results of calculations. The usual way to do such verification - is a comparison of results of calculations for specially selected model problems with the results obtained with the help of other methods. A comparison with the exact analytical solutions is the most preferred one. In the theory of elasticity there are a lot of exact analytical solutions, but in the framework of the nonlinear theory of elasticity at large strains the number of exact analytical solutions is small. Typically, such solutions are known either for incompressible materials [1,2] or for materials whose mechanical properties are described by potentials of a special form, e. g. John potential (harmonic) [3]. In this connection, a semi-inverse method for solution of nonlinear elasticity problems is of interest. The method consists in the following, that a certain class of motions of a continuous medium is set up in the form of relationships between the coordinates of a particle in initial and final states containing unknown functions. By substituting these relationships into the equations and boundary conditions of the boundary value problem we obtain new equations and boundary conditions, which are less complicated than the initial ones. For example, the initial equations are the differential equations in partial derivatives, but after the substituting we obtain ordinary differential equations. The, the researchers aim is to find solutions to these new equations.

There are many publications dedicated to semi-inverse method, for example [4-11]. This method does not always give an opportunity to get an exact analytical solution of the problem. However, even in cases, where such a decision can not be found, the semi-inverse method can simplify the task greatly.

In this paper, the semi-inverse method is applied to solving of Lame problems for hollow cylinder at large deformations. Generally, this problem for compressible hyperelastic materials has no exact analytical solution, but it can be reduced to a boundary value problem for an ordinary second-order nonlinear differential equation, and in some cases - to the Cauchy problem. A numerical solution of this problem can be used as a test one for finite element calculations carried out in three-dimensional statement. Some results of such verification (finite element calculations were performed using the Fidesys CAE-system) are presented. The calculations were performed for the case where the mechanical properties of the cylinder material are described by defining relations for compressible Mooney-Rivlin material.

2. Problem statement

The static statement of nonlinear elasticity problem for finite deformations is formulated in the coordinates of the initial (non-deformed) state and includes the following equations.
Equilibrium equation (in the absence of mass forces)

$$\overset{0}{\nabla}\cdot \mathbf{P} = 0 \tag{1}$$

Constitutive relations

$$\mathbf{P} = \frac{dW}{d\mathbf{\Psi}} \tag{2}$$

Kinematic relations

$$\mathbf{\Psi} = \overset{0}{\nabla}\mathbf{R} \tag{3}$$

In the equations (1)-(3) $\mathbf{R}$ – the radius vector of a particle in a deformed state; $\mathbf{\Psi}$ - deformation gradient; W - elastic potential (strain potential energy density); $\mathbf{P}$ - the first Piola-Kirchhoff stress tensor, which is related with the true stress tensor $\boldsymbol{\sigma}$ in the following way:

$$\boldsymbol{\sigma} = J^{-1}\mathbf{\Psi}^{T}\cdot \mathbf{P} \tag{4}$$

here $J = \det \mathbf{\Psi}$; $J-1$ is the relative volume change.

The elastic potential for a compressible Mooney-Rivlin material is defined as follows [12]:

$$W = C_1\bar{I}_1 + C_2\bar{I}_2 + D(J-1)^2 \tag{5}$$

Here C_1. C_2. D are material constants; $\bar{I}_1 = J^{-2/3}I_1$; $\bar{I}_2 = J^{-4/3}I_2$; I_1. I_2 - invariants of the Green deformation tensor $\mathbf{G} = \mathbf{\Psi}\cdot\mathbf{\Psi}^{T}$.

Boundary conditions at a given pressure p at the boundary can be written as follows

$$\mathbf{n}\cdot\mathbf{P} = -pJ\mathbf{n}\cdot\left(\mathbf{\Psi}^{-1}\right)^{T} \tag{6}$$

A model problem of deformation of a hollow circular cylinder has been solved, a predetermined pressure was applied to the inner border of the cylinder, the outer boundary is free of loads and cylinder bases are fixed, so that displacements in the direction of the axis of the cylinder are equal to zero.

3. Approaches to problem solving

Finite-element solution was prepared using the Fidesys CAE-system [13]. In this system, a finite element method based on the Galerkin method [14-17] is implemented. A system of nonlinear algebraic equations, obtained with the help of the finite element approximation, is solved using the Newton-Kantorovich method.

In solving the problem using semi-inverse method, a cylindrical coordinate system was used. Let's denote coordinates of the particle in the undeformed state in this coordinate system with (r,z,φ) , and coordinates of the particle in the final state in the system - with (R,Z,Φ). Considering the symmetry of the problem and fixation conditions, the solution can be found in the following form

$$R = R(r),\ Z = z,\ \Phi = \varphi \tag{7}$$

Here $R(r)$ - unknown function.

Substitution of expression (7) in the kinematic relations (3) gives

$$\Psi_{rr} = R'(r),\ \Psi_{\varphi\varphi} = R(r)/r,\ \Psi_{zz} = 1 \tag{8}$$

The off-diagonal components of the deformation gradient are equal to zero.

The expressions for the components of the first Piola stress tensor **P** are determined from (2) as follows:

$$P_{rr} = \frac{\partial W}{\partial \Psi_{rr}},\ P_{\varphi\varphi} = \frac{\partial W}{\partial \Psi_{\varphi\varphi}},\ P_{zz} = \frac{\partial W}{\partial \Psi_{zz}} \tag{9}$$

The off-diagonal components of this tensor are equal to zero due to the isotropy of material. Considering (8), one can see that all the components of the tensor **P** in this problem will depend on the r only. Therefore, the vector equilibrium equation (1) reduces to a single scalar equation:

$$\frac{dP_{rr}}{dr} + \frac{1}{r}(P_{rr} - P_{\varphi\varphi}) = 0 \tag{10}$$

The other two equilibrium equations are completed identically.

Equation (10) in view of (9) and (8) is converted to an ordinary differential equation of the second order with respect to $R(r)$. The boundary conditions for this equation are written on the basis of the boundary condition (6). By denoting a and b the radii of the outer and inner cylinder, respectively, these boundary conditions can be written in the form

$$P_{rr}(a) = 0,\ P_{rr}(b) = -p\Psi_{\varphi\varphi}(b) \tag{11}$$

Along with the boundary value problem for a differential equation obtained from (10) with the boundary conditions (11), we can consider the Cauchy problem for the same equation with the given values of $R(a)$ and $R'(a)$. From a mechanical point of view, this means that instead of setting up the pressure on the inner boundary, radius variation of the outer cylinder under the influence of this pressure should be given. The value of $R'(a)$ is determined from the no-load conditions at the external boundary. For this, the equation $P_{rr}(a)=0$ is solved with respect to $\Psi_{rr}(a)$ for a given value of $\Psi_{\varphi\varphi}(a)=R(a)/a$. The obtained value $\Psi_{rr}(a)$ according to (8) is equal to $R'(a)$.

Let's note that to solve the Cauchy problem than the boundary value problem. From the solution of the Cauchy problem one can define all the characteristics of the stress-strain state at $b \le r \le a$ and, in particular, to find the pressure p at the inner boundary (at $r=b$). By solving this problem for several values $R(a)$, one can draw a table of dependence of the pressure p at the inner boundary against the outer cylinder radius after deformation. According to this table (for example, using interpolation) one can approximately find characteristics of the stress-strain state for a given pressure p. An accuracy depends on the pitch of the table: the smaller the step is, the more accurate solution will be obtained.

4. Calculation results

Finite element solution was obtained for a cylinder with a relation of the inner radius to the outer one of $b/a=0.6$ with a height of $h=2a$. Material parameters specification: $C_2=0$, the D value is varied from $2C_1$ to $64C_1$. The calculations were carried out for three values of the internal pressure: $p/C_1=0.25$, 0.5 and 0.7. An unstructured mesh was used, the number of nodes - 8736, the number of elements - 1092.

Some results of the problem solution with the help of finite element method are shown in Table 1, and the results of solution of this problem using the semi-inverse method are shown in Table 2. The tables show the value of true hoop stress $\sigma_{\varphi\varphi}$ on the inner boundary of the cylinder for different values of the internal pressure p and the material constant D. The stress is referred to the material constant C_1.

Table 1. True hoop stress $\sigma_{\varphi\varphi}$ on the inner boundary of the cylinder for different values of the internal pressure p and constant of material D. Finite element solution. All quantities are referred to C_1.

D/C_1	p/C_1		
	0.25	0.5	0.7
2	0.56	1.48	3.29
4	0.56	1.45	2.93
8	0.56	1.43	2.79
16	0.55	1.41	2.71
32	0.55	1.38	2.62
64	0.53	1.33	2.49

Table 2. True hoop stress $\sigma_{\varphi\varphi}$ on the inner boundary of the cylinder for different values of the internal pressure p and constant of material D. Solution obtained via semi-inverse method. All quantities are referred to C_1.

D/C_1	p/C_1		
	0.25	0.5	0.7
2	0.63	1.65	3.67
4	0.63	1.63	3.30
8	0.64	1.63	3.18

16	0.64	1.62	3.13
32	0.64	1.62	3.11
64	0.64	1.62	3.09

By comparing the results shown in Tables 1 and 2, it can be seen that the difference between the finite element solution and the solution obtained via semi-inverse method amounts to 10 - 20% of these solutions. Evidently, this difference is explained by the use of a large enough finite element mesh.

5. Conclusion

Thus, the article offered the option of applying of semi-inverse method to the solution of the Lame problem for large deformations. Numerical results are obtained for a compressible Mooney-Rivlin material. The results can be used for verification of finite element calculations, in particular the calculations carried out using the Fidesys CAE-system.

Acknowledgments

The research for this article was performed in FIDESYS LLC and financially supported by the Russian Ministry of Education and Science (project No.~14.579.21.0112, project ID RFMEFI57915X0112).

References

[1] Truesdell C 1972 *A First Course in Rational Continuum Mechanics* (Baltimore: The Johns Hopkins University)
[2] Lurie A I 1990 *Nonlinear Theory of Elasticity* (Amsterdam, North-Holland)
[3] John F 1966 Plane elastic waves of finite amplitude. Hadamard materials and harmonic materials *Comm. Pure and Appl. Math.* **19** 309–41
[4] Zelenina A A, Zubov L M 2008 First integrals of the equilibrium equations in one-dimensional problems of nonlinear elasticity theory *Doklady Phys.* **53** 379–82
[5] Aron M 2006 On certain deformation classes of compressible Hencky materials *Math. Mech. Solids* **11** 467-78
[6] Arrigo D J, Hill J M 1996 Transformations and equation reduction in finite elasticity. II: Plane stress and axially symmetric deformations *Math. Mech. Solids* **1** 177–92.
[7] De Pascalis R, Destrade M, Saccomandi G 2007 The stress field in a pulled cork and some subtle points in the semi-inverse method of nonlinear elasticity *Proc. Roy. Soc. A.* **463** 2945-59.
[8] Fu Y B, Lin Y P 2002 A WKB analysis of the buckling of an everted neo-Hookean cylindrical tube *Math. Mech. Solids* **7** 483-501.
[9] Hill J M 1973 Partial solutions of finite elasticity – axially symmetric deformations *Zeit. Angew. Math. Phys.* **24** 409–18.
[10] Hill J M 1973 Partial solutions of finite elasticity – plane deformations *Zeit. Angew. Math. Phys.* **24** 401–8.
[11] Walton J R, Wilber J P 2004 Deformations of neo-Hookean elastic wedge revisited *Math. Mech. Solids* **9** 307-27.
[12] Ogden R W 1997 *Non-Linear Elastic Deformations* (New York: Dover Publications).
[13] Fidesys LLC official website. Available online: http://cae-fidesys.com
[14] Zienkiewicz O C, Taylor R L 2000 *The finite element method* vol 1 *The basis* (Oxford: Butterworth-Heinemann)
[15] Zienkiewicz O C, Taylor R L 2000 *The finite element method* vol 2 *Solid mechanics* (Oxford: Butterworth-Heinemann)
[16] Levin V A, Vershinin A V 2008 Non-stationary plane problem of the successive origination of

stress concentrators in a loaded body. Finite deformations and their superposition *Comm. Numer. Meth. Eng.* **24** 2229–39.

[17] Levin V A, Zingerman K M, Vershinin A V, Freiman E I, Yangirova A V 2013 Numerical analysis of the stress concentration near holes originating in previously loaded viscoelastic bodies at finite strains *Int. J. Solids Struct.* **50** 3119–35.

Institute of Physics
Dirac House, Temple Back
Bristol BS1 6BE UK

ISSN: 1757-8981
ISBN 978-1-5108-3524-5

CURRAN ASSOCIATES INC.
proceedings.com